SPACE TECHNOLOGY AND APPLICATIONS INTERNATIONAL FORUM—1998

SPACE TECHNOLOGY & APPLICATIONS INTERNATIONAL FORUM (STAIF - 98)

January 25-29, 1998, Albuquerque, NM

1ST CONFERENCE ON GLOBAL VIRTUAL PRESENCE

1ST CONFERENCE ON ORBITAL TRANSFER VEHICLES

2ND CONFERENCE ON APPLICATIONS OF THERMOPHYSICS IN MICROGRAVITY

3RD CONFERENCE ON COMMERCIAL DEVELOPMENT OF SPACE

3RD CONFERENCE ON NEXT GENERATION LAUNCH SYSTEMS

15TH SYMPOSIUM ON SPACE NUCLEAR POWER AND PROPULSION

Cosponsored by:

NATIONAL AERONAUTICS AND SPACE ADMINISTRATION
Headquarters
Field Centers

UNITED STATES DEPARTMENT OF ENERGY
Headquarters
Los Alamos National Laboratory
Sandia National Laboratories

UNITED STATES AIR FORCE
Air Force Research Laboratory

In cooperation with:

AMERICAN ASTRONAUTICS SOCIETY

AMERICAN INSTITUTE OF AERONAUTICS & ASTRONAUTICS
National & Local Section

AMERICAN INSTITUTE OF CHEMICAL ENGINEERS
Heat Transfer and Energy Conversion Division

AMERICAN NUCLEAR SOCIETY,
Trinity Section

AMERICAN SOCIETY OF MECHANICAL ENGINEERS
Nuclear Engineering Division
Heat Transfer Division

AMERICAN SOCIETY FOR TESTING & MATERIALS
Committee E-10 on Nuclear Technology and Applications

INSTITUTE OF ELECTRICAL AND ELECTRONICS ENGINEERS, INC.
Nuclear and Plasma Sciences Society

INTERNATIONAL ASTRONAUTICAL FEDERATION

NEW MEXICO ACADEMY OF SCIENCE

NASA NATIONAL SPACE GRANT COLLEGE AND FELLOWSHIP PROGRAM
New Mexico Space Grant Consortium

PROFESSIONAL AEROSPACE CONTRACTORS ASSOCIATION

Organized by:

INSTITUTE FOR SPACE AND NUCLEAR POWER STUDIES
School of Engineering, University of New Mexico
Albuquerque, New Mexico
Webb Address: http://www-chne.unm.edu/isnps/isnps.htm
(505) 277-0446

SPACE TECHNOLOGY AND APPLICATIONS INTERNATIONAL FORUM—1998

1st Conference on Global Virtual Presence
1st Conference on Orbital Transfer Vehicles
2nd Conference on Applications of Thermophysics in Microgravity
3rd Conference on Commercial Development of Space
3rd Conference on Next Generation Launch Systems
15th Symposium on Space Nuclear Power and Propulsion

Albuquerque, NM January 1998

PART ONE

EDITOR
Mohamed S. El-Genk
*Institute for Space and Nuclear Power Studies,
University of New Mexico*

American Institute of Physics

AIP CONFERENCE PROCEEDINGS 420

Woodbury, New York

Editor:

Mohamed S. El-Genk
Institute for Space and Nuclear Power Studies
The University of New Mexico
Farris Engineering Center, Room 239
Albuquerque, NM 87131-1341
EMAIL: mgenk@unm.edu

Authorization to photocopy items for internal or personal use, beyond the free copying permitted under the 1978 U.S. Copyright Law (see statement below), is granted by the American Institute of Physics for users registered with the Copyright Clearance Center (CCC) Transactional Reporting Service, provided that the base fee of $10.00 per copy is paid directly to CCC, 222 Rosewood Drive, Danvers, MA 01923. For those organizations that have been granted a photocopy license by CCC, a separate system of payment has been arranged. The fee code for users of the Transactional Reporting Service is: 1-56396-747-2/ 98 /$10.00.

© 1998 American Institute of Physics

Individual readers of this volume and nonprofit libraries, acting for them, are permitted to make fair use of the material in it, such as copying an article for use in teaching or research. Permission is granted to quote from this volume in scientific work with the customary acknowledgment of the source. To reprint a figure, table, or other excerpt requires the consent of one of the original authors and notification to AIP. Republication or systematic or multiple reproduction of any material in this volume is permitted only under license from AIP. Address inquiries to Office of Rights and Permissions, 500 Sunnyside Boulevard, Woodbury, NY 11797-2999; phone: 516-576-2268; fax: 516-576-2499; e-mail: rights@aip.org.

L.C. Catalog Card No. 97-77866
ISSN 0094-243X

Casebound:			Paperback:	
ISBN	1-56396-779-0	Part One	1-56396-783-9	Part One
	1-56396-780-4	Part Two	1-56396-784-7	Part Two
	1-56396-781-2	Part Three	1-56396-785-5	Part Three
	1-56396-747-2	Set	1-56396-786-3	Set

DOE CONF- 980103

Printed in the United States of America

CONTENTS

Introduction xxv
Publications (Part One) xxvii
Organizing Committee (Part One) xxviii

PART ONE

PLENARY SESSIONS

Breakthrough Propulsion Physics Workshop Preliminary Results 3
    M. G. Millis

FIRST CONFERENCE ON GLOBAL VIRTUAL PRESENCE

[A1] ACTIVE REMOTE SENSING

Superresolution Techniques for Active Remote Sensing 15
    R. K. Mehra and B. Ravichandran
A Tunable MWIR Laser Remote Sensor for Chemical Vapor Detection 21
    T. L. Bunn, P. M. Noblett, and W. D. Otting
Geoscience Laser Altimeter System—Stellar Reference System 27
    P. S. Millar and J. M. Sirota
The Geoscience Laser Altimeter System (GLAS) 33
    J. B. Abshire, J. C. Smith, and B. E. Schutz

[A2] LONG-TERM MONITORING AND CHANGE DETECTION - I

256×256 GaN Ultraviolet Imaging Arrays 39
    Z. D. Huang, D. B. Mott, and P. K. Shu
High Performance ZnSSe Ultraviolet Photoconductors .... (NA)*
    Z. C. Huang, D. B. Mott, and P. K. Shu
Split-Geometry Detectors, Our Eyes in Space 45
    R. J. Martineau, F. Peters, A. Burgess, C. Kotecki, S. Manthripragada, J. Godfrey,
    D. Krebs, D. Mott, P. Shu, J. Z. Shi, and K. Hu
CdZnTe Technology for Gamma Ray Detectors 55
    C. Stahle, J. Shi, P. Shu, S. Barthelmy, A. Parsons, and S. Snodgrass

[A3] LONG-TERM MONITORING AND CHANGE DETECTION - II

Large Aspect Ratio Detector Array for Space Applications 61
    A. Ewin, M. Fortin, P. Shu, N. C. Das, S. Aslam, W. Smith, and J. McAdoo
The Next Generation of Monolithic Infrared Detector Arrays 67
    A. M. Joshi, M. Jhabvala, and P. Shu
Overview of the New Millennium Program's Earth Orbiter #1 Mission .... (NA)*
    B. Cramer

(NA)* indicates that the paper was not available at time of press

Wide-Field High-Performance Geosynchronous Imaging... 76
    H. J. Wood, D. Jenstrom, M. Wilson, S. Hinkal, and F. Kirchman
The New Millennium Program's Earth Orbiter #1 Advanced Land Imager........................ (NA)*
    S. G. Ungar

[A4] SPACE LASER TECHNOLOGY - I

Laser Diode Pumped Solid State Lasers for Space Applications.................................... 77
    R. L. Byer
Lasers and Space Optical Systems Study.. 80
    C. Giuliano and A. L. Annaballi
Semiconductor Lasers for NASA Applications... 85
    S. Forouhar, M. G. Young, S. Keo, L. Davis, R. Muller, and P. Maker
High-Efficiency Laser Development for Space Applications.. 91
    C. Muller, C. Denman, M. Gregg, and L. DeHainaut
Solid State Laser Technology and Atmospheric Sensing Applications 95
    J. C. Barnes

[A5] SPACE LASER TECHNOLOGY - II

Enabling Technologies for a Global Tropospheric Wind Lidar System 101
    B. M. Gentry
Diode-Pumped Solid State Lasers for Space-Based Applications................................ 107
    R. S. Afzal
Semiconductor Diode Laser Development for Space Communications......................... (NA)*
    M. Gregg
Solar Pumped IBr Laser .... (NA)*
    G. Hager
Rigorous Derivation of Laser Resonator Dynamics Based on Generalized
'Fox-Li' Integral Equation... 112
    E. J. Bochove

[A6] SATELLITE AUTONOMY AND SATELLITE CONSTELLATION TECHNOLOGY

An Algorithm for Enhanced Formation Flying of Satellites in Low Earth Orbit................. 119
    D. C. Folta and D. A. Quinn
Autonomous Soft-Landing Technique onto the Moon in SELENE Project 127
    S. Hikida and H. Itagaki
Autonomous Failure Detection, Identification and Fault-Tolerant Estimation
for Spacecraft Guidance, Navigation and Control... 134
    R. Mehra, C. Rago, S. Seereeram, and D. S. Bayard
Technology Initiatives for the Autonomous Guidance, Navigation, and Control
of Single and Multiple Satellites... 141
    J. Croft, J. Deily, K. Hartman, and D. Weidow
Nonlinear Predictive Control for Spacecraft Trajectory Guidance, Navigation
and Control... 147
    R. K. Mehra, S. Seereeram, J. T. Wen, and D. S. Bayard
Design of an Integrated Satellite (INT-SAT) using Advanced Semiconductor Technology .. 153
    A. M. Joshi

(NA)* indicates that the paper was not available at time of press

[A7] LIGHTWEIGHT STRUCTURES AND OPTICAL SYSTEMS - I

Lightweight, Highly Integrated Optical Systems for the New Millennium Program 159
    M. Schwalm, D. Wang, M. Curcio, D. H. Rodgers, and P. M. Beauchamp

Lightweight Composite Reflectors for Space Optics 166
    B. E. Williams, S. R. McNeal, and R. M. Ono

A Critical Review of Ultralightweight Composite Mirror Technology 173
    E. P. Kasl and D. A. Crowe

Issues to be Addressed in the Design and Fabrication of Ultralight Meter Class Mirrors 179
    G. C. Krumweide, J. Dyer, E. P. Kasl, and G. V. Mehle

Advanced Lightweight Optics Development for Space Applications 185
    J. W. Bilbro

[A8] LIGHTWEIGHT STRUCTURES AND OPTICAL SYSTEMS - II

The Infrared Telescope Technology Testbed: Design, Fabrication and Testing
of an Ultra-lightweight Cryogenic Beryllium Telescope .... (NA)*
    D. R. Coulter, S. A. Macenka, and M. T. Stier

Research on the Problem of High-Precision Deployment for Large-Aperture
Space-Based Science Instruments 188
    M. S. Lake, L. D. Peterson, M. R. Hachkowski, J. D. Hinkle, and L. R. Hardaway

Inflatation-Deployed Space Cameras—A Vision .... (NA)*
    J. B. Breckinridge, A. Meinel, and M. Meinel

High Precision Sensing and Control Technology .... (NA)*
    R. A. Laskin

Advanced Lighweight Mirrors Using Glass Membranes with Active Rigid Support .... (NA)*
    J. Burge, R. Angel, B. Martin, S. Miller, P. Hinz, D. Sandler, D. Bruns, and D. Tenerelli

[A9] PAYLOAD SUPPORT TECHNOLOGIES

Functional and Life Test Data for a Two-Stage Stirling Cycle Mechanical Cryocooler
for Space Applications 199
    H. Carrington, W. J. Gully, W. K. Kiehl, S. Banks, E. James, and S. Castles

A Mechanical Cooler for Dual-Temperature Applications 205
    W. Gully, H. Carrington, W. Kiehl, and K. Byrne

Directions in US Air Force Space Power Technology for Global Virtual Presence 211
    D. Keener, K. Reinhardt, C. Mayberry, D. Radzykewycz, C. Donet, D. Marvin, and C. Hill

Kinematic Path Planning for Space-Based Robotics 223
    S. Seereeram and J. T. Wen

[A10] COMMUNICATIONS TECHNOLOGIES

The Globalstar Satellite Cellular Communication System Design and Status 229
    F. J. Dietrich

Telepresence by High-Data Rate Satellite Communications 235
    D. J. Hoder, M. J. Zernic, P. G. Mallasch, K. Bhasin, D. E. Brooks,
    and D. R. Beering

(NA)* indicates that the paper was not available at time of press

NASA's Wireless Augmented Reality Prototype (WARP)..................................... **236**
    M. Agan, L.-A. Voisinet, and A. Devereaux
High-Rate Optical Communications Links for Virtual Presence in Space.......................... **243**
    J. Lesh, K. Wilson, J. Sandusky, M. Jeganathan, H. Hemmati, S. Monacos,
    N. Page, and A. Biswas
Key Issues on the Constellation Design Optimization for NGSO Satellite Systems................. **249**
    A. W. Wang

[A11] KNOWLEDGE ON DEMAND AND DATA FUSION

Onboard Science Data Analysis: Opportunities, Benefits, and Effects on Mission Design............ **(NA)***
    P. Stolorz
Enabling Information-on-Demand with a Regional Validation Center............................ **(NA)***
    R. F. Cromp
Smart Optical Sensing at the Phillips Laboratory... **(NA)***
    L. McMackin
Polarization and Hyperspectral Information Fusion for Remote Sensing from Space............... **(NA)***
    R. L. Grotbeck

FIRST CONFERENCE ON ORBITAL TRANSFER VEHICLES

[B1] OTV MISSIONS AND APPLICATIONS

Space Transportation: Historical Development, Barriers to Progress, and the Way Ahead.......... **(NA)***
    E. Cady
Impact of Future Launch Systems on Solar Orbit Transfer Vehicle Payoffs....................... **(NA)***
    T. M. Miller and P. E. Frye
ROTV as Part of a Reusable Space Architecture... **257**
    T. H. Phillips, R. O'Leary, and F. Widman
Characterization of the Dielectric Properties of the Propellants MON and MMH.................. **258**
    A. A. M. Delil

[B2] OTV SYSTEMS

Radiation Effects in Low-Thrust Orbit Transfers... **264**
    J. E. Pollard
Reusable Solar Thermal Orbital Transfer Vehicle for MEO and GEO Bound Payloads............. **(NA)***
    M. Bangham, E. Cady, and D. Luther
Non-Toxic Propulsion for Spaceplane "Pop-up" Upper Stages.................................. **270**
    J. B. Eckmann, R. L. Wiswell, and E. Haberman
Status and Design Concepts for the Hydrogen On-Orbit Storage and Supply Experiment.......... **276**
    D. J. Chato, M. Van Dyke, J. C. Batty, and S. Schick

[B3] SOLAR CONCENTRATOR TECHNOLOGIES

Naval Research Laboratory Solar Concentrator Program...................................... **282**
    I. Sokolsky and M. A. Brown
Ultralight Inflatable Fresnel Lens Solar Concentrators....................................... **288**
    M. J. O'Neill and M. F. Piszczor

(NA)* indicates that the paper was not available at time of press

Concentrator Array Testing and Integration Verification of an Inflatable Collector .... (NA)*
    J. P. Paxton, J. C. Pearson, P. A. Gierow, and M. R. Holmes
Spline Radial Panel Concentrator Designs for ISUS Systems .... (NA)*
    G. Borell, D. Wilson, B. Drawdy, and C. Haeuber
ISUS Engine Ground Demonstration Solar Concentrator Design and Testing .... (NA)*
    J. Laxton, C. Lawandowski, D. Wilson, and G. Borell
A High-Efficiency Refractive Secondary Solar Concentrator for High Temperature
Solar Thermal Applications .... (NA)*
    M. F. Piszczor and R. P. Macosko

[B4] ELECTRIC PROPULSION/ENERGY CONVERSION TECHNOLOGIES

Development of an Advanced Pulsed Plasma Thruster (PPT) and Plume
Diagnostic Experiment (PDE) for Demonstration on the MightySat II.I Spacecraft 295
    T. Peterson, E. Pencil, L. Arrington, W. A. Hoskins, N. J. Meckel, J. R. LeDuc,
    D. R. Bromaghim, J. Malak, J. J. Blandino, and B. Moore
Comparison of Chemical and Electric Propulsion Concepts for Orbit Transfer
of Space Based Radar Systems .... (NA)*
    R. M. Salasovich and R. A. Spores
The Electric Propulsion Space Experiment (ESEX)—A Demonstration of High Power
Arcjets for Orbit Transfer Applications 302
    D. R. Bromaghim, R. M. Salasovich, J. R. LeDuc, and L. K. Johnson
Performance Evaluation of Thermionic Converter with Macro-Grooved Emitter
and a Smooth Collector 308
    M. S. El-Genk and Y. Momozaki
A Rigorous Approach for Predicting Thermionic Power Conversion Performance 317
    A. C. Marshall
Conductively Coupled Multi-cell TFE with Electric Heating Pretest Ability 318
    Y. V. Nikolaev, R. Ya. Kucherov, S. A. Eryomin, O. L. Izhvanov, V. U. Korolev,
    N. V. Lapochkin, D. L. Tsetshladze, T. A. Lechtenberg, and L. L. Begg

[B5] GROUND DEMONSTRATION AND TESTING

Re-START: The Second Operational Test of the String Thermionic Assembly
Research Testbed 324
    F. J. Wyant, D. Luchau, and T. D. McCarson
Experimental Testing of a Foam/Multilayer Insulation (FMLI) Thermal Control
System (TCS) for Use on a Cryogenic Upper Stage 331
    L. J. Hastings and J. J. Martin
STUSTD LH2 Storage and Feed System Test Program 342
    A. D. Olsen
Integrated Solar Upper Stage (ISUS) Engine Ground Demonstration (EGD) 348
    C. T. Kudija and P. E. Frye

[B6] OTV SPACE DEMONSTRATION AND EXPERIMENTS

Solar Orbital Transfer Vehicle (SOTV) Development Program .... (NA)*
    K. E. Dayton and E. C. Cady
Solar Thermal Propulsion Shooting Star Experiment .... (NA)*
    L. A. Curtis and R. G. Toelle

(NA)* indicates that the paper was not available at time of press

Propulsive Small Expendable Deployer System (ProSEDS) Space Demonstration.................. 354
   L. Johnson and J. Balance
Upper Stage Technology Demonstration Off of a Minuteman II Booster .... (NA)*
   D. R. Perkins

[B7] SOLAR RECEIVER TECHNOLOGIES

Thermionic Converters for Ground Demonstration Testing...................................... 359
   G. J. Talbot, W. D. Ramsey, and E. L. James
JSUS Solar Thermal Thruster and Its Integration with Thermionic Power Converter............ 364
   M. Shimizu, K. Eguchi, K. Itoh, H. Sato, T. Fujii, K-i Okamoto, and T. Igarashi
MULTI-FOIL© Insulation for High Temperature Applications; Summary
of RACCET and EGD Experience for the ISUS Program....................................... 370
   G. Miskolczy, J. Burchfield, and J. Malloy
Advanced High Temperature Insulation Systems for Solar Thermal Rockets..................... (NA)*
   J. Malloy
Testing of a Receiver-Absorber-Converter (RAC) for the Integrated Solar
Upper Stage (ISUS) Program 375
   K. O. Westerman and B. J. Miles
Interdiffusion of Rhenium and Poco Graphite... 381
   J. Li and R. H. Zee

SECOND CONFERENCE ON APPLICATIONS OF THERMOPHYSICS IN MICROGRAVITY

[C1] MICROGRAVITY THERMOPHYSICS FUTURE RESEARCH DEVELOPMENT

Real-Time X-ray Transmission Microscopy for Fundamental Studies of Solidification:
Al—Al$_2$Au Eutectic 389
   P. A. Curreri, W. F. Kaukler, and S. Sen
High Temperature Electrostatic Sample Levitator as a Future Containerless
Materials Processing Facility in Space.. 397
   W-K Rhim
Current and Future Experiments in Microgravity Fundamental Physics .... (NA)*
   U. E. Israelsson and M. C. Lee
The Microgravity Applications Promotion Programme of the European Space Agency, ESA 399
   H. U. Walter, R. Binot, E. Kufner, and O. Minster

[C2] COMPUTATION AND MODELING OF MICROGRAVITY THERMAL-FLUID PHENOMENA AND MATERIALS PROCESSING

Numerical Investigations of Buoyancy Effect on Thermocapillary Migration of Bubbles........... (NA)*
   A. Esmaeeli and V. Arpaci
Stability of a Capillary Surface in a Rectangular Container................................. 404
   M. M. Weislogel and K. C. Hsieh
Hydrodynamic Dryout in Two-Phase Flows: Observations of Low Bond Number Systems 413
   M. M. Weislogel and J. B. McQuillen

(NA)* indicates that the paper was not available at time of press

Applicability of Theoretical Models to the Evaluation of Thermal Conductivity
in Microgravity 422
    G. Passerini, F. Polonara, and F. Gugliermetti
Thermocapillary Flow Near a Corner in a Thin Liquid Layer 429
    R. Balasubramaniam
Gravitational Effects on Diamond and Carbon Fiber Fabrication 435
    E. B. Kennel and C. Tang

[C3] FLUID DYNAMICS AND HEAT TRANSFER IN MICROGRAVITY

Effects of Gravity on Capillary Motion of Fluid 440
    S. H. Chan, T. R. Shen, G. D. Proffitt, and B. Singh
Heat Transfer in Microgravity for Bounded Fluids with Moderate Prandtl Numbers .... (NA)*
    F. Stella, F. Gugliermetti, and L. Ottaviani
Experimental Study and Modeling of the Effect of Low-Level Impurities
on the Operation of the Constrained Vapor Bubble 446
    J. Huang, M. Karthikeyan, J. Plawsky, and P. C. Wayner, Jr.
Bubble Flow in Reduced Gravity 452
    T. L. Brower
A Theoretical Approach to the Evaluation of Gravity Influence on Heat and Mass
Transfer in Liquids... 458
    G. Latini, G. Passerini, F. Polonara, and G. Galli

[C4] SPACECRAFT THERMAL CONTROL

Experimental Investigation of Reducing Startup Time on Capillary Pumped
Loop with EHD Assistance 464
    B. Mo, M. M. Ohadi, S. V. Dessiatoun, J. H. Kim, and K. Cheung
A Study of the Fundamental Operations of a Capillary Driven Heat Transfer Device
in Both Normal and Low Gravity, Part I. Liquid Slug Formation in Low Gravity 471
    J. S. Allen, K. P. Hallinan, and J. Lekan
Testing and Evaluation of Small Cavitating Venturis with Water at Low Inlet Subcooling 479
    S. G. Liou, I. Y. Chen, and J. S. Sheu
Thermal Design and Analysis of the Spartan Lite Spacecraft 488
    J. Tolson and D. F. Powers
Improved Cooler Performance Using Spectrally Selective Thermal Coatings 495
    D. Neuberger, N. Ackerman, and G. Harris

[C5] MICROGRAVITY TWO-PHASE FLOW

Design, Fabrication and Testing of an SP-100-Like Phase Separator
in a Microgravity Environment 501
    S. Bragg, D. L. Lockridge, D. Dorsey, J. Fuller, M. Ellis, and F. R. Best
Void Fraction Measurements by Quick Acting Valves and Capacitance Measurements 505
    J. H. Chang and F. R. Best
Loop Heat Pipe Flight Experiment 511
    W. B. Bienert
Sensors and Components for Aerospace Thermal Control, Life Sciences,
and Propellant Systems 514
    A. A. M. Delil, A. Pauw, R. G. H. M. Voeten, and P. van Put

(NA)* indicates that the paper was not available at time of press

PART TWO

THIRD CONFERENCE ON COMMERCIAL DEVELOPMENT OF SPACE

[D1] UNDERSTANDING IMPEDIMENTS TO THE COMMERCIAL USE OF SPACE

The Utilization of Space as a Tool for Applied Research H. J. Sprenger	525
Identifying the Impediments to Technology Transfer to Industry—A Case Study B. Kronberg	531
Commercial Space Three Problems—Three Solutions D. A. Rossi	(NA)*
A Canadian Viewpoint on the Challenges of Space Station Commercialization G. Frappier, P. Gregory, R. Herring, and D. McCabe	534
Mindset Impediments to the Commercialization of Space R. C. Hill	535

[D2] SIGNIFICANT HIGHLIGHTS IN PRODUCTION DEVELOPMENT USING THE ATTRIBUTES OF SPACE

Space Transfer Services as a Precursor to Space Business Parks D. V. Smitherman, Jr.	538
Investigating the Nucleation and Growth of Zeolite Crystals in Space A. Sacco, Jr., N. Bac, J. Warzywoda, I. Guray, M. Marceau, T. L. Sacco, and L. M. Whalen	544
A Transparent Furnace as a Commercial Research Facility C. Watson, C. Lundquist, F. Wessling, and R. Naumann	551
Commercialization of a Demodulating Camera System Originally Developed for Microgravity Combustion Diagnostics M. Linne, C. Fisher, and N. Middleton	556
Radiation Hardening of InP Solar Cells for Space Applications M. F. Vilela, A. Freundlich, C. Monier, F. Newman, and L. Aguilar	563

[D3] COMMERCIAL APPLICATIONS IN SPACE AGRICULTURE AND ENVIRONMENTAL SCIENCE

Production of Potato Minitubers Using Advanced Environmental Control Technologies Developed for Growing Plants in Space R. G. Britt	569
Impact of Accelerated Plant Growth on Seed Variety Development E. Christopherson	574
Space Flight Research Leading to the Development of Enhanced Plant Products: Results from STS-94 L. S. Stodieck, A. Hoehn, and G. Heyenga	578
Commercial Products Developed from Plant Oils Produced in Microgravity N. A. Draeger	586
Capabilities of the Commercial Plant Biotechnology Facility to Support Plant Activities on the International Space Station W. Zhou and R. J. Bula	593
Mid-IR Interband Cascade Lasers for Remote Chemical Sensing Applications D. Zhang, C.-H. Lin, R. Q. Yang, B. H. Yang, S. S. Pei, J. Harper, and M. B. Weimer	603

(NA)* indicates that the paper was not available at time of press

[D4] ADVANCES AND RECENT COMMERCIAL ACCOMPLISHMENTS IN BIOPRODUCT DEVELOPMENT IN SPACE

Investigation of PACE™ Software and VeriFax's Impairoscope Devices
for Quantitatively Measuring the Effects of Stress 609
    G. W. Morgenthaler, G. R. Nuñez, A. M. Botello, J. Soto, R. Shrairman, and A. Landau
Autonomous Biological System—A Unique Method of Conducting Long Duration
Space Flight Experiments for Pharmaceutical and Gravitational Biology Research 616
    G. A. Anderson, T. K. MacCallum, J. E. Poynter, and D. Klaus
A Model System for Studies on Bone Matrix Formation by Osteogenic Cells in Microgravity 622
    T. M. Quinton, H. K. Fattaey, F. Motaffaf, and T. C. Johnson
Benefits Attained from Space Flight in Pre-Clinical Evaluation of Candidate Drugs 627
    L. S. Stodieck, T. Bateman, R. Ayers, V. Ferguson, and S. Simske
Antibiotic Production in Space 633
    D. Klaus, R. Brown, and K. Cierpik
The Use of Space as a Model for Spinal Cord Regeneration Experiments .... (NA)*
    K. I. Clark

[D5] TELE-SCIENCE TOOLS AND TECHNOLOGY FOR IMPROVING SPACE AND TERRESTRIAL LABORATORY RESEARCH CAPABILITIES

Utilization of Commercial Communication Systems for Space Based
Research Applications 638
    C. Overmyer and C. Thompson
COMCAP: a Cost-Effective Commercial Capsule System for Microgravity
Research in Low Earth Orbit .... (NA)*
    J. M. Cassanto and J. K. von der Lippe
Single Board Controller for Spaceflight Payloads 643
    V. Strength
Advanced Imaging Microscope Tools Applied to Microgravity Research Investigations 647
    L. Peterson, J. Samson, D. Conrad, and K. Clark

[D6] PRODUCTS APPROACHING THE MARKET PLACE RESULTING FROM MICROGRAVITY RESEARCH

A Hot Electrons-Based Wide Spectrum On-Orbit Optical Calibration Source 648
    D. Starikov, I. Berichev, N. Medelci, E. Kim, Y. Wang, and A. Bensaoula
Influence of Growth Transients on Interface and Composition Uniformity of Ultra
Thin In(As,P) and (In,Al,Ga)As Epilayers Grown by Chemical Beam Epitaxy 654
    F. Newman, L. Aguilar, A. Freundlich, M. F. Vilela, and C. Monier
Lunar Regolith Thin Films: Vacuum Evaporation and Properties 660
    A. Freundlich, T. Kubricht, and A. Ignatiev
Boron Carbon Nitride Materials for Tribological and High Temperature Device Applications 666
    N. Badi, A. Tempez, D. Starikov, V. Zomorrodian, N. Medelci, A. Bensaoula, J. Kulik,
    S. Lee, S. S. Perry, V. P. Ageev, S. V. Garnov, M. V. Ugarov, S. M. Klimentov, V. N. Tokarev,
    K. Waters, and A. Schultz
Conducting and Interfacial Properties of Epitaxial SVO Films 672
    D. L. Ritums, N. J. Wu, X. Chen, D. Liu, and A. Ignatiev

(NA)* indicates that the paper was not available at time of press

[D7] SPACE POWER

Simplified Proton Exchange Membrane Fuel Cells for Space and Terrestrial Applications 679
    H. P. Dhar, K. A. Lewinski, and V. K. Tripathi

Magnetic Bearing Development for Support of Satellite Flywheels 685
    A. Palazzolo, M. Li, A. Kenny, S. Lei, D. Havelka, and A. Kascak

Flywheel Magnetic Bearing Field Simulation with Motion Induced Eddy Currents .... (NA)*
    D. Havelka, A. Palazzolo, A. Kenny, and A. Kascak

Improvement and Optimization of $InAs_xP_{1-x}InP$ Multi Quantum Well Solar Cells 693
    L. Aguilar, F. Newman, I. Serdiukova, C. Monier, M. F. Vilela, A. Freundlich, A. Delaney, and S. Street

Molecular Beam Epitaxy of InP Single Junction and $InP/In_{0.53}Ga_{0.47}As$ Monolithically Integrated Tandem Solar Cells using Solid Phosphorous Source Material 698
    A. Delaney, K. Chin, S. Street, F. Newman, L. Aguilar, A. Ignatiev, C. Monier, M. Velela, and A. Freundlich

[D8] SPACE-SMALL BUSINESS INNOVATIVE RESEARCH HIGHLIGHTS: 1996–1997—I

Applying Dynamic Control To Crystallization in Space 703
    L. Arnowitz

High Temperature Transparent Furnace Development.. 711
    S. C. Bates, K. S. Knight, and D. W. Yoel

Microscopic and Macroscopic Modeling of Layer Growth Kinetics and Morphology in Vapor Deposition Processing.. 718
    P. J. Stout, S. A. Lowry, and A. Krishnan

Concentration of Atomic Oxygen in Low Earth Orbit and in the Laboratory for Use in High Quality Oxide Thin Film Growth 724
    J. A. Schultz, K. Eipers-Smith, K. Waters, S. Schultz, M. Sterling, D. Starikov, A. Bensaoula, T. Minton, and D. J. Garton

[D9] SPACE-SMALL BUSINESS INNOVATIVE RESEARCH HIGHLIGHTS: 1996–1997—II

Benzoporphyrin Derivative and Light-Emitting Diode for Use in Photodynamic Therapy: Applications of Space Light-Emitting Diode Technology............................. 729
    H. T. Whelan, J. M. Houle, D. M. Bajic, M. H. Schmidt, K. W. Reichert, II, and G. A. Meyer

Verifax: Biometric Instruments Measuring Neuromuscular Disorders/Performance Impairments 736
    G. W. Morgenthaler, R. Shrairman, and A. Landau

Novel Approaches to Calibration of High Temperature Furnaces............................. 743
    L. Vujisic and S. Motakef

Microgravity Processing of Polymets for NLO Applications 749
    D. J. Trantolo, J. D. Gresser, Y. Y. Hsu, R. L. White, and D. L. Wise

[D10] MICROGRAVITY SCIENCES— RESULTS TO COMMERCIALIZATION

Effect of Gravity on the Combustion Synthesis of Porous Materials 755
    D. A. Pacas, J. J. Moore, F. Schowengerdt, and T. W. Wolanski

(NA)* indicates that the paper was not available at time of press

Microgravity Effects on Magnetotactic Bacteria 761
    J. E. Urban
Resistance to Chemical Disinfection Under Conditions of Microgravity 765
    G. L. Marchin
**Commercialization of Microgravity Experimental Results Through Improvements
to Ceramic Fabrication Process** .... (NA)*
    R. Sood
**Development of THM Crystal Growth Technology to Produce Commercial
Terrestrial Semiconductors and to Aid Microgravity Experimental Design** 769
    R. F. Redden, N. Audet, R. P. Bult, J. R. Butler, and P. W. Nasmyth
Measurement of Soret Coefficients of Crude Oil in Microgravity 773
    D. Hart, J-C. Legros, and F. Montel

[D11] COMMERCIAL OPPORTUNITIES UTILIZING
THE INTERNATIONAL SPACE STATION

Space Station Access with VentureStar™ 779
    W. Faulconer and D. Randolph
Commercial Opportunities Utilizing the International Space Station 786
    M. E. Kearney, P. Mongan, C. M. Overmyer, and K. Jackson
**Boeing's Mir Pathfinder Project and its Implications for Commercializing International
Space Station** .... (NA)*
    R. J. Foss and T. Johnston
Commercial Biotechnology Processing on International Space Station 793
    M. S. Deuser, J. C. Vellinger, J. R. Hardin, and M. L. Lewis
**Commercial Opportunities for the On-Orbit Use of the Extreme Temperature
Translation Furnace** 798
    M. W. Riley, S. Noojin, and J. E. Smith, Jr.
**Commercial Life Sciences Prospects for the International Space Station-Building
on Lessons Learned** 803
    G. W. Morgenthaler and J. Berryman

[D12] SPACE LIFE SCIENCES
AND THE PUBLIC GOOD

Measuring the Returns to NASA Life Sciences Research and Development 810
    H. R. Hertzfeld
Preparing the Way, Space Life Sciences Outreach 816
    D. Atchison and R. A. Grymes
Fundamental Gravitational Research and Emerging Technologies: The Connection 822
    M. D. Ross, J. Smith, K. Montgomery, R. Cheng, and S. Linton
Earth-Based Applications of Space Biosensor Technologies .... (NA)*
    J. W. Hines
NASA's Telemedicine Testbeds: Commercial Benefit 829
    C. R. Doarn and R. Whitten
Commercialization of Regenerative Life Support Systems 835
    M. Flynn and D. Bubenheim

[D13] COMMERCIAL APPLICATIONS
OF COMBUSTION IN SPACE

**Opportunities for Microgravity Research in the Development of Practical
Catalytic Combustion Systems** 840
    R. J. Kee and L. Raja

(NA)* indicates that the paper was not available at time of press

High-Strength Diamond Drills Produced by Combustion Synthesis .... (NA)*
    B. Radtke

Diode-Laser-Based Combustion Sensing .... (NA)*
    D. Kane

Commercial Production of Heavy Metal Fluoride Glass Fiber in Space 847
    D. S. Tucker, G. L. Workman, and G. A. Smith

Modeling Study of Water-Mist as Flame Suppressant 852
    R. Srivastava, J. T. McKinnon, and J. R. Butz

Commercial Potential of a Burner Developed for Investigating Laminar
and Turbulent Premixed Flames in μg 858
    R. K. Cheng, M. R. Johnson, and L. W. Kostiuk

THIRD CONFERENCE ON NEXT GENERATION LAUNCH SYSTEMS

[E1] X-33 UPDATE

X-33 Flight Assurance Status .... (NA)*
    C. Meade

VentureStar™—A Revolutionary Space Transportation Launch System 867
    R. I. Baumgartner

Systems Integration Challenges for the X-33 Program .... (NA)*
    J. Laube

A Primer on the Use of Commercial Off-the-Shelf Software in the Development
of Low Cost Launch Control Systems 875
    S. M. Tobin and R. Summerford

[E2] MILITARY SPACEPLANE

Feasibility of Atmospheric Skipping for Reusable Launch Vehicles .... (NA)*
    R. H. Moszée

Atmospheric Considerations for Skipping Spaceplane Trajectories 881
    D. Stapleton, T. Galati, and F. McDougall

Mini-Spaceplane Center Wingbox Technology Demonstrator .... (NA)*
    E. R. Anselmo and A. Del Mundo

Developing the Military Spaceplane—From Concept to Hardware 887
    P. L. Klevatt and W. A. Gaubatz

The Military Spaceplane Integration Technology Testbed .... (NA)*
    C. Johnston

[E3] COMMERCIALIZATION OF LAUNCH SYSTEMS

The Commercial Implications of the EELV Program 888
    S. E. Sasso

SeaLaunch Program .... (NA)*
    J. Stenovec

The Commercial Atlas IIAR Program 893
    R. L. Hauser, Jr.

Transportation Requirements for the Fast Freight Market .... (NA)*
    D. G. Andrews, M. J. Dunn, and M. Rubeck

(NA)* indicates that the paper was not available at time of press

Development of a Quick-Reaction Commercial Launch Site at the Cape
Canaveral Air Station... 899
    R. L. Schuiling and E. A. O'Connor
Economic Trade-offs for Spaceplanes Over a Large Range of Projected Traffic 905
    J. P. Penn and C. A. Lindley

[E4] INTERNATIONAL PROGRAMS

The European Space Agency's FESTIP Initiative 921
    D. Burleson
Russian Aluminum-Lithium Alloys for Advanced Reusable Spacecraft 926
    R. O. Charette, B. G. Leonard, W. F. Bozich, and D. A. Deamer
Current Status of the H-II Orbiting Plane-Experimental (HOPE-X) Development 937
    T. Fukui, K. Miho, and E. Nakano
Comparison of Basic Launch Vehicles of Leading Space Countries 943
    Y. G. Korotkiy
The Need and Processes for International Cooperation in Future Human
Space Exploration.. (NA)*
    W. H. Siegfried

[E5] COST AND OPERATIONS

Insuring RLV Transportation Services 949
    J. S. Greenberg
A Survey of Space Cost Models 956
    W. T. Harwick
A Novel Methodology for Estimating Upper Limits of Major Cost Drivers for
Profitable Conceptual Launch System Architectures... 962
    R. E. Rhodes and R. J. Byrd
Researching the Cost of Launch Operations .... (NA)*
    A. Matthews

[E6] LESSONS LEARNED

DC-XA Auxiliary Propulsion System Lessons Learned.. (NA)*
    K. E. Dayton
Delta Clipper Lessons Learned for Increased Operability in Reusable Space Vehicles.............. 969
    R. O. Charette, D. A. Steinmeyer, and R. R. Smiljanic
The Benefits of X-Vehicles—DC-XA and Beyond .... (NA)*
    B. G. Leonard, R. O. Charette, D. A. Deamer, and P. W. Ferguson
Lessons Learned and Results of the DC-XA Program .... (NA)*
    D. L. Dumbacher
Lessons Learned from LASRE Development... (NA)*
    P. Best

[E7] ADVANCED/NOVEL CONCEPTS

Highly Reusable Space Transportation: Advanced Concepts and Technologies
that may Achieve $100/Lb to Low Earth Orbit.. (NA)*
    J. C. Mankins
Very High Thrust-to-Weight Rocket Engines... 979
    J. F. Glass, B. D. Goracke, and D. J. H. Levack
Magnetohydrodynamic Propulsion Using On-Board Sources 985
    J. A. Martin

(NA)* indicates that the paper was not available at time of press

Very Advanced HRST Exploiting Off-Board Beamed Power.................................... (NA)*
    L. N. Myrabo

An Advanced Military Space Plane Design Concept .... (NA)*
    R. L. Chase, L. E. McKinney, H. D. Froning, Jr., P. Czysz, R. Boyd, and M. Lewis

[E8] ENTREPRENEUR INITIATIVES

Incrementally Developing a Cultural and Regulatory Infrastructure for Reusable Launch Vehicles.. 991
    R. Simberg

Space Transportation: Historical Development, Barriers to Progress and the Way Ahead .... (NA)*
    M. S. Kelly

Commercial Space Development Needs Cheap Launchers.. 995
    J. W. Benson

OASIS: A Global Commercial Space Transportation System .... (NA)*
    L. Ortega, T. Gregory, and J. R. Briarton

An Affordable RBCC-Powered 2-Stage Small Orbital Payload Transportation Systems Concept Based on Test-Proven Hardware 999
    W. J. D. Escher

[E9] LAUNCH VEHICLE TECHNOLOGIES—I

Experimental Investigation of a Graphite-Composite Intertank Section for a Reusable Launch Vehicle.. 1007
    J. W. Sawyer and H. Bush

Thermo-Mechanical Evaluation of Carbon–Carbon Primary Structure for SSTO Vehicles 1020
    H. C. Croop, H. B. Lowndes III, S. E. Hahn, and C. A. Barthel

Development of Actively Cooled Panels for Advanced Propulsion Systems...................... 1027
    B. K. Hauber

Mechanical Attachments for Flexible Blanket TPS 1033
    C. W. Newquist, D. M. Anderson, M. W. Shorey, and K. S. Preedy

[E10] LAUNCH VEHICLE TECHNOLOGIES—II

Single Stage and Thrust Augmented Reusable Launch Vehicle Stability and Performance Study... 1039
    P. A. Tanck and K. B. Steadman

Design, Fabrication and Test of a Liquid Hydrogen Titanium Honeycomb Cryogenic Test Tank for Use as a Reusable Launch Vehicle Main Propellant Tank........................ 1045
    P. B. Stickler and P. C. Keller

Thermal Structures Technology Development for a Reusable Launch Vehicle's Cryogenic Propellant Tanks .... (NA)*
    T. F. Johnson

Next-Generation Thermal Protection System Materials for Reusable Launch Vehicles.............. (NA)*
    S. Heng and B. E. Williams

[E11] LAUNCH VEHICLE PROPULSION

The Next Generation in Rocket Engines—The RD-180 1051
    R. N. Ford, W. E. Pipes, III, and J. F. Josef

The RS-68 LOX Hydrogen Engine for EELV .... (NA)*
    B. Beckman

(NA)* indicates that the paper was not available at time of press

Integrated Powerhead Demonstration Full Flow Cycle Development 1056
    J. M. Jones, J. T. Nichols, W. F. Sack, W. D. Boyce, and W. A. Hayes
50K Expander Cycle Engine Demonstration 1062
    A. M. Sutton, S. D. Peery, and A. B. Minick
Investigation of Next Generation Engines for Reusable Launch Vehicle 1066
    A. Konno, K. Kishimoto, and M. Atsumi

PART THREE

FIFTEENTH SYMPOSIUM ON SPACE NUCLEAR POWER AND PROPULSION

[F1] POWER BEAMING—I

Review of Direct-Drive Laser Space Propulsion Concepts 1073
    C. Phipps
Flight Experiments and Evolutionary Development of a Laser Propelled,
Trans-Atmospheric Vehicle .... (NA)*
    L. N. Myrabo, F. B. Mead, Jr., and D. G. Messitt
Nuclear Pumped Lasers for Space Power Beaming 1081
    M. Petra and G. H. Miley
Use of a Powered Photon Beam for Space Applications 1087
    H. Takahashi, Y. An, and Y. Yamazaki
The SABER Microwave-Powered Helicopter Project and Related WPT Research
at the University of Alaska Fairbanks 1092
    J. Hawkins, S. Houston, M. Hatfield, and W. Brown

[F2] ADVANCED RADIOISOTOPE POWER SOURCES

Advanced Converter Technology Evaluation and Selection for ARPS 1098
    J. F. Mondt, M. L. Underwood, and B. J. Nesmith
Small Universal Thermalpile Power Source for Space and Terrestrial Applications .... (NA)*
    J. Bass and N. Elsner
Recommended Design and Fabrication Sequence of AMTEC Test Assembly 1107
    A. Schock, V. Kumar, H. Noravian, and C. T. Or
The Development of a Milliwatt–Level Radioisotope Power Source 1119
    D. C. Bugby and T. R. McBirney
Microminiature Thermionic Converter Demonstration Testing and Results .... (NA)*
    F. J. Wyant

[F3] MISSIONS

High Risk Low Cost Mars Missions Scenarios 1125
    B. N. Cassenti and R. W. Bass
MITEE: A New Nuclear Engine Concept for Ultra Fast, Lightweight Solar System
Exploration Missions 1131
    J. Powell, J. Paniagua, H. Ludewig, G. Maise, and M. Todosow
Reducing the Risk to Mars: The Gas Core Nuclear Rocket 1138
    S. D. Howe, B. De Volder, L. Thode, and D. Zerkle
Interplanetary Missions with the GDM Propulsion System 1145
    T. Kammash and W. Emrich, Jr.

(NA)* indicates that the paper was not available at time of press

Fast-Transit Jovian Moon Probe Using HPS Reactor .... (NA)*
    M. Berte

**Reusable Launch Vehicles, Enabling Technology for the Development of Advanced
Upper Stages and Payloads** 1151
    J. D. Metzger

[F4] RADIOISOTOPE POWER SYSTEMS AND LESSONS FROM CASSINI

Lessons Learned from RTG Programs 1157
    R. M. Reinstrom and R. D. Cockfield

Production of ^{238}PuO$_2$ Heat Sources for the Cassini Mission 1163
    T. G. George and E. M. Foltyn

**Application of the Cobalt Based Superalloy Haynes Alloy 25 (L605)
in the Fabrication of Future Radioisotope Power Systems** 1167
    D. P. Kramer, J. R. McDougal, J. D. Ruhkamp, D. C. McNeil, F. A. Koehler,
    R. A. Booher, and E. I. Howell

The Cassini Project: Lessons Learned Through Operations 1173
    E. D. McCormick

[F5] AFFORDABLE FISSION POWER AND PROPULSION SYSTEMS

Launch Approval Considerations for Space Nuclear Power Systems 1179
    D. Skinner and J. M. Phillips

Championing Programs within the Air Force 1185
    F. G. Kennedy

Heatpipe Power System and Heatpipe Bimodal System Development Status 1189
    M. G. Houts, D. I. Poston, and W. J. Emrich, Jr.

Engineering Design Aspects of the Heatpipe Power System 1196
    B. Capell and M. Berte

Balance of Plant Options for the Heatpipe Bimodal System 1200
    M. Berte and B. Capell

The Human Mars Mission: Transportation Assessment 1206
    L. Kos

[F6] POWER BEAMING—II

Ground-Based Laser Propulsion for Orbital Debris Removal 1212
    J. W. Campbell and C. R. Taylor

Summary of Twenty-First Century Power Needs and Supply Options 1219
    D. R. Criswell

**Toward Space Solar Power: Wireless Energy Transmission Experiments Past,
Present, and Future** 1225
    F. E. Little, J. O. McSpadden, K. Chang, and N. Kaya

**The Potential for Space Solar Power Using Innovative Concepts
and New Technologies** .... (NA)*
    J. C. Mankins

The Invisible Extension Cord 1234
    S. Gunn

(NA)* indicates that the paper was not available at time of press

[F7] SAFETY AND LAUNCH APPROVAL

Response of the GPHS/RTG System to Potential Launch Accident Environments..................	1245
M. Mukunda	
Steady-State Temperature Predictions for General Purpose Heat Source in Vacuum	1257
C. T. Or and E. A. Skrabek	
Deconvolution of Variability and Uncertainty in the Cassini Safety Analysis	1263
F. J. Kampas and S. Loughin	
Lessons Learned from the Galileo and Ulysses Flight Safety Review Experience	1269
G. L. Bennett	
Cassini Nuclear Risk Analysis with SPARRC..	1275
C. T. Ha and N. A. Deane	

[F8] ADVANCED PROPULSION CONCEPTS

Nuclear Modules for Space Electric Propulsion..	1281
F. C. Difilippo	
Aneutronic Fusion Propulsion for Earth-to-Orbit and Beyond...................................	1289
H. D. Froning Jr. and R. W. Bussard	
International Space Station Technology Demonstrations	1295
A. C. Holt	
Requirements for a Common Nuclear Propulsion and Power Reactor for Human Exploration Missions to Mars..	1301
R. L. Cataldo and S. K. Borowski	

[F9] RADIOISOTOPE POWER SYSTEMS, SAFETY AND LAUNCH APPROVAL

Recycle of Scrap Plutonium-238 Oxide Fuel to Support Future Radioisotope Applications	1307
L. D. Schulte, G. L. Silver, G. M. Purdy, G. D. Jarvinen, K. Ramsey, J. Espinoza, and G. H. Rinehart	
Overview of Advanced Technologies for Stabilization of ^{238}Pu-Contaminated Waste................	1314
K. B. Ramsey, E. M. Foltyn, and J. M. Heslop	
Radioisotope Thermoelectric Generator/Thin Fragment Impact Test............................	1321
M. A. H. Reimus and J. E. Hinckley	
Light-Weight Radioisotope Heater Unit (LWRHU) Impact Tests	1329
M. A. H. Reimus, G. H. Rinehart, A. Herrera, B. Lopez, C. Lynch, and P. Moniz	

[F10] FUSION SYSTEMS—I

Physics Basis for the Gasdynamic Mirror (GDM) Fusion Rocket	1338
T. Kammash and W. Emrich, Jr.	
System/Subsystem Engineering Interface Considerations and R&D Requirements for IEF/QED Engine Systems	1344
R. W. Bussard and H. D. Froning, Jr.	
Magnetized Target Fusion for Energy and Propulsion	(NA)*
K. Schoenberg and R. Siemon	
An Overview of the Star Thrust Experiment	1352
K. Miller, J. Slough, and A. Hoffman	
Muon-Catalyzed Fusion for Space Propulsion, and a Compressed Target for Producing and Collecting Anti-Protons	1359
H. Takahashi and A. Yu	

(NA)* indicates that the paper was not available at time of press

[F11] FUSION SYSTEMS—II

Antiproton-Catalyzed Microfission/Fusion Propulsion Systems for Exploration
of the Outer Solar System and Beyond 1365
    G. Gaidos, J. Laiho, R. A. Lewis, G. A. Smith, B. Dundore, J. Fulmer, and S. Chakrabarti

Scaling of the Inertial Electrostatic Confinement (IEC) for Near-Term Thrusters
and Future Fusion Propulsion 1373
    G. Miley, B. Bromley, B. Jurczyk, R. Stubbers, J. DeMora, L. Chacon, and Y. Gu

Nuclear Device-Pushed Magnetic Sails (MagOrion) .... (NA)*
    D. Andrews and R. Zubrin

A Fusion-Driven Gas Core Nuclear Rocket 1377
    T. Kammash and T. Godfroy

[F12] ENERGY CONVERSION—I

Performance Tuned Radioisotope Thermophotovoltaic Space Power System 1385
    W. E. Horne, M. D. Morgan, and S. B. Saban

The Quantum Efficiency of InGaAsSb Thermophotovoltaic Diodes 1394
    R. U. Martinelli, D. Z. Garbuzov, H. Lee, N. Morris, T. Odubanjo, G. C. Taylor,
    and J. C. Connolly

A Novel Approach for the Improvement of Open Circuit Voltage and Fill Factor
of InGaAsSb/GaSb Thermophotovoltaic Cells 1400
    D. Z. Garbuzov, R. U. Martinelli, V. Khalfin, H. Lee, N. A. Morris, G. C. Taylor,
    J. C. Connolly, G. W. Charache, and D. M. DePoy

A Review of Recent Thermophotovoltaic Energy Conversion Technology Development
at NASA Lewis Research Center 1410
    D. M. Wilt and D. L. Chubb

Thermophotovoltaic Technology Development at Essential Research, Inc. 1417
    L. M. Garverick, N. S. Fatemi, and P. P. Jenkins

[F13] SPACE SAFETY AND RELIABILITY

The Energy Interaction Model: **A Promising New Methodology for Projecting
GPHS-RTG Cladding Failures, Release Amounts & Respirable Release Fractions
for Postulated Pre-Launch, Launch, and Post-Reentry Earth Impact Accidents** 1423
    J. R. Coleman, J. A. Sholtis, Jr., and W. H. McCulloch

Nondestructive Inspection of General Purpose Heat Source (GPHS) Fueled
Clad Girth Welds 1429
    M. A. H. Reimus, T. G. George, C. Lynch, M. Padilla, P. Moniz, A. Guerrero,
    M. W. Moyer, and A. Placr

Verified All-Regime Model for Reactivity Changes in Topaz-2 System .... (NA)*
    N. N. Ponomarev-Syepnoi, Y. A. Nechaev, K. A. Pavlov, B. S. Stepennov,
    and I. M. Khazanovich

Real-Time Monitoring Results During Transportation of the Cassini
Radioisotope Thermoelectric Generators (RTGs) Using the Radiotope
Thermoelectric Generator Transportation System (RTGTS) .... (NA)*
    B. Pugh, C. Barklay, and R. Miller

[F14] BREAKTHROUGH PHYSICS—I

Precision Tests of Einstein's Weak Equivalence Principle for Antimatter 1435
    R. A. Lewis, G. A. Smith, F. M. Huber, and E. W. Messerschmid

(NA)* indicates that the paper was not available at time of press

An Electromagnetic Basis for Inertia and Gravitation: What are the Implications for 21st Century Physics and Technology? 1443
    B. Haisch and A. Rueda
Space Coupling by Specially Conditioned Electromagnetic Fields 1449
    H. D. Froning, Jr. and T. W. Barrett
Propulsion Using the Electron Spiral Toroid 1455
    C. Seward

[F15] ENERGY CONVERSION—II

Parametric Analyses of Vapor-Anode, Multitube AMTEC Cells for Pluto/Express Mission 1461
    M. S. El-Genk and J.-M. Tournier
Experimental Uncertainties in Vacuum Tests of PX-Series AMTEC Cells 1471
    L. Huang and M. S. El-Genk
PX-5 AMTEC Cell Development 1479
    R. K. Sievers, J. R. Rasmussen, C. A. Borkowski, T. J. Hendricks, and J. E. Pantolin
Design and Fabrication of Multi-Cell AMTEC Power Systems for Space Applications 1486
    M. E. Carlson, J. C. Giglio, and R. K. Sievers
Development and Experimental Validation of a SINDA/FLUINT Thermal/Fluid/Electrical Model of a Multi-Tube AMTEC Cell 1491
    T. J. Hendricks, C. A. Borkowski, and C. Huang

[F16] BREAKTHROUGH PHYSICS—II

Interstellar Travel by Means of Wormhole Induction Propulsion (WHIP) 1502
    E. W. Davis
A Propulsion—Mass Tensor Coupling in Relativistic Rocket Motion 1509
    H. H. Brito
Conceptual Design of Space Drive Propulsion System 1516
    Y. Minami
Survey and Critical Review of Recent Innovative Energy Conversion Technologies 1527
    P. G. Bailey, T. Grotz, and J. J. Hurtak
Twisting Gravity: An Interstellar Propulsion System Utilizing a New Theory of Gravity 1535
    A. A. Larson

[F17] ENERGY CONVERSION—III

Performance Testing and Analysis Results of AMTEC Cells for Space Applications 1542
    C. A. Borkowski, A. Barkan, T. J. Hendricks, J. Rasmussen, and R. K. Sievers
Radiation Conduction Model for Multitube AMTEC Cells 1552
    J-M. Tournier and M. S. El-Genk
Cascaded Space Solar Power System with High Temperature Cs-Ba Thermionic Converter and AMTEC 1565
    A. Ya. Ender, I. N. Kolyshkin, V. I. Kuznetsov, E. V. Yakovlev, and D. V. Paramonov
Irreversible Thermodynamics of AMTEC Devices 1571
    A. M. Strauss and S. W. Peterson
AMTEC Performance and Evaluation Analysis Model (APEAM); Comparison with Test Results of PX-4C, PX-5A, and PX-3A Cells 1576
    J.-M. Tournier and M. S. El-Genk

[F18] ENERGY CONVERSION—IV

Optimization of Liquid-Return Artery in a Vapor-Anode, Multitube AMTEC 1586
    M. S. El-Genk and J-M. Tournier

(NA)* indicates that the paper was not available at time of press

Sodium Vapor Flow Regimes and Pressure Losses on Cathode Side of Multitube AMTEC Cell.. 1595
    J-M. Tournier and M. S. El-Genk

Lifetime Modeling of TiN Electrodes for AMTEC Cells... 1607
    M. A. Ryan, R. M. Williams, M. L. Homer, W. M. Phillips, L. Lara, and J. Miller

Vacuum Testing of High Efficiency Multi-Base Tube AMTEC Cells: February 1997—October 1997.. 1613
    J. M. Merrill, M. J. Schuller, and L. Huang

Hydro-Capillary Thermal to Electric Energy Conversion Device... 1621
    A. P. Sorokin, V. S. Egorov, V. G. Lisitsa, V. G. Maltsev, O. A. Olenchenko,
    A. G. Portianoy, E. N. Serdun, and M. M. Trevgoda

[F19] ENERGY CONVERSION—V

New Directions in Materials for Thermomagnetic Cooling... 1628
    A. Migliori, F. Freibert, T. W. Darling, J. L. Sarrao, S. A. Trugman, and E. Moshopoulou

Improved Materials for Thermoelectric Conversion (Generation)... 1634
    Z. Dashevsky, D. Rabinovich, I. Drabkin, and V. Korotaev

Design Peculiarities Prospects of the Use of Small-Size Low Power RTG for Investigating Planets and Small Bodies of the Solar System... 1641
    A. A. Pustovalov, V. V. Gusev, and M. I. Pankin

Development of High Efficiency Thermoelectric Generators Using Advanced Thermoelectric Materials.. 1647
    T. Caillat, J-P. Fleurial, and A. Borshchevsky

Simulation of a Thermoelectric Element Using B-Spline Collocation Methods..................... 1652
    S. W. Peterson and A. M. Strauss

Conference Program.. 1659
Author Index... A1

(NA)* indicates that the paper was not available at time of press

INTRODUCTION

The 1998 Space Technology and Applications International Forum (STAIF-98) is the largest of our annual meetings ever, in terms of the number of technical and plenary sessions in the program and the number of technical papers accepted for presentation at the six hosted conferences. Such growth confirms the success of the STAIF format in fostering timely dissemination and exchange of technical information among the attendees of the hosted conferences. These conferences cover a broad spectrum of space science and technology that spans the range from basic research, such as thermophysics in microgravity and breakthrough physics for propulsion, to the most recent advances in space power and propulsion, space exploration and commercialization, and next generation launch systems. STAIF provides an excellent opportunity for professional interaction among members of academia, industry, and government, program managers, and developers of space technologies in the US and abroad. STAIF continues to place strong emphases on international participation and collaboration. This is reflected in the membership of the various STAIF committees, the participation in the organization of hosted conferences, the invited speakers at the Plenary Session IV: *View from Abroad*, and by the many international papers accepted for presentation in the various technical sessions.

STAIF-98 is hosting six technical conferences that share the common interest in space exploration, technology, and commercialization and this year's theme: **PROGRESS IN EXPANDING THE SPACE FRONTIER**. These conferences are:

- **1st Conference on Global Virtual Presence**
   Chair: *Christine Anderson*, USAF Research Laboratory
   Co-Chair: *Peter Ulrich*, NASA Headquarters
- **1st Conference on Orbital Transfer Vehicles**
   Chair: *Michael Jacox*, USAF Research Laboratory
   Co-Chair: *Saroj Patel*, NASA Marshall Space Flight Center
- **2nd Conference on Applications of Thermophysics in Microgravity**
   Chair: *Rodney Herring*, Canadian Space Agency
   Co-Chair: *Liya Regel*, Clarkson University
- **3rd Conference on Commercial Development of Space**
   Chair: *Mark Nall*, NASA Marshall Space Flight Center
   Co-Chair: *Frederick R. Best,* Texas A & M University;
- **3rd Conference on Next Generation Launch Systems Technology**
   Chair: *Ken Verderame*, USAF Research Laboratory
   Co-Chair: *William Gaubatz*, The Boeing Company
- **15th Symposium on Space Nuclear Power and Propulsion**
   Chair: *Michael G. Houts*, Los Alamos National Laboratory
   Co-Chair: *William J. Emrich*, NASA Marshall Space Flight Center.

The STAIF-98 program features four plenary sessions, two on Monday morning and one each on Tuesday and Wednesday mornings, which are attended by all attendees. At these plenary sessions, prominent national and international speakers from governmental agencies, industry and academia are invited to address timely topics of importance that relate the STAIF-98 theme. The Monday morning Plenary sessions are *"Views from the Top"* and *"Views from the Frontier"*. The Tuesday morning Plenary Session is *"Views of the Future"* and the Wednesday Plenary Session is *"Views from Abroad"*. We are grateful to the speakers of the plenary sessions, the session Chairs and Co-Chairs, and the members of the Steering and Advisory Committees for their effort in developing the meeting's theme and organizing these sessions.

The outreach component of STAIF-98 on Monday morning includes a space design competition *"Manned Spacecraft for Mars Mission for the year 2011"*, for Secondary School students from throughout the State of New Mexico. In addition, a featured presentation will be given by Marc

Millis, of NASA Lewis Research Center on *"Interstellar Travel Challenge & Opportunities"*. More than 100 students, 30 teachers, and parents are expected to participate in this year's activities. The Outreach program of STAIF-98 is co-sponsored by NASA's New Mexico Space Grant Consortium, the American Nuclear Society Trinity Section, and the University of New Mexico's Institute for Space and Nuclear Power Studies (ISNPS). The student Space Design Competition winners and their supervising teachers are recognized and receive awards at the STAIF-98 Luncheon. We wish to express special thanks to the organizers of these events, Irene El-Genk, West Mesa High School in Albuquerque, NM; Steve Seiffert, Consultant, Albuquerque, NM; Maureen Alaburda of ISNPS, and the members of the ISNPS's Education Outreach Advisory Board. Our thanks and congratulations to the organizers of this successful outreach program.

The technical program of STAIF-98 offers a over 340 paper presentations, shared among the six hosted conferences, in 72 topical sessions and two special sessions: *"The Cassini Launch and Mission to Saturn"* and *"Nuclear Power and Propulsion for Exploration Missions"*. We are very grateful to the program chairs and co-chairs and the members of technical committees of the hosted conferences, members of the executive technical program committee, and the chairs and co-chairs of the technical sessions for all their hard work and dedication in organizing this year's technical program. We wish to express our appreciation and thanks to the authors for their contributions to the technical program and to the STAIF-98 Proceedings.

I am pleased to introduce the three-volume proceedings which contains the full manuscripts of papers presented at STAIF-98. We are grateful to the authors for their cooperation and commitment to contribute to a clear and technically sound publication. Special thanks are due to the ISNPS staff: Mary Bragg, STAIF Administrative Chair, and Maureen Alaburda and Carolyn Marcum, Administrative Co-Chairs, for their dedication, hard work, and commitment in preparing and compiling the entries to the program, editing and preparation of the final manuscripts, and for their effective communication with the session organizers, members of various committees, and the authors, and for the coordination of the many administrative functions of the meeting.

On behalf of the STAIF-98 Steering, Advisory, Technical Program and Executive Committees, I also wish to extend our thanks to the sponsoring organizations, national societies and participating organizations from government, industry, national laboratories, US Air Force, NASA Centers, universities, and international members for their input and contributions to this year's program. We also wish to acknowledge the contribution of the exhibitors for their timely and informative displays on the latest in space technology, which are an integral part of the annual meeting. Without the commitment, dedication, and contribution of all aforementioned individuals, members of various STAIF committees, contributing and participating organizations, exhibitors, and organizers of various events, this year's meeting would not have been possible.

My final and heartfelt thanks go to the families of the members of the ISNPS staff and to my family for their understanding, patience and continued encouragement and support through the demanding task of organizing this year's events. Last but not least, special thanks are due to the University of New Mexico, its School of Engineering, and the Department of Chemical and Nuclear Engineering for their continued support.

Mohamed S. El-Genk
Regents' Professor and Director, ISNPS
STAIF Technical and Publication Chairman

PUBLICATIONS

<u>Available from UNM's Institute for Space and Nuclear Power Studies</u> (Add $10 for shipping and handling within the U.S., $25 outside the U.S.)

Transactions of the 2nd - 5th Symposia (1985 - 1989)...$10.00 (each)
Transactions of the 6th Symposium (1989).. $15.00

<u>Available from the American Institute of Physics, c/o AIDC, P. O. Box 20, Williston, VT 05495, Phone 1-800-809-2247</u> (Add $3.75 for shipping and handling; $1.00 for each additional book.)

Proceedings of the 8th Symposium (1991) (3-vol. hardback set), ISBN # 0-88318-838-4 AIP Conference Proceedings #217 .... $175.00
Proceedings of the 9th Symposium (1992) (3-vol. hardback set), ISBN # 1-56396-027-3 AIP Conference Proceedings #246 .... $225.00
Proceedings of the 10th Symposium (1993) (3-vol. hardback set), ISBN # 156396-137-7 AIP Conference Proceedings #271 .... $275.00
Proceedings of the 11th Symposium (1994) (3-vol. hardback set), ISBN # 1-56396-305-1 AIP Conference Proceedings #301 .... $285.00
A Critical Review of Space Nuclear Power and Propulsion (1984-1993) (Anniversary Issue), ISBN # 1-56396-3175 .... $ 75.00
Proceedings of the 12th Symposium on Space Nuclear Power and Propulsion, Conference on Alternative Power from Space, and Conference on Accelerator-Driven Transmutation Technologies and Applications (1995) (2-vol. hardback set) ISBN # 1-56396-427-9 AIP Conference Proceedings # 324 .... $225.00
Proceedings of the 1st Conference on NASA Centers for Commercial Development of Space (1-vol. hardback book), ISBN # 1-56396-431-7 and AIP Conference Proceedings # 325 .... $125.00
Proceedings of the Space Technology and Applications International Forum (STAIF-96): 1st Conference on Commercial Development of Space; 1st Conference on Next Generation Launch Systems, 2nd Spacecraft Thermal Control Symposium, and 13th Symposium on Space Nuclear Power and Propulsion (1996) (3-vol. hardback set), ISBN # 1-56396-562-3 AIP Conference Proceedings # 361............... $275.00
Proceedings of the Space Technology and Applications Internatioal Forum (STAIF-97): 1st Conference on Future Science and Earth Science Missions; 1st Conference on Synergistic Power and Propulsion Systems Technology; 1st Conference on Applications of Thermophysics in Microgravity; 2nd Conference on Commercial Development of Space; 2nd Conference on Next Generation Launch Systems; 14th Symposium on Space Nuclear Power and Propulsion (1997) (3-vol. Hardback set), ISBN # 1-56396-679-4 AIP Conference Proceedings # 387 .... $295.00

<u>Publications available from Orbit Book Company, P. O. Box 9542, Melbourne, FL 32902-9542, Phone: (407) 724-9542</u>

Space Nuclear Power Systems (1984 - 1988) ....$125.00 (each) or $500.00 (set)
Space Nuclear Power Systems (1989) $149.50

ORGANIZING COMMITTEE

FORUM HONORARY CHAIR
Honorable Steven H. Schiff, United States House of Representatives (R, NM)

FORUM HONORARY CO-CHAIR
Honorable Pete V. Domenici, United States Senate (R, NM)

FORUM GENERAL CHAIR
Ricardo de Bastos, Orbital Sciences Corporation, Polomas, MD

FORUM TECHNICAL AND PUBLICATION CHAIR
Mohamed S. El-Genk, Institute for Space & Nuclear Power Studies (ISNPS), University of New Mexico

FORUM ADMINISTRATIVE CHAIR AND CO-CHAIRS

Mary J. Bragg, Chair	Maureen Alaburda, Co-Chair	Carolyn Marcum, Co-Chair
ISNPS, University of New Mexico	ISNPS, University of New Mexico	ISNPS, University of New Mexico

FORUM EDUCATION OUTREACH

Irene L. El-Genk	Steve Seiffert	Maureen Alaburda
West Mesa High School	Consultant	ISNPS
Albuquerque, NM	Albuquerque, NM	University of New Mexico

1ST CONFERENCE ON GLOBAL VIRTUAL PRESENCE
PROGRAM CHAIR: Christine M. Anderson, Air Force Research Laboratory
PROGRAM CO-CHAIR: Peter Ulrich, NASA Headquarters

1ST CONFERENCE ON ORBITAL TRANSFER VEHICLES
PROGRAM CHAIR: Mike Jacox, Air Force Research Laboratory
PROGRAM CO-CHAIR: Saroj Patel, NASA Marshall Space Flight Center

2ND CONFERENCE ON APPLICATIONS OF THERMOPHYSICS IN MICROGRAVITY
PROGRAM CHAIR: Rodney Herring, Canadian Space Agency
PROGRAM CO-CHAIR: Liya Regel, Clarkson University

3RD CONFERENCE ON COMMERCIAL DEVELOPMENT OF SPACE
PROGRAM CHAIR: Mark Nall, NASA Marshall Space Flight Center
PROGRAM CO-CHAIR: Frederick R. Best, Texas A&M University

3RD CONFERENCE ON NEXT GENERATION LAUNCH SYSTEMS
PROGRAM CHAIR: Kenneth Verderame, Air Force Research Laboratory
PROGRAM CO-CHAIR: William A. Gaubatz, The Boeing Company

15TH SYMPOSIUM ON SPACE NUCLEAR POWER AND PROPULSION
PROGRAM CHAIR: Michael G. Houts, Los Alamos National Laboratory
PROGRAM CO-CHAIR: William J. Emrich, NASA Marshall Space Flight Center

STEERING COMMITTEE

Honorable Steven H. Schiff, Chair
United States House of Representatives (R, NM)

Honorable Pete V. Domenici, Co-Chair
United States Senate (R, NM)

Christine M. Anderson
Director, Space Technology Directorate
Air Force Research Laboratory

Richard W. Davis
Commander, Wright Laboratory
Department of the Air Force

William C. Gordon
Vice President for Academic
  Affairs/Provost
University of New Mexico

Noel Hinners
Vice President-Flight Systems
Lockheed Martin Astronautics

Charles F. Marvin
Vice President & General Manager
McDonnell Douglas Aerospace, Huntsville

Shigeaki Nomura
Technical Special Advisory
NASDA

K. Lee Peddicord
Associate Vice Chancellor
Texas A&M University

Ian Pryke
Head of European Space Agency
European Space Agency, Washington

R. Rhoads Stephenson
Director, Technology and Applications
Jet Propulsion Laboratory

Earl Wahlquist
Deputy Associate Director
Office of Engineering & Tech. Dev.
U. S. Department of Energy

Raymond P. Whitten
Program Manager
NASA Headquarters

Joan Woodard
Vice President, Energy & Environment
Sandia National Laboratories

ADVISORY COMMITTEE

Mohamed S. El-Genk, Chair
University of New Mexico

Samit K. Bhattacharyya
Argonne National Laboratory
Dan Bland
SPACEHAB, Inc.
David Boyle
Texas A&M University
Joseph L. Cecchi
University of New Mexico
Beverly Cook
U. S. Department of Energy
Donald M. Ernst
Thermacore
William Gaubatz
The Boeing Company
Jean-Claude Gauthier
Commissariat a l'Energie Atomique
Robert Haslett
Northrop Grumman Corp.
Richard Hemler
Martin Marietta Astro Space
Mark D. Hoover
Lovelace Respiratory Research Inst.
Charles Hogge
Air Force Research Laboratory

Michael Houts
Los Alamos National Laboratory
James Kee
Booz-Allen & Hamilton, Inc.
Mary Kicza
NASA Goddard Space Flight Center
Michael Jacox
Air Force Research Laboratory
Thomas Lechtenberg
General Atomics
Mary Martinek
Congressman Schiff's Office
George H. Miley
University of Illinois
Robert Mucica
The Boeing Company
Nestor Ortiz
Sandia National Laboratories
Gary Payton
NASA Headquarters
Ian Pryke
European Space Agency
Harrison Schmitt
Consultant

Joseph A. Sholtis, Jr.
Private Engineering/Safety Consultant
L. Kevin Slimak
Air Force Research Laboratory
R. Joseph Sovie
NASA Lewis Research Center
Jess Sponable
Air Force Research Laboratory
Frank V. Thome
Team Specialty Products
Sadayuki Tsuchiya
NASDA
Giulio Varsi
Jet Propulsion Laboratory
Atsutaro Watanabe
NASDA
Kurt Westerman
Babcock & Wilcox Company
Raymond P. Whitten
NASA Headquarters
David M. Woodall
University of Idaho
Richard A. Zavadowski
Nuclear Fuel Services, Inc.

EXECUTIVE TECHNICAL PROGRAM COMMITTEE

Mohamed S. El-Genk, Chair
University of New Mexico

Christine M. Anderson
Air Force Research Laboratory
Frederick R. Best
Texas A&M University
Beverly Cook
U. S. Department of Energy
Mohamed S. El-Genk
University of New Mexico
William J. Emrich
NASA Marshall Space Flight Center
William A. Gaubatz
The Boeing Company

Rodney Herring
Canadian Space Agency
Michael G. Houts
Los Alamos National Laboratory
Mike Jacox
Air Force Research Laboratory
Mary Kicza
NASA Goddard Space Flight Center
Mark Nall
NASA Marshall Space Flight Center
Saroj Patel
NASA Marshall Space Flight Center

Ian Pryke
European Space Agency
Liya Regel
Clarkson University
R. Joseph Sovie
NASA Lewis Research Center
Peter Ulrich
NASA Headquarters
Giulio Varsi
Jet Propulsion Laboratory
Kenneth Verderame
Air Force Research Laboratory

1ST CONFERENCE ON GLOBAL VIRTUAL PRESENCE

Christine Anderson, Program Chair
Air Force Research Laboratory

Peter Ulrich, Program Co-Chair
NASA Headquarters

Steve Alejandro
Air Force Research Laboratory
Leon Alkalaj
Jet Propulsion Laboratory
Thomas Brackey
Hughes Space & Communication Co.
Rich Carreras
Air Force Research Laboratory
Ramon P. DePaula
NASA Headquarters

Don Edberg
McDonnell Douglas Aerospace
Janet Fender
Air Force Research Laboratory
Michael W. Fitzmaurice
NASA Goddard Space Flight Center
Robert J. Hayduk
NASA Headquarters
Gordon I. Johnston
NASA Headquarters
Jesse Leitner
Air Force Research Laboratory

Barry D. Meredith
NASA Langley Research Center
Mel Montemerlo
NASA Headquarters
Cliff Muller
Air Force Research Laboratory
Tim Murphy
Air Force Research Laboratory
John O'Hair
Air Force Research Laboratory

1ST CONFERENCE ON ORBITAL TRANSFER VEHICLES

Michael Jacox, Program Chair
Air Force Research Laboratory

Saroj Patel, Program Co-Chair
NASA Marshall Space Flight Center

George Borell
Harris Corporation
Dave Byers
TRW Space & Technology Division
Ed Cady
McDonnell Douglas Aerospace
Jim Calogeras
NASA Lewis Research Center
Leslie Curtis
NASA Marshall Space Flight Center

Alan Darby
The Boeing Company
Keith Dayton
The Boeing Company
Patrick Frye
The Boeing Company
Lara Medoff
SMC/XRT
Jay Polk
Jet Propulsion Laboratory

Robert L. Sackheim
TRW Space & Technology Division
Jim Shoji
The Boeing Company
Ron Spores
Air Force Research Laboratory
Kurt Westerman
Babcock & Wilcox Company

2ND CONFERENCE ON APPLICATIONS OF THERMOPHYSICS IN MICROGRAVITY

Rodney Herring, Program Chair
Canadian Space Agency

Liya Regel, Program Co-Chair
Clarkson University

Frederick R. Best
Texas A&M University
Ad Delil
National Aerospace Laboratory NLR
J. C. Duh
NASA Lewis Research Center
Cindy Edgerton
Swales Aerospace

Mohamed S. El-Genk
University of New Mexico
Olivier Lebaigue
French Atomic Energy Commission
Marianna Long
University of Alabama/Birmingham
Thomas R. Reinarts
United Technologies

Ted D. Swanson
NASA Goddard Space Flight Center
Eugene Trinh
Jet Propulsion Laboratory
Peter Wayner
Rensselaer Polytechnic Institute

3RD CONFERENCE ON COMMERCIAL DEVELOPMENT OF SPACE

Mark Nall, Program Chair
NASA Marshall Space Flight Center

Frederick R. Best, Co-Chair
Texas A&M University

Rose Allen
NASA Marshall Space Flight Center
Ray Bula
University of Wisconsin
Jim Calogeras
NASA Lewis Research Center
Vita Cevenini
NASA Headquarters
David Conrad
ERIM
Jim Fountain
The Boeing Company
Ed Gabris
NASA Headquarters

Alex Ignatiev
University of Houston
Ron Ignatius
Quantum Devices, Inc.
Terry Johnson
Kansas State University
Michael Kearney
SPACEHAB, Inc.
George Morgenthaler
University of Colorado
William Powell
NASA Marshall Space Flight Center

Muriel Ross
NASA Ames Research Center
David A. Rossi
SPACEHAB, Inc.
Frank Schowengerdt
Colorado School of Mines
Edward Sloot
Guigne Technologies Limited
Helen Stinson
NASA Marshall Space Flight Center
Joan Vernikos
NASA Headquarters
Raymond P. Whitten
NASA Headquarters

3RD CONFERENCE ON NEXT GENERATION LAUNCH SYSTEMS

Ken Verderame, Program Chair
Air Force Research Laboratory

William A. Gaubatz, Program Co-Chair
McDonnell Douglas Aerospace

Dana Andrews
The Boeing Company
John Anttonen
Air Force Research Laboratory
Gene Austin
NASA Marshall Space Flight Center
William Bozich
McDonnell Douglas Aerospace
Robert G. Brengle
The Boeing Company
Jackie O. Bunting
Lockheed Martin Astronautics

Chris Clay
USAF Wright Laboratory
Steve Creech
NASA Marshall Space Flight Center
Mel Eisman
RAND Corporation
B. David Goracke
The Boeing Company
Joe Hamaker
NASA Marshall Space Flight Center
Joe T. Howell
NASA Marshall Space Flight Center

Tom Ingersoll
Universal Space Lines
Jay Penn
Aerospace Corporation
Dave Perkins
Air Force Research Laboratory
Wayne Pritz
Air Force Research Laboratory
Rene Rey
Allied Signal Technical Services Corp.
John Skratt
ECON, Inc.

15TH SYMPOSIUM ON SPACE NUCLEAR POWER AND PROPULSION

Michael G. Houts, Program Chair
Los Alamos National Laboratory

William J. Emrich, Program Co-Chair
NASA Marshall Space Flight Center

C. Eugene Athon
Nuclear Fuel Services, Inc.
H. Sterling Bailey
Advanced Data Concepts, Inc.
Eric W. Baker
US Air Force
Chad Barklay
EG&G Mound Applied Technologies
Lester Begg
General Atomics
Gary L. Bennett
Metaspace Enterprises
Samit K. Bhattacharyya
Argonne National Laboratory
Stanley K. Borowski
NASA Lewis Research Center
Whitt Brantley
NASA Marshall Space Flight Center
Robert G. Brengle
The Boeing Company
Edward J. Britt
Space Power, Inc.
Thierry Caillat
Jet Propulsion Laboratory
Jonathan Campbell
NASA Marshall Space Flight Center
Mario Carelli
Westinghouse Electric
Ken Chidester
Los Alamos National Laboratory
John Connolly
NASA Johnson Space Center
Beverly Cook
U. S. Department of Energy
David Criswell
University of Houston
Tim A. Frazier
U. S. Department of Energy
Janell W. Hales
Westinghouse Hanford Company
Lisa Herrera
U. S. Department of Energy
Mark D. Hoover
Lovelace Respiratory Research Institute

W. E. Horne
EDTEK, Inc.
Steven D. Howe
Los Alamos National Laboratory
Maribeth E. Hunt
The Boeing Company
Thomas K. Hunt
Advanced Modular Power Systems
Oleg Izhvanov
Scientific Industrial Association "Lutch"
Allan T. Josloff
Martin Marietta Astro Space
Kent Joosten
NASA Johnson Space Center
Terry Kammash
University of Michigan
James Kee
Booz-Allen & Hamilton, Inc.
William C. Kincaide
Teledyne Energy Systems
Donald B. King
Sandia National Laboratories
Arvind Kumar
University of Missouri-Rolla
Roger X. Lenard
Sandia National Laboratories
John W. Lyver, IV
NASA Headquarters
Francois Linet
Commissariat a l'Energie Atomique
Albert C. Marshall
Defense Special Weapons Agency
Edward F. Mastal
U. S. Department of Energy
George May
NASA Stennis Space Center
Clay Mayberry
Air Force Research Laboratory
Tim McKelvey
Field Command, DNA
Frank Mead, Jr.
Air Force Research Laboratory
John Merrill
Nichols Research Corporation

John D. Metzger
Northrop Grumman Corp.
Jim Meyers
McDonnell Douglas Aerospace
George H. Miley
University of Illinois
Roger Miller
EG&G Mound Applied Technologies
Marc G. Millis
NASA Lewis Research Center
Jack F. Mondt
Jet Propulsion Laboratory
Dan Mulder
Sandia National Laboratories
Ray Nelson
Field Command, DNA
Yuri Nikolaev
Scientific Industrial Assoc., "Lutch"
Dmitry V. Paramonov
Westinghouse Electric
Randy C. Parsley
Pratt & Whitney
David I. Poston
Los Alamos National Laboratory
Eric Proust
Commissariat a l'Energie Atomique
Lyle L. Rutger
U. S. Department of Energy
Amy Ryan
Jet Propulsion Laboratory
Michael J. Schuller
Texas A&M University
Gene E. Schwarze
NASA Lewis Research Center
Stephen L. Seiffert
Consultant
Gerald A. Smith
Penn State University
Jean-Michel Tournier
University of New Mexico
Ben Wernsman
Westinghouse Electric
Robert M. Zubrin
Pioneer Astronautics

CONTRIBUTING ORGANIZATIONS

Air Force Research Laboratory
NASA Headquarters and Field Centers
Sandia National Laboratories
U. S. Department of Energy

SCHREIBER-SPENCE AWARD

<u>1997-1998 AWARD COMMITTEE</u>: **Jack Stocky** (Chair), Jet Propulsion Laboratory; **Paul Regeon**, White House, Office of Science & Technology Policy; **Jess Sponable**, Air Force Research Laboratory; and **Peter Ulrich**, NASA Headquarters

MANUEL LUJAN, JR. STUDENT PAPER AWARD

<u>AWARD COMMITTEE</u>: **Michael G. Houts** (Chair), Los Alamos National Laboratory; **Ray Bula**, University of Wisconsin; **Thomas R. Reinarts**, United Technologies, Inc.; **Fred Shair**, Jet Propulsion Laboratory; and **Dale Shell**, Air Force Research Laboratory

OUTSTANDING PAPER AWARD

<u>AWARD COMMITTEE</u>: **John D. Metzger** (Chair), Northrop Grumman Corp.; **John Anttonen**, Air Force Research Laboratory; **David Conrad,** Environmental Research Institute of Michigan; **Gene Ungar**, NASA Johnson Space Center; **Giulio Varsi**, Jet Propulsion Laboratory; and **Bob Wiley**, Booz-Allen & Hamilton.

PLENARY SESSIONS

BREAKTHROUGH PROPULSION PHYSICS WORKSHOP PRELIMINARY RESULTS

Marc G. Millis
NASA Lewis Research Center
21000 Brookpark Rd., MS 60-4, Cleveland, OH 44135
(216) 977-7535, Fax (216) 977-7545
marc.g.millis@lerc.nasa.gov

Abstract

In August, 1997, a NASA workshop was held to assess the prospects emerging from physics that might lead to creating the ultimate breakthroughs in space transportation: propulsion that requires no propellant mass, attaining the maximum transit speeds physically possible, and breakthrough methods of energy production to power such devices. Because these propulsion goals are presumably far from fruition, a special emphasis was to identify affordable, near-term, and credible research that could make measurable progress toward these propulsion goals. Experiments and theories were discussed regarding the coupling of gravity and electromagnetism, vacuum fluctuation energy, warp drives and wormholes, and superluminal quantum tunneling. Preliminary results of this workshop are presented, along with the status of the Breakthrough Propulsion Physics program that conducted this workshop.

INTRODUCTION

The objective of the NASA Breakthrough Propulsion Physics Program is to make measurable and credible progress toward the seemingly long range goal of creating propulsion breakthroughs. One of the first major milestones of the program was to convene a workshop with established physicists, government researchers and select innovators to jointly examine new theories and phenomena from scientific literature that have reawakened consideration that such breakthroughs may be achievable. Preliminary results of the workshop are presented along with the status of the program. This program, managed by Lewis Research Center, is funded out of the "Advanced Space Transportation Plan" (ASTP) managed by Marshall Space Flight Center (MSFC).

WORKSHOP GOALS

The purpose of the workshop was to examine emerging physics in the context of seeking *propulsion* breakthroughs. It is desired to channel the continuing advancements in science toward answering the fundamental questions of how to propel a spacecraft farther, faster, and more efficiently. Specifically, these goals are:

(1) MASS: Discover new propulsion methods that eliminate or dramatically reduce the need for propellant. This implies discovering fundamentally new ways to create motion, presumably by manipulating inertia, gravity, or by any other interactions between matter, fields, and spacetime.
(2) SPEED: Discover how to attain the ultimate achievable transit speeds to dramatically reduce travel times. This implies discovering a means to move a vehicle at or near the actual maximum speed limit for motion through space or through the motion of spacetime itself.
(3) ENERGY: Discover fundamentally new modes of onboard energy generation to power these propulsion devices. This third goal is included since the first two breakthroughs could require breakthroughs in energy generation, and since the physics underlying the propulsion goals is closely linked to energy physics.

To make near-term and measurable progress toward these ambitions, the workshop sought to produce a list of next-step, incremental research approaches. Specifically, this means identifying research tasks that are of relatively short duration and that address the immediate questions raised by the emerging physics and program goals.

WORKSHOP METHODS

Three major elements were used at the workshop; a plenary sequence of invited presentations to review emerging physics, a poster paper segment to provide thought-provoking ideas, and breakout sessions to produce a list of candidate next-step research tasks. The first day of the workshop and the morning of the second day featured the invited presentations. This included having opening remarks from Congressman Dennis Kucinich. The breakout sessions followed the invited presentations and were completed by noon of the third day. Summaries of the breakout groups were presented in a plenary session on the afternoon of the third day. The posters were on display throughout the entire workshop.

To keep the number of participants to a manageable number and to provide a constructive mix of physicists, government researchers, and thought-provoking innovators, an invitation-only format was used. Attendance was limited to about 90 participants due to the breakout sessions, where a maximum of 15 participants per each of the six groups was desired. For the overviews of emerging physics, established physicist were invited, including some with constructively pessimistic viewpoints. To foster collaboration between NASA and other government labs, several government researchers were invited. To provoke thought and discussion, poster papers were invited from individuals who had previously submitted materials to NASA on this subject. In total, 84 participants attended the workshop, including 16 from universities; 28 from industry; 11 from government labs including Los Alamos, Oak Ridge, Fermi, Brookhaven and the Air Force Research Labs at Edwards and Kirtland; 17 from NASA including from Lewis, Langley, Marshall, Johnson, and the Jet Propulsion Laboratory; and 12 students.

Since this workshop dealt with seeking breakthroughs in science, it asked participants to be *visionary*. Admittedly, these breakthroughs may turn out to be impossible, but progress is not made by conceding defeat. For the sake of promoting progress, participants were asked to entertain, for the duration of the workshop, the notion that these breakthroughs are indeed achievable. Simultaneously, however, this workshop looked for sound and tangible research approaches. Therefore, participants were also asked to be *credible* -- credible progress toward incredible possibilities.

To provide a list of next-step research tasks, the participants were divided into six breakout groups. Each of the three goals mentioned on page 1 were addressed by two of the six groups. Each group was led by a facilitator through a process designed to elicit a large number of creative ideas and then to evolve these ideas into candidate next-step research tasks. This process consisted of the following sequence:

(1) VISION: Participants were asked to assume a priori that the physics breakthroughs needed to create practical interstellar travel were achievable. They were asked to imagine that they were far enough into the future where these breakthroughs have been realized, and were asked to suggest ways that such feats were accomplished.
(2) ISSUES: Participants were asked to identify the critical unknowns and make-or-break issues associated with the ideas from Step-1, and to identify any curious effects (confirmed or unconfirmed) that may support the goals.
(3) NEXT STEPS: Participants were asked to propose the next-step experiments, theoretical analyses, or further theoretical developments that would be needed to resolve the issues identified in Step-2. This included transforming *objections* into research *objectives*.
(4) EVALUATION: The ideas from Step-3 were scored using a spreadsheet containing the evaluation criteria shown below. This scoring was more of an experiment of the evaluation process than it was a genuine attempt at scoring the task ideas. As expected, none of the groups were able to score all of the task ideas they generated. However, each group selected a few representative tasks for presentation in the closing plenary session.

- Relevance To Program:
    - Directness to Program (Must seek propulsion relevant advances in physics)
    - Magnitude of Potential Gains for Goal #1 (Mass) + Goal #2 (Speed) + Goal #3 (Energy)
- Readiness:
    - Level of Progress Achieved To Date (using scientific method levels as status metric)

- Testability (ease of empirical testing)
  (Note: experiments are considered closer than theory to becoming technology)
- Credibility: (As reflected by peer reviewed scientific literature)
- Minimum Credibility Criteria for Approaches Not in Peer Reviewed Literature *
  - Fits credible data
  - Advantageous to propulsion or power goals
  - Discriminating test suggested
- Research Task Factors:
  - Level of Progress to be Achieved Upon Completion of Task (using scientific method levels as metric)
  - Breadth of Work (experiment, theory, and/or comparative study)
  - Triage (will it be done anyway or must this program support it?)
  - Lineage (will it lead to further relevant advancements?)
  - Time Required to Complete Task (reciprocal scoring factor)
  - Funding Required (reciprocal scoring factor)
  - Probability of Successful Task Completion (credentials and realism of proposal) *

* Due to time limitations during the workshop, these criteria were not used.

INVITED PRESENTATIONS

The invited presentations, from established physicists, covered many of the relevant areas of emerging physics. The intent of these presentations was to provide credible overviews of where we stand today in physics and introduce the unknowns and unresolved issues. Below is a short synopsis of these presentations in the order that they were presented. Where a related or equivalent work is available in the open literature, a reference is cited.

(1) Lawrence Krauss (Case Western Reserve University, Cleveland OH), *"Propellantless Propulsion: The Most Inefficient Way to Fly?"*: The physics behind manipulating spacetime for propellantless propulsion is reviewed. Even considering the possibilities of new physics, it is shown that such achievements would be probably impossible in practice. (Krauss 1995 and Pfenning 1997)

(2) Harold Puthoff (Institute for Advanced Studies at Austin, TX), *"Can the Vacuum be Engineered for Spaceflight Applications?: Overview of Theory and Experiments"*: Discusses the phenomena of electromagnetic vacuum fluctuations, its empirical evidence, existing applications, and implications for breakthrough space propulsion and power. (Puthoff 1989, Cole 1993, and Haisch 1994)

(3) Raymond Chiao (University of California at Berkeley, CA), and A. Y. Steinberg, *"Quantum Optical Studies of Tunneling Times and Superluminality"*: Presents the experimental methods and results of measuring the tunneling time of a photon to cross a photonic band-gap tunnel barrier, where an effective tunneling speed of 1.7 times the speed of light is measured. The author concludes, however, that *information* did not travel faster than light. (Chiao 1994)

(4) John G. Cramer (University of Washington, Seattle, WA), *"Quantum Nonlocality and Possible Superluminal Effects"*: Discusses the quantum mechanics of *nonlocality*, which is often used to speculate about superluminal effects. Examples and unknowns are highlighted, including the transactional interpretation. (Cramer 86)

(5) Ronald J. Koczor, and David Noever (NASA MSFC, Huntsville, AL), *"Experiments on the Possible Interaction of Rotating Type II YBCO Ceramic Superconductors and the Local Gravity Field"*: Presents the status and interim results of experiments being conducted at MSFC to investigate claims of gravity effects in the vicinity of rotating superconductors in strong magnetic fields (Podkletnov 1992). Only static measurements have been completed to date, with inconclusive results (change of less than 2 parts in 10^8 of the normal gravitational acceleration) (Li 1997). Work continues toward measurements with rotating superconductors.

(6) Robert Forward (Forward Unlimited, Clinton, WA), *"Apparent Endless Extraction of Energy from the Vacuum by Cyclic Manipulation of Casimir Cavity Dimensions"*: Proposes a conceptual energy extraction method using cyclic dimensional changes of irregular Casimir cavities. This concept uses data on the energy densities of electromagnetic quantum fluctuations within irregular Casimir cavities, where portions

of the data plots are double-valued (Ambjørn 1983 and Forward 1984).

(7) Bernhard Haisch (Lockheed Palo Alto CA) and A. Rueda, *"The Zero-Point Field and the NASA Challenge to Create the Space Drive"*: Proposes that the Newtonian F=ma equation can be derived from Maxwell's equations as applied to the electromagnetic vacuum fluctuations (Haisch 1994). The effective momentum of the vacuum fluctuations is speculated to be a possible basis of propulsion. The author also notes that "negative mass" is incompatible with this theory.

(8) Alfonso Rueda (California State University, Long Beach, CA) and B. Haisch, *"Inertial Mass as Reaction of the Vacuum to Accelerated Motion"*: Presents further investigations of the theories linking inertia to vacuum fluctuations. (Haisch 1994).

(9) Daniel C. Cole (IBM Microelectronics, Essex Junction, VT), *"Calculations on Electromagnetic Zero-Point Contributions to Mass and Perspectives"*: Challenges the theories linking inertia to vacuum fluctuations, but supports the idea of vacuum energy exchange (Cole 1993).

(10) Peter W. Milonni (Los Alamos National Labs, Los Alamos NM), *"Casimir Effect: Evidence and Implications"*: Reviews the empirical evidence used to support vacuum fluctuation theories, showing that source fields and macroscopic manifestations of intermolecular forces can also explain observed phenomena. (Milonni 1994).

(11) Hüseyin Yilmaz (Electro-Optics Technology Center, Winchester, MA), *"The New Theory of Gravitation and the Fifth Test"*: Compares the Einstein and Yilmaz theories with data of the gravitational perturbations on Mercury by the other planets. (Yilmaz 92).

(12) Arkady Kheyfets (Dept. of Mathematics, N. Carolina State Univ., Raleigh, NC) and Warner A. Miller, *"Hyper-Fast Interstellar Travel via a Modification of Spacetime Geometry"*: Addresses the key features and obstacles confronting the Alcubierre warp bubble and the Krasnikov tube for hyper-fast travel. Issues include the casual structure, weak and null-energy conditions as well as the violation of chronology protection. It is suggested that quantum gravity may be a more productive avenue to further investigate these possibilities and issues. (Alcubierre 94, Krasnikov 95, and Pfenning 97)

(13) Frank J. Tipler, III (Tulane University, New Orleans, LA), *"Ultrarelativistic Rockets and the Ultimate Future of the Universe"*: Presents a case that no new scientific discoveries are required to traverse the galaxy, provided that antiproton annihilation rockets are used and virtual (computerized) humans do the traveling.

(14) George Miley (University of Illinois, Urbana IL), *"Possible Evidence of Anomalous Energy Effects in H/D-Loaded Solids -- Low Energy Nuclear Reactions"*: Presents empirical evidence of excess energy, radiation emission, and transmutations of elements from experiments involving lattices loaded with deuterium by various methods.

POSTER PAPERS

To provide imaginative material to help provoke discussion and research ideas, poster papers were solicited for the workshop. Selected posters were on display throughout the workshop, and are listed below, alphabetically by author. In those cases where a related or equivalent work is available in the open literature, a reference is cited. Some posters were based on peer reviewed publications while others were more adventurous and less rigorous. In pioneering work it can be difficult to distinguish between the 'crazy' ideas that will one day evolve into breakthroughs, and the more numerous, genuinely crazy ideas. Even though many ideas proposed for this subject are likely to be incorrect, they can still be useful by provoking other, more viable, ideas. It was in this spirit that ideas beyond the conventional were invited for poster papers.

- D. Alexander (MSE Technology Applications, Inc.), *"Replication of an Experiment Which Produced Anomalous Excess Energy."*
- C. Asaro (Molecudyne Research), *"Special Relativity with Complex Speeds."* (Asaro 1996).
- J. D. Baxter (Student, ITT Technical Institute, Salt Lake City), *"A Plan For Exceeding The Light Barrier."*
- E. W. Davis (National Institute for Discovery Science), *"Wormhole Induction Propulsion (WHIP)."*
- K. J. Davis (Grad Student, Rutgers University), *"Study of M-Theoretic Alcubierre Type Warp Drives."*
- S. Dinowitz (Underwriters Laboratories, Inc.), *"Michelson-Morley on the Space Shuttle: A Possible Experiment to Test a Field Distortion Theory."* (Dinowitz 1996).

- G. F. Erickson (Los Alamos National Lab), *"QED Casimir Force Electrical Power Supply."*
- R. Forward (Forward Unlimited), *"Observational Search for Negative Matter in Intergalactic Voids."* (da Costa 1996 and Bondi 1957).
- H. D. Froning Jr. (Flight Unlimited), *"Experiments to Explore Space Coupling by Means of Specially Conditioned Electromagnetic Fields."* (Froning 1997).
- U. Gat and P. Carpenter (Oak Ridge National Lab), *"Nuclear Isomer Decay: A Possibility for Breakthrough Space Propulsion."*
- J. G. Hartley (IBM Microelectronics), *"Possible Experimental Test of Wheeler-Feynman Absorber Theory."* (Heron 1974 and Cramer 1986).
- N. W. Kantor (Integram), R. D. Eagleton and M. N. Kaplan, *"Determination of Existence of the Vacuum Structure."* (Eagleton 1983).
- M. N. Kaplan (Motorotor), R. D. Eagleton and N. W. Kantor, *"Force Field Propulsion."*
- G. Landis (Ohio Aerospace Institute), *"An Alcubierre Drive Using Cosmic String."*
- J. Maclay (Microfabrications Applications Lab, Univ. of Illinois), M. Serry, B. R. Ilic, P. Neuzil, and D. Czaplewski, *"Use of AFM (Automatic Force Microscope) Methods to Measure Variations in Vacuum Energy Density and Vacuum Forces in Microfabricated Structures."* (Serry 1995).
- G. L. Matloff (New York Univ.), *"The Zero-Point Energy (ZPE) Laser and Interstellar Travel."*
- M. G. Millis (NASA Lewis), *"The Challenge to Create the Space Drive."* (Millis 1997b).
- J. M. Niedra, *"Vacuum Fluctuations, Connectivity and Superluminal Physics for Interstellar Travel."*
- H. Ringermacher (GE Corp. Research and Development Ctr.), B. Cassenti and D. J. Leopold, *"Search for Effects of an Electrostatic Field on Clocks in the Frame of Reference of Charged Particles."* (Ringermacher 1994).
- J. J. Roser, *"Laboratory Scale Vacuum Energy Extraction Modeled on Weak Nuclear Force Reactions in a Spinning Black Hole System."*
- F. Rounds, *"Anomalous Weight Behavior in $YBa_2Cu_3O_7$ Compounds at Low Temperature."*
- D. K. Sen (Grad Student, Dept. of Phys., Univ. of Washington), *"Recent Results Concerning the Properties and Structure of the Electromagnetic Vacuum."*
- C. Seward (Electron Power Systems, Inc.), *"Propulsion and Energy Generation Using the Electron Spiral Toroid."* (Seward 1996).
- G. Sobczak (Grad Student, Astronomy Dept., Harvard Univ.), *"The Modified Casimir Force in a Uniformly Accelerating Reference Frame and in a Gravitational Field."*
- J. V. Vargas and D. G. Torr (Univ. of South Carolina), *"Lurking Breakthrough Physics."* (Vargas 1991).
- E. L. Wall (Institute for Basic Research), *"A First Tangible Step in the Quest for Hyperluminal Space Travel."*
- C. K. Whitney (Tufts Univ., Electro-Optics Technology Ctr.), *"Challenging the Speed of Light."*
- J. F. Woodward (Dept. of Physics, California State Univ.), *"Mach's Principle and Impulse Engines: Toward a Viable Physics of Star Trek?"* (Woodward 1992 and 1994).
- C. A. Yost (Electric Spacecraft Journal), *"Electric Field Propulsion Concepts from Independent Researchers."*
- E. Zampino (NASA Lewis), *"Can a Hyperspace Really Exist?"*

CANDIDATE NEXT-STEP RESEARCH APPROACHES

Based on the invited presentations, poster papers, and the ideas generated during the breakout sessions, several next-step research approaches were identified and are presented next. These are arranged according to the three program goals and highlight the intriguing phenomena and theories, critical issues, and candidate next-step approaches for each program goal. Regarding the specific task ideas generated during the breakout sessions, about 80 task ideas were collected. These have not yet been fully reviewed, but many are integrated into the discussions below.

Toward Goal 1 - Eliminating Propellant Mass

It is known that gravity, electromagnetism and spacetime are coupled phenomena. Evidence includes the bending of light, the red-shifting of light, and the slowing of time in a gravitational field as illustrated in Figure 1. This coupling is most prominently described by general relativity (Misner 1973). Given this coupling and our technological proficiency for electromagnetics, it has been speculated that it may become possible to use electromagnetic technology to manipulate inertia, gravity, or spacetime to induce propulsive forces (Millis 1997b). Another phenomena of interest is the Casimir Effect, Figure 2, where closely spaced plates are forced together, presumably by vacuum fluctuations (Lamoreaux 1997). One explanation is that this force is the net radiation pressure of the virtual vacuum fluctuation photons, where the pressure is greater outside the plates than within, since wavelengths larger than the plate separation are excluded. The force is inversely proportional to the 4th power of the distance. Even though this effect can be explained by various theories (Milonni 1994), the idea that the vacuum might create these forces leads to speculations that an *asymmetric* vacuum effect, if possible, could lead to a propulsive effect (Millis 1997b). There are many unsolved issues regarding these speculations, including whether these phenomena can lead to controllable net-force effects and whether such effects can be created, even in principle, without violating conservation of momentum and energy (Millis 1997b).

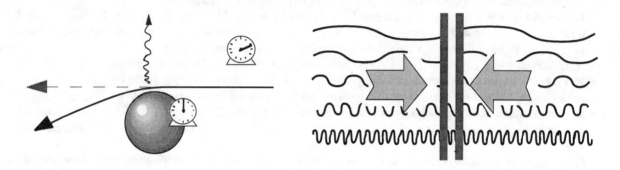

FIGURE 1. Coupling of Gravity, Spacetime, and Electromagnetism.

FIGURE 2. The Casimir Effect.

Although it is presently unknown if such propellantless propulsion can be achieved, several theories have emerged that provide additional research paths. It should be noted that all of these theories are too new to have either been confirmed or discounted, but their potential utility warrants consideration. This includes negative mass propulsion (Bondi 1957), theories that suggest that inertia and gravity are affected by vacuum fluctuations (Puthoff 1989 and Haisch 1994), and numerous other theories about the coupling between matter, electromagnetism, and spacetime (Dinowitz 1996, Froning 1997, Ringermacher 1994, Vargas 1991, Woodward, 1992, and Yilmaz 1992). Another recent development, which has yet to be credibly confirmed or discounted, is where anomalous weight changes are observed over spinning superconductors (Podkletnov 1992).

During the workshop these possibilities were discussed with an emphasis on experimental verification. A poster by Forward suggested a search for evidence of negative mass based on recent astronomical data (da Costa 1996). The posters of Dinowitz, Froning, Ringermacher, and Woodward all offered experiments to test their theories. Several experiments were suggested to test the theories linking inertia to vacuum fluctuations, including experiments described in existing literature (Forward 1996). And interest was expressed in continuing the experiments to test the claims of weight changes over spinning superconductors (Li 1997).

Toward Goal 2 - Achieving the Ultimate Transit Speed

Special relativity states that the speed of light is an upper limit for the motion of matter through spacetime. Recently, however, theories using the formalism of general relativity have suggested that this limit can be *circumvented* by altering *spacetime* itself. This includes "wormhole" and "warp drive" theories. A wormhole is a shortcut created through spacetime (Morris 1988 and Visser 1995), as illustrated in Figure 3, where a region of spacetime is warped to create a shorter path between two points. A warp drive involves the expansion and contraction of spacetime to propel a region of spacetime faster than light (Alcubierre 1994). Figure 4 illustrates the Alcubierre warp drive, showing the opposing regions of expanding and contracting spacetime that propel the center region.

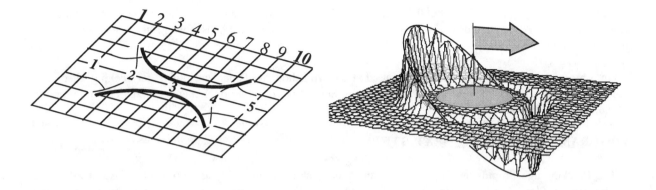

FIGURE 3. Wormholes - Spacetime Shortcuts.   FIGURE 4. The Alcubierre Warp Drive.

It has also been suggested that the light speed limit could be exceeded if velocities could take on imaginary values (Asaro 1996). In addition, there are theories for "nonlocality" from quantum physics that suggest potentially superluminal effects (Cramer 1986). These theories not only present challenging physics problems, but are intriguing from the point of view of future space travel. Do these theories represent genuinely possible physical effects, or are they merely mathematical curiosities?

In addition to theories, there are some intriguing experimental effects. Photons have been measured to tunnel across a photonic band-gap barrier at 1.7 times the speed of light (Chiao 1994). Even though the author concludes that *information* did not travel faster than light, the results are intriguing. During the workshop several suggestions were made to conduct similar experiments using matter rather than photons to unambiguously test the *information* transfer rate. In addition, recent experiments of the rest mass of the electron antineutrino have measured an imaginary value (Stoeffl 1995). Even though this result is attributed to possible errors, an imaginary mass value is a signature characteristic of a tachyon. Tachyons are hypothetical faster-than-light particles. In the workshop it was suggested to revisit this and other similar data to determine if this can be credibly interpreted as evidence of tachyons. It was also pointed out that other experiments have been suggested to search for evidence of tachyons (Chiao 1996).

The notion of faster-than-light travel evokes many critical issues. These were summarized in the presentation by Kheyfets. Issues include causality violations, the requirement for negative energy, and the requirement for enormous energy densities to create the superluminal effects. Suggestions were made during the workshop for a number of theoretical approaches to address these issues, including the use of quantum gravity to study the wormhole and warp drive concepts.

Toward Goal 3 - Discovering New Modes of Energy Production

Since the first two breakthroughs could require breakthroughs in energy generation, and since the physics underlying the propulsion goals is closely linked to energy physics, it is also of interest to discover fundamentally new modes of energy generation. The principle phenomena of interest for this category is, again, the vacuum fluctuations. It has been theorized that this energy can be extracted without violating conservation of energy or any thermodynamic laws (Forward 1984, Cole 1993). It is still unknown if this vacuum energy exists as predicted, how much energy might be available to extract, and what the secondary consequences would be of extracting vacuum energy.

At the workshop, the techniques described in the posters of Maclay (Serry 1995), Erickson, and Sen were suggested to investigate the energy extraction concepts described in the poster of Erickson and by the presentation by Forward. These techniques involve the use of micromechanical structures. Not only are micromechanical structures an emerging technology, but the dimensions of such structures are similar to the dimensions required for Casimir effects. Also, should any viable device be engineered, these methods hold promise for high-volume manufacturing.

On a more conventional vein, ideas were raised at the workshop by Tipler and others for seeking alternative methods of antimatter production. Also, the poster by Seward presented a novel energy storage device involving toroidal plasmas (Seward 1996).

PROGRAM STATUS AND NEXT STEPS

The Breakthrough Propulsion Physics Program was established in 1996 as part of the "Advanced Space Transportation Plan" (ASTP) managed by MSFC (Millis 1997a). A government Steering Group containing volunteers from various NASA centers, DoD and DoE laboratories and led by the Lewis Research Center, has been established to guide this program. This includes the development of the research solicitation and selection criteria used during the workshop, and which are being refined for future evaluations of research proposals.

The first major milestone of the program was this kick-off workshop. A publicly available Conference Proceedings is being assembled to fully document the results. The list of task ideas generated during the workshop will be used to advocate for funding to support this research.

Once funded, this program plans to use an annual "NASA Research Announcement" (NRA) to solicit and support research tasks. This solicitation will be open to academia, industry, government labs, and NASA centers. Selection will be via a peer review process led by the Steering Group and using the prioritization criteria to provide an initial ranking. Because it is too early to focus on a given approach, it is anticipated that multiple, different approaches will be supported from the top ranking candidates. Proposed tasks should be of relatively short duration (1-3yrs), modest cost ($50 to $150K), and traceable to at least one of the three program goals.

The next step of the program, with or without funding, is to reopen participation to the broader set of government, university and industry researchers. An internet site has been established as a first step to begin this collaboration (http://www.lerc.nasa.gov/WWW/bpp/). Also, a limited access site is envisioned to contain works in progress, more in-depth annotated bibliographies, and allow on-line discussions. Access will be limited to a Contributor Network selected by the Steering Group. The process for nominating and selecting Contributor Network members has not yet been established.

CONCLUSIONS

New theories and laboratory-scale effects have emerged in the scientific literature which provide new approaches to seeking major propulsion breakthroughs. During the recent workshop, many of these new approaches were reviewed, and about 80 specific research task ideas were generated for making progress toward propulsion breakthroughs. A peer review system has been drafted that can be used to rank these and other future proposals.

Acknowledgments

Special thanks is owed to the Lewis Research Center *volunteers* who helped make this workshop a success: Obasi Akan, Sheila Bailey, Michael Binder, David Chato, Dane Elliott-Lewis, Cynthia Forman, James Giomini, Jon Goldsby, Scott Graham, Al Juhasz, Geoffrey Landis, Grace Scales, Gary Scott Williamson, Natalie Woods, Ed Zampino, and especially to Joe Hemminger for arranging the equipment and volunteers for the breakout sessions. Special thanks is also owed to the *volunteer* members of the BPP Steering Group, especially to Frank Mead. In addition, thanks is owed to the Honorable Dennis Kucinich for his opening remarks, and to the NYMA Inc. staff for smoothly handling the workshop logistics; Dr. Richard Ziegfeld, Linda Oliver, and John Toma.

References

Ambjørn, J. and Wolfram, S. (1983) "Properties of the Vacuum, 1. Mechanical and Thermodynamic, and Properties of the Vacuum, 2. Electrodynamic," *Annals of Physics*, 147:1-56.

Alcubierre, M. (1994) "The Warp Drive: Hyper-fast Travel Within General Relativity," *Classical and Quantum Gravity*, 11:L73-L77.

Asaro, C. (1996) "Complex Speeds and Special Relativity," *Am. J. Phys.*, 64(4):412-429.

Bondi, H. (1957) "Negative Mass in General Relativity," *Reviews of Modern Physics*, 29:423-428.

Chiao, R. Y., Steinberg, A. M., and Kwiat, P. G. (1994) "The Photonic Tunneling Time and the Superluminal Propagation of Wave Packets," *Proc. of the Adriatico Workshop on Quantum Interferometry*, DeMartini, Denardo, and Zeilinger, eds., World Scientific, Singapore, p. 258.

Chiao, R. Y., Kozhekin, A. E., and Kurizki G. (1996) "Tachyonlike Excitations in Inverted Two-Level Media," *Phys. Rev. Lett.* 77:1254

Cole, D. and Puthoff, H. (1993) "Extracting Energy and Heat from the Vacuum," *Phys Rev E*, 48:1562-1565.

Cramer, J. G. (1986) "The Transactional Interpretation of Quantum Mechanics," *Reviews of Modern Physics*, A. Phys. Soc., 58:647-688.

da Costa, L. N., Freudling, W., Wegner, G., Giovanelli, R., Haynes, M. P., and Salzer, J. J. (1996) "The Mass Distribution in the Nearby Universe," *Astrophysical Journal Letters*, 468: L5-L8 and Plate L1

Dinowitz, S. (1996) "Field Distortion Theory," *Physics Essays*, 9:393-418.

Eagleton, R. D. and Kaplan, M. N. (1983) "The Radial Magnetic Field Homopolar Motor," *Am. J. Phys*, 56:858.

Forward, R. L. (1984) "Extracting Electrical Energy from the Vacuum by Cohesion of Charged Foliated Conductors," *Physical Review B*, 15 AUG. 1984 B30:1700-1702.

Forward, R. L. (1996) "Mass Modification Experiment Definition Study," Report # PL-TR-96-3004, Phillips Lab, Edwards AFB, CA.

Froning, H. D. and Barrett, T. W. (1997) "Inertial Reduction and Possible Impulsion by Conditioning Electromagnetic Fields," AIAA 97-3170, 33rd AIAA/ASME/SAE/ASEE Joint Propulsion Conference.

Haisch, B., Rueda, A., and Puthoff, H. E. (1994) "Inertia as a Zero-Point Field Lorentz Force," *Physical Review A*, 49:678-694.

Heron, M. L. and Pegg, D. T. (1974) "A Proposed Experiment on Absorber Theory," *J. Phys A*, 7:1965-1969.

Krasnikov, S. V. (1995) "Hyper-Fast Interstellar Travel in General Relativity," *gr-qc*, 9511068.

Krauss, L. M. (1995) *The Physics of Star Trek*, Basic Books, NY.

Lamoreaux, S. K. (1997) "Demonstration of the Casimir Force in the 0.6 to 6 μm Range," *Phys. Rev. Letters*, 78:5-8.

Li, N., Noever, D., Robertson, T., Koczor, R., and Brantley, W. (1997) "Static Test for a Gravitational Force Coupled to Type II YBCO Superconductors," *Physica C*, (to be published Sept-Oct, 97).

Millis, M. (1997a) "Breakthrough Propulsion Physics Research Program," NASA TM 107381, Lewis Research Center. Also at: http://www.lerc.nasa.gov/WWW/bpp/.

Millis, M. (1997b) "Challenge to Create the Space Drive," *Journal of Propulsion and Power*, 13:577-582.

Milonni, P. W. (1994) *The Quantum Vacuum*, Academic Press, San Diego, CA.

Misner C. W., Thorne, K. W., and Wheeler, J. A. (1973) *Gravitation*, W. H. Freeman and Company, NY.

Morris, M. and Thorne, K. (1988) "Wormholes in Spacetime and Their Use for Interstellar Travel: A Tool for Teaching General Relativity," *American Journal of Physics*, 56:395-412.

Pfenning, M., Ford, L. (1997) "The Unphysical Nature of Warp Drive," *gr-qc*, 9702026.

Podkletnov, E. and Nieminen, R. (1992) "A Possibility of Gravitational Force Shielding by Bulk $YBa_2Cu_3O_{7-x}$ Superconductor," *Physica C*, C203:441-444.

Puthoff, H. E. (1989) "Gravity as a zero-point-fluctuation force," *Phys Rev A*, 39:2333-2342.

Ringermacher, H. (1994) "An Electrodynamic Connection," *Classical and Quantum Gravity*, 11:2383-2394.

Serry, F. M., Walliser, D., Maclay, G. J. (1995) "The Anharmonic Casimir Oscillator," *J. Microelectromechanical Systems*, 4:193.

Seward, D. C. (1996) "Energy Storage System," US Patent 5,589,727.

Stoeffl, W. and Decman D. J. (1995) "Anomalous Structure in the Beta Decay of Gaseous Molecular Tritium," *Physical Review Letters*, 75:3237-3240.

Vargas, J. (1991) "On the Geometrization of Electrodynamics," *Foundations of Physics*, 21:379-401.

Visser, M. (1995) *Lorentzian Wormholes - From Einstein to Hawking*, AIP Press, Woodbury, NY.

Woodward, J. F. (1992) "A Stationary Apparent Weight Shift From a Transient Machian Mass Fluctuation," *Foundations of Physics Letters*, 5:425-442.

Woodward, J. F. (1994) "Method for Transiently Altering the Mass of an Object to Facilitate Their Transport or Change their Stationary Apparent Weights," US Patent 5,280,864.

Yilmaz, H. (1992), "Toward a Field Theory of Gravitation," *Il Nuovo Cimento*, 107B:941-960.

FIRST CONFERENCE ON GLOBAL VIRTUAL PRESENCE

SUPERRESOLUTION TECHNIQUES FOR ACTIVE REMOTE SENSING

R. K. Mehra and B. Ravichandran
Scientific Systems Company Inc.
500 West Cummings Park, Suite 3000
Woburn, MA 01801
(781) 933 5355

Abstract

Superresolution offers the potential to improve system performance by increasing the resolution. We have developed a suite of five algorithms, applicable to 1–D and 2–D and to the discrete and distributed cases. We also have shown specific applications of superresolution for air-to-ground surveillance, data resolution enhancement, SAR ATR and FOPEN ATR. We are developing algorithms tuned for remote sensing based on a mosaic type approach.

INTRODUCTION

Resolution is a fundamental limitation to any processing based on SAR data. Conventional radar imaging techniques, in general, make use of the FFT to determine the spatial location of a target from its scattered field. The resolution of these images is limited by the bandwidth of the interrogating radar system and the aspect angle sector over which the target is observed. In such cases, *superresolution* offers the potential to improve system performance by increasing the resolution. As observed by Luttrell (Luttrell 1990), *Superresolution is the process of increasing the effective bandwidth of an image (or time series) by introducing collateral data to augment the dataset; thus the Rayleigh resolution imposed by the size of the dataset is overcome by the introduction of the collateral data*

SUPERRESOLUTION ALGORITHMS

Scientific Systems has developed a suite of superresolution algorithms (Mehra et al. 1996), based on complex exponential models, AR, ARMA, and state space models, includes five classes of superresolution algorithms using techniques from the fields of system identification, signal processing, and modern control theory. They are:

1. Stochastic Realization Algorithm (SRA) for multiple time series modeling and spectral estimation using state space models (Mehra et al. 1996, Akaike 1974, Faurre 1976, and Rao Arun 1992).

2. 2-D superresolution algorithm for complex exponential signals using a matrix pencil (MP) approach (Mehra et al. 1996 and Hua 1992).

3. SRA–BE: 2–D superresolution algorithm using a combination of 1–D SRA, Bandwidth Extrapolation, and FFT.

4. Deterministic Stochastic Realization Algorithm (DSRA) for multi-input multi-output (MIMO) stochastic state space models using extensions of the SRA and subspace methods (Viberg 1994).

5. A fast superresolution algorithm (Gough 1994 and Li Stoica 1996) called RELAX that makes use of separate FFT computations to iteratively estimate the parameters of all the sinusoidal components in the data.

Scientific Systems has also extended the SRA based algorithm to obtain statistically optimal Global Maximum Likelihood Estimates (GMLE) using Adaptive Partitioned Random Search (APRS) (Mehra et al. 1996).

The unique and innovative aspects of SRA, MP, and DSRA supplemented by GMLE are that they have been shown both theoretically and empirically to be superior to other algorithms for 1D and 2D spectral estimation, multiple time series forecasting and transfer function estimation from noisy data. The comparison with other methods have been done on both simulated and real data (Mehra et al. 1996 and Mehra et al. 1992). The comparisons involve test problems from the fields of System Identification, time series analysis, and spectral estimation (Kay 1988). Figure 1 shows a comparison of techniques on a sample benchmark problem reported in the spectral estimation literature based on estimating two closely spaced complex sinusoids. Most current techniques report performances that coincide with the Cramer-Rao bound for SNR greater than $3db$. Thus, an investigation using GMLE estimation and APRS on cases with varying SNR was run. The experiments, based on 100 Monte Carlo runs, show that APRS performs as well the techniques reported in the current literature, and is able to detect these two frequencies at noise levels less than $3db$. Further, when constraints on the search space are included in the search the SRA-APRS approaches the Cramer-Rao bound.

Also, Compared to the Fast Fourier Transform (FFT), SRA can improve resolution by a factor of 4 to 100 depending on SNR in terms of separating closely spaced frequencies. It is shown in (Mehra et al. 1992) that the SRA also reduces false alarm rates and detects very weak signals in radar signal processing thereby improving target detection and classification accuracies in high clutter environments.

APPLICATIONS OF SUPERRESOLUTION ALGORITHMS

Specific applications of SRA, MP, and DSRA have been demonstrated for:

1. *Air-to-ground surveillance* of targets using Doppler radar for the classification of tanks and trucks for UAV radar (Mehra et al. 1992) and MTI radar, and for the classification of tanks, trucks, and SAMs for Joint STARS SAR (Mehra et al. 1996).

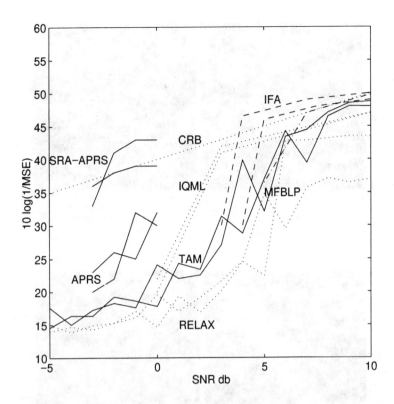

FIGURE 1. Comparison of the Cramer Rao Bound (CRB) on the Benchmark Problem.

2. *Resolution enhancement* using MP for Synthetic Aperture Radar (SAR) imagery for Joint STARS radar and Norden's MMRS radar (Mehra et al. 1996). Examples of four fold resolution enhancement on real SAR data provided by Norden is shown for a discrete targets and and for a distributed target in Figure 2.

3. Superresolution has been applied in conjunction with a *SAR target recognition* system based on Flexible Template Matching (Mehra et al. 1995) and, as shown in Table 1, for 2:1 superresolution the classification rate is very close to the high resolution case.

TABLE 1. Classification with and without Superresolution.

	Target Classification	
FFT Resolution	Standard Processing	Superresolved to 1 m
1 m	158/180 : 88%	
2 m	141/180 : 78%	157/180 : 87%
3 m	126/180 : 70%	136/180 : 76%

4. Superresolution has been applied in conjunction with a *FOPEN target recognition* system based on Fisher's Linear Discriminants (Mehra et al. 1997) and the results show that classification with superresolution has significant improvements over classification without superresolution.

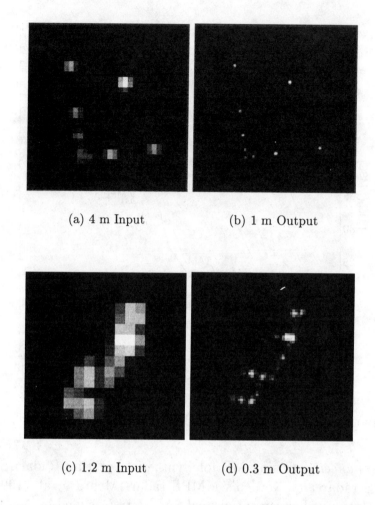

FIGURE 2. Example of Discrete Target Superresolution: (a) 4 m Input to Superresolution Algorithm (b) 1 m Output from Superresolution Algorithm. Example of Distributed Target Superresolution: (c) 1.2 m Input to Superresolution Algorithm (d) 0.3 m Output from Superresolution Algorithm.

APPLICATIONS OF SUPERRESOLUTION TO REMOTE SENSING

NASA's Mission to Planet Earth (MTPE) includes the Space borne Imaging Radar C/X Band Synthetic Aperture Radar (SIR-C/X-SAR) to monitor the earth's surface. Data from SIR-C/X-SAR will be used for the study and understanding of vegetation type, extent and deforestation; soil moisture content; ocean dynamics, wave and surface wind speeds and directions; volcanism and tectonic activity; and soil erosion and desertification. SIR-C/X-SAR is the next step in a series of space-borne imaging radars, beginning with SEASAT (1978), SIR-A (1981), Germany's Microwave Remote Sensing Experiment (1983), and SIR-B (1984). It is a precursor to the Earth Observing System (EOS) imaging radar system planned for the end of the decade. Also, data from SIR-C/X-SAR will be used to develop automatic techniques for extracting information from radar image data, in preparation for the EOS SAR mission.

Thus, in context of a remote sensing application such as NASA MTPE, the applicability of superresolution includes the areas of Earth and Space Science Research and Applications and Commercial Remote Sensing. Specific benefits of using superresolution will include:

1. *Improvements in radar resolution* using signal processing software instead of additional system hardware: the resolution of the SIR-C/X-SAR radar is 10 - 200 meters. 2:1 superresolution will produce resolution of up to 5 meters, and 3:1 will produce resolution of up to 3.3 meters.

2. The capability for superresolution allows the operator to collect data at lower resolutions. Lower resolutions allow (a) *reductions in transmission bandwidth*. Data from the SIR-C-C/X-SAR is recorded and relayed to the ground via data link. Superresolution will allow efficient use of this link. (b) *the radar can scan larger areas*.

3. Superresolution detects very weak signals thereby improving the performance of *target and region detection and classification*.

We are in the process of developing and testing superresolution algorithms for remote sensing. For testing and evaluation, we are using SAR images from RADARSAT (McCann 1995). Compared to superresolution and ATR, one of the main issues in context of super resolution and remote sensing is that the images are much larger. Thus, we are developing algorithms using a *mosaic* type approach (Degraaf 1997). In essence, a mosaic approach partitions a large image into tiles. These tiles may or may not be overlapping. Then, superresolution is performed on each one of these tiles, and the superresolved tiles are grouped back together into a mosaic.

SUMMARY AND CONCLUSIONS

Superresolution offers the potential to improve system performance by increasing the resolution. We have developed a suite of five algorithms, applicable to 1–D and 2–D and to the discrete and distributed cases. We also have shown specific applications of superresolution for air-to-ground surveillance, data resolution enhancement, SAR ATR and FOPEN ATR. We are developing algorithms tuned for remote sensing based on a mosaic type approach.

Acknowledgments

Support for this work came from Air Force contracts F19628-94-C-0005 and F19628-96-C-0065.

References

Akaike H. (1974) "Stochastic Theory of Minimal Realization", *IEEE Transactions on Automatic Control*, 19(6).

DeGraaf S. R (1997) "Sar Imaging Via Modern 2D Spectral Estimation Techniques", *IEEE Transactions on Image Processing*, To Appear.

Faurre P. L, (1976) "Stochastic Realization Algorithms", *Academic Press*, New York.

Gough P. T (1994) "A Fast Spectral Estimation Algorithm Based on the FFT", *IEEE Transactions on Signal Processing*, 42(6):1317–1322.

Hua Y. (1992) "Estimating Two-Dimensional Frequencies by Matrix Enhancement and Matrix Pencil", *IEEE Transactions on Signal Processing*, 40(9):2267–2280.

Kay S. M. (1988) "Modern Spectral Estimation", Prentice Hall. New York.

Li J. and P. Stoica (1996) "Efficient Mixed-Spectrum Estimation with Applications to Target Feature Extraction", *IEEE Transactions on Signal Processing*, 44(2):281–295.

Luttrell S. P. (1990) "A Bayesian Derivation of an Iterative Auto Focus/Super Resolution Algorithm", *Inverse Problems*, 6:975–996.

McCann S. (1995) "Radarsat Product Specification. Technical Report RZ-SP-50-5313, Macdonald Dettwiler, Richmond B.C., Canada.

Mehra R. K, S. Mahmood, and O. Sarfaty (1992) "System Indentification Based Target Classification in a High Clutter Environment", Technical Report SSCI-1158-Final, Scientific Systems Company, Woburn, MA.

Mehra R. K and B. Ravichandran (1996) "A System Identification Based Target Classification Technique in a High Clutter Environment. Technical Report 1174 Final, Scientific Systems Company Inc., Woburn, MA.

Mehra R. K, B. Ravichandran, A. K. Bhattacharjya, and M. Greenspan (1995) "Automatic Target Recognition for Joint STARS using Flexible Template Matching", Technical Report 1180 Final, Scientific Systems Company Inc., Woburn, MA.

Mehra R. K and B. Ravichandran (1997) "FOPEN Radar ATR using Superresolution and Eigentemplates", Technical Report 1211 Final, Scientific Systems Company Inc., Woburn, MA.

Rao B. D and K. S. Arun (1992) "Model Based Processing of Signals: A State Space Approach", *Proceedings of the IEEE*, 80(2):283–309.

Viberg M. (1994) "Subspace Methods in System Identification", In *IFAC Symposium on System Identification*, Copenhagen, Denmark.

A TUNABLE MWIR LASER REMOTE SENSOR FOR CHEMICAL VAPOR DETECTION

Thomas L. Bunn, Patricia M. Noblett and William D. Otting
Boeing Defense and Space Group
Rocketdyne Division
6633 Canoga Avenue
Canoga Park, CA 91309-7922
818-586-2720

Abstract

The Air Force vision for Global Virtual Presence suggests a need for active remote sensing systems that provide both global coverage and the ability to detect multiple gaseous chemical species at low concentration from a significant standoff distance. The system will need to have acceptable weight, volume, and power characteristics, as well as a long operating lifetime for integration with various surveillance platforms. Laser based remote sensing systems utilizing the differential absorption lidar (DIAL) technique are promising for long range chemical sensing applications. Recent advancements in pulsed, diode pumped solid state laser (DPSSL) technology and in tunable optical parametric oscillators (OPO) make broadly tunable laser transmitters possible for the DIAL system. Also the characteristic narrow spectral bandwidth of these laser devices provides high measurement sensitivity and spectral selectivity with the potential to avoid interfering species. Rocketdyne has built and tested a tunable, midwave infrared (MWIR) DIAL system using DPSSL/OPO technology. The key to the system is a novel tuning and line narrowing technology developed for the OPO. The tuning system can quickly adjust to the desired wavelength and precisely locate a narrow spectral feature of interest. Once the spectral feature is located, a rapid dither tuning technique is employed. The laser pulses are tuned "on" and "off" the spectral resonance of a molecule with precise and repeatable performance as required to make the DIAL measurement. To date, the breadboard system has been tested by measuring methane, ethane, and sulfur dioxide in a calibrated gas cell at a range of 60 meters.

INTRODUCTION

Lasers are ideal light sources for optical remote sensing. The laser's well collimated and intense light beam propagates a considerable distance through the atmosphere. This coupled with the spectral purity and broadly tunable output wavelength of the DPSSL/OPO system permits remote spectroscopic chemical analysis of a broad range of chemical species. The DPSSL/OPO laser also provides very short pulse times enabling range resolved concentration determination based on time-of-flight measurements. The most sensitive method for sensing atmospheric properties remotely is differential absorption lidar or DIAL (Baumgartner 1978a, 1978b). The operation of a DIAL based system is illustrated in Figure 1. DIAL systems measure gas concentrations remotely using the difference in light absorption at two distinct wavelengths. Absorption cross sections at target molecular resonance wavelengths are large enough to cause measurable attenuation, even at low concentration levels. As depicted as in Figure 1, the "on" resonance laser

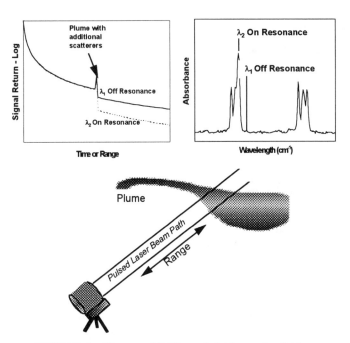

FIGURE 1. Illustrated Differential Absorption Lidar.

pulse, λ_2, is attenuated by the target molecular species in the gas plume in accordance with the Beer-Lambert law. The "off" resonance laser pulse, λ_1, is not absorbed by the molecular species, and therefore, is used as a reference. This reference signal represents the expected return assuming no target gas is present and thus allows for normalizing the signal at λ_2 for atmospheric effects along the optical path such as backscatter and atmospheric transmission or visibility effects. The difference between the two signals is due solely to the absorption contribution of the target molecular species. The "on" and "off" laser pulses are closely spaced temporally to assure that both pulses travel along the same optical path.

TRANSMITTER DEVELOPMENT

Unlike conventional spectrometers where the spectral separation is done at the receiver, in a tunable laser DIAL system, the spectral separation is accomplished at the transmitter. The receiver only needs to collect photons onto any sensitive detector. The key technology for DIAL, therefore is the tunable transmitter.

A design for a breadboard transmitter was developed using a Nd:YAG pump laser and an extracavity LiNbO$_3$ OPO. The pump laser consists of two diode pumped Nd:YAG heads arranged in a linear cavity. A Q-switch provides 45 nsec pulses at a pulse repetition frequency of 250 Hz. The laser provides linearly polarized light with a spectral linewidth of < 0.1 cm^{-1} and a nominal average power of 6 watts. A schematic diagram of the breadboard transmitter is shown in Figure 2.

Pump Laser Parameter	Design or Performance
• Diodes	
• Pulsed	• 200 μs
• Duty Cycle	• 5 %
• YAG Rod Length	• 6 cm
• Cavity Length	• 1.45 m
• Beam Size	• 900 mm x 500 mm (hxv) FWHM
• Beam Quality	• M2 = 1.15
• Pulse Rep Rate	• 250 Hz
• Pulse Duration	• 45 nsec (Q-switched)
• Pulse Energy	• 24 mJ/pulse
• Average Power	• 6 W
• Etalon	• 3.3 cm^{-1} FSR
• Linewidth	• 0.08 cm^{-1}

Ring OPO Parameter	Design or Performance
• Crystal Length	• 40 mm (2 crystals)
• Pump Threshold	• 10 mJ w/o etalon
	• 16 mJ w/etalon
• Conversion Efficiency	• ~ 15%
• Wavelength Range	• Signal - 1.43 to 1.61 μm
	• Idler - 4.16 to 3.15 μm
• Linewidth	• without etalon - 3.6 cm^{-1}
	• with etalon - 0.1 cm^{-1}
• E/O Tuning Range	• ± 40 nm

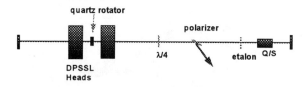

FIGURE 2. Breadboard Pump Laser.

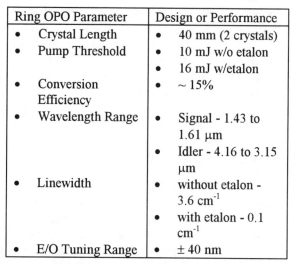

FIGURE 3. OPO incorporating singly resonant ring cavity design.

The optical parametric oscillator (OPO) is a singly resonant ring design. Figure 3 illustrates the OPO design featuring a matched LiNbO$_3$ crystal pair. The opposite orientations of the crystals' extraordinary axes compensate for birefringence walk-off. A half-wave retardation waveplate located in the resonated wave (ω_{signal}) allows for adjusting the cavity round-trip loss. This important adjustment affects the threshold condition for the OPO and is used to prevent "over-coupling" which occurs when the intracavity ω_{signal} intensity becomes much larger than that of the pump laser, ω_{pump}. An intracavity etalon is used to narrow the spectral bandwidth of the signal. This correspondingly reduces the linewidth of the idler beam as well. The net effect of the etalon is to force gain within a narrow spectral region providing for an output linewidth of 0.1 cm^{-1}.

Tuning System Characterization

The wavelengths of the signal and idler are determined by the phase matching of the pump, signal, and idler. The phase matching for a uniaxial crystal results from satisfying the equations for the conservation of energy and momentum. OPO tuning is effected by changing the orientation of the index ellipsoid of the crystal causing a change in the index of refraction at the pump, signal, and idler wavelengths. New signal and idler waves will build out of the noise when phased matched. There are several ways to change the index ellipsoid of the crystal. In the present OPO design, the orientation of the ellipsoid is adjusted two ways:

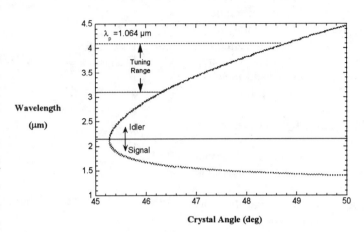

FIGURE 4. LiNbO$_3$ OPO Phase Matching Tuning Curve.

- Angle tuning - physically rotating the crystal
- Electro-optic tuning - applying an electric field across the crystal.

Coarse tuning of the OPO is accomplished by precisely counter-rotating the crystals with respect to one another. The phase matching curve or tuning curve for the LiNbO$_3$ OPO is shown in Figure 4. The tuning range of the idler, λ_{idler}, is 3.15 - 4.16 µm. Many molecules have absorptions in this spectral range and this design easily facilitates tuning from one chemical species to another.

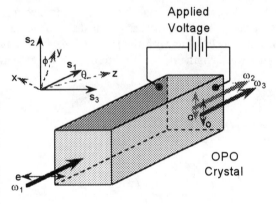

FIGURE 5. Electro-optic Tuning of LiNbO$_3$.

The index ellipsoid can be altered much faster using electro-optic techniques. A high voltage electric field applied perpendicular to the crystal faces is used to tune the output over a narrow bandwidth. The high voltage pulses rapidly shift the wavelength over a narrow (~3cm^{-1}) range. This technique is used to shift the wavelength "on" and "off" the narrow molecular transitions in a dithering fashion as required for DIAL. Figure 5 illustrates the electro-optic tuning technique. The electro-optic tuning characteristics for the LiNbO$_3$ OPO were measured as a function of electric field strength and crystal angle. Figure 6 shows the measured change in wavelength as a function of applied field strength. A tuning range of ± 40 nm is achieved for the idler.

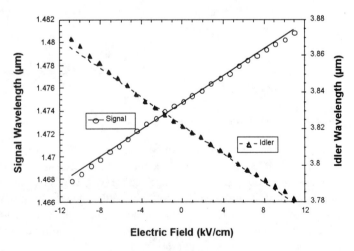

FIGURE 6. Electro-Optic Wavelength Tuning as a Function of Electric Field Strength.

Figure 7 illustrates similar electro-optic tuning data for a number of mechanical angle tuned orientations, showing that the electro-optic tuning technique is effective across the tuning range of the OPO and that, combined with mechanical angle tuning, continuous narrow band radiation is available from 3 - 4 μm.

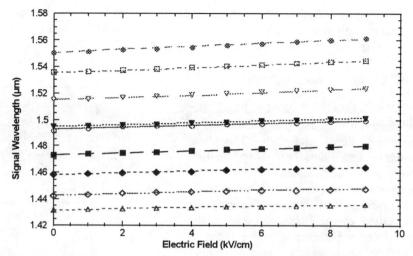

FIGURE 7. Electro-optic Tuning Characteristics, Parameterized at Nine Different Crystal Angles.

DIAL Tuning Implementation

The transmitter's output characteristics for DIAL operation are depicted in Figure 8. The broad curve represents the intrinsic gain curve of the OPO with no etalon in the resonator, typically 4 cm^{-1} wide. The shaded curve represents the Airy transmission function for the etalon, in this case, with a free spectral range (FSR) of 3.55 cm^{-1}. The etalon forces the OPO gain to the narrow transmission window providing for a narrow output linewidth. To dither tune "on" and "off" a molecular absorption resonance, the broad OPO gain curve is electro-optically tuned alternately between two fixed etalon transmission peaks. In effect, the laser is made to mode hop on the etalon transmission peaks by applying an ac voltage function synchronous with the laser firing. The pulse-to-pulse output wavelength alternates between the "on" resonance wavelength and the "off" resonance wavelength

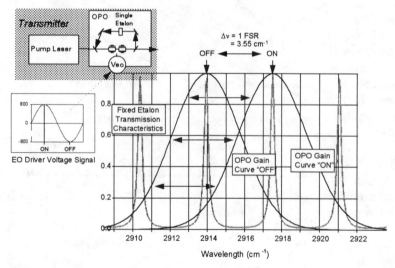

FIGURE 8. DIAL Dither Tuning System.

Breadboard System Description

A photograph of the laboratory testbed is shown in Figure 9 with a system block diagram shown in Figure 10. The system consists of the pump laser and OPO, gas cell diagnostics, optics and receiver, signal processing, and laboratory support equipment. The gas cell diagnostic consists of two Molectron pyroelectric detectors and a "reference" cell filled with a subject gas of known concentration. The "null" detector (without the gas cell) provides shot-to-shot pulse energy information so that laser intensity fluctuations between the outgoing "on" and "off" resonance pulses can be normalized. The reference cell (2.5 cm in length) has CaF$_2$ windows followed by a Molectron detector. The reference detector measures the relative transmission of the on and off resonance pulses through the reference cell serving two functions. First, the reference arm provides quick and positive spectral feature identification. Secondly, the relative absorption between the on and off resonance pulses measured by the reference detector is ratioed to the relative absorption measured by the "sample" detector thereby normalizing for pulse-to-pulse frequency jitter. This eliminates the need of knowing the exact absorption cross sections as a function of wavelength.

The optics and receiver subsystem enable coalignment of the transmitted beam and collection optical path. The outgoing pulse illuminates a target region, with the divergence of the transmitter selected to match the illuminated target area to the receiver subsystem field-of-view (FOV) at the target range. The f/2, 12" diameter parabolic mirror collects the backscattered energy and focuses it onto a single-element, liquid nitrogen cooled, 0.5 mm diameter indium antimonide IR detector at the prime focus. The image of this detector area projected through the receiver, results in approximately a 5 cm spot at the target 60 meters away.

FIGURE 9. MWIR Breadboard DIAL System.

The detector output is sent directly to a Stanford Instruments Boxcar Averager. The integrated pulse is sent to a signal processing computer programmed with LabView. The LabView program controls the firing of the pump laser and the generation of the OPO electro-optic waveform synchronous with the pump laser firing. It also stores raw data, process this data stream, and stores the resulting computed data.

The buildup of the system breadboard was implemented in two stages as indicated in Figure 10. Initially, gas cell measurements were conducted with bench-top gas cells by simply adding a third gas cell/detector as the sample leg. The bench-top cells were 2.5 cm in length also with CaF_2 windows. The bench-top tests measured the relative transmission, "on" vs "off" resonance, for each of the cells.

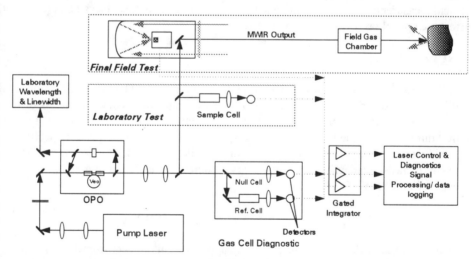

FIGURE 10. Breadboard DIAL System Block Diagram.

This configuration enabled the development of the tuning techniques, the development of a DIAL algorithm and statistical processing routines, as well as making initial concentration measurements.

For the field tests, the bench-top sample cell was removed and a field cell was added, as shown in Figure 10. The laser is directed outside the laboratory and propagated across the neighboring test range. The field gas cell was located approximately 60 m downrange. The field cell was 0.64 m in length and hermetic with CaF_2 windows. This cell was charged with a known concentration of test gases in balance air. Direct assessment of the measurement performance capability of the breadboard system was ascertained during testing.

RESULTS OF INITIAL TESTING

The tuning techniques for locating spectral feature and dithering on and off resonance were developed during the initial laboratory tests. Dither tuning was found to be repeatable at 250 Hz PRF. Once steady-state "on" resonance operation was attained, no drift in laser frequency was observed. The measured output linewidth of the laser OPO was 0.10 cm^{-1}. Signal processing routines were developed during the initial gas measurement tests. Gas cell tests with methane and ethane were conducted with concentration-pathlength products ranging from 50 to 2000 ppm-m. As demonstrated in Figure 11 for methane, the DIAL concentration measurements agreed well with the actual concentration over the entire cell pressure range. The standard deviations of the mean was typically less than 3 ppm-m for 10 seconds of averaged data. This standard deviation is expected to improve as further noise reduction schemes are explored and implemented. Sulfur dioxide has a smaller absorption cross section in the MWIR but gave similar results relative to the absorbance. For the field test, the necessary optics and telescope were integrated with the system as illustrated in Figure 10. The return signal back to the collection optics was provided by reflection off a Lambertian target located beyond the gas cell. The system provided a strong return signal for a variety of target materials. Following optical alignment of the system, gas cell tests similar to the bench top tests were conducted using methane, ethane, and sulfur dioxide. Initial testing identified a zero offset in the data sets corresponding to about 5% differential return. The offset was attributed to wavelength sensitive beam splitters used in the diagnostics. Using an offset correction technique, the results from the field measurement tests were found to be similar to the laboratory measurements. Work is currently ongoing to correct the optical layout of the diagnostics.

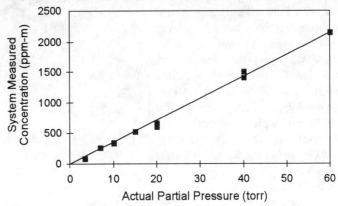

FIGURE 11. Measured methane concentration versus known gas cell pressure.

CONCLUSIONS

The DPSSL/OPO system is promising for a DIAL application. The tuning flexibility of the OPO allows tuning to multiple chemical species with IR spectra, as was demonstrated by measurements of methane, ethane, and sulfur dioxide using the current breadboard system. Additionally, achieving an output linewidth of 0.1 cm^{-1} across the entire tuning range of the OPO provides the spectral selectivity necessary for spectroscopic applications. Finally, the rapid electro-optic dither tuning system demonstrated the feasibility of making the DIAL on and off resonance wavelength measurements using a single laser source. The dither tuning system operated precisely and was repeatable at a PRF of 250 Hz.

Acknowledgments

The authors acknowledge the conscientious support on this project from Dr. Jay Fox of the Army Night Vision and Electronics Sensors Directorate and Cynthia Swim of the Army Chemical and Biological Defense Command.

References

Baumgartner, R.A. and R.L. Byer, (1978a) "Remote SO$_2$ Measurements at 4 µm with a Continuously Tunable Source," in *Optics Letters*, **2**(6):163-165.

Baumgartner, R.A. and R.L. Byer, (1978b) "Continuously Tunable IR Lidar with Applications to Remote Measurements of SO2 and CH4," *Applied Optics* **17**(22):3555-3561.

GEOSCIENCE LASER ALTIMETER SYSTEM - STELLAR REFERENCE SYSTEM

Pamela S. Millar
NASA Goddard Space Flight Center
Laser Remote Sensing Branch, Code 924
Greenbelt, MD, 20771
301-286-3793

J. Marcos Sirota
University of Maryland at Baltimore County
5401 Wilkens Ave
Baltimore, MD, 21228-5398
301-286-8242

Abstract

GLAS is an EOS space-based laser altimeter being developed to profile the height of the Earth's ice sheets with ~15 cm single shot accuracy from space under NASA's Mission to Planet Earth (MTPE). The primary science goal of GLAS is to determine if the ice sheets are increasing or diminishing for climate change modeling. This is achieved by measuring the ice sheet heights over Greenland and Antarctica to 1.5 cm/yr over 100 km x 100 km areas by crossover analysis (Zwally 1994). This measurement performance requires the instrument to determine the pointing of the laser beam to ~5 urad (1 arcsecond), 1-sigma, with respect to the inertial reference frame. The GLAS design incorporates a stellar reference system (SRS) to relate the laser beam pointing angle to the star field with this accuracy. This is the first time a spaceborne laser altimeter is measuring pointing to such high accuracy. The design for the stellar reference system combines an attitude determination system (ADS) with a laser reference system (LRS) to meet this requirement. The SRS approach and expected performance are described in this paper.

INTRODUCTION

The Geoscience Laser Altimeter System (GLAS) is a space-based lidar being developed to monitor changes in the mass balance of the Earth's polar ice sheets (Thomas *et al.* 1985). GLAS is part of NASA's Earth Observing System (Schutz 1995), and is being designed to launch into a 600 km circular polar orbit in the year 2001, for continuous operation over 3 years. The orbit's 94 degree inclination has been selected to allow good coverage and profile patterns over the ice sheets of Greenland and Antarctica. The GLAS mission uses a small dedicated spacecraft, which is required to have a very stable nadir and zenith pointing platform.

Accurate knowledge of the laser beam's pointing angle (in the far field) is critical since pointing the laser beam away from nadir biases the altimetry measurements (Gardner 1992, Bufton *et al.* 1991). This error is a function of the distance of the laser centroid off nadir multiplied by the orbit altitude and the tangent of the slope angle of the terrain. Most of the ice sheet surface slopes are less than 1° resulting in pointing knowledge bias of only 5 cm with 5 urad accuracy, and overall single shot height accuracy of ~15 cm. However, over a 3° surface slope pointing knowledge to ~5 urad is the largest error source (15 cm) in achieving 20 cm height accuracy. The GLAS design incorporates a stellar reference system (SRS) to relate the laser beam pointing angle to the star field to an accuracy of ~5 urad. The stellar reference system combines an attitude determination system (ADS) operating at 10 Hz coupled to a 40 Hz laser reference system (LRS) to perform this task.

STELLAR REFERENCE SYSTEM DESCRIPTION

The simplest approach for measuring the pointing of the GLAS laser beam with respect to the star field is to couple the laser directly into a star camera. Unfortunately this approach is not feasible with current star camera technology, regarding repetition rate (there are no cameras faster than 10 Hz), and angular resolution (the typical 8 x 8 degree field of view (FOV) would require centroiding to 1/60th of a pixel). Upgrades to star camera technology are being studied for a next generation (GLAS-II) SRS system. For these reasons an approach using a separate sensor to measure the pointing of the sampled laser beam at 40 Hz is baselined.

The overall approach for the stellar reference system is shown in figure 1, with GLAS SRS parameters listed in table 1. The GLAS laser beam is coupled into the LRS along with collimated reference sources from the ADS components. All optical beams are spatially superimposed into the telescope but are offset in angle to avoid image overlap on the array focal plane. The ADS measures the pointing of the instrument platform with respect to the star field while the LRS samples the laser beam and measures its alignment with respect to the components of the ADS.

In the SRS approach a small fraction of the GLAS laser beam is folded into the laser reference sensor's FOV with two lateral transfer retroreflectors (LTR's). LTR's have the same characteristics as retro-reflectors, they preserve the parallelism between the input and output beams as long as the alignment integrity between the three faces are maintained. The optic as a whole can move (within several degrees) and not affect the parallelism between the input and output beams. The first LTR encountered by the outgoing laser beam is constructed with three fused silica faces, the output face is anti-reflective (AR) coated while the other two have reflective coatings. The AR coatings on the output face ensures that at least 99% of the laser energy gets transmitted to Earth surface. The next LTR, constructed of highly reflective coated Zerodur, relays the laser beam into the laser reference telescope where it is imaged for each laser shot fired.

The LRS consists of a (5 x 5 mrad) narrow FOV camera operating at 40 Hz frame rate. The camera includes a telescope, a charge coupled device (CCD) array, and ancillary electronics for imaging and computing centroid locations of the GLAS laser beam and other fiducial reference images. There are 2 collimated (50 urad) reference beam sources attached to both the star camera housing and gyro housing. The optical beams generated by these sources are coupled into the off-axis telescope with LTR's to observe the alignment of the star camera and gyro housings. The alignment and stability between the reference sources and the star camera CCD array and gyro spin axis will influence how well the pointing of the GLAS laser beam can be determined, these error components are accounted for in the LRS error budget. The LRS will also be able to occasionally image stars which will enable a boresight check between the LRS and the larger FOV (8° x 8°) star camera. An annulus of 3.8 cm thickness with ~120 cm^2 total area is available for collecting star light. The relative movement between the far field pattern of the GLAS laser beam, reference sources from the ADS instruments, and an occasional star will be determined. This data combined with the processed ADS data will yield the pointing of the laser beam in inertial space. GPS and satellite laser ranging data, from the ground to a retroreflector target on the instrument, are used to determine the center of gravity (CG) orbit position of the spacecraft to ~5 cm. Knowledge of the laser footprint location on the ground is then determined by combining the pointing knowledge of the outgoing laser beam with the position of the spacecraft CG.

PERFORMANCE

The performance of the SRS is influenced by many factors such as frequency of the boresight check between the star camera and the laser reference camera, post-launch calibration, and stability of individual components on the optical bench. The approach taken by the SRS is to measure as many references as possible, as frequently as possible while staying within GLAS mission constraints. The design allows for movement of the individual components after launch and in-orbit. In this section we have defined an SRS error budget allotment, calibration strategy, and determined star observation statistics and expected performance of the laser reference camera.

Error Budget Allotment

The error budget allotment of the SRS is shown in table 2. The overall system error in determining the laser pointing is estimated to be 7.2 urad, 1- sigma, (1.5 arcseconds) at 40 Hz. This value may be improved to 5 urad after post flight calibrations are performed. The ADS error component is derived by combining the performance specifications from a Litton HRG gyro and a Hughes Danbury HD-1003 star camera with Kalman filtering. Unmeasured errors take into account any thermal distortions and high frequency (> 10 Hz) vibrations of the bench and components on the bench.

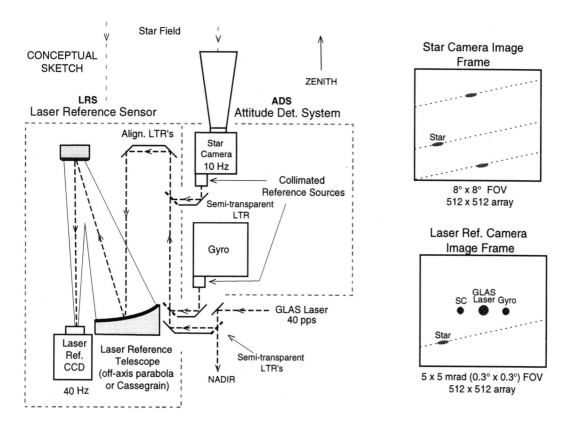

FIGURE 1. GLAS Stellar Reference System Conceptual Approach.

TABLE 1. Laser Reference Sensor System Parameters.

GLAS LASER BEAM PARAMETERS	
divergence	100 urad
beam diameter, $1/e^2$	5 cm
energy @ 1064 nm	70 mJ (altimetry channel)
spatial profile	~ TEM_{00}
rep. rate	40 pps
LATERAL TRANSFER RETROREFLECTORS	
stability	0.2 urad for Zerodur
	1.9 urad for Fused Silica
	(semi-transparent LTR's)
Parallelism between entrance and exit	≤ 2 minutes of arc
STAR CAMERA: (COMMERCIAL)	
Field of view (FOV)	8° x 8°
Frame rate	10 Hz
Single star accuracy/star	3.5 urad
GYRO: (COMMERCIAL)	
Update rate	30 Hz
Scale factor	100 ppm (@ 30 Hz)
Drift Rate Bias (1-sigma)	6.45×10^{-03} urad/sec
Random Walk Rate Noise (1-sigma)	1.55×10^{-04} urad/(sec)$^{3/2}$
LASER REFERENCE CAMERA	
Field of view	5 x 5 mrad
Update rate	40 Hz
Sensitivity	+8 Magnitude Stars

Details of the LRS error component are listed in table 3. This subsystem is the core instrument of SRS, it combines the laser reference camera (CCD array plus telescope) with coupling optics and collimated reference sources. The values listed for the lateral transfer retroreflectors and collimated reference sources are based on finite element calculations assuming thermal gradients are limited to ±1°C. A description of the the laser reference camera performance is described in the following section.

Laser Reference Camera

To enable multiple co-alignment checks per orbit between the LRS and ADS, the laser reference camera (LRC) is required to detect stars of +8 magnitude, with centroid resolution of ~2 urad while operating at 40 frames per second (the laser pulse rate). The LRC is also required to determine the ranging laser image and star camera and gyro fiducial image centroids to a resolution of at least 1 urad.

The laser reference camera is formed by the laser reference sensor and its associated telescope. The sensor is a 512 x 512 pixel array. At 40 Hz frame rate the integration and read out time for the imaging array is < 25 msec, which is much shorter than typical times used by star trackers which operate at frame rates ≤ 10 Hz. The shorter integration time is compensated by using a large optical collecting aperture and a dual frame transfer CCD array with four independently clocked outputs. The frame transfer architecture, with a very fast frame shift, permits use of most of the 25 msec for integration. The multiple outputs decrease the pixel read out rate, which translates into lower read out noise.

TABLE 2. GLAS SRS Error Budget Allotment.

SYSTEM	ERROR MAGNITUDE (urad, 1 sigma)	FREQUENCY (Hz)	REFERENCE
Attitude Determination System	4.7	10	ADS ref. sources to star field
Laser Reference Sensor (LRS)	4.9	40	Laser beam to ADS ref. sources
Unmeasured Errors	2.2	10	Bench vibrations & Instrument angular disturbances
Stellar Reference System	7.2	40	Laser Beam to Star Field

TABLE 3. GLAS SRS Laser Reference Sensor (LRS) Error Budget Allotment.

ERROR TYPE	ERROR (urad)	COMPONENT DESCRIPTION
Camera Centroid Resolution:	0.97	Laser image
	0.97	Star camera reference image
	0.97	Gryo reference image
	1.94	Star image
Telescope Distortion:	1.45	Off-axis parabola
Coupling Optics Stability:	1.94	Star Camera LTR, Fused Silica (semi-transparent)
	1.94	Gyro LTR, Fused Silica (semi-transparent)
	1.94	Output Glas laser LTR, Fused Silica (semi-transparent)
	0.19	LTR obscuring Telescope, Zerodur (all reflective)
	1.45	Star Camera collimated reference beam stability
	1.45	Gyro collimated reference beam stability
Total LRS RSS Error	4.9	(1-sigma)

The telescope is an F/10 off-axis optical system with 16 cm aperature and 8 cm central obscuration (caused by the LTR). The smearing produced by the orbital motion corresponds to 2.3 pixels along track. Under these conditions, the noise equivalent angle (NEA) in the along-track direction for a star of magnitude +8 in the sensor becomes 1.2 urad. Readout noise, dark current, and photon noise are included in this value. When accounting for estimated centroiding errors, optical error, and defocusing, a total RSS value of 1.9 urad per frame per star is obtained. Given the small FOV, statistics indicate that for most frames only one star will be visible in the field, therefore it will not be possible to average over multiple stars as is typically done by star cameras. However, an average star track in our camera FOV should produce approximately 100 images, which would yield a factor of 10 reduction in the overall error if averaged over the whole track. Sweep time for the 5 mrad FOV is approximately 4 sec. Averaging star images over the track over these periods will be valid for boresight verification assuming that the boresight between the two cameras does not vary while performing the measurement. The validity of this assumption will be checked during engineering model testing and analysis.

The laser ranging spot, as shown in figure 1, is imaged on a region of interest of 10 x 10 pixels wide. The NEA in this case is determined essentially by photon noise. The largest contribution to centroiding error is pixel response non-uniformity. The errors accrued in determining the location of the centroid of the ranging laser image and ADS fiducial images are much lower since the optical intensity of these sources are greater than dim stars. Preliminary calculations show that, accounting for all contributions, the required resolution (1 urad or 0.2 arcsec) per shot can be met.

On-Orbit Calibration

There are two approaches being investigated for an end to end calibration of the laser beam pointing angle determination post launch. One approach uses an aircraft underflying the GLAS satellite. In this case a wide FOV camera viewing nadir will image the ground where geo-located optical beacons are used as markers. When the GLAS laser illuminates the ground viewed by the aircraft camera the markers will be used to determine the geo-located position of the GLAS laser beam footprint. This calibration will be performed every 6 months for a period of 8 days.

The other calibration approach takes place while the instrument is orbiting over Earth's oceans, a roll and pitch maneuver is performed to determine the range while normal to the ocean surface. True nadir can then be determined using knowledge of the ocean surface. This technique has so far been proven over flat lakes (Hofton, *et al.* 1996) but still needs to be confirmed over ocean. This calibration will be performed over the equator while the sky is clear of cloud cover.

Laser Reference Sensor Star Observation Statistics

Star observation statistics were calculated for the GLAS orbit for +2 through +6 and +8 magnitude stars and 5 mrad x 5 mrad field of view for the laser reference sensor and 8 degree x 8 degree field of view for the star camera respectively. For the laser reference sensor field of view the average time between star observations is ten minutes over the South Pole and nine minutes over the North Pole. On average it takes about sixteen minutes to traverse either polar region and there is only one star in the field of view at any time.

For the wider 8 degree x 8 degree field of view star camera, which can detect stars up to +6 magnitude, the average time between observing stars is one minute for both North and South polar regions with an average of about three stars at any given time.

There are clear advantages in being able to detect up to 8th magnitude stars in terms of number of stars per orbit and average time between observations. Fortunately there are more stars located over the poles where it is most important for the GLAS measurement of the ice sheet heights.

SUMMARY

The GLAS stellar reference system takes laser altimetry to a more accurate level (15-20 cm per measurement) which is a function of the slope angle of the terrain. We have developed an approach for relating the laser pointing angle to the inertial reference frame given the limits of current technology. The estimated pre-launch SRS performance is close to the mission goal at 7.2 urad. It is likely with the planned post-launch calibrations that the SRS will reach 5 urad accuracy while operating on-orbit.

Acknowledgments

This work is being funded by NASA's EOS Mission to Planet Earth. The authors would like to thank the GLAS SRS team with special thanks to Joe Garrick, Paul Hannan, and Jim Lyons for their contributions on the SRS instrument development.

References

Bufton, J.L., Garvin, J.B., Cavanaugh, J.F., Ramos-Izquierdo, L., Clem, T.D. and Krabill, W.B., (1991) "Airborne lidar for profiling of surface topography," *Opt. Engr.*, 30:72.

Gardner, C.S., (1992) "Ranging Performance of Satellite Laser Altimeters," *IEEE Trans. Geosc. Rem. Sensing*, 30: 1061.

Hofton, M. A., Ridgway, J. R., Minster, J. B., Williams, N., Bufton, J. L., Blair, J. B., and Rabine, D. L., (1996) "Repeat Airborne topographic survey of Long Valley, California, Using Scanning Laser Altimeters," EOS Trans., 77:S262.

Schutz, B. E., (1995) "GLAS Geoscience Laser Altimeter System," *MTPE/EOS Reference Handbook*, NP-215, 132.

Thomas, R. H., Bindschadler, R. A., Cameron, R. L., Carsey, F. D., Holt, B., Hughes, T. J., Swithinbank, C. W. M., Whillans, I. M., and Zwally, H. J., (1985) "Satellite Remote Sensing for Ice Sheet Research," in *Nasa Tech. Memo. 86233*, 17.

Zwally, H. J., (1994) "Detection of Change in Antarctica," in *Antarctic Science: Global Concerns*, Gotthilf Hempel, Ed., Springer-Verlag Berlin Heidelberg, 126-143.

THE GEOSCIENCE LASER ALTIMETER SYSTEM (GLAS)

James B. Abshire and James C. Smith
NASA - Goddard Space Flight Center
Laser Remote Sensing Branch, Code 924
Greenbelt MD 20771, USA
(301) 286-2611   Fax: (301) 286-0213

Bob E. Schutz
University of Texas at Austin
Center for Space Research, WRW 402
Austin, TX 78712, USA
(512) 471-4267

Abstract

GLAS is a space-based lidar designed for NASA's Mission to Planet Earth (MTPE) Laser Altimeter Mission (LAM). The GLAS instrument will precisely measure the heights of the polar ice sheets, to profile areas of the Earth's land topography, and to profile the structure of clouds and aerosols on a global scale. The LAM mission utilizes a small dedicated spacecraft in a polar orbit at 598 km altitude with an inclination of 94 degrees. GLAS is being developed to launch in 2001 and to operate continuously for a minimum of 3 years with a goal of 5 years.

INTRODUCTION

The most important measurements for GLAS are to determine the seasonal and annual changes in the heights of the ice sheets which cover Greenland and Antarctica (NASA 1995). For these surface lidar measurements, GLAS will measure the distance to the ice sheet at the laser wavelength of 1064 nm with an accuracy of 10 cm per laser shot. When combined with 10 cm orbit uncertainty, this results in measurement of the ice sheet topography with 20 cm accuracy. Subsequent data processing will allow a determination of 1.5 cm-level changes/year over surface areas of 200x 200 km. Seasonal and annual fluctuations in ice sheet thickness will be determined by comparing successive GLAS measurement sets. The planned series of 3 such missions will monitor ice sheet heights over 15 years, and information gained should greatly improve understanding of ice sheet dynamics and its contribution to the issue of global sea-level rise.

GLAS will also measure atmospheric backscatter at both 1064 and 532 nm to determine the vertical distributions of clouds and aerosols below its flight path. The GLAS 1064 nm atmospheric lidar measurements will provide unambiguous estimates of profiles of cloud height and vertical structure. The more sensitive 532 nm measurements will be used to determine of the height distributions of thin clouds and aerosols. With data averaging these can be used to determine the height of the planetary boundary layer. The lidar measurements of the vertical aerosol structure over a global scale will help improve understanding of aerosol-climate effects.

The GLAS instrument design utilizes three diode-pumped Q-switched ND:YAG lasers, a 100 cm diameter Beryllium telescope, a narrow (~ 25 pm) optical filter at 532 nm, Si APD detectors for both 1064 and 532 nm, and a subsystem to measure the laser pointing angles to the arc-second level. The measurement performance of the altimeter and lidar have been estimated by using analysis and simulations. The mission and the instrument and subsystem designs are briefly discussed in this paper.

MISSION OVERVIEW

GLAS will be launched into a 598 km circular polar orbit in summer of 2001. The orbit's 94 degree inclination has been selected to allow good coverage and profile patterns over the ice sheets of Greenland and Antarctica as well as data comparison with other Mission to Planet Earth (MTPE) instruments. The spacecraft will be a commercial spacecraft derivative which will determine its orbit altitude from an on-board geodetic quality GPS receiver. The GLAS instrument design goal is to operate continuously for 3 years with a 5 year goal. The repeated laser altimeter measurements over the polar regions should permit assessment of changes in the polar ice sheet volume on time scales longer than 1 month. The MTPE plan is that two subsequent follow-on laser altimeter missions will be used to span the 15 year overall measurement time frame.

Surface Measurements

Lidar allows precise surface height and atmospheric measurements to be made from aircraft and spacecraft [2-5]. The GLAS instrument combines a 10 cm precision surface lidar with a dual wavelength cloud and aerosol lidar. The 1064 nm surface and cloud channel will measure the range to ice and snow surfaces, using a 60 m diameter spot with 175 m spot spacing. Over surfaces where slopes are < 3 deg., every laser measurement should have 10 cm resolution. The orbit and measurements for GLAS are summarized in Table 1, and the error budget for the ice measurements is summarized in Table 2. When over land, GLAS will profile the topographic heights, which will allow the Earth's topography to be referenced, for the first time, to a common global grid with m-level accuracy. Analysis of the GLAS echo waveforms will also allow remote measurements of the roughness or slopes of the terrain and the vertical distributions of tree and other vegetation illuminated by the laser beam.

TABLE 1. GLAS Orbit and Measurements.

Orbit altitude	598 km
Orbit inclination	94 degree
Orbit repeat tracks	1 km every 183 days
Ground track spacing	15 km at equator 2.5 km at 80 deg latitude
Post-Processing pointing	² 3 arcsec knowledge (all axes)

Position requirements:
   Radial orbit height
   Along-track
	
	Post-processing < 5 cm
	< 20 cm

Measurement orientation	Fixed, nadir viewing
Laser firing rate	40 pps

Measurements wavelengths:
   Surface & cloud tops
   Atmospheric aerosols
	
	1064 nm
	532 nm

Spot diameter on ground	60-70 m
Along-track spot separation	170 m

TABLE 2. Ice Altimetry Measurement Error Budget.

Source	Error Type	Magnitude
Instrument	a. Single shot accuracy (3 deg. surface features)	10 cm
	b. Range bias	> 5 cm
	c. Radial orbit uncertainty	5 cm
	d. Pointing uncertainty (1 arcsec, 3 deg. surface)	18 cm
	e. Clock synchronization (>1 usec)	1 cm
Spacecraft	Distance uncertainty from S/C POD to GLAS zero ref. point	0.5 cm
Environment	Atmospheric error (10 mbar error, 0.23 cm/mbar)	2 cm
	RSS error	~20 cm

Atmospheric Measurements

GLAS also incorporates a dual wavelength atmospheric lidar which shares the transmitter, receiver telescope and the 1064 nm detector with the surface lidar. The instrument specifications are summarized in Table 3. The 1064 nm lidar measurements will be used to profile the heights of clouds and dense aerosols with 75 m vertical and 175 m horizontal resolution. The 532 nm lidar receiver will use a narrow band optical filter at 532 nm and a set of photon counting detectors to measure the vertical distribution of optically thin aerosols during both day and night. The will accumulate 20-40 consecutive measurements, producing an average lidar profile at a 1 Hz rate with 75 m vertical and 3.5-7 km along-track resolution.

TABLE 3. GLAS Instrument Specifications.

Laser Type	ND:YAG slab, 3 stage Q-switched, Diode pumped
Number of lasers	3 each, one operated at any time
Laser firing rate	40 pps
Laser pulse width	4-6 nsec
Laser Divergence angle	100-120 urad
Telescope diameter	100 cm
Detector types	Si APD
Mass	300 kg
Power	300 W average
Duty cycle	100%
Data rate	~ 300 kbps
Physical size	~ 100 x 100 x 80 cm
Thermal control	Radiators with heaters, heat pipes

Laser Design

The GLAS transmitter is a frequency doubled ND:YAG laser [6,7], which is pulsed continuously at a 40 Hz rate. The GLAS laser specifications are summarized in Table 4. The 100 urad laser beam divergence produces laser footprints on the surface which are 60 m diameter and separated along-track by 175 m. GLAS incorporates 3 identical laser transmitters. One laser is used at any one time, a second is needed to meet the lifetime goal, and the third is available as a spare. There is one transmit path, and the second and third lasers are optically selected via flip mirror assemblies. The ND:YAG lasers are Q-switched, diode-pumped, conductively cooled and emit ~ 5 nsec wide pulses in a TEMoo beam. The transmitted pulse energy is ~ 75 mJ at 1064 nm and 30 mJ at 532 nm. The present laser design uses a few mJ energy Q-switched oscillator, followed by two double-pass zig-zag slab laser amplifiers.

TABLE 4. GLAS Laser Specifications.

Number of Flight Lasers	3
Energy/pulse (1064 nm)	75 mJ
(532 nm)	35 mJ
Pulse Repetition rate	40 Hz
Pulse Shape/Width Stability	± 5%
Pulse width (FHWM)	4 - 6 nsec
Beam divergence	100-120 µrad
Beam Profile	Nominal Gaussian
Beam pointing stability*	± 10 µrad
Mass: (total for 3 lasers)	38 kg
Size: (each laser)	< 20 x 20 x 10 cm
Power	< 100 W @ 28 V
Efficiency	> 6% wall plug @ 28 V
Lifetime goal	5 years continuos operation
Total laser shots in 5 years	6.3×10^9
Target laser lifetime	3.15×10^9 shots/laser

* - depends on SRS measurement capability

Laser Pointing Angle Determination

Pointing the laser beam away from nadir biases the surface lidar measurements, and accurate knowledge of the laser beam's pointing angle is critical [8] for the Laser Altimetry Mission. Over surfaces with 3 deg. slopes, knowledge to ~ 5 urad is required to achieve 10 cm height accuracy. GLAS uses a stellar reference system to determine the pointing angle of each laser firing to inertial space.

The stellar reference system approach uses a star camera oriented toward local zenith. The far field pattern of the laser beam is measured on every laser firing with a separate zenith-viewing laser reference camera. This sub-system folds a small fraction of the outgoing laser beam into the camera with two cube corner assemblies. The first cube corner, which has a transparent entry face, folds a small fraction of the laser beam angle into the second cube corner, which directs it into the laser reference camera. The laser reference camera digitizes the laser far field pattern along with several alignment markers. These are optical signals from the alignment references of the star camera and gyroscope, which are also folded into the laser reference camera. Optically measuring the laser image relative to the star camera and gyroscope references permits the laser's firing angle to the determined relative to inertial space in subsequent ground-based data processing. The laser reference camera can also view stars which pass through its field of view, which permits alignment biases between the star camera and laser reference camera to be determined with sub-arcsecond precision.

Receiver Design

The laser backscatter from clouds, aerosols and the surface are collected by the receiver telescope. The GLAS telescope is an all Beryllium Cassegrain design, with a diameter of 100 cm and a field-of-view of 300 urad. The 1064 nm detectors are a silicon avalanche photodiodes with ~ 200 MHz bandwidth. One serves as the prime detector and the backup unit can be switched into the path via a flip mirror assembly. The 1064 nm optical receiver is a larger higher performance version of the MOLA [9] receiver.

The 1064 nm surface echo is spread in time due to the slope and roughness of the terrain surface [10]. An electronics timing unit is used to measure the time of flight of the laser pulse. In contrast to conventional surface lidar receivers, the GLAS electronics receiver is an all digital approach, which digitizes and record both the transmit and echo signals at a 1 GHz rate. This permits post detection search algorithms to search the digitized window, find the largest energy echo signal, and to calculate the transmit and echo pulse energies and centroid occurrence times. This receiver approach permits a closer approximation of a maximum likelihood receiver for range, and results in better performance than simpler threshold, pulse width and energy measurements used on MOLA [11]. The 1064 nm cloud lidar electronics utilize filter the echo waveform with 2 MHz (75 m) resolution, for the lowest 30 km in range. The lidar reflections at 1064 nm are reported for every laser firing.

The 532 nm lidar receiver uses movable mirror assembly, which permits this receiver channel to align to the laser beam with a narrower 150 urad field of view. A narrow band 532 nm etalon filter is also used to reject background light. The etalon has a bandwidth of ~ 25 pm, and its center wavelength is tunable to track the laser frequency. The lidar signals from the filter are distributed via beam splitters to eight photon counting detectors. These operate in the Geiger mode and have > 50% photon counting efficiency. Using eight detectors permits > 80 MHz photon counting rates from the 532 nm receiver, which allows photon counting even with bright clouds in the telescope's field of view. At 532 nm, the sum of 40 individual lidar profiles measured by the photon counting detectors are reported once per second.

The measurement capabilities of the altimeter and lidar have been estimated by using analysis and simulations. Both the altimeter and lidar have > 3 dB performance margins. The details of the mission, instrument requirements, design and results of analysis and simulations will be presented.

Acknowledgments

We would like to acknowledge the talented contributions of the entire GLAS Instrument, GLAS Science and Laser Altimeter Mission teams who are making GLAS a reality.

References

Abshire, J.B., S.S. Manizade, W. H. Schaefer, R.K. Zimmerman, J.S. Chitwood and J.C. Caldwell (1991) "Design and Performance of the Receiver for the Mars Observer Laser Altimeter," *Technical Digest* - CLEO'91, Paper CFI4, *Opt. Soc. Amer*, Washington DC.

Afzal, R.S. (1998) "Diode-Pumped Solid State Lasers for Space-Based Applications," in Proceedings of *Space Technology and International Forum-98*, CONF. 98-0103, M. S. El-Genk, Editor, American Institute of Physics, New York.

Afzal, R.S. and M.D. Selker (1994) "Design Considerations for the GLAS Laser Transmitter," Processings, 1994 International Laser Radar Conference (ILRC), Sendai Japan.

Bufton, J.L. (1989) "Laser Altimetry Measurements from Aircraft and Spacecraft," Proc. IEEE, 77: 463.

Bufton, J.L., J.B. Garvin, J.F. Cavanaugh, L. Ramos-Izquierdo, T.D. Clem and W.B. Krabill (1991) "Airborne Lidar for Profiling of Surface Topography," *Opt. Engr.*, 30: 72.

Gardner, C.S. (1992) " Ranging Performance of Satellite Laser Altimeters," IEEE Trans. *Geosc. Rem. Sensing*, 30: 1061.

McGarry, J. F., L.K. Pacini, J.B. Abshire, and J.B. Blair (1991) "Design and Performance of an Autonomous Tracking System for the Mars Observer Laser Altimeter Receiver," *Technical Digest* - CLEO'91, paper CThR27, *Opt. Soc. Amer.*, Washington DC.

Millar, P.S. and J. M. Sirota (1998) "The Geoscience Laser Altimeter System Stellar Reference System", in Proceedings of *Space Technology and International Forum-98*, CONF. 98-0103, M. S. El-Genk, Editor, American Institute of Physics, New York.

NASA Headquarters (1995) "Geoscience Laser Altimeter System," *1995 MTPE/EOS Reference Handbook*, Washington DC, 132.

Ramos-Izquierdo, L., J.L. Bufton and P.Hayes (1994) "Optical System Design and Integration of the Mars Observer Laser Altimeter," *Appl. Optics*, 33: 307.

Zuber, M.T., D.E. Smith, S.C. Solomon, D.O. Muhleman, J.W. Head, J.B. Garvin, J.B. Abshire and J.L. Bufton (1992) "The Mars Observer Laser Altimeter Investigation, " *J. Geo. Res.*, 97: 7781.

256x256 GaN ULTRAVIOLET IMAGING ARRAY

Z.C. Huang, D.B. Mott and P.K. Shu
NASA/Goddard Space Flight Center,
Solid State Device Development Branch, Code 718.1,
Greenbelt, MD 20771
301-286-7020, Fax: 301-286-1672

Abstract

We have successfully developed a 256x256 photoconductive GaN area imaging array using a metal-semiconductor-metal structure. The array, with its pixels (30 µm^2) indium bump bonded to a readout integrated circuit, showed a 90% of yield. The spectral response of the array is consistent with the measured spectral response of the single photoconductors fabricated elsewhere on the source wafer.

INTRODUCTION

There has been a strong demand for compact solar blind solid-state ultraviolet (UV) photodetectors for high temperature operations in both private sectors and government. The use of III-V nitrides for photodetector applications is expected to yield high responsivities with low dark currents over a wide range of temperature operations. GaN and aluminum nitride (AlN) have direct bandgaps of 3.4 and 6.2 eV, respectively, with corresponding cutoff wavelengths of 365 and 200 nm. Since they are miscible with each other and form a complete series of aluminum gallium nitride (AlGaN) alloys, it should be possible to develop detectors with wavelength cutoffs anywhere in this range.

During the past three years, prototype nitride-based single element photodetectors and linear arrays have been fabricated and studied by several groups (Khan 1992, Kung 1995, Chen 1995, Huang 1996, and 1997). It is commonly reported that the detector performances varied from sample to sample depending on the material quality. Although these materials show great promise, there is still a great deal of work to be done to develop their manufacture so that their properties reach their full potential. Despite the need for continued research, these materials have reached sufficient maturity to warrant a vigorous device development effort to create UV image sensors out of these materials. Device development of detectors from these materials will require extensive effort over several years. In this work, we report for the first time our success in making large area photoconductive arrays on GaN using metal-semiconductor-metal (MSM) structures. The spectral response of the array is consistent with those reported on the element GaN photoconductors.

METHODS AND RESULTS

Semi-insulating undoped GaN materials were grown by metalorganic chemical vapor deposition (MOCVD) at 1100°C (Wickenden 1995). The MSM 256x256 area array with each

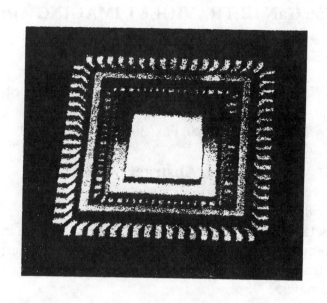

30 μm

FIGURE 1. The GaN Area Photoconductive Array.

pixel size of 30 μm² was made by conventional lift-off technique. The part of the array is shown in Fig.1. The ohmic contacts were made by evaporating 200A Ti, and then 300A Al followed by 1200A Au, and annealing at 450°C for 5 minutes in a N_2 environment. After the fabrication, each array was sliced and bump bonded to a Lockheed Martin Fairchild Systems LT9601 Readout Integrated Circuit (see Fig.1). Due to the insulating nature of the sapphire substrate, it is impossible to collect the signal through the substrate as it is done in Si-based detector arrays. Instead, the incident light was illuminated on the GaN array through the back side, i.e. sapphire side, and the readout system was bonded through the front side. No absorption loss was observed by the buffer layer as long as it is not too thick (in our sample, the thickness of buffer layer is 500A). Fig.2 shows a comparison of the spectral responses between the front and the back illuminations from a single element detector. The responsivity through the front illumination is even smaller than that through the back illumination, presumably due to the absorption and/or reflection by the GaN surface.

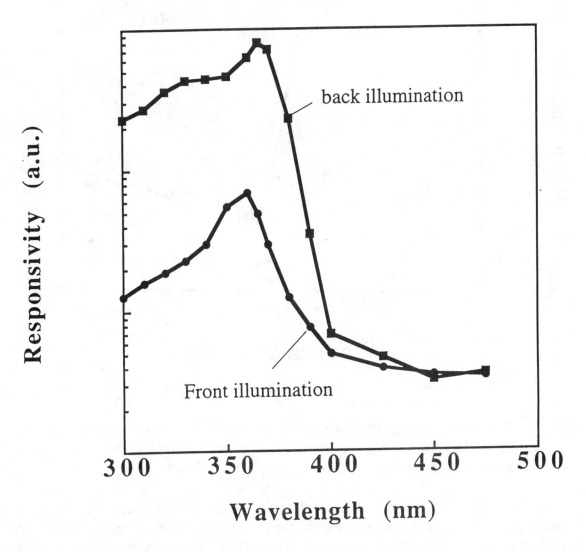

FIGURE 2. The Comparison of the Spectral Response from a Single Element Detector Between the Front and the Back Illumination.

Fig.3(a) shows a mean spectral response from our first array under an operation bias of 5 V. The response from an unmated individual pixel located elsewhere on the starting wafer showed a similar result. Since the starting material has a pretty high conductivity, the visible response in

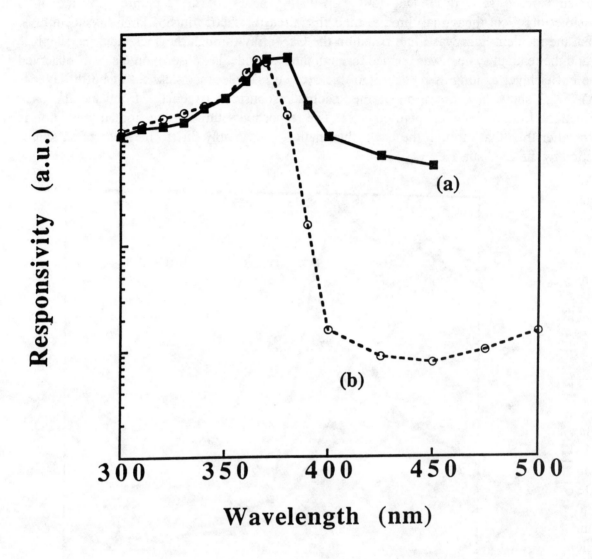

FIGURE 3. (a) The Mean Spectral Response of the GaN Array Under a Bias of 5V Through Back Illumination. (b) The Spectral Response from a Test Pixel of the GaN Array in the Second Production Run.

Fig.3(a) is due to large leakage current, rather than a result of light leakage to the silicon readout chip, because we used an extra layer to protect the readout chip from the light. Also shown (in Fig.3(b)) is the spectral response of a test pixel from a GaN wafer in the second production run, indicating much improved visible light rejection. These results are consistent with the previous reported spectral responses from the GaN single element (Huang 1997) and linear detectors (Huang 1996).

The success of GaN area array makes it possible to replace the current UV detection devices, such as photomultiplier tubes and silicon-based sensors in Earth observation applications.

SUMMARY

In conclusion, we have reported for the first time the fabrication of large area GaN array using a metal-semiconductor-metal structure through back illumination. The device consists of a 256x256 array of 30 μm square GaN photoconductors indium bump bonded to a Lockheed Martin Fairchild Systems LT 9601 Readout Integrated Circuit. About 90% of the pixel are responsive. The spectral response showed a good visible light rejection.

Acknowledgments

We would like to thank Dr. D.K. Wickenden for sample preparation and valuable discussions.

References

Khan M.A., Kuznia J. N., Olson D. T., Van Hove J. M., Blasingame M., and Reitz L. F., (1992) "High-responsivity photoconductive ultraviolet sensors based on insulating singe-crystal GaN epilayers," Appl. Phys. Lett., 60: 2917

Kung P., Zhang X., Walker D., Saxler A., Piotroski J., Rogalski A. and Razeghi M., (1995) "Kinetics of Photoconductivity in N-Type GaN Photodetector," Appl. Phys. Lett., 67: 3792

Chen Q., Khan M. A., Sun C. J. and Yang J. W., (1995)"Visible-Blind Ultraviolet Photodetectors Based on GaN p-n Junctions," Electron. Lett., 31: 1781

Huang Z. C., Chen J. C. and Mott D. B. (1997) "Improved Performance of GaN Ultraviolet Detectors," J. Electron. Mater., 26: 330

Huang Z.C., Chen J.C., Mott D.B. and Shu P.K., (1996) "High Performance GaN Linear Array," Electron. Lett., 32: 1324

Wickenden A.E., Wickenden D.K. and Kistenmacher T.J., (1995) "The Effect of Thermal Annealing on GaN Nucleation Layers Deposited on (0001) Sapphire by MOCVD," J. Appl. Phys., 75: 5367

SPLIT-GEOMETRY DETECTORS, OUR EYES IN SPACE

R. J. Martineau, F. Peters, A. Burgess,
C. Kotecki, S. Manthripragada, J. Godfrey
D. Krebs, D. Mott, and P. Shu
NASA Goddard Space Flight Center
Code 718, Greenbelt, MD 20771
(301) 286-8006

J. Z. Shi and K. Hu
Hughes STX Corp.
Lanham, MD 20706
(301) 286-7944

Abstract

Infrared detectors have projected our ability to explore our planet and our solar system far beyond the spatial, temporal, and spectral limitations of our natural vision. As such, they are our eyes in space, constantly searching the heavens, and sending back information about the origin, constitution, and dynamics of planetary atmospheres, and other processes of interest. Their ability to do this effectively depends on their sensitivity. Today, long wave PC (photoconductive) HgCdTe detectors are the detectors of choice for applications requiring high sensitivity at long wavelengths and elevated temperature. However, planetary exploration and space surveillance of the earth's climatic condition are presently still limited by the sensitivity of available detectors. This paper will describe detectors developed at Goddard to provide enhanced performance for applications such as the CIRS/Cassini mission to Saturn and Titan, and the GOES weather satellite. Specifically, this paper will show theoretically and experimentally how detectors of split-geometry design can be exploited to increase detector resistance, responsivity, and detectivity, while decreasing 1/f noise and power dissipation. Photomicrographs of split-geometry detectors will be shown, and data demonstrating theoretical split-geometry design advantages will be presented for flight arrays built for the CIRS/Cassini mission, and for advanced detectors for GOES.

INTRODUCTION

First generation PC (photoconductive) infrared detectors have been largely supplanted by second generation PV (photovoltaic) detectors because the latter are easier to build in large numbers, have negligible power dissipation, and have very low 1/f noise and superior linearity. In addition, because their high impedance allows efficient coupling to multiplexers, lead count is significantly reduced, allowing very large PV focal plane arrays to be built and packaged. However, PC detectors are still the detectors of choice for applications requiring high sensitivity at long wavelengths and high temperature, since they outperform PV detectors under these circumstances. This has led to dual-mode focal plane array architectures for such programs as MODIS (Moderate Resolution Imaging Spectrometer), (Dowler 1993), AIRS (Atmospheric Infrared Sounder), (Morse 1995), CIRS (Composite Infrared Spectrometer), (Kunde 1996), and GOES (Geostationary Operational Environmental Satellite), (Hursen 1996 and Murphy 1996) where PV is used for shorter wavelengths, and PC for the longest wavelengths. For such missions, instrument capability is still limited by the sensitivity of available long wave PC detectors. It is the purpose of this paper to show theoretically and experimentally how split-geometry detectors can be used to advantage to enhance the performance of long wave PC detectors for such application. In order to accomplish this, we will present case examples taken from detector development programs for CIRS/Cassini flight arrays and advanced detectors for GOES. But first we will describe some potential advantages afforded by the split-geometry design, and present the theoretical basis for those advantages.

POTENTIAL ADVANTAGES OF SPLIT-GEOMETRY DETECTORS

Typically photoconductive detectors have square or rectangular active areas with electrical leads at either end as shown in figure 1.

(a) Square Detector   (b) Rectangular Detector

FIGURE 1. Typical Photoconductive Detectors with a) Square, and b) Rectangular Active Areas.

If slits are introduced into the active area to narrow and lengthen the current path between electrodes, the device is known as a split-geometry or serpentine detector. A wide variety of such devices are possible. Figure 2a illustrates a split-square detector with a metal current-return cap. Figure 2b illustrates a split-square detector with a slit extending to the center of the detector and no end cap, and figure 2c depicts a split-circle detector.

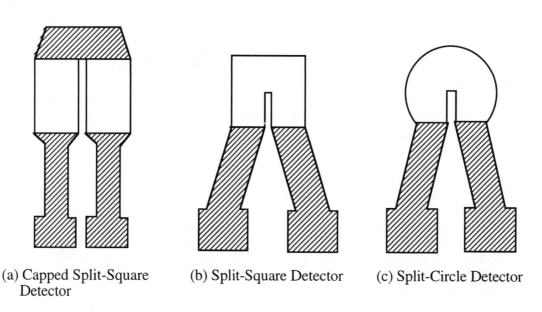

(a) Capped Split-Square   (b) Split-Square Detector   (c) Split-Circle Detector
    Detector

FIGURE 2. Split Detectors with a) Capped Split-Square, b) Split-Square, and c) Split-Circle Active Areas.

If one compares figures 1a and 2a, one immediately sees that the split-square detector will have a resistance 4 times that of the square detector since the ratio of the squares in the current path is 4 to 1. If the PC detector is not in sweepout, and is limited by thermal gr-noise, one also concludes that the responsivity and noise of the split-square detector will be twice that of the square detector when operated at constant power dissipation, since both responsivity and noise will go like the square root of resistance. This means that it will be easier for a split-square detector to meet a high responsivity requirement at low bias power than it will be for a detector of traditional design. The split-square detector will also make it easier for detector resistance to dominate lead resistance, and for detector noise to dominate preamp noise. In addition, if a program calls for close butting of arrays of different wavelengths, or of PV and PC arrays as was required for CIRS, the split-geometry approach is the natural, and perhaps the only way to go.

While the CIRS detectors had active areas 200 μm on an edge, and were not operated in heavy sweepout, GOES long wave sounder detectors are much smaller, having nominal active areas of 50 μm x 50 μm, and are more susceptible to carrier sweepout. Sweepout is especially troublesome for GOES detectors, since the high responsivity requirement is generally met by increasing detector bias which aggravates sweepout. Thus it is important to examine split-geometry detector operation under sweepout conditions if the advantages of small split-geometry detectors are to be properly understood.

For present purposes we are only interested in obtaining a rough understanding of the functional dependence of detector performance on bias and geometry. To this end we examine a simplified model of a detector with a slit that extends the entire length of the active area to a metal shorting cap. This detector is easier to model than the split-square and split-circle detectors of figures 2b and 2c, yet reveals the mechanisms at work in harder-to-model split-geometry designs.

When we do this, the detectivity, D^*_λ, can be written in terms of the standard expressions for responsivity, R_λ, and total noise per root Hz, Vn, i.e.,

$$D^*_\lambda = \frac{R_\lambda}{V_N} \cdot \sqrt{A}, \tag{1}$$

where V_N is the rms sum of the 1/f noise, Vfn, the gr-noise, Vgr, and the preamp noise, Va. The only difference is that the bulk lifetime, τ_b, in the standard expressions for R_λ and Vgr gets replaced with an effective lifetime, τ_e, given by:

$$\tau_e = \tau_b \left[1 + \frac{\tau_b}{\tau_d} \cdot \left(Exp\left(-\frac{\tau_d}{\tau_b}\right) - 1 \right) \right], \tag{2}$$

where τ_d is the time it takes minority carriers to drift across the length of the device in the applied field. That is:

$$\tau_d = \frac{l^2}{\sqrt{P_D R_D} \cdot \mu_a}, \tag{3}$$

where l is the length of the current path, P_D is the power dissipation, R_D is detector resistance, and μ_a, the ambipolar mobility, is essentially the hole mobility in n-type material.

It is important to notice that the effective lifetime is essentially the bulk lifetime at low bias, and can become very small at high bias, tending to $\tau_d/2$ when the bias is so high that $\frac{\tau_d}{\tau_b} \ll 1$. When this occurs the device is said to be operating in heavy sweepout, a mode of operation which should be avoided, as we shall see.

The usual expressions for R_λ and V_N, with τ_b replaced by τ_e, are given by:

$$R_\lambda = \frac{\eta \cdot \lambda_c}{h \cdot c} \cdot \left(\frac{\sqrt{P_D R_D} \cdot \tau_e}{n \cdot l \cdot w \cdot t} \right), \tag{4}$$

and

$$V_N = \sqrt{V_{fn}^2 + V_{gr}^2 + V_a^2}. \tag{5}$$

The V_{gr} noise and 1/f noise are in turn given by the expressions

$$V_{gr} = 2 \cdot \tau_e \cdot \left(\frac{\sqrt{P_D R_D}}{n \cdot t \cdot \sqrt{l \cdot w}} \right) \cdot \sqrt{\frac{n \cdot p}{n + p}} \cdot \sqrt{\frac{t}{\tau_e}}, \tag{6}$$

and

$$V_{fn} = \beta \cdot \frac{\sqrt{P_D R_D}}{\sqrt{f}}. \tag{7}$$

In the above, λ_c is the cutoff wavelength, n and p are carrier concentrations, $l \cdot w \cdot t$ is the detector volume, A its area, η its quantum efficiency, f is the noise frequency, and β is a dimensionless, process dependent parameter.

Examination of the expressions for R_λ, thermal Vgr, Vfn, and D^*_λ for negligible preamp noise indicates the following explicit dependence on $\sqrt{P_D R_D}$ and τ_e:

$$\begin{aligned}
& R_\lambda \propto \sqrt{P_D R_D}, \quad V_{gr} \propto \sqrt{P_D R_D}, \quad V_{fn} \propto \sqrt{P_D R_D}, \quad D^*_\lambda \propto Const, \\
\\
& R_\lambda \propto \tau_e, \quad V_{gr} \propto \sqrt{\tau_e}, \quad V_{fn} \propto Const, \quad D^*_\lambda \propto \sqrt{\tau_e}, \quad \text{(if Vgr dominates).}
\end{aligned} \tag{8}$$

For heavy sweepout, we have seen that:

$$\tau_e \to \frac{1}{2}\tau_d \propto \frac{1}{\sqrt{P_D R_D}}. \tag{9}$$

Thus for sweepout, the dependence on $\sqrt{P_D R_D}$ and τ_e can be combined to give:

$$R_\lambda \propto Const, \quad V_{gr} \propto \left(P_D R_D\right)^{1/4}, \quad V_{fn} \propto \sqrt{P_D R_D},$$

$$D^*_\lambda \propto \left(P_D R_D\right)^{-1/4} \text{ (if Vgr dominates).}$$
(10)

Therefore at moderate bias, i.e. $\sqrt{P_D R_D}$, responsivity keeps up with noise and D^*_λ remains constant, but as bias increases and the device goes into sweepout, the responsivity saturates while noise continues to increase, resulting in decreasing D^*_λ. If one reduces bias for devices under sweepout, one would therefore expect to achieve higher D^*_λ until the Vgr and 1/f noise sources drop below the preamp noise. When this happens, responsivity decreases faster than the total noise, resulting in decreasing D^*_λ. Thus D^*_λ as a function of bias has a maximum at low bias in a region between heavy sweepout and preamp-noise dominance. The maximum is larger for split-geometry detectors than for standard detectors because R_λ has a stronger dependence on τ_e than does the noise, and τ_e improves with lower bias.

The 1/f knee, the frequency at which the 1/f noise is equal to the white noise, is given by the expression:

$$f_{knee} = \frac{(\beta \cdot V_b)^2}{V_{gr}^2 + V_a^2}.$$
(11)

When the gr-noise dominates the preamp noise, fknee has no explicit dependence on $\sqrt{P_D R_D}$ but is proportional to $1/\tau_e$. Therefore under heavy sweepout fknee is proportional to $\sqrt{P_D R_D}$. This means that the 1/f knee of a device operated in sweepout will decrease when the bias is lowered.

Finally we notice that no benefit accrues to use of high bulk lifetime materials or of processes that preserve that lifetime if the device is operated under heavy sweepout. This follows from the fact that τ_b is replaced by τ_e which becomes proportional to $1/\sqrt{P_D R_D}$ at high bias. In this case we have seen that D^*_λ degrades with increasing bias. However, if we reduce bias, D^*_λ becomes proportional to $\sqrt{\tau_e}$ which approaches $\sqrt{\tau_b}$. Therefore if one operates the device at low bias, one can realize the benefits of using good processes and materials having high bulk lifetimes.

LOW SWEEPOUT, SPLIT-GEOMETRY DETECTORS FOR GOES

Sweepout is an important consideration for GOES Long Wave Sounder (LWS) detectors since the active area is small, the drift time is short, and both responsivity and D^*_λ specs are high. To address these problems we designed and fabricated split-square and split-circle detectors, examples of which are shown in figures 3 and 4 (Martineau 1996).

 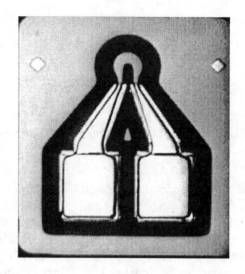

FIGURE 3. A LWS Detector of Split-Square Design

FIGURE 4. A LWS Detector of Split-Circle Design.

LWS detector requirements are specified at 40 Hz and 100 Hz for 102 K operation, with most of the screen testing being done at 77 K for ease, and at 40 Hz to address the harder spec. D^*_λ normally falls by a factor of 2 to 3 for 102 K operation.

Screen test data for a split-square detector is shown in figure 5 (Martineau 1996). Usually we test split-geometry detectors at two biases, 0.25 and 0.50 mA. This device was tested at 77 K from 0.1 mA to 2.0 mA to obtain the bias dependence of the data. The resistance was measured at 134 ohms, and the cutoff wavelength was 17.2 µm. The total noise measured at 0.1 mA and 40 Hz with a 1.8 nV/√Hz preamp was 2.7 nV/√Hz.

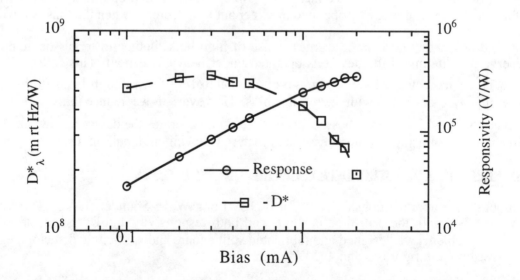

FIGURE 5. Measured D^*_λ and Responsivity as a Function of Bias for a Split-Square Detector.

We see from figure 5 that even a split-geometry detector will experience sweepout as bias increases, leading to sublinear responsivity. However, the degree of sweepout is significantly reduced relative to a detector of standard geometry since the current path has been greatly increased. At lower biases the preamp noise becomes comparable to the other noise sources, and doesn't decrease with decreasing bias as the other noise sources and the responsivity do, leading to lower D^*_λ at lower biases. Thus dominance of preamp noise at low bias and sweepout at high bias produces a maximum in the D^*_λ curve at intermediate bias levels as predicted by theory.

Variable temperature data for a split-circle detector is presented in figures 6 and 7 (Martineau 1997). The resistance of this device is 188 Ω, and its cutoff wavelength is 15.7 μm at 77 K. Data was taken at two

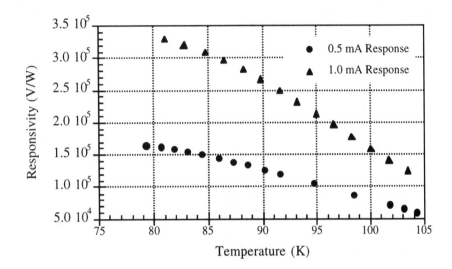

FIGURE 6. Responsivity vs. Temperature for a Split-Circle Detector.

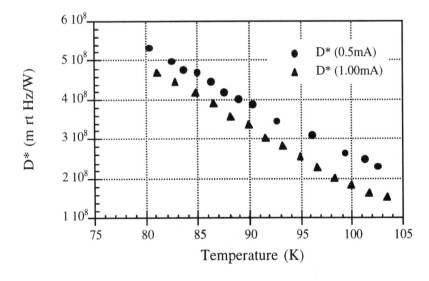

FIGURE 7. Detectivity vs. Temperature for the Same Split-Circle Detector.

biases, 0.5 mA and 1.0 mA, and noise was measured at 40 Hz. From figure 7, we notice that D^*_λ is higher at lower bias as might be expected for a split-geometry device. This device is not yet A/R coated, but it exceeds the D^*_λ spec of 2×10^8 m $\sqrt{Hz}$/W at 40 Hz and 102 K, and the responsivity spec of 60,000 V/W. With A/R coating, these values would increase by about 25%.

BUTTABLE, SPLIT-GEOMETRY FLIGHT ARRAYS FOR CIRS

The Solid State Device Development Branch at Goddard Space Flight Center was tasked with delivering 10 element PC HgCdTe flight arrays to the CIRS/Cassini program for detection of IR radiation in the 9.1 to 16.7 μm spectral band. The arrays were required to be buttable within 500 μm to 10 element PV arrays provided by CE-Saclay-France for detection of radiation in the 7.1 to 9.1 μm band. Together, the arrays would populate the focal plane of the mid-IR Michelson interferometer of the CIRS instrument.

The active area for CIRS detectors was specified as 200 μm x 200 μm nominal, so sweepout was not a major consideration. However, array buttability, and low power dissipation with high detector resistance and responsivity dictated a split-geometry detector approach. An example of the type of buttable, split-geometry detector array we designed, fabricated, qualified, and delivered is shown in figure 8. Notice that the bonding pads are all on one side of the array, allowing close butting of arrays.

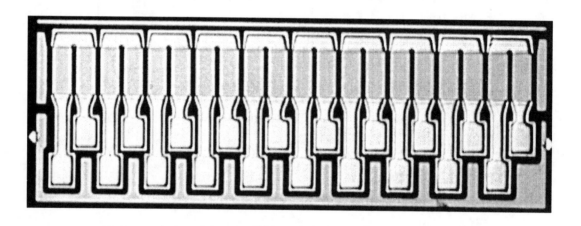

FIGURE 8. A Buttable 10-element CIRS Array of Split-Geometry Design.

CIRS arrays were tested at 36 Hz with a preamp having an input referred noise at 36 Hz of 1.5 nV/$\sqrt{Hz}$ and a gain of 1×10^4. Table 1 shows the performance of a flight array at 80 K, its intended operating temperature. Notice the long cutoff wavelength of 18.7 μm needed to meet the peak response at 16.2 μm. Due to the split-geometry configuration, the array was able to exceed the D^*_λ spec of 2.2×10^8 m $\sqrt{Hz}$/W, and the responsivity spec of 10,000 V/W while dissipating less than 0.5 mW per element as required. Figure 9 shows a histogram by element number of peak responsivity and D^*_λ. Excellent uniformity is observed without the use of individual bias resistors. The array was also tested and met program requirements for linearity of response, crosstalk, spectral uniformity, and radiation hardness.

TABLE 1. CIRS Flight Array Performance at 80 K.

Det.	R (Ω)	λ Cutoff (μm)	I bias (mA)	Signal (V)	Vnoise 36 Hz (V/rtHz)	G factor	Peak Resp (V/W)	D^*_λ Peak (m rtHz/W)
1	104	18.5	2.15	9.95E-05	1.03E-08	0.484	1.33E+04	2.59E+8
2	101	18.6	2.18	9.11E-05	1.01E-08	0.464	1.27E+04	2.51E+8
3	99	18.6	2.20	9.21E-05	9.41E-09	0.460	1.29E+04	2.74E+8
4	97	18.7	2.22	8.93E-05	9.12E-09	0.460	1.25E+04	2.74E+8
5	96	18.7	2.23	9.33E-05	9.16E-09	0.462	1.30E+04	2.85E+8
6	96	18.7	2.23	9.19E-05	1.01E-08	0.468	1.27E+04	2.50E+8
7	95	18.7	2.24	9.30E-05	9.04E-09	0.476	1.26E+04	2.79E+8
8	91	18.7	2.29	9.16E-05	9.29E-09	0.474	1.25E+04	2.68E+8
9	92	18.7	2.28	9.11E-05	8.71E-09	0.474	1.24E+04	2.85E+8
10	91	18.7	2.29	9.30E-05	7.99E-09	0.492	1.22E+04	3.05E+8

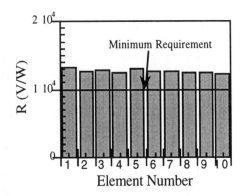
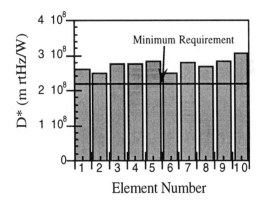

FIGURE 9 Peak Responsivity and Detectivity at 80K Versus Element Number.

CONCLUSIONS

Although first generation PC detectors have their limitations, they are still the detectors of choice for a variety of programs requiring high performance at long wavelengths. Even so, program requirements are often limited by the performance of available detectors. We have shown that split-geometry PC detectors can be used to advantage to provide enhanced performance for such applications. Specifically, using examples from GOES and CIRS, we have shown how split-geometry detector designs can address buttability requirements, and can provide higher detector resistance, responsivity, and detectivity at low operating frequencies, while assuring low power dissipation.

Acknowledgments

The authors thank NOAA and Marty Davis, GOES Project Manager, for their interest in the GOES detector development effort, and Virgil Kunde and Richard Barney for their support of the CIRS detector development effort. We gratefully acknowledge their funding, without which these efforts could not have proceeded.

References

Dowler, M. et. al. (1993) "Broad Spectrum, Multi-Band Focal Plane Arrays for the Moderate Resolution Imaging Spectroradiometer (MODIS)," in Proc. IRIS Detector, 2: 113-122.

Morse, P., Miller, C., Chahine, M., O'Callaghan, F., Aumann, H., Karnik, A. (1995) "Candidate Future Atmospheric Sounder for the Converged U. S. Meteorological System," in Proc. SPIE, 2553: 329-343.

Kunde, V. et. al. (1996) "Cassini Infrared Fourier Spectroscopic Investigation," in Proc. SPIE, 2803: 162-177.

Hursen, K. and Ross, R. (1996) "The GOES Imager: Overview and Evolutionary Development," in Proc. SPIE, 2812: 160-173.

Murphy, J. and Hinkal S. (1996) "GOES Sounder Overview," in Proc. SPIE, 2812: 174-181.

Martineau, R. et. al. (1996) "High Performance HgCdTe Infrared Detectors for the GOES Long-Wave Sounder," in *GOES-8 and Beyond*, Washwell, E., Editor, Proc. SPIE, 2812: 490-500.

Martineau, R. et. al. (1997) "PC Detector Passivation for High Performance," in *Infrared Spaceborne Remote Sensing V*, Strojnik, M. and Anderson, B., Editors, Proc. SPIE, 3122: 392-398.

CDZNTE TECHNOLOGY FOR GAMMA RAY DETECTORS

Carl Stahle, Jack Shi, Peter Shu, Scott Barthelmy, Ann Parsons, and Steve Snodgrass
NASA Goddard Space Flight Center
Greenbelt, MD 20771
(301) 286-0968, 286-7944, 286-5191, 286-3106, 286-1107, 286-8029

Abstract

CdZnTe detector technology has been developed at NASA Goddard for imaging and spectroscopy applications in hard x-ray and gamma ray astronomy. A CdZnTe strip detector array with capabilities for arc second imaging and spectroscopy has been built as a prototype for a space flight gamma ray burst instrument. CdZnTe detectors also have applications for medical imaging, environmental protection, transportation safety, nuclear safeguards and safety, nuclear non-proliferation, and national security. This can be accomplished from space and also from portable detectors on earth. One of the great advantages of CdZnTe is that the detectors can be operated at room temperature which eliminates the need for cryogenic cooling. CdZnTe detectors have good energy resolution (3.6 keV at 60 keV) and excellent spatial resolution (< 100 microns). NASA Goddard has developed the fabrication technology to make a variety of planar, strip, and pixel detectors and integrated these detectors to high density electronics. We have built a 2 x 2 and a large area (60 cm^2, 36 detectors) 6 x 6 strip detector array. This paper will summarize the CdZnTe detector fabrication and packaging technology developed at Goddard.

INTRODUCTION

CdZnTe (CZT) is an optimum technology for hard x-ray astronomy as well as ground-based applications. Large area and large volume spectrometers are of interest for medical imaging, environmental protection, transportation safety, nuclear safeguards, nuclear non-proliferation, and national security. The many applications of these detectors include the monitoring of air emissions and liquid effluents escaping from nuclear and fossil fuel plants, new high sensitivity x-ray monitoring devices at airport baggage checks, real time dosimetry, inexpensive continuous monitoring and early leak detection of radioactive waste disposal containers, and portable monitoring devices for nuclear safeguards and treaty verification. These novel devices also promise to revolutionize medical imaging devices by reducing the exposure of the patients to x-rays and improving the quality of the images.

A CZT strip detector array is being developed at NASA Goddard Space Flight Center for a proposed gamma ray Burst Arc Second Imaging and Spectroscopy (BASIS) space flight mission to accurately locate gamma ray bursts, determine their distance scale, and measure the physical characteristics of the emission region. The energy range of interest is 10 - 150 keV. A CZT strip detector array with less than 100 μm spatial resolution combined with a coded aperture mask will allow a gamma ray burst to be positioned on the sky to a few arc second accuracy. This is an advance of more than an order of magnitude in angular resolution compared to the best previous instrumentation in hard x-ray and gamma ray astronomy. In addition, the good energy resolution of the CZT strip detector will be used to measure the spectral characteristics of the burst. Previous work has shown the good imaging and spectral resolution for this CZT strip detector (Bartlett et al. 1996 and Kurczynski et al. 1996).

A large area CZT detector array is needed to achieve the high sensitivity required to detect gamma ray and hard x-ray sources with the coded aperture mask. A double sided 2 x 2 array (6.5 cm^2, 4 detectors) and a 6 x 6 array (60 cm^2, 36 detectors) has been assembled and integrated to electronics. Details on the performance of the 2 x 2 array can be found elsewhere (Parsons et al. 1997 and Palmer et al. 1997). The configuration of the 6 x 6 strip detector array places challenging fabrication requirements on the detectors. Since 6 strips are connected in series to make 1 long (7.6 cm) strip, the leakage current in the individual strips of each detector needs to be small to have low leakage current noise from the long strip. Our goal for the fabrication run was to have an average strip leakage current < 500 pA/detector strip. In order to achieve the science goal of > 90% active area in the array, > 99% of the strips are required to be active after integration to the electronics. This yield of > 99% includes strips which are not open or shorted due to fabrication defects or wire bond failures. The focus of this paper will be on the fabrication and testing of a large number of strip detectors to achieve these goals. Critical issues to be addressed include fabricating and testing metal contacts with low leakage current and with excellent wire bonding yield, achieving high yield for good strips, surface cleaning and passivation.

FABRICATION OF CZT STRIP DETECTORS

The materials used in this work were discriminator grade 1.5 cm x 1.5 cm x 0.2 cm $Cd_{0.9}Zn_{0.1}Te$ purchased from eV Products. The two dimensional CZT strip detectors were produced by having strips on each side of the detector orthogonal to each other. The strip detector was patterned with 100 μm pitch, 50 μm wide metal strips. The fabrication process consisted of the following steps: mechanical polishing for producing a smooth starting surface, solvent cleaning to remove any debris from the polishing, chemical etching for removing damage from the polished surface and to ensure a fresh surface, photolithography to develop the strip pattern, metalization and lift-off. For the second side process, the procedure was the same except the strips on the finished side were protected with photoresist. After both sides of the detector were fabricated, the detectors were passivated to increase the interstrip resistance and annealed to improve the adhesion of the metal to the CZT. Figure 1 shows a closeup of a finished CZT strip detector. The strip pitch is 100 μm with 50 μm wide strips. The bonding pads are staggered with a length of 400 μm and a width of 120 μm. The guard ring is 450 μm wide.

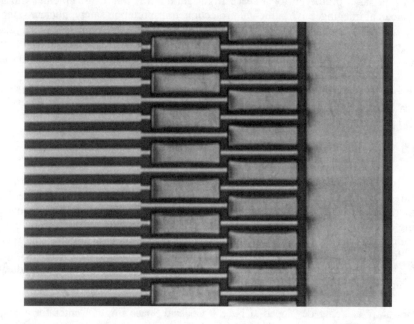

FIGURE 1. Closeup View of the CZT Strip Detector.

Mechanical Polishing

The surfaces of both sides of the CZT were mechanically polished with 0.5 μm Al_2O_3 to produce a smooth surface for patterning and metal deposition. The samples were mounted on a jig with crystal bonding wax at 90 - 100 °C and then were polished on a DF 200 pad (nylon pad) for 30 minutes at 60 rpm. The sample was demounted from the jig and cleaned by soaking in acetone, methanol, and deionized water.

Solvent Cleaning

The polished CZT samples were soaked in acetone for 2 minutes and then the surfaces were cleaned by a soft swab in acetone to remove any residues from mechanical polishing. The sample was rinsed in methanol, water and blown dry by nitrogen. At this point, the surface is inspected with a microscope to evaluate the effectiveness of the cleaning. Keeping the surfaces clean is critical for achieving a high yield of good strips.

Chemical Etching

The samples were etched in 5% bromine in ethylene glycol for six minutes followed by rinses in ethylene glycol and water to remove damage from the mechanical polish and to create a good chemical surface for the metal contact. Approximately 6 μm of CZT was removed in this process. One problem encountered from this step was that some defects in the surface of the CZT etched at a slower rate creating bumps on the CZT surface. These "defect bumps" broke the photoresist during the masking step and gave shorted strips after the metal was deposited. A chemical etch was used after photolithography and before metal deposition to improve the adhesion of the platinum metal to CZT and to aid in the lift-off process. This additional etching removed between 4 - 6 μm of CZT in the metal film area.

Photolithography

A thin layer of photoresist (~2.7 μm) was spun on the CZT surface at 4000 rpm followed by a 90 °C bake for 30 minutes. The fine pitch strip pattern on the CZT surface was created by a mask aligner with an exposure time in the range from 10 to 20 seconds. Alignment marks were made on the mask to aid in centering the pattern on the detector. The strips on each side could be aligned to one another with the aid of a backside camera. After a 5 minute chlorobenzene soak and 5 minute 90 °C bake, the pattern was developed in a chemical solution for 90 seconds.

Metalization

A layer of platinum and gold was sputter deposited for the metal strips and bonding pads. The platinum film was used to achieve both good adhesion and low leakage current. The platinum film thickness was 0.09 μm with a deposition rate of 0.045 μm/min. The total thickness of the strips was 0.2 μm, and the total thickness of the bonding pads was 2.5 μm. The metal thickness of the strips was kept thin to minimize the absorption of low energy (< 20 keV) x-rays in the metal. Thick gold was deposited on the bonding pads to minimize the damage to the CZT during wire bonding. After metal deposition, the metal pattern was finished by lift-off with an acetone soak and spray.

Surface passivation

After both sides of the detector were patterned with metal strips, the detector was cleaned in acetone, methanol, and water and passivated to improve the interstrip resistance and long term stability. The interstrip resistance typically increased from a few megaohms to greater than gigaohms after surface passivation.

Annealing

After passivation, the detector was annealed in vacuum at 100 °C for 10 hours to improve the adhesion of the metal to CZT for wire bonding. The time has not been optimized. We have observed that the strip leakage current increased after annealing for the platinum contact but decreased for the electroless gold contact. In addition, the interstrip resistance was increased. More work is needed to understand the effects of annealing on contacts and surfaces.

Electrical Testing and Final Cleaning

The strip detectors were electrically tested for strip leakage current and interstrip resistance in a probe station. The detectors were then cleaned in acetone, methanol, and water and an oxygen plasma. The main purpose of the oxygen plasma clean was to remove organics from the metal bonding pads so the wire bonds would stick easier with less ultrasonic power. Following a final surface passivation, the detectors were visually inspected and ready for mounting in an array.

EVALUATION OF CZT STRIP DETECTORS

The three methods used to evaluate and qualify CZT strip detectors for an array were visual inspection, electrical testing, and wire bonding tests. The visual inspection detected open strips and shorted strips as well as cracks, voids, and other defects in the detector which might impair detector performance. The electrical tests measured the leakage current from individual strips through the bulk or volume of the detector and the interstrip leakage current. An interstrip resistance was calculated from the interstrip leakage current. The wire bonding tests were done on a small number of samples to determine the typical wire bonding yield on the metal bonding pads. In addition, pull tests were done on all the wire bonds of these samples to test the adhesion of the wire to the metal pad and the adhesion of the metal pads to the CZT.

Visual Inspection

In a visual inspection of 44 CZT strip detectors fabricated for the 2 x 2 and 6 x 6 arrays, there were 38 open strips out of 11,176 total strips or 0.34%. The open or broken strips were caused by voids or pipes at the surface of the CZT. For a void across or along a strip, the photoresist was too thick in the void so the photoresist would not be completely exposed. Subsequently, the metal strip would be broken in this area during the lift-off process. A void across a strip was much more destructive than a void along a strip.

For the same 44 strip detectors, there were 94 shorted strips out of a total of 11,264 strips and guard rings or 0.83%. The majority of shorts were between two strips or bonding pads. These strips were shorted either by voids, defect bumps or a metal liftoff problem where the metal did not break off between strips. The defect bumps would break the photoresist between the strips when the detector was in contact with the mask which allowed metal to be deposited between the strips. The photoresist metal liftoff problem was solved for later detector samples by making sure the trenches for the metal films were deep enough so the metal lifted off easily and cleanly. For this set of 44 samples, the total yield of good strips (not open or shorted) was 98.8%. Although this is good for a first try at processing a large number of detectors, we need to improve. If CZT material without voids is used, we believe the fabrication yield for good strips will be close to 100%.

Electrical Tests

The electrical tests for the CZT strip detectors provided the first indication of the quality of our metal contacts and surfaces. We measured the strip leakage current though the detector and between strips. Details of these measurements have been published previously (Stahle et al. 1996). The average strip leakage current and interstrip resistance was calculated from the average of measurements on 4 strips across one side of the detector. The interstrip resistance is calculated from the leakage current between strips and provides an indication of the leakage currents between strips if there are voltage differences between strips. For 49 strip detectors, the average of the strip leakage current was 332 ± 232 pA, and the interstrip resistance was 5.4 ± 4 GΩ. The variation in strip leakage current and interstrip resistance between samples is large and not understood. Possible explanations include variations in our metal contacts and surfaces or material differences between CZT samples. Since all the detectors were fabricated with the same process and many were fabricated in the same batch, we can not identify an obvious processing variation. More work is needed to understand these variations. The average total strip leakage current for 6 strips would be 2 nA. For 2 nA of leakage current and a typical integration time of 2 μs, the expected leakage current or shot noise is 158 electrons rms. This is significantly less than the expected capacitance noise of 320 electrons (25 pF capacitance load from strips) expected from the high density electronics readout chip. With passive cooling to -20 °C for a spaceflight instrument, the leakage current would probably decrease by a factor of 10 to 100 rendering the shot noise negligible. From the geometry of the strip detector, our average interstrip resistance is calculated to be 1.4×10^{12} ohms/square. This is an important number for a variety of other electrode configurations for CZT detectors where interelectrode resistance is critical for a uniform electric fields and low leakage current betweeen electrodes. Our fabrication process will produce a high surface resistance for a variety of CZT electrode geometries.

Wire Bonding Tests

One of the critical tests for evaluating the CZT detector fabrication process was the wire bond tests. Our goal was to achieve a wire bonding yield > 99.5%. From our final fabrication process used to produce the 44 detectors for the 2 x 2 and 6 x 6 arrays, we prepared 4 strip detectors for wire bond tests. Two samples were used to test the "A" side or the gamma ray incident side of the detector, and two samples were tested on the other side or "B" side. For each side, 512 total wire bonds were attempted. An automatic ultrasonic wedge wire bonder was used with 25 μm diameter aluminum (99.99%) bonding wire. We have previously found that aluminum bonding wire with 1% silicon alloy (which is commonly used for wirebonding silicon chips) caused cratering in the CZT material under the bonding pads. We believe the cratering was due to the increased hardness of the wire and higher ultrasonic power required to achieve a satisfactory bond. To minimize stress in the CZT material, the first bond was placed on the CZT bonding pad and the second bond on the ceramic bridge. The second bond experiences a significant pull force immediately after the bond is made when the machine cycles through the wire tear-off portion of the cycle. Pull tests were done on the wire bonds to determine the strength of the bonds. The results from these tests were very good. On the "A" side, we achieved 100% wire bonding yield and 0 bond peels on the pull tests. The wire broke before the bond peeled which is the best result. For the "B" side, we achieved 99.6% wire bonding yield (2 bonds peeled during bonding) and had 7 bonds peel during the pull test. However, all of the 7 bond peels resulted from a force between 2.5 and 4.6 grams which exceeds the MIL SPEC standard of 1.5 grams. For both sides with a total of 1024 wire bonds, we achieved a 99.8% wire bonding yield. We found that the "B" was more difficult to bond, and the ultrasonic power had to be increased by 20% to get the bonds to stick. This is probably the reason we had some bond peeling because the higher ultrasonic power caused greater damage to the CZT surface underneath the metal bonding pad. We have found in previous work that the CZT surface can be broken if the ultrasonic power is too high. CZT is a much more brittle material compared to Si. Improved surface cleaning of the "B" side before wire bonding should alleviate this problem. The "B" side of the detector is fabricated first so it is susceptible to surface contamination while the "A" side of the detector is processed. We believe the keys to successful wire bonding to CZT are to use an ultrasonic wedge bonder with soft aluminum wire, low ultrasonic power, excellent metal adhesion, thick metal pads (2-3 µm) to minimize damage to the CZT, and a large bonding pad area so the force and power to apply the bond are spread over a larger metal pad area. In other words, the energy coupled into the CZT should be minimized. Finally, the length of the bonding pad should be made long enough to have space for two wire bonds in case rebonding is necessary from a failed first attempt.

CONCLUSION

We have developed a robust fabrication process for CZT strip detectors to make the large number of detectors needed for a large area array. The strip detectors can be fabricated with a high yield of good strips, low leakage current, and high interstrip resistance. In addition, the strips can be connected to high density electronics with wire bonds with a yield close to 100%. A 6 x 6 strip detector array has been assembled and is shown in Figure 2. The array consists of 36 double sided CdZnTe strip detectors, resistor and capacitor arrays, and high density analog readout electronics.

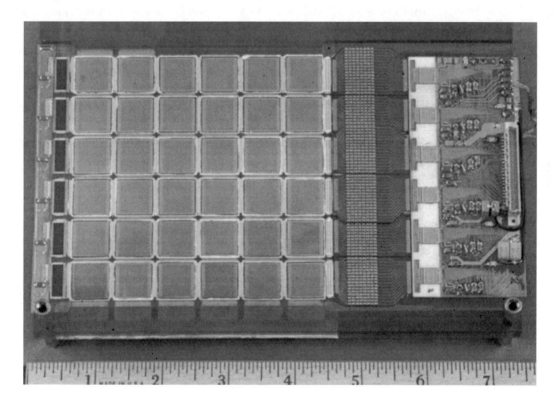

FIGURE 2. 6 x 6 CZT Strip Detector Array.

Acknowledgments

The authors acknowledge the excellent wire bonding and assembly work for the arrays by John Lehtonen and Kathy Mach of the Johns Hopkins Applied Physics Laboratories. The authors also acknowledge the work of Carol Sappington and Frank Peters in the earlier wire bonding tests, Andre Burgess in assisting with mechanical polishing of the CZT material, and Jim Odom with assembly of the electronics for the arrays.

References

Bartlett, L. M., C. M. Stahle, D. Palmer, L. M. Barbier, S. D. Barthelmy, F. Birsa, N. Gehrels, J. F. Krizmanic, P. Kurczynski, J. Odom, A. M. Parsons, C. Sappington, P. Shu, B. J. Teegarden, and J. Tueller (1996) "CdZnTe Strip Detectors for Astrophysical Arc Second Imaging and Spectroscopy: Detector Performance and Radiation Effects," in *Gamma-Ray and Cosmic-Ray Detectors, Techniques, and Missions,* B. D. Ramsey and T. A. Parnell, eds, CONF-2806, SPIE, Bellingham, WA, 2806: 616-628.

Kurczynski, P., J. F. Krizmanic, C. M. Stahle, A. Parsons, D. M. Palmer, L. M. Barbier, L. M. Bartlett, S. D. Barthelmy, F. Birsa, N. Gehrels, J. Odom, C. Hanchak, P. Shu, B. J. Teegarden, and J. Tueller (1997) "CZT Strip Detectors for Imaging and Spectroscopy: Collimated Beam and ASIC Readout Experiments," *IEEE Trans. Nuclear Science*, 44:1011-1016.

Palmer, D. M., A. M. Parsons, P. Kurczynski, L. M. Barbier, S. D. Barthelmy, L. M. Bartlett, E. Fenimore, N. A. Gehrels, J. F. Krizmanic, D. C. Mancini, C. M. Stahle, J. Tueller, and B. J. Teegarden (1997) "Arcsecond Source Positions with a Prototype BASIS Imaging System," in *EUV, X-Ray, and Gamma-Ray Instrumentation for Astronomy VIII*, O. H. Siegmund and M. A. Gummin, eds, CONF-3114, SPIE, Bellingham, WA, 3114.

Parsons, A., D. M. Palmer, P. Kurczynski, L. Barbier, S. Barthelmy, L. Bartlett, N. Gehrels, J. Krizmanic, C. M. Stahle, J. Tueller, and B. Teegarden (1997) "Position Resolution Performance of Prototype Segmented CdZnTe Arrays," in *EUV, X-Ray, and Gamma-Ray Instrumentation for Astronomy VIII*, O. H. Siegmund and M. A. Gummin, eds, CONF-3114, SPIE, Bellingham, WA, 3114.

Stahle, C. M., A. M. Parsons, L. M. Bartlett, P. Kurczynski, J. F. Krizmanic, L. M. Barbier, S. D. Barthelmy, F. Birsa, N. Gehrels, J. Odom, D. Palmer, C. Sappington, P. Shu, B. J. Teegarden, and J. Tueller (1996) "CdZnTe Strip Detector for Arc Second Imaging and Spectroscopy," in *Hard X-Ray/Gamma-Ray and Neutron Optics, Sensors, and Applications*, CONF-2859, R. B. Hoover and F. P. Doty, eds, SPIE, Bellingham, WA, 2859:74-84.

LARGE ASPECT RATIO DETECTOR ARRAY FOR SPACE APPLICATIONS

A. Ewin, M. Fortin and P. Shu
NASA, GSFC, Solid State Device
Branch, Greenbelt, MD 20771
(301)-286-5374

N. C. Das and S. Aslam
Hughes STX Corp., 7701 Greenbelt
Road, Greenbelt, MD 20770
(301)-286-1330

W. Smith
Orbital Science Corp., Greenway
Center, Greenbelt, MD 20770
(301)-286-2820

J. McAdoo
Electro-Optics & Control Branch,
NASA/LARC, Hampton, VA 23681
(757)-864-1640

Abstract

At the Goddard Space Flight Center (GSFC) Detector Development Laboratory (DDL), we have custom designed and fabricated a chip which consists of an interleaved linear array of 1 x 512 pinned photo-diode (PD) sensing pixels, which have dimensions 200 µm x 6 µm, together with a parallel detector register and a serial readout register. The registers use a triple polysilicon CCD design. This paper presents some aspects of the design, processing and operating techniques used for optimizing the full well capacity of the odd shaped photo-diodes, efficiency of photo-diode to CCD transfer and the charge handling capability of the various CCD/output structures. By using a pinning implant in the PD area together with a larger transfer gate voltage the desired photo-diode signal level of 5 Me$^-$ was achieved.

INTRODUCTION

Space sensor applications require the use of CCD imagers having exceptional performance in terms of full well capacity and charge transfer efficiency (Geary 1990). An additional requirement often is large photo-diode detecting pixels in order to attain higher resolution of low-light level distant objects within the desired dynamic range.

At GSFC we have custom designed a 1 x 512 large photo-diode pixel array for the Gas and Aerosol Measuring Sensorcraft (GAMS) project which will measure various atmospheric species in the upper troposphere using solar occultation. Because of the size constraints of the miniaturized grating spectrometer, long and narrow (200 µm x 6µm) photo-sensing elements with a 10 µm periodicity were used in the design of the CCD imaging sensor. The photo-diode was designed to be 200 µm long in order that enough signal could be generated from the low light levels being detected. There are many problems associated with charge detection by long and narrow photo-sensing elements, namely: (i) inefficient charge transfer from the photo-diode to the CCD channel (Banghart 1991), (ii) full well capacity of the photo-diode (Janesick 1989) and (iii) spilling of signal charge to the adjacent pixels (Kuriyama 1991). Therefore, the ideal detector is one which has a maximum full well signal at low light levels and minimum noise but which also has exceptionally good charge transfer of all signal levels. We have made an effort to address some of these issues which are reported in the following sections.

DESIGN ASPECTS

Specifications of the Image Sensor

Table 1 shows the specifications of the CCD image sensor for GAMS. The sensor consists of a 1 x 536 linear array composed of a parallel detector register and a serial readout register. The parallel register consists of 512 optically active pixels for detection of light and 24 pixels for determining the dark current or reference electrical signal levels. Twelve of these 24 inactive pixels are located on each side of the optically active pixels. The serial readout register consists of 536 pixels and includes four output amplifiers.

TABLE 1. Specifications of GAMS CCD image sensor

number of active pixels	512
number of inactive pixels	24
pixel size	200 µm x 6 µm
pixel arrangement	linear
pitch of pixels	10 µm
number of outputs	4 (2 redundant)
chip size	8 mm x 3 mm

Modes of Operation

The GAMS CCD sensor array supports the following basic modes of operation:

1. Integration mode.
2. Readout modes where half (268) pixels are clocked to one output amplifier and half to the other amplifier or where all the pixels are clocked to either output amplifier.

Array Layout

Because of the limiting minimum geometry dimensions in polysilicon processing, the array was designed with two serial CCD readout registers, see Figure 1. The top register transfers the even pixels while the bottom transfers the odd. The two registers can be operated either simultaneously or independently to transfer charge to one of two output structures associated with each register. All four output structures on each imaging sensor chip are identical, two of which provide redundancy for the operation of the device.

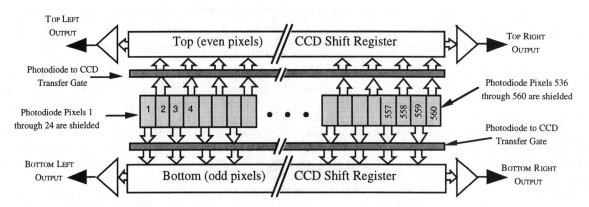

FIGURE 1. Schematic Layout of the GAMS Sensor Chip

Photo-diode/CCD Pixel Full Well

The pixel elements in the GAMS sensors have a 10 µm periodicity. The 4 µm minimum width channel-stops between pixels allows the actual width of each photo-diode element to be 6 µm. Generally, this does not present a problem in the fabrication of low signal level detectors. The technological challenge for the GAMS device comes from seeking a very large signal from a photo-diode element of such narrow width.

The lowest expected capacitance in the photo-diode is 0.8 fF/μm^2, corresponding to approximately 4800 electrons/μm^2. A 200 µm x 6 µm photo-diode pixel should provide a full well in excess of 5 Me$^-$. Optimizations in the processing technology has markedly increased the photo-diode's capacitance to give in excess of 8 Me$^-$ full well. It is expected that this signal level will increase with further refinements in the fabrication process.

The CCD pixels are required to hold a minimum of 15Me⁻. Since each CCD register receives signal charge from alternate photo-diode elements, each 3-phase CCD pixel is 20 μm long (in the direction of charge transfer). The two registers are identical with 500 μm wide channels which gives each CCD pixel a total area of 10,000 μm^2. Typical CCD structures fabricated in the GSFC DDL exhibit at least 2000 electrons/μm^2 for a 3 phase pixel area. Thus the GAMS CCD registers are designed to provide a pixel full well of 20 Me⁻.

FABRICATION PROCESS

The photodiode/CCD array consists of 13 mask levels including selective implants for the photo-diode area to achieve boron surface pinning to the substrate as well as additional phosphorous implants to adjust the charge handling capacity of the photo-diode.

The sensors are fabricated from a 20-50 Ω-cm <100> p-type epitaxial on a p$^+$ wafer. The thin gate dielectric consists of 500 Å of silicon dioxide and 700 Å of silicon nitride. A n-buried channel is then formed by a phosphorous implant at 180KeV and dose of 1.6×10^{12} /cm^2 followed by high temperature annealing. A shallow boron implant was done to adjust the threshold voltage of the transfer gate.

Three layers of polysilicon, architectured using Reactive Ion Etching (RIE) in a SF_6+O_2 plasma, were used for phases ø1, ø2 and ø3 of the CCD storage resistor respectively. The RIE process was optimised (Aslam 1997) in such a way that the polysilicon electrodes are etched very selectively to silicon nitride. The cross-section of one pixel of of the device is shown in Fig.2.

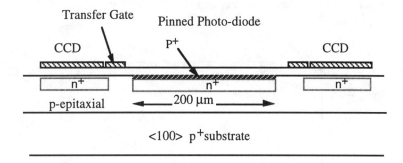

FIGURE 2. Cross Section of One Pixel

To improve the photo-diode charge handling capacity the entire surface of the photo-diode receives a shallow p-type implant shutting the surface to the guard bands (Burkey 1984). This ensures full diode depletion. Phosphorous was also implanted into the photo-diode to enhance the charge handling capacity. Finally the CCD register and all non-imaging areas are shielded with a layer of polyimide, low temperature SiO_2 and aluminum exposing only the photosites. The chip was diced and mounted on a 40 pin ceramic package and taken for testing. A photograph of the processed chip is shown in Figure 3.

FIGURE 3. Photograph of the Processed GAMS Chip

RESULTS AND DISCUSSION

Various biases and clock voltages used for testing of the sensor arrays are shown in Table 2. The output of the serial register is a source follower operated in the constant gain region (0.7).

Table 2. Bias and Clock Voltages used in Testing

Function	Voltage levels
Output Drain (OD)	24.0 V
Reset Drain (RD)	17.0 V
Output Gate (OG)	2.0 V
Summing Well (SW)	-4.0 to 6.0 V
Reset Gate (RG)	-3.0 to 7.0 V
Phase 1 (ø1)	-5.5 to 4.5 V
Phase 2 (ø2)	-4.5 to 5.0 V
Phase 3 (ø3)	-5.5 to 5.0 V
Transfer Gate (TG)	-5.0 V

The bias levels are experimentally selected to optimize the charge transfer efficiency and then to maximize the full well (signal output) vs. light power input. Of particular interest is the dependency of the full well signal as a function of the phosphorous doping in the photo-diode. This implant determines the photo-diode capacitance and a plot of full well capacity vs. the phosphorous dose is shown in Figure 4.

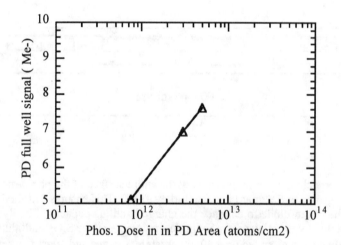

FIGURE 4. Photo-diode Full Well Signal as a function of Phosphorous Dose

For this application the goal was to optimize the full well of the photo-diode and this was achieved with a phosphorous dose of 2×10^{12} /cm^2 at an ion implantation energy of 180 KeV. The charge handling density of the photo-diode was 5000 e$^-$/µm^2 at this dose. The clock voltage levels on the transfer gate between the photo-diode and CCD were also found to have profound effect on the maximum signal level that could be collected in the photo-diode. This effect is illustrated in Figure 5.

It should be noted that since we are interested in large full wells, it was necessary to deliberately desensitize the integrating mode capacitor to ensure we did not have extraordinarily large output voltage swings. A variety of output structures were simultaneously fabricated ranging in sensitivity from 0.1 to 0.6 µV/e$^-$. At 6M electrons this would give us a maximum operating range of 0.6 to 3.6 V output. We settled on a device that gave a node sensitivity ranging from 0.15 to 0.20 µV/e$^-$ (about 1.0 V output swing).

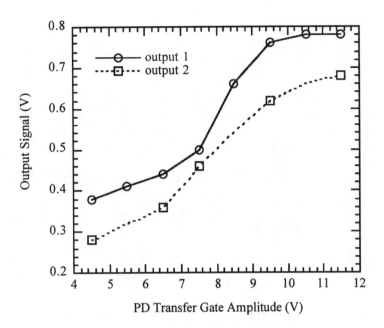

FIGURE 5. Photo-diode Output Signal as a function of Photo-diode Gate Amplitude

CONCLUSIONS

A unique pinned photo-diode /CCD device has been designed, fabricated and evaluated for use as the focal plane array in GAMS spectrometer. Large aspect ratio photo-diodes capable of charge storage in excess of 5 million electrons were successfully integrated with a CCD serial shift register. The effect of photo-diode doping profile and various operating voltages was investigated and results presented.

Acknowledgments

This work was performed under a contract from NASA/ Langley Research Center, VA. The authors wish to thank S. Babu, Hal Isenberg and Joe Santos for providing invaluable technical support.

References

Geary, J. C., et al. (1990), Development of a 2048X2048 Imager for Scientific Applications, Proc. SPIE Conf. on CCD and Solid State Sensors, Santa Clara, CA, 38.

Banghart, E. K., et al. (1991), A Model for Charge Transfer in Buried Channel Charge Coupled Devices at Low Temperatures, IEEE Trans. ED-38: 1168.

Janesick, J., et al. (1989), SPIE, Optical Sensors and Electronic Photography, CA, 1071.

Kuriyama, T., et al. (1991), IEEE Electron Devices, ED-38: 949.

Aslam, S., N. C. Das and P. Shu (1997), SPIE, Process, Equipment, and Materials Control in Integrated Circuit Manufacturing III, 3213: 73.

Burkey, B. C., et al. (1984), Proc. of EIDM, 28.

THE NEXT GENERATION OF MONOLITHIC INFRARED DETECTOR ARRAYS

Abhay M. Joshi
Discovery Semiconductors, Inc.
186 Princeton-Hightstown Road
Cranbury, NJ 08512
Phone: (609) 275-0011

Murzy Jhabvala & Peter Shu
NASA Goddard Space Flight Center
Greenbelt Road, Mail Code 718
Greenbelt, MD 20771
Phone: (301) 286-5232

Abstract

Recently NASA announced an initiative "X-2000" whose ultimate goal is "satellite-on-a-chip." We propose advanced, Monolithic InGaAs-on-silicon Short Wave Infrared (SWIR), and Monolithic InSb-on-silicon Medium Wave Infrared (MWIR) detector arrays for X-2000. To achieve "satellite-on-a-chip" goal, it is imperative to have a fully integrated Focal Plane Array on one wafer substrate, preferably silicon. A (512 x 512) High Resolution Focal Plane Array structure is proposed that employs a technique to selectively epitaxially grow InGaAs and InSb photodetectors on silicon substrate. Growth of lattice mis-matched InGaAs and InSb layers on silicon substrate is improved by limiting the growth area, selecting appropriately oriented wafers, as well as controlling the growth conditions, and thus, the RoA product, an important parameter of the photodetectors will be enhanced. At 150 K, the expected RoA is 300 to 400 Kohm- sq. cm for InGaAs and 500 to 1,000 Ohm - sq. cm. for InSb. A new CMOS readout circuit will be developed from the present Discovery Semiconductors' 1D and 2D FPA readout circuits for the (512 x 512) High Resolution Monolithic InGaAs/InSb-on-Silicon SWIR/MWIR FPA. It has four readout methods suitable for a wide range of RoA product of the IR detectors. The silicon readout circuit will be fabricated by 0.6 um CMOS process technology for obtaining a higher fill factor, readout speed, signal injection, and readout efficiency of the FPA. The above approach addresses the application requirements of NASA's X-2000 initiative and solves some problems inherent to Hybrid Focal Plane Arrays.

INTRODUCTION

Microsatellites are fast becoming important scientific and commercial realities. However, most satellites that fall in this class are still fairly large (~50 Kg, ~0.5 m). One of the major obstacles in reducing these parameters further is the lack of integration of all the satellite's functions. Normally, most satellites are constructed from physically separate sub-systems, each of which is composed of some combination of circuit boards and components (Payne 1992, Priedhorsky 1989). This partitioned approach wastes valuable space and weight by increasing the demands on structure and power resources of the satellite. This is the problem that Discovery Semiconductors and NASA Goddard Space Flight Center wish to address.

We propose to develop the next generation of microsatellites which overwhelmingly use semiconductor technology to reduce the weight of the satellites, ultimately achieving the "satellite-on-a-chip" objective. One of the first devices that we have targeted to utilize this concept and technology is Monolithic InGaAs and InSb-on-Silicon Photodiode Arrays.

(512 x 512) HIGH RESOLUTION, MONOLITHIC InGaAs-ON-SILICON (SWIR) and MONOLITHIC InSb-ON-SILICON (MWIR) FOCAL PLANE ARRAYS

According to the NASA's application, a high image resolution system that eliminates the need for scanning mechanisms is needed, and it should cover the 1 -3 μm SWIR and 3 - 5 μm MWIR wavelength band. The Infrared Focal Plane Array (IR FPA) is an important component, and it should combine detection, integration, processing, and self-scanning signal readout of the IR imaging system so as to make it light weight, utilize less power, and exhibit high reliability.

For realizing NASA's X-2000 requirements, the IR FPA in the imaging system must have following features:

1. Higher operation temperature requiring minimal or no active cooling. The idea is that passive radiation cooling from space (≥ 105K) will be utilized, thus, eliminating any need for active cooling such as Sterling coolers.
2. Two dimensional (Staring Type) eliminating the mechanical scanning structure.
3. Larger format (more sensitive elements) for getting higher space resolution for the imaging system.
4. Should be fully integrated on one wafer substrate, preferably silicon.

WHY DEVELOP MONOLITHIC InGaAs-ON-SILICON AND MONOLITHIC InSb-ON-SILICON FOCAL PLANE ARRAYS?

There are several reasons for developing this advanced technology:

1) <u>To Achieve Satellite-On-A-Chip</u>:

If NASA wants to demonstrate a satellite-on-a-chip by year 2006, it has to focus on integrating the Imaging Sensors with other peripherals all on one semiconductor wafer. Imaging sensors allow the spacecraft to take pictures of the scene of interest, whether it is the Martian surface or the Earth's surface. Imaging sensors are vital to our explorations, both for Inter-Planetary expeditions as well as for the Mission to Planet Earth.

2) <u>Difficulties facing Hybrid FPAs</u>:

With the increase in the size and pixel density for IR FPAs, the interconnection reliability of hybrid FPAs is a major problem. Also the yield and the operability of the large-area, high density hybrid IR FPA is another problem. For the hybrid InSb FPAs, the detector substrate has to be thinned down to 10 μms or less for acceptable quantum efficiency. This thinning leads to breakages, especially during Indium bump bonding and hence, to reduced yields.

3) <u>Advantages of Monolithic FPAs</u>:

- Compatibility with the Silicon VLSI Process: Like the Schottky-barrier FPA, the Monolithic InGaAs and InSb-on-Si FPA will be fabricated by a well established silicon VLSI process, therefore, at the present time it represents the most advanced technology for fabricating large-area, high-density FPAs.
- High Operability, Reliability, and Yield: The sensitive elements of the monolithic InSb-on-Si FPA will be "<u>selectively</u>" grown in input side of Si readout circuit. This will result in a high operability, reliability and yield.

For these reasons, the high resolution IR FPAs for NASA's future needs should use the monolithic integration technology.

METHODOLOGY FOR MONOLITHIC InSb-on-Si DETECTOR ARRAY FABRICATION

The different steps required to manufacture Monolithic InSb-on-silicon FPA are illustrated in Figs. 1(a) thru (e) and are listed below:

a) Fabrication of silicon CMOS readout circuitry, except Aluminium metalization..
b) Etching of silicon dioxide and silicon for selective epitaxial growth on the silicon substrate.
c) Selective epitaxial growth of GaAs and InSb through opened windows (wells) in silicon.
d) Fabrication of InSb PIN Diodes.
e) Final metalization of InSb devices to the CMOS readout circuitry.

In this entire process, we will be growing and fabricating photovoltaic InSb p-i-n diodes in etched "wells" in the silicon substrate using Discovery Semiconductors patented technique of "selective epitaxy". The proposed InSb device structure is shown in Table 1.

Fig. (a) Fabrication of silicon CMOS circuitry, except metalization.

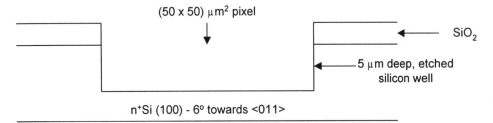

Fig. (b) Etching of silicon well for GaAs/InSb growth.

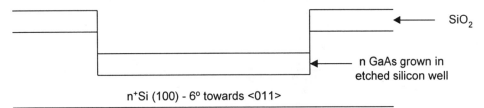

Fig. (c) Selective epitaxial growth of GaAs buffer on silicon.

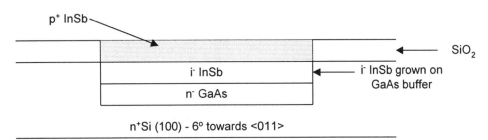

Fig. (d) Selective epitaxial growth of i InSb & p+ InSb on GaAs buffer.

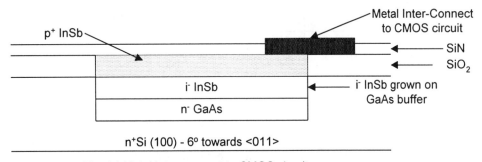

Fig. (e) Metal inter-connect to CMOS circuitry.

FIGURE 1. (a) - (e): Methodology for Achieving Monolithic Integration of InSb-on-Silicon Detectors with the Silicon Readout Circuitry Using the Technique of "Selective Epitaxy".

TABLE 1. Epitaxial Growth Layers for Monolithic InSb-on-Silicon.

Layer	Lattice Constant (A)	Thickness (µm)
Silicon Substrate (n+)	5.43	N/A
GaAs Buffer (n)	5.63	1
InSb (i)	6.47	3
InSb (p+)	6.47	1

<u>Novel Capability of Discovery Semiconductors : Selective Epitaxial Growth of III-V Compound Materials on Silicon Substrates</u>

Discovery has developed the technology of growing high quality GaAs - InGaAs materials <u>selectively</u> on a silicon substrate (Joshi 1997) and has successfully demonstrated the selective epitaxy of InGaAs-on-silicon that has resulted in device quality InGaAs layers on silicon. The purpose of these selective area depositions is to provide an accessible sink to threading dislocations (Joshi 1997 and Fitzgerald 1989). Small diameter deposits should have a lower dislocation density and, hence, better electrical performance.

Etch Pit Density (EPD) measurements reveal (100 - 1000X) reduction in the dislocation density for 50 µm diameter growth areas as compared to areas where MBE growth was done over the entire silicon substrate. Table 2 shows the average etch pit density data for the different diameter InGaAs dots selectively grown on silicon. This reduction in the dislocation density by selective epitaxy is critical to make monolithic, low leakage, high yield InSb-on-silicon FPAs. As can be seen in the Table II below, reduction of the growth area significantly reduces the dislocation density. This reduction is by a factor of up to three orders of magnitude!

TABLE 2. Etch Pit Density Measurements of InGaAs-on-Ssilicon.

Specimen (Grown on mis-oriented silicon)	EPD/cm^2
Blanket InGaAs	1.1×10^8/cm^2
500 µm dot InGaAs	1.4×10^6/cm^2
300 µm dot InGaAs	4.4×10^5/cm^2
50 µm dot InGaAs	5.1×10^5/cm^2

We have processed five different 1.3 µm InGaAs-on-silicon p-i-n diode sizes varying from (40 x 40) µm^2 to (200 x 200) µm^2 in steps of 40 µm (Joshi 1997). We measured I-V and C-V characteristics of these different diode sizes at room temperature. The following Table 3 shows the InGaAs-on-silicon Photodiode DC characteristics.

TABLE 3. 1.3 µm InGaAs-on-silicon Photodiode DC Characteristics at Room Temp.

Diode Area A (cm^2)	Reverse Bias (mV)	Leakage Current (A)	Junction Resistance Ro (KΩ)	RoA (Ω-cm^2)	Capacitance (pF)
1.6e-5	10	1.39e-8	1823	29.2	16.6
6.4e-5	10	4.80e-8	714.7	45.7	25.0
1.44e-4	10	5.30e-8	270.2	38.9	34.0
2.56e-4	10	7.74e-8	146.3	37.5	47.3
4.0e-4	10	1.37e-7	72.3	28.9	61.9

The process used is limited-area (selective) epitaxy. We have shown that in general both grading layers and epitaxy on reduced areas should lead to reduced defect densities in highly mismatched semiconductor systems (Joshi 1997, Fitzgerald 1989). Discovery Semiconductors has developed a technology which has used the growth of InGaAs in small areas to form InGaAs photodetectors on Si (Joshi 1997). The detectors worked well due to the decrease in threading dislocations density from the reduced-area growth.

The process utilizes GaAs-on-Si growth first, known to produce many threading dislocations in the GaAs layer. However, a graded-composition InGaAs layer was grown after the GaAs layer, which creates gradual mismatch stress that can move the threading dislocations at the surface of the GaAs, pushing them to the edge of the pattern (see Fig. 2). Thus, the graded InGaAs serves as a structural tool: the InGaAs relaxes via the motion of the threading dislocations inherited from the GaAs.

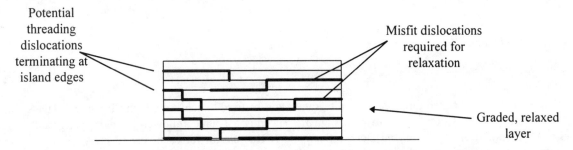

FIGURE 2. Schematic Revealing the Ability of the Graded InGaAs Layer to Move the Threading Dislocations which are Mobile from the GaAs/Si to the Edge of the Pattern, Relaxing the InGaAs Layer in the Process.

The combination of a graded-compositional layer and epitaxially growing the layer in a small island area as shown above in Fig. 2 results in many of the threading dislocations, which are moving to relax the InGaAs, exiting the InGaAs at the pattern edge. Thus, we can expect that the relaxed InGaAs-on-silicon, and the PINs made out of it, to have a lower threading dislocation density than GaAs-on-Si. Indeed, based on some of the work by Prof. Fitzgerald at MIT, a group at Xerox has shown that graded InGaAs on GaAs/Si does decrease the dislocation density, as compared to the GaAs/Si without the graded InGaAs (Fitzgerald 1989). Our results from Phase II SBIR with the NASA Goddard agree with our expectations and their results, since we were able to obtain working InGaAs-on-Silicon PIN Detectors using this process (Joshi 1997).

As can be seen from the Table 3 above, (80 x 80) μm^2 size diode with an area of 6.4e-5 cm^2 exhibited a superior RoA product indicating that the selective epitaxial growth area of (80 x 80) μm^2 or less produces the fewest number of threading dislocations, and thus, a lower leakage current, and higher RoA product. The (40 x 40) μm^2 area diode showed poorer RoA due to processing difficulties and errors. However, we plan to remove these processing errors so that smaller size detectors can be fabricated. Furthermore, smaller area diodes show smaller capacitance. A lower capacitance is needed for a photodetector to exhibit a decent speed by reducing its RC time constant.

DESIGN FOR CMOS READOUT CIRCUIT

The design parameters for CMOS readout circuit are as follows:

1. Large current input range for suiting a wide range of RoA (1 to 1,000 $\Omega\text{-cm}^2$) and operating temperatures (77 to 200K) for InGaAs/InSb-on-Si photodetectors.
2. Multiple readout methods for suiting different pixel density, and different integration times of the FPA.
3. Reducing the effect of parasitic capacitance for enhancing the operating speed of the device, and injection and readout efficiencies of the photodetector.

CMOS Process Consideration

We will use 0.6 μm CMOS technology to fabricate the IR FPAs in order to obtain:

1. High Fill Factor

The 0.6 μm CMOS process technology will permit the unit CMOS readout circuit of each pixel to occupy smaller space and hence, achieve a higher fill factor. We expect a fill factor of about 45% for the Monolithic InGaAs/InSb-on-Si FPAs for an individual pixel size of (50 x 50) μm^2.

2. High Speed

The large format of the proposed (512 x 512) FPA will require a high readout speed for the CMOS circuitry. Thus, 0.6 μm CMOS technology will enable us to design a high speed device compared to 1.2 or 2 μm CMOS technology.

Readout Circuit Configuration

1. Unit Circuit

A simple input circuit called direct detector integration will be used to achieve a high fill factor for the FPA. The schematic for unit pixel circuit is shown in Fig. 3. There is a source follower for each pixel, and a current integration amplifier and a Correlated Double Sampling (CDS) circuit for each row of the FPA. This circuit is designed for enhancing the signal injection efficiency and overcoming the effect of the bus parasitic parameters, and thus, improve the bandwidth.

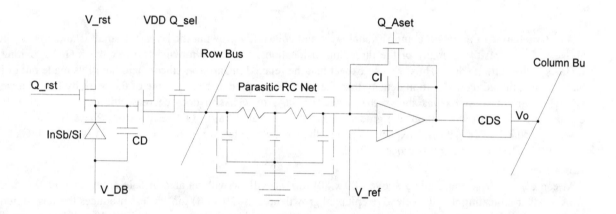

FIGURE 3. Unit Pixel Circuit of the (512 x 512) High Resolution, Monolithic InSb-on-Si Focal Plane Array.

2. Circuit Configuration

If the dark current of the photodetector is higher (or RoA is lower), the integration time is shorter. If the cut-off wavelength is longer, the dark current is higher (or RoA is lower) for the same operating temperature. For suiting different integration times of the FPA due to different RoA's of the photodetector, a new readout circuit configuration is proposed. Fig. 4. shows the block diagram of the circuit. The unit pixel circuit is same as the circuit shown in Fig. 3. The random row and column shift registers realize different readout methods for the FPA.

3. Readout Method

There are four readout methods in the proposed (512 x 512) High Resolution, InGaAs/InSb-on-Si Focal Plane Array. They are:

1) **Conventional Frame Time Integration Readout:** The integration time of each pixel is the Frame Time of the FPA. The two random shift registers are shifted sequentially. This method is suitable for a device in which the dark current of each photodetector is very low, or the integration time is very long.

2) **Pixel Interlacing Readout:** For a (512 x 512) image, it can be separated into 4 fields of (256 x 256), 16 fields of (128 x 128) or 64 fields of (64 x 64), or 512 fields of (1 x 1) images. The image of (512 x 512) FPA is readout one interlacing pixel by one interlacing pixel. Fig. 5. is an example for readout of 4 x (256 x 256). We can restore the 512 x 512 image with the pixel interlacing fields from the video signal. This readout method suits FPAs in

which the integration time is limited by the dark current of the photodetector. For example, if the maximum readout speed of the FPA is 10 MHz, we can use interlacing of 16 x (128 x 128) for 2 ms integration time for each pixel, and 4 x (256 x 256) interlacing for 8 ms integration time.

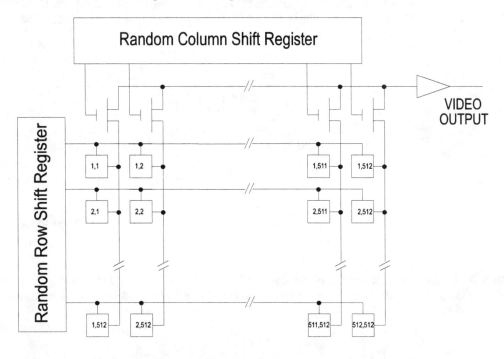

FIGURE 4. The Block Diagram of the (512 x 512) High Resolution, InGaAs/InSb-on-Si Focal Plane Array. Each square represents 1 pixel. The numbers in the square denote the address (location) for that particular pixel. For e.g., 1, 512 is 1st column, 512nd row.

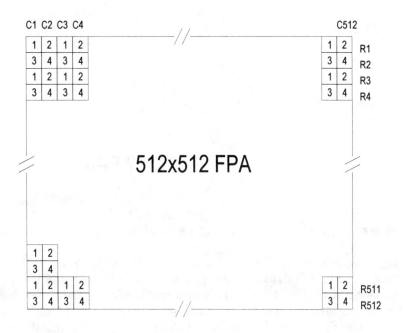

FIGURE 5. An Example of Pixel Interlacing Readout: 4 x (256 x 256). The (512 x 512) image is formed partitioning it in 4 areas of (256 x 256)s.

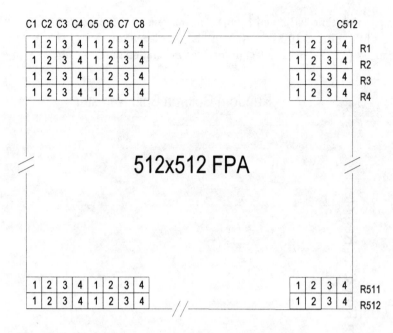

FIGURE 6. An example of Line Interlacing Readout - 4:1. The (512 x 512) imaging area is again partitioned in 4 parts and the final image is restored by combining (inter-lacing) video signal from these 4 areas.

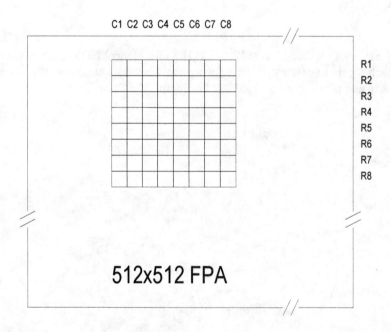

FIGURE 7. An example of Random Readout of (8 x 8) pixel area.

3) **Line Interlacing Readout:** Like the Pixel Interlacing readout, the (512 x 512) image can be separated into 2 fields of 256 x 512 (2:1), 4 fields of 128 x 512 (4:1) or 8 fields of 64 x 512 (8:1), ..., 512 fields of 1 x 512 (512:1) images. The image of (512 x 512) FPA is readout one interlacing line by one interlacing line. Fig. 6 is an example for interlaced readout of (4:1). Similarly, we can restore the 512 x 512 image with the line interlacing fields from the video signal. *This readout method also suits FPAs in which the integration time is limited by the dark current of the photodetector.* The advantage of this method is that the column amplifier and CDS are not idle.

4) Random Readout: In this 4th and last readout method, we can only readout an interesting pixel or area of the FPA by programming the row and column random shift register. Fig. 7 shows an example for random readout, where only an area of (8 x 8) pixels is readout.

SUMMARY

To summarize, we are proposing to design and fabricate, Monolithic, InGaAs-on-Silicon and InSb-on-Silicon SWIR/MWIR FPAs for NASA's X-2000 initiative. Discovery Semiconductors will use its proprietary and patented "selective epitaxial" growth of III-V compound semiconductors on silicon to achieve the above goal.

Acknowledgment

The innovations described in this paper have mainly resulted from the past 30 months of work on a Phase II SBIR funded by NASA Goddard Space Flight Center. We sincerely thank NASA Goddard for funding this research activity.

References

Fitzgerald, E. A, G. P. Watson, E. P. Proano, D. G. Ast, P. D. Kirchner, G. D. Pettit, and J. M. Woodall, (1989) "Nucleation Mechanisms and the Elimination of Misfit Dislocations at Mismatched Interfaces by Reduction in Growth Area," *J. of Appl. Phys.*, **65**(6) : 2220-2237.

Joshi, A. M, R. Brown, E. A. Fitzgerald, X. Wang, S. Ting, and M. Bulsara, (1997) "Monolithic InGaAs-on-Silicon Short Wave Infrared Detector Array," *Proc. SPIE*, 2999 : 211-224.

Payne, R. A. Jr. (1992) "Applications of the Petite Amateur Navy Satellite (PANSAT)," "*Master's Thesis, Naval PostGraduate School*," Monterey, CA, Sep. 1992.

Priedhorsky, W. C. (1989) "Low-cost Small Satellites for Astrophysical Missions," *NASA workshop on High-Energy Astrophysics*, Dec. 1989.

WIDE-FIELD HIGH-PERFORMANCE GEOSYNCHRONOUS IMAGING

H. John Wood, Del Jenstrom, Mark Wilson, Sanford Hinkal and Frank Kirchman
NASA Goddard Space Flight Center
Mail Code 551
Greenbelt, MD 20771
301-286-8278

Extended Abstract

The NASA Mission to Planet Earth (MTPE) Program and the National Oceanographic and Atmospheric Administration (NOAA) are sponsoring the Advanced Geosynchronous Studies (AGS) to develop technologies and system concepts for Earth observation from geosynchronous orbit. This series of studies is intended to benefit both MTPE science and the NOAA GOES Program. Within the AGS program, advanced imager trade studies have investigated two candidate concepts for near-term advanced geosynchronous imagers. One concept uses a scan mirror to direct the line of sight from a 3-axis stabilized platform. Another eliminates the need for a scan mirror by using an agile spacecraft bus to scan the entire instrument. The purpose of this paper is to discuss the optical design trades and system issues encountered in evaluating the two scanning approaches.

The imager design started with a look at first principles: what is the most efficient way to image the Earth in those numerous spectral bands of interest to MTPE scientists and NOAA weather forecasters. Optical design trades included rotating filter wheels and dispersive grating instruments. The design converged on a bandpass filter instrument using four focal planes to cover the spectral range 0.45 to 13.0 micrometers.

The first imager design uses a small agile spacecraft supporting an afocal optical telescope. Dichroic beamsplitters feed refractive objectives to four focal planes. The detectors are a series of long linear and rectangular arrays which are scanned in a raster fashion over the 17 degree Earth image. The use of the spacecraft attitude control system to raster the imager field-of-view (FOV) back and forth over the Earth eliminates the need for a scan mirror. However, the price paid is significant energy and time required to reverse the spacecraft slew motions at the end of each scan line. Hence, it is desired to minimize the number of scan lines needed to cover the full Earth disk. This desire, coupled with the ground coverage requirements, drives the telescope design to a 1.6 degree square FOV to provide full Earth disk coverage in less than 12 swaths. The telescope design to accommodate the FOV and image quality requirements is a 30 cm aperture three-element off-axis anastigmat. The size and mass of the imager instrument that result from this optical configuration are larger than desired. But spacecraft reaction wheel torque and power requirements to raster the imager FOV are achievable using existing spacecraft technology. However, launch mass and cost are higher than desired.

In the second high-level trade study, the AGS imager team is looking at incorporating a scan mirror and having the satellite three-axis stabilized. The use of the scan mirror eliminates the long turn-around times of the spacecraft scanning approach, allowing for faster Earth coverage. Thus the field of view of the afocal telescope can be reduced by half while still satisfying ground coverage requirements. The optical design of the reduced field afocal telescope is being studied to shrink its size and improve its performance. Both a three-mirror Cassegrain afocal and a two-mirror pair of confocal paraboloids are being considered. With either telescope, the size, mass, and power requirements of this imager are significantly less than those of the first imager design.

Both imager designs appear to be feasible and both meet envisioned MTPE and NOAA geosynchronous imaging needs. The AGS imager team is continuing to explore the optical trade space to further optimize imager designs.

LASER DIODE PUMPED SOLID STATE LASERS
FOR SPACE APPLICATIONS

Robert L. Byer
Professor Applied Physics
Stanford University
Stanford, CA 94305-4085
(650) 723 0226

ABSTRACT

Laser Diode pumped solid state lasers are now the most coherent laser source available with linewidths less than 1Hz. The recent demonstration of cw amplification in LD pumped Nd:YAG opens the possibility of preserving coherence while extending power levels from tens of watts to kilowatts over the next five years. Continued improvements in power, efficiency, reliability at lower cost will open future space based applications such as laser communications, laser based precision ranging, laser remote sensing, and laser atmospheric wind sensing

INTRODUCTION

Laser diode pumped solid state lasers are efficient, coherent sources of optical radiation that offer the potential for scaling to high average power levels while preserving the coherence necessary for space based optical communications, interferometry and remote sensing.

The early work in laser diode (LD) pumped solid state lasers was motivated by the need for a coherent optical source that could meet the requirements for global wind sensing from a satellite platform in earth orbit. The early work led to the invention and demonstration of the essential components for coherent wind sensing. These components were a local oscillator, an amplifier and a coherent detection system based on beam propagation and mixing in single mode optical fiber. Wind sensing using a LD pumped Nd:YAG laser first successfully demonstrated at Stanford in 1987.

THE Nd:YAG LOCAL OSCILLATOR

The laser local oscillator was the NonPlanar Ring Oscillator (NPRO) Nd:YAG monolithic ring resonator invented by Byer and Kane. This single crystal laser oscillator, isolated from laboratory noise by its millimeter dimensions and packinging, produced output of a few milliwatts with a free running linewidth of less than 10kHz. Over the past decade the power of the NPRO has been in creased to the 1W level while the bandwidth has been decreased to less than 1Hz relative stability and to 10^{-14} or 1Hz absolute stability when locked to molecular I_2 lines at the 532nm harmonic of the 1064nm Nd:YAG transition. This laser local oscillator has become the "quartz crystal" of the laser industry and is now an essential element in many applications that require a highly stable coherent laser source.

The NPRO local oscillator is tunable over a 30GHz range at a rate of 10Hz. It is tunable over a 100MHz range at a rate of greater than 100kHz. Further, the NPRO operates in a single frequency and in a single spatial mode and is available as a fiber coupled source at both 1064 and 1321nm wavelength. The intrinsic coherence of the NPRO has been investigated by many groups. It is known to be shot noise, or quantum noise limited, at frequencies greater than 1MHz. Recent work has extended the stability of the laser in both amplitude and frequency by filtering with a high finesse interferometer following the laser source.

The ability to phase lock NPROs in optical phase over a frequency offset range from 1Hz to greater than 1GHz is of importance for space applications including coherent communication, laser ranging, and in the future coherent optical receiver arrays for optical astronomy. Optical phaselocking is relatively straightforward for the LD pumped Nd:YAG source because of its intrinsic low noise noise.

LASER AMPLIFICATION

The first laser amplifier used for coherent laser radar and the detection of wind velocity by Doppler shift was based on a flashlamp pumped Nd:YAG slab geometry laser. Demonstrated by Kane, this pulse laser operated at 10Hz repetition rate with 60dB of optical gain. It was the power amplifier used in coherent wind detection.

The advent of LD pumping using high power 20W LD bars enabled laser amplification on a continuous wave basis in 1997. Early experiments demonstrated amplification of Nd:YAG power from 40W to 64W of output using a LD pumped Nd:YAG slab laser with a 64mm length gain region. The quantum noise characteristics of this amplifier are being investigated for applications to the Laser Interferometer Gravitational Wave project or LIGO project managed for the NSF by Caltech and MIT. Recent advances in cw laser amplification has demonstrated 10W of output power from a 600mW NPRO source. This diffraction limited laser source has been selected as the laser oscillator for the LIGO project.

Preliminary measurements indicate that the 10W cw laser amplifier is stable in beam pointing, amplitude fluctuations and spectral noise characteristics. The design meets the demand that the laser source have a "soft" failure mode over its operational lifetime and that it might be "repaired" while operational by selective substitution of spare laser diodes
for those pump diodes that have failed.

This recent progress is promising in that optical communications technology requirements for space based laser communications plans for 5W of laser power at 1064nm by 2000. 10W of laser power is expected at 532nm by the year 2010. Commercial development of LD pumped Nd:YAG sources will certainly exceed these "aggressive" expectations.

FIBER BASED DETECTOR DEVELOPMENT

In the early coherent wind detection experiment the use of single mode optical fibers was implemented. The results were an easy to align receiver with the heterodyne mixing accomplished in single mode optical fibers prior to illumination of the detector. These benefits are important in that they relieve the optical systems engineer from difficult alignment and control aspects associated with bulk optics. In detector development over the past decade, the InGaAs detectors at 1064 nm are now exceeding operational quantum efficiencies of 90% at power levels on the detector of greater than 100mW. Work is underway to design and test 1W to 10W detectors for the LIGO project. This detector capability is important for heterodyne detection as the local oscillator power can be increased to improve the detector sensitivity.

FUTURE PROPECTS FOR LASER DIODE PUMPED LASERS

The next generation of LD pumped solid state laser will be based on the ion Yb doped into the YAG crystal. Yb oscillates at 1030nm in the near infrared but at a projected wall plug efficiency of 20%. While not as efficient at the LD pump source at 40%, the gain in spatial and spectral mode brightness of the LD pumped solid state laser is a clear system advantage in many applications.

Recent work has led to the demonstration of LD pumped Yb:YAG lasers at continuous wave output power levels of 450W. Work is underway to extend the power levels to greater than 1kW, the power required for an Earth to Jupiter communications link. Conceptual designs of LD pumped Yb:YAG lasers point to the possibility of 10kW cw lasers of dimensions 4x4x8cm. If employed as power amplifiers, these lasers could be used in wide bandwidth ground to space communications links or even for ground to satellite platform power transmission applications.

The wavelength of the Yb:YAG laser can be frequency doubled to 514nm which in turn can be optically locked to the I_2 molecule for absolute frequency stabilization. Prospects are encouraging for an absolute frequency stability of 10^{-16} with a frequency tuning range of 3THz. This opens the possibility for interferometry for both alignment and control of co-orbiting satellites and for distance measurements with unprecedented precision. Recent research has demonstrated single frequency operation of LD pumped Yb:YAG laser oscillators with output at both the 1030nm and 514nm wavelength regions. Early experiments demonstrated 150mW of continuous wave output at 515nm from the LD pumped Yb:YAG laser source.

The wide gain bandwidth of the Yb:YAG laser offers the possibility of modelocked operation for short pulse generation. To date pulse widths of 350 fsec have been demonstrated for LD pumped Yb:YAG. Such short pulses may prove useful for local ranging in space-based construction applications.

CONCLUSION

The progress in solid state laser sources over the past decade has been extraordinary in both coherence and in power. A backward look projected into the future gives the following trends: the power output available from commercial LD pumped solid state lasers increased by a factor of 10 every two years with 20W being delivered in 1998. The cost of the lasers is relatively constant as laser diode prices fall to offset the increase in power demanded by users. The coherence of the LD pumped solid state lasers has reached unprecedented values of one part in 10^{14} but is expected to be improved by two orders of magnitude over the next five years. This bodes well for laser based interferometer experiments such as the proposed LISA program aimed at the detection and measurement of gravitational wave radiation over a 5×10^6 km baseline Michelson interferometer.

LASERS AND SPACE OPTICAL SYSTEMS STUDY

Concetto Giuliano and Angela L. Annaballi
Air Force Research Laboratory/Phillips Laboratory
Kirtland Air Force Base NM 87117-5776
(505) 846-1683, (505) 846-2603

Abstract

The Air Force and other government organizations have considered the application of space-based lasers since the early 1970s. Recent studies have identified the enormous potential of lasers and optical systems in space to support the Full-Spectrum Dominance envisioned by the Joint Chiefs of Staff in "Joint Vision 2010." The Air Force Research Laboratory has undertaken the LAsers and Space Optical Systems (LASSOS) Study to examine in detail how space lasers and optics (defined as any laser system based in space or any terrestrial-based laser whose beam transits space) could best be used to satisfy this critical need. This twelve-month study will identify promising technology concepts for space laser/optic systems, develop system concepts based on these technologies with special emphasis on systems capable of performing multiple missions, assess how well these systems can accomplish operational tasks in a quantitative manner, and design technology development roadmaps for selected concepts. Since work on the study had commenced only days before the publication deadline, this manuscript is necessarily limited to a description of the background, motivation, and organization of the study. The "Concept Definition" phase of the study is scheduled to be completed by the time of the STAIF conference. By that time, study participants will have identified key concepts that best satisfy criteria for timely and cost-effective augmentation of combat capability. A final report, which will be made available to authorized recipients, will be written after completion of the study in August 1998.

INTRODUCTION

The dramatically changed geo-political situation following the end of the Cold War has motivated several "futures" studies having the goal of assessing what directions the United States Air Force should take as we enter the 21st century. One of the common themes of these studies is the vision of global presence -- the ability to gather information about and/or project power to any location on the globe. Another common theme is the awareness of the profoundly changed nature of the threat -- most notably the proliferation of ballistic missiles and weapons of mass destruction among a number of political entities which may not hesitate to commit acts of terror against or threaten the vital interests of the United States or its allies. These studies have concluded that the unique capabilities envisioned for space laser systems, defined as systems which include space-based lasers or which employ laser beams that transit space, raise the promise of both achieving global presence and developing a credible defense against the post Cold War threat. SPACECAST 2020, a 1994 Air University study, identified the space-based high energy laser system as one of two leading concepts "clearly ahead of the rest . . . because it could fulfill a variety of important force application and space defense missions, and its optical system could also provide a surveillance capability" (Air University, 1994). New World Vistas (NWV), an Air Force Scientific Advisory Board (SAB) study, noted that a postulated future "energy-frugal, cost-efficient Global Precision Optical Weapon (GPOW) would have a clearly dominant role in warfare" (Air Force Scientific Advisory Board, 1995). In addition, NWV foresees space lasers enabling a future global "Virtual Presence," defined as the ability "to transmit interactive presence to distant points of the globe at the speed of light" (Air Force Scientific Advisory Board, 1995). Lt Col Mark Rogers concluded, in a comprehensive Air University paper surveying a wide range of space laser concepts, that "the operational enhancement value of lasers in space is firmly established" (Rogers, 1997). General Richard Paul, Commander of Air Force Research Laboratory, has noted the recent maturing of lasers and optics technologies as a result of Department of Defense programs (Paul, 1997). Based on this observation and the conclusions of the futures studies, General Paul directed the Air Force Research Laboratory (AFRL) to carry out the LASSOS study. General

John Piotrowski (USAF, Ret.), former Air Force Vice Chief of Staff and Commander-in-Chief of North American Air Defense Command and U.S. Space Command, is the study director. Participating organizations, besides AFRL, include Air Force Major Commands, Space and Missile Systems Center, Space Warfare Center, Ballistic Missile Defense Organization (BMDO), Air War College (AWC), Department of Energy laboratories, and Federally Funded Research and Development Centers. SAB members will advise the study members throughout their work.

PURPOSE

The primary objective of the LASSOS study is the identification of promising applications of space lasers and their associated optical systems to military missions. Once these concepts are identified, the study members will evaluate them for technical feasibility and cost. Since cost is an increasingly important consideration due to continuing deficit-reduction efforts, the study team will concentrate on developing system concepts capable of performing more than one mission. For those concepts that the study members assess to be most promising (considering operational utility, technical feasibility, and cost) the team will create development plans for essential new technologies or novel applications of existing technologies. These plans may include advanced technology demonstrations (ATDs) -- operationally realistic field experiments designed to demonstrate a system's readiness to be produced and fielded. Results of the LASSOS study will also be used as inputs to the existing Air Force mission planning and acquisition processes. The goal is to develop and validate the utility and fieldability of new ways to augment warfighter capabilities and options.

STUDY APPROACH

The study is divided into two phases: a "Concept Definition" phase and a "Concept Evaluation" phase. During the Concept Definition phase, study participants will compile an extensive list of space laser concepts and conduct an initial assessment to determine which concepts merit more detailed investigation. In addition to hardware and software, the laser "system" includes mission planning, system deployment, operation, and operational results assessment.

The study team is organized as depicted in Figure 1. Descriptions of the individual panel functions are contained in the following text.

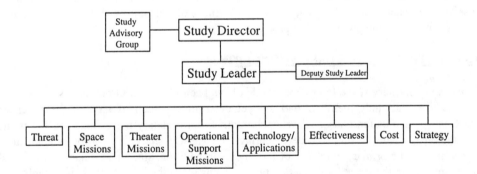

FIGURE 1. LASSOS Study Organization.

It has been observed that technical solutions to military problems have come about in two ways. One way, sometimes termed "requirements pull," results when warfighters identify what is needed to perform their mission better. The second stimulus to technical solutions, sometimes called "technology push," takes place when technologists identify military applications for technologies they have developed through their laboratory or theoretical investigations. The LASSOS study is organized in such a way as to capitalize on both of these methods of stimulating ideas. Study participants are divided into panels of approximately a dozen members each. Field-grade officers from the major user commands (Air Combat Command, Air Force Space Command, Air Mobility Command, Air Force Materiel Command, and Air Force Special Operations Command) lead Mission Panels. These panels, which also include technologists from AFRL and other government research institutions, will review stated military requirements found in a number of planning documents (such as Mission Needs Statements, Operational Requirements Documents, and Mission Area Plans), and look for space laser concepts which could address these needs. The goal is to achieve a synergistic interaction among the warfighters, who are the customer for the LASSOS study, and the technology developers. To further promote this interaction, members of operational units will describe many of their missions in detail to the study participants, allowing the technologists to generate system concepts inspired by their renewed or enhanced understanding of the warfighter's task. A separate Technology Panel, composed entirely of technologists, will focus primarily on proposing military uses for the latest technical developments. All participants will review previous studies, which have already identified a number of space laser concepts. However, they will not be constrained either by the results of previous studies or the statements in existing requirements documents.

The output of the Concept Definition phase, scheduled for January 1998, will be a shorter list of the most promising system concepts. These concepts will have passed an initial screening, to ensure mission relevance, technical feasibility, and cost practicality, and will have had top-level Concepts of Operations, Measures of Effectiveness, and Measures of Performance defined. The study will then go into the Concept Evaluation phase, which will be primarily the work of the Technology Panel, Cost Panel, and Effectiveness Panel. They will further refine the system concepts and assess them in terms of expected performance, technical risk, and life-cycle cost. Finally, the technology development roadmaps will be constructed. This phase is scheduled to be completed in August 1998, with the publication of a final report and briefing.

Throughout the study, an AWC Strategy Panel will consider how the proposed space laser concepts could support national security policy by providing a credible, reliable, and non-entangling tool for the National Command Authority. The Strategy Panel will also examine the potential impact of proposed space laser systems on existing agendas and treaties

STUDY GOALS

The goal of this study is to determine if the unique capabilities of space lasers enable cost-effective ways to provide substantial, preferably near-term, enhancements to the capabilities of the warfighters and peacekeepers to do their jobs. It is expected that the study will identify at least one such concept that will earn the strong advocacy of the users themselves. Through such advocacy, earlier deployment of the improved capability might be achieved through the Air Force planning and programming process.

PREVIOUSLY IDENTIFIED SPACE LASER CONCEPTS

This section lists some military space laser concepts that have been identified in previous studies. At the time of the publication deadline, the study members had not yet determined which, if any, of these concepts merited inclusion in the Concept Evaluation phase. This representative, but certainly not exhaustive, list is included here only to illustrate the types of ideas being considered during the LASSOS study. Non-military applications will also be considered, but are not the primary emphasis of the study. In addition, study participants will consider whether the optical system supporting the space laser can do other missions when the laser is not being employed. However, these non-laser applications will not be studied in detail.

Space-Based Laser Target Designator (SBLTD)

A laser target designator illuminates a target with light at a wavelength that can be detected by a sensor on a precision-guided munition. Most Americans are familiar with these laser-guided weapons from news coverage of the 1991 Gulf War. This concept involves a space-based laser designed to illuminate targets. The advantage of such as system is that our soldiers could be out of harm's way as they employ long-range precision weapons, perhaps laser-guided cruise missiles.

Space-Based Battlefield Illuminator

Similar to the SBLTD concept, this idea involves flood-illuminating a larger region with laser light in order to enhance the capability of reconnaissance sensors or night-vision devices.

Laser Alignment and Docking System

Some concepts for assembling large space-based systems involve launching several relatively small modules, which automatically connect together on-orbit. One advantage of these concepts is that new modules could replace older systems as technology matures. Laser alignment and docking systems could be used to provide accurate range and closure rate information when connecting spacecraft.

Active Remote Sensing for Battle Damage Assessment, Treaty Verification, and Weather Forecasting

Researchers have suggested that space-based differential absorption lidars could be used to probe the atmosphere around a target to detect propellants, chemical warfare agents, or other substances that may provide information about the results of an attack. Remote sensing might also be used to survey a battle area before troops move in. In addition, such sensors could be used in peacetime to monitor suspected chemical, and perhaps biological, production and weapon storage facilities as part of international treaty verification. Improved wind measurement using Doppler lidar could greatly improve the accuracy of weather forecasting, with commensurate enhancement of operational capability.

Space Debris Cataloging and Removal

According to New World Vistas, there are about 300,000 pieces of man-made space debris in orbit, which threaten the safety of manned and unmanned spacecraft (Air Force Scientific Advisory Board, 1995). Some of these items are too small to be detected by ground-based systems. Conceivably, space-based laser trackers and rangers could be used to update and maintain the space-debris catalog. Some have suggested that high-power lasers could be used to deorbit space debris by vaporizing some of its material, creating a thrust in the right direction.

Active Imaging of Satellites

Space Command keeps track of the status of earth-orbiting satellites as part of its space surveillance function. Some satellites are too dim to be observed using solar illumination. Plus, the sun illuminates satellites only during limited periods of time. Laser illumination from either ground-based or space-based systems could aid the space surveillance mission by providing enough signal for high resolution orbital ephemeris data, imagery or other types of signatures.

Laser Communications

Modern warfare is increasing in tempo. Efforts to stay ahead of the enemy lead the warfighters to require immediate access to information on demand. Such requirements lead to a need for extremely high bandwidth communication systems, which motivate interest in laser communications. Other advantages of laser communications include smaller space aperture requirements and lower probability of intercept by unauthorized recipients.

Laser Rocket Propulsion

Some researchers have suggested that it may be cost-effective to boost a satellite from low-earth orbit to geosynchronous orbit by means of ion engines powered by ground-based lasers, compared to the conventional method of lifting heavy chemical rocket engines and fuel.

Power Beaming

Concepts for using ground-based lasers to recharge satellite batteries during eclipse have been advanced. Conversely, beaming power from space (collected using solar panels and/or large inflatable solar concentrators) to remote locations on the earth is another suggestion considered in previous studies.

Laser Weapons

Ground and space-based laser weapons have been investigated for about 20 years by the Defense Advanced Research Projects Agency, Strategic Defense Initiative Organization, BMDO, and the Armed Services. Their advantages include their continual readiness for employment and speed-of-light delivery. In a ballistic missile defense role, they could destroy the missiles in the boost phase, thereby causing the debris to fall well short of the intended target, perhaps on the aggressor's territory. Treaties presently restrict placing operational anti-ballistic missile systems in space, but these concepts remain under investigation. If the main ballistic missile threat continues increasingly to be perceived as coming from terrorists or other "rogue" political entities, such a system may become politically acceptable in the future. Counterair and anti-satellite missions have also been suggested for space laser systems.

CONCLUSION

At the time of the publication deadline for these proceedings, the Air Force had just initiated a comprehensive study to identify and evaluate space laser system concepts that are operationally effective, technically feasible, and affordable. After August 1998, a final report will be available to authorized recipients.

Acknowledgments

The authors would like to acknowledge Major General Richard R. Paul, Commander of the Air Force Research Laboratory, for commencing the study. We also acknowledge Colonel Robert A. Duryea, Director Lasers and Imaging Directorate, Ms. Christine M. Anderson, Director Space Technology Directorate, and Colonel (Retired) William A. Byrne, former Director Plans and Programs Directorate for their support of this study. Additionally, we would like to thank the Office of Aerospace Studies for their help in the planning phase of this project.

References

Air Force Scientific Advisory Board (1995) *New World Vistas: Air and Space Power for the 21st Century – Directed Energy Volume,* Washington, DC.

Air University (1994) "Operational Analysis," in *Spacecast 2020,* Maxwell Air Force Base, AL, Volume 4.

Paul, Maj. Gen. Richard R. (1997) Memorandum, Air Force Research Laboratory, Wright-Patterson Air Force Base, OH, 24 June 1997.

Rogers, Lt. Col. Mark E. (1997), *Lasers in Space – Speed of Light Technology Meeting the Warfighter's Needs,* AU/AWC/RWP162/97-04, Air University, Maxwell Air Force Base, AL.

SEMICONDUCTOR LASERS FOR NASA APPLICATIONS

S. Forouhar, M.G. Young, S. Keo, L. Davis, R. Muller, P. Maker
Jet Propulsion Laboratory, California Institute of Technology
4800 Oak Grove Drive
Pasadena, CA 91109
818-354-4967

Abstract

Lasers play a key role in a wide variety of applications relevant to NASA missions and instruments. These include such diverse areas as spectroscopy, communications, signal processing, metrology and micro-instruments. The unique characteristics of lasers including high optical power in a narrow wavelength range, small beam divergence, phase coherence, and high-frequency modulation, make lasers ideal sources for many of these applications. However, the large size, high power requirements, and relative fragility of conventional gas and solid-state lasers have limited their usefulness for many spaceborne instruments. With the advent of semiconductor lasers, it is now possible to fabricate much smaller, higher efficiency lasers and laser instruments, making them practical choices for a wide variety of applications in space missions. Semiconductor lasers also offer orders-of-magnitude increases in communication bandwidth which enables tremendous enhancements in spacecraft to ground and inter-instrument data transmission rates. Further reduction in power and weight can be gained through the monolithic integration of lasers with other optical and electrical components. Integrated architectures also offer advantages in robustness and system simplicity. The implementation of semiconductor lasers and integrated optics can impact a wide variety of systems capabilities. The following describes few applications of semiconductor lasers that are of significant interest to NASA.

SPECTROSCOPY

Near-Infrared tunable diode lasers (TDLs) have properties that make them well suited for detection of trace gases by optical absorption (Sandstrom 1996, Sonnenfroh 1996, and Tai 1992). Many molecular species have absorption bands in the 1-5 µm wavelength region where III-V compound semiconductor diode lasers operate. Single frequency diode lasers can be temperature or current tuned so that an output wavelength coincident with a specific gas-absorption line can be obtained. Detection techniques based on modulation of laser wavelength yield sensitivities for measuring one part per million (ppm) absorption or smaller. Compatibility with fiberoptics makes the techniques attractive for certain remote-sensing applications, where a single laser source might probe gas concentrations in different locations accessed by fiberoptics. Their sensitivity, and ability to quickly discriminate gases make them ideal for life support applications in the Shuttle and for the Space Station. Unlike the mass spectrometers and gas chromatography that are usually employed in these applications, TDLs require no pumps or other moving parts and are much less susceptible to poisoning by reactive gases (hydrocarbons, sulfur gases, acids). They do therefore provide enhanced reliability as well as lower mass and power consumption. With their sensitivity, low power consumption, low mass, and compact size, instruments based on these near-infrared TDLs also have numerous commercial applications too. They could be used in toxic gas monitoring (workplace detection or industrial site monitoring), for medical applications such as breath analysis, mine safety monitors (methane and carbon monoxide detection), monitoring of pollutants in stack gases, and on-line monitors of combustion or chemical processes.

Spectroscopic applications typically require single wavelength operation with some degree of tunability. Highly-stable single-longitudinal-mode operation may be met by a distributed feedback

(DFB) or distributed Bragg reflector (DBR) laser cavity. Both types of devices require the incorporation of a submicron lithographically defined grating buried within the laser structure which is accomplished using state-of-the-art fabrication and epitaxial techniques. DFB and DBR lasers have been commercially developed in the Indium Phosphide (InP) material system at the fiber-optic wavelengths of 1.3 and 1.55 µm by a host of companies. However, the availability of semiconductor lasers at specific wavelengths suitable for spectroscopy applications is extremely limited. Figure 1 shows a schematic of a DFB laser fabricated at JPL for incorporation into spectroscopy instruments. Operating wavelengths of 1.37 µm, 1.43 µm and 2.04 µm for the detection of water, carbon dioxide, and their isotopes have been demonstrated. These devices are currently being used for atmospheric trace gas measurements in JPL's aircraft programs, and are being implemented in the Mars Volatiles and Climate Surveyor (MVACS) instrument as part of the Mars '98 flight (Paige 1995).

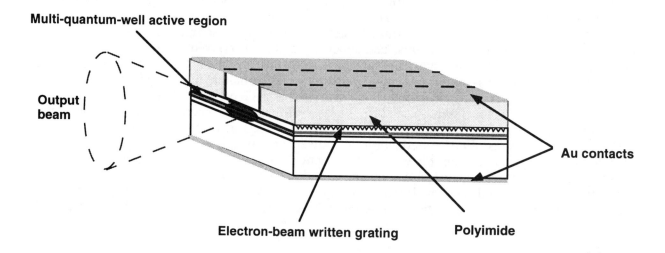

FIGURE 1. Schematic Diagram of a DFB Laser. Typical Dimensions are 350 µm Long and 100 µm Wide.

Single-mode semiconductor lasers that can produce continuous wave (CW) output at wavelengths between 2.1 and 5 microns at temperatures above 200 K would provide further opportunities to develop much smaller instruments for spectroscopy. JPL, in collaboration with several laboratories and universities is developing Gallium Antimonide (GaSb) based single-mode semiconductor lasers in the 2 to 5 µm wavelength range that can operate at these temperatures. To obtain single-mode operation in this material system, the laterally coupled DFB design recently developed at JPL's Microdevices Laboratory will be used. The laterally coupled design is best suited to fabricate DFB lasers in material systems for which epitaxial regrowth embedding the grating structure is extremely difficult.

COMMUNICATIONS

In general, laser communication systems employ either fiber-optic or free-space transmission channels. Fiber-optic communications require laser wavelengths matched to the minimum loss or dispersion wavelength of the transmitting fiber. The evolution of such technologies based on low-loss fiber transmission at 1.3 and 1.55 µm has benefited from intensive commercial development. While free-space communications has no inherent wavelength restrictions based on transmission properties, the development of solid-state laser sources has been dominated by the commercial availability of high-quality Yttrium Aluminum Garnet (YAG) lasers at 1.064 μm. The development of diode laser sources has been driven by the desired integration with electronic and photonic device development, for which GaAs- (850 nm) and InP- (1.3 μm, 1.55 μm) based systems have flourished. In general, NASA's communication applications do not require excursions outside these well-developed wavelengths and materials systems, but the premium on size, mass and system robustness associated with long-duration space missions can place special requirements on the long-term reliability and stability of the lasers. For example, the enhanced reliability under high power (over 1 W) and high temperature operation (to 200° C) of InGaAs/GaAs lasers (900–1000 nm) has made them strong candidates for space-based communications.

One very important application of photonic components and semiconductor lasers in particular, is their use in millimeter wave telecommunication systems within the spacecraft. Within this frame of work, photonic components can enable a reduction in spacecraft size and mass by replacing Radio Frequency (RF) components and subsystems with equivalent optical architecture and front end signal processing at low power levels. The key candidates for this technology insertion are the Ka-band photonic phase array telecommunication systems and fiber optic spacecraft buses for digital/RF signal distribution. To achieve the anticipated pay-off of reduction in volume, mass, and power consumption, in addition to enhanced performance through increased data rate and more design flexibility, the performance of the existing photonic components such as photodetectors and electrooptic modulators as well as the semiconductor lasers need to be improved.

High power (>100 mW), low relative intensity noise (RIN<-165 dBc), and high efficiency lasers required for these optical RF systems are not commercially available. Currently, the only lasers with acceptable low RIN and output power are diode pumped solid-state lasers. However, these devices are bulky and provide poor power conversion efficiencies when compared with semiconductor lasers. The availability of high power reliable semiconductor lasers with good RIN characteristics is essential to the realization of high performance optical signal distribution systems within the spacecraft.

METROLOGY

The stellar interferometry missions require laser sources suitable for high precision interferometric metrology. Interferometric metrology systems are used to form optical trusses for structure definition and stabilization and for measurement of the stellar interferometer baseline. For these applications the lasers need to be high-power, extremely frequency stable, and have a very narrow line-width, i.e. long coherence length, so that an interferometric measurement with 1-10 km path length differences between the spacecraft comprising the stellar interferometer can be performed. In addition to performance, the choice of technology used to implement the metrology laser sources must also take into account the ruggedness, reliability, size, weight, and power consumption requirements associated with space qualification and deployment on future small and inexpensive spacecraft. Though semiconductor lasers are by far the simplest, and most reliable type of lasers available, high power (>100 mW) narrow-line-width (< 20 KHz) semiconductor lasers have not been available to date for these applications.

Conventional DFB semiconductor lasers with uniform gratings or phase shifted gratings have been utilized in high speed communication systems. Although phase shifted DFB lasers have a high probability of single-mode operation, their stability is reduced significantly at higher output powers due to frequency fluctuations caused by the variation of refractive index with carrier density inside the laser cavity. The typical line-width of a single-mode DFB semiconductor laser (determined by both intensity and frequency fluctuations) is ~50 MHz, which is too broad as light sources for metrology applications. One promising approach pursued at JPL, is to achieve stable single-mode operation at high output powers using a corrugation pitch modulated grating DFB laser structure. In this approach, the variation of refractive index with carrier density is compensated through a phase arranging region within the laser cavity, thereby increasing modal stability at high output powers. High power, narrow-line-width semiconductor lasers make available compact, and robust lasers sources for space interferometry applications at significantly lower costs than conventional laser systems.

MICRO-SENSORS AND MICRO-INSTRUMENTS

The semiconductor diode laser as a discrete device or when integrated with optics, electronics, or microsensors offers unique opportunities in the miniaturization of scientific instruments and sensors. Such applications are only now beginning to be identified, and an ever-increasing proliferation of such opportunities can be anticipated to surface over the next decade. While there may be commonality between some generic components for commercial and defense applications and NASA needs, in many cases the applications will be unique to NASA's space science mission. The following are some examples where diode lasers play an important role in the miniaturization of scientific instruments.

Laser Doppler Anemometry

Laser Doppler Anemometry is based on the principle of measuring the Doppler shift of light scattered from particles carried by currents of wind or fluids. Laser Doppler anemometers (LDA) are the most accurate instruments available for measuring wind in harsh environments. This accuracy is intrinsic to the technique, in which wind speed is deduced from a direct measurement of the velocity of particles entrained in the flow. In sharp contrast to thermal anemometers, the calibration of an LDA depends only on the geometry of the device and the accurate measurement of the Doppler frequency. The measurement of wind at a fixed distance from the sensor is a crucial advantage of the LDA technique, particularly on the Martian surface where the wind can come from any direction and sunlight causes errors in thermal anemometers. Most wind sensors measure the mechanical and thermal interaction of the wind with a sensor. The sensing therefore disturbs the flow. The LDA measures the undisturbed free-stream flow away from the sensor, leading to measurements which are more accurate and easier to interpret. In spite of these advantages, state-of-the-art LDA instruments are too large and fragile to survive landing and deployment on Mars' surface, given the hard landing schemes adopted for small Mars landers. Existing LDAs are not small enough, stable enough, or rugged enough for Mars. Therefore, miniaturization of the existing LDAs will provide a major advance in anemometer technology for microinstruments and microspacecrafts, enabling the use of this powerful technique on Mars landers for in situ measurement of wind.

A unique micro-laser Doppler anemometry (μLDA) design in which all of the components of a conventional LDA are integrated onto a single optoelectronic integrated circuit is shown in figure 2. The μLDA consists of a DFB laser, a wave guide, and diffractive optics. Instead of splitting a single beam, as in the conventional LDA configuration, light from the DFB laser propagates through a waveguide, and is coupled out of the plane by diffractive optical elements patterned onto the chip. The coherent beams emitted by opposite ends of the laser are focused onto a measurement volume above the plane of the chip, with the focal length and the diffraction angle determined by the device geometry. The Doppler shift of the scattered light is then measured by a large area photodetector which is also integrated on the same chip.

Diode Laser Pumped Helium Magnetometers

Analysis has shown that the magnetic field equivalent noise of a helium magnetometer can decrease by a factor of 100 using laser pumping with radiation at 1.083 μm instead of conventional helium lamp pumping. Replacing the gas discharge lamp with a semiconductor laser can lead to further instrument simplifications. The laser pump source need not to be co-located in the sensor with the absorption cell but could be located in the spacecraft bus and optical fiber could be used to deliver pumping power to the absorption cell. The 1.083 μm wavelength radiation can be achieved by InGaAs/GaAs lasers, a material system which has been developed for fiber-optic communications (as pump sources for fiber amplifiers). The commercial application requirement for InGaAs/GaAs lasers has focused only at 980 nm, with no regard for spectral purity and tunability both of which are required for replacing helium gas discharge lamps.

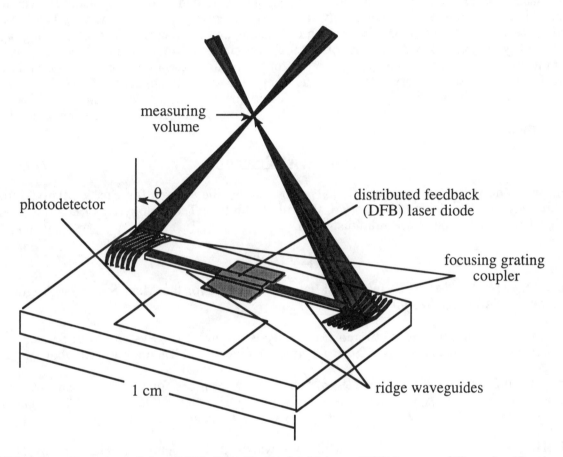

FIGURE 2. Drawing of the μLDA (Not To Scale) Using a DFB Laser and Focusing Grating Couplers to Create the Two Beam Measurement.

Fiber Optic Rotational Sensor

Spacecraft rotation can be sensitively detected with a fiber-optic rotation sensor (FORS), which utilizes the interference of light beams counter propagating in an optical fiber loop. For highest sensitivity, a very low-coherence source is required. The best choice for this application has been found to be super luminescent diode at 1.3 μm or 1.55 μm wavelength. While such devices are currently available commercially, questions of reliability, power level, and spectral quality remain. Rotation sensing could also benefit from the simplicity and robustness of a fully integrated technology, which is possible through monolithic fabrication of semiconductor diode sources, waveguide channels, and processing circuitry.

Other possible applications for in-situ microsensors are also being identified. Compact, robust microsensors are expected to play a major role in future robotic missions, both in autonomous rover vehicles testing the plantery environment as a precursor to human exploration, and in robotic exploration of harsh environments on or off planet Earth which preclude direct human access.

CONCLUSIONS

Considerable development is still required before the benefits of semiconductor lasers and integrated optics can be fully realized in space-based systems. Extensive industrial development of diode lasers has been driven by a handful of high-volume commercial applications, which has resulted in devices operating at a relatively small set of fixed wavelengths. New wavelength ranges need to be accessed, additional spectral and temporal characteristics achieved, and specialized integrated systems developed for many of NASA's missions and instruments. Given the diversity of applications and technologies for semiconductor lasers, a coherent development plan is in place at the Microdevices Laboratory (MDL) of JPL to provide optimal support for NASA's short- and long-term needs. In addition to ensuring the development of lasers for specific instruments or missions, this plan offers the flexibility to move rapidly in new directions when a new area of need is identified.

Acknowledgments

The work described in this paper was performed by the Center for Space Microelectronics Technology, Jet Propulsion Laboratory, California Institute of Technology under contract with the National Aeronautic and Space Administration.

References

Paige, D.A. (1995) "MVACS- Integrated Payload Proposal for the Mars Surveyor Program '98 Lander," *NASA AO No. 95-OSS-3*.

Sandstrom, Lars (1996) "Near-Infrared Gas Analysis Spectroscopy Systems with Semiconductor Lasers," in *Technical report No. 299L*, Chalmers University of Technology, Goteborg, Sweden

Sonnenfroh, D. and Mark G. Allen (1996) "Diode Laser Sensors for Combustor and Aeroengine Emissions Testing: Applications to CO, CO2, OH, and NO," *19th AIAA Advanced Measurement and Ground Testing Technology Conference*, New Orleans, LA.

Tai, H., K. Yamamoto, M. Uchida, S. Osawa, and K. Uehara (1992) "Long Distance Simultaneous Detection of Methane and Acetylene by Using Diode Lasers Coupled with Optical Fibers," *IEEE Photonics Technology Letters*, (4) 7.

HIGH-EFFICIENCY LASER DEVELOPMENT FOR SPACE APPLICATIONS

Clifford Muller, Craig Denman,
Michael Gregg and Linda DeHainaut
Laser Systems Division, Directed Energy Directorate
Air Force Research Laboratory
Kirtland Air Force Base, NM 87117-5776
(505) 846-4026, 846-4601, 846-8992, 846-4020

Abstract

High-efficiency lasers are critical to space-based systems where power consumption severely constrains systems. This paper will review Air Force Research Laboratory/Phillips Laboratory (AFRL/PL) laser development programs which provide high-efficiency electric lasers for space-based communications and solid-state lasers for space-based clocks.

INTRODUCTION

The Air Force Research Laboratory/Phillips Laboratory explores, develops and applies laser technology to meet Air Force and national objectives. Recently, laser technology development pursued at Phillips Laboratory has focused on air and ground applications. Air Force studies such as Spacecast 2020, AF 2025, and New World Vistas have contributed to the Air Force's vision of global awareness. The Air Force initiative in Global Virtual Presence is dependent upon the capability to detect threats and interact globally on a real-time basis. Laser technologies can provide or enhance space-based surveillance, geopositioning, target designation, remote sensing and communication missions, helping to make Global Virtual Presence a reality. The Phillips Laboratory's Laser Systems Division is formulating a laser technology development program which addresses space as well as air and ground requirements. Semiconductor laser and solid-state laser technologies are being developed for future space applications.

SPACE-BASED LASER COMMUNICATIONS

The High Power Semiconductor Laser Technology (HPSLT) program develops power scaling technology for high-power, coherent laser sources using diode lasers and diode laser arrays. As the technology matures, it offers advantages over alternative laser sources; semiconductor lasers are:

- extremely efficient -- 60% wallplug efficiency demonstrated,
- wavelength flexible -- 0.5 to 5.0 microns demonstrated,
- compact, light weight -- 1 cubic foot laser head due FY98,
- arrays are scalable to high powers -- 200 Watts incoherent demonstrated,
- arrays offer graceful degradation for limited access systems
- mass producible -- low cost,
- low voltage requirements -- battery-operated, portable.

These advantages make diode lasers attractive for space-based laser systems which must be light weight, compact, and efficient.

Diode lasers are particularly suited for space-based laser communications, and AFRL/PL has a well-defined program to develop semiconductor lasers for lasercom. Since an established commercial base for communications at 1.55 µm exists, the program approaches lasercom from two system perspectives - a high-power diode laser pump (emitting at 980 nm) for Er-doped fiber transmitters emitting at 1.55 µm and a direct diode, high-power transmitter emitting at 1.55 µm. Table 1 outlines the current laser technology development program for space laser communications.

TABLE 1. Development Goals for Space-Based Applications.

Near-Term Goals	Mid-Term Goals
<div align="center">**High-Power 980 nm Pump**</div>	
• 5 W CW, 980 nm, diffraction limited, through a single-mode fiber	• 10 W CW, 980 nm, diffraction limted, through a single-mode fiber • 5 W average power, 980 nm, diffraction limited, modulated at 10 Gbps
• Reliability: 50K hrs at 1 Watt	• Reliability: 100K hrs at 10 Watts
<div align="center">**Coupling to Single-Mode Fibers**</div>	
• Compact, rugged, efficient coupling system	• 1 W out of single-mode fiber, 80% efficiency
<div align="center">**1.55 μm Transmitter Source**</div>	
• 1 Watt, CW, 1.55 μm through a single-mode fiber	• 3 Watts, diffraction limited 1.55 μm source
<div align="center">**Fiber Lasers/Amplifiers**</div>	
• 100 Watt fiber laser at 1.06 μm	• High-power fiber array concept

Global Virtual Presence will require the handling of large amounts of information making it necessary to increase laser communication data rates. Higher data rates can be achieved by increasing modulation rates and transmitter output power, and the most effective way to increase the output power from Er-doped fiber amplifiers is to increase the pump power. One of the AFRL/PL lasercom program goals is to demonstrate 10 Watts, diffraction limited power at 980 nm through a single-mode fiber. This is a factor of ten increase over commercially available 980 nm diode lasers. Broad-area, flared amplifiers have been developed; and 6 Watts of continuous wave, diffraction limited laser output power has been demonstrated with a slope efficiency of 75 percent.

Laser reliability continues to be an issue since logistics and space environments place more stringent reliability requirements on space-based systems. On the average, 980 nm diode lasers currently have an operating lifetime of 100,000 hours at 100 mWatts and 3,000 to 5,000 hours at 1 Watt. The Laser Systems Division is beginning development to extend the lifetime to 100,000 hours at 10 Watts. Our near-term goal is to extend the diode laser lifetime to 50,000 hours at 1 Watt within one to two years.

In addition to providing higher power, reliable diode lasers, efficiently coupling pump power into fibers is another way to increase transmitter power and improve overall system efficiency. As in all space-based system parts, coupling devices must be compact and rugged. The AFRL/PL is pursuing development of a compact, rugged coupling system capable of delivering pump power into the fiber with 80 percent efficiency. The coupling system is based upon a single-surface anamorphic lens made in GaP to directly focus the laser output into a single-mode fiber. Coupling efficiencies as high as 67 percent (350 mWatts out of the fiber) have been measured, and a rugged prototype package device has been fabricated with 60 percent demonstrated efficiency. The prototype package dimensions are on the order of 25 cm^2.

While pumping Er-doped fibers with higher-power pumps allows increased communication rates, a high-power diode laser operating at 1.55 μm would be a more efficient transmitter. It is estimated that replacing a fiber source with a direct diode laser could increase transmitter efficiency by two orders of magnitude. Therefore, AFRL/PL is pursuing development of a 1.55 μm source with the near-term goal of achieving moderate power and a mid-term goal of increasing the coherent output power to 3 Watts, diffraction limited, a factor of 10 increase over commercially available diodes. Over 4 Watts, CW, incoherent power at room-temperature at 1.55 μm has been demonstrated from a broad-area device.

FIBER LASERS/AMPLIFIERS

Since fiber lasers and amplifiers are compact, highly efficient and scalable to higher powers, they can play a crucial role in space-based systems where packaging, thermal management, and power consumption are issues. The AFRL/PL has a new fiber laser/amplifier program to investigate the feasibility of high-power fiber laser arrays. Computer models indicate that it is possible to achieve 100 Watts from one single-mode fiber. Combining fibers into a coherent array has the potential to provide a light weight, compact, high-power laser for space-based applications. The near-term project goal includes the demonstration of a 50 Watt fiber laser operating at 1.06 μm for use as an Er-doped fiber pump. The mid-term goal is investigation of fiber array architectures.

LASER CLOCKS

The AFRL/PL has recently begun a five-year program to develop a Nd:YAG laser clock based on a cavity-enhanced saturated absorption on the 5th-overtone resonance of carbon dioxide ($2v_1 + 3v_3$ located at 1.064507 μm). This transition offers a narrow natural linewidth, 170 Hz, for the absolute frequency stabilization of a new generation of laser clocks.

The potential of a CO_2 laser clock can be put in perspective by considering the stability achieved using the strongly absorbing transition in iodine and the second harmonic of Nd:YAG. On a Doppler-broadened linewidth of about 850 MHz a relative stability of $dv/v \approx 10^{-11}$ was achieved by A. Arie, et al., (1993). Similarly, J. Hall, et al., (1995) of University of Colorado (JILA) demonstrated a stability of about 10^{-13} on a 1 MHz iodine doppler-free linewidth obtained at very low pressure.

Over the past year, AFRL/PL researchers conducted experiments using two Nd:YAG NPRO lasers stabilized to adjacent axial modes of a commercially available high finesse optical cavity. They achieved a 700 mHz linewidth beatnote between the two lasers and less than a 3 Hz linewidth beatnote stabilization at 30 Watts (TEM_{00} polarized CW) by injection locking a high-power slave to the frequency stabilized low-power NPRO. Additionally, to prepare for future doppler-free spectroscopy/laser stabilization experiments, the doppler broadened spectroscopy of CO_2 has been completed to better establish the wavelength of the 5th-overtone resonance and estimate the pressure broadening coefficient. Because of the many rotational-vibrational resonances available in CO_2, a beatnote between two lasers locked to adjacent resonances could be used for down conversion of the optical frequency standards for clock time-keeping purposes. Unfortunately, only one resonance is accessible to Nd:YAG; therefore, we are looking at other lasing media. One based on a more tunable dopant such as Yb may be more desirable.

CONCLUSION

The AFRL/PL is expanding its research to laser development for space-based applications. High-efficiency semiconductor lasers are being developed for space laser communications. A fiber laser/amplifier program is being pursued to provide high-power sources for applications where packaging and thermal management are critical. Development is expanding to solid-state lasers, and the first development project is a high-precision laser clock. As space-based applications are more clearly understood, AFRL/PL laser programs will be expanded to meet system requirements.

Acknowledgements

This work is being performed by the Laser Systems Division, Directed Energy Directorate, Air Force Research Laboratory.

References

Arie, A. and R. L. Byer (1993) "Frequency Stabilization of the 1064-nm Nd:YAG Lasers to Doppler-Broadened Lines of Iodine," in *Applied Optics*, 20 December 1993, 32 (36).

Hall, J. L. and M. L. Eickoff (1995) "Optical Frequency Standard at 532 nm," in *IEEE Transactions on Instrumentation and Measurement*, April 1995. 44 (2).

SOLID STATE LASER TECHNOLOGY AND ATMOSPHERIC SENSING APPLICATIONS

James C. Barnes
NASA Langley Research Center
Mail Stop 474
Hampton, VA 23681
(757) 864-1637 and 864-7174

Abstract

Future NASA space missions to enable global monitoring of the earth's atmosphere will include measurements of ozone, water vapor, aerosols and clouds as well as global wind velocity. These important elements and parameters effect climate change, atmospheric chemistry and dynamics, atmospheric transport and, in general, the health of the planet. NASA **Di**fferential **A**bsorption **L**idar (DIAL) and backscatter lidar techniques are leading candidates for active remote sensing of molecular constituents and atmospheric phenomena from advanced high-altitude aircraft and space platforms. DIAL and lidar systems operating from space have the capability to completely map the earth's atmosphere for these and other molecules as well as for tracking atmospheric phenomena such as winds and volcanic dust transport. This paper provides an overview of NASA Langley's development of advanced solid state lasers, harmonic generators, and wave mixing techniques aimed at providing the broad range of wavelengths necessary to meet measurement goals of NASA's Mission To Planet Earth (MTPE) Enterprise and potentially for applications in the Space Science Enterprise.

INTRODUCTION

NASA is developing new solid state laser and nonlinear optics technologies to meet energy, efficiency, wavelength, pulsewidth, linewidth, and reliability requirements for advanced DIAL and lidar instruments. These development efforts will provide significant improvements in efficiency and reduction in the size and weight of laser-based sensors having the capability to measure ozone, water vapor, aerosols and clouds, and wind velocity with high vertical and horizontal resolution and high accuracy. These atmospheric constituents are very important with respect to global climate change: Measurements of water vapor is recognized as a key strategic step towards meeting the MTPE objective of making "observations to monitor, describe, and understand seasonal-to-interannual climate variability, with the aim of improving skill in long-range weather forecasting and seasonal climate predictions." Water vapor is not as well characterized at the higher altitudes, hence, the gap in our knowledge base is considerable (Ismail 1989). Aerosols, especially those arising from combustion of fossil fuels and biomass burning, reduce the Earth's surface temperature by reflecting a portion of the incident solar radiation back to space, countering some of the effects of increased greenhouse gas concentrations. Clouds also reflect solar radiation, and are considered the major uncertainty in understanding global climate change. Ozone has received increasingly more attention in recent years as a result of its reported depletion in the stratosphere and its importance to absorption of harmful ultraviolet solar radiation. Instruments that can measure these four atmospheric constituents with high spatial resolution and accuracy will make important contributions to understanding the mechanism of global climate change (Browell 1994 and Grant 1989). Global climate change scientists, atmospheric dynamics researchers, and numerous agencies such as NASA, NOAA, DOD, and DOE greatly desire continuous, global measurements of tropospheric wind with good horizontal and vertical coverage and resolution, i.e. a horizontal vector accuracy of about 1 m/s (Kavaya 1996). Descriptions of the development of advanced solid state laser light sources for NASA DIAL and lidar measurements of ozone, water vapor, aerosols, clouds and wind velocity are discussed in detail below.

Laser Materials Modeling and Spectroscopy

A quantum mechanical laser materials model is used by NASA researchers to guide the search for new laser materials that can meet the stringent and unique space-based DIAL and lidar laser requirements; these include very specific wavelengths, narrow linewidths, high energy output with high efficiency, high repetition rates, and long operational lifetimes with high reliability. Purely experimental approaches for the development of new laser materials result in impractical developmental cost and in time consuming, high risk research and development efforts. Consequently, a quantum mechanical modeling and spectroscopy approach is used by NASA. The NASA quantum mechanical model is used to predict specific wavelengths, transition rates, and thermal occupation factors of new laser materials. By using existing spectroscopic data for known laser materials that are similar to the new materials being investigated, the quantum mechanical model can be validated. The validation of the model through the existing data on similar materials lends credence to the model's extrapolation of the optical and laser characteristics of the new material. The next step in this approach is the spectroscopic evaluation of the laser materials using spectroscopic samples which are relatively inexpensive compared to laser grade samples. The evaluation tests the predictions of the quantum mechanical model with regards to material parameters such as emission wavelength, fluorescence lifetime, and transition strength. The evaluation also provides spectroscopic data with which to fine tune the results, if necessary. Finally, based on the model results, a laser material composition is chosen to demonstrate laser performance (Jani et al. 1994 and Hart et al. 1996). If warranted, the material is further developed to laser grade quality to meet the particular DIAL system's requirements.

2-μm, Ho:Tm:YLF Laser System for Wind Velocity Measurements

An energetic, low divergence, narrow linewidth all solid-state laser system in the eye-safe region is required as a transmitter for coherent wind lidar measurements from ground, air, and space-borne platforms. Development of a diode-pumped, 2-μm, Ho:Tm:YLF laser system has been initiated at NASA Langley Research Center (LaRC) as it emits radiation in the eye-safe region and can achieve efficient diode pumped operation using commercially available laser diodes. An extensive theoretical model (Barnes et al. 1996 and Barnes et al. 1996) has been established at NASA/LaRC and was used to identify the laser material for optimum gain at 2-μm. The Ho:Tm:YLF crystal, operating at 2.05 μm, was selected as the laser gain medium based on the model. This group previously used a flashlamp pumped oscillator and five diode pumped amplifiers to generate 700 mJ at 1 Hz (Williams-Byrd et al. 1997). Since then, an injection-seeded diode pumped oscillator has been developed and used in the laser system. The pulse repetition rate has been increased to 10 Hz. Amplifier gain at 10 Hz is measured to be comparable to that at 1 Hz. Performance of the system has been fully characterized. A schematic of the injection-seeded Ho:Tm:YLF laser system is shown in Fig. 1. It consists of a CW master oscillator, power oscillator and amplifiers. The diode pumped Q-switched power oscillator utilizes a three meter long, figure-eight ring resonator configuration and is injection-seeded by a CW microchip master oscillator. This combination

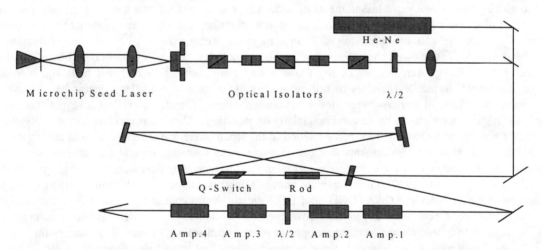

FIGURE 1. Schematic of Injection-Seeded Ho:Tm:YLF Laser System.

provides a single frequency, high beam quality extraction beam for the amplifiers. The amplifier chain comprises four diode pumped amplifiers which produce a gain of about 9 in the single pass configuration used for this work.

Ti:Sapphire Laser Transmitter for Airborne Water Vapor DIAL Instrument

In the mid-to-late 1980's, NASA Langley researchers and several industry partners pioneered $Ti:Al_2O_3$ (titanium-doped sapphire) laser material development and laser pumping and injection seeding techniques

(Brockman et al. 1986; Barnes et al. 1988; Rines et al. 1989). The material quality was improved by an order of magnitude, allowing the researchers to demonstrate the superior performance of tunable (from 0.66 μm to 1.1 μm) Ti:Al$_2$O$_3$ lasers over conventional tunable dye lasers. With this, Ti:Al$_2$O$_3$ became a viable laser device with a multitude of potential uses including medical, industrial, communication, military, and scientific applications. We make a conservative estimate of 40 - 50 million dollars of commercial Ti:Al$_2$O$_3$ sales over the past 5 years based on published reports indicating ~25 million dollars of sales through 1992 (Kales 1992 and Kales 1993).

With the advantage of having performed the earlier work, NASA laser engineers designed and built a fully autonomous tunable laser system (TLS), based on Ti:Al$_2$O$_3$ laser technology, for NASA's Lidar Atmospheric Sensing Experiment (LASE). Since 1994, the LASE Instrument has flown on over 26 successful water vapor measurement missions on NASA's high altitude ER-2 aircraft without a single failure of the Ti:Al$_2$O$_3$ laser (Browell 1995). The TLS was designed to operate in a double-pulse mode at 5Hz, with energy outputs up to 150mJ per pulse with tunable wavelengths of 813 to 819nm and with 99% of the output energy within a spectral interval of 1.06 pm. The Ti:Sapphire power oscillator uses a frequency-doubled Nd:YAG laser as the pump source and a single mode diode laser as a injection seeder for the Ti:Al$_2$O$_3$ laser (J. Barnes et al. 1993; N. Barnes et al. 1993; J. Barnes et al. 1993). The LASE instrument was recently reconfigured to fly on a NASA P-3 Aircraft and participated in the Southern Great Plains Hydrology Experiment (SGP97) in the summer of 1997. The SGP97 is a joint NASA, DoE, NOAA, NSF, USDA, NRC of Canada, and Oklahoma Mesonet interdisciplinary investigation designed to establish that retrieval algorithms for surface soil moisture developed at high spatial resolutions for ground and airborne sensors can be extended to the coarser resolutions expected from satellites (Jackson et al. 1995).

Tunable Cr:LiSAF and Nd:Garnet Development for Space-Based Water Vapor DIAL

As a precursor to space flight deployment, NASA researchers and contractors are currently developing a water vapor DIAL laser transmitter for remotely piloted vehicle deployments. The baseline laser transmitter is based on the promising new laser material Cr:LiSrAlF$_6$. Cr:LiCaLiF$_6$ and Cr:LiSrGaF$_6$, two similar materials with slightly different wavelength and thermal properties are also being investigated for future applications. These materials are attractive because of they have broad absorption (~550nm-750nm) and emission (~750nm-1000nm) bands and because they can be heavily doped with the active, Cr, ion. The broad tuning range resulting from the broad emission allows for multiple NASA DIAL applications. Tropospheric water vapor measurements can be addressed at wavelengths around 815nm and 940nm, frequency tripling of 900nm yields wavelengths that address atmospheric ozone and the residual 450 and 900nm wavelengths would provide aerosol and cloud measurements. The broad pump absorption bands of the materials peak in the 670 - 690nm region but extend into the 750nm region where well developed AlGaAs laser diode arrays (LDA's) can be used to pump heavily doped materials (Payne et al. 1992). NASA has supported the development of advanced visible 675nm InGaP laser diode array technology to optimize efficiency in pumping the Cr lasers. LDA pumping increases the wall-plug efficiency and operational lifetime over current flashlamp pumping, both of which are critical parameters for NASA space missions. The researchers used 3 each six bar stacks of the new 675nm LDA's that operate with 360W output in 200μ s pulses to demonstrate the first ever 675nm, LDA side-pumped Cr:LiSrAlF$_6$ laser oscillator (Johnson et al. 1996). The oscillator produces 33mJ of energy in the normal mode and has 3mJ of q-switched energy. The oscillator has been used to measure hard targets and water vapor in a joint NASA/Los Alamos National Laboratory demonstration (Early et al. 1996). Because of the relatively short upper state lifetime of these materials (approximately 65 to 70μs), they require high peak power pumping in order to operate efficiently. Investigations are underway to identify host with longer lifetimes and to improve both the efficiency and the reliability of the 675nm LDA pumps when operating at short pulsewidth and high power.

Compositional tuning of Nd:Garnet laser emission to match strong water vapor absorption features in the 940 nm wavelength region is also being pursued by NASA researchers for water vapor DIAL applications. Garnet laser materials are pumpable by well developed, less costly, AlGaAs laser diode array technology and promise to provide the efficiency, reliability, and affordability required for future space-based DIAL missions. The emission peak of the garnet is tuned by adjusting the fractional composition of Ga in Nd:YGAG, a garnet laser material. NASA researchers have demonstrated tuning of the fluorescence emission peak of the garnet, Nd:YGAG, laser transition from 935nm to 945nm. In this wavelength region, the water absorption lines are 20 times stronger than the lines currently used at ~ 815nm, hence, these wavelengths are attractive for measuring low water vapor concentrations in the stratosphere and upper troposphere. Work remains to demonstrate a compositionally tuned garnet laser that perfectly overlaps a desired water absorption feature while suppressing the strong laser gain found at around 1060nm. Other garnets materials and new resonator optics designs are being investigated to solve these technology tall-poles.

UV Transmitter for Ozone DIAL

Development of ultraviolet laser transmitter technology for space-based ozone DIAL has recently begun at NASA Langley Research Center. Atmospheric models predict the transmitter energy output must be about 500mJ at wavelengths around 305nm and 315nm and it must operate at a minimum 5 Hertz, double-pulsed rate. The current NASA design, for demonstration of the UV laser transmitter technology, will split the frequency doubled output of a Nd:YLF laser into three 524nm beam lines. One beam line will be used to pump an optical parametric oscillator (OPO). The OPO output is tunable between 740nm and 820nm and it will be amplified in a titanium-doped sapphire laser amplifier, which will be pumped by a second 524nm beam. The output of the titanium-doped sapphire laser amplifier and the third 524nm beam will be simultaneously transmitted into a BBO sum frequency mixer to generate the tunable ultraviolet output. Initial demonstrations will be in the 100mJ energy range but with designs to show scalability to 500mJ or more. The technology tall poles for this development include eliminating optical damage to the ultraviolet optics and mixing crystal, providing the precise timing required for an efficient mixing process, and obtaining good beam quality in the OPO and amplifier output.

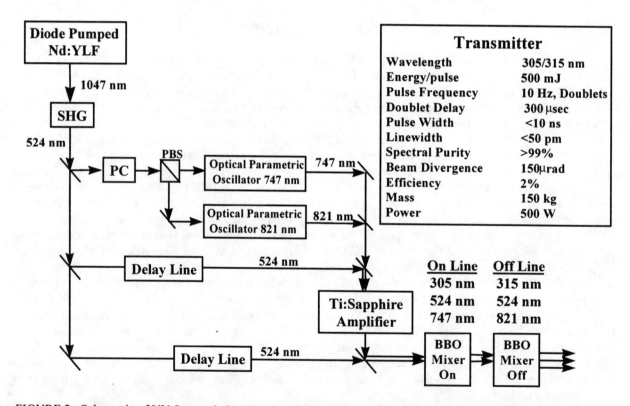

FIGURE 2. Schematic of UV System being Developed for Spaceborne DIAL Ozone Measurements.

Acknowledgments

The laser and DIAL research are supported by NASA Headquarter's Codes X, S and Y. The author recognizes the contributions to this paper by Norman Barnes, William Edwards, Larry Petway, Christyl Johnson, Julie Williams-Byrd, Upendra Singh, Waverly Marsh, Jirong Yu, Mark Storm, Brian Walsh, and Elka Ertur.

References

Barnes, J. C., N. P. Barnes, and G. E. Miller (1988) "Master Oscillator Power Amplifier Performance of Ti:Al$_2$O$_3$," *IEEE J. Quantum Electron.*, 24, (6):1029-1038.

Barnes, J. C., W. C. Edwards, L. Petway, and L. G. Wang (1993) "NASA Lidar Atmospheric Sensing Experiment's Titanium-doped Sapphire Tunable Laser System," *OSA Optical Remote Sensing of the Atmosphere Technical Digest*, 5:459-565.

Barnes, J. C., N. P. Barnes, L. G. Wang, and W. C. Edwards (1993) "Injection Seeding II: Ti:Al$_2$O$_3$ Experiments," *IEEE Journal of Quantum Electronics*, 29, (10):2683-2692.

Barnes, N. P., and J. C. Barnes (1993) "Injection Seeding I: Theory," *IEEE Journal of Quantum Electronics*, 29 (10):2670-2683.

Barnes, N. P., Filer, E. D., Morrison, C. D., and Lee, C. J. (1996) "Ho:Tm Lasers I: Theoretical", *IEEE J. Quant. Elect.* QE-32, 92-103.

Barnes N. P., Rodriguez, W. J., and Walsh, B. M. (1996) "Ho:Tm:YLF Laser AmplifiersC, *J. Opt. Soc. Am. B*, 13, (12):2872-2882.

Brockman, P., C. H. Bair, J. C. Barnes, R. V. Hess, and E. V. Browell (1986) "Pulsed Injection Control of Titanium-Doped Sapphire Laser, " *Opt. Lett.*, 11 (11):712-714.

Browell, E. V. (1994) "Remote Sensing of Trace Gases from Satellites and Aircraft," in *Chemistry of the Atmosphere: The Impact on Global Change*, edited by J. Calvert, Blackwell Scientific Publications, 121-134.

Browell, E. V. and S. Ismail (1995) "First Lidar Measurements of Water Vapor and Aerosols from a High-Altitude Aircraft," *Proceeds of the OSA Optical Remote Sensing of the Atmosphere Conference*, 2:212-214.

Early, J.W., C. S. Lester, N. J., Cockroft, C.C. Johnson, D. J. Reichle, and D. W. Mordaunt "Dual-Rod Cr:LiSAF Oscillator-Amplifier for Remote Sensing Applications," *Trends in Optics and Photonics on Advanced Solid State Lasers*, 1:126 -129.

Grant, W. B., ed. (1989) *Ozone Measuring Instruments for the Stratosphere*, Optical Soc. Am., 438.

Grant, W. B., et al. (1994) "Aerosol-Associated Changes in Tropical Stratospheric Ozone Following the Eruption of Mount Pinatubo," *J. Geophys.Res.*, 99:8197- 8211.

Hart, D. W., M. G. Jani, and N. P. Barnes (1996) "Room-Temperature Lasing of End-Pumped Ho:Lu$_3$Al$_5$O$_{12}$," *Opt. Lett.*, 21 (10):728-730.

Ismail, S. and E. V. Browell (1989) "Airborne and Spaceborne Lidar Measurements of Water Vapor Profiles: A Sensitivity Analysis," *Appl. Opt.*, 28:3603-3615.

Jani, M. G., N. P. Barnes, K. E. Murray, and R. L. Hutcherson (1994) "Diode-Pumped Ho:Tm:Lu$_3$Al$_5$O$_{12}$ Room Temperature Laser," *OSA Proceedings on Advanced Solid State Lasers*, 20:109-111.

Jackson, T. J., LeVine, D. M., Swift, C. T., Schmugge, T. J., and Schiebe, F. R. (1995) "Large Area Mapping of Soil Moisture Using the ESTAR Passive Microwave Radiometer in Washita '92," *Remote Sensing Environment*, 53:27-37.

Johnson, C. C., D. J. Reichle, N. P. Barnes, G. J. Quarles, J. W. Early, and N. J. Cockroft (1996) "High Energy Diode Side Pumped Cr:LiSAF Laser," *Trends in Optics and Photonics on Advanced Solid State Lasers,* 1:120 - 125.

Kales, D. (1992) "Review and Forecast of Laser Markets:1992," *Laser Focus World,* 28 (1):56-58.

Kales, D. (1992) "Laser Suppliers Lower Forecast for 1992," *Laser Focus World,* 28 (8):57-60.

Kales, D. (1993) "Review and Forecast of Laser Markets: 1993," *Laser Focus World,* 29 (1):70-88.

Kavaya, M. J. (1996) "Novel Technology for Satellite Based Wind Sensing," *AIAA Space Programs and Technologies Conference,* AIAA 96-4276:1-3.

Payne, S. A., W. F. Krupke, et al. (1992) "752nm Wing-Pumped Cr:LiSAF Laser," *IEEE Journal of Quantum Electronics,* 28 (4):1188-1196.

Rines, G. A., P. F. Moulton, and J. Harrison (1989) "Narrowband, High Energy Ti:Al_2O_3 Lidar Transmitter for Spacecraft Sensing," *OSA Proceedings on Tunable Solid State Lasers,* 5:2-8.

Williams-Byrd, J. A., Singh, U. N., Barnes, N. P., Lockard, G. E., Modlin, E. A. and Yu, J. (1997) "Room-Temperature, Diode-Pumped Ho:Tm:YLF Laser Amplifiers Generating 700 mJ at 2 μm", *OSA Trends in Optics and Photonics Series,* Advance Solid State Lasers, 10:199-201.

ENABLING TECHNOLOGIES FOR A GLOBAL TROPOSPHERIC WIND LIDAR SYSTEM

Bruce M. Gentry
NASA Goddard Space Flight Center,
Laboratory for Atmospheres, Code 912
Greenbelt, MD 20771
Phone: 301/286-6842, Fax 301/286-1762
gentry@agnes.gsfc.nasa.gov

Abstract

Remote measurement of tropospheric winds is recognized as one of the most important measurements for the atmospheric sciences. Winds can be measured as a function of altitude from satellites using lidar (LIght Detection And Ranging) systems which sense the Doppler shifted frequency of the laser signal backscattered from the atmosphere. NASA has had a long term commitment to developing the technologies required to enable global wind sensing. This paper describes recent developments in the Doppler lidar key technology areas.

INTRODUCTION

Research has established the importance of global wind measurements for large scale improvements in numerical weather prediction (Baker 1995). In addition, global wind measurements provide data that are fundamental to the understanding and prediction of global climate change. These tasks are closely linked with the goals of the NASA Mission to Planet Earth (MTPE) and NASA Global Climate Change programs. NASA has been actively involved in the development of a satellite based global wind observing system to meet the needs of the atmospheric science community.

The development of a global wind measurement capability is identified as a key science objective of NASA's New Millennium Program. The New Millennium Program is a unique collaboration of industry, university and government technology partners brought together to focus on development of advanced spacecraft and instrument technologies for the next generation satellites. Space flight opportunities are available to demonstrate advanced concepts, flight test new enabling technologies and reduce risk associated with those new technologies for future science and operational systems. A path to develop a global wind sensing capability using lidar was identified early in the New Millennium Program. The key technologies and a roadmap to develop those technologies was adopted that leads to an early demonstration flight. The New Millennium roadmap includes technology paths for both a Coherent Doppler lidar approach and a Direct Detection Doppler lidar approach. This paper will discuss several of the key lidar technologies identified in that roadmap with an emphasis on developments in the areas of space qualified solid state lasers, tunable local oscillators, and tunable high spectral resolution filters.

DOPPLER LIDAR

The wind profile can be measured with a spaceborne lidar system that senses the Doppler shift in the frequency of the laser signal backscattered from the atmosphere. A conceptual diagram of the spaceborne Doppler lidar is shown in Figure 1. The Doppler lidar transmitter is a high power, narrowband, pulsed laser directed to the atmosphere at a nadir angle ❏ (typically ❏ = 30 to 45❏). A fraction of the laser energy will be backscattered by aerosols and molecules and collected by a telescope. The collected signal is recorded as a function of range then processed by the detection system which measures the Doppler shift in the laser frequency produced by the motion of the atmosphere. The measured Doppler shift can be used to determine the component of the wind along the line of sight of the laser. A scanning system is used with the Doppler lidar to obtain multiple look angles (e.g. forward and backward views) of each sample volume. The horizontal components of the wind can be obtained by combining line-of-sight velocities from the multiple views.

The Doppler shifted laser frequency is measured using either coherent detection or direct detection techniques. A coherent detection Doppler lidar measures the shift by beating the atmospheric return backscattered from aerosols with a local oscillator to produce a beat frequency which is proportional to the Doppler shift. The key technologies are a high energy pulsed narrow spectral linewidth laser transmitter, a precise local oscillator, and a diffraction limited telescope and scanning system. A direct detection Doppler lidar measures the Doppler shifted frequency by observing the frequency of the backscattered signal through a high spectral resolution optical filter. The direct detection lidar can process the signal return from either aerosols or molecules. The molecular return is thermally broadened by the random motion of the molecules which spreads out width the molecular spectrum to about 600 m/sec in equivalent Doppler velocity units. The spectral spreading of the molecular signal makes a precise measurement of the wind more difficult but this disadvantage is offset by the fact that the molecular signal is always present while the aerosol return is highly variable in altitude as well as globally. In addition the backscattered signal can be greatly increased by operating at shorter wavelengths in the ultraviolet because of the $\bullet^{-4}$ dependence of the backscatter coefficient. The key technologies for the direct detection Doppler lidar are the high energy pulsed laser transmitter, the high spectral resolution filters and sensitive photon counting detectors. Lightweight, large aperture telescopes and scanning systems are also required although these systems do not require diffraction limited performance as the telescope performs more like a "light bucket" in the direct detection lidar.

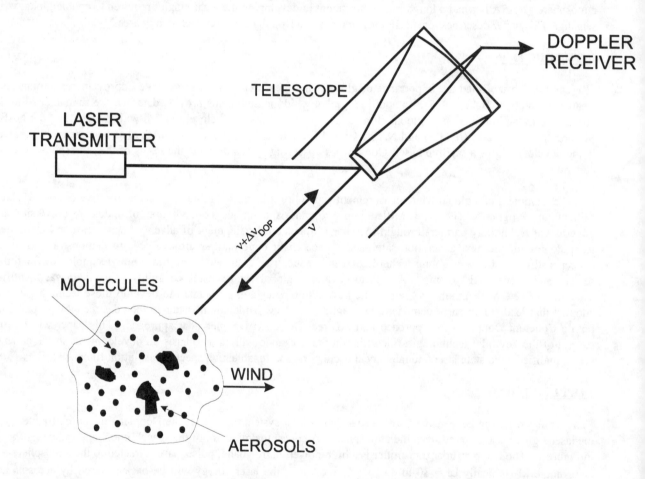

FIGURE 1. Conceptual Diagram of A Spaceborne Doppler Wind Lidar System.

KEY TECHNOLOGIES

The technological demands for a spaceborne wind lidar system are quite challenging and push the limits of the laser and optical technologies employed. The scale of this challenge is brought home when it is realized that measurement of the wind speed to the meter/second level requires a measurement of frequency to a few parts in 10^9

from a spaceborne platform. Early development efforts concentrated on coherent Doppler lidar using heterodyne detection and CO_2 lasers at a wavelength of 10 μm (Huffaker 1978). The Laser Atmospheric Wind Sounder (LAWS), a proposed facility instrument for the Earth Observing System, utilized this approach (NASA 1987). More recently, advances in solid state lasers and other lidar technologies has led to the development of a number of promising new approaches using both coherent (Henderson 1993) and direct detection Doppler lidar (Korb 1992 and Abreu 1992). The use of efficient diode pumped solid state laser transmitters offers the possibility to fly compact, robust wind lidar systems that will operate within the resource envelopes of the next generation of small spacecraft.

Within the NASA New Millennium Program Marshall Space Flight Center has taken the lead in the development of advanced coherent Doppler lidar technologies. This includes the development of high energy 2 μm solid state lasers as well as novel diffraction limited telescopes and scanners. Work carried out at and Langley Research Center has advanced the state of the art in narrowband high energy 2 μm lasers. Recently a breadboard diode pumped Tm:Ho:YLF laser system operating single mode with 700 mJ per pulse energy at 1 Hz repetition rate has been reported (Singh 1997). This laser is a master oscillator power amplifier design which is injection seeded for single frequency operation. The oscillator generates 50 mJ at 10 Hz with a pulse length of 200-400 ns. A chain of amplifiers is used to increase the laser energy to the requisite level. The design goal for the 2μm coherent Doppler lidar transmitter is 500 mJ per pulse at a repetition rate of 10 Hz. A highly stable local oscillator is also required for the coherent Doppler lidar. A widely tunable, stable, narrow linewidth Tm:Ho:YLF laser is being developed for this purpose at JPL (Menzies 1997). The local oscillator must be tunable to offset the large Doppler shift introduced by the spacecraft motion. For example, with a conically scanned lidar system operating with a nadir angle of 30°, the 7.5 km/sec velocity of the spacecraft can introduce Doppler shifts of +/- 4GHz. In order to reduce the dynamic range required from the photomixing detector and associated electronics a frequency agile local oscillator is needed which can meet the wide tuning requirements while maintaining single frequency operation.

Goddard Space Flight Center is developing a direct detection Doppler lidar method, the edge technique, as part of the New Millennium Program. The edge technique can be used to make high accuracy wind measurements using well developed Nd:YAG laser technology (Korb 1997). It can be employed at the YAG fundamental wavelength of 1064 nm to measure winds using the aerosol backscattered signal or at the third harmonic wavelength of 355 nm to measure winds using the molecular backscattered signal. An optimum approach is a combined system which incorporates both the 1064 nm aerosol and the 355 nm molecular techniques. This approach requires a single laser transmitting both wavelengths, a common telescope with high reflectivity at both wavelengths, and two receivers: one for the molecular Doppler measurement and one for the aerosol.

The laser energy required for a combined aerosol/molecular direct detection wind lidar system would be 1-3 J per pulse at repetition rates of 10-30 Hz. Single frequency operation is required but the laser linewidth can be many times the equivalent measurement accuracy without a significant penalty in performance (Korb 1992). Recently Goddard has flown diode-pumped space qualified Nd:YAG lasers in the Mars Orbiter Laser Altimeter (MOLA), an instrument on Mars Global Surveyor (Afzal 1994), and on two Shuttle missions as part of the Shuttle Laser Altimeter (SLA) experiment. The MOLA laser is diode pumped, compact, and generates 40 mJ per pulse at 10 Hz in a conductively cooled design. Goddard is continuing to develop this laser technology for future EOS space altimetry and lidar missions including the Geoscience Laser Altimeter (GLAS) (Afzal 1997). Recent developments in high brightness Nd:YAG laser designs have been reported by TRW, a New Millennium industry partner for solid state lasers (St.Pierre 1993). The TRW laser breadboard is a diode pumped master oscillator power amplifier design utilizing phase conjugation to provide high power and excellent beam quality. The reported laser energy is 1 J at 100 Hz, 100 W average power, with a beam quality of 1.1 times diffraction limit. The high peak power and good beam quality make this laser ideal for efficient third harmonic generation. The optical efficiency of this laser is 22% and the overall electrical to optical efficiency is reported to be 9.4% (excluding cooling).

An additional important technology for direct detection Doppler lidar is the high spectral resolution optical filter used for the Doppler frequency shift measurement. Fabry Perot etalons have been successfully employed in ground based lidar measurements and have the required resolution and high throughput for spaceborne wind lidar measurements. Piezoelectric tuning of the etalon cavity can be used to precisely tune the center frequency of the bandpass. Capacitance sensors can be used as a reference in a servo loop to control the etalon gap and plate parallelism to parts in 10^{10} (Hicks 1984). The capacitance sensors and piezo tuning etalon control system can also be used to introduce a precise frequency offset to perform spacecraft Doppler motion compensation in a fashion analogous to the frequency agile local oscillator used in the coherent Doppler lidar.

Goddard and Langley are also actively involved in the development of the other key technologies for direct detection lidar including solid state lasers, filters, detectors and advanced telescope and optical scanner designs. These same technologies enable significant improvements in climate science with a high efficiency cloud and aerosol lidar and advanced DIAL lidars for profiling of water vapor and ozone. The technology synergy of the direct detection lidar approaches indicate that a multifunctional lidar capable of measuring winds, radiative properties of clouds and aerosols as well as DIAL measurement of constituent molecules may be possible.

Acknowledgements

Much of the work reported here has been funded by the NASA Office of Mission to Planet Earth and by the NASA Cross Cutting Technology Initiative under the auspices of the Advanced Technology and Mission Studies Division of the Office of Space Science. The author wishes to acknowledge the work of the many scientists and technologists who have contributed to the Doppler lidar technology base summarized here. Particular thanks to Drs. Larry Korb, Upendra Singh, Robert Menzies and Michael Kavaya for contributing material presented in this paper.

References

Abreu, V.J., J.E. Barnes and P.B. Hayes, (1992) "Observations of winds with an incoherent lidar detector", Appl. Opt., 31:4509-4514.

Afzal, R.S. (1994)"The Mars Observer Laser Altimeter: Laser Transmitter", Appl. Opt., **33**:3184.

Afzal, R. S , A. W. Yu, W. A. Mamakos, (1997) "The GLAS Laser Transmitter Breadboard", OSA TOPS, Vol. 10, C. R. Pollack and W. R. Bosenberg Eds., (10) 102.

Baker, W., G. D. Emmitt, F. Robertson, R. Atlas, J. Molinari, D. Bowdle, J. Paegle, R. M. Hardesty, R. Menzies,T. Krishnamurti, R. Brown, M.J. Post, J. Anderson, A. Lorenc and J. McElroy (1995) "Lidar-Measured Winds from Space: A Key Component for Weather and Climate Prediction", Bull. Amer. Meteor. Soc., 76:869-888.

Henderson, S. W., P. J. M. Suni, C.P. Hale, S. M. Hannon, J.R. McGee, D.L. Bruns and E.H. Yuen, (1993)."Coherent Laser Radar at 2 Om using Solid State Lasers", IEEE Trans.Geosci. and Remote Sens., 31:4-15.

Hicks, T.R., N.K. Reay and P.D. Atherton, (1984) "The application of capacitance micrometry to the control of Fabry-Perot etalons", J.Phys.E:Sci. Instrum, 17: 49-55.

Huffaker, R. M.(1978) "Feasibility study of satellite-borne lidar global wind monitoring system", NOAA Tech Memo, ERL WPL-37.

Korb, C.L., B.M. Gentry and C.Y. Weng, (1992) "Edge Technique Theory and Application to the Measurement of Atmospheric Winds", Appl. Opt., 31:4202-4213.

Korb, C.L., B.M. Gentry and S.X. Li, (1997) "Edge Technique Wind Measurements with High Vertical Resolution", Appl. Opt., 36:5976-5983.

Menzies, R.T., H. Hemmati and C.Esproles, (1997)"Frequency Agile Tm,Ho:YLF Local Oscillator for a Scanning Doppler Wind Lidar In Earth Orbit", Proceedings 9th Conference on Coherent Laser Radar, Linkoping, Sweden, June 23-27, 62-64.

NASA, (1987)"Laser Atmospheric Wind Sounder. EOS Instrument Panel Report, Vol IIg. Satellite Doppler Lidar Wind Measuring System", NASA Rep MSFC-MOSD-146.

Singh, U. N., J.A.Williams-Byrd, N.P.Barnes, J.Yu, G.E.Lockard and E.A.Modlin, (1997) "Diode-Pumped 2Om Lidar Transmitter for Wind Measurements", Proceedings 9th Conference on Coherent Laser Radar, Linkoping, Sweden, June 23-27, 66-69.

St Pierre, R. J., H. Injeyan, R.C. Hilyard, M.E. Weber, J. G. Berg, M.G. Wickham, C.S. Hoefer and J.P Machan, (1993) "One Joule Per Pulse, 100 Watt, Diode Pumped, Near Diffraction Limited, Phase Conjugated Nd:YAG Master Oscillator Power Amplifier", OSA Proceedings on Advanced Solid State Lasers, A. Fano and T.Y. Fan eds., 15:2-8.

DIODE-PUMPED SOLID STATE LASERS FOR SPACE-BASED APPLICATIONS

Robert S. Afzal
Code 924, NASA-GSFC
Greenbelt, MD 20771
(301) 286-5669 (V) - 1761 (F)
Internet - Robert.Afzal@gsfc.nasa.gov

Abstract

NASA is embarking on a new era of laser remote sensing instruments from space. This paper focus' specifically on the laser technology involved in two present NASA missions. The first is a current mission, the Mars Orbiter Laser Altimeter and the second is the Geoscience Laser Altimeter System scheduled to launch in 2001. The laser transmitters in these space-based remote sensing instruments are discussed in the context of requirements of these two missions.

INTRODUCTION

Recently there has been great interest at NASA in laser remote sensing instruments for Earth orbit and planetary science missions. These instruments typically use diode-pumped solid state lasers for the laser transmitter. The mission specifications and constraints of space qualification, which are mission specific, place strict requirements on the design and operation of the laser. Although a laser can be built in the laboratory to meet performance specifications relatively routinely, the mission constraints demand unique options and compromises to be made in the materials used and the design in order to ensure the success of the mission. Presently, the best laser architecture for a light weight, rugged, high peak power and efficient transmitter is a diode laser pumped Nd:YAG laser. Diode lasers can often obviate the need for water cooling, reduce the size and weight of the laser, increase the electrical to optical efficiency, system reliability, and lifetime. Two distinct laser designs are described in this paper and represent benchmarks on the state of the technology.

MARS ORBITER LASER ALTIMETER: LASER TRANSMITTER

On November 5, 1996, NASA launched the Mars Global Surveyor (MGS). MGS is intended to recover as much of the science data possible from the earlier loss of the Mars Observer spacecraft. One of the science instruments on MGS is the Mars Orbiter Laser Altimeter (MOLA). MOLA's primary mission is to gather Martian topographic data on a 0.2° x 0.2° grid with 30 m vertical accuracy and short baseline (Å 100 km) topographic profiles with a < 2 m vertical accuracy (Zuber 1992). The MOLA laser transmitter (Afzal 1994) utilizes a diode-laser pumped Nd:YAG laser which when a previous version was flown on ill-fated Mars Observer, was the first diode-laser pumped solid state laser to operate in space. The laser was designed and built by McDonnell-Douglas Aerospace, St. Louis, MO, with the diode pump arrays provided as government furnished equipment. A photograph of the laser is shown in figure 1.

The MOLA laser transmitter is required to deliver >35 mJ pulses at 10 Hz, through 0.6 billion shots (Å 2 Earth years - 1 Martian year). The laser weighs < 5.5 Kg and is 3.3% efficient from 28 V prime power to light out. Some of the environmental conditions that the instrument must survive uncompromised are, launch vibrations both random and sinusoidal, pyrotechnic shock and launch sound levels. The instrument capabilities must not be degraded under the conditions of thermal cycling (between -5 and 40 °C) and must withstand radiation dosages expected during the mission. The laser is designed to operate in vacuum and great care and effort was expended to make the laser vacuum compatible.

For this mission not only does the laser transmitter (LT) need to survive launch and the journey to Mars, but had to meet the specifications summarized in Table 1. The MOLA laser design is a side-pumped zig-zag slab, crossed porro resonator, power oscillator. It's very robust and can operate over 30 K while providing > 80% of its peak energy. To date the MOLA laser has operated in-space on four separate occasions over the 10 month cruise from Earth to Mars and accumulated over 500,000 shots with no indications of degradation. MGS will enter its final mapping orbit in March, 1998 where MOLA will then be operated for continuous mapping of the planet.

Table 1

Parameter	Weight (Kg)	Available Power (W)	Pulse Energy (mj)	Pulse Width (ns)	Repetition Rate (Hz)	Lifetime (shots)
Specifications	6.26	15	40	10	10	6×10^8
Actual	5.38	14.7	49	8.4	10	TBD

GEOSCIENCE LASER ALTIMETER SYSTEM: LASER TRANSMITTER

A future laser remote sensing mission will be the Geoscience Laser Altimeter System (GLAS) scheduled to launch in 2001 (Abshire 1994). GLAS will be a satellite laser altimeter whose primary mission is the global monitoring of the Earth's ice sheet mass balance. GLAS will also use a lidar for global monitoring of cirrus cloud heights. The laser transmitter will have the following performance characteristics: pulse energy - 100 mJ @ 1 µm, 50 mJ @ 0.5 µm, repetition rate - 40 Hz, pulse-width 4 - 6 ns, beam divergence - 95 µrad, beam profile - nominally Gaussian, >5% electrical efficiency, with a > 3 billion shot lifetime. In addition the laser components must be tolerant of the orbital radiation environment. The GLAS laser will generally have an order of magnitude higher performance than MOLA and will represent the next generation of space-based remote sensing laser transmitters.

One of the more challenging aspects of the performance requirements is to simultaneously generate 4 ns, 150 mJ, near diffraction limited beam quality pulses efficiently with long-life using a rugged architecture. This performance regime is not easily accessible to the Q-switched laser due to gain and cavity-length constraints and Q-switched, cavity-dumped lasers typically require complex high-voltage switching stages which hold little promise in lasting the mission duration. Performance of high peak power lasers capable of generating 4 ns pulses are also typically limited by laser induced damage of the optical components. We believe presently, that the development of a master-oscillator power-amplifier (MOPA) design is the most promising technique for meeting the transmitter performance objectives. In addition to meeting the performance requirements the laser must be rugged, reliable and capable of long term operation in the space environment over temperature ranges expected by the spacecraft.

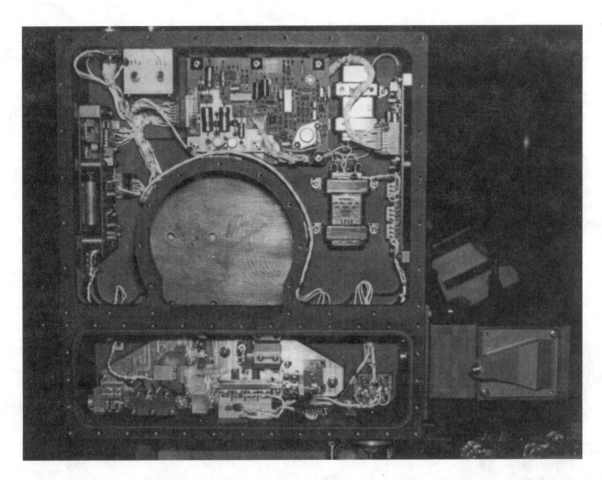

FIGURE 1. Photograph of the MOLA Laser Transmitter with Covers Off. For Scale, the Main Box is about 30 cm x 30 cm in Size.

We have demonstrated a low energy, short pulse with high beam quality oscillator (Afzal 1995) in conjunction with double-pass amplifier stages which preliminarily meet the performance goals of the GLAS transmitter (Afzal 1997). The laser is designed as follows and a schematic is in figure 2.

A single 100 W Q-cw diode bar pumped oscillator slab laser generates 2 mJ, 4.5 ns near diffraction limited ($M^2 < 1.1$) pulses at 40 Hz. The pulses are expanded by a 2x telescope, then amplified by a double-pass preamplifier stage pumped by 8, 100 W bars. After passing through the preamplifier, the pulse is 20 mJ with an M^2 Å 1.5. This stage utilizes a polarization coupled double pass zig-zag slab with a porro prism for beam symmetrization. The beam next enters a power amplifier pumped by 44, 100 W bars after another 2x expansion. The 20 mJ pulses are amplified to 150 mJ after a double pass with an M^2 Å 2. The peak laser fluence in the final amplifier is Å 4 J/cm^2. Table 2 shows the operational requirements and the lasers' maximum capability. Figure 3 shows the output energy and prime power draw as a function of diode drive current.

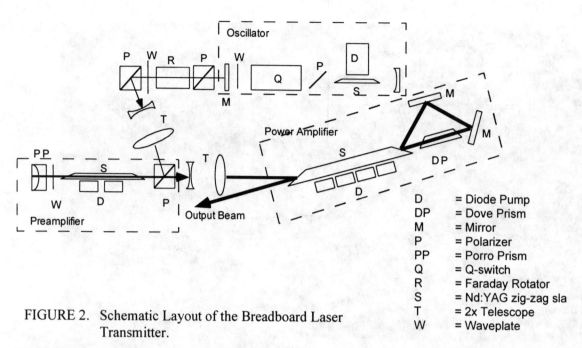

FIGURE 2. Schematic Layout of the Breadboard Laser Transmitter.

Table 2

Parameter	Weight (Kg)	Available Power (W)	Pulse Energy (mj)	Pulse Width (ns)	Repetition Rate (Hz)	Lifetime (shots)
Operation	Å 10	100	110	< 6	40	> 3 Billion
Maximum	Å 10	120	150	5	40	TBD

We are now in the process of mechanical ruggedization and environmental testing of the laser in preparation fro flight laser delivery in 2000

CONCLUSIONS

Diode-pumped solid state lasers are not only becoming more common in space, but are beginning to be considered "operational." MOLA and GLAS are two pertinent examples of where the technology is and the direction it's going.

Acknowledgments

This work was funded by the NASA EOS Chemistry and Special Flights project office. The author thanks A. Yu, B. Mamakos, J. Abshire, J. Dallas, A. Lukemire, M. Selker, M. Stephen, and M. Krainak for helpful discussions and technical assistance. Also the author thanks the MOLA team at McDonnell-Douglas for their hard work and perseverance.

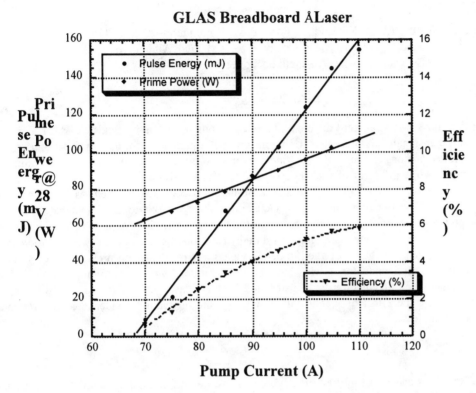

FIGURE 3. Pulse Energy and Prime Power Draw of the Laser vs. Diode Drive Current. Also Plotted is the Laser Efficiency from Spacecraft 28 V Prime Power.

References

Abshire, J.B., J.C. Smith and B.E. Schutz, (1994), "Geoscience Laser Altimeter System (GLAS)", *Proceedings 17th International Laser Radar Conference*, July 25 - 29, Sendai, Japan, Paper no. 26D5, 215.

Afzal, R. S., A. W. Yu and W. A. Mamakos, (1997), "The GLAS Laser Transmitter Breadboard," *OSA Trends in Optics and Photonics on Advanced Solid State Lasers,* C. R. Pollock and W. R. Bosenberg, eds. (Optical Society of America, Washington DC, (10), 102.

Afzal, R. S., M. D. Selker, (1995) "A simple high efficiency, TEM_{00}, diode laser pumped, Q-Switched Laser," March 1, 1995, *Opt. Lett.*,(20), 5: 46.

Afzal, R. S., (1994), "Mars Observer Laser Altimeter: Laser Transmitter," *Applied Optics*, (33), 15, 20 May, 1994, 3184.

Zuber, M. T., D. E. Smith, S. C. Solomon, D. O. Muhlman, J. W. Head, J. B. Garvin, J. B. Abshire, and J. L. Bufton; (1992) "The Mars Observer Laser Altimeter Investigation", *J. Geophys. Res.*, (May 25, 1992, (97), E5: 7781-7797.

RIGOROUS DERIVATION OF LASER RESONATOR DYNAMICS BASED ON GENERALIZED 'FOX-LI' INTEGRAL EQUATION

Erik J. Bochove
Phillips Laboratory/LIDA, Kirtland Air Force Base, Albuquerque, NM 87115
and Center for High Technology Materials, University of New Mexico, Albuquerque, NM 87131
Tel: (505) 846-5937, e-mail: bochovee@plk.af.mil

Abstract

A new formulation for deriving 'exact' coherent differential rate equations in the cavity fields of laser resonators of arbitrary geometry and dimensionality is presented. Using an integral equation as formal starting point, e.g. a Fox-Li expression, optical rate equations are derived following a rigorous procedure that guarantees preservation of the resonator modal structure. While the form of such rate equations is generally not unique, all obtained in this manner are equivalent in the sense that the same dynamical information is contained in them as in the original integral equation. The method bridges two previously unconnected disciplines in resonator optics, so that the results of one discipline can be applied to the other, presenting new opportunities for vastly increased power in the concise theoretical description of laser cavities of almost any kind. As a demonstration, the example of two longitudinally-coupled Fabry-Perot lasers separated by a gap is studied in detail, yielding agreement with previous models in the weak interaction limit, but for semiconductor lasers, having comparatively strong coupling, the numerical corrections are very significant. The rate equation for a distributed feedback laser is derived in a second example, producing a simple expression for the effect of spatial hole burning on the linewidth. Evidence of experimental consistency is cited in both examples.

INTRODUCTION

The modeling of laser dynamics can be simplified by use of a first-order differential equation to describe the temporal evolution of each active resonator mode. In such a description all references to spatial coordinates is removed. The possibly standard procedure of systematic derivation of rate equations, the coupled mode approach, is based on expansion of the electric field variables as a linear combination of resonator modal functions, as applied by Spencer & Lamb (1972) and Marcuse (1985) to study the dynamics of two longitudinally-coupled lasers. However, since individual-cavity modal functions can not satisfy exactly the true physical boundary conditions connecting interior and exterior fields (at the mirror locations), the cavity fields can not be accurately represented by such expansions. Another problem is that the calculation of coupling coefficients is tedious and must normally be performed numerically, while in some cases hundreds of terms per coefficient must be included.

An alternative approach (Lang & Yariv 1985) starts from a set of rigorous homogeneous modal equations of plane-wave fields, at frequency ω, written as,

$$\tilde{\Gamma}(\omega)\tilde{E}(\omega) = 0 \quad , \tag{1}$$

where $\tilde{E}(\omega)$ is a column vector in the electric fields, and $\tilde{\Gamma}(\omega)$ is a square matrix having elements derived by application of the boundary conditions to the solutions of the wave equation. Following an analytical approximation procedure, (1) is linearized to first order about a reference frequency, ω_0 (e.g. the operating frequency of the laser). By an inverse Fourier transformation a set of first-order differential equations in the time-dependent fields is obtained, in which the relevant quantity is the coupling matrix, $\tilde{K}(\omega_0)$. Although (1) may quite generally be rigorously derived, and is fully consistent with the physical boundary conditions, that rigor is seriously curtailed by the described linearization procedure. In addition, ambiguity is introduced by the fact that $\tilde{K}(\omega_0)$ is physically non-unique, in part caused by the lack of uniqueness of $\tilde{\Gamma}(\omega)$. While the offered criterion of selecting the procedure yielding the best fit of the linearized matrix to $\tilde{\Gamma}(\omega)$ is likely valid, a clear guide for its implementation has not been given.

The approach taken here differs in that its starting point is an integral equation (or set of such equations), which for the sake of definiteness may be identified with the known Fox-Li equation (Fox & Li 1961) describing the field's evolution over a round-trip in the resonator, and second, the rate equation is derived from this equation by a rigorous procedure. The former assumption lends a physical flavor, but it is due to the mathematical rigor of the derivation of the rate equation that the normal mode solutions of the rate equation are truly those of the actual resonator. Even though a given integral equation is usually consistent with an infinite number of rate equations, the important point is that all formulations of the latter that are obtained in this manner are physically equivalent - any physically consequential differences between them are introduced only at the point at which approximations are applied. As that last step is not performed blindly, but is applied directly on the rate equation at the end of the derivation, a worker will be guided by his or her physical understanding of the application at hand. The last faculty is also needed in choosing sometimes between various alternative versions of rate equations, for which simple inspection generally suffices in determining the most suitable.

FORMAL SOLUTION AND SOME GENERAL PROPERTIES

In this section a given general integral equation in the cavity fields, specified by a linear operator $\hat{\Lambda}$, is formally transformed into an equivalent differential-integral version wherein $\ln \hat{\Lambda}$ plays the role of the coupling operator.

The Differential Rate Equation

Starting from Maxwell's equations describing the fields of a quite arbitrary resonator at fixed frequency, ω, a set of equations of formally simple form is obtained, $\widetilde{E}(\omega) = \widetilde{A}(\omega)\widetilde{E}(\omega)$ (viz. Eq.(1)), where $\widetilde{A}(\omega)$ is a square matrix operator, and $\widetilde{E}(\omega)$ the field vector. This notation allows for any spatial dependence (e.g. transverse, in a fixed reference plane). Multiplying by a factor $\widetilde{T}(\omega) \equiv \exp(-i\omega\tau)$ (containing the constant τ to be determined in the applications) we have $\widetilde{T}(\omega)\widetilde{E}(\omega)\exp(-i\omega\tau) = \widetilde{\Lambda}(\omega)\widetilde{E}(\omega)$, where $\widetilde{\Lambda}(\omega) \equiv \widetilde{T}(\omega)\widetilde{A}(\omega)$. Fourier-transforming, and for generality including an injected (or stochastic noise) field, $E_{in}(t)$, we have the following *integral equation*:

$$E(t+\tau) = \hat{\Lambda}E(t) + E_{in}(t), \tag{2}$$

where $\hat{O}E(t) \equiv FT^{-1}[\widetilde{O}(\omega)\widetilde{E}(\omega)]$. τ is formally arbitrary, but in practice must be defined with care, as seen below. One can show from (2) that $E(t)$ also satisfies a differential *rate equation*:

$$\dot{E}(t) = \hat{K}E(t) + \tau^{-1}(\hat{\Lambda} - \hat{T})^{-1} \ln(\hat{\Lambda}\hat{T}^{-1})E_{in}(t), \tag{3a}$$

where,

$$\hat{K} \equiv \tau^{-1} \ln \hat{\Lambda} \tag{3b}$$

will be called the coupling operator. The matrix operator $\hat{\Lambda}$ can be the cavity Green function, or Fox-Li round-trip operator, although this identification is not essential.

In spite of its formal form, (3) can be readily applied, as shown in two examples below. For 3-dimensional cavities it is generally useful to expand the various field components in terms of bi-orthogonal sets of basis functions that are eigenfunctions of $\hat{\Lambda}$ and its transpose (Siegman 1987).

In cases where (2) represents a set of $N > 1$ equations, then other forms than (3) are possible, e.g. as obtained by grouping the fields. For two groups, those in one are treated as injected fields for one set of equations, and those in the other group as injected fields in the other set of equations. The resulting formulations, although rigorously equivalent, are not all equally useful. We found, however, that determining the best solution seems rarely to present an obstacle.

Perturbation Solution for Coupling Matrix

In many applications, $\hat{\Lambda} = \hat{\Lambda}_0 + \hat{\lambda}$, where $\hat{\lambda} \ll \hat{\Lambda}_0$. Choosing a representation of bi-orthogonal eigenfunctions of Λ_0 (Siegman 1987), then to first-order (e.g. by using Cauchy's theorem):

$$K_{nm}\tau \equiv \ln \Lambda|_{nm} = \ln \gamma_{0n} \delta_{nm} + \frac{\ln(\gamma_{0n}/\gamma_{0m})}{\gamma_{0n} - \gamma_{0m}} \lambda_{nm}, \quad (4)$$

where γ_{0j} are eigenvalues of $\hat{\Lambda}_0$. Eq.(4) applies to all values of dimensionality. Below, this expression is used to derive approximate results for the coupled-cavity laser.

APPLICATIONS

Two applications will be presented. The first, that of the coupled-cavity laser, is included mainly for its "historical" interest, (Spencer-Lamb 1972, Marcuse 1985, Lang-Yariv 1985), while the second application to a distributed feedback (DFB) laser illustrates the efficacy of the method to a system of greater current concern.

Coupled-Cavity Laser: Exact Solution

Consider two, for convenience identical, Fabry-Perot lasers, separated by a gap, d. Assuming counter-propagating plane-wave solutions at frequency ω, the modal equation is $\widetilde{E}(\omega) = \widetilde{A}(\omega)\widetilde{E}(\omega)$ where $\widetilde{E}(\omega)$ is a 2-component vector in the cavity fields (Fig.1). $\widetilde{A}(\omega)$ has components $\widetilde{A}_{11} = \widetilde{A}_{22} = gr_b S_{11} e^{i\omega\tau_c}$, $\widetilde{A}_{12} = \widetilde{A}_{21} = gr_b S_{12} e^{i\omega\tau_c}$ where g is the round-trip amplitude gain factor inside either cavity, $\tau_c = 2nL/c$ the individual cavity round-trip time, r_b, r_f are facet amplitude reflectivities, and S_{ij} is the gap "scattering matrix", given by

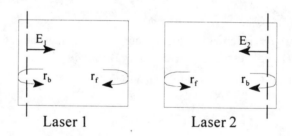

FIGURE 1. Illustrating Application to Symmetric Coupled-Cavity Laser.

$S_{11} = S_{22} = r_f - e^{2ik_0 d} r_f (1 - r_f^2)/(1 - e^{2ik_0 d} r_f^2)$, and $S_{12} = S_{21} = e^{ik_0 d} r_f /(1 - e^{2ik_0 d} r_f^2)$.

Identifying $\tau = \tau_c$, then $\widetilde{\Lambda}(\omega) = \widetilde{A}(\omega) e^{-i\omega\tau_c}$. The exact solution for the coupling matrix is calculated from (3), as

$$\widetilde{K}_{11} = \frac{1}{2\tau_c} \ln\left[\frac{g^2 r_b^2 (e^{2ik_0 d} - r_f^2)}{e^{2ik_0 d} r_f^2 - 1}\right] + 2im\pi/\tau_c, \quad \widetilde{K}_{12} = \frac{1}{2\tau_c} \ln\left[\frac{(e^{2ik_0 d} + r_f)(e^{2ik_0 d} r_f - 1)}{(e^{2ik_0 d} - r_f)(e^{2ik_0 d} r_f + 1)}\right], \quad (5)$$

with $\widetilde{K}_{22} = \widetilde{K}_{11}$, $\widetilde{K}_{21} = \widetilde{K}_{12}$, $\pm m = 0,1,2,...$, and $k_0 = n_0 \omega/c$ is the gap wave vector with index, n_0. The gain-clamping condition is $gr_b = 1$. As discussed, alternative physically equivalent formulations are possible, but it is not difficult to be convinced that (5) represents indeed the best of all solutions. Such exact expressions have not been derived before. Below we obtain approximate expressions and these with Marcuse's "heuristic" expression.

Coupled-Cavity Laser: Perturbation Solution

We compare (5) with the result of the perturbation formula, (4), in which the unperturbed matrix is obtained by neglecting the coupling terms between the lasers. This gives: $\widetilde{K}_{ij}^{pert} = \ln\gamma_0 \delta_{ij} + \lambda_{ij}/\gamma_0$, where, $\gamma_0 = gr_f r_b$ is the degenerate unperturbed eigenvalue, and $\lambda_{ij} = S_{ij} - r_f \delta_{ij}$. Explicitly,

$$\widetilde{K}_{11}^{pert} = \ln(gr_f r_b) - \frac{e^{2ik0d} r_f \kappa_i}{1 - e^{2ik0d} r_f^2}, \quad \widetilde{K}_{12}^{pert} = \frac{e^{ik0d} \kappa_i}{1 - e^{2ik0d} r_f^2}, \tag{6}$$

where $\kappa_i = \tau_c^{-1}(1 - r_f^2)/r_f$ is identified as the injection rate. Marcuse gave an expression for $\widetilde{K}_{12}$ that is r_f times the above, with (seemingly) incorrect phase. Since $r_f \approx 1$, the former difference is insignificant. Our $\widetilde{K}_{12}$ agrees with Lang's result (R. Lang 1982) for injection locking, allowing for the denominator describing multiple-pass effects of the external cavity. No corresponding expression for the diagonal element has been derived before that given above. It is emphasized that all elements of the coupling matrix must be known in order to have a useful theory.

TABLE 1. Comparison of Exact and Perturbation Values of $\widetilde{K}_{ij}$ for GaAs Lasers.

Mode freq. (Ghz) - symmetry	K_{11}	K_{11}^{pert}	K_{12}	K_{12}^{pert}	ω_r/ω_{rs}	γ_r/γ_{rs}
128 - e	-0.54i	-.47-.7i	-.76i	-.65-1.1i	0	3.7
438 - o	.58i	-.6+.7i	-.82i	.78-1.1i	0	0.46
562 - e	-1.4i	-1.6-1.9i	1.5i	1.8+.23i	3.9	2.2
679 - o	-0.5i	-.47-.7i	.71i	.65+1.1i	0	.46
989 - e	.67i	-.6+.7i	.82i	-.78+1.1i	0	3.7

In Table 1 are compared exact numerical values of $\widetilde{K}_{ij}$ of a few modes with perturbation theory, using representative numerical values for two GaAs lasers having high-reflectance coated rear facets ($r_b^2 = 0.9$) to enhance coupling. We chose m=0 in (5). The gap is equal to the optical length of the lasers; a "line-broadening" factor, $\alpha \equiv -\Delta n'/\Delta n'' = 3$, was assumed. The mode frequencies, along with their symmetry, are listed in the first column. $\widetilde{K}$ is given in units of τ_c^{-1}. Notable are the large differences between the values computed by the exact and approximate theories, being of the order of τ_c^{-1} in magnitude, which is a convincing demonstration of the failure of approximate theoretical models to systems like these (viz. Agrawal 1985,1986).

To demonstrate a more useful application, we computed the relaxation oscillation frequency and damping rate of each mode from the solution of the complete set of linearized rate equations, including those for carriers, and listed their values in the last two columns of Table 1. The ratios are relative to the corresponding solitary laser value, pumped at the same level. When confirmed, the third mode would have a greatly enhanced modulation bandwidth, and since this should be a practically realizable system on a monolithically integrated chip, results like these would be valuable in designs of future laser optical communications systems.

The modulation bandwidth enhancement predicted above appears to be consistent with reported experimental observations (Vahala et al 1985).

Linewidth of Distributed Feedback Laser

In a distributed feedback (DFB) semiconductor laser end-reflectors are replaced by periodic variations in the medium, usually in the form of a corrugated layer outside the active layer, but making contact with the optical mode, creating continuous coupling between right- and left-propagating fields through backward Bragg scattering. Such devices are becoming increasingly important spectrally pure sources in the IR spectral region. Here, the calculation method of this paper is applied to model the effect of spatial hole burning on the linewidth of a DFB laser.

The field round-trip condition for a single-section, uniform grating, DFB laser of length, L, is written in the form $\widetilde{E}(\omega) = \widetilde{A}(\kappa,\zeta)\widetilde{E}(\omega)$, where (Kogelnik & Shank 1972):

$$\widetilde{A}(\kappa,\zeta) = \frac{\zeta - \sqrt{\kappa^2+\zeta^2}}{\zeta + \sqrt{\kappa^2+\zeta^2}} e^{2\sqrt{\kappa^2+\zeta^2}L} \tag{7}$$

in terms of the DFB coupling constant, κ, and the complex parameter, $\zeta = (1-i\alpha)\gamma - in_0\Delta\omega/c$, where γ is the gain (treated as uniform over the cavity), α is defined above, n_0 is the refractive index, and $\Delta\omega = \omega - ck_G/n_0$ is the detuning from the grating frequency. The steady-state requires

$$\widetilde{A}(\kappa,\zeta) = 1. \tag{8}$$

The coupling function is given by $\widetilde{K}(\kappa,\zeta) = -i\omega + \tau^{-1}\ln\widetilde{A}(\kappa,\zeta)$. An instructive feature of this example is that the "time" τ must be determined carefully, due to the strong frequency dependence of the Bragg effect. The optimum choice for τ is the *complex* value making $\widetilde{K}$ extremum at the operating frequency, ω_L, or:

$$\tau = -i\widetilde{A}^{-1}\left.\frac{d\widetilde{A}}{d\omega}\right|_{\omega=\omega_L} = \left.\frac{\tau_c(\zeta L - 1)}{\sqrt{\kappa^2+\zeta^2}L}\right|_{\omega=\omega_L}, \tag{9}$$

where ω_L, and the gain, γ, are obtained from the solutions of (8). Eq.(9) shows that $\tau \to \tau_c \equiv 2nL/c$ as $\Delta\omega$ increases with increasing mode-order satisfying (8).

Expressing the cavity field as $E(t) = \sqrt{S+s(t)}\, e^{-i\omega_L t + i\varphi(t)}$, then the rate equations for fluctuating phase, normalized power, are obtained from (3), yielding, together with an equation for carrier number,

$$\dot{\varphi}(t) = K_{N,i}n(t) + K_{S,i}s(t) - K_{\omega,i}\dot{\varphi}(t) + \frac{1}{\sqrt{S}}f_i(t),$$
$$\dot{s}(t) = 2S[K_{N,r}n(t) + K_{S,r}s(t) - K_{\omega,r}\dot{\varphi}(t)] + 2\sqrt{S}f_r(t), \quad \dot{n}(t) = -\Gamma_e n(t) - Gs(t) \tag{10}$$

where subscripts r and i indicate real and imaginary parts, subscripts N,S,ω indicate partial derivatives, capital letters represent steady-state values, $G = 2c\gamma/n_0$, $\Gamma_e = \tau_e^{-1} + 2c\gamma_N S/n_0$, $\gamma_N = \partial\gamma/\partial N$, and $f(t) = f_r(t) + if_i(t)$ is a Langevin force describing SE noise, satisfying $<f_i(t)f_r(t')> = 0$, $<f_i(t)f_i(t')> = <f_r(t)f_r(t')> = \frac{1}{2}R_s\delta(t-t')$, where R_s is the spontaneous emission rate. The standing wave character of the field has the effect of "burning" a weak grating into the medium, which we model by taking κ to be a function of S (normalized to equal the photon-number). Thus, from the above, $K_\omega = -i(1+n_0K_\zeta/c) = 0$, $K_\kappa = c(\kappa^2L^2 + \zeta L)/[n_0\kappa L(\zeta L - 1)]$. The full linewidth at half-maximum of a Lorentzian line shape is given by $\Delta\omega_L = \lim_{\upsilon\to 0}<|\upsilon\widetilde{\varphi}(\upsilon)|^2>/2\pi$, which yields:

$$\Delta\omega_L = (1+\alpha_{eff}^2)\frac{R_{sp}}{4\pi S}, \tag{11a}$$

where

$$\alpha_{eff} = \frac{2\alpha c^2\gamma\gamma_N + \Gamma_e\kappa_S K_{\kappa i}}{2c^2\gamma\gamma_N - \Gamma_e\kappa_S K_{\kappa r}} \tag{11b}$$

is an "effective" α-parameter, defined in analogy with the normally appearing factor $1+\alpha^2$ (Henry 1982). Plots of $\Delta\omega_L$ are shown in Fig.(2), obtained assuming the expression $\kappa = \kappa_0(1+S/S_0)$ and using the following values:

$\kappa_0 = 1080$, $\kappa_0 L = 0.4$, $\alpha = 5$, $\gamma_N = 2.610^{-5}$, $n_0 = 3.45$, $G_N = 4500$, $\tau_e = 210^{-9}$, all in SI units. The figure shows power re-broadening in all three cases, while the power-independent linewidth is very nearly vanishing, as indicated by the fact that the continuation of the high-bias asymptote nearly passes through the origin, in agreement with experimental facts (Wenzel et al 1991, and Nakajina et al 1991). It is also seen that its slope is strongly affected by spatial hole burning.

SUMMARY

We have demonstrated a way of deriving coherent rate equations for optical resonators which retain the complete modal structure of the resonator, or, equivalently, are consistent with Maxwell's equations. The method is proposed as a tool in describing laser dynamics when the optical rate equations are supplemented by suitable dynamic rate equations for the medium. Two examples of applications were studied in detail, in which simple rigorous solutions were found to problems which were not previously solved. Other applications of this method are foreseeable to external optical feedback in lasers, injection locking, dynamics of broad-area semi-conductor lasers, phased-array lasers, other types of distributed feedback lasers, lasers in unstable resonator configurations, and lasers containing phase-conjugating mirrors, or receiving external feedback (Bochove 1997).

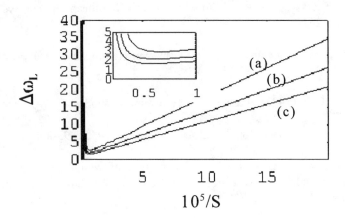

FIGURE 2. Linewidth as Function of Inverse Internal Field Power for Single-Section DFB Laser (Dimensionless units). $S_0 = (a) 2.10^5$, $(b) 4.10^5$, $(c) 2.10^7$.

Acknowledgments

The author thanks John McInerney for pointing out to him the need for critical evaluation of past rate equation models, and Greg Dente for many relevant discussions on linewidth issues.

References

Agrawal, G.P., "Generalized Rate Equations and Modulation Characteristics of External Cavity Semiconductor Lasers," *J. Appl. Phys.*, 56: 3110-3115 (1984); "Coupled-Cavity Semiconductor Lasers Under Current Modulation: Small-Signal Analysis," *IEEE J. Quant. Electron.*, QE-21: 255-263 (1985).

Bochove, E. J. (1997) "Semiconductor Laser with Phase-Conjugate Optical Feedback," *Phys. Rev. A*, 55: 3891-3899.

Fox, A. G. and T. Li (1961) "Resonant Modes in a Maser Interferometer," *Bell System Tech. J.*, 40: 453-488.

Henry, C. (1982) "Theory of the Linewidth of Semiconductor Lasers," *IEEE J. Quant. Electron*, QE-18: 259-264.

Kogelnik, H., and C. V. Shank (1972) "Coupled Wave Theory of Distributed Feedback Lasers," *J. Appl. Phys.* 43: 2327-2335.

Lang, R (1982) "Injection Locking Properties of a Semiconductor Laser," *IEEE J. Quant. Electron.*, QE-18: 976-983.

Lang, R.J. and A. Yariv (1985) "Local-Field Rate Equations for Coupled Optical Resonators," *Phys. Rev. A*, 34: 2038-2043.

Marcuse, D. (1985) "Coupled Mode Theory of Optical Resonant Cavities," *IEEE J. Quantum Electron.*, QE-21, 1819-1826.

Marcuse, D. (1986) "Coupling Coefficients of Coupled Laser Cavities," *IEEE J. Quant. Electron.*, QE-22: 223-226.

Nakajima, H. and J. C. Bouley (1991) "Observation of Power Dependent Linewidth Enhancement Factor in 1.55 µm Strained Quantum Well Lasers," *Electron. Lett.* 27: 1840-1841.

Siegman, A.E. (1989) "Excess Spontaneous Emission in Non-Hermitean Optical Systems. I. Laser Amplifiers," *Phys. Rev. A*, 39: 1253-1263.

Siegman, A.E. (1989) "Excess Spontaneous Emission in Non-Hermitean Optical Systems. II. Laser Oscillators," *Phys. Rev. A*, 39: 1264-1258.

Spencer, M.B., and W.E. Lamb (1972) "Laser with a Transmitting Window," *Phys. Rev. A*, 5: 884-892.

Spencer, M.B. (1972) "Theory of Two Coupled Lasers," *Phys. Rev. A*, 5: 893-898.

Vahala, K, J. Paslaski, and A. Yariv (1985) "Observation of Modulation Speed Enhancement, Frequency Modulation Suppression, and Phase Noise Reduction by Detuned Loading in a Coupled-Cavity Semiconductor Laser," *Appl. Phys. Lett.*, 46: 1025-1027.

Wenzel, H., H. J. Wunsche, and U. Bandelow (1991) "Linewidth Rebroadening in Semiconductor Lasers Due to Lateral Spatial Holeburning," *Electron. Lett.*, 27: 2301-2302.

AN ALGORITHM FOR ENHANCED FORMATION FLYING OF SATELLITES IN LOW EARTH ORBIT

David C. Folta and David A. Quinn
Guidance, Navigation, & Control Center
NASA/Goddard Space Flight Center
Greenbelt, MD. 20771
(301) 286-6082

Abstract

With scientific objectives for Earth observation programs becoming more ambitious and spacecraft becoming more autonomous, the need for innovative technical approaches on the feasibility of achieving and maintaining formations of spacecraft has come to the forefront. The trend to develop small low-cost spacecraft has led many scientists to recognize the advantage of flying several spacecraft in formation to achieve the correlated instrument measurements formerly possible only by flying many instruments on a single large platform. Yet, formation flying imposes additional complications on orbit maintenance, especially when each spacecraft has its own orbit requirements. However, advances in automation and technology proposed by the Goddard Space Flight Center (GSFC) allow more of the burden in maneuver planning and execution to be placed onboard the spacecraft, mitigating some of the associated operational concerns. The purpose of this paper is to present GSFC's Guidance, Navigation, and Control Center's (GNCC) algorithm for Formation Flying of the low earth orbiting spacecraft that is part of the New Millennium Program (NMP). This system will be implemented as a close-loop flight code onboard the NMP Earth Orbiter-1 (EO-1) spacecraft. Results of this development can be used to determine the appropriateness of formation flying for a particular case as well as operational impacts. Simulation results using this algorithm integrated in an autonomous 'fuzzy logic' control system called AutoCon™ are presented.

INTRODUCTION

Advanced navigation and attitude control systems, especially for closed-loop autonomy, have recently become a driver on formation design and has driven a need for new technology. The Guidance, Navigation, and Control Center (GNCC) at GSFC is developing and implementing an enhanced formation flying autonomous system which has a positive change on the operations while minimizing impacts to onboard software and hardware requirements, software development, and integration. As a backbone to the GNCC's Precision Image & Orbit Navigation Experiment for Earth Remote Sensing (PIONEERS) program, this algorithm will be flown as a new technology onboard the NMP EO-1 spacecraft. The NMP EO-1 mission

FIGURE 1. Formation Flying Spacecraft.

has a major success criteria of successful completion of numerous paired scene observations to validate the technological advanced mappers. To enable the paired scene process, the EO-1 spacecraft must fly over the current groundtrack of Landsat-7 within 3 km. Also, in order to maintain a safety aspect, the minimum nominal along-track separation will be one minute. The tolerance of this requirement is dictated by the relationship of the +/- 3 km groundtrack to an along-track tolerance on the one minute separation.

Mechanics and Definitions of Formation Flying

Formation flying involves position maintenance of multiple spacecraft relative to measured separation errors. It involves the use of an active control scheme to maintain the relative positions of the spacecraft. Optimally, this process will be performed autonomously onboard the spacecraft and is called Enhanced Formation Flying, such as that which will be implemented by GSFC for the New Millennium EO-1 mission. An example of the orbit dynamics of formation flying is shown in Figure 1.

An overview of formation flying using a two spacecraft example is presented here. If two spacecraft are placed in the same orbital plane and at the same altitude with a small initial separation angle they will be similarly affected by atmospheric drag and by the Geopotential field of the Earth, provided that they have identical ballistic properties. As long as the separation angle is small enough that atmospheric density and gravitational perturbations can be considered constant, the relative separation will remain the same. If the spacecraft are separated in the radial direction, and the respective ballistic properties are different, their orbit velocities are also different, and one spacecraft (the formation flyer) will appear to drift relative to the other (reference flyer). The drifting is most apparent in the along-track (orbital velocity) direction. The radial separation can be operationally planned or induced by differential decay rates caused by environmental perturbations. The concept of formation flying is based on the constructive use of the differential decay rates as a direct function of differential ballistic properties between a reference and a free-flying spacecraft.

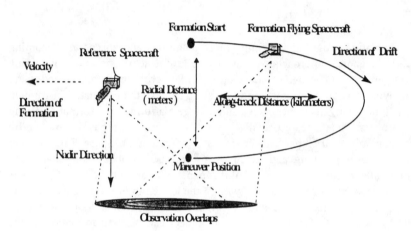

FIGURE 2. Formation Flying Example.

In general, the ballistic properties of spacecraft will not be identical due to differences in effective frontal areas and masses. These differences result from varied instrument location and system engineering considerations such as solar array size to meet power requirements of the instruments. The spacecraft masses will vary due to the spacecraft and instrument design. The measurement of the ballistic properties is defined in terms of ballistic coefficient (B_c) calculated using mass(m), frontal area(a), and coefficient of drag (c_d), where $B_c = m/(a*c_d)$ and the C_d is assumed to be equal to a constant value, usually that of a sphere, 2.2. Since the B_c is a measure of the spacecraft's ability to overcome the effects of drag, the spacecraft with the smaller B_c will decay faster than the spacecraft with the larger B_c. This differential decay rate as a result of differential B_c will cause the spacecraft to be at slightly different altitudes at the same time. As previously mentioned, they will therefore have different velocities and will appear to drift with respect to each other. As the orbits decay at different rates, the relative radial and along-track position of the formation flyer can be predicted and controlled with respect to a reference. The direction of the drift, either upwards or downwards and away from or towards the reference is solely dependent upon the differential B_c. The drifting is most apparent in the along track direction as shown in Figure 1. Maneuvers are generally positioned at the end of each formation cycle to reestablish the formation.

Basic Formation Types

Formations are defined as Close, Ideal, and Dynamic. The Close Formation involves two spacecraft kept to a close separation ranging from tens of meters to less than one kilometer in the orbital velocity (along track) direction. This results in a time separation on the order of hundredths of seconds and would be applicable to a congruent observation requirement. Ideal Formations require a separation in both the along-track direction and in the normal to this vector (crosstrack). This formation has a constant alongtrack separation ranging from greater than 1 km to

100's of km (derived from seconds to minutes in temporal coincidence). The ideal formation is appropriate when constant time gaps between observations are required. While this formation is similar to the close formation in that the relative separations are held constant or minimized, it also has a cross-track component. The cross-track separation enables the formation flyer to meet the viewing requirements along the same ground track of the reference. The third formation defined herein is call a Dynamic formation. It begins with an ideal formation in which the separation is held constant. The effect of formation flying with a ballistic coefficient difference is assessed by allowing the formation flyer to drift with respect to the reference flyer. This drift can be hundreds of kilometers or held to a smaller value by active control. This formation type will be used to meet EO-1 requirements. Figure 2 presents these formations. An example would be approximately 450km alongtrack and 30 km crosstrack at the initial conditions.

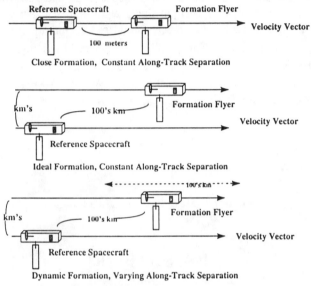

FIGURE 3. General Formation Flying Types.

Formation flying techniques are be used to meet a variety of mission separation requirements. When the mission requirements call for a tightly controlled separation (kilometer range), whether the overall separation is small or large, frequent control becomes necessary. Loosely controlled formations can also be established where along track separations on the order of thousands of kilometers are maintained by routine groundtrack maneuvers. Formations of spacecraft are identified using tight or loose control methods. While some separations may seem exceedingly large, they are determined by the science requirement to view coincident sites or a communication requirement of a ground station to view only one spacecraft at a time. For large separations, one must consider the rotation of the Earth if the formation is used to meet concurrent or sequential imaging of the same locations on the ground. Therefore, relative cross track separations are used to follow the reference groundtrack for any temporal requirement.

AUTONOMOUS CLOSED LOOP 3-AXIS NAVIGATION CONTROL ALGORITHM

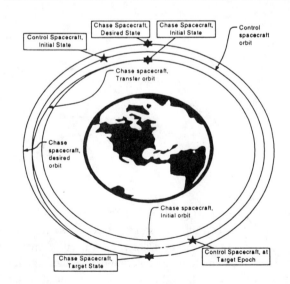

FIGURE 4. Formation Orbital Parameters.

This section provides an overview of the algorithm developed by the authors. The algorithm is based to a large extent on mathematics derived in Battin. These mathematics have been adapted to the formation flying regime. A patent rights application was submitted to the GSFC patent counsel by the authors for the application of Autonomous Closed Loop 3-Axis Navigation Control Of Spacecraft. This patent arose from the design and application of algorithms which will be implemented into the New Millennium Earth Observing-1 (EO-1) spacecraft as part of new technology growth. This patent will allow full closed-loop maneuver autonomy onboard any spacecraft rather that the tedious and costly operational activity associated with ground based operations and control. The application to other missions is unlimited and can be used to fully explore the NASA mandate of faster, better, cheaper, spacecraft.

Algorithm Description

This algorithm for formation flying solves this problem by combining Lambert's two point boundary value problem and the use of the 'C*' guidance and navigation matrix application of the f and g series discussed by Battin[10] with the use of an autonomous fuzzy logic system called AutoCon ™. The algorithm enables the spacecraft to execute complex 3-axis orbital maneuvers autonomously. Figure 3 illustrates the basic sets of information required for formation targeting as it is incorporated into AutoCon™. The algorithm is suited for multiple burns scenarios but is explained in a two-burn approach for clarity. The formation flying problem involves two spacecraft orbiting the Earth. One spacecraft, referred to as the control spacecraft, orbits without performing any formation flying maneuvers. The second spacecraft is the chase spacecraft. It monitors the control spacecraft, and performs maneuvers to maintain the desired formation phasing. The goal of the formation flying algorithm is to perform maneuvers to move the chase spacecraft along a specified trajectory, called the transfer orbit, from its initial state $S_0 = (r_0, v_0)$ at a given time t_0 to a target state $S_t = (r_t, v_t)$ at a later time t_t. This goal is accomplished by finding the state the spacecraft would have at the current time in order to achieve the target state at the target epoch without maneuvering. This new state is called the desired state $S_d = (r_d, v_d)$; it is the target state propagated backwards in time from the target epoch to the epoch of the initial state. The difference between the initial state and the desired state is:

$$\delta S = \begin{pmatrix} \delta r \\ \delta v \end{pmatrix} = \begin{pmatrix} r_0 - r_d \\ v_0 - v_d \end{pmatrix} \tag{1}$$

Then, following the derivation of the state transition matrix given in Battin, the relevant state transition matrix submatrices are:

$$R(t_t) = \frac{|r_d|}{\mu}(1-F)[(r_t - r_d)v_d^T - (v_t - v_d)r_d^T] + \frac{C}{\mu}[v_t, v_d^T] + G[I] \tag{2}$$

$$\tilde{R}(t_t) = \frac{|r_t|}{\mu}[(v_t - v_d)(v_t - v_d)^T] + \frac{1}{|r_t|^3}[|r_t|(1-F)r_t r_d^T + C v_t r_d^T] + F[I] \tag{3}$$

The Expressions for F, G, and C are derived from the universal variable formulation, as discussed in chapter 9 of Battin. From these submatrices, the C* matrix is computed as follows:

$$R^*(t_0) = -R^T(t_t) \tag{4}$$

$$V^*(t_0) = \tilde{R}^T(t_t) \tag{5}$$

$$C^*(t_0) = V^*(t_0)[R^*(t_0)]^{-1} \tag{6}$$

The Expression for the impulsive maneuver follows immediately:

$$\Delta v = C^*(t_0)\delta r - \delta v \tag{7}$$

Keplerian and Non-Keplerian Transfer Orbits

Having established both actual and desired states of a spacecraft's location using GNCC's GPS filter algorithms, all that is needed is a means of autonomously zeroing the difference between the two states. Given two Keplerian trajectories and a chronologically defined maneuver window, a reference non-Keplerian trajectory may be determined which will smoothly transport the spacecraft from its position on the first Keplerian path at the beginning of the maneuver window to a desired position on the second Keplerian path at the conclusion of the maneuver window. Control points on the reference trajectory in Figure 4 are calculated at regular time intervals consistent with the ability of the spacecraft to receive and process position data, fire its thrusters, and account for the effects of each firing.

At each step in the process, the next control point on the reference path is examined and back-computed along a Keplerian path to determine small differences between spacecraft position and velocity on the reference path and determine which Keplerian path would intersect the reference path at the next control point. These differences are then fed into a system of linearized state transition matrices to determine the incremental ΔV required to get the spacecraft to the next control position on the reference trajectory. At the conclusion of the maneuver window, a final burn is required to match the velocity required to maintain the new Keplerian trajectory.

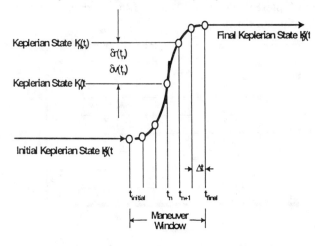

FIGURE 5. Non-Keplerian Reference Trajectory During Maneuver Window.

In the case of the EO-1 mission, the desired orbital trajectory will have the EO-1 spacecraft trace out the same ground-track as Landsat-7 separated by 1 minute in time. Since the injection orbit of EO-1 may be expected to be close to the desired orbit, both the true and the desired trajectories of EO-1 are expected to intersect at a single point. A long, iterative window requiring many small burns will not be necessary and ΔV maneuvers are expected to be executed at a single firing. The generalized approach does not, however, require that the two Keplerian (pre- and post- window) paths intersect, nor does it require that the intervening reference trajectory be Keplerian. The approach is therefore a means of executing any smooth non-Keplerian trajectory which will get a spacecraft from anywhere to anywhere, limited only by time, fuel, and spacecraft capabilities.

Algorithm Control

The algorithm needs input data for the current spacecraft state, the target state, and the desired state. These data are provided by the algorithm driver. The driver takes the current state of the control spacecraft, calculates its orbital period. It then propagates this state for a user specified fraction of the period. This propagation provides the location of the control spacecraft at the target epoch. User specified offsets are applied to this state to create the target state. The target state is then propagated back to the epoch of the initial state, producing the desired state. This procedure creates the required inputs to the GSFC algorithm.

Establishing the desired state of a spacecraft's location is as varied as spacecraft missions themselves. Autonomous orbit control of a single spacecraft requires that a known control regime be established by the ground consistent with mission parameters. That data must then be provided to the spacecraft. When orbital perturbations carry the spacecraft close to any of the established boundaries, the spacecraft reacts (via maneuver) to maintain itself within its error box. Once an error box is provided to the spacecraft, no further ground interaction is required. Enhanced formation flying takes the next step up the technological ladder by permitting the spacecraft themselves to establish where their own control boxes should be. This requires cooperation between all the members of the formation, and therefore a depth of communication between all the individual satellites that is not practical (or in some cases even possible) from the ground. This may occur through cooperative "agreement" by controllers of all the spacecraft in the formation or by maintaining a relative position from a designated 'lead', or by some hybrid of these two methods.

AutoCon Simulations

The following results are taken directly from the AutoCon™ ground system which utilizes the GSFC algorithm. The results are divided into formation flying of two spacecraft to maintain either a close or a dynamic. The initial conditions where derived from the orbit elements for the Landsat7 mission which has a sun-synchronous orbit with a mean local time of 10:00 and a ground track repeat of 233 orbits. The results show formation flying evolution and the effect on the mission groundtrack requirements. Evolution figures are presented in a control spacecraft rotating coordinate system with the radial direction the difference in radius magnitude and the alongtrack direction the arc between the position vectors.

Close Formations

The first two figures present the maintenance of a formation that has a 10 meters radial separation only. Figure 5 presents the formation evolution in radial and separation distances for a period of 90 days. To re-initialize this orbit, two maneuvers are used in a Hohmann like transfer. The first ΔV to re-establish the 10m radial separation position by using the algorithm targeting method with a ½ orbit period and the second ΔV by using the same method with a .01 orbit period to adjust the velocity components. Figure 2 presents the ground track of these orbits. The initial orbital condition placed the ground track at the "0" error location for convenience.

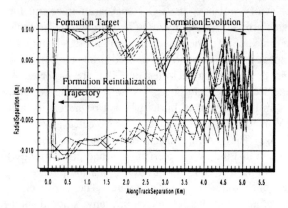

FIGURE 6. Close Formation Radial and Alongtrack Separation.

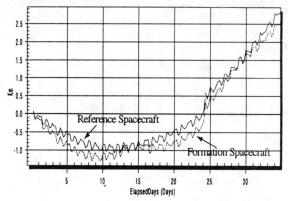

FIGURE 7. Close Formation Relative Groundtrack

Close Formations with Groundtrack Maneuvers

Figures 7 and 8 present results of starting with an initial alongtrack separation of 0 meters and an initial radial separation of 20 meters and then targeting to a 10 m radial and 0 AlongTrack separation whenever a maneuver occurs from either spacecraft. Therefore, the first maneuver of the formation flyer is to adjust to both the groundtrack of the control spacecraft after its maneuver and to re-establish the initial formation parameters. Figure 7 presents the results when groundtrack maneuvers have occurred for the control spacecraft. As seen in Figure 8, a groundtrack maneuver takes place slightly before the time when the alongtrack separation is near zero. The smaller

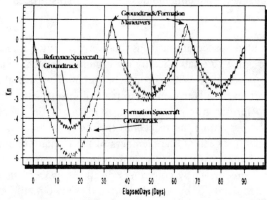

FIGURE 8. Close Formation Groundtrack with Maneuvers.

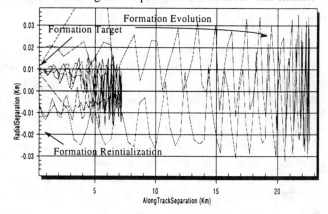

FIGURE 9. Close Formation Evolution.

parabola represent the maintenance of the formation to the 10m radial separation. The formation evolution in radial and separation distances are presented for a period of 90 days.

Close Formations with Inclination Maneuvers

The next set of figures present the simulation results of allowing the control spacecraft to perform an 0.5 degree inclination maneuver and targeting the formation flyer to not only change its orbit inclination, but to also re-establish the initial condition for the formation. Figures 9 and 10 present results of starting with an initial alongtrack separation of 0 meters and an initial radial separation of 20 meters and then targeting to a 10 m radial and 0 alongtrack separation whenever an inclination maneuver occurs from the control spacecraft. In this scenario, the formation evolution was approximately one-fourth completed, since an inclination maneuver of the control spacecraft was performed after 5 days from initial epoch.

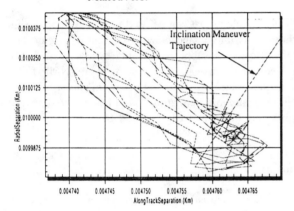

FIGURE 10. Close Formation with Inclination Maneuvers.

As seen in the figure, the trajectory of the formation flyer reinitialized at a radial and alongtrack separation of 10m and 0 m with an inclination change of 0.5 degrees. This maneuver was not a combined maneuver as optimization was not performed. The inclination maneuvers were performed at the descending node for both spacecraft. Ironically, due to a typo in the control script, the formation was to be maintained at each 5 minute integration step though-out the next 85 days of the simulation. This meant that the algorithm was targeting to the 10m and 0 m separations every 5 minutes. Figure 10 presents the formation evolution for this 'oversight' in the script. The radial separation was maintained to +/- 5.25×10^{-5} m and 3.0×10^{-5} m in the alongtrack separation.

Dynamic Formations

The next simulation consist of maintaining a dynamic formation where the formation flying spacecraft was in a different orbit plane with an alongtrack separation on the order of 450 km. To simulate this, the initial state of the control spacecraft was propagated backward for 1 minute (450 km at 7.5 km/s) and to maintain the groundtrack requirement, the right ascension of ascending node adjusted to account for a one minute Earth rotation. Figure 11 and 12 present the formation evolution in the radial versus alongtrack and crosstrack versus alongtrack separation for several days. As seen, the effect of the perturbations on the orbit elements has an immediate effect in the osculating orbital elements. This results in a very large radial separation approaching +/- 1km. A crosstrack of +/- 30km was anticipated since that is the effect of the node difference. As the formation evolved , a maneuver was required to re-established the formation at the initial separation of 0m alongtrack and 30km crosstrack at a radial separation of

FIGURE 11. Close Formation Constant Control.

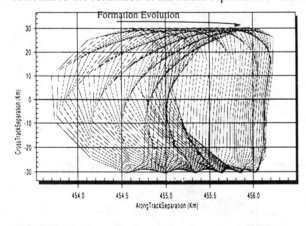

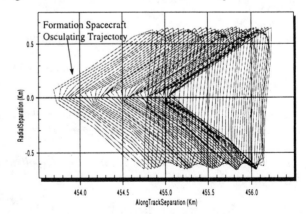

FIGURE 12. Dynamic Formation Crosstrack Evolution.   FIGURE 13. Dynamic Formation Evolution.

10m. Figure 13 presents the trajectory of the formation flyer. The figure shows the radial separation change from about -500m to a +10m separation and a 450km alongtrack separation to a 0km separation. After this state was targeted, a maneuver was performed to maintain a formation similar to the close formation. Figure 14 presents the crosstrack separations for this simulation.

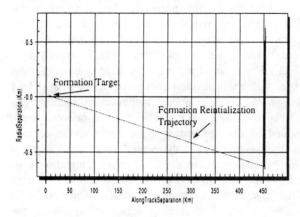

FIGURE 14. Dynamic Formation Reinitialization.

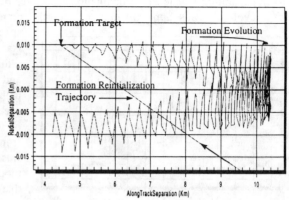

FIGURE 15. Dynamic Formation Evolution Post Maneuver.

CONCLUSIONS

We have shown that our formation flying algorithm is a feasible technology that can be used in a close-loop design to meet science and mission requirements of Low Earth Orbiting missions in the New Millennium Program and Mission To Planet Earth Program. The algorithm is very robust in that it supports not only benign groundtrack control, but demanding 3-D control for inclination and non-Keplerian transfers. To best meet the NMP EO-1 requirements, this innovative technology will be flown onboard the spacecraft which launches in May 1999. The algorithms are being integrated into AutoCon™ for both ground support validation and closed-loop onboard autonomy. The application to other NMP or MTPE programs is unlimited and can be used to fully explore the NASA mandate of faster, better, cheaper, spacecraft.

Acknowledgments

The authors would like to thank Dr. Mike Griffin (former NASA Chief Scientist) for his astrodyanmic expertise bestowed during a graduate class which lead to the formulation of this algorithm design.

References

Battin, R. (1987) *An Introduction to the Mathematics and Methods of Astrodynamics*, AIAA Education Series, Chapters 9 and 11.

Carter, R. (1997) Internal memo of EO-1 Requirements, NASA GSFC Greenbelt, MD.

Folta, D.C. (1992) "Considerations on Formation Flying Separations For Earth Observing Satellite Missions", Proceedings of the AAS 92-144 Spaceflight Mechanics Meeting, Colorado Springs, Co.

Folta, D.C. and L. Newman (1996) "Foundations of Formation Flying For Mission To Planet Earth and New Millennium, AIAA 96-3645, AIAA Guidance, Navigation, and Control Conference and Exhibit, San Diego, CA

Quinn, D.A. and D. C. Folta (1996) *Patent Rights Application and Derivations of Autonomous Closed Loop 3-Axis Navigation Control Of EO-1.*

Scolese, C.J., D. Folta, and F. Bordi (1991) "Field of View Location and Formation Flying For Polar Orbiting Missions", AAS 91-170 Spaceflight Mechanics Meeting, February, 1991, Houston, TX.

Sperling, R. (1997) *AutoCon User's Guide,* AI Solutions, Inc., Greenbelt, MD. 20770

AUTONOMOUS SOFT-LANDING TECHNIQUE ONTO THE MOON IN SELENE PROJECT

Sumio Hikida and Haruaki Itagaki
National Space Development Agency of Japan (NASDA)
Tsukuba Space Center, 2-1-1, Sengen,
Tsukuba-city, Ibaraki 305, Japan
phone: +81 (298) 52-2407, fax: +81 (298) 52-2407

Abstract

NASDA and ISAS have investigated the experimental lunar lander which is based on Japanese own technologies in SELENE project. In this project, the spacecraft consists of three parts; an orbiter, a relay satellite, and experimental lunar lander. This experimental lander is planned to acquire the basic soft-landing technology. For landing, this experimental lander has to do "powered descent and moon soft-landing" autonomously. But soft-landing onto the moon is the first experience in Japan. This paper describes the outline of concepts and technological problems of this experimental lunar lander and introduces some unique characteristics under consideration, such as autonomous obstacle avoiding control.

INTRODUCTION

The lunar exploration project named SELENE (SELenological and ENgineering Explore) is now investigated in Japan. SELENE is the first NASDA/ISAS joint lunar project and planned to be launched by H-IIA rocket around 2003. The major objectives of SELENE are to obtain scientific data of the lunar origin and evolution, and develop the technologies for the future lunar exploration. For these objectives, a spacecraft in SELENE consists of three parts; an orbiter, a relay satellite and an experimental lunar lander. FIGURE 1 shows SELENE system. The propulsion module in this figure operates as a lunar lander after separation from the mission module.

The major purposes of the lunar lander are acquisition of the basic soft-landing technique for future activities on the moon and demonstration of soft-landing based on Japanese own technologies. This is the first experience in Japan. But, in consideration, the lander has a quite challenging technique that has never accomplished in the world; the autonomous obstacle avoiding control. In this paper, following sections describe the outline of the lunar lander system, and concept of the landing sequence and method.

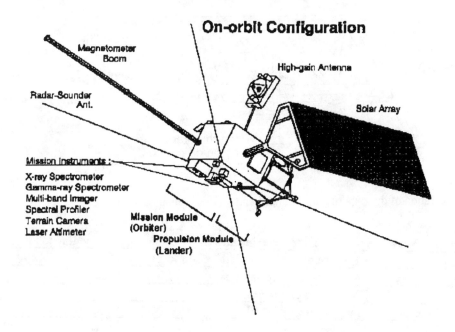

FIGURE 1. System Configuration of SELENE (A relay satellite is not in this figure.)

THE OUTLINE OF THE LUNAR LANDER SYSTEM

The configuration of the lander and its mission summary is shown in FIGURE 2 and TABLE 1, respectively. And key technologies is shown in TABLE 2.

As mentioned before, the major objective of this lander is the demonstration of soft-landing based on Japanese technologies. These technologies include the autonomous obstacle avoiding control. It is expected to be effective to raise the landing reliability and to make it possible to land on various points on the moon, such as the far side of the moon, the polar region and the other points where the geographical features are unknown or with much undulations. The technique is also expected applying to the mission that requires landing at the particular points on the surface.

For addition of the function without much increase of propellant, hovering is not adopted. Detection and avoidance of obstacles are accomplished during descent. As a result, the available time for the control is only as long as 100 seconds. This is why simple logic is adopted. In this purpose, the following methods are selected.

(1) Detection of obstacles with shades on the lunar surface using the image taken by one onboard camera.
(2) Avoidance accomplished by open-loop control

In addition, the lander has unique characteristics such as the followings in order to lead the project to success without huge development cost&risk. In other words, the numbers and the qualities of new components to be developed are restricted.

(1) The lander has a only narrow range (3km) altimeter.: Wide range (more than 10km) altimeters were adopted in Apollo missions and Surveyor missions.
(2) The lander accomplishes the control of thrust value only by ON/OFF of the thrusters which have constant thrust.: Throttling engines have been adopted in previous soft-landing missions, such as Apollo etc.

Of course, the difficulties in the development of software increase. But the conceptual study shows our concept has feasibility.

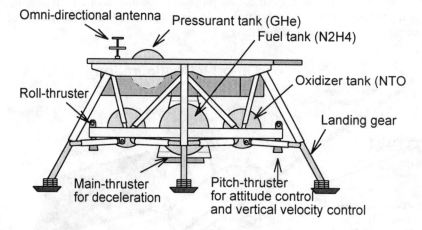

FIGURE 2. Configuration of the Experimental Lunar Lander (Conceptual design)

TABLE 1. Mission Summary of the Experimental Lunar Lander (Corresponding to FIGURE 2.)

Dimension	1.5m(height) x 2.4m(width) x2.4m(depth)
Weight	480 kg on the circular lunar orbit, 216 kg on the lunar surface
Landing site	Maria area on the near side
Mission equipment	radio source for VLBI

TABLE 2 Key Technologies in the Experimental Lunar Lander Development

Item	Explanation
Guidance and landing operations	Velocity=0 in all components, and vertical attitude must be attained finally (altitude=0) with all the error sources concerning orbit determination, navigation and geographical features of the Moon. It needs to select and change proper sensors, actuators and guidance method which meets the altitude and other conditions, and to stabilize closed control loop in order to perform descent and obstacles-avoidance successfully.
Landing sensors (altimeter and speedometer)	They are currently installed in the airplanes, however, they must be improved for the space use, taking the environment around the Moon and the influence of the thrusters into account.

Thrust power control	Thrust control is necessary for soft-landing with powered descent.
	Considering various error factors, the thrust power should be controlled to 30% of the full power when the lander is descending in uniform velocity.
	There are 2 candidates for the power control methods by;
	  (1) variable power engines controlled by throttling, and
	  (2) clustered low-power engines controlled by ON/OFF switching.

Tracking and control	NASDA has no experience to operate spacecraft which is farther than geostationary orbit. The error concerning orbit determination will become the largest factor of the positioning error until the landing sensors catch the lunar surface, so the orbit determination should be as precise as possible in order to correct the course in the final landing operation phase after the landing sensors are available.
Shock absorption	To avoid thrust-gas rebound and regolith raising, the engines are cut-off 2m(TBD) above the ground and the lander falls freely after that. Therefore, some shock absorbing mechanism, which absorbs not only vertical but horizontal components of the velocity energy, is required to protect the components and payloads on board.
Obstacles detection and avoiding control	To avoid falling down and collision with rocks on landing, obstacles avoidance control is assumed to be done. Installing the function, onboard processing mechanism using simple and reliable logic should be necessary because the available time for the control is only as long as 100 seconds.
Thermal control and power supply on the lunar surface	On the Moon, day and night changes every 14 (earth-)days. So the thermal control system that meets both the high temperature in the day and low in the night is needed to accomplish missions on the moon. In the long night, the temperature falls down to -170 deg. Under the condition, the total amount of power for one night survival is estimated to be quite large, even if the mission power looks small. Then it is unavoidable that the weight of its battery become very heavy. So, it is desirable to adopt effective thermal control and power supply.
System lightening	On the soft-landing from the circular lunar orbit, the weight of the propellant necessary for the operation is just the same as that of lander on the surface itself. Namely, the increase of the hardware causes the same amount of the propellant increase from a viewpoint of weight. So, lightening effort is more important and effective for such a landing missions.
System verification on the ground	Some of the critical technologies shown above should be tested on the ground before the experimental lander is developed, if possible. It is also important to develop the technique for the system verification in advance.

FLIGHT SEQUENCE

The lander is carried by the orbiter onto an lunar polar orbit whose altitude is 100km. On the orbit, the lander separates from the orbiter. After separation, the lander transfers to 15x100km elliptical orbit. At the perilune of the orbit, the lander performs powered descent and soft-landing autonomously all the phase after the separation. After successful landing, the radio source for VLBI is activated and works for more than 2 months, to observe the vibration and gravity field of the moon.

Landing process of the lunar lander is shown in Figure 3.

The continuous injection of a NTO/N2H4 main thruster from the perilune of the 15x100km elliptical orbit makes the horizontal velocity relative to the lunar surface nearly zero.

The horizontal velocity needs to be adjusted to nearly zero before the lander reaches 3km altitude. The altitude sensor available only below 3km makes the lander achieve the condition without direct information of altitude. Although navigation errors are accumulated until the lander reaches 3 km altitude, the analysis of errors shows the feasibility by targeting 5 km as the nominal altitude where horizontal velocity becomes zero.

The vertical descent phase following the above phase needs constant vertical velocity for absorbing errors. It is achieved by on/off control of N2H4 thrusters whose thrust are constant. During this phase, it detects obstacles such as craters using the images of the surface below, and performs avoidance maneuver autonomously not to land on them . The thrusters are cut off at the point 2m above the surface.

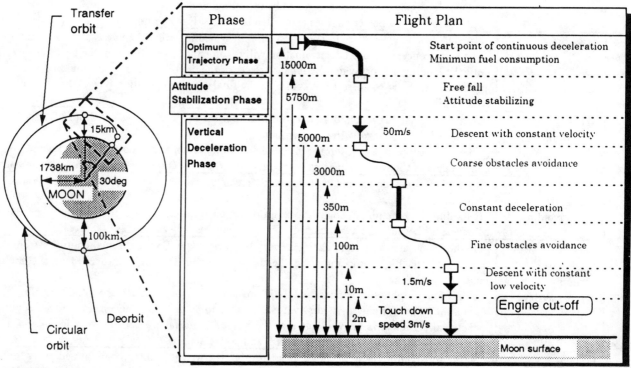

FIGURE 3. Landing Process

OBSTACLES AVOIDANCE SYSTEM

During the final approach to the landing site on the moon, it will be designed to land avoiding craters (lager than 2m diameter) for safe landing, especially for absorbing the touch down impact, and prevention of fall down. This autonomous obstacle avoiding control system is executed by onboard computer which provides real time procedure. In order to detect obstacles, the image of CCD camera is more suitable than other methods (Aoki 1996); for example digital elevation map taken by RADAR or LIDAR, because it doesn't need any complicated hardware. However its calculation load shall be considered. In other words, it is necessary to investigate the simple logic for detecting obstacles with CCD images. Several process for image detection have been trade off, then following method is considered.

Obstacles Detection

There are some kinds of methods about images recognition; stereo vision, optical flow methods, shape-from-shading, and other methods by using brightness. In this lander, calculation load is taken into account, because the lander has only a little time until touch down. Meanwhile it should be paid attention to that it's necessary only to detect obstacles, not to reappear exact terrain. In this case, calculation time could be enough short. And then following method have possibility under these condition.

Combination of Simple Shape-from-shading and Moment of Brightness

Shape-from-shading is known as a method to reappear an original shape, but here it must be simplified enough because of adopting the calculation capacity of electronics of the lunar lander. Shape must be steep around the edge of crater. If a simple shape-from-shading is applied to suitable sized mosaic divided from original picture image, it can be estimated steep of the slope.

On the other hand, it's considered that gradation of brightness corresponds to roughness of terrain. Local quadratic moment of brightness (S^2ij) can be applied to gradation of brightness in order to estimate roughness.

$$S^2ij = dij^2 + (mij - m)^2 .$$

Where, Sij is small area around pixcel(i,j), mij is local mean of brightness (mean of brightness in Sij), m is :mean or median of all pixcels, and dij^2 is local dispersion of brightness.

Taking logical AND of above each estimations, landing spot will be targeted.

Conceptual flow chart and estimation of calculation are shown in FIGURE 4 and TABEL 3, respectively. This table says, for reduction of calculation time, the simple estimation logic is necessary to investigate.

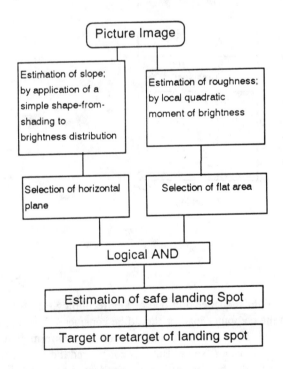

FIGURE 4. Flow Chart of Calculation

TABLE 3. Estimation of Calculation[*1]

Function	Contents of processing	time
Estimation of slope	Shape-from-shading	10.5sec
	Fitting by method of least squares	negligible
Estimation of roughness	Calculation of moment of brightness	1.1sec
Estimation of landing spot	Logical AND	negligible
Target or retarget	Calculation of the center of maximum circumscribed of obstacles	0.5sec

*1:Assuming to use R3000 CPU

Avoiding Maneuver

In FIGURE 2, the CCD images are taken at 3000m and 100m altitude. After detection of obstacles, the lunar lander is led to a suitable landing site within 80m circle by open loop control. This movement is accomplished by attitude rotation around pitch or yaw axis. Under the condition of the engine configuration shown in FIGURE 5, a simulation of maneuver is solved. Its results are shown in FIGURE 6. In this simulation, thrust is controlled by ON/OFF switching of 150N thrusters to achieve constant vertical velocity, and 50N thrusters are available for attitude control. This simulation shows a feasibility for this control of the lunar lander.

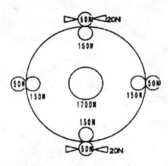

FIGURE 5. Engine and Thrusters Configuration in the Simulation.

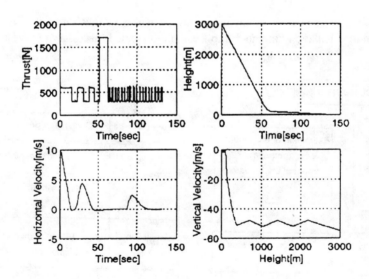

FIGURE 6. Result of the Simulation about Avoiding Maneuver.

CONCLUSIONS

The basic concept of the experimental lunar lander system, flight sequence and one of key technologies; obstacle avoiding control, are described. In the view of a reduction of the computer load, it should be considered how easily the terrestrial characteristics is determined. To realize the concept of obstacle avoiding control, more information is necessary; for example the correct reflection model of the lunar surface and so on. But it is considered the conceptual that study makes almost sure the feasibility of the lander. Currently, further studies are conducted for the launch around 2003.

Acknowledgment

This work is a result of the conceptual study of the SELENE project. The authors wish to thank the engineers who supported this project under the companies: Mitsubishi Electric Co., NEC Co., and Toshiba Co.

Reference

Aoki,H. ,S.Ishikawa,E.Namura, and K.Yamanaka (1996) "Critical Technology of MOON Soft-Landing," in *echnical Report of IEICE*,SANE96-24 (1996-06):11-18.

AUTONOMOUS FAILURE DETECTION, IDENTIFICATION AND FAULT-TOLERANT ESTIMATION FOR SPACECRAFT GUIDANCE, NAVIGATION AND CONTROL

R. Mehra, C. Rago, S. Seereeram
Scientific Systems Company
500 West Cummings Park, Suite 3000
Woburn, MA 01801
781 933-5355

David S. Bayard
Jet Propulsion Laboratory
4800 Oak Grove Drive, M/S 198-326
Pasadena, CA 91109
818 354-8208

Abstract

In this paper, we propose a novel approach for Failure Detection and Identification (FDI) in nonlinear systems based on the Interacting Multiple Model (IMM) Extended Kalman Filter (EKF) approach. In the nonlinear-system FDI application, the main idea consists of representing each failure mode by a model and combining the outputs of EKF's based on different models in a near-optimal way. This IMM-FDI filter provides not only failure detection and identification but also a near-optimal estimate of the system state (even during a failure). The approach has been applied successfully to a problem of spacecraft autonomy for the detection and identification of sensor (gyro, star tracker) and actuator failures. The results of this application show that IMM-EKF detects and identifies failures much more rapidly and reliably than the multi-hypothesis EKF. Furthermore, it handles satisfactorily both permanent and transient failures.

INTRODUCTION

Future space missions call for unprecedented levels of autonomy, reliability and precision. Spacecraft like other engineering systems will encounter unexpected failures and environmental disturbances. Failure detection, identification and protection are an essential component of future spacecraft, whether earth-orbiting or inter-planetary, with increased autonomy requirements at reduced cost. Current methods for spacecraft operation call for significant investment in ground support and telemetry capability. Spacecraft data is continually produced and relayed via downlink to ground stations where trained personnel collect, store and analyze to determine the operational status of the satellite mission. This is a complex, time-consuming and personnel-intensive procedure. Additionally, space operations require the availability of mission "experts" to deal with anomalous situations. Any loss of ground control effectively results in satellites flying out of control. With recent growth in communications and remote sensing requirements/infrastructure, the numbers of in-orbit satellite missions are expected to increase significantly. In order to support these missions, future satellite and ground monitoring systems necessitate increasing use of automated data processing in order to extract relevant information from data feeds, and reduce the complex and burdensome task for human operators and satellite engineers. For inter-planetary craft, significant hardware redundancy – together with the associated increase in cost, payload and fuel requirements – has been necessary to reliably ensure mission success.

Scientific Systems has developed analytic techniques for Failure Detection, Identification and Compensation (FDIC) for S/C guidance and control systems. In particular, we recently developed the Interacting Multiple Model (IMM) filter, which proves to be an efficient, fast-reacting detector for mode switching among various failures, as well as providing fault-tolerant attitude estimation for guidance and control purposes. Together with other techniques such as Multi-hypothesis Extended Kalman Filtering, Failure Detection Filters, Signature and Residual Analysis and Classification/Detection Algorithms, SSC is developing an overall Health-Monitoring system architecture for spacecraft and other complex engineering systems. In this paper, we present recent results of applying the IMM technique for identifying GNC component failures, and fault-tolerant estimation of spacecraft attitude using gyros (IRUs) and star cameras (SRUs).

The spacecraft FDI problem is one of a vast class of problems where the model representing the dynamics of the system can switch (unpredictably) from one model to another. Failures of components (gradual or sudden), statistical changes in noise parameters, variations in physical structure (damage to the spacecraft), etc. can all render standard control techniques unreliable. The use of a single model (a "compromise model") in these situations usually gives very poor estimates, and these estimates are incomplete in the sense that they lack information relevant to the model in effect. This is particularly important in the failure detection case. In the IMM approach (Blom and Bar-Shalom 1988, Bar-Shalom and Li 1993, 1995), this problem is solved by having M filters running in parallel (Figure 1) incorporating a hidden Markov switching mechanism among the models. For the FDI problem, each failure hypothesis corresponds to a different model within the IMM, i.e. each model is constructed to map a failure hypothesis into the dynamics of the system.

A related approach based on Multiple Hypothesis EKF, or MH-EKF was presented in Mehra *et al*, 1995, where several EKF's are run in parallel, and a hard-switching decision is made at each sampling time based on the innovation sequences and likelihood functions of each filter. A common problem encountered with this approach is the delay in detection due to build up of the likelihood functions for active hypotheses and divergence of individual EKF states for the inactive hypotheses. The IMM approach proposed here overcomes this limitation of MH-EKF and provides a more reliable state estimate as well as rapid detection of failures.

THE INTERACTING MULTIPLE MODEL (IMM) FILTER

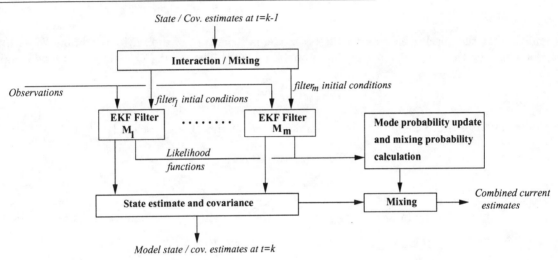

FIGURE 1. The Interacting Multiple Model Filter.

Assume that the system dynamics can be represented by a set of M models:

$$\mathbf{x}_{k+1} = \mathbf{F}_k^j \mathbf{x}_k + \Gamma_k^j \nu_k^j \qquad j = 1 \cdots M , \tag{1}$$

$$z_k = H_k^j \mathbf{x}_k + \omega_k^j , \tag{2}$$

where M^j represents the j'th model, and the system can "jump" unpredictably among models. Given a whole measurement set Z^k at time k $Z^k = \{z_i\}_{i=1}^{k}$ the optimal state estimate at time k, $\hat{\mathbf{x}}_k$ is the conditional expected value $E\{\mathbf{x}_k|Z^k\}$. Using the total probability theorem, the posterior probability density function (pdf) of the state conditioned on the measurement set is given by:

$$p\left(\mathbf{x}_k|Z^k\right) = \sum_{j=1}^{M} p\left(\mathbf{x}_k|Z^k, M_k^j\right) p\left(M_k^j|Z^k\right) = \sum_{j=1}^{M} p\left(\mathbf{x}_k|z_k, Z^{k-1}, M_k^j\right) \mu_k^j , \tag{3}$$

where μ_k^j is model M_j posterior probability at time k. The posterior pdf of the state given model M_k^j can be rewritten as:

$$p\left(\mathbf{x}_k|z_k, Z^{k-1}, M_k^j\right) = \frac{p\left(z_k|\mathbf{x}_k, Z^{k-1}, M_k^j\right)}{p\left(z_k|Z^{k-1}, M_k^j\right)} p\left(\mathbf{x}_k|Z^{k-1}, M_k^j\right) . \tag{4}$$

Using (2), conditioned on the model and the state the measurement z_k does not depend on the past measurements Z^{k-1}. The prior pdf of the state given the past measurements and the actual model can be written as:

$$p\left(\mathbf{x}_k|Z^{k-1}, M_k^j\right) = \sum_{i=1}^{M} p\left(\mathbf{x}_k|Z^{k-1}, M_k^j, M_{k-1}^i\right) p\left(M_{k-1}^i|Z^{k-1}, M_k^j\right). \quad (5)$$

Assuming that all the past through $k-1$ can be summarized by the set of M states $\hat{\mathbf{x}}_{k-1}^l$ and the corresponding covariances (P_{k-1}^l), (5) can be written as:

$$p\left(\mathbf{x}_k|Z^{k-1}, M_k^j\right) \approx \sum_{i=1}^{M} p\left(\mathbf{x}_k|M_k^j, M_{k-1}^i, \hat{\mathbf{x}}_{k-1}^i, P_{k-1}^i\right) \mu_{k-1}^{i|j}, \quad (6)$$

where $\mu_{k-1}^{i|j}$ are called the *mixing probabilities*. Assuming that each element of the sum is Gaussian, then the prior pdf is a Gaussian mixture and can be approximated (via moment matching) by a single Gaussian:

$$p\left(\mathbf{x}_k|Z^{(k-1)}, M_k^j\right) \approx \mathcal{N}\left(\mathbf{x}_k; E\{\mathbf{x}_k|M_k^j, \sum_{i=1}^{M} \hat{\mathbf{x}}_{k-1}^i \mu_{k-1}^{i|j}\}, cov_{mix}[\cdot]\right). \quad (7)$$

This implies that the initial condition for filter j model is a mixture of the initial conditions of each filter ($\hat{\mathbf{x}}_{k-1}^i$) weighted by the corresponding mixing probabilities $\mu^{i|j}$. The mixing probabilities can be computed as follows:

$$\mu_{k-1}^{i|j} = P\left(M_{k-1}^i|M_k^j, Z^{k-1}\right) = \frac{1}{\sum_{l=1}^{M} p^{lj} \mu_{k-1}^l} p^{ij} \mu_{k-1}^i, \quad (8)$$

where p^{ij} is the *prior* transition probability from model M^i at time $k-1$ to model M^j at time k, i.e.:

$$p^{ij} = P\left(M_k = M^j|M_{k-1} = M^i\right) \quad ; \quad \sum_j p^{ij} = 1. \quad (9)$$

The values of p_{ij} are design parameters, selected based on *a priori* reliability information about the sensors and actuators. Implicitly, the IMM uses an embedded Markov switching mechanism characterized by $[p^{ij}]$.

SPACECRAFT FAILURE DETECTION AND IDENTIFICATION

In this paper we will use the spacecraft attitude dynamics model described in (Mehra et al., 1994, 1995). Two sensors are used for attitude determination: the Stellar Reference Unit (SRU) or star camera and the Inertial Reference Unit (IRU) or gyro. The star camera operates by locating known celestial reference stars in the camera's field of view, and comparing this position against a stored star catalog. The output of the star camera is a measurement of the attitude (quaternion). Star camera and gyro data are available at different rates, and are used depending on the required mode of navigation. The data provided by these sensors are processed by the estimation filter (IMM-FDI) to provide an estimate of the attitude and also detect any failures in the sensors or the thrusters.

The dynamics of the spacecraft are based on the Euler rigid-body equations:

$$\mathbf{J}\dot{\omega}(t) = \tau(t) - \omega(t) \times \mathbf{J}\omega(t), \qquad \dot{q}(t) = \frac{1}{2}\mathbf{\Omega}(t)q(t), \quad (10)$$

where ω is the S/C angular velocity, q is the attitude quaternion, τ is the total external torques, and $\mathbf{J}$ is the S/C inertia. The attitude propagation is determined by the operator:

$$\mathbf{\Omega} = \begin{bmatrix} 0 & \omega(3) & -\omega(2) & \omega(1) \\ -\omega(3) & 0 & \omega(1) & \omega(2) \\ \omega(2) & -\omega(1) & 0 & \omega(3) \\ -\omega(1) & -\omega(2) & -\omega(3) & 0 \end{bmatrix}. \quad (11)$$

In the IMM filter, each model M^j uses a standard Extended Kalman Filter (see, for example, Mehra et al, 1995). with the nominal $\mathbf{J}$, and $\mathbf{\Omega}$ based on the latest estimates ($\hat{\mathbf{\Omega}}(k)$). Measurement noise for the gyros is assumed to be additive white Gaussian noise. For the star camera, white Gaussian noise is composed with the true quaternion. For a small-noise approximation, the SRU measurement noise can be assumed additive and decoupled. The measurement equation is then given by

$$z(t) = h(x(t), n(t)) \approx \mathbf{H}x(t) + n(t) . \qquad (12)$$

The EKF(s) use a noisy measurement of the torque, τ_m, consisting of the true value of τ plus white Gaussian noise ($\sigma^2 = 10^{-4}$ (N/m)^2). The inertia matrix used to model the spacecraft is given by $\mathbf{J} = \mathrm{diag}(18.4, 18.2, 6.8)$ (N/m).

Spacecraft FDI Simulation

Figure 2 depicts the failure scenario used for the simulations. Each of the failures is explained in detail below. In all cases, the performance of the IMM-EKF filter is also compared to that obtained using a single

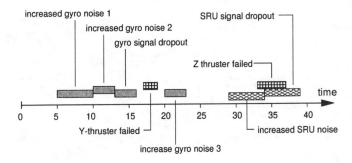

FIGURE 2. Failure Sequence used for FDI Simulation.

EKF (designed for nominal operation). Various component failures are described in the following list:

Actuator failures: There are two actuator (thruster) failures, during which the torque produced by the thrusters falls to zero, whereas the avionics' torque signal remains at its command level.

Gyro failures: The gyro output is sampled at 100 Hz. Measurement noise is additive white Gaussian noise, with a variance (for each axis) of $\sigma_w^2 = 1.2 \times 10^{-9}$ (rad/sec)^2 nominal, and $\sigma_w^2 = 1.2 \times 10^{-7}$ (rad/sec)^2 (factor of 100) in failure mode. We consider two kinds of gyro failures: increased noise and no-signal failures. The simulation includes three increased-noise gyro failures: measurement noise (variance) 50 times larger than the nominal, measurement noise 100 times larger than nominal, and measurement noise 400 times larger than nominal. The failure model is held fixed at 100 times larger than nominal, representing a mismatch in failure model variances. A few seconds after each failure, the back-up gyro is ready and the system recovers from the failure.

SRU failures: The SRU output is sampled at 0.5 Hz. A random quaternion (obtained from a white Gaussian random angle with variance σ_α^2 and a uniformly distributed direction vector) is composed with the true attitude quaternion to produce the measurement. The variance of the noise (angle) in this case is given by $\sigma_\alpha^2 = 2.5 \times 10^{-7}$ (rad)^2 nominal, and $\sigma_\alpha^2 = 2.5 \times 10^{-5}$ (rad)^2 (factor of 100) in failure mode. As before, both increased noise and signal dropout failures are considered.

Because of the difference in the sampling rates between the gyro and the SRU, two parallel IMM-FDI filters are used: IMM-FDI^I considers gyro and actuator failures, while IMM-FDI^{II} considers SRU failures. Each filter is run at the maximal update rate corresponding to the relevant sensors. The IMM-FDI^{II} filter must be coasted during the times for which no SRU measurements are available. IMM-FDI^I uses the models: H_0 nominal operation, H_1 noisy gyro, H_2 gyro signal dropout, H_3 X-thruster failure, H_4 Y-thruster failure, and H_5 Z-thruster failure. IMM-FDI^{II} uses the models: H_0 nominal operation, H_1 noisy SRU, and H_2 SRU signal dropout.

Simulation Results

The IMM-FDI filters previously described were applied to the fault scenario above. Analysis of the posterior probabilities of hypothesis H_0 (nominal operation) for the IMM-FDI^I clearly shows four failure periods (Figures 3(a) and 3(b)). Posterior probabilities corresponding to the different hypotheses ($H_0 \ldots H_5$) indicate when the system has suffered the failures (represented by 0-1 and 1-0 switchings). Furthermore, failures are identified unambiguously and almost instantaneously. Minimal processing is needed to classify unique events by combining the outputs of the dual IMM-FDI filter. Because of the low sampling rate for the SRU, there is a "natural" delay (until the first measurement arrives) in detecting the failures (as well as in recovering from the failure).

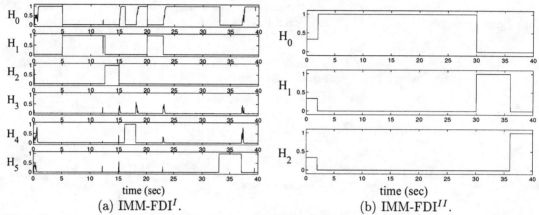

(a) IMM-FDI^I.   (b) IMM-FDI^{II}.
FIGURE 3. Posterior Probabilities for Dual-Rate Filters.

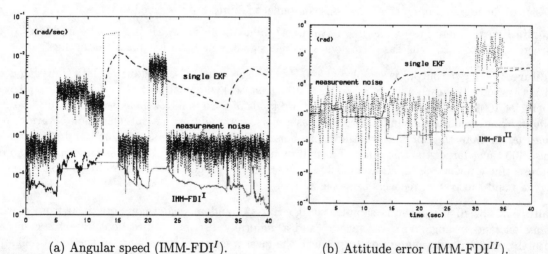

(a) Angular speed (IMM-FDI^I).   (b) Attitude error (IMM-FDI^{II}).
FIGURE 4. Comparison of Estimation Errors between Single EKF and IMM-EKF.

The attractiveness on the IMM-FDI approach is highlighted when one considers its dual-purpose utility as fault-tolerant estimators. For the simulation shown above, the spacecraft rate and attitude estimation errors computed by the IMM-FDI are consistently orders of magnitude less than that of a single extended Kalman filter designed around nominal operating parameters (Figures 4(a) and 4(b)). It is clear that the IMM-FDI^I outperforms the single EKF during the failure periods by at least an order of magnitude. The transients after a failure are also shorter for the IMM-FDI^I filter. Both IMM-FDI filters perform similarly during the non-SRU-failure periods. However, during the SRU failure period IMM-FDI^{II} outperforms the IMM-FDI^{I}. It should also be mentioned that all the failures are quickly detected (within 2 to 3 samples) and unambiguously identified, which is much better than the performance of the MH-EFK filters (Mehra et al. 1995).

CONCLUSIONS AND FUTURE WORK

This paper presents a new method for FDI using the IMM approach. Preliminary studies have indicated superior performance of the IMM-EKF for rapid failure detections, unambiguous identification, and robust attitude estimation in the event of a failure (fault-tolerant attitude estimation). The IMM-FDI filter outperforms a single EKF by an order of magnitude in attitude and rate estimation errors. Detection and identification delays are significantly shorter than those from the likelihood function MH-EKF approach. The IMM-FDI filter is able to detect sequential failures such as a gyro failure followed by a star camera failure or vice versa. Ongoing effort will expand the IMM-FDI approach to include other kinds of system failures and enhance its robustness to uncertainty in the system models.

Acknowledgment

This research was supported through contract NAS7-1383 from NASA Jet Propulsion Laboratory (JPL) and the support of Drs. Bayard and Hadaegh from JPL is gratefully acknowledged.

References

Blom, H.A.P., and Y. Bar-Shalom (1988) "The Interacting Multiple Model Algorithm for Systems with Markovian Switching Coefficients," *IEEE Transactions on Automatic Control*, **33(8)**:780–783, August 1988.

Brown, G.M. (1993) "Cassini AACS Functional Failure Modes and Effects Analysis," Preliminary version, Jet Propulsion Laboratory, February 1993.

Bar-Shalom, Y., and X.R. Li (1993) "*Estimation and Tracking: Principles, Techniques and Software,*" Artech House, Inc., 1993.

Bar-Shalom, Y. and X.R. Li (1995) "*Multitarget-Multisensor Tracking: Principles and Techniques,*" YBS Publishing, 1995.

Pell *et al.* (1996) "A Remote Agent Prototype for Spacecraft Autonomy," in SPIE Proceedings Vol. 2810, "*Space Sciencecraft Control and Tracking in the New Millennium,*" August, 1996.

Mehra, R.K., and J. Peschon (1971) "An Innovations Approach to Fault Detection and Diagnosis in Dynamics Systems," *Automatica*, **7**:637-640, Pergamon Press, 1971.

Mehra, R.K., and S. Seereeram (1994) "Neural Nets and Adaptive Reconfigurable Spacecraft Guidance and Control under Failure Conditions," Technical Report 93-1-09.03-5355, Scientific Systems Company Inc., July 1994.

Mehra, R.K., S. Seereeram, D. Bayard, and F. Hadaegh (1995) "Adaptive Kalman Filtering, Failure Detection and Identification for Spacecraft Attitude Estimation," in *IEEE Conference on Control Applications*, Albany, NY, pp 176–181, September 1995.

Rauch, H.E. (1994) "Intelligent Fault Diagnosis and Control Reconfiguration," *IEEE Control Systems*, pp 6–12, June, 1994.

Willsky, A.S. (1976) "A Survey of Design Methods for Failure Detection," *Automatica*, pp 601–611, November, 1976.

TECHNOLOGY INITIATIVES FOR THE AUTONOMOUS GUIDANCE, NAVIGATION, AND CONTROL OF SINGLE AND MULTIPLE SATELLITES

John Croft, John Deily, Kathy Hartman, and David Weidow
NASA Goddard Space Flight Center
Greenbelt, Maryland 20771
(301) 286-3239, 286-2934, 286-5696, and 286-5711

Abstract

In the twenty-first century, NASA envisions frequent low-cost missions to explore the solar system, observe the universe, and study our planet. To realize NASA's goal, the Guidance, Navigation, and Control Center (GNCC) at the Goddard Space Flight Center sponsors technology programs that enhance spacecraft performance, streamline processes and ultimately enable cheaper science. Our technology programs encompass control system architectures, sensor and actuator components, electronic systems, design and development of algorithms, embedded systems and space vehicle autonomy. Through collaboration with government, universities, non-profit organizations, and industry, the GNCC incrementally develops key technologies that conquer NASA's challenges. This paper presents an overview of several innovative technology initiatives for the autonomous guidance, navigation, and control (GN&C) of satellites.

INTRODUCTION

Enterprises created by NASA's Strategic Plan include Centers of excellence for Space Science, Mission to Planet Earth, Human Exploration and Development of Space, Aeronautics, and Space Technology. Each Enterprise has a roadmap, or activities plan, which demonstrates how they accomplish NASA's mission. Common to all the challenging Enterprise roadmaps are instrument, spacecraft, and operations technologies. Part of the Space Technology Enterprise, the GNCC, in partnership with the other Enterprises, is tasked with developing and verifying enabling, cutting-edge technologies for future space science, exploration, and commercial missions. The Space Technology Enterprise also tasks the GNCC with identifying and maturing high-risk, high-payoff advanced concepts that enable revolutionary space activities (Goldin 1996). Reducing satellite operating costs while simultaneously enabling new science is a key NASA goal. GNCC technology programs are structured to meet this goal by automating spacecraft and ground systems.

TECHNOLOGY PROGRAM OVERVIEW

Autonomous systems can improve science data collection and greatly reduce the need for continuous satellite monitoring in centralized ground facilities. To reduce satellite monitoring requirements, future "smart" satellites will incorporate autonomous GN&C systems and spacecraft health-monitoring systems such as anomaly detection and fault recovery. Science data collection will be improved by automating onboard data processing with automatic replanning and goal-directed resource management and control. System autonomy is crucial to meeting the challenge of many, low-cost, science missions as budgets decrease.

The GNCC has an extensive history of supporting NASA and non-NASA missions with diverse and challenging GN&C requirements. GNCC technology programs are dedicated to developing key technologies that reduce overall mission support costs and enable science using autonomous spacecraft and ground systems. Technology needs are determined by reviewing Enterprise roadmaps and extrapolating for specific GN&C technologies. GNCC technologies include the use of the Global Positioning System (GPS) for spacecraft navigation, time and attitude as well as for autonomous on-orbit control of formation-flying satellites. Other technology programs focus on automating ground systems, developing image-aided attitude sensors and autonomous star trackers.

Use of the Global Positioning System

The GNCC is utilizing the Department of Defense (DOD) NAVSTAR Global Positioning System (GPS) for the autonomous guidance, navigation, and control of satellites. GPS can be used for accurate, onboard real-time orbit, and attitude determination, and time distribution (Bauer 1995). The use of GPS will nearly eliminate the need for and attitude determination, and time distribution (Bauer 1995). The use of GPS will nearly eliminate the need for dedicated operations personnel who perform routine orbit and attitude determination and spacecraft time maintenance. A reduction in flight hardware on a spacecraft occurs when GPS is used for attitude and timing, e.g., sensors and oscillators. A complete elimination of tracking support for navigation is an advantage. GPS offers a similar cost reduction in overall spacecraft, and systems development. Combining the use of GPS with integrated inertial navigation systems results in improved performance over the individual systems, and offers a more robust navigation service. GPS enables new mission concepts like formation flying by providing a radio frequency (RF) communications link for relative navigation control and data transfer between satellites. GPS also enables an array of potential options for enhanced spacecraft autonomy; formation flying/coordinated platforms; closed-loop attitude navigation and control; constellation control of multiple satellites; rendezvous and proximity operations (Figure 1).

 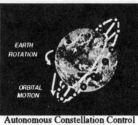  

Closed-Loop Navigation   Autonomous Constellation Control   Autonomous Formation Flying   GPS-Based Navigation, Attitude, Time

Figure 1. GPS-Based Mission Concepts.

GPS is becoming more attractive to NASA satellite missions now that the system is fully operational and with the promise that continued civilian use is guaranteed by the DOD. The establishment of a second civilian frequency will improve accuracy and system reliability for users. The NASA Mission to Planet Earth (MTPE) Enterprise is using GPS for many future missions and is planning formation flying of multiple satellites for concurrent science data. Other NASA science Enterprises are planning the use of multiple autonomous satellites.

Receiver Development
- Orbit, Time & Attitude
- New Algorithms
- Robust Hardware

Testing
- Receiver Performance
- Subsystem Validation
- Spacecraft End-to-End Tests

Flight Experiments
- Validation of New Algorithms, New Hardware, & New Techniques

Figure 2. GSFC GNC GPS Areas of Expertise.

Despite the increased interest in GPS for satellite operations, the technology maturity level, and receiver availability are far from adequate. For these reasons, the GNCC has been working many years to advance the state-of-the-art for command and control of satellites using GPS, and to promote rapid development and deployment of the technology to enable future missions.

The GNCC collaborates with industry and other government agencies in advancing GPS technology to enable future missions. Areas of expertise include algorithm and receiver development; end-to-end testing using a world-class GPS facility at the Goddard Space Flight Center; and conducting flight experiments (Figure 2).

The GNCC has conducted many GPS experiments, and has new technology initiatives underway. Recent experiments in 1996 include the GPS Attitude Determination and Control System (GADACS) on the Shuttle/Spartan flight; GPS attitude software on the REX-II satellite which performed closed-loop attitude control; and attitude software on the Shuttle/GANE flight. Planned for 1997 are the GADFLY experiment on the SSTI/Lewis satellite for attitude, navigation, and timing; the AMSAT highly elliptical orbit which will operate above the GPS constellation; and attitude determination software on the SSTI/Clark satellite. Other work includes developing a low-cost GPS receiver for the Shuttle/Spartan-lite satellite series; developing an integrated inertial navigation systems with GPS (SIGI) for a 1998 Shuttle flight; an antenna pointing system using GPS for the Shuttle Spacehab payload. Another critical piece of work includes helping industry test their GPS receivers in the GNCC test facility prior to launch. More information on GNCC GPS activities can be found at the at the following internet addresses: http://fdd.gsfc.nasa.gov and http://www710.gsfc.nasa.gov.

Future GPS work includes improving algorithms and miniaturizing hardware; enhancing receiver robustness; adding transmit/receive capability for relative navigation/formation flying applications; and including Wide Area Augmentation System compatibility.

Formation Flying and Autonomous Control Architecture Environment

NASA Enterprises are recognizing the advantages of flying multiple satellites in coordinated virtual platforms and constellations to accomplish science objectives. The GNCC developed a ground-based system to autonomously plan orbit maneuvers while minimizing analyst intervention. Much of the analyst's time is spent optimizing the maneuver while considering operational constraints: shadow coverage, operations staffing, maneuver location, station coverage. The autonomous control system (AutoCon-G) uses fuzzy logic to resolve multiple conflicting constraints encountered when planning orbit maneuvers. AutoCon-G also employs natural language scripting to represent control algorithms allowing updates without complete software changes. An operational version of AutoCon-G was delivered to the Mission to Planet Earth (MTPE) Earth Observing System (EOS)-AM1 Project. AutoCon-G will be used for automated mission design and maneuver planning to control the evolution of the EOS-AM1 groundtrack. This system will save approximately fifty percent of a mission analyst's time per year in routine operations.

Concurrently the basic AutoCon ground system has been modified for formation flying concepts on the New Millennium Program (NMP) Earth Orbiter (EO)-1 mission (Bauer 1997). A flight version, AutoCon-F, is under development and is the integrating onboard architecture for the Goddard closed-loop orbit control algorithm and other algorithms from partners including the Jet Propulsion Laboratory, and Phillips Laboratory. The first build of the EO-1 formation flying flight system will be delivered in January 1998, and validated after launch, scheduled for May 1999. All formation flying algorithms will be validated during the second year of operations.

The AutoCon system and planned enhanced formation flying experiments on the NMP EO-1 mission are key technologies which lower total mission risk associated with autonomous orbit control systems, reduce operations cost, and improve science data collection for future NASA missions. Enhanced formation flying and space vehicle autonomy will revolutionize space and Earth science missions and enable many small, inexpensive spacecraft to fly in formation and gather concurrent science data. The AutoCon system enables increased onboard autonomy, automated constellation control, and "lights-out" operations. The general formation flying algorithms developed are applicable to numerous mission types including deep space interferometry, constellations, and virtual platforms.

Automation of Ground Based Routine Flight Dynamics Functions

Recognizing that spacecraft autonomy will not always offer a viable solution to cost reduction, alternative means of reducing costs associated with ground based flight dynamics product generation functions are being researched. A joint team of government, academic, and commercial entities have come together under the Flight Dynamics Automation Studies Program with the goal of eliminating the costs associated with ground-based routine flight dynamics functions. The overall technical approach of the Automation Studies Program highlights 4 basic technology focus areas being addressed to attain this goal. These focus areas consist of (1) the application of existing technologies (e.g., expert systems), (2) the application of state of the art technologies (e.g., data mining, genetic algorithms), (3) the development of fundamentally new approaches to flight dynamics product generation (abandoning long supported but inherently labor intensive approaches), and (4) the development of flight dynamics "smart" tools which promote, facilitate, and contribute to non-human operations. In support of the program, the Automation Studies Testbed has been established as an extension to the University of Maryland Flight Dynamics and Control Laboratory (FDCL). The testbed provides facilities which support the development, validation, and demonstration of new techniques and technologies. The FDCL consists of an operational component (currently supporting SAMPEX flight dynamics operations) and a testbed component which allows parallel operations to be conducted. This facility allows new techniques and technologies to be demonstrated and validated against operational benchmarks to obtain a quantitative understanding of the impact on operations costs associated with routine flight dynamics operations. Several initiatives are now underway through the Automation Studies Program within each of the technology focus areas identified above.

Bowie State University, in collaboration with the Flight Dynamics Division and Computer Sciences Corporation, is developing an expert system framework to perform flight dynamics product generation functions. Supported by a rulebase developed by Flight Dynamics Division engineers, the framework is capable of automatically performing tracking data processing, orbit determination, and planning product generation functions. Input data, intermediate products, and final products are evaluated for correctness by the framework and corrective action is taken when necessary to address anomalies identified in the performance of these functions. Because of the flexibility inherent within an expert system approach, the framework, through it's accompanying rulebase, can be as sophisticated or as simple as required. In either case, when a scenario is encountered which the framework is unable to resolve on its own, the responsible flight dynamics engineer is notified (via phone call and/or email) of the situation and asked to provide assistance. The system is currently being exercised within the FDCL Automation Studies Testbed and is expected to be deployed for use by the TRACE, Landsat-7, and EOS-AM1 spacecraft during the next year.

Towson State University and Computer Sciences Corporation are conducting fundamental research in the application of the general method of data mining and specific techniques of neural networks for pattern recognition and genetic algorithms for data correction in an attempt to improve the ability of an autonomous system to monitor and control its performance. Currently, these techniques are being applied to tracking data evaluation and reporting functions in support of the development of an early warning system. This system will be capable of notifying flight dynamics engineers of possible anomalies within tracking data prior to the use of that data in the orbit determination process. In addition, the ability of these techniques to support functions such as tracking data filtering and enhanced quality assurance functions to evaluate the relationships between tracking data quality and the resulting orbit determination solution quality are being investigated for use.

AI Solutions, Inc., in collaboration with the Flight Dynamics Division and the University of Maryland are conducting research related to the adaptation of neural network controllers to Kalman filters in an attempt to create a self-tuning Kalman filter. The self-tuning Kalman filter will form the basis for an autonomous orbit determination server process expected to execute continuously within the flight dynamics product generation system. This system will process tracking data and/or propagate orbit solutions to maintain up-to-date and accurate orbital state information for the spacecraft. As such, the orbit determination function is performed continuously within the system. Client processes, such as a flight dynamics planning product generation function, would query the server for orbital state information at the current time or some requested time as required. Such an approach constitutes a significant deviation from traditional approaches currently in use by the Flight Dynamics Division.

IA/GNC - Image-Aided Guidance, Navigation and Control (*Real-time GN&C using a 2-d image*)

IA/GNC promises to be a low cost supplement to attitude control systems for coarse and fine pointing Earth-referenced satellites. Using two-dimensional images from the spacecraft's science instrument or a secondary low-cost camera, IA/GNC systems generate attitude control and image stabilization information that enhances pointing performance and ultimately permits autonomous onboard geo-referencing and geometric rendering of data.

GN&C packages for Earth referenced missions have classically been designed to meet coarse pointing requirements with horizon sensors and gyro units, and fine pointing requirements with star trackers and gyro units. Course systems which navigate by looking at the Earth often have marginal performance because of horizon variations caused by the atmosphere. Costly fine pointing systems are stellar based and require precise knowledge of orbital position. In either case, spacecraft attitude is computed on-board and Earth-referenced instrument pointing is deduced after the fact using orbital position knowledge which locates the Earth relative to the stars.

For Earth resource missions like NASA's Landsat 7, scientists must post-process instrument data to render geometric corrections (correcting the shape of the pixels) and geo-registration (tagging the pixels in longitude and latitude.) IA/GNC promises Earth-referenced attitude determination and control in the fine pointing category with real-time, on-board geometric correction and geo-registration of the data, alleviating the need for time and labor intensive post processing.

IA/GNC relies heavily on image correlation tracking (ICT) techniques which compute the translational offset between an instantaneous image and a stored reference image derived from a previous image or a model. For an Earth pointing instrument, the offset between an instantaneous image and a reference image is determined by cross-correlating between the images to produce a correlation surface. Sub-pixel interpolation of this new surface produces a robust estimate of the translational offset between the reference image and the instantaneous image. The offsets derived from ICT can be recorded and used in the data geometric correction and geo-registration process, or used to close the loop in a servo system to point or stabilize an instrument or spacecraft.

The programs contained within this initiative explore the various uses of IA/GNC techniques for airborne and spaceborne scientific platforms. Our long-range goal is to develop the knowledge, expertise, and technological means (software, hardware, etc.) required to apply this technology to a wide range of aerospace applications.

Three programs highlight our dash into IA/GNC: The first is a collaboration with the Johns Hopkins Applied Physics Lab which takes advantage of the Ballistic Missile Defense Organization's Midcourse Space Experiment (MSX) to test Earth referenced remote sensing ICT algorithms. The second is an internal GSFC collaboration to develop and implement an ICT-based geo-registration and instrument pointing system on a remotely piloted vehicle (RPV), the third is a collaboration with the Army to develop an image-aided yaw correction algorithm for NASA's EO-1 mission.

Future programs include the development of an in-house IA/GNC lab where algorithms, software builds, and hardware components will be developed, tested, and perfected. The lab will also spawn an ICT software kernel that will be implemented wherever needed, and eventually, a stand-alone ICT hardware component for applications where usable instrument data is unavailable or nonexistent.

Precision Guidance Initiative - Advanced Autonomous Star Sensors

State-of-the-art commercial star sensors do not meet the needs of many NASA "next generation" spacecraft and instruments. Precision guidance missions such Next Generation Space Telescope will require sub-arc-second star sensors that are a factor of ten lower in weight and power than commercially available units. On the average, commercial units weigh 15 pounds, attain five to ten arc-second accuracy, and use roughly ten watts of power. Our objective is to employ advances in microelectronics and processors, detectors, and algorithms to allow star sensor designers to develop a one pound, two watt, sub-arc-second star sensor which autonomously identifies the star field in a 'stars in, quaternion out' stellar compass mode.

Present day star sensors are designed to image a portion of the sky and report the location of any stars that are tracked. Among other factors, the accuracy of these devices is historically limited by three constants:
- An eight degree square field of view
- The capability to track 5 stars simultaneously
- A CCD imager with 512 pixel square imaging area.

These choices have led to a limiting accuracy on the order of 5 arc seconds per star per frame. In addition, the burden of attitude initialization, estimation, star catalog development and manipulation, and attitude filtering are typically performed by the spacecraft's processor or the ground.

The GNCC has a multiphase, multi-year program to inject revolutionary technologies into current star sensor designs and to enhance the capabilities of existing sensors to the levels required by the next generation of spacecraft and instruments. The program's four phases were developed to provide both a roadmap and incremental measurement tool for star sensor enhancement. NASA designs and financial resources are being merged with industry to provide partnerships to attain our goal.

Phase one, a star sensor enhancement study by multiple vendors of commercial star sensors, is near completion. Four star sensor vendors were asked to examine their commercial products and processes to identify improvements that would allow them to meet these performance goals:

Pointing Accuracy	0.20 arc-sec	(1 sigma)
NEA	0.10 arc-sec	(1 sigma)
Update Rate	40 Hz	
Power	2 W	
Weight	0.5 kg	

In the second phase, the GNCC will identify aspects of the reports that can readily be explored with industry partners to produce tools needed to meet the robust goals targeted in phase one. These tools may come in the form of hardware, software, or even further studies. Once identified and categorized, specific areas of exploration will be determined. In phase three, the GNCC will seek partnerships with industry experts to share resources and capital to allow the exploration of enhanced star sensor capabilities. Throughout the program, incremental enhancements will be incorporated and tested to provide a measurement tool to verify adherence to the overall goals and to enable the benefits from this plan to be utilized prior to the final product development. In phase four, these enhanced products will become available to the aerospace community.

CONCLUSIONS

NASA's success will depend on its ability to successfully integrate cutting-edge technologies into satellite programs to meet the space challenges of the twenty-first century. The use of multiple, small, autonomous satellites, some in formations and constellations, is repeated throughout several of the NASA Enterprise roadmaps. The GNCC is an organization with a coordinated, well-planned program dedicated to advancing key technologies critical to meeting NASA mission challenges. The top priority of the GNCC is developing technologies supporting autonomous GN&C systems, and "lights-out" operation of satellites. This paper highlighted only a few key autonomous guidance, navigation and control technologies, and program elements, which enable Earth and space science while reducing spacecraft systems and development time, and operations costs.

Acknowledgments

The authors wish to recognize the dedication and hard work of all the GNCC technologists who make the overall technology program a great success at the Goddard Space Flight Center. A special thanks to Steve Hammers and Albin Hawkins of the Hammers Company, Darrel Conway and Robert Sperling of AI Solutions, David Folta, John Bristow, and Dave Quinn of the GSFC, for their work on the enhanced formation flying technology on the NMP EO-1 mission. Glenn Lightsey, Roger Hart, and Eleanor Ketchum are mentioned for their outstanding work with GPS, and Martin Houghton for his contributions to IA/GNC, and Robert Spagnuolo for his work with advanced star trackers. And to the members of the GNCC management team, and especially Frank Bauer, Head of the GNCC, for their guidance, and continued promotion of the entire technology program.

References

Bauer, F. H., et al (1997) "Satellite Formation Flying Using an Innovative Autonomous Control System (AUTOCON) Environment," *1997 AIAA Guidance, Navigation, and Control Collection of Technical Papers,* GNC, AFM, and MST Conference and Exhibit, New Orleans, Louisiana, Part 2, 657-666.

Bauer, F. H., et al (1995) "The GPS Attitude Determination Flyer (GADFLY): A Space-Qualified GPS Attitude Receiver on the SSTI Lewis Spacecraft," *ION GPS-95 Proceedings,* ION GPS-94 Conf., Salt Lake City, Utah

Goldin, D. S., et al (1996) *NASA Strategic Plan,* National Aeronautics and Space Administration, Washington, DC, 18-19.

NONLINEAR PREDICTIVE CONTROL FOR SPACECRAFT TRAJECTORY GUIDANCE, NAVIGATION AND CONTROL

R.K. Mehra, S. Seereeram	John T. Wen	David S. Bayard
Scientific Systems Company	Rensselaer Polytechnic Inst.	Jet Propulsion Laboratory
500 West Cummings Park, Suite 3000	110 8th Street, CII 8213	4800 Oak Grove Dr. 198-326
Woburn, MA 01801	Troy, NY 12180	Pasadena, CA 91109
(617) 933-5355	(518) 276-8744	818 354-8208

Abstract

A class of iterative methods has recently been proposed for the path planning for a variety of fully and under-actuated mechanical systems, including robots and spacecraft. These methods all involve the basic idea of warping an initial path iteratively to an acceptable final path by using a Newton-type algorithm. Once a path is found off-line, a feedback controller can then be used to follow the path. A modification of the off-line methods transforms them directly into a nonlinear predictive feedback controller. with guaranteed closed-loop asymptotic stability when the system model is known, and certain robustness when the model information is imperfect. This method represents a special class of model predictive control (MPC), since the control action at each time instance is determined based on the future predicted trajectory. Preliminary results are presented illustrating the application of this nonlinear controller to three-axis attitude control of fully-actuated and under-actuated spacecraft.

INTRODUCTION

This paper presents a new methodology for Model Predictive Control (MPC) based on a function-space optimization of control inputs. A class of iterative methods has recently been proposed (Sontag and Lin 1992, Divelbiss et al 1993) for planning and control of a variety of fully and under-actuated mechanical systems. The key idea of this approach is to transform the problem into a root finding problem which can be solved using the Newton method. Algorithmically, an initial path is iteratively modified to an acceptable final path by using a best-step Newton update. This method converges to a feasible path provided that a singular path (a path about which the linearized system is uncontrollable) is not encountered during the iterations. The appeal of this method lies in its generality and ability to include inequality constraints (e.g., saturation limits of actuators).

Once an initial path is found off-line, a feedback controller can then be used to follow the path. In Lizarralde and Wen (1995), a modification of these off-line methods was introduced to transform them directly into a feedback controller using a receding horizon concept. The main idea is to couple the iteration variable to the actual time, thus the control is executed during the path iteration, <u>before</u> the convergence. This scheme can guarantee the closed loop asymptotic stability when the system model is known, and possesses certain robustness when the model information is imperfect. Furthermore, by using interior penalty functions, inequality constraints can be incorporated into the algorithm. This type of feedback control, termed Function-space Model Predictive Control (F-MPC), is a special class of model predictive control, since the control action at each time instance is determined based on the future predicted trajectory. However, in contrast with the standard MPC schemes where an optimization problem needs to be solved at each control time interval, only a small number (possibly *one*) of Newton steps are required, with a fixed amount of computation.

This technique was first developed and successfully applied in the context of off-line motion planning for robotic manipulators and mobile robots (Divelbiss et al 1993). Further development of this technique resulted in its application to on-line planning and feedback control for motion planning. This technique has been developed and successfully implemented for kinematically redundant manipulators and automated guidance of wheeled vehicles. Further study investigated and developed theoretical convergence and robustness aspects of this approach (Lizarralde and Wen 1995 and Lizarralde et al 1997).

For spacecraft (S/C) attitude dynamics, the F-MPC method is eminently suited for predictive control. Using model information to predict and adjust the control input, in the presence of constraints, fits naturally into the framework of the function-space optimization technique. Aspects such as representation, planning and control on the orientation manifold ($SO(3)$) were handled successfully for robotic planning, and are an integral part of S/C attitude control.

In the following sections, we describe the F-MPC strategy for autonomous S/C attitude control. Several scenarios were considered, representing assumptions on the models used for modeling the dynamics and control input. The basic ideas of the F-MPC method for solving this class of nonlinear control problems are described, together with. the S/C dynamic models and simplifying assumptions used in this study. Selected preliminary results are given. Some conclusions and directions for future work are indicated in the last section.

FUNCTION-SPACE MODEL PREDICTIVE CONTROL

Consider a time-invariant nonlinear system, affine in the control:

$$\dot{x}(t) = f(x) + g(x)u(t), \tag{1}$$

where f and g are smooth, $x \in \mathbf{R}^n$ and $u \in \mathbf{R}^m$. Full state measurement and global controllability are assumed. The precision pointing task can be stated as follows. Given Equation (1), find an input control history over the prediction horizon T, $\underline{u} \in \mathcal{U}$ where $\underline{u} \triangleq (u(t), t \in [0,T))$, such that $x(T) = x_d$, which minimizes $J(\underline{x}, \underline{u}, T)$, subject to applicable constraints on $u(t)$ and $x(t)$. For a slewing task, we desire $\underline{u} \in \mathcal{U}$ such that $\underline{x} \triangleq (x(t), t \in [0,T]) = \underline{x}_d$. In other words, the entire trajectory (or part thereof) is specified.

$J(\cdot)$ corresponds to a weighted combination of control effort (minimum fuel), or execution time (minimum time). F-MPC optimizes the state trajectory by choosing the input history $\underline{u}$ over a pre-defined prediction horizon T. If T is the entire task duration, then $\underline{u}$ represents an open-loop optimal control. MPC applies this strategy over sub-intervals of the complete task time, accounting for differences in actual vs. predicted performance by recomputing the segmented control histories at set intervals as the task evolves.

Main Idea

For the system given in Equation (1), with initial configuration $x(0) = x_0$, desired final configuration x_d, and applicable control set $\mathcal{U}$, write the end-point map

$$x(T) = \int_0^T (f(x) + g(x)u)\,dt + x(0) \triangleq F(u, x_0, T). \tag{2}$$

Define an error term (precision pointing task) as follows

$$e(\underline{u}) \triangleq x(T) - x_d. \tag{3}$$

We desire a control $\underline{u}$ which will result in zero error, i.e. $e(u(\tau)) \to 0$ as a function of an iteration variable $\tau \to \infty$. Global controllability means that for any x_0 and x_d, there is at least one solution to Equation (3). In other words, the control problem has been converted into a nonlinear root-finding problem, i.e. solve

$$e(\underline{u}, x_0, T) = 0. \tag{4}$$

By differentiating the endpoint error map (Equation (2)), it follows that:

$$\frac{de}{d\tau}(\tau) = \nabla_{\underline{u}} F(\underline{u}) \frac{du}{d\tau}(\tau) \triangleq D(\underline{u}) \frac{du}{d\tau}, \tag{5}$$

where $D(\underline{u}) \triangleq \nabla_{\underline{u}} F(\underline{u})$ is the gradient (Fréchet derivative) of the endpoint map (Equation (2)). Equation (5) suggests the following procedure: choose $\underline{u}$ to satisfy

$$\frac{du}{d\tau} = -\alpha D^{\dagger}(\underline{u}) e(\underline{u}(\tau), x_0, T), \tag{6}$$

where $D^\dagger(\underline{u})$ denotes the Moore-Penrose pseudo-inverse of $D(\underline{u})$, and $\alpha > 0$. For full-rank $D(\underline{u})$, the error term follows an exponential decay:

$$\frac{de}{d\tau}(\tau) = -\alpha e(\tau),\tag{7}$$

with a prescribed decay rate α.

Equation (6) can be implemented in iterative form using a discretized basis for the control function space $\mathcal{U}$ according to the following update law:

$$\underline{u}_{k+1} = \underline{u}_k - \lambda D^\dagger(\underline{u}_k)e(\underline{u}_k).\tag{8}$$

By combining this update with a line search procedure to find λ a best-step Newton-type algorithm is obtained. The gradient $D(\underline{u}(k))$ can be computed from the system Equation (1) linearized about the path corresponding to $\underline{u}(k)$.

A sufficient condition for convergence of $e(\tau)$ to zero as $\tau \to \infty$ is that $D(\underline{u})$ is full-rank for all τ. This is equivalent to the control input history $\underline{u}$ being nonsingular at every iteration. In practice, the possibility of encountering singular controls is "generically rare" and in the unlikely case that one is encountered, a generic loop can be appended to render $D(\underline{u}(k))$ non-singular. Another technique (Wen and Popa 1995) is to augment the update law using a basis which spans the null space of D (i.e. $D\tilde{D} = 0$):

$$\underline{u}_{k+1} = \underline{u}_k - \lambda D^\dagger(\underline{u}_k)e(\underline{u}_k) + \tilde{D}(\underline{u}_k)\beta.\tag{9}$$

We can exploit the freedom in choosing β to avoid singularities, an ensure norm bounded-ness of the control. A more detailed discussion of singularity considerations, and classes of problems for which the non-singularity of $D(\underline{u})$ is guaranteed can be found in Wen and Popa (1995).

SPACECRAFT ATTITUDE CONTROL PROBLEM

To investigate the F-MPC technique for spacecraft attitude control, we selected tasks representative of future spacecraft (earth-orbiting as well as inter-planetary) missions. A recognized mission-critical control problem is that of precision pointing and attitude stabilization for spacecraft (S/C). Precision pointing is used in many tasks, including mosaic'ing for survey of a planetary surface, orientation of medium and high-gain antennae for uplink/downlink communication, and precision trajectory guidance and control during maneuvers such as orbit insertion, orbit raising/repositioning and interplanetary cruise guidance. These represent challenging attitude control problems for the current open and closed-loop techniques.

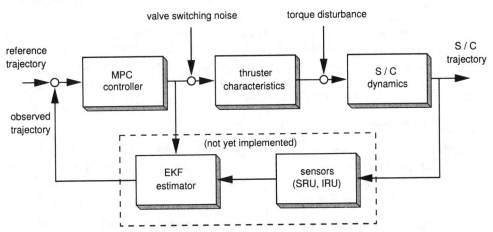

FIGURE 1. Schematic of F-MPC Implementation for Spacecraft Control.

Preliminary studies (Kissel 1995) for the Pluto mission precision pointing task concluded that specifications could not be met by traditional closed-loop set-point control. Specifically, the proposed gyros (IRU) were of too low quality to be used for tight closed-loop control during imaging tasks. Low update rate of the star camera (SRU) suggests a predictive control scheme for maintaining a desired attitude trajectory. Adaptive estimation and nonlinear control schemes were suggested as a means of achieving better closed-loop performance. The key advantage of predictive control is the closed-loop control of the trajectory along a pre-determined attitude profile, with control optimized for precise tracking. Incorporation of an Extended Kalman Filter (EKF) for fusion of sensor measurements from various sensors, at multiple time rates, is needed in order to use the maximum information available from a given set of hardware. Optimization of the control input with respect to the overall trajectory is critical to meeting fuel budgets on extended missions.

Attitude Dynamics

For this study, we use a standard rigid-body model for attitude dynamics. For simplicity, take the principal axes as the coordinate axes of the S/C's reference frame. In this case the inertia tensor reduces to its diagonal form, and we have:

$$J\frac{d\omega}{dt} = \tau - \omega \times J\omega , \qquad (10)$$

where τ is the total applied torque acting on the S/C – control, disturbance and environmental. (This model does not include momentum-conserving forces, such as the effect of reaction wheels.) Assign the angular velocity ω and orientation (unit quaternion) q as the state vector. Actuator or thruster torques are computed using $\tau_{\text{ctrl}} = \sum_{i=1}^{n} r_i \times F_i$ where F_i is the thrust produced by the i'th thruster located at a displacement of r_i from the S/C reference frame origin.

For simulation and control purposes, the Euler symmetric parameterization (unit quaternion) is widely used in aerospace studies. This has desirable properties in terms of representation singularities, computational complexity and propagation equations. Orientation kinematics are concisely represented by the following model:

$$\frac{dq}{dt} = \frac{1}{2}\Omega(\omega)q ; \quad \Omega = \begin{bmatrix} 0 & \omega_z & -\omega_y & \omega_x \\ -\omega_z & 0 & \omega_x & \omega_y \\ \omega_y & -\omega_x & 0 & \omega_z \\ -\omega_x & -\omega_y & -\omega_z & 0 \end{bmatrix}. \qquad (11)$$

Control Objective

The overall S/C trajectory is nominally specified in terms of an inertial frame (Frame 0). Let $^0A_S(t)$ represent the true trajectory (attitude) of the S/C in Frame 0, and let $^0A_d(t)$ be the desired trajectory (attitude) of the S/C in the inertial frame. The general control objective can then be stated as follows: Given $^0A_d(t)$ and $\omega_d(t)$, design a control strategy, using observed values for S/C orientation and rate, which yields $^0A_S(t) \to {}^0A_d(t)$ and $\omega(t) \to \omega_d(t)$ as $t \to \infty$. Practically, a nominal performance specification in terms of trajectory tracking accuracy is also to be met, i.e. the trajectory tracking error $\|(^0_\bullet A_S(t) - {}^0A_d(t))\|$ should fall below a given accuracy ϵ within a finite time.

Thruster (Actuator) Models

In this study, micro-thruster actuators are modeled after gas-jet systems (eg. cold-gas or Hydrazine units) commonly used for fine pointing control. By properly positioning jet couples, it is possible to produce symmetric rotational torques centered about the S/C's principal axes. We assume that offset forces have been transformed to principal axis torques as necessary.

Thruster actuation from a controller standpoint can be modeled as a pulse-width modulated signal. Control signals are mapped into valve on/off commands by varying the duty cycle and switching frequency. Thruster rise and fall times, valve command delays, and switching constraints imposed by the actuator mechanism are incorporated into the thruster models in order to achieve optimal predictive control performance.

Un-modeled Effects

A number of assumptions are made in terms of un-modeled effects: S/C flexible structure analysis; propellant slosh; spin-rate disturbance due to mass relocation aboard S/C (including reaction-wheel and other momentum-conserving forces); mass/inertia adjustments due to propellant usage; and explicit electromagnetic disturbances due to nearby planetary masses. Due to their complexity, these effects are deferred for later incorporation into the model as warranted. It should be noted, however, that for future miniature S/C missions (e.g. image acquisition pointing/slewing for Pluto Fast Fly-by), many of the above effects may be second order, higher, or even non-existent for the control tasks of interest.

SAMPLE RESULTS

Scientific System's F-MPC algorithms have been tested successfully on precision pointing tasks using gas-jet thrusters. Preliminary results indicate robustness to imperfect S/C model information as well as good disturbance rejection of on-board and external force/torque disturbances. Figure 2 (a) shows the effective thrust computed to execute a 90 degree sprint turn efficiently and smoothly to the commanded attitude. The almost linear rotation (Figure 2(b)) with virtually no extraneous motion is indicative of near-optimality of the trajectory. The regulation error (Figure 2(c)) can be made arbitrarily small by choice of termination criterion, subject to the minimum impulse limitations of the actuation hardware.

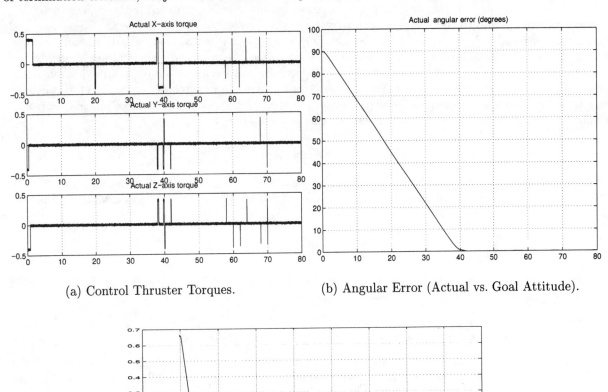

(a) Control Thruster Torques.   (b) Angular Error (Actual vs. Goal Attitude).

(c) Regulation Angular Error (deg).

FIGURE 2. 90 deg. Sprint Turn using RCS Thrusters.

One key advantage of our F-MPC feedback stabilization algorithm is that it can handle certain well-known difficult cases that arise in S/C trajectory control. For example, if one actuator fails, conventional feedback algorithms can no longer position a spacecraft to arbitrary orientation while our *unmodified* F-MPC algorithm remains viable. Simulations have demonstrated re-orientation tasks in the case of under-actuation (such as two-axis torquers). Other applicable scenarios include the use of reaction wheels or control moment gyros as torque actuators. Reaction wheel momentum management and gyro stops present unique control problems. Rather than the current method of using RCS thrusters to maintain attitude, dump momentum or re-center the gyro, the overall problem can be addressed by our F-MPC algorithm *without* the additional thruster control.

CONCLUSIONS AND ANTICIPATED BENEFITS

This paper presented a new formulation of nonlinear predictive control based on a function-space optimization of the control input. A feedback form has been developed, and applied successfully to a variety of mechanical systems. This paper indicates the nature of results of its application to the spacecraft attitude control problem. Ongoing work focuses on developing robustness and convergence criteria for the F-MPC algorithm, and quantifying its benefits for closed-loop predictive control for spacecraft trajectory optimization and control. Anticipated benefits of nonlinear MPC include: robust/fault-tolerant controllers for global trajectory and attitude control of spacecraft; optimal use of fuel/energy resulting in economies in power/storage requirements; comprehensive or centralized control architectures for a variety of spacecraft systems.

Preliminary problems will address spacecraft actuation systems such as gimbaled main engines (delta-V), reaction control micro-thrusters (RCS), and/or reaction wheel assemblies (RWA). Specific mission-critical operations, such as orbit insertion, precision pointing and stabilization, spin-stabilization, sun or earth acquisition, prediction adjustment and trajectory optimization are candidate applications of predictive control techniques. Other technologies such as Solar Electric Propulsion are expected to provide challenging problems in Thrust Vector Control and Autonomous Management. Commercial interests which will benefit from this technology include remote-sensing, global communications, and multi-variable process control in chemical, automotive, power generation, manufacturing and metallurgy industries.

Acknowledgment

This work was performed under a NASA Small Business Innovative Research award (Contract NAS8-40718). The authors would like to thank Fred Hadeagh and Glen Kissel at JPL for helpful technical discussions on spacecraft control requirements.

References

Kissel, G.J. (1995) "Precision Pointing for the Pluto Mission Spacecraft," in *Proceedings of the 18th Annual AAS Guidance and Control Conference*, February 1995.

Lizarralde, F. and J.T. Wen (1995) "Attitude Control Without Angular Velocity Measurement: A Passivity Approach," in *IEEE Trans. Aut. Contr.*, (accepted for publication) 1995.

Lizarralde, F., J.T. Wen, and L. Hsu (1997) "Control of an Underactuated Mechanical System using Path Space Iteration," in *Proceedings of the 1997 American Control Conference*, Alburqueque, New Mexico, 1997.

Popa, D. and J.T. Wen (1995) "Identification of Singularity Points for Non-Holonomic Motion Planning Algorithms," Submitted to the 1996 IFAC Conference, June 1996.

Seereeram, S., A. Divelbiss and J.T. Wen (1993) "A Global Approach to Kinematic Path Planning for Robots with Holonomic and Non-Holonomic Constraints," in *Proceedings of the 1993 IMA Workshop on Robotics*, Minneapolis, MN, January 1993.

Sontag, E.D. and Y. Lin (1992) "Gradient Techniques for Systems with No Drift," in *Proceedings of 1992 Conference on Signals and Systems*, 1992.

DESIGN OF AN INTEGRATED SATELLITE (INT-SAT) USING ADVANCED SEMICONDUCTOR TECHNOLOGY

Abhay M. Joshi
Discovery Semiconductors, Inc.
186 Princeton-Hightstown Road
Bldg. 3A, Box 1
Cranbury, NJ 08512
Phone: (609) 275-0011

Abstract

Microsatellites are fast becoming important scientific and commercial realities. However, most satellites that fall in this class are still fairly large (~50 Kg, ~0.5 m). One of the major obstacles in reducing these parameters further is the lack of integration of all the satellite's functions. Normally, most satellites are constructed from physically separate sub-systems, each of which is composed of some combination of circuit boards and components. This partitioned approach wastes valuable space and weight directly by increasing demands on structure and power resources of the satellite (Payne 1992 and Priedhorsky 1989). This is the problem that Discovery Semiconductors wishes to address. We propose to design and construct an entirely integrated satellite (INT-SAT) that contains all the electronics of a fully functional satellite condensed into one module. This module, which would be very compact and light-weight, would be constructed from specially designed 4 to 8 inches silicon wafer size chips. Through the creative use of semiconductor processing and proprietary crystal growth techniques, all the components required for a state of the art satellite can be implemented on the silicon chips. The INT-SAT's condensed electronics module will be compact and light weight, and will occupy only 200 cubic centimeters, and will weigh less than 500 grams. Since the electronics are the heart of almost every mission, this integration will produce drastic savings in weight and size, enabling the reduction of satellite mass to well below 10 Kg.

INTRODUCTION

Discovery Semiconductors proposes to develop the next generation of microsatellites which overwhelmingly use semiconductor technology to reduce the weight of the satellites. Discovery Semiconductors wants to build "light and miniature" microsatellites ranging from $2 million to $20 million that can be easily and cheaply launched (Joshi 1994). Launch costs in the United States are approximately $10,000 per pound of payload. Hence, a satellite weighing under 10 Kg (22 lbs) will cost less than $220,000 to launch.

Currently, Discovery Semiconductors is working in 3 research areas in the field of semiconductors & opto-electronics integrated circuits: *(1) Monolithic InGaAs-on-Silicon, Optically Resonant, Infrared Photodetector Arrays, (2) Monolithic InGaAs-on-Silicon, Optical Interconnects for Massively Parallel Computing, and (3) Development of an InP Based Integrated High-speed Photoreceivers.* These basic technologies lay the foundation for achieving the goal of a completely integrated satellite (INT-SAT).

While the goal of our activities is the total integration of all satellite functions, eventually into a single device (satellite-on-a-chip), we recognize that progress towards this goal will be gradual and have useful intermediate results. Therefore, we propose:

1. As a first step, the creation of an INT-SAT 'platform' containing communications, avionics, power, and processing functions, comprised of a stack of chip modules,
2. The second step would be to develop a spectroscopic / imaging remote sensing instrument-on a-chip-module, and demonstrate integration with the INT-SAT platform.
3. Once the instrument and the bus are constructed, they could be flight tested on the Space Shuttle using the Get Away Special (GAS) or other such program.
4. The final step would be to combine the bus / instrument with the required peripheral components (solar panels, optics, batteries) to create a complete, free-flying INT-SAT.

Applications of INT-SAT Technology

1) Resource Mapping
2) Weather Tracking
3) Mobile Communications
4) Navigation Aids
5) Planetary Probes
6) Incorporation into conventional satellites

Advantages of INT-SATs over Conventional Satellite Designs

1) Reduced Weight, Size, and Power
2) Reduced Launch Costs / More Launch Opportunities
3) Improved Reliability
4) Modular Design
5) Smaller Mission Risk

DESCRIPTION OF THE INT-SAT MODULE

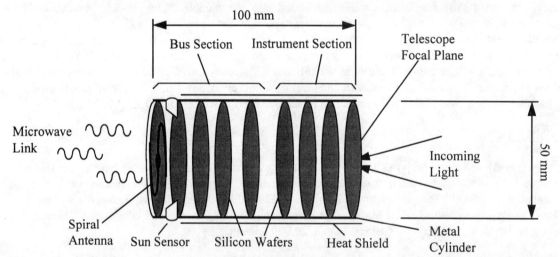

FIGURE 1. The Integrated Instrument - Satellite Bus Architecture for Discovery Semiconductors' Integrated Satellite (INT-SAT).

The primary goal of the INT-SAT is to integrate the various functions of a satellite on a few semiconductor chips (or wafers). This is the genesis of a "satellite-on-a-chip" (Joshi 1994). Our concept for doing this is shown in Fig. 1. There are multiple silicon wafers supported by sapphire mounts. Each wafer could contain any combination of silicon CMOS electronics signal processing circuitry, InGaAs, InP or Gallium Arsenide (GaAs) opto-electronics such as photodetectors and lasers, metallic patterning, and even micro-mechanical devices. The sapphire wafer mounts would be bonded together and mounted inside a light-weight metal cylinder for shielding. The resulting cylindrical structure will be strong (because of its shape), stiff (from the use of advanced materials), yet lightweight and compact.

The multiple wafer design shown in Fig. 1 above is modular in nature. All of the basic satellite platform functionality would be placed on wafers in the bus section of the INT-SAT, which leaves the instrument section free for adaptation to different missions. *The idea is to have an architecture made of a stack of wafers so that each wafer can have certain functionality*. If there are changes to be made to a particular wafer's design, then that wafer can be replaced fairly easily with another. In addition, if higher reliability is needed, parallel wafers having same functionality could be installed to provide back-up capability. *Thus, during a launch, if one wafer is lost, the other one with the same functionality can get the mission done. This is the beauty of Discovery Semiconductors' INT-SAT architecture.* Furthermore, in a semiconductor foundry, usually a batch of 25 to 50 identical wafers is routinely manufactured. Thus, manufacturing multiple wafers of the same functionality does not increase the production costs.

Inter Module Communication on INT-SAT

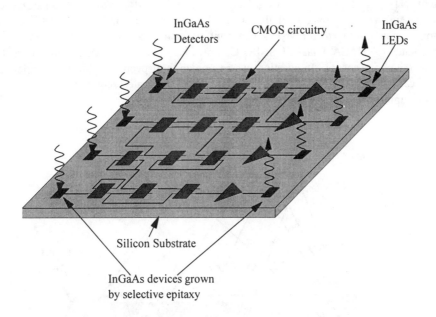

FIGURE 2. Schematic of Discovery Semiconductors' Patented Monolithic InGaAs-on-Silicon LED/Detector Optical Interconnect Integrated Circuit.

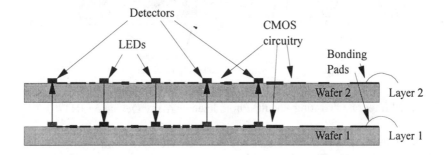

FIGURE 3. Proximity Focused 3-Dimensional Integration Architecture for INT-SAT. The Data Transfer Between 2 Wafers is Done Using Optical Links Between the Detectors and Light Emitting Diodes (LEDs).

Communication between the wafers will be accomplished in two ways: optically and electrically. The optical connections will be made by placing InGaAs Light Emitting Diodes (LEDs) and photodiodes on adjacent wafers. These optical interconnects will be used for digital data transfer between different semiconductor wafers. Discovery Semiconductors has already exhibited InGaAs-on-Silicon Optical Interconnects for USAF Rome Laboratories SBIR. A block diagram of a basic device interconnect, Fig. 2, can be used to illustrate this. Detectors formed from InGaAs on silicon, operating at wavelengths in the 1100 nm to 1600 nm band, serve as inputs to a generic signal processing unit. The use of InGaAs material and wavelength band provides compatibility with fiber optics, good detector performance, and through chip transparency. The circuitry would be implemented using silicon CMOS technology. This allows the processor to be nearly unlimited in size, power, and complexity, providing for great design flexibility. The outputs of the processor could drive LEDs or lasers formed from the same InGaAs on silicon as the detectors. This capability of growing optical sources on the same chip as the processor is the unique feature that enables the entire system to function. *Fig. 3 shows the optical information exchange between 2 adjacent semiconductor wafers on an INT-SAT.*

SPECTROSCOPIC IMAGING INSTRUMENT

The architecture of INT-SAT is specifically designed to support many different mission instruments, but our initial focus is for remote sensing applications, especially for visible, near infrared, and short wave infrared imaging of the Earth. InGaAs-on-Silicon photodetector arrays can capture images at wavelengths from 0.3 to 2.6 microns [Discovery Semiconductors' NASA Phase II SBIR, Contract Number NAS5-32809], by using both silicon and InGaAs materials as the photosensors. In addition, we recognize that spectrographic information is of great use in remote sensing applications. Therefore, we propose to develop an integrated remote sensing spectroscopic imaging instrument as our first application on Discovery's INT-SAT.

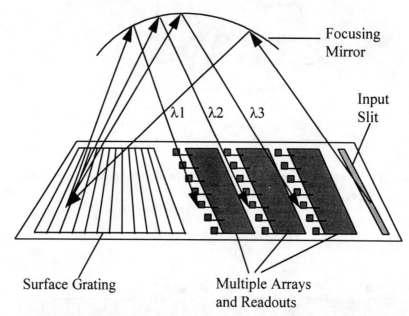

FIGURE 4. The Proposed Integrated Grating - Multi-Photodetector Array Device on INT-SAT.

The basic concept to be developed is to make both the photodetector array and the optical surface grating on the same silicon substrate (See Fig. 4). The silicon chip is divided into three areas: the detector array, the readout electronics, and the grating area. The incoming light would be directed onto the surface grating, which would be metallized, and therefore, reflective. The light would diffract off the grating and be focused by a curved focus mirror (which would be part of a neighboring wafer mount) onto the photodetector array. In the multi-array device shown, there are several arrays, and they are placed parallel to the grating grooves at specific locations. This arrangement results in each array capturing a row of pixels at a specific wavelength, each array registering a different wavelength. This is best suited for spectral window imaging. "Spectral window imaging" means the capture of a complete image using two (or more) predefined wavelengths. The data set would be of the form $I(x,y,\lambda_i)$ where i=1,2...n. This device is best suited for use in a spectral line scan camera. The mechanical scanning needed to produce a complete image could be accomplished by using the motion of the satellite in its orbit (Pushbroom Technique).

COMPLETE, FREE-FLYING, INTEGRATED SATELLITE (INT-SAT)

Now that the INT-SAT's modular architecture has been described, we wish to show how it would be used. The following paragraphs give a picture of a complete satellite application of the INT-SAT module. The conceptual drawing of the satellite is shown in Fig. 5.

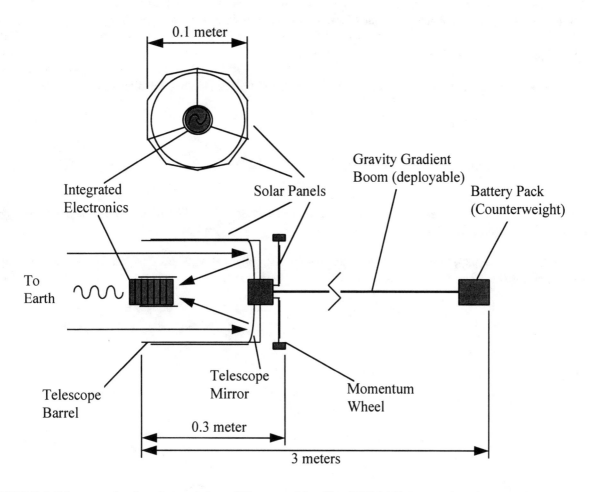

FIGURE 5. Discovery Semiconductors' Overall Integrated Satellite (INT-SAT) Layout (Version V.1) for the Visible, Near-Infrared, and Short Wave-Infrared Spectroscopic Imaging.

The satellite would be maintained in an Earth pointing orientation in a low Earth polar orbit. The INT-SAT's <u>bus - instrument module</u> (Integrated Electronics) is placed at the Earth side of the satellite. The remote sensing telescope forms the bulk of the satellite. The optical design would be either a simple Newtonian or a Schmitt telescope, which provide the correct Earth-facing geometry and have a single mirror. The barrel of the telescope is a ten-sided polygonal solid, covered on the exterior faces with solar cells. The roll control momentum wheel is placed behind the telescope mirror. The outer surface of the wheel is also covered with solar cells. A long, extensible gravity gradient boom extends from the rear of the satellite. This would carry the battery pack, which would serve as the counterbalancing mass. Once deployed, the gravity gradient effect would serve to point the satellite towards the Earth, thereby passively controlling the pitch and yaw. Note that all the components besides the bus / instrument (Integrated Electronics) are passive.

The entire satellite would be very small, occupying only ~5000 cubic centimeters. The INT-SAT's bus - instrument (Integrated Electronics) module itself only occupies ~200 cc, and will probably weigh less than 500 grams. This shows how integration will shrink the payload size & weight.

SUMMARY

Discovery Semiconductors proposes to develop the next generation of microsatellites which overwhelmingly use semiconductor technology to reduce the weight of the satellites. The innovation includes design of an innovative, compact, light-weight microsatellite called INT-SAT (Integrated Satellite). The INT-SAT is composed of multiple semiconductor wafer-scale integrated circuit/system modules. Each semiconductor wafer module performs a specific satellite function, so that together they form a complete satellite.

Acknowledgment

The author thanks his colleague and friend, Dr. Frank J. Effenberger for helpful discussions in preparing this paper.

References

Joshi, A. M. (1994) "Startup to Develop Satellite-on-a-Chip," *Military & Aerospace Electronics*, Feb. 1994, Vol. 5(2) : 1.

Payne, R. A. Jr. (1992) "Applications of the Petite Amateur Navy Satellite (PANSAT)," in "*Master's Thesis, Naval PostGraduate School*," Sep. 1992, Monterey, CA.

Priedhorsky, W. C. (1989) "Low-cost Small Satellites for Astrophysical Missions," in *NASA Workshop on High-Energy Astrophysics, Taos, NM*, Conference 891276, Dec. 10-14, 1989.

LIGHT WEIGHT, HIGHLY INTEGRATED OPTICAL SYSTEMS FOR THE NEW MILLENNIUM PROGRAM

Mark Schwalm, Dexter Wang and Michael Curcio
SSG, Inc.
65 Jonspin Road
Wilmington, MA 01887
(978) 694-9991

David H. Rodgers and Patricia M. Beauchamp
Jet Propulsion Laboratory
1800 Oak Grove Drive
Pasadena, CA 91107
(818) 354-4321

Abstract

Over the past four years SSG has been working with the support of JPL to develop light weight, high performance optical systems for earth orbit and deep space missions. These programs, which include both ground demonstrations and flight instruments, achieve the improved weight and performance through a combination of two main features: multifunctional design and silicon carbide optics and structures. Multifunctional designs include aperture sharing telescopes and spectrometers that not only perform science, but can be used for optical navigation and laser communication. SiC, in monolithic form for optics and in composite form for fracture-tough structures, enables very light weight systems that are passively athermal to cryogenic temperatures. Ground demonstrations of several of these technologies have led to the upcoming flight demonstrations on DS-1 and EO-1, as well as candidate telescopes for other technology flights. These advances are currently being extended to the next level of multifunctionality in a sensorcraft demonstration, in which a SiC telescope performs science, optical navigation, and optical communication, while the structure serves as both the telescope bench and spacecraft bus. This paper presents a general discussion of multifunctional designs and SiC optical systems in the context of recent, present, and future work.

INTRODUCTION

The desire to reduce weight and improve performance of space optical systems is not new, but NASA's mandate of Faster, Better, Cheaper, has dictated a dramatic shift in space system design over the past few years. Spacecraft budgets are being reduced by an order of magnitude, system size and weight are being cut to accommodate smaller, cheaper launch vehicles, and schedules are being shrunk to three years and below from program initiation to launch—all while greater performance is required. Such spacecraft and satellite pressures flow down to the instrument level, and small, incremental improvements are insufficient to meet these needs. Significant improvements in optical system technology are required, but such improvements require new technologies and their inherent risks.

NASA has therefore undertaken the New Millennium program, a series of deep space and earth orbiting flights designed to prove out new technologies and give them the required flight heritage to be applied to science and operational systems. Within the New Millennium program, as well as through Small Business Innovated Research (SBIR) programs and other technology demonstrations, SSG has been working with the support of JPL to develop state of the art optical systems that are lighter and smaller than existing systems but provide greatly improved performance. These efforts began as technology studies and ground demonstrations, but have since led to flight instruments.

Two separate innovations have enabled such improvements: multifunctional design and SiC optics and structures. These two innovations are discussed in detail below in the context of several current or recent ground and flight demonstrations.

MULTIFUNCTIONAL DESIGN

A single optical system that can perform multiple functions can take the place of two or more others, provide redundancy where none existed, and gather more science data within a given package. SSG's goal has been to develop optical systems that provide infrared, visible, and ultraviolet imagery and/or spectrometry in a single system, and to design the system such that these instruments can be used not only for science gathering, but also for optical navigation and laser communications. Two recent instruments—one a flight instrument, and the other an SBIR phase II ground demonstration—combine several of these various functions.

MICAS

The Miniature Integrated Camera And Spectrometer (MICAS), shown in Figures 1 and 2, is the scientific instrument to be flown on the New Millennium program's Deep Space-1 flight, a spacecraft which will perform a near-earth asteroid and comet fly-by. The instrument's complete opto-mechanical system was completed by SSG and delivered to JPL earlier this year, where it has undergone integration and environmental testing (thermal, vibration, and shock), and is currently being interfaced to the spacecraft. DS-1 is scheduled for launch in July, 1998.

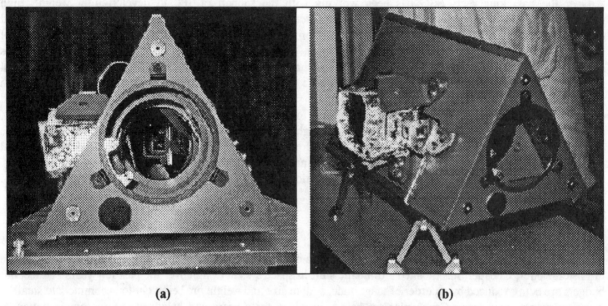

FIGURE 1. The Miniature Integrated Camera and Spectrometer (MICAS) to be Flown on the New Millennium Program's Deep Space-1 Flight, Shown in Front (a) and Isometric (b) Views.

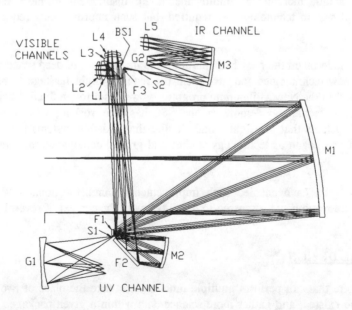

FIGURE 2. The MICAS Optical Schematic.

The MICAS instrument, based on a JPL optical design, is multifunctional in that it has two visible imaging channels, an infrared spectrometer, and an ultraviolet spectrometer, and the imaging channels will perform both scientific data collection and optical navigation. As such, this single instrument replaces what typically would have been four higher cost, heavier instruments. For example, MICAS performs the same job as the Voyager instruments, but weighs less than 5 kg, including both the infrared and optical radiators, whereas Voyager's instruments weighed more than 200 kg. Such a design also allows the use of smaller launch vehicles for further cost savings. MICAS is an all-SiC telescope with 10 cm aperture shared fore-optics that feed the four separate channels, and the entire package is the size of a bread box: 220 mm x 240 mm x 380 mm. The telescope was completed in 16 months at SSG and integrated in 5 months at JPL.

Dual-Use Lasercom Telescope

Another example of multifunctional design is the all-SiC optical system shown in Figures 3 and 4, which was designed and built under a phase II SBIR program with JPL to demonstrate a telescope that functions as both a 20:1 afocal laser communication transmitter/receiver and a remote sensing telescope front end. The 30 cm aperture telescope has a 20:1 afocal magnification, a 1 mrad field of view, and weighs 6.5 kg. Current SSG developments can further reduce this weight to below 4 kg. A fold mirror is located at the accessible exit pupil of the afocal telescope so that a fast steering mirror can be used for the laser communications beam acquisition and stabilization. In the dual-use design, a beam splitter separates the laser transmitter channel from the 1 mrad field-of-view visible camera/lasercom receive channel. Such a design is also consistent with further back-end channels for spectroscopy and/or altimetry.

Combining a lasercom transmitter/receiver with a visible imaging telescope not only eliminates the redundancy of having two telescopes with similar performance requirements, it can also provide improved performance through the higher data rate and reduced aperture size of optical communications. This benefit is especially true for deep space missions such as Pluto or Neptune flights, where, for a given power, laser communications can provide over 10 times the data rate at half the mass with smaller aperture sizes than RF communications. As optical communication technology proliferates, dual-use optical telescopes can provide significant cost, weight, and performance benefits.

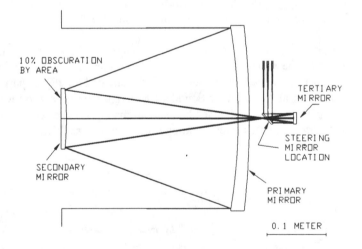

FIGURE 3. The dual-Use Telescope For Lasercom and Remote Sensing.

FIGURE 4. The Dual-Use Lasercom Optical Schematic

SILICON CARBIDE OPTICAL SYSTEMS

The second innovation that has enabled light weight, high performance systems is the use of silicon carbide optics and structures. This section addresses two aspects of SiC optical systems: the benefits of SiC mirror substrates; and the advantages of combining SiC optics with SiC structures to create passively athermal systems to cyrogenic temperatures.

The Benefits of SiC Mirror Substrates

SiC has been widely recognized as an attractive material for mirror substrates due to its high specific stiffness, excellent polishability, and superior thermal stability to visible levels over wide temperature ranges and gradients (Paquin 1995, Tobin 1995). Its modulus (E) of 455 GPa and density (ρ) of 3.2 gm/cm^3 yield a specific stiffness (E/ρ) 5.5 times greater than aluminum and 4.5 times greater than low expansion glass. And its thermal conductivity (K) of 165 W/mK and coefficient of thermal expansion (CTE, or α) of 2.2×10^{-6}/K yield a thermal stability (K/α) 7 times greater than aluminum and greater than or similar to low expansion glass, depending on the glass's CTE tolerance. Compared to beryllium, SiC has a roughly equal specific stiffness and a thermal stability that is three times greater. These two benefits enable mirrors designs that are extremely light weight and resistant to deformation in the presence of thermal loads and gradients.

SiC also supports an excellent optical finish. Hot pressed SiC has been polished bare to better than 5 angstroms rms surface roughness, while the castable SiC has been measured at better than 15 angstroms rms. Flats and spheres are typically finished in the bare SiC, and are amenable to all typical reflectivity overcoats. Aspheres are typically finished in a thin silicon cladding applied to the optical surface. This silicon is softer than bare SiC for quicker material removal during polishing, and can be single point diamond machined. Silicon is an excellent CTE match to SiC; in testing, bimetallic effects have been shown to be negligible to visible levels even to cryogenic temperatures.

Light-weighting SiC mirrors can be accomplished in several manners, and the exact method is an integral part of the trade between different SiC types. The hot pressed SiC provides the higher material properties but is limited in the complex geometries that can be formed. Castable SiC has a slightly lower modulus, but can be fabricated in more efficient, complex honeycomb designs. Typically, in small aperture systems such as MICAS, hot pressed SiC mirrors can be made stiff enough to survive processing and gravity release with thin cross-sections alone. This simple solution is both very light weight and low in cost and risk. Such hot pressed mirrors for visible level optical systems are gaining flight qualification on both the MICAS instrument and on the larger aperture Advanced Land Imager for the New Millennium Earth Orbiting-1 flight.

For larger aperture systems, more complex light-weighting designs are required, and castable SiC typically becomes the material of choice. One example of a telescope incorporating such castable mirrors is a ground demonstration telescope currently being fabricated for JPL as part of a technology demonstration for the Integrated Multispectral Atmospheric Sounder (IMAS), a next generation weather sounder. This system, shown in Figure 5, is in the final stages of assembly, and alignment is expected to begin in mid September. Once aligned, the system will undergo launch level vibration testing, as well as thermal vacuum testing to 135K.

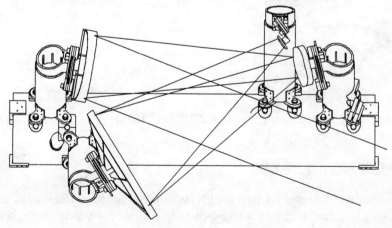

FIGURE 5. The IMAS Technology Demo Telescope Currently in Assembly and Environmental Testing.

All-SiC Optical Systems for Passive Athermal Performance

Although SiC provides excellent weight and performance gains for mirror substrates, the greatest advantage SiC offers is not at the component level but at the system level. When SiC mirrors are combined with a SiC structure, the mirrors and structure have **matching CTEs over all temperatures**, and thus create a passively athermal system. Such athermal behavior significantly reduces design complexity, weight, testing, and integration. For example, alignment of cryogenic systems can be performed at room temperature, with no need for further alignment or focus adjustment. Such benefits, in turn, translate into a lower risk, lower cost, shorter schedule instrument, and for this reason, the passive athermal design has been a driving factor in the various instruments discussed here.

Optics and structures expand and contract over temperature, and if the expansion of the structure does not match that of the optics, misalignment occurs, degrading the optical performance. Solutions to this problem have included active alignment at temperature, active temperature control, CTE compensation with dissimilar materials, very low expansion materials, "cryo-nulling" (i.e. misaligning at room temperature to obtain correct operating temperature performance), or some combination thereof. Each of these solutions is limited by either its effective temperature range or its complexity. In contrast, a single material optical system that offers passive athermal performance over all temperatures does not suffer such limitations and offers numerous benefits. For example, unlike graphite epoxy, SiC has a zero coefficient of moisture expansion (CME) and does not suffer from outgassing. And compared to an all-aluminum design, SiC offers superior specific stiffness and thermal stability.

The MICAS instrument is an excellent example of SiC's advantages. During instrument design, the all-SiC system obviated the need to design for thermal mismatch. Risk was low because there were no active mechanisms, temperature restrictions, or CTE balancing by design. Alignment was performed at room temperature even though the instrument's operating temperature was 135K. Subsequent thermal vacuum testing at SSG with a retro-sphere at the focal plane location was successful: the instrument performance remained in specification and any changes were on the order of the noise level of the test (<0.1λ P-V WFE). The results are shown in Table 1. Furthermore, there was no observable performance change even in the presence of large thermal gradients (>20K) during the uncontrolled rapid cool-down—a benefit very relevant to earth-orbiting missions that have varying, cyclical thermal loads. The instrument was then delivered to JPL and the retro-sphere replaced with the focal plane. When the system was thermal vacuum tested to 135K at JPL, the focal plane was in correct focus and no alignment was required, providing significant cost and time savings during integration. Such athermal performance is a critical enabling technology for deep space missions, where the weight, cost, and risk of active focus adjustment would have major impact.

TABLE 1. Results of MICAS Thermal Vacuum Testing ($\lambda = 0.633$ μm)

Temperature	Peak-to-Valley Wavefront Error	RMS Wavefront Error
20K	$0.70\ \lambda$	$0.13\ \lambda$
-137K	$0.79\ \lambda$	$0.16\ \lambda$

A final example of the athermal performance of an all-SiC optical system is the dual-use lasercom telescope, shown in Figures 3 and 4. This SBIR ground demonstration telescope has a composite SiC structure and monolithic SiC mirrors. This system was thermal vacuum tested (the first test of such a composite SiC structure system), and the results are shown in Figure 6. The system was thermally matched to within 0.06λ rms wavefront error from 290K to 220K, even with the sensitive tolerances of an f/1.7 primary mirror.

THE SENSORCRAFT CONCEPT

What is the next goal? With JPL's support, SSG is in the early stages of developing a silicon carbide sensorcraft. A sensorcraft is a microspacecraft in which the sensor (in this case an optical system) and the spacecraft are not separable entities fabricated independently and then assembled together, but instead a union of

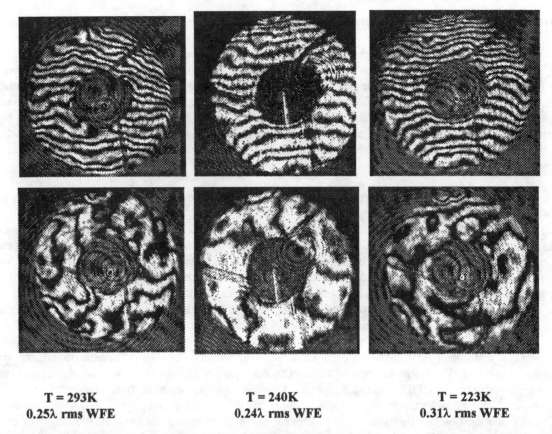

| T = 293K | T = 240K | T = 223K |
| 0.25λ rms WFE | 0.24λ rms WFE | 0.31λ rms WFE |

FIGURE 6. Double-Pass Interferograms of the Dual-Use Lasercom Telescope Show Thermal Stability to Within 0.06λ rms Wavefront Error from Room Temperature to 223K.

the two, so that the instrument and spacecraft share a common structure. Furthermore, the optical system performs several roles—science (imaging and/or spectroscopy), laser communications, optical navigation, and altimetry—in order to eliminate the need for separate subsystems. Such a concept eliminates redundancy at the structural level, combining spacecraft bus and optical bench into one structure, and at the system level, using one optical instrument in place of several systems.

SSG investigated this concept under a phase I SBIR with JPL, and is currently developing and fabricating a ground demonstration sensorcraft structure and optical system under a phase II SBIR. Microspacecraft such as the sensorcraft being developed under this effort are a key element in NASA's vision for a low cost, virtual presence in space. Such microspacecraft weigh on the order of 10 kg, and thus can be deployed singly or in groups on small launch vehicles for a fraction of the cost of even today's small spacecraft. Current mission designs include both earth orbiting missions for surface and atmospheric monitoring and measurement, and deep space exploratory missions.

Clearly there are technical challenges associated with realizing the benefits of a sensorcraft. First, the weight gains that result from combining separate telescope structures will only be appreciable if the optical system itself is a significant percentage of the entire spacecraft weight. This requirement is met with current microspacecraft concepts, which incorporate a 0.3 meter aperture telescope onto a spacecraft that is not much larger—roughly 40 cm in diameter and less than twice the length of the telescope itself. Second, combining the structures eliminates the ability to place mechanical and thermal isolation between the optical instrument and spacecraft subsystems. This requirement significantly increases the loads that the telescope must accommodate, and strongly points towards the thermal and mechanical stability of SiC as an enabling technology. And finally, the multiple roles of the optical system place restrictions on the simultaneity of functions during operation, since, for example, the single optical system is limited in its ability to point in one direction for gathering science and at the same time point in another direction to transmit or receive data. This challenge, too, can be solved by proper mission sequencing.

CONCLUSIONS

The upcoming New Millennium flights of MICAS and the Advanced Land Imager will provide the needed flight heritage for hot pressed SiC multifunctional optical systems, and the ground demonstrations of composite SiC structures provided by the lasercom telescope and the IMAS demo telescope will provide the basis for a similar flight validation for composite SiC structures in the near future. Other technologies such as the dual-use lasercom transmitter/receiver and the sensorcraft concept are still in the ground demonstration phase, but if successful, these technologies will follow a similar path towards operational flight maturity. A light weight, passively athermal cryogenic SiC spectrometer is one of several enabling technologies for JPL's IMAS atmospheric sounder, conceived as a candidate for the NPOESS weather satellites.

As these technologies complete the progression from risky new ideas to flight proven realities, further applications of passively athermal SiC systems can play a role in upcoming space remote sensing and exploration. For example, athermal performance over a very large temperature range would allow such optical systems to function without thermal control on the surface of Mars or other solar system bodies, where the local ambient temperature will fluctuate greatly. SiC optical systems may also play a significant role in space based interferometry and the Next Generation Space Telescope. In general, SiC and multifunctional designs will be among key new innovations that fuel the trend towards lighter, cheaper, high performance space instruments.

Acknowledgments

Besides the co-authors from JPL, SSG would like to thank the following for their support of these technical developments: Dr. James Lesh and JPL's Optical Communications group for their sponsorship of the dual-use lasercom telescope; Ross Jones and JPL's microspacecraft group for their sponsorship of the sensorcraft instrument; Gun-Shing Chen from JPL for his support on PICS/MICAS and other programs; and NASA/Goddard and Lincoln Laboratories for their sponsorship of the EO-1 advanced land imager flight telescope.

References

Paquin, Roger A. (1995), "Materials for Mirror Systems: An Overview," in *Silicon Carbide Materials for Optics and Precision Structures*, SPIE Proceedings Vol. 2543: 2-11.

Tobin, E., M. Magida, S. Kishner, and M. Krim (1995) "Design, Fabrication, and Test of a Meter-Class Reaction Bonded SiC Mirror Blank," in *Silicon Carbide Materials for Optics and Precision Structures*, SPIE Proceedings Vol. 2543: 12-21.

LIGHTWEIGHT COMPOSITE REFLECTORS FOR SPACE OPTICS

Brian E. Williams and Shawn R. McNeal
Ultramet
12173 Montague Street
Pacoima, CA 91331
818-899-0236

Russell M. Ono
Ono Laboratories
5104-H Airport Drive
Ontario, CA 91761
909-390-0238

Abstract

The primary goal of this work was to advance the state of the art in lightweight, high optical quality reflectors for space- and Earth-based telescopes. This was accomplished through the combination of a precision silicon carbide (SiC) reflector surface and a high specific strength, low-mass SiC structural support. Reducing the mass of components launched into space can lead to substantial cost savings, but an even greater benefit of lightweight reflectors for both space- and Earth-based optics applications is the fact that they require far less complex and less expensive positioning systems. While Ultramet is not the first company to produce SiC by chemical vapor deposition (CVD) for reflector surfaces, it is the first to propose and demonstrate a lightweight, open-cell SiC structural foam that can support a thin layer of the highly desirable polished SiC reflector material. SiC foam provides a substantial structural and mass advantage over conventional honeycomb supports and alternative finned structures. The result is a reflector component that meets or exceeds the optical properties of current high-quality glass, ceramic, and metal reflectors while maintaining a substantially lower areal density.

INTRODUCTION

The operation of precision reflectors for extended periods in space presents significant challenges with regard to materials. In addition to conventional reflector considerations such as surface figure, thermal expansion and distortion, strength, and durability, materials for on-orbit reflector panels require low areal density, resistance to atomic oxygen and solar radiation, and the ability to withstand temperatures as low as 141 K.

The optical performance of any mirror depends on the optical properties of its surface, the type of material used, and the design of the substrate. The substrate is responsible for maintaining the precise geometric surface over the lifetime of the system, regardless of environment, thermal variations or gradients, mechanical stresses, etc.

Critical properties for space-based reflector materials are density, thermal expansion coefficient (CTE), thermal conductivity, elastic modulus, natural frequency and inertial loading parameter, thermal stress parameter, and thermal distortion parameter. The last of these is especially important for reflectors. High values for each property are desirable, except density and thermal expansion (Towell 1994).

OPTICAL MATERIALS

In high-performance optical applications, the traditional mirror substrate materials (Pyrex and other glasses) have largely been replaced by several high-quality grades of low-expansion fused silica. These materials have the desired low density and CTE, but also relatively low thermal conductivity and elastic modulus. Thermal cycling-induced hysteresis, or failure of a material to return to its original dimensions, has been shown to result in permanent surface figure changes in these glasses following spatially nonuniform heating (Jacobs 1987), as indicated by interferometric and dilatometric analysis. Unlike isotropic thermal cycling, nonuniform heating can result in permanent surface figure change because the uneven distortion does not completely relax upon cooling.

Work has also been directed at developing improved metal mirrors using substrates of aluminum, beryllium, copper, molybdenum, nickel, and other metals. Although aluminum and copper have high thermal conductivity, they exhibit low inertial loading and high thermal expansion. Beryllium has a very high inertial loading parameter and is an excellent thermal conductor, while having moderate strength and reasonably good oxidation resistance at low to moderate temperatures. However, it has a relatively high CTE, and its optical performance has been erratic and unpredictable due to instabilities in the material. Beryllium is also expensive, in part due to costs associated with the toxicity of its oxide. Silicon exhibits generally good properties for reflector applications, except for key disadvantages of low inertial loading parameter, unavailability in large sizes at a practical cost, and lack of a suitable, compatible lightweight support structure. Fabrication of a silicon foam support, similar to the silicon carbide foam used in the current work, may be possible and has been investigated at Ultramet under internal R&D funding.

SILICON CARBIDE FOR OPTICAL APPLICATIONS

Over the past decade, there has been growing interest in the production of bulk silicon carbide (SiC) for optical applications. Most bulk SiC is fabricated by sintering and/or hot-pressing SiC powders. However, these forms of SiC are usually porous materials that do not produce a good optical surface on polishing unless they are silicon-clad. Silicon cladding improves polishability but results in a multiphase material with variable CTE, which leads to greater thermal distortion compared to monolithic SiC produced by chemical vapor deposition (CVD). CVD process technology, in which the desired material is constructed at the atomic level, allows the fabrication of SiC that is fully dense, isotropic, near-net shape, and highly polishable. The mechanical properties of SiC are superior to those of the glasses, and although it has a higher CTE than the glasses, SiC also has a higher thermal conductivity, leading to a comparable thermal distortion parameter.

CVD SiC is therefore a good candidate material for optical (especially large, lightweight) reflectors. Although SiC is highly polishable (surface figure $\leq \lambda/12$, where $\lambda=0.6328$ μm, the wavelength of midrange visible light; surface finish or roughness ≤ 0.1 nm rms), its high hardness has previously led to high grinding and polishing costs. Through a codevelopment effort with Ono Laboratories, which performed optical grinding and polishing in the current effort, a proprietary process was developed to increase the speed of grinding and polishing and thereby substantially reduce cost. Once polished, the hard CVD SiC mirror surface is very resistant to scratches and damage from debris impact.

PREVIOUS WORK AT ULTRAMET (PHASE I)

In Phase I of this work, Ultramet demonstrated the feasibility of the CVD SiC foam support/CVD SiC reflector faceplate technology through the fabrication of a lightweight (4 kg/m^2 areal density), all-ceramic composite mirror structure composed of an open-cell SiC foam substrate onto which a SiC mirror surface was deposited by CVD. The 15.24-cm (6-in) diameter × 182.88-cm (72-in) radius of curvature mirror was polished to a surface finish of <1 nm rms and a surface slope error of <1.2 μm. One such mirror structure was subjected to 200 thermal cycles from 273 to 393 K (0 to 120°C) at Ono Laboratories, with interferogram measurements taken after every 20 cycles. The specimen was then subjected to 10 cycles from 73 to 393 K (-200 to +120°C), after which a single interferogram measurement was conducted. The surface slope error did not exceed 2 μm at any time during the 273-393 K cycling and averaged only 1.2 μm, well within the <3.6 μm specification for the project. The single interferogram measurement taken after the cryogenic testing showed that no permanent deformation occurred. In addition, the thermal shock resistance and compatibility of the composite mirror components were also demonstrated.

CURRENT EFFORT (PHASE II)

The current, Phase II effort concentrated on optimizing the mechanical and optical properties of the same CVD SiC foam/SiC faceplate composite mirror structure through optimization of the fabrication and grinding/polishing processes, and scaling up the entire process for components up to 45.72 cm (18 in) diameter. The effort proved to be a continual struggle between minimizing mass and producing a component that could survive the grinding process and exhibit the desired optical characteristics, with the end result being a very encouraging reflector material.

Although scaleup of the CVD SiC foam and SiC faceplate fabrication processes from 10.16-cm (4-in) to 25.4-cm (10-in) diameter proved to be a challenge, the subsequent scaleup to 45.72-cm (18-in) diameter was rapid, and further scaleup to 1-m (39.37-in) diameter is expected to be reasonably straightforward in commercial application. The process allows extremely stiff, lightweight, and directly contoured optical structures to be fabricated quickly and efficiently from abundant, inexpensive, and non-strategic raw materials.

MIRROR STRUCTURE FABRICATION

The baseline composite mirror structure, which remained unchanged throughout the effort, was a CVD SiC foam structural support, a CVD SiC reflector coating, and a proprietary interlayer between them that was encapsulated with CVD SiC. The CTEs of all three materials were closely matched. The interlayer served as a very effective closeout layer for the foam, allowing for the deposition of a smooth and relatively thin (0.762 mm, or 0.030 in) CVD SiC faceplate surface without any print-through of the foam structure. As the CVD SiC reflector surface was being deposited on the interlayer, the SiC completely encapsulated the interlayer, leaving none exposed, and fully locked the interlayer to the SiC foam support. A virtually all-SiC structure is thus created, and the testing performed so far supports this claim. Optimization of the composite mirror structure began with the 10.16-cm (4-in) diameter size, followed by 25.4-cm (10-in) and 45.72-cm (18-in) diameter parts.

The bulk density of the SiC foam used in this project was 400 kg/m^3 (0.4 g/cm^3), which was shown to provide good structural support and stability for the CVD SiC mirror surface up to 45.72 cm (18 in) diameter. Prior knowledge of the thermal and mechanical properties of the CVD SiC and SiC foam for various densities and porosities allowed for more straightforward modeling of the foam configuration required in Phase II for the fabrication of scaled-up reflectors. It was decided to increase the density of the SiC foam to 400 kg/m^3 (0.4 g/cm^3) in Phase II, compared to the 200 kg/m^3 (0.2 g/cm^3) material used in Phase I, in order to provide increased strength, rather than attempting a sandwich panel structure with two faceplates. The reason for this was that adding a faceplate on the back of the reflector would increase the total weight by about two-thirds, whereas doubling the SiC foam density increases the total weight by only one-third. In addition, it is critical that the foam provide adequate support for the SiC faceplate right at the interface between the two, in order to withstand localized stresses resulting from the grinding process. Increasing the strength of the foam, by increasing its density, enhances support at the interface, which becomes increasingly important with increasing reflector size.

GRINDING/POLISHING PROCESS OPTIMIZATION

Establishing the desired surface figure often takes more than 90% of the total machining time, with the remainder used for meeting surface finish (roughness) requirements. Altering the surface figure specifications of a reflector will thus have the greatest impact on machining cost.

SiC coating thicknesses ranging from 0.254 to 0.762 mm (0.010-0.030 in) were applied to various flat substrates in order to determine the minimum amount required for successful polishing. Approximately 0.127 mm (0.005 in) of the SiC surface was removed during polishing, depending on the as-deposited roughness. It was determined that a minimum SiC coating thickness of 0.508 mm (0.020 in) is required to withstand the stresses induced during machining, given the SiC foam and proprietary interlayer materials used to support the SiC faceplate.

Following deposition, any small nodules that were present on the SiC surface were removed. The mirror surface was then rough ground with 220-mm diamond such that no surface low points existed, followed by further grinding with successively finer abrasive (30-mm and 9-mm diamond). Final polishing was performed with 1-mm diamond.

10-cm (4-in) REFLECTOR DEVELOPMENT AND TESTING

A total of 38 components were fabricated to varying degrees of completion during process optimization for the 10.16-cm (4-in) diameter reflector size. At first, flat specimens were sent to Ono Laboratories for polishing to the surface figure target specification of <3 μm rms. Approximately 0.127 mm (0.005 in) of SiC was generally removed from the faceplate during grinding and polishing of this reflector size. A flat, polished 10.16-cm (4-in) diameter × 1.27-cm (0.5-in) thick SiC composite reflector was then delivered to NASA for evaluation. The part had a good surface figure of 1 μm, but the areal density of 1.1 g/cm^2 was higher than desired. The surface finish was ≈0.5 nm rms, and the part had several small pits that likely could have been removed by polishing; however, it was decided that no chances should be taken of damaging the reflector, considering the good surface figure that was achieved. The relatively high weight resulted from a problem that was subsequently minimized: the foam support was open at the edges, which allowed the SiC reactant gases to infiltrate and deposit around the perimeter and resulted in a significant weight increase. For every 1 g of SiC deposited on the faceplate, 2 g was being applied to the foam.

Following optimization of the flat reflectors, 10.16-cm (4-in) diameter × 182.88-cm (72-in) radius of curvature reflectors were fabricated and polished to <2 fringes, which was well within the 3.9 fringes required to meet the target 3-μm rms surface figure extrapolated over a 1-meter diameter. Areal densities in the range of 0.60-0.99 g/cm^2 were achieved, depending on foam density and faceplate thickness. A single optimized curved reflector of this size was delivered to NASA, while additional parts were used for nondestructive evaluation (NDE) and thermal/optical testing. The deliverable part sent to NASA was a 10.16-cm (4-in) diameter × 1.27-cm (0.5-in) thick SiC composite reflector with a surface figure of 2.5 μm rms and a surface finish of ≈0.5 nm rms. The total weight was 81.9 g, corresponding to an areal density of 0.99 g/cm^2.

Cryogenic testing was performed at Ono Laboratories on two optimized 10.16-cm (4-in) diameter reflectors (2-fringe surface figure), as well as on an unpolished substrate of the same size. The test setup was initially designed so that the reflector would be suspended over a 77 K (-196°C) liquid nitrogen bath, in order to provide an indication of cryogenic thermal shock behavior. However, the unpolished reflector test piece performed so well in initial testing that all of the components were instead subjected to much more stringent conditions. All three specimens were literally immersed in liquid nitrogen for a number of minutes, after which they were removed from the bath and allowed to thaw at room temperature for approximately ten minutes. No microcracking of the reflectors was found, and interferogram analysis showed that no permanent surface figure change had occurred. A second, even more demanding test was then conducted by immersing the reflectors in the liquid nitrogen bath on edge such that only half of the reflector was frozen, creating a >150 K (150°C) thermal gradient across the reflector faceplate. Again, no damage or distortion was observed.

The same three specimens were then exposed to 30 minutes of isothermal exposure at 473 K (200°C) in a static air furnace. No microcracking was found through dye penetrant examination. Interferogram measurements were taken on the two polished reflectors, and both exhibited approximately 3 fringes, a slightly higher number than the as-polished condition of approximately 2 fringes, but still within the target of 3.9 fringes (i.e., 3-μm surface figure).

In-situ interferogram measurement of a 10.16-cm (4-in) diameter reflector, from ambient temperature to 423 K (150°C), was then conducted at Ono Laboratories. The reflector was placed in a holder at the center of an oven, and interferogram measurement was then conducted through a quartz window in the oven door. The surface figure of the reflector did not deviate by more than 1 fringe during testing, well within the target allowable deviation of 3 μm. These data were supported in repeated testing of the reflector. These results are very important, as they establish the thermal compatibility of the constituents of the composite reflector structure.

25-cm (10-in) REFLECTOR DEVELOPMENT AND TESTING

A total of 32 components were fabricated to varying degrees of completion during development of the 25.4-cm (10-in) diameter, 182.88-cm (72-in) radius of curvature reflector size. As expected, scaleup of the SiC foam support material was straightforward. In an initial optimization effort, 25.4-cm (10-in) diameter × 1.27-cm (0.5-in) thick graphite disks were successfully coated with SiC to a thickness of 0.889 ±0.0762 mm (0.035 ±0.003 in). The first 25-cm diameter SiC foam substrate (with a proprietary interlayer) was then coated with the CVD SiC faceplate material, again 0.889 mm (0.035") thick. The SiC coating on this first foam substrate was found to be unacceptable, however, because of two problems: significant coating thickness nonuniformity (up to 40%) across the faceplate diameter, and a SiC "ridge" that developed around the faceplate perimeter. A 25-cm diameter, partially ground reflector was delivered to NASA at this stage in development for evaluation.

Although subsequent CVD SiC runs produced coating thicknesses that were more uniform than that achieved on the first attempt, they were still not axially symmetric, which caused problems during the grinding process. Attempts to grind parts with this thickness variation were unsuccessful because only the thick portion of the reflector surface came into contact with the grinding surface at the outset of the grinding operation, which induced stresses in the faceplate that often led to microcracking. To improve the uniformity of the SiC faceplate, the CVD work station was retrofitted with a substrate rotation device so that the SiC precursor gases flowed more uniformly over the part. Substrate rotation can sometimes induce unusual flow patterns, but in this case the rotation promoted much greater coating thickness symmetry.

During Phase I SiC composite reflector development at Ultramet, for reflector sizes ranging from 2.54 to 15.24 cm (1 to 6 in) diameter, typical SiC microhardnesses had ranged from 3000 to 3200 VHN, resulting in a good combination of polishability, scratch resistance, and fracture toughness. However, the brittle behavior of many of the early 25.4-cm (10-in) diameter parts indicated that the larger reflector sizes may require a somewhat lower hardness level. As the hardness is reduced, the scratch resistance also decreases and, if the hardness is low enough, polishability can also decrease. However, reducing the hardness increases the fracture toughness, which is a property that will undoubtedly become more critical with each increase in reflector size.

A small number of 2.54-cm (1-in) square graphite test pieces were coated with CVD SiC to establish the deposition conditions resulting in coating hardnesses as low as 2700 VHN. Varying the CVD SiC faceplate microhardness over the 2700-3500 VHN range resulted in no substantial change in either the polishability of the material or the ability to achieve the desired surface figure and finish. It was therefore decided that a lower hardness of 2700-2900 VHN would be targeted for the larger (≥25-cm diameter) reflectors, which when combined with other processing improvements led to the successful fabrication of the larger reflectors. The grinding and polishing process was also improved for the 25.4-cm (10-in) diameter reflectors through use of a lower load per unit area than had been used for the 10.16-cm (4-in) diameter size, which increased processing time but was less destructive to the reflector.

A preliminary evaluation of reflector behavior in a vibrating environment was performed using a small, 60-Hz vibrator attached to an aluminum plate. An unpolished 25.4-cm (10-in) diameter × 182.88-cm (72-in) radius of curvature SiC foam/proprietary interlayer/SiC faceplate reflector was adhesively bonded to the aluminum plate, then vibrated for 30 minutes. Dye penetrant testing conducted before and after the vibration test revealed no defects.

An optimized 25.4-cm (10-in) diameter reflector was then fabricated, ground and polished, and delivered to NASA, with the following properties: 25.4-cm (10-in) diameter × 1.524-cm (0.6-in) thick at edge (flat backside); 182.88-cm (72-in) radius of curvature; 567.5 g total mass; 1.1 g/cm^2 areal density; <1 nm rms surface finish; and <1.2 μm surface slope error. Before delivery to NASA, Ono Laboratories performed *in-situ* interferogram testing on this component over the temperature range from ambient to 473 K (200°C) in air. As targeted, the surface figure did not deviate by more than 3 μm during thermal cycling.

Although the deliverable 25.4-cm (10-in) diameter reflector met the target optical properties, both the optical figure and surface finish could be substantially improved with further polishing, which could not be performed in this project due to schedule and budgetary constraints. The perimeter was left unpolished to eliminate the possibility of initiating radial cracks from the edge during the grinding process. It is now believed that cracks will not occur during grinding, and therefore, under internal R&D funding, Ultramet and Ono Laboratories have since fabricated and polished an additional reflector of the same size, radius of curvature, and surface finish, but with an improved areal density of 0.85 g/cm^2 and a surface figure of $\lambda/14$ at $\lambda=0.5640$ μm.

45-cm (18-in) REFLECTOR DEVELOPMENT

As with scaleup from 10.16-cm (4-in) to 25.4-cm (10-in) diameter, scaleup of the SiC foam support material to the 45.72-cm (18-in) diameter size was readily achieved. With regard to CVD SiC faceplate deposition, many of the processing improvements made during 25-cm reflector development were applicable to the 45-cm parts as well, and good quality (uniform and smooth) SiC faceplate coatings were readily achieved on the 45-cm diameter SiC foam substrates. A total of six 3.302-cm (1.3-in) thick SiC foam supports were fabricated, of which the proprietary interlayer was applied to two, which were then coated with 0.127-mm (0.050-in) thick CVD SiC faceplates.

A single unpolished 45.72-cm (18-in) diameter SiC foam/graphite interlayer/SiC faceplate reflector was delivered to NASA, with the following properties: 45.72-cm (18-in) diameter × 3.302-cm (1.3-in) thick at edge (flat backside); 182.88-cm (72-in) radius of curvature; 3459.7 g total mass (prior to grinding/polishing); and 2.1 g/cm^2 areal density. This reflector was left unpolished due to schedule and budgetary constraints. The relatively high areal density of this component is due in part to the fact that the SiC faceplate surface was unpolished; the mass would be lowered if the SiC faceplate were ground and polished. A greater mass contribution is due to the fact that a relatively thick (3.302 cm) SiC foam support was required to accommodate the 182.88-cm (72-in) radius of curvature over the 45.72-cm

(18-in) diameter in a part having a flat backside. A substantial reduction in mass would be realized in both the 45.72-cm (18-in) and 25.4-cm (10-in) diameter reflector sizes if the back surface were made convex.

CONCLUSIONS

The primary objective of this work, namely to advance the state of the art in lightweight precision reflectors, was clearly achieved. A composite reflector was developed, up to 45.72 cm (18 in) diameter, composed of a stiff and lightweight CVD SiC open-cell foam support, a graphite interlayer, and a fully dense CVD SiC faceplate. In the 10.16-cm (4-in) and 25.4-cm (10-in) diameter sizes, components were fabricated and polished to well within the desired surface figure of <3.0 μm rms (extrapolated over a 1-m diameter) and surface slope error of ≤3.0 μm during thermal cycling to 473 K (200°C). For both sizes, the areal densities achieved were close to 1.0 g/cm^2, which is slightly higher than the target of <0.8 g/cm^2 (≤8.0 kg/m^2). However, a 25.4-cm (10-in) diameter component has since been fabricated by Ultramet and Ono Laboratories with an areal density of 0.85 g/cm^2.

In-situ evaluation of surface figure under cryogenic thermal cycling conditions was beyond the time and budgetary resources of the project, but no thermal cycling hysteresis was found after actual immersion in liquid nitrogen at 77 K (-196°C). Use of standard dye penetrant testing was found to be an effective means of locating flaws in the CVD SiC faceplate, while acoustic microimaging was successfully used to evaluate the bond between the graphite/CVD SiC faceplate and the SiC foam support.

Although scaleup from the 10.16-cm (4-in) diameter to 25.4-cm (10-in) diameter reflectors was quite a challenge, subsequent scaleup of the fabrication process to the 45.72-cm (18-in) diameter size was achieved significantly more quickly. Demonstration of the grinding/polishing process at the 45-cm diameter size remains to be performed. However, Ono Laboratories is prepared to polish parts of this size if follow-on funding can be obtained. The coating uniformity and texture of the 45-cm diameter parts is similar to that of the 25-cm components, and it is therefore believed that the larger size can be polished successfully.

One of the first tasks to be performed in follow-on work will be to evaluate the thermal stability of the SiC composite mirror structure under cryogenic conditions. An additional area requiring future investigation is an evaluation of possible joining techniques for the formation of a segmented reflector structure, and for bonding the reflectors to substructures. Ultramet has extensive experience in the area of ceramic-to-ceramic and ceramic-to-metal joining, typically by developing interlayers that can be deposited by CVD.

Acknowledgments

This work was performed under contract NAS1-19972 for NASA Langley Research Center (Hampton, VA), funded through the NASA Small Business Innovation Research (SBIR) program. Ultramet would like to thank the late NASA project manager, Timothy J. Towell, for his extensive input, interest, and support over the course of the project; his enthusiasm will be missed. Special thanks also go to Russell Ono of Ono Laboratories, whose invaluable efforts in grinding, polishing, and testing the reflectors contributed greatly to the success of the project. A more detailed description of this work is available elsewhere (Williams 1997).

References

Jacobs, S.F. *et al.* (1987) "Surface Figure Changes Due to Thermal Cycling Hysteresis," *Applied Optics* 26, 4438.

Schneider, S.J., ed. (1991) *Engineered Materials Handbook*, Vol. 4: *Ceramics and Glasses*, ASM International, Metals Park, OH.

Towell, T.J. (1994) Personal communication, NASA Langley Research Center, Hampton, VA.

Williams, B.E. (1997) "Lightweight SiC Composite Reflectors for Space Optics, Phase II," Final Report, Contract NAS1-19972, NASA Langley Research Center, Hampton, VA.

A CRITICAL REVIEW OF ULTRALIGHTWEIGHT COMPOSITE MIRROR TECHNOLOGY

Eldon P. Kasl and David A. Crowe
Composite Optics, Inc.
9617 Distribution Avenue
San Diego, CA 92121
(619) 621-5700

Abstract

The objective of this paper is to provide an overview of ultralightweight composite mirror technology. The overview includes a description of the technology, differences between traditional and composite designs, significant industry-wide demonstrations of the technology based on available literature (see References section), and a projection for future applications. The emergence of composite designs provides exciting potential for nontraditional, accurate, lightweight, stable, stiff, and high strength composite mirrors, such as those shown in Figure 1. This evolving technology promises significant improvement in reducing weight, cost and cycle time for future infrared, visible, and x-ray systems. Customers currently embracing composite mirror technology for radiometric use are already reaping substantial system performance benefits. Other customers interested in lidar, IR, visible, and grazing incidence x-ray applications are eagerly awaiting successful completion of current technology development and demonstration efforts.

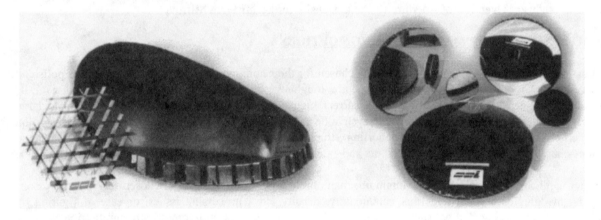

FIGURE 1.   MLS Reflector   NASA SBIR Mirrors

INTRODUCTION OF TECHNOLOGY & DESCRIPTION

The objective of this technology is to develop lightweight (5-10 kg/m^2), large (0.5-2 meter), dimensionally stable mirrors for aircraft and space optical systems. High performance optical systems have historically relied upon glass or metallic materials to provide optimum mirror designs based on trades for stiffness, weight, low, mid and high frequency figure error, CTE, dimensional stability, and thermal conductivity requirements. These optical systems require metering structures, optical benches, baffles and other structures with demanding stiffness, weight, and stability requirements. In most cases the design, behavior, and mounting designs of these structures are dominated by the supported optical components (mirrors, detectors, focal planes, etc.) The existing state-of-the-art areal density for a primary mirror assembly for diffraction limited visible systems is 25-30 kg/m^2 as shown in Figure 2. On the other hand, hundreds of composite reflectors with areal densities ranging from 1 to 3 kg/m^2 have been fabricated for microwave and millimeter wave applications. By combining the technical performance features of composite optical bench and reflector technologies with precision optical fabrication methods, the significant weight savings, excellent thermal stability characteristics, and reasonable cost afforded by this mirror technology will enable development of the next generation precision lightweight optical instruments.

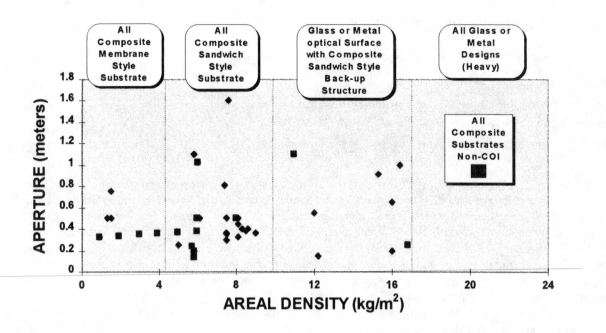

FIGURE 2. Industry Overview of Aperture vs. Areal Density for Lightweight Mirrors

TRADITIONAL MIRROR MATERIAL DESIGN OPTIONS

In most cases, glass or glass ceramics have been chosen for their homogeneity, stability, low CTE and polishing characteristics. Glass' low tensile strength, fracture toughness, and modulus of elasticity coupled with its moderately high density lead to the industry's current higher areal densities, particularly in medium to large aperture substrates. These properties force the system design and areal density to meet demanding 1G-distortion design criteria, increasing mount and supporting structure weight and yielding relatively low fundamental frequency, and in some cases requiring actuators and extensive damping provisions in active or steered systems.

In other applications, metals such as aluminum, titanium, and beryllium have been chosen for various reasons including modulus of elasticity, thermal conductivity, density, cost, processing issues, or ease of figuring by grinding, diamond turning, and polishing. Use of these materials can result in improved structural efficiency over glass in most cases, but exhibit high coefficients of thermal expansion (CTE). In many cases, the high CTE of metals requires elaborate thermal control systems to keep figure degradation within acceptable levels. The CTE problem is further exacerbated by thermal gradient sources such as transient heating or partial illumination conditions that occur during system operation. Dealing with such issues complicates the design and significantly impact cost.

In many cases, the structural efficiency (i.e., bending stiffness to weight ratio) of glass and metal can be improved moderately by lightweighting (i.e., pocketing) the back of the substrate. Pocketing removes underutilized material and results in a rib stiffened faceplate. Pocketing's effectiveness at lightweighting and creating a homogenous substrate also presents major technical issues based on the material's properties, equipment capabilities, grinding pressure, etc. The material's stiffness, strength, CTE properties, and machining characteristics affect the fabrication sequence. Equally as important are the equipment's ability to access all necessary areas to minimize radii of rib and faceplate intersections, minimize rib thickness, and control tolerances. The structural efficiency of designs for either glass or metals can be significantly improved with a sandwich style design (i.e., two facesheets separated by a core). Joining a lightweighted core to the facesheets in a predictable manner is difficult and requires specialized and reliable processes. Joining techniques include low and high temperature fusion and frit bonding for glass and brazing or soldering for metal. These sandwich technologies have increased the number of business opportunities for these materials. However, most of the aforementioned limitations still persist.

COMPOSITE MIRROR DESIGN OPTIONS

A third material candidate that is worthy of consideration is carbon fiber reinforced composite. Composite materials offer an impressive combination of material properties and processing options perhaps unmatched by any other material. These properties and processes include:

1. A near-zero CTE over a wide temperature range via proper fiber and resin selection.

2. High structural efficiency that surpasses beryllium via tailorable thin-ply manufacturing and design methods.

3. High reflectivity aluminum, silver, gold or multilayer coatings.

4. Low surface roughness that may be suitable for x-ray applications via coating and/or polishing processes.

5. High strength via proper fiber selection.

6. High fracture toughness via toughened, space-qualified matrix (i.e., resin) selection.

7. Design and manufacturing flexibility that permits the synergistic use of a composite support structure, a glass facesheet, and state-of-the-art glass figuring equipment.

In conjunction with this combination of benefits comes a dimensional sensitivity to changes in moisture content. Understanding the effects of moisture on laminates is essential. The deleterious effects of moisture on mirror figure can be minimized by proper design, material selection, sealants, and dry fabrication practices. Nitrogen gas or space environments are ideal for composite mirrors.

In the early 1990s, the application of low moisture uptake resins such as the cyanate esters to space hardware programs has dramatically reduced the epoxy distortion issues experienced in the 1980s. Current developments with siloxane resin systems and sealing (i.e., moisture barrier) technologies look promising and will further improve upon the cyanate performance or possibly eliminate moisture diffusion entirely.

SIGNIFICANT DEMONSTRATIONS OF THE TECHNOLOGY

The use of composites for mirrors began approximately twenty years ago. Over 75 composite mirrors have been fabricated and tested in the United States and Europe. The majority of the mirrors have been produced by COI. Most of the mirrors have been developed as technology demonstrators for microwave, infrared, and visible applications.

The mirror substrates developed have either been all composite substrates or hybrid designs consisting of all composite substrates with glass or metal clad facesheets. Mirrors developed for visible applications have typically been hybrid designs and have relied on traditional optical finishing techniques to achieve the appropriate figure. Examples of the all composite and hybrid mirror technology demonstrators include:

1. Composite substrates that were replicated or vapor deposited with highly reflective coatings (Figures 1 and 4).

2. A nickel-plated composite substrate that was subsequently replicated (Figure 1).

3. A nickel-plated composite substrate that was subsequently superpolished.

4. Composite substrates with a ground and polished glass facesheet supported via flexures (Figure 5).

5. A composite substrate with a diamond-turned integral aluminum surface (Figure 6).

6. A composite substrate that supports ground and polished glass facesheet (Figure 7).

7. A composite core that supports two ground and polished glass facesheets (Figure 7).

FIGURE 4. ISUS Solar Concentrator Mirror Facets
FIGURE 5. High Altitude Large Optics

FIGURE 6. MIT Mirror & Detail
FIGURE 7. GOES Mirror Composite Optics Study

A plot of aperture vs. areal density for the aforementioned mirrors and many others that have been produced is provided in Figure 8. Note that for 0.5 - 1.6 meter class all composite mirrors, areal densities have been as low as 1.5 kg/m^2 but are typically in the range of 7-9 kg/m^2. The hybrid passive designs with apertures as large as 1 meter exhibited 12-16 kg/m^2 areal density. The active surface hybrid mirror's areal density without the actuators was slightly heavier at 16.5 kg/m^2. The passive areal density ranges established by the data in Figure 8 form the guidelines for the all composite and hybrid performance projections appearing in Figure 1. Summary comments pertaining to various performance features are as follows.

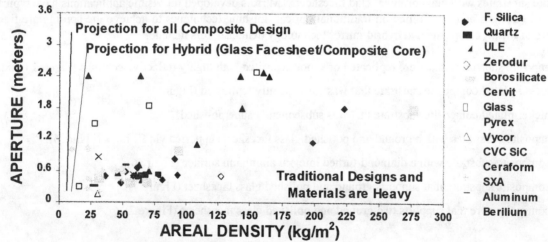

FIGURE 8. Aperture vs. Areal Density

Mirror Quantity vs. Profile Type

Most of the surface profiles of the mirrors have been plano or spherical. Other profiles produced include torroidal and parabolic both on-axis and offset.

Figure vs. Areal Density

The figure vs. areal density data indicate that the hybrid (i.e., heavier) mirrors have exhibited the best visible-quality ambient figure. However, one of the more recently produced all composite mirrors (Peck and Raquel 1994) has approached the ambient figure exhibited by the hybrid mirrors.

Figure vs. Temperature

The figure vs. temperature data indicate that several of the recently produced all composite designs and one of the recent hybrid designs have exhibited exceptional figure stability over a broad temperature range. These data demonstrate the benefits of near-zero CTE and proper design. The all composite designs had plano and spherical profiles with fundamental frequencies in excess of 2200 Hz. The hybrid mirror was a plano.

Figure vs. Aperture

The figure vs. aperture data indicate that most of mirrors fabricated have approximately 0.5 meter apertures with 1-3 μm RMS figure. Two visible-quality ($\lambda/20$ and $\lambda/50$ at .633 μm) meter class mirrors and one 1.6 meter aperture offset parabola with 4.5 μm RMS figure have also been produced. A few smaller aperture mirrors exhibited excellent figure at ambient temperature. In general, one of the traditional thoughts in the industry -- the larger the aperture, the larger the figure error -- is not substantiated by this data.

Figure vs. Year

The figure vs. time data indicate that the figure of all composite mirrors has generally become more precise in recent years. The impressive figure of the hybrid design has been consistent over the years. Recently, however, meeting this figure quality with lower areal density hybrid design concepts has presented new technical challenges.

PROJECTIONS FOR THE FUTURE

Many different applications will benefit from this technology in the near future. The biggest beneficiaries include pointing and fast scanning optical systems. Examples include optics for laser communications, surveillance instruments, remote sensing, and telescopes for space and airborne applications. Other potential applications on earth include machinery for high volume precision fabrication processes. Current research efforts at COI are supporting research and development of products for such applications.

CONCLUSIONS

Many ultralightweight mirrors have been fabricated with composite materials. Large aperture (0.5 to 1.6 meter) all composite mirrors have demonstrated 7-9 kg/m^2 areal density, figure accuracies of a few microns RMS, excellent figure stability over temperature, and high fundamental frequency. Large aperture (1 meter) hybrid composite mirrors have demonstrated 12-16 kg/m^2 areal density and figure accuracies of $\lambda/20$ and $\lambda/50$ at .633 μm.

Acknowledgments

The authors would like to acknowledge the advice and assistance provided by Dr. Eri J. Cohen, JPL. We also wish to acknowledge the support of NASA's Small Business Innovative Research Program.

References

Abt, B., Gunter Helwig, and Dietmar Scheulen (1990) "Composite Technology for Lightweight Optical Mirrors," SPIE, Advanced Technology Optical Telescopes IV, 1990, :1236.

Coulter, D. (1992) "Survey of Lightweight Mirror," Memo #355-2-92-020 to M. Lou, JPL, 11 February 1992.

Crowe, D.A. (1997) "Lightweight Mirror Summary," Interoffice Memo to Eldon Kasl, COI, 9 May 1997.

Eastman Kodak Company (1992) "Composite Optic Technology Development Program Final Report" developed by EKC and COI for Goddard Space Flight Center, Rochester, NY 14653-8123, 21 December 1992.

Ehmann, D., Wilfried Becker, and Hubert Salmen (1994) " Aspects of Structural Design Optimization for High Precision Reflectors and Mirrors," 1994.

Freeland, R.E. P.M. McElroy, and R.D. Johnston (1989) "Technical Approach For the Development of Structural Composite Mirrors," SPIE, Active Telescope Systems, 1989, :1114.

Gormican, J.P., S. Kulick, E.P. Kasl, and L.B. Abplanalp (1993) "Dimensional Stable, Graphite-Fiber Reinforced Composite Mirror Technology," SPIE, Space Astronomical Telescopes and Instruments II, Orlando, 1993, :1945.

Helms, R. G., C.P. Porter, Y.C. Wu, C.P. Kuo, R.N. Miyake, and P.M. McElroy (1989) "Lightweight Composite Mirror Analysis and Testing," SPIE, Active Telescope Systems, 1989, :1114.

Jean-Michel, Lamarre (1996) "PRONAOS," Facsimile to Eri Cohen, JPL, 6 May 1996.

Kasl, Eldon, and Zvi Friedman (1996) "Lightweight, Thermally Stable, High Resolution, Composite Optics," Phase II Final Report, April 1996.

Peck, S. and F.C. Raquel (1994) "Composite Scan Mirror," 39th International SAMPE Symposium, 11-14 April 1994.

Petersson, Mikael "Development of Extremely Accurate Space Reflectors," 41st International SAMPE Symposium and Exhibition, Material & Process Challenges: Aging Systems, Affordability, and Alternate Applications, :41(1 of 2).

Rapp, D. (1991) "Precision Composite Mirror Panel Development Task," Final Report, JPL Report No. D-9204, 20 December 1991.

Rapp, D. (1992) "Precision Composite Mirror Panel Development Task," Revised Final Report Draft, JPL, November 1992.

Stern, Theodore (1997) "Graphite Fiber Reinforced Composite Submillimeter Reflectors," SBIR Phase II Final Report, February 1997.

Sultana, J.A. and S. E. Forman (1990) "A Graphite/Epoxy Composite Mirror for Beam Steering Applications," Optical Engineering, November 1990, :29(11).

Tompkins, S.S., J.G. Funk, D.E. Bowles, T.W. Towell, and J.W. Connell (1992) "Composite Materials for Precision Space Reflector Panels," SPIE, Design of Optical Instruments 1992, :1690.

Tompkins, S., D.E. Bowles, J.G. Funk, J.A. Lavoie, and T.W. Towell (1990) "The Development of Composite Materials for Spacecraft Precision Reflector Panels," Composite Materials for Optical and Electro-Optical Instruments, Orlando, 1990.

ISSUES TO BE ADDRESSED IN THE DESIGN AND FABRICATION OF ULTRALIGHTWEIGHT, METER CLASS OPTICS

Gary C. Krumweide
Director, Advanced Programs
Composite Optics, Incorporated
9617 Distribution Avenue
San Diego, CA 92121
619/621-5700

Abstract

There is a growing need for large aperture, ultralightweight, deployable optics (mirrors) for various science, military and commercial compact satellites (Pleimann 1997). This paper will examine the engineering and manufacturing considerations that must be addressed in order to satisfy the requirements for these sought after optics. In order to limit the scope of this paper, only Graphite Fiber Reinforced/Polymer Matrix Composites (GFR/PMC) will be under consideration because of the potential to satisfy ultralightweight mirror requirements (5 - 10 Kg/m^2) (Kasl 1997). The requirements associated with specular mirror concepts that Composite Optics, Incorporated (COI) has proposed to Air Force Research Laboratory (AFRL/VS) and NASA Langley Research Center for visible range optics and LIDAR (light bucket) optics, respectively, will also be our interest. Moreover, it is the intent of this paper to illustrate how COI's proposed design/manufacturing concepts for visible and LIDAR optics have evolved based on overcoming, or working around, material constraints and/or undesirable characteristics associated with GFR/PMC.

INTRODUCTION

Given the need for ultralightweight, meter class mirrors (optics) for various science, military, and commercial compact satellite applications, COI has set out to address several challenging issues in the production of such mirrors. The feasibility of producing visible range mirrors can be attributed to recent technology advances in materials, innovative design/fabrication approaches, moisture barrier development and mirror post-machining processes. These recent technological advances and several supporting Small Business Innovative Research (SBIR) Programs (see Table 1) allowed COI to propose a mirror design/fabrication concept that provided a "workaround" solution to the following issues:

1. How to employ an anistropic laminate of GFR/PMC in a mirror or mirror substrate that requires micron-level dimensional stability.
2. How to employ a hygroscopic laminate of GFR/PMC in a mirror or mirror substrate that requires micron-level dimensional stability.
3. How to match the Coefficient of Thermal Expansion (CTE) of glass (necessary to produce visible range [$\lambda/40$ RMS] optical surfaces) with GFR/PMC (needed as a lightweight substrate).
4. How to assure GFR/PMC will not microcrack.
5. How to meet 5 - 10 Kg/m^2 weight requirements.
6. How to obtain/maintain figure ($\lambda/40$ RMS visible optics and $\lambda/2$ RMS LIDAR optics) during assembly of the optics.
7. How to mount optics without distortion or distortions of operation.
8. How to assure long-term stability of optics.
9. How to produce ultralightweight optics that are dynamically stable.
10. How to produce large, ultralightweight optics for visible range and LIDAR economically.

These types of issues have been a concern for many years when using GFR/PMC for dimensionally stable structures and will be addressed in the ensuing discussion (Krumweide 1988).

In order to illustrate how a specific issue (above) is being addressed for the two mirrors (i.e., visible, LIDAR), a design concept for each will be presented and the various features of each reviewed for their function and/or resolution of the issue at hand.

TABLE 1. COI SBIRs Supporting Mirror Development Technology.

NAS8-39826	1994	Phase I	NASA/Jet Propulsion Laboratory	Graphite Fiber Reinforced Composite Submillimeter Reflectors
NAS8-40170	1993	Phase II	NASA Marshall Space Flight Center	Thermally Stable, Large Aperture, High Resolution Optics
NAS8-40511	1994	Phase I	NASA/Marshall Space Flight Center	Fiber Print Thru Avoidance and Stability Enhancement Using Carbon Fiber Composites for Grazing Incidence X-Ray Optics
NAS7-1259	1994	Phase I	NASA/Jet Propulsion Laboratory	Graphite Fiber Reinforced Composite Submillimeter Reflectors
NAS8-97202	1997	Phase I	NASA/Marshall Space Flight Center	Lightweight Carbon Fiber Composite Mirror Fabrication Using Advanced Core Technology
NAS8-97151	1997	Phase I	NASA/Marshall Space Flight Center	Barrier Coatings That Minimize Hygroscopic Change Of Ultra High Modulus Carbon Reinforced Cyanate Ester Resins
NAS8-97274	1997	Phase II	NASA/Marshall Space Flight Center	Fiber Print Through for Grazing Incidence X-Ray Mirrors
F29601-97-C-0084	1997		Phillips Laboratory	Development of Lightweight Visible Range, Meter Class Optics Using PM/CFC) Materials

DISCUSSION

The visible range optics and LIDAR optics will be discussed separately in order to understand their differences (i.e., design/fabrication/assembly).

Visible Range Optics

The design concept for this mirror is mostly driven by the high resolution requirement defined by the AFRL/VS. Table 2 lists the current SBIR Phase I mirror requirements provided by AFRL/VS (Pleimann 1997). The surface roughness of 10-20 angstroms figure of $\lambda/40$ RMS and areal density of 5-8 Kg/m^2 suggests a hybrid design of ULE glass and GFR/PMC be considered.

TABLE 2. SBIR Phase I Mirror Requirement.

Definition	Requirement
Size	0.25m diameter substrate
Shape	Spherical
Radius of Curvature	2m
Areal Density (Kg/m^2)	5-8
Surface Roughness (angstroms)	10-20
Figure	$\lambda/40$ rms
Wavefront	$\lambda/20$ rms
Spectrum	Visible
Temperature Range (K)	±5° from room temperature
Survivable Thermal Range (K)	268° to 328°
Testing	Optical axis in vertical position
Launch Environment	Space Shuttle
First Mode	>80 Hz w/free-free mount
Scalability	Multiple meter class

Figure 1, which COI calls its Fixed Third Surface Mirror Design, has been selected by the AFRL/VS for evaluation for Ultra-LITE (a sparse aperture, deployable, ground-based test program). The mirror is 0.25m diameter and has a 2m radius of curvature (spherical).

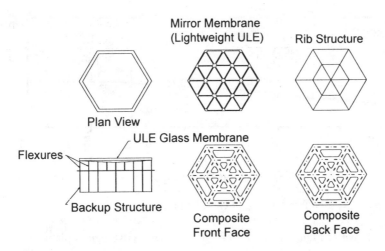

FIGURE 1. Fixed Third Surface Mirror Design.

The lightweighted ULE glass membrane, with its isogrid pattern, mounts on flexures (located at each node) protruding out of the GFR/PMC back-up structure. This back-up structure consists of GFR/PMC facesheets (laminate) and GFR/PMC ribs (laminate). The ribs form a similar isogrid pattern that matches the isogrid pattern of the lightweighted ULE glass membrane. Recesses, waterjet machined when the basic isogrid node is machined (by Waterjet Technologies, Inc.), provides a bonded structural interface between the ULE glass and the dowel-like flexures. The flexures are uni-directional GFR/PMC. The lockout position of each flexure, by "wick bonding" at both the top and bottom skin, is its only faying surface with the back-up structure. Given this basic design approach, the issues presented in the introduction will now be addressed.

Anistropic laminate workaround: Because through-the-thickness CTE properties of a GFR/PMC are approximately 36 ppm/°C, while its in-plane properties for a candidate material are approximately -0.30 ± 0.2 ppm/°C, construction of the back-up structure design has to "null out" the effect of the high through-the-thickness CTE. Since micron stability is desired, this CTE of 36 ppm/°C with operational temperature variations from room temperature ±5°C is undesirable. By using thin ribs (>0.75mm thick) that bond to the inside surfaces of the facesheets and not bonding the ribs at the node points (where they cross one another in the isogrid), the distorting effect of through-the-thickness is eliminated. It should be noted that the symmetry of the back-up structure design and manufacturing approach assures greater dimensional stability of this structure.

Hygroscopic laminate workaround: The effects of moisture ingress and egress (CME) on GFR/PMC is, by 2 or 3 orders of magnitude, a greater problem than CTE. This is true even when Polycyanate Ester resins are used; which have one-third lower (CME) than Epoxy resins. For this design, COI is employing, a moisture barrier to eliminate the need to do all testing of the mirror assembly in a vacuum, or a very low relativity humidity (RH) environment. Otherwise, the mirror would ingress the moisture typical of an assembly environment (30 - 60% RH) and egress the moisture when operating in a dry space environment (0% RH). Figure errors would result unless the mirror is final figured and operates in a dry (0% RH) environment.

COI moisture barriering methods have been used successfully. Figure 2 shows moisture barriered telescope assembly coupons fabricated by COI in 1985 (Krumweide 1989). Most recent testing of moisture barriers applied to coupons indicates even better moisture barriering results.

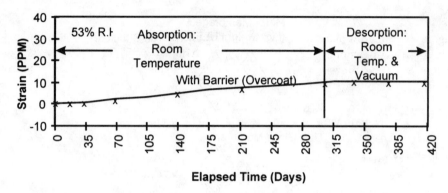

FIGURE 2. Moisture Barrier on .04 inch thick P75/930 Gr/E laminate.

For the back-up structure shown in Figure 1, a 100% moisture barrier coverage is planned after the structure is assembled.

CTE mismatch workaround: COI has been very successful in tuning a GFR/PMC laminate CTE to a near zero CTE – meaning a CTE of 0.0±0.024 ppm/°C is possible, limited mostly by the measurement accuracy of our Laser Optical Comparator (LOC). A workaround method that is appropriate for the Fixed Third Surface Mirror, shown in Figure 1, is to measure the lightweighted ULE glass membrane "side-by-side" with the moisture barriered back-up structure using a tilt mirror across the two parts; a typical measurement technique for the LOC. Any substantial difference in CTE can be verified and corrected.

Another workaround plan is the use of flexures. These dowel-like flexures will mitigate any CTE mismatches that are built into the assembled mirror or result from long-term exposure to the space environment (i.e., radiation effects).

Microcracking workaround: COI incorporates fiber/resin combinations necessary to prevent microcracking (Krumweide 1991).

Weight requirement of 5-10 Kg/m^2: This is the most difficult requirement to satisfy because the lightweighted ULE glass is over half the weight of the design. The glass is waterjet machined to reduce its basic skin thickness to under 2mm which is optional for this post-machining process and for maintaining surface figure of the mirror.

The back-up structure has been designed to remove all the unnecessary facesheet material (top and bottom cutouts) and still provide the desired stiffness. Note: Relatively large spacing between facesheets.

Surface figure of $\lambda/40$ rms: Four steps are necessary to assure this requirement is achievable. The first is to support the ULE glass membrane on flexures at isogrid node points. The second is to machine the lightweight ULE glass to a fraction of a wave λ rms ($\approx \lambda/2$). The third is to bond flexures only to the membrane and, yet, unattached (unbonded) to the back-up structure until the original (or near original) surface accuracy is adjusted into the ULE glass membrane via the axial movement of the flexures. Here adjusters are attached to each flexure and systematically activated while surface figure is monitored using a laser interferometer. The fourth step is to post machine the lightweighted ULE glass.

Mounting distortions: The use of a Fixed Third Surface Mirror design is of importance here. The flexures will compensate for any CTE mismatch type of distortions associated with mounting. It is assumed that a 3-point (determinate) mount system will be employed.

Long-term stability: As mentioned previously, the use of flexures is to mitigate long-term distortion due to radiation effects on the polymer matrix.

Dynamic stability: The diameter of the individual flexure to provide adequate stiffness for 1 "G" sag and produce a frequency of approximately 250 Hz will also provide enough flexibility to compensate for a differential CTE between the glass and GFR/PMC (0.04 ppm/°C) and maintain figure.

Economical design: By using flat laminate construction techniques for the back-up structure, the design is very economical. That is, no expensive tooling (PDMO) is required. Also, the capability of cutting enough parts from a single flat laminate to make several mirrors is possible. This design requires no master mold, for mirror figure, as is necessary in a replicated mirror figuring processes.

Thus, an ultralightweight visible optic, as illustrated in Figure 1, may be possible using GFR/PMC if attention is given to the undesirable characteristics of this GRF/PMC material.

LIDAR (Lightweight) Optics

Figure 3 depicts the Fixed Third Surface Mirror being proposed for a NASA LaRC LIDAR Mirror concept. The similarity to the AFRL/VS mirror is not coincidental but is intended to get some synergism from these two programs. COI proposed to utilize the same back-up structure for both mirrors. So the primary difference is that the ULE glass membrane for the AFRL/VS mirror is replaced with a thin GFR/PMC membrane for the NASA LaRC mirror. The thin membrane is the same material and layup used for the flat facesheet laminates in the back-up structure. Another minor difference is that a circular doubler (doughnut-shaped) made from the same membrane (outer trim of basic laminate) is bonded to the top of each flexure and to the outer surface of the membrane itself. Injection of a "wicking adhesive" into the faying surface between a fully moisture barriered membrane and fully moisture barriered flexures, bonds the flexures to the membrane and its doubler. Given this basic design approach, the issues presented in the introduction will now be addressed.

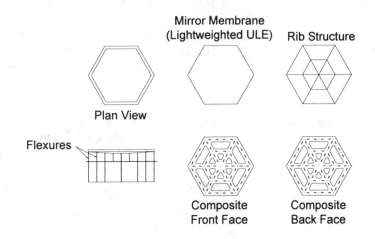

FIGURE 3. Fixed Third Surface Mirror Design

Anistropic laminate workaround: Same as for AFRL/VS Mirror.

Hygroscopic laminate workaround: Same as for AFRL/VS Mirror.

CTE mismatch workaround: Same as for AFRL/VS mirror except GFR/PMC membrane is measured side-by-side with back-up structure using the LOC. CTE mismatch is expected to be minimal because the materials used in fabricating the membrane and in the fabrication of the back-up structure are essentially the same.

Microcracking workaround: Same as for AFRL/VS mirror.

Weight requirement-of-4-5 Kg/m^2: The lightweight GFR/PMC membrane allows this type of weight reduction over a glass membrane to be achieved.

Surface figure of $\lambda/2$ rms: Several very important factors bear on achieving this requirement. The first is symmetry of the membrane. This includes the incorporation of "rotate and fold" lay-up methods, moisture barrier on both sides of membrane (100%), and use of the replication adhesive on back surface as is used when front surface when replicated. It should be noted that a Fixed Third Surface Mirror easily allows this manufacturing procedure. The second factor is the use of the replication approach in providing an acceptable smooth, specular surface that, for the most part, does replicate the quantity of glass master mold surface.

The third factor is the use of adjusters attached to the flexures that adjust the membrane to near its original surface figure (when released from the glass mold). The flexures are then locked out with adhesive as described for the AFRL/VS mirror. The use of a laser interferometer is also applicable here.

Mounting distortion: Same as for AFRL/VS mirror.

Long term stability: Same as for AFRL/VS mirror.

Dynamic stability: Same as for AFRL/VS mirror.

Economical design: Same as for AFRL/VS mirror although, the cost of replication adds to the mirrors final cost as does the need for a glass master mold to do this replication on.

As with to the AFRL/VS mirror, it is very possible that a Fixed Third Surface Mirror for LIDAR application can be designed and fabricated if attention is given to the undesirable characteristics of the GFR/PMC material.

CONCLUSIONS

The ability of a Fixed Third Surface Mirror to meet visible range optics and LIDAR (light bucket) requirements is going to be demonstrated soon. Design and Manufacturing methods are to be incorporated into the mirror to workaround the undesirable characteristics of graphite fiber reinforced polymer matrix composite materials.

Acknowledgments

The author would like to thank Messrs. Eldon Kasl, Jack Dyer, and Greg Mehle for their support in the preparation of this paper. We also wish to acknowledge the support of the Air Force Phillips Laboratory and NASA Small Business Innovative Research programs.

References

Kasl, Eldon P. and David E. Crowe, (1997) "A Critical Review of Ultralightweight Composite Mirror Technology," SPIE, Critical Review of Optical Science 1997; CR67.

Krumweide, Gary C. and David N. Chamberlin, (1988) "Adaptation and Innovation in High Modulus Graphite/Epoxy Composite Design: Notes on Recent Developments", SPIE, O-E LASE, Los Angeles, CA, January 1988.

Krumweide, Gary C., Ed A. Derby, and David N. Chamberlin, (1989) "The Performance of Effective Moisture Barriers for Graphite/Epoxy Instrument Structures," SAMPE, Atlantic City, NJ 1989.

Krumweide, Gary C. and Richard A. Brand, (1991) "Attacking Dimensional Instability Problems in Graphite/Epoxy Structures, " Composite Design, Manufacture and Application, ICCM/8, Honolulu, HI.

Pleimann, Mark E. (1997) "Lightweight Meter Class Optics," IEEE, Aspen, Colorado, March 1998.

ADVANCED LIGHTWEIGHT OPTICS DEVELOPMENT FOR SPACE APPLICATIONS

James W. Bilbro
Optics & Radio Frequency Division
NASA/Marshall Space Flight Center
Huntsville, AL 35812
(205)544-3467

Abstract

A considerable amount of effort over the past year has been devoted to exploring ultra-lightweight optics for two specific NASA programs, the Next Generation Space Telescope (NGST), and the High Throughput X-ray Spectrometer (HTXS). Experimental investigations have been undertaken in a variety of materials including glass, composites, nickel, beryllium, Carbon fiber reinforced Silicon Carbide (CSiC), Reaction Bonded Silicon Carbide, Chemical Vapor Deposited Silicon Carbide, and Silicon. Overall results of these investigations will be summarized, and specific details will be provided concerning the in-house development of ultra-lightweight nickel replication for both grazing incidence and normal incidence optics. This will include x-ray test results of the grazing incidence optic and cryogenic test results of the normal incidence optic. The status of two 1.5 meter diameter demonstration mirrors for NGST will also be presented. These two demonstrations are aimed at establishing the capability to manufacture and test mirrors that have an areal density of 15 kilograms per square meter. Efforts in thin membrane mirrors and Fresnel lenses will also be briefly discussed.

INTRODUCTION

NASA's George C. Marshall Space Flight Center (MSFC) has been involved in the development of advanced optics for both grazing and normal incidence for over 30 years, beginning with experimental work for the proposed Large Space Telescope in the late 60's and continuing through the development and flight of the Apollo Telescope Mount, the SO-56 X-ray Telescope, the High Energy Astrophysics Observatory (HEAO) Series (which included the Einstein Observatory), and the Hubbell Space Telescope. Most recently, MSFC has had the responsibility for the development of the Advanced X-ray Astrophysics Facility - Imaging (AXAF-I) which is due to be launched in 1998. MSFC began concentrating in the development of electro-formed nickel mirrors for the Advanced X-ray Astrophysics Facility - Spectroscopy (AXAF-S) in 1992 and in 1994 began to focus on producing lightweight x-ray mirrors. The effort was expanded in 1996 to begin work in the development of extreme lightweight normal incidence mirrors in support of the Goddard Space Flight Center's study of the Next Generation Space Telescope (NGST). The overall activity now includes work in silicon carbide, composites, glass, beryllium, alumina, silicon, nickel and plastics. The effort is largely divided into two major categories, grazing incidence and normal incidence, although there is additional activity in the development of large deployable Fresnel lenses and diffractive optics for micro-miniature optics. The following sections will cover these areas in more detail.

GRAZING INCIDENCE OPTICS

The purpose of this effort is aimed at the development of very lightweight, high resolution mirrors that can be produced rapidly and cheaply.

The primary in-house concentration in grazing incidence optics is in the development of ultra-lightweight electro-formed nickel Wolter Type I optics. The primary and secondary mirrors are respectively paraboloid and hyperboloids of revolution and are manufactured as a single piece. The process starts with the manufacture of a mandrel made of aluminum which has been machined into near net shape. The mandrel is then diamond turned, coated with kanigen (electroless nickel), diamond turned again for final figure, and then polished. The mandrel is then coated with gold and the nickel shell electro-formed over the coated mandrel. The shell is then cryogenically separated from the mandrel. The investigations are currently concentrated on determining the minimum shell thickness that can be successfully separated from the mandrel without the shell being deformed either due to separation stresses or subsequent handling stresses. A number of techniques are being examined including using reinforcing the shell by electro-forming rings onto the shell after it has reached a certain thickness. The use of special shell support structures for used during the separation process are also being investigated.

A number of other materials are being investigated under various contracts. These can be divided into two basic approaches. One approach is aimed at producing a shell which directly replicates the optical surface in the process of generating the shell and the second approach is to produce a structural support shell and then bond the support shell to an optical surface which has been produced separately. The first approach is what is used for the electro-formed nickel optic described above. This "Direct Replication Process" (DRP) is being investigated for Chemical Vapor Deposition (CVD) Silicon Carbide (SiC) under a contract with Morton International. This effort involves the deposition of Silicon Carbide on to a polished mandrel and then separating the shell from the mandrel in a manner similar to that done for the electro-formed nickel. Two main areas under investigation for this process. The first involves finding suitable mandrels that can be polished to the surface finish and accuracy required for the optic and that at the same time withstand the hostile environment associated with the CVD process. The second involves the development of uniform passivation techniques that will permit accurate replication of the mandrel optical surface and that will permit separation of the shell from the mandrel without adversely affecting the optical surface of the shell or of the mandrel. This investigation is a collaborative undertaking with the normal incidence program. A second approach to the DRP involves the use of vacuum deposition of Beryllium. Investigation of this approach has just begun. A number of other techniques involving plasma spray forming of materials such as alumina are being investigated jointly with the Smithsonian Astrophysical Observatory. The second approach involves the production of two mandrels, one which will be used to produce the optical surface and one which is slightly oversized to produce the support shell. This investigation involves examining different materials for use in support shells as well as examining different methods for bonding (or otherwise producing) the optical surface (such as epoxy replication). For this approach, again Morton International is investigating the production of Silicon Carbide shells. In this case the optical surface is not an issue and the primary goal is to produce a shell which will maintain the roundness and axial figure necessary to permit bonding of the optical surface. The dimensional tolerances in this case are considerably relaxed form the previous case in that what is desired here is to maintain a reasonably constant (and narrow) epoxy bond gap so that as the epoxy shrinks during curing it will not distort the optical surface. The use of composite materials for shells is being investigated under a Phase II Small Business Innovative Research (SBIR) contract with Composite Optics Incorporated. An alumina support shell is being fabricated by Coors.

NORMAL INCIDENCE OPTICS

This effort is focused on supporting the Goddard Space Flight Center's NGST study. The goal of the program is to demonstrate the ability to produce an ultra-lightweight primary mirror that can be adaptively corrected or phased at an areal density of 15 kg/m2 including mirror, back plane and actuators. This translates into a mirror areal density of 5-7 kg/m2 - a significant step in the production of lightweight mirrors. The program has two primary components, one aimed at the production of two 1.5m diameter demonstration mirrors and the other aimed at developing alternative mirror materials in smaller 0.5m diameter demonstrations. The latter effort makes use of both project funds and funds from the Small Innovative Research program. The two 1.5m mirror demonstrations were competitively won by the University of Arizona and Composite Optics Incorporated (COI). The University of Arizona concept involves the production of a 2mm thick glass shell supported by a series of actuators against a composite back plane structure. The COI approach involves the production of a lightweighted 10mm thick glass face sheet bonded to a composite support structure which is the supported by a series of actuators against a composite back plane structure. These mirrors are to be completed by the summer of 1998. In parallel, the other component of the program is developing a number of smaller mirror demonstrations which if successful will result in a second 1.5m diameter demonstration in the 2001 time frame. A request for proposal is currently being prepared for the demonstration of a 0.5m Beryllium mirror. The contract is expected to be awarded in the November, 1997 time frame. As in the case of the grazing incidence program, work is underway in-house on Nickel replication and at Morton International on CVD SiC replication. Small study contracts have recently been awarded to IABG, Munich Germany to investigation into Carbon fiber reinforced Silicon Carbide (CSiC) and Schafer Inc. for Silicon mirror investigations. The SBIR program has efforts at Xicera for Ceraform and Ceracore Reaction Bonded Silicon Carbide, SSG for CSiC, Ultramet for Pyrolized Graphite Composite and

COI for Composites. An additional SBIR effort is underway at Waterjet Tech. Inc., involving lightweighting techniques employing water jet cutting.

THIN MEMBRANE OPTICS

Three separate efforts are involved in the area of thin membrane optics, one area involves the development of techniques for spin casting of Polymide films on convex mandrels to produce membranes that would take a concave shape in a zero g environment, a second area which is to start in October, 1997 is the investigation of techniques for the production of large diameter (2m) Fresnel lenses for the detection of Cosmic Ray Showers. The third area involves the production of diamond turned mandrels for use in spin casting of polymide Fresnel lenses and the direct generation of Polymide film Fresnel lenses using diamond turning.

SUMMARY

Investigations in advanced ultra-lightweight optics have been undertaken for both normal incidence and grazing incidence mirrors as well as for Fresnel lenses. These effort will lead to low cost, lightweight, high performance mirrors systems which will enable a new class of space based missions in the post 2000 era.

Acknowledgments

Material for this paper was provided by William D. Jones, John Redmon Jr., Edward V. Montgomery IV, Michael D. Watson and Vinson B. Huegele.

References

The NGST (Next Generation Space Telescope) Study Team (1997) "Visiting a Time When Galaxies Were Young," *The Next Generation Space Telescope,* H.S. Stockman, The Association of Universities for Research in Astronomy, Inc. Pages: 1-163.

RESEARCH ON THE PROBLEM OF HIGH-PRECISION DEPLOYMENT FOR LARGE-APERTURE SPACE-BASED SCIENCE INSTRUMENTS

Mark S. Lake
NASA Langley Research Center
Hampton, VA 23681
757-864-3114

Lee D. Peterson, M. Roman Hachkowski,
Jason D. Hinkle, and Lisa R. Hardaway
University of Colorado
Boulder, CO 80309
303-492-1743

Abstract

The present paper summarizes results from an ongoing research program conducted jointly by the University of Colorado and NASA Langley Research Center since 1994. This program has resulted in general guidelines for the design of high-precision deployment mechanisms, and tests of prototype deployable structures incorporating these mechanisms have shown *micro*dynamically stable behavior (i.e., dimensional stability to parts per *million*). These advancements have resulted from the identification of numerous heretofore unknown microdynamic and micromechanical response phenomena, and the development of new test techniques and instrumentation systems to interrogate these phenomena. In addition, recent tests have begun to interrogate *nano*mechanical response of materials and joints and have been used to develop an understanding of nonlinear *nano*dynamic behavior in *micro*dynamically stable structures. The ultimate goal of these efforts is to enable *nano*-precision active control of *micro*-precision deployable structures (i.e., active control to a resolution of parts per *billion*).

INTRODUCTION

NASA's Office of Space Science has recently inaugurated the Astronomical Search for Origins and Planetary Systems (Origins) Program. Between the years 2005 and 2010, this program will launch a series of extraordinary new astrophysical science instruments including the Next Generation Space Telescope (NGST, Fig. 1(a)). Also, NASA's Office of Mission to Planet Earth is studying a class of non-imaging-quality deployable telescopes for future atmospheric science missions (Fig. 1(b)). These so-called "lidar" (light direction and ranging) telescopes are to be used for the measurement of climatically important variables in the atmosphere such as water vapor, aerosol and clouds, and certain chemical constituents such as ozone. The success of these revolutionary science instruments depends on the development of many new materials and structures technologies including sub-micron-stable, actively controllable deployable structures (Peterson et al. 1996).

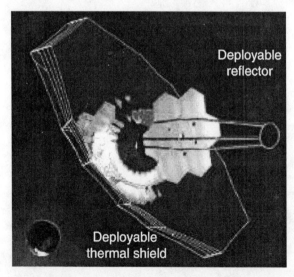
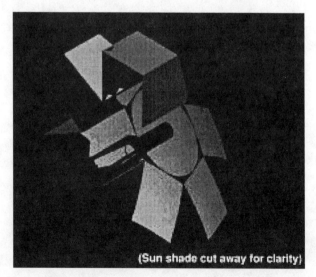

(a) Next Generation Space Telescope Concept.   (b) Deployable Lidar Telescope Concept.

FIGURE 1. Future NASA Deployable Telescope Concepts.

Since 1994, the University of Colorado and NASA Langley Research Center have been jointly conducting a research program to advance the state-of-the-art of high-precision deployable structures. The broad goals of this program are to enable the design of deployable structures that exhibit sufficient dimensional stability to function as metering structures for optical instruments, and to minimize the complexity and cost of such structures by maximizing passive dimensional precision and minimizing the use of active control. The specific goals of this program are: 1) to characterize experimentally and to model analytically microdynamic instabilities in deployable structures that arise from deployment mechanisms and material nonlinearities; 2) to characterize experimentally and to model analytically the micromechanical response within deployment mechanisms that gives rise to dimensional instabilities; and 3) to develop design rules for deployable structures to minimize dimensional instabilities, and to maximize deployment precision and post-deployment stability.

The present paper provides an overview of this ongoing research program. Specifically, techniques for conducting dynamic-response testing of precision deployable structures and results from some tests are presented. Dominant nonlinear microdynamic response effects derived from these tests are discussed, and the affects of these microdynamic responses on the design of active shape control systems are outlined. In addition, techniques for conducting micromechanical-response tests on deployment mechanisms and results of these tests are presented, and general guidelines for designing high-precision deployment mechanisms are derived. Finally, an overview of current research thrusts is presented.

PASSIVE PRECISION VERSUS ACTIVE CONTROL OF DEPLOYABLE STRUCTURES

The design of any lightweight precision deployable structure must involve a trade between passive precision and active control. Before explaining the issues that must be considered in performing an active-versus-passive system trade, it is useful to define explicitly two aspects to the dimensional precision of a deployed structure:

1. **Deployment Precision**: The error in the final deployed shape of a structure as compared to its ground-measured shape.

2. **Post-Deployment Stability**: The variation in the deployed shape of a structure in response to on-orbit thermal and mechanical loads.

For moderate-precision applications (i.e., precision up to parts per hundred thousand), it is possible to satisfy the deployment precision and post-deployment stability requirements passively through proper mechanical design of the deployment mechanisms. However, high-precision applications will require on-orbit active control to meet dimensional stability requirements. To minimize the cost and complexity of active control systems, it is prudent to characterize carefully the passive deployment precision and post-deployment stability of the structure and to establish adequate, but not overly conservative, requirements for active control. For example, characterizing the passive deployment precision of a structure determines the need and requirements for an on-orbit quasi-static active control. Similarly, characterizing the post-deployment stability of a structure defines the need and requirements for high-bandwidth active control.

Without a clear understanding of deployment precision and post-deployment stability, it is difficult to establish reasonable requirements for on-orbit active-control systems. Uncertainty in these requirements may lead to increased complexity (and cost) in the development of an on-orbit active-control system. Furthermore, uncertainty in the requirements leads to increased system-development risk because estimates for deployment precision and post-deployment stability requirements that are thought to be conservative might, in fact, neglect effects, such as high-bandwidth microdynamics, that could substantially diminish the performance of the active system.

Thus, attaining a better understanding of passive deployment precision and post-deployment stability is necessary for the design of efficient (i.e., not overly conservative) and low-cost active control systems. While it is clear that active control cannot or should not be eliminated entirely, experience to date indicates that accurate characterization of passive deployment precision and post-deployment stability can lead to improvements in passive performance and can enable the use of efficient and low-cost active control systems. The present research program can be viewed as an effort to enable the design of precision deployable structures which are very low in cost because they incorporate a reasonable combination of passive precision and active control.

MICRODYNAMICS OF PRECISION DEPLOYABLE STRUCTURES

The word "microdynamics" is ordinarily associated with a broad class of nonlinear structural dynamic phenomena with response magnitudes at or below the microstrain level (i.e., 10^{-6} times a characteristic dimension of the structure). By definition, microdynamics are nonlinear and difficult to model analytically. Furthermore, experience has shown that the inherent uncertainties associated with microdynamics leads directly to uncertainty in the post-deployment stability of a structure.

Perhaps the first question which one might ask regarding microdynamics is: "How can nonlinearities arise when motion becomes small?" This question follows from the fact that the nonlinear terms in the general governing equations of solid mechanics become negligible as deflections become small. However, these governing equations consider only geometric and material nonlinear effects which, by definition, are negligible for small motions.

For sub-microstrain levels of motion, the usual assumptions used for classical mechanics models of structures and materials are questionable. Generally, classical mechanics models exclude inhomogeneous material effects (e.g., intercrystalline slippage in metals, or fiber-matrix debonding in composite materials) and non-conservative interface mechanics (e.g., sliding, yielding, and creep that happens in friction interfaces within the structure). Experience to date indicates that non-conservative interface mechanics within deployment mechanisms dominates the microdynamics of deployable structures at the tens of microstrain level down to the nanostrain level. It is currently theorized that inhomogeneous material effects dominate the microdynamics of structures at sub-nanostrain levels. In any event, it is important to realize that the usual assumptions for classical mechanics models fail to include the effects which dominate the microdynamic response of deployable structures. However, it is incorrect to infer from this observation that microdynamic response is purely random and unpredictable.

Dominant Microdynamic-Response Effects

While our understanding and observations of microdynamic nonlinearities are incomplete, there are several effects which we currently know can affect both deployment precision and post-deployment stability. The principal effects of microdynamics which have been observed to date are the phenomena referred to as "microlurch" and the "equilibrium zone" (Warren 1996). Microlurch is a residual change in the shape of a deployed structure which occurs after a dynamic transient motion, and the equilibrium zone is the variability of a deployed structure's shape due to random microlurch response. Hence, equilibrium zone is a measure of the deployment stability of the structure.

Microlurch has been seen to occur in response to dynamic forces, such as impulses, and it is believed to occur in response to quasi-static thermal loading, although no such occurrences have been measured to date. Current models predict that microlurch is caused by a release of strain energy that is built up due to frictional interactions within the joints and latches of a deployed structure. Microlurch behavior can be either random or progressive depending on the state of the structure prior to loading. Specifically, it has been observed that just after deployment, microlurches tend to be progressive such that successive microlurches accumulate into a net, quasi-stable, shape change of the structure. In other words, it appears that structures tend to change their shape slightly after deployment to relax internal stresses built up by the deployment process. After structures have undergone a post-deployment, progressive microlurch, subsequent microlurching responses tend to be random but bounded in magnitude. The boundary of this random microlurch response is the equilibrium zone of the structure.

Repeated deployment and stowage tests of precision deployable structures have shown that deployment precision might be an order of magnitude worse than post-deployment stability once the structure has microlurched to within its equilibrium zone. The most important implication of these observations is that the equilibrium zone geometry, and NOT the initial deployed shape, determines the post-deployment stability. Furthermore, data suggest that it might be possible to design deployed structures with post-deployment stabilities to better than one part per million of the overall dimension. However, it is also important to realize that all data to date have been derived from ground tests, and the effects of gravity on the deployment repeatability and post-deployment stability of precision structures is currently unknown.

Affect of Microdynamics on the Problem of Active Control

The existence of microlurching and an equilibrium zone, as opposed to a single equilibrium geometry, establishes the need for quasi-static shape control and possibly high-bandwidth active control for applications in which precision requirements are smaller than the size of the equilibrium zone itself. To complicate the implimentation of quasi-static shape control systems, the possibility of repeated microlurching of the structure implies that quasi-static shape adjustment might have to be repeated after fairly short intervals of time.

To complicate the implementation of high-bandwidth active shape control systems, the sudden release of strain energy during a microlurch has the potential to excite high frequency structural dynamics which can fall outside the practical bandwidth of an active control system. It is perhaps this aspect of microdynamics which is most troubling to the developer of active shape-control systems. For example, future space-based interferometers require nanometer stability up to at least 1,000 Hz, but the nominal active control bandwidth is typically limited to several hundred Hertz. If a microlurch occurs during the formation of an image or during an astrometric measurement, it is likely that the instrument would need to be realigned and the measurement began again. Persistent events of this type would directly limit mission science data.

Microdynamic Test Methodology

Microdynamic testing is unlike conventional dynamic testing in that microdynamic-response phenomena possess random attributes that can be easily confused with random noise in the test setup. The fundamental challenges to performing accurate microdynamic tests are: 1) to control the environmental influences that are often neglected in macroscopic tests but which are significant systematic errors at microscopic or nanoscopic levels; 2) to resolve the microdynamic mechanics with "multiple witness" sensors with nanometer and nanostrain resolution; and 3) to isolate particular mechanical effects of the test apparatus from the mechanical effects of interest.

At the University of Colorado, a new facility and associated test methodologies have been recently developed which enable nanometer-resolution measurements to be made of microdynamic phenomena. The "Thermal Acoustic Stabilization Chamber" (TASC) is a 3-meter cube enclosing an optics table mounted on vibration isolation legs. The walls of the chamber are shielded for passive acoustic attenuation, and are insulated by approximately 10 inches of styrofoam for thermal isolation. Tests in the chamber typically achieve 6 micro-g vibration levels and hour-long milli-degree Kelvin thermal stability. Additionally, the facility includes high sensitivity, lightweight accelerometers, a nanometer resolution laser interferometer, and picostrain resolution strain gauges. In particular, the laser interferometer measurement path is enclosed to reduce the effect of variations in the index of refraction in air. The net result is approximately 2.5 nanometers of resolution for sampling rates ranging from quasi-static to many 10's of kiloHertz. Such a performance level was previously only possible in a vacuum.

Microdynamic Testing of a Prototype Deployable Telescope Metering Truss

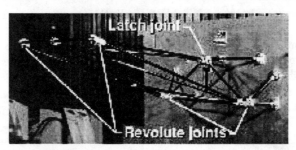

FIGURE 2. Deployable Telescope Metering Truss.

Warren (1996) presents results from tests of a deployable telescope metering truss (Fig. 2) which incorporates four precision revolute (i.e., hinge) joints and one end-of-deployment latch joint. This test article represents a portion of a metering truss that could support one reflector panel in a segmented telescope mirror (e.g., one of the six perimeter panels shown in either Fig. 1(a) or 1(b)). The metering truss is approximately 1.2 m in length, and exhibits deployment precision of approximately 20 microns in the position of the outboard nodes. Results from microdynamic testing indicate that the structure microlurches up to approximately 10 microns in response to transient disturbances subsequent to deployment. However, the structure consistently settles into an equilibrium zone of approximately 2 microns in size after progressive microlurching from its initial deployed condition.

Stick-Slip Friction Model of Microlurching

Analytical results presented from a simple two-degree-of-freedom model suggest that microlurching is an artifact of stick-slip instability due to load transfer through friction (Warren 1996). Since both the revolute joints and the latch joint possess mechanical interfaces at which load is partially transferred through friction, it is plausible that these mechanisms are responsible for microlurching in the deployable telescope metering truss.

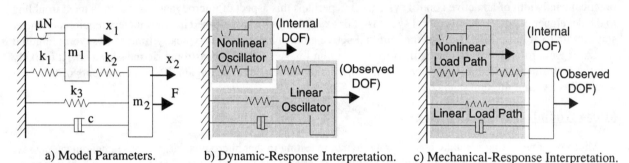

a) Model Parameters. b) Dynamic-Response Interpretation. c) Mechanical-Response Interpretation.

FIGURE 3. Simple System Model from Warren (1996).

The simplified model used by Warren is shown schematically in Fig. 3 with model parameters described in Fig. 3a. In these analyses, the simplified model was used to simulate a single degree of freedom of the dynamic response of the deployable telescope metering truss shown in Fig. 2 (i.e., vertical motion of an outboard joint). The parameters of the simplified model represent modal parameters of the truss (e.g., modal masses and stiffnesses) rather than physical parameters of the truss (e.g., member masses and stiffnesses). For these dynamic-response analyses, the model can be interpreted to represent a linear oscillator coupled with a nonlinear oscillator as shown in Fig. 3b. The nonlinear oscillator exhibits stick-slip instabilities due to the Coulombic friction element. These instabilities give rise to microlurch motion in the total system response.

The analyses presented by Warren (1996) using this model closely agree with his extensive experimental data on the microlurching response of the deployable telescope metering truss shown in Fig. 2. These analyses reinforced laboratory findings that microlurching is random and possibly chaotic (i.e., deterministic but highly dependent on initial conditions), and dependent on excitation energy. The correlation between experimental and analytical results strongly supports the notion that microlurching is caused by stick-slip interactions within the structure.

MICROMECHANICS OF DEPLOYMENT MECHANISMS

As a consequence of lessons learned through microdynamic testing of deployable structures, new high-precision deployment mechanisms have been developed, and substantial micromechanical response experiments have been performed. Specifically, numerous tests have been conducted to quantify the hysteretic response of these deployment mechanisms under load cycling, and efforts have been made to correlate these measured response results with analytical results from models of frictional interactions within the deployment mechanisms.

Precision Revolute Joint Concept

Early in this research program, and with very little a priori knowledge of the relationship between nonlinear response behavior in joints and dimensional instabilities in structures, NASA Langley Research Center developed a revolute (i.e., hinge) joint concept with the hope that it would function well as a precision deployment mechanism (Lake et al. 1996). The joint concept was based on the simple premise that nonlinear mechanical response within deployment mechanisms is the fundamental limitation to post-deployment dimensional stability in a mechanically deployable structure, and therefore, a precision joint is one which responds LINEARLY to load cycling.

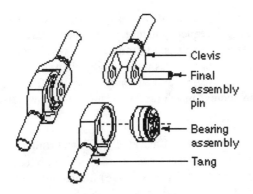
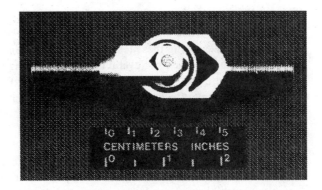

(a) Linear Revolute Joint Diagram. (b) Linear Revolute Joint Photograph.

FIGURE 4. Linear Revolute Joint.

The revolute joint concept (Fig. 4) represents a substantial departure from conventional pin-clevis joint designs in that it incorporates a precision, preloaded angular-contact ball bearing to allow rotation instead of a simple pin. The bearing is internally preloaded to eliminate freeplay, and the bearing diameter is maximized to minimize stiffness changes due to nonlinear interface conditions between the balls and the races. Early tests of the joint indicated that its load-cycle response is very nearly linear with less than 2% hysteresis under quasi-static load cycling.

Hysteretic Response Testing

Substantial efforts have been made to characterize the hysteretic-response behavior of the precision revolute joint, and to correlate this behavior with predictions from analytical models. Three separate studies have investigated the hysteretic load-cycling response of the precision revolute joints. Lake et al. (1996, 1997) presented results of extensional load-cycle tests conducted over a wide range of load-cycle magnitudes. Bullock (1996) presented results of extensional and rotational load-cycle tests conducted at low load-cycle magnitudes.

Instrumentation and Test Procedures. In each of these experimental programs, great care was taken to minimize noise and hysteresis in the load and displacement instrumentation in order to ensure accurate characterization of the hysteretic response of the specimens. In particular, fiber-optic displacement transducers were used instead of electro-mechanical displacement transducers because the fiber-optic instruments exhibit no hysteresis in their response. Also, the load cells used were calibrated prior to testing to ensure that their hysteretic response was insignificant.

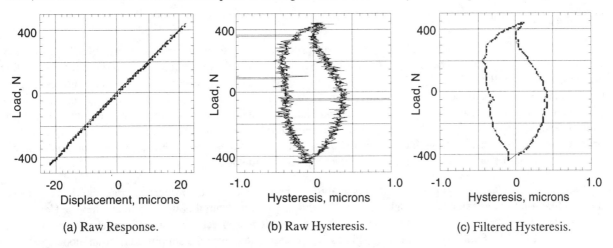

(a) Raw Response. (b) Raw Hysteresis. (c) Filtered Hysteresis.

FIGURE 5. Typical Load-Displacement Response from Precision Revolute Joint.

A typical load-displacement response from these tests is presented in Fig. 5 (Lake et al. 1997). The raw (i.e., unfiltered) response is presented in Fig. 5(a), and the corresponding raw hysteretic response (derived by subtracting the best-fit straight line from the total response) is presented in Fig. 5(b). These data illustrate that the displacement data possess more noise than the load data, and this noise is of a high-frequency nature. Consequently, substantial efforts were made during these tests to filter the displacement data channels and to improve the resolution of the displacement measurements. Several numerical filtering and averaging procedures were employed during these tests and the results are presented in Fig. 5(c). Comparing Figs. 5(b) and 5(c), it can be seen that the data filtering and averaging procedures effectively reduce displacement data noise by more than an order of magnitude leaving a filtered displacement measurement resolution of approximately 25 nm.

General Results. Results presented in the three references (Lake et al. 1996, Lake et al. 1997, and Bullock 1996) from extensional load-cycle testing of the joint indicate that hysteretic response varies dramatically with load-cycle magnitude and internal preload within the joint. For example, data from Bullock (1996) indicate that the joint exhibits negligible hysteresis in response to low-magnitude load cycling (i.e., below 24 N (5 lb_f) in load-cycle magnitude and approximately 200 nm of total deflection response). Data from Lake et al. (1997) indicate the hysteretic loss of the joint increases monotonically with load-cycle magnitude, and can reach values of 1 to 2% for some specimens. Together, these results indicate that friction-induced micro slippage between the internal mechanical components of the joint essentially vanishes at low load-cycle magnitudes. Therefore, within the limitations of the present tests, the joint exhibits perfect elastic response to quasi-static load-cycling at these low load-cycle magnitudes. The broader implication of this result is that structures incorporating these joints should be elastic and dimensionally stable if disturbance forces are small.

Data from Lake et al. (1997) also indicate that hysteretic energy loss occurs in BOTH the angular-contact bearing that allows rotation of the joint, and the press-fit pin that is used to affect final assembly of the joint (see Fig. 4(a)). This result might be counter-intuitive due to the fact that the press-fit pin is designed to be a highly preloaded interface that would commonly be expected to exhibit a perfectly elastic response to load cycling. Based on these data, it appears likely that nonlinear microdynamic response would even be a problem in heavily preloaded structures.

In addition to extensional load-cycle testing, Bullock conducted rotational load-cycle testing of the precision revolute joint. The motivation for these tests was to understand the nature of the hysteretic response of the joints under rotational motion, and to study the effects that this hysteretic response might have on deployment repeatablity of structures incorporating the joint. The primary result of Bullock's rotational load-cycle tests is that, unlike the case of extensional load cycling, the joint exhibits measurable hysteresis for all load-cycle magnitudes during rotational load cycling. In other words, the joint fails to be perfectly elastic in response to rotational load-cycling and structures incorporating these joints should be expected to exhibit some degree of dimensional uncertainty between successive deployments (i.e., error in deployment repeatability).

Hysteretic Response Analyses

Hachkowski (1996) correlated Bullock's rotational-response data with an analysis of the frictional resistance of the bearing to rotation. Hachkowski's analysis is based on the Todd-Johnson tribological friction model of rolling resistance within ball bearings (Todd and Johnson 1987), and includes the effects of Coulombic micro-slippage between the bearing components and material hysteretic damping. His analysis is developed using nondimensional parameters to describe bearing preload, geometry, and material properties, a formulation that facilitates parametric analysis and design optimization. The significant contribution of Hachkowski's work is that he correlated analytical predictions of rolling resistance with experimental data, and demonstrated that the error between the predicted and actual friction torques is within the uncertainty of the properties of the bearings.

More recently, Hachkowski has extended his analysis to predict the effects of micro-slippage on the response of the revolute joint under extensional load cycling. Preliminary results from this work indicate that the Coulombic micro-slippage between the bearing components leads to hysteretic loss effects similar to those found experimentally by Lake and Bullock. Furthermore, Hachkowski has demonstrated that even the simplified friction model presented in Fig. 3 predicts, qualitatively, the gross hysteretic-response behavior seen in the precision revolute joint.

The general conclusions of analyses to date are that: 1) hysteretic losses within the revolute joint are caused by localized friction-induced micro-slippage between the interfaces of the various mechanical components within the joint; and 2) these effects can be represented qualitatively FOR MOST, IF NOT ALL, PRECISION DEPLOYMENT MECHANISMS by the dual-load-path model suggested in Fig. 3.

DESIGNING HIGH-PRECISION DEPLOYMENT MECHANISMS

The general conclusions from Hachkowski's analyses can be applied to develop general rules for the design of high-precision deployment mechanisms. Specifically, since hysteretic response under load cycling is caused by load transfer through friction at interfaces between *load-carrying* components within the mechanism, it follows that a general principal to be applied in the design of high-precision deployment mechanisms is to design the load-carrying mechanical interfaces such that minimal load is transferred through friction. To understand how this design requirement might be met in practical designs consider first how friction develops at an interface.

Load Transfer Across Mechanical Interfaces

Under the action of externally applied loads, internal load paths develop between the various mechanical components of a mechanism. As load is transmitted between adjacent mechanical components, interface stresses and localized material deformations are developed. The interface stresses and deformations can be decomposed into normal and tangential (i.e., shear) components as depicted in Fig. 6. Load transfer through normal stress at the interface involves no friction and is thus elastic. However, load transfer through shear stress at the interface involves friction, and can thus be hysteretic (i.e., *inelastic*) if slippage occurs.

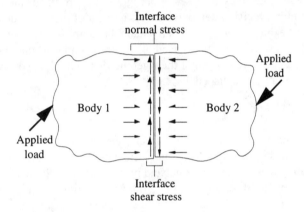

FIGURE 6. Interface Loads Between Two Mechanical Components Within a Deployment Mechanism.

The simplified model presented in Fig 3 to explain microlurching response can also be used to illustrate this dual-load-path behavior across an interface (Fig. 3(c)). This model indicates how an externally applied load, F, can be divided into elastic (i.e., normal) and inelastic (i.e., tangential) load path components for load transfer across the interface. The elastic load path involves only a single elastic element, k_3, which transmits the normal component of the externally applied load. The inelastic load path includes an elastic element, k_2, which transmits the shear component of the externally applied load from the point of application to the interface. The interface portion of the shear load path is represented by a second elastic element, k_1, in parallel with a friction element, μN. This subdivision of the shear load at the interface accounts for the fact that some of the shear load (i.e., in areas where slippage *does not occur*) is transferred across the interface elastically, whereas some of the shear load (i.e., in areas where slippage *does occur*) is transferred across the interface inelastically.

General Design Rule

To minimize the potential for inelastic load transfer, and thus to minimize the possibility of interface slippage, it is desirable to minimize shear stresses at the interfaces between mechanical components within the mechanism. A number of factors affect the interface stress distribution between two elastic components in contact, and understanding which of these factors lead to the development of shear stresses at the interface is a critical step towards developing guidelines for the design of mechanisms which minimize interface shear stresses. However, a general rule that can be applied is:

to minimize interface shear stresses and the possibility of micro-slippage, minimize locally the tangential stiffness in the region of the interface (i.e., k_2 in Fig. 3).

In the case of the precision revolute joint presented in Fig. 4, the numerous interfaces between the balls and the races within the bearings exhibit minimal local tangential stiffness by virtue of the fact that the balls are kinematically free to roll. Hence, ball bearings are inherently good mechanisms for high precision applications.

Conforming Versus Non-Conforming Interfaces

Practically, this general rule discourages the use of "conforming" interfaces (i.e., interfaces whose geometries match over a relatively large area). Conforming interfaces involve large contact regions within which it would be very difficult to reduce or eliminate local shear stiffness. Conversely, non-conforming interfaces (e.g., a ball bearing against a race) involve relatively small regions of contact, thus lending themselves to local elastic tailoring of the structural design adjacent to the contact region to reduce shear stiffness.

CURRENT RESEARCH THRUSTS

During the past three years, the present research program has resulted in substantial advancements in understanding the problem of high-precision deployment. As a result of these advancements, it is now possible to design deployable structures that exhibit *micro*dynamically stable behavior (i.e., dimensional stability to the level of parts per *million*). To further advance our understanding of the nonlinear microdynamics of these precision deployable structures and to enable active control of their geometries to the level of parts per *billion*, a number of new research thrusts have been recently established.

Inhomogeneous Material Effects

As mentioned previously, inhomogeneous material effects, such as intercrystalline slippage in metals, or fiber-matrix debonding in composite materials might have a significant effect on nanostrain-level response in precision structures. To begin investigating these effects, thermal and mechanical stability tests have been conducted recently on a low-CTE composite strut developed for application to optical metering structures. These tests were performed in the TASC facility at the University of Colorado, and were designed to isolate nanostrain-level nonlinear response similar to microlurch. Specifically, the response of the strut to gradually varying thermal and mechanical loads was interrogated for occurrence of sudden high-frequency events such as acoustic emissions.

During these tests various combinations of axial and thermal loading were applied to the strut and its extensional response was measured to nanometers of resolution. For extensional stress up to 5 psi, temperature changes up to 20° C, and extensional strains up to 250 nanostrain, the test specimen showed benign monotonic response to variations in loads with no evidence of the acoustic emissions.

The Effects of Interface Preload on Micro-Slippage and Load-Cycle Hysteresis

A common misconception regarding the design of deployment mechanisms is that preload prevents slippage in a joint or latch if the load does not exceed the Coulombic limit. In fact, recent test results have shown that this notion is not the case, and that significant presliding slippage (also known as microslip) occurs in preloaded interfaces. Preload therefore does not mitigate microdynamic effects.

To investigate the phenomenon of microslippage across a preloaded interface, and to develop a better understanding of the relationship between interface preload and microdynamic response effects such as microlurch, a new test apparatus has been constructed recently at the University of Colorado. The objective of this apparatus is to test the nanostrain-level response of an interface across which load is being transferred through friction. To date, no data have been found in the literature to describe such behavior, and the present test is viewed as a benchmark effort in this area.

Systematic Methodology for Spatial Localization of Microdynamic Events

An effort is currently underway to develop test and analysis techniques to isolate the "origin" of microdynamic events within deployable structures (i.e., the specific joint within the structure which is microslipping). In support of this effort, microdynamic tests will be conducted on a prototype deployable reflector truss and a deployable interferometer boom. The microdynamic response of these structures to a variety of static and dynamic thermal and mechanical loads are to be studied at nanometer resolution. To isolate the origin of microdynamic events, the measured response will be compared with a predicted response generated using linear models. An important feature of this approach is that it also includes the "microdynamic error" in a mathematical form which is convenient for control system design.

On-Orbit Microdynamic Experiments

One of the most critical questions in microdynamic research is the effect of gravity (in both joints and materials) on the microdynamic response of structures. To answer this question, the Micron Accuracy Deployment Experiments (MADE) facility is being developed. Intended as a permanent facility on the International Space Station, MADE is scheduled for installation in 2002. It is being developed as part of NASA's Engineering Research and Technology (ERT) program. The broad goal of MADE is to provide a capability to perform multiple, long-duration microdynamic experiments on-orbit, and to correlate the on-orbit test results with ground measurements to determine the effects of gravity on microdynamic behavior.

Defining Passive Limits for Deployment Repeatability and Post-Deployment Stability

NASA Langley Research Center is currently planning the design, fabrication, and testing of a 2.1-m-diameter deployable lidar telescope mirror test article for ground microdynamic testing (Fig. 1(b)). The test article will incorporate recent advancements in precision deployment mechanisms, and recent advancements in low-areal-density and low-CTE composite reflector panels and structures. The test article is intended to help define the practical limits of passive deployment repeatability and post-deployment stability. The rationale for selecting a lidar telescope application as the focus of this test article design effort is that the dimensional precision requirements for lidar telescopes are on the order of parts per million - a level that might be attainable without the use of active control. Current plans are for the test article to be completed and the microdynamic testing to begin during 1998.

SUMMARY

Prior to the present research program, experience indicated that traditional deployment mechanisms inherently limit the dimensional stability of deployable structures to the level of approximately one millimeter over a one meter span (i.e., dimensional stability to parts per *thousand*.) The present research program has resulted in general guidelines for the design of high-precision deployment mechanisms, and tests of prototype deployable structures incorporating these mechanisms have shown *micro*dynamically stable behavior (i.e., dimensional stability to parts per *million*). These advancements have resulted from the identification of numerous heretofore unknown microdynamic and micromechanical response phenomena, and the development of new test techniques and instrumentation systems to investigate these phenomena. In addition, tests have been begun recently to investigate the *nano*mechanical response of materials and joints and to develop an understanding of the nonlinear *nano*dynamic behavior of *micro*dynamically stable structures. The ultimate goal of these efforts is to enable *nano*-precision active control of *micro*-precision deployable structures (i.e., active control to a resolution of parts per *billion*).

Acknowledgment

The work presented herein and conducted at the University of Colorado has been jointly funded by the NASA Langley Research Center (Grant No. NAG1-1840) and the Jet Propulsion Laboratory (Contract No. 960896).

References

Bullock, Steven J. (1996), "Identification of the Nonlinear Micron-Level Mechanics of Joints for Deployable Precision Space Structures," Ph.D. Dissertation, University of Colorado.

Hachkowski, M. R. (1996), "Friction Model of a Revolute Joint for a Precision Deployable Spacecraft Structure," presented at the 37th AIAA/ASME/ASCE/AHS/ASC Structures, Structural Dynamics, and Materials Conference, Salt Lake City, UT, April 15-17, 1996, AIAA Paper No. 96-1331.

Lake, Mark S. et al. (1996), "A Revolute Joint with Linear Load-Displacement Response for Precision Deployable Structures," presented at the 37th AIAA/ASME/ASCE/AHS/ASC Structures, Structural Dynamics, and Materials Conference, Salt Lake City, UT, April 15-17, 1996, AIAA Paper No. 96-1500.

Lake, Mark S. et al. (1997), "Experimental Characterization of Hysteresis in a Revolute Joint for Precision Deployable Structures," presented at the 38th AIAA/ASME/ASCE/AHS/ASC Structures, Structural Dynamics, and Materials Conference, Kissimmee, FL, April 7-10, 1997, AIAA Paper No. 97-1379.

Peterson, L. D. et al. (1996), "Micron Accurate Deployable Antenna and Sensor Technology for New-Millennium-Era Spacecraft," presented at the 1996 IEEE Aerospace Applications Conference, Snowmass, CO, February, 1996.

Todd, M.J. and Johnson, K.L. (1987); "A Model for Coulomb Torque Hysteresis in Ball Bearings," *International Journal of Mechanical Science*, 29, 1987, pp. 339-354.

Warren, Peter A. (1996), "Sub-Micron Non-Linear Shape Mechanics of Precision Deployable Structures," Ph.D. Dissertation, University of Colorado.

FUNCTIONAL AND LIFE TEST DATA FOR A TWO-STAGE STIRLING CYCLE MECHANICAL CRYOCOOLER FOR SPACE APPLICATIONS

H. Carrington, W. J. Gully, and W. K. Kiehl
Ball Aerospace Systems Division
P.O. Box 1062
Boulder, CO 80306-1062
(303) 939-5416

S. Banks, E. James, and S. Castles
Cryogenics, Propulsion, and Fluid Systems Branch
NASA/Goddard Space Flight Center
Greenbelt, MD 20771
(301) 286-5405

Abstract

We have been developing a long-life mechanical cryocooler for space applications since 1991. We started with a "clean sheet of paper" with this cooler to customize the best current technology for space applications. The space requirements include reliability, power efficiency, mass, system compatibility, and the ability to withstand adverse environmental conditions. We have delivered an engineering model and a flight prototype under the current contract. Each unit was typically preceded by several breadboard versions used to explore alternate configurations. We are reporting in this paper on test data for the flight prototype cooler, delivered from Ball to the Goddard Space Flight Center (GSFC) at the end of 1996. Verification tests on this cooler were performed at Ball. The performance tests were confirmed and expanded upon at GSFC, and the cooler is now in for a period of extended running to evaluate its long-term performance. We are building a version of this cooler for the High Resolution Dynamics Limb Sounder (HIRDLS) program.

INTRODUCTION

There is a need for mechanical cryocoolers that can provide modest cooling in a satellite for an extended period of time. This is a complex task because the low temperatures they produce are incompatible with the ordinary kinds of lubrication that make long life in a mechanism possible. We discuss here an approach where the cooler employs unlubricated mechanisms in which the running clearances in the linearly oscillating components are completely supported by flexing diaphragm springs.

The task is to develop a cryocooler package that addresses the concerns of a flight system. The cooler must have high reliability and low mass. It must not rub or produce vibration or EMI. It must be able to reject heat in a vacuum environment, and draw power from the typical dc space battery. It must be autonomous, easy to interface to, and capable of withstanding the usual launch loads and space radiation environments. Building on the early work of the RAL group (Bradshaw 1990), we re-evaluated each aspect of the design in light of the new technical requirements and created a series of coolers to test and evaluate the results. The final product is a 15 kg system consisting of a twin compressor, a two-stage displacer/counterweight, and a set of control electronics. The mechanical components are shown in Figure 1. This particular cooler is thermodynamically optimized to provide 0.4 W at 30 K for less than 75 W of input power, and was delivered to the Goddard Space Flight Center at the end of 1996. We have spun off single-stage and three-stage versions of this cooler that are tailored to other refrigeration requirements. We are currently building several flight coolers for the HIRDLS program based on our single-stage design.

We are now partway through the testing cycle. We had evaluated the cooler's performance in a series of environmental tests at Ball (Berry 1996), and the thermal performance has been explored over a wide range at GSFC. We are now beginning the time-consuming task of at least verifying that this one cooler can run for a long period of time. We begin the discussion with a description of the unit under test, describe the verification tests, and provide the status of the extended running. We will finish with a few observations on our experience with the flight coolers.

SYSTEM DESCRIPTION

Reliability is a primary concern, but it has been difficult to quantify because the mechanisms in our cooler have no antecedents. Our approach has been to analyze every critical feature in an effort to screen out all possible technical failures. To this end, we have tried to build a cooler where all motions are elastic and without wear in the hope that its operation is transparent.

FIGURE 1. The 30 K Cryocooler System Consists of a Twin Compressor, a Two-Stage Displacer, and a set of Autonomous Control Electronics. The Cooler Is Mounted in a Six-Axis Vibration Dynamometer.

Our twin compressor provides a balanced method of providing the pressure wave for the Stirling thermodynamic cycle. Each flexing component is designed for virtually infinite fatigue life, which is necessary because the mechanism accumulates over 3 million cycles every day. We also have used a very minimum of non-metallics to lessen the internal contamination problem. However, the key feature of our cooler is its provisions for non-contacting operation. Our moving armatures are on diaphragm flexure springs, which are capable of supporting the piston in its clearance gap. But maintaining this clearance dynamically at full power is difficult, and we had to learn how to do this through testing and verification. To carry out this program, we added research sensors that enabled us to monitor the radial positions of the pistons while the cooler was in operation. With this capability we were able to determine the construction and alignment details necessary to eliminate the contact. The special feedthroughs for these sensors are visible on the ends of both the compressor and displacer in Figure 1. Finally, we discuss a few details of our interface to the system. By alternating stiff mounting lugs and aluminum skirts, we can provide both accurate force feedback and adequate thermal coupling for heat rejection. The compressor constitutes about 7 of the 15 kg of system mass.

Our displacer is consciously a miniature of our compressor mechanism. It incorporates all of the features described above, including the diaphragm spring support of the armature, and the internal sensors for monitoring the gaps in the close tolerance clearance seals. The difference between them is that the displacer piston is much longer and has to be cantilevered into the cold end of the machine. Our approach has been to accommodate this by making the displacer as light as possible. We departed from customary displacer design by off loading the regenerator to the stationary part of the cold finger. We then developed an extremely lightweight, hollow, thin wall piston, which dropped the cantilevered mass by a factor of five. We can now maintain the gap clearance even when the armature is horizontal in a gravitational field. As an added bonus of the fixed regenerator design, the now external regenerator stiffens the cold finger and affords excellent thermal contact between the working fluid and the load. The displacer accounts for approximately 4 kg of the 15 kg system mass.

Since the beginning of the program, our control electronics evolved along with the mechanical hardware as a key part of the system. The electronics for our protoflight system weighs approximately 4 kg and fits in a package that is approximately 200 mm x 200 mm x 150 mm. It is divided into three sections: a power section that conditions the dc power input and provides the drive waveforms, an analog section that supports the various sensors, and a digital section that has the control processor that runs the cooler. With the cooler at rest, the electronics draws a quiescent power of only 14 W. It can be built with flight-quality parts, and in that condition it is immune to the SEU and total dose radiation threats expected for a typical low Earth orbit for five years.

We use a Harris RTX microprocessor, which executes a firmware code written in FORTH. It runs continuously in a loop where it reads the sensor information, calculates a proper response, updates the drive waveforms in the fallow part of the ping/pong RAM, and toggles it into the circuit that sends it repetitively to the cooler when it is ready. The cooler accepts a 14-byte control command and transmits an 18-byte status via an RS-422 telemetry link. The control command sets parameters and flags for the autonomously operating processor. The controller provides a number of sophisticated features. It can control temperature by adjusting the stroke of the armatures. And it can balance the strokes of the opposing components to minimize the axial output vibration as measured in the mount. In combined temperature and vibration control, it will coordinate the changes to simultaneously pursue both goals.

VERIFICATION TESTS AT BALL

The cooler went through a series of verification tests at Ball before it was delivered to GSFC. Some of these tests were described in a previous report (Berry 1996). The cooler successfully met the requirements listed in Table 1, with two minor exceptions. These anomalies were in two subsections of different tests, and are discussed below. In each case we have implemented corrective action and expect to verify compliance by the end of 1997.

TABLE 1. Verification Tests for the Phase IV GSFC Cooler, 1996.

Test	Requirements
Thermal performance	Control at 30 K, carry .3 W @ 30 K for 75 W, all orientations, cool down time
Non-contact	No touch/verify with stiction and internal sensors all orientations
EMI, emissions	CE01, CE03, RE02, RE04
EMI, susceptibilities	CS01, CS02, RS02, taken in full temperature and vibration control mode
Vibration (shaker)	14.1 Grms random all axes, 15G swept sine all axes
Exported vibration	< 234 mN output combined compressor and displacer, each axis
Thermal vacuum	8 cycles, 233 K to 323 K, cool down and operate 2 times per cycle

The first anomaly occurred in our test for conducted susceptibility. In a narrow frequency band around 100 Hz, the 5 Vrms voltage applied to the power lines coupled through to the motor drives in a way that caused a slight increase in the compressor output vibration. Our solution was to implement a dc/dc converter on the input for complete isolation from the raw 28 Vdc power line. The second anomaly apparently occurred during random vibration. After the test we were able to detect a shift in the alignment of the displacer. During our subsequent investigation, the alignment re-established itself, indicating a shift in the alignment of one of our subassemblies. We have addressed the problem by replacing the assembly with a single part.

PERFORMANCE TESTING AT GSFC

At GSFC the cooler was installed in a large environmental chamber to study the cooler behavior over a range of ambient temperatures in a vacuum environment. The cooler was characterized as a function of frequency, stroke, and heat load at each temperature. It is important to characterize these trends, because a cryocooler has to be sized to meet its requirements over its expected operating range of parameters.

In Figure 2 we show an example of our results for 36 Hz operation and 293 K ambient, for two different operating strokes of the cooler. (The compressor and displacer have the same stroke.) Reading from the interpolation line marked 30 K, and for the case of 90% stroke, we see that the cooler can sink 0.48 W for an input power of 77 W from the 28 Vdc supply. Of the 77 W, approximately 55 W are transmitted to the compressor. We have performed similar tests for a wide range of operating frequencies and ambient temperatures. In order to condense the data, we focus on a key task and note the input power and heat lift while maintaining 30 K at a given stroke. The results for a range of ambient temperatures are tabulated in Table 2. In general, the power goes down with temperature, but the net cooling is complex because it is the difference between the gross cooling and the parasitic losses, both of which are dropping with temperature. The result, using the specific power as a figure of merit, is that the efficiency first rises and then drops with decreasing ambient temperature.

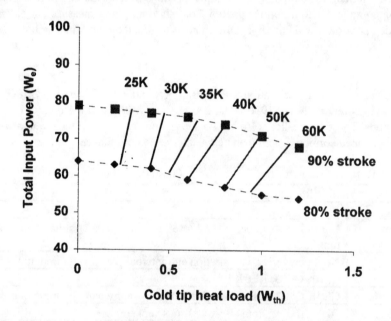

FIGURE 2. The Thermal Performance of the Two-Stage Cooler at 293 K and 36 Hz as a Function of Stroke.

TABLE 2. Characteristics of the GSFC Cooler at 30 K and 36 Hz as a Function of Ambient Temperature.

Ambient Temperature (K)	Stroke (%)	Input Power (Watts$_e$)	Heat Lift (Watts$_{th}$)	Specific Power (W$_e$/W$_{th}$)
293	90	77	.48	160
	80	62	.39	159
273	90	70	.57	123
	80	57	.42	135
253	90	68	.40	170
	80	56	.28	200

We can provide a similar table for the performance of the cooler over a range of operating frequencies. In this case, we hold the ambient temperature to 273 K. It is clear from Table 3 that the user can increase the cooling capacity by about 50% by changing the operating frequency from 32 to 39 Hz. We find the highest efficiencies at the highest frequencies because the input power, which is roughly proportional to the frequency, only goes up by about 20%.

TABLE 3. Performance at 30 K and at an Ambient Temperature of 273 K as a Function of Frequency.

Operating Frequency (Hz)	Stroke (%)	Input Power (Watts$_e$)	Heat Lift (Watts$_{th}$)	Specific Power (W_e/W_{th})
32	90	63.6	.46	138
	80	53.2	.33	160
36	90	70	.57	123
	80	57	.42	135
39	90	78.2	.63	124
	80	62.7	.45	139

EXTENDED RUNNING TEST

After the performance characterization, we simply ran the cooler in order to accumulate operating hours. We collect hours at the rate of 730 hours a month, and have been operating continuously since the end of June. We have made a number of changes to support the cooler in this long-term effort. The cooler is operating in the same large chamber, but we now provide a separate vacuum for the cold tip so we can continue running even when the main chamber is open. We power the cooler with an uninterruptable power supply, and provide a protection circuit that shuts off the cold tip heater when the cooler stops. We intend to monitor the cooler on a weekly basis, and will perform a full set of load curves every 5000 hours. Since this is not an accelerated test, we will have to continue with the test for five years to meet our goal.

FIGURE 3. The Ball Cooler Mounted Within the Thermal Radiator Panel Used on the HIRDLS Instrument.

HIRDLS SYSTEM INSIGHTS

We are preparing several coolers for use in an Earth observation NASA program. The cooler will differ from the one described above in that it has a single-stage displacer, and it will not include the internal sensors that we used to study non-contacting. Several interesting integration issues are illustrated in Figure 3, which shows our cooler submerged into the plane of the radiator panel. We have found that this configuration benefits both the cooler and the system. Active vibration control requires fidelity in the force feedback, and this configuration provides a stiff and symmetric mount. We also take advantage of the isolation between the radiator and the rest of the system to further attenuate the vibration. Finally, we find that this compact configuration is adequate to reject the heat from our mechanisms.

SUMMARY

Ball and NASA GSFC have been developing a cryocooler for use in space systems. The cooler has been designed specifically to meet environmental and compatibility requirements. We have characterized the cooler, and are now running the cooler to establish its lifetime.

Acknowledgments

We thank Dan Berry, who has led the development of the interface on the HIRDLS program, for his many contributions. The cryocooler work has been funded by the Goddard Space Flight Center.

References

Berry, D., H. Carrington, W. J. Gully, M. Luebbert, and M. Hubbard (1996) "System Test Performance for the Ball Two-Stage Stirling Cycle Cryocooler ," in *Proc. of the 9th International Cryocooler Conference,* Plenum, 69-77.

Bradshaw, T. W., and A. Orlowska (1990) "Closed Cycle Cooler for Temperatures Below 30 K," in *Cryogenics* 20:246-248.

A MECHANICAL COOLER FOR DUAL-TEMPERATURE APPLICATIONS

W. Gully, H. Carrington, and W. Kiehl
Ball Aerospace Systems Division
P.O. Box 1062
Boulder, CO 80306-1062
(303) 939-5416

Kevin Byrne
U.S. Air Force Research Laboratory
Space Vehicle Technologies
Albuquerque, NM 87117
(505) 846-2686

Abstract

Ball Aerospace has been developing Stirling cycle mechanical cryocoolers specifically for space applications. These coolers are special in that they are designed from the beginning for power efficiency, high reliability, and compatibility with sensitive instruments. We have delivered several of these coolers to NASA Goddard Space Flight Center, and are currently assembling one for the High Resolution Dynamics Limb Sounder (HIRDLS) program. In our current research effort, funded by the Ballistic Missile Defense Organization (BMDO), we are tailoring our basic design to new requirements from the Air Force Research Laboratory and its customers. We describe our success in optimizing a cooler to efficiently provide refrigeration at two different temperatures simultaneously. This two-temperature application requires 0.4 W of cooling at 35 K, and 0.6 W of cooling at 60 K. We have met these requirements with an input power of approximately 70 W from a dc source with a breadboard version of the cooler. We expect to deliver the protoflight version of this cooler to the Air Force Research Laboratory in January 1998.

INTRODUCTION

We have gone to extraordinary lengths with these coolers to ensure the lifetime and reliability of our mechansim. We have adopted a Stirling refrigeration cycle approach that employs flexure spring supported linear compressors and displacers, which have no wearing parts (Bradshaw 1990). We have built a series of these coolers with funding from the Goddard Space Flight Center (GSFC), the most recent of which we describe in another paper in this proceedings, "Functional and Life Test Data for a Two-Stage Stirling Cycle Mechanical Cryocooler."

The dual-temperature cooler discussed in this paper is an outgrowth of the GSFC cooler, which we will quickly review. The GSFC cooler was specifically optimized for power efficiency while delivering 0.3 W at 30 K. It consisted of a compressor, a two-stage displacer, and a microprocessor based set of control electronics. The complete system (including electronics) weighed less than 15 kg, was autonomous, and operated from a 28 Vdc power source. A key feature of the GSFC program was the development of noncontacting and nonwearing clearance seals in all parts of the cooler. We did this by employing unique instrumentation for monitoring the gaps in an operating cooler, which helped us put in place the fabrication and alignment capabilities necessary to eliminate contact. The cooler was also designed to withstand the space environment, to control EMI and exported vibration, and to provide heat rejection in a vacuum environment.

We built a new cooler based on the GSFC design specifically to meet the Air Force Research Laboratory dual-cooling requirement. We now wanted to deliver 0.4 W at 35 K as well as 0.6 W at 60 K. It would be clearly advantageous to the system if we could meet both loads with a single cooler because of the system mass and power penalties associated with two coolers. Our approach was to develop a new cold head but to make use of our existing compressor. However, this meant that we had to improve our thermodynamic efficiency because we intended to get more refrigeration out of the same pumping source.

THEORETICAL APPROACH

We have arrived at a rather efficient cooler, and would like to provide some insight into how that came about. Once we had selected the right expander configuration, we were able to size it properly with an analytical model (Urieli 1984) that had been perfected during the development of the GSFC cooler. But we settled on the initial configuration by using two fundamental design principles. The first is the Second Law of Thermodynamics, which states that more work is required to lift heat from a lower temperature. The second is an appreciation for the economy of scale, which reflects the fact that the gross refrigeration of an expander is proportional to its volume while the parasitic losses are proportional to its surface area.

We had already rejected an approach using two separate coolers because of the increase in size and complexity. We felt that an approach with a single compressor and two expanders had the same problem as well as an additional inefficiency due to the two smaller expanders. So we chose an approach with a single expander.

The simplest single expander configuration would have a single expansion space at the cold tip. We would connect the tip directly to the heat source at the lower temperature, and connect the tip to the warmer heat source through a thermal strap sized to accommodate the temperature drop at the anticipated heat flow rate. However, this configuration has an unnecessarily high input power since both loads end up being refrigerated at the lower temperature. There is a variant in which the strap is connected to an intermediate temperature on the cold finger. Heat inflow there suppresses some of the original parasitics for a modest improvement in efficiency. But the degree of improvement depends upon how far apart the temperatures are, and on the provisions in the cold finger for thermal contact at the point of attachment. In our case this approach was not competitive.

Theoretically, the most efficient approach would be to use a multistage cooler with an expansion stage at the temperature of each of the heat sources. The traditional way to implement this with a single expander is to use a stepped displacer piston, which leads to a second toroidal expansion space part way up the cold finger. This nested approach offers an additional benefit in that the upper stage also intercepts the parasitic heat load that would have flowed from ambient temperature to the colder stage. Since this constitutes the largest part of the heat load on the cold tip, intercepting it is the most significant factor in improving the efficiency of the machine. In fact, this turned out to be so important that it suggested a third expansion stage solely to intercept the bulk of the parasitic losses at an even higher temperature.

FIGURE 1. Three-Stage Breadboard Stirling Displacer to Provide Refrigeration at 35 K at the Cold Tip and at 60 K at the Copper Flange Approximately 40 mm Behind the Cold Tip.

We therefore had a choice between an enhanced two-stage and a three-stage expander for the dual temperature application. At this point we performed a detailed quantitative analysis of their respective performances in order to decide between them. We started with our model of the well-characterized GSFC two-stage 30 K cooler. This cooler had ample capacity at the cold tip, but could only cool to 100 K at the upper stage. We would have to increase its gross refrigeration by well over a watt to enable the upper stage to cool to 60 K *and* have 0.6 W left over for the external load. We modeled two departures from this starting point. In the first we enlarged the upper stage to increase its gross cooling. In the second we interposed a third expansion stage between the old upper stage and room temperature to cut down on the internal parasitics. We found that the latter approach made much better use of the available fixed compressor displacement and resulted in the most efficient cooler. We designed and built our breadboard expander based on the results of these calculations.

HARDWARE ISSUES

The theoretical advantage of a multistage cooler can be lost in the imperfect performance of the hardware when the design is realized, especially in the miniature size required by these relatively low-capacity coolers. It is therefore important to address a few critical design details associated with the stepped stages in the expander.

The main difficulty in a multistage cooler is how to provide for the cold seal between expansion stages. It is essential to get the bulk of the gas to flow through the interstage regenerators and not let it leak past the expander piston. This gets more difficult when both ends of the seal are at low temperatures because the gas gets more dense. We are able to use noncontacting clearance seals because of our fixed regenerator design. In the flexure spring approach to Stirling cycle displacers, the displacer piston is cantilevered from a support at the warm end. To use a clearance seal, the mechanism must be able to maintain the small gap. In our fixed regenerator design, we started with a lightweight, thin wall, all-metal piston, and cantilever it out on a shaft rigid enough to support the anticipated side loads, including the gravitational sag. Finally, we stabilized its lateral motion by careful fabrication and alignment of the seals. In our experience these simple seals eliminate the variability in the low-temperature performance often found with a contacting seal.

The next issue is the internal heat exchange. A temperature drop between the load and the expansion stage is inefficient because it forces the cooler to run colder than necessary. Although less efficient, in small coolers it is possible to get by without an explicit heat exchanger. But because of our fixed regenerator design, we have been able to explicitly add flow heat exchangers at both refrigerating stages. We were able to do this because the fixed regenerator geometry has the regenerator on the outside of the assembly (Berry 1995), which gives an external heat exchanger ready access to the gas flow. In our testing we found that these heat exchangers were necessary for meeting our performance requirement.

BREADBOARD PERFORMANCE

We now discuss the thermodynamic performance of the breadboard version of our "two-temperature" cooler. We will discuss the special attributes of a breadboard, and present the load lines for both cooling stages. We will then indicate the general trend in the off baseline performance to suggest how the cooler could be used in other applications.

In our terminology a breadboard is a simplified set of hardware that is only meant to be used to demonstrate the thermodynamic performance of a design. To make one we typically bolt our latest thermal design onto an outdated expander drive mechanism. But this in turn restricts us to an earlier generation of our drive electronics. Therefore, some allowance has to be made for the difference between the actually measured breadboard power, and the power we would project for the final flight system. For clarity we will always report the actual power draw from the 28 Vdc supply for our breadboard control electronics, even though this number is typically about 7 W higher than would be required by the protoflight system. The difference is mainly due to extra diagnostic circuitry not present in the streamlined flight version of our electronics.

We start with the cooler performance in its "baseline" configuration. This means that the cooler charge pressure is set to 10^6 Pa, and it is operating at an ambient temperature of 295-300 K with forced air cooling. We are also operating at a frequency of 36 Hz, and are using a 90% stroke amplitude for each component. We define 100% stroke to be the distance between mechanical stops, and 90% stroke to be the maximum amount safely available for use. We distinguish between the lower cold stage at the very tip of the expander and the upper cold stage, which is at the step about 40 mm back from the tip (Figure 1).

We begin with a load line for each stage in turn, with the applied load on the other stage held constant. The upper stage load line is shown in Figure 2. The lower cold stage is held at a fixed heater load of 0.4 W. Note the decrease in input power with increasing upper stage load, which illustrates that the load relieves the expander of the task of operating at a low temperature. This measured performance is in quite good agreement with our numerical model predictions. Note that with flightlike electronics, the input power at the requirement would be only 65 W of power from a 28 Vdc supply.

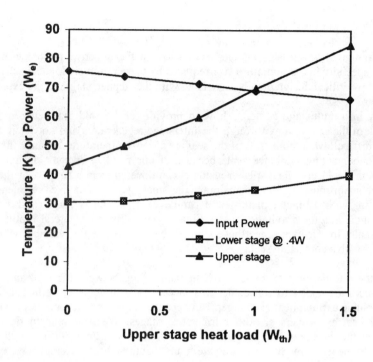

FIGURE 2. Load Line for the Upper Cold Stage of the Three-Stage Breadboard Cooler. It Provides More Than 0.6 W at 60 K *and* 0.4 W at 35 K for 72 W of 28 Vdc Input Power.

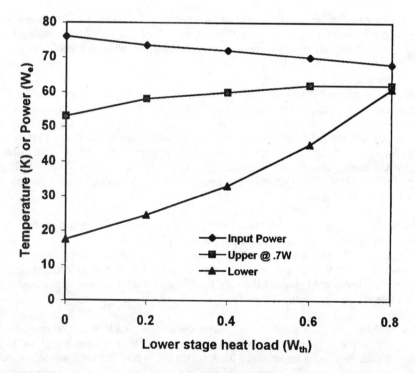

FIGURE 3. Load Line for the Lower Cold Stage of the Three-Stage Breadboard Cooler.

We plot the load line for the cold tip in Figure 3. In this case we have applied a constant 0.7 W to the upper cold stage at the same time. The cooler again demonstrates that it handily meets its design requirements.

We now turn to the cross coupling between the stages. In an ideal case, the stages would be independent of one another. That would simplify temperature control because "make-up" heater power added in an attempt to maintain the temperature of one stage would not interfere with the temperature at the other stage. We illustrate our measured performance in Figure 4. Each data point in the figure represents the temperatures of two stages at two applied heat loads. The stages will be independent to the extent that the lines (shifts in temperature) are either horizontal or vertical in the plot, which is the case for our cooler near its design point. The independence occurs because an applied load on one stage causes two effects on the other stage that happen to cancel each other out. As one would expect, some of the heat applied to one stage is indirectly shared with the other stage and would tend to warm it up. But the applied heat also warms up the heated stage significantly, which lowers the density of the gas in the surrounding region. This frees up working gas, resulting in a small increase in the cooling capacity at every stage in the expander. It so happens in our cooler that this increase in cooling capacity is just large enough to cancel the indirect heating, which leads to the results shown in Figure 4.

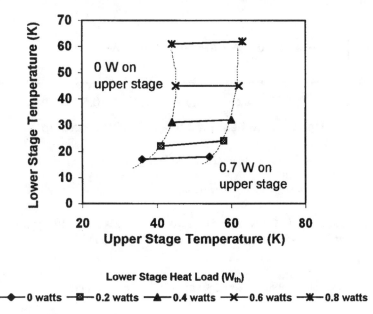

FIGURE 4. Simultaneous Load Curves on the Upper and Lower Stages Illustrating Cross Coupling.

In Figures 1 through 4 we presented the performance of the cooler while we were operating it in our baseline configuration. We can get somewhat more heat lift if we raise the charge pressure or the operating frequency, but at the expense of increased input power. The change with frequency is typically more relevant because it is available even after the cooler is sealed. In Figure 5 we present a set of curves like the one shown in Figure 2, but for a range of operating frequencies. The results indicate that we can increase the performance by perhaps 100 mW at each stage, although we are meeting with diminishing returns for the input power.

FUTURE WORK

We are currently fabricating the protoflight version of this cooler that meets the full set of flight requirements. It will consist of this new expander, our compressor, and the flightlike version of our electronics. After a complete set of verification tests at Ball, we will deliver this unit to the U.S. Air Force Research Laboratory for further testing.

Acknowledgments

We thank Steve Castles, Ed James, and Stuart Banks at GSFC for their support and assistance during the development of these coolers. We thank Dave Glaister and Dave Curran of the Aerospace Corporation for their contributions to the midstage cooling concepts. And finally, our thanks to Larry Crawford for his foresight, encouragement, and continued support, which have enabled us to achieve this successful outcome. This work has been funded by BMDO.

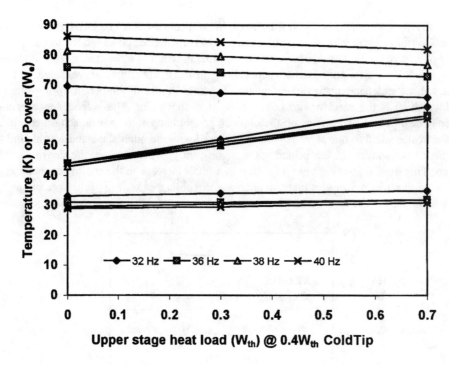

FIGURE 5. The Heat Lift Increases With Frequency Near the Design Point.

References

Berry, D., H. Carrington, and W. J. Gully (1995) "Two Stage Cryogenic Refrigerator for High Reliability Applications," in *Proc. of the 1995 Cryogenic Engineering Conference, Advances in Cryogenic Engineering,* 41A:1585-1593.

Bradshaw, T. W., and A. Orlowska (1990) "Closed Cycle Cooler for Temperatures Below 30 K," in *Cryogenics* 30:246-248.

Urieli, I., and D. M. Berchowitz (1984) "Stirling Cycle Engine Analysis," Adam Hilger, Bristol.

DIRECTIONS IN US AIR FORCE SPACE POWER TECHNOLOGY FOR GLOBAL VIRTUAL PRESENCE

David Keener, Kitt Reinhardt, Clay Mayberry,
Dan Radzykewycz, and Chuck Donet
US Air Force Research Laboratory, AFRL/VSDV
3550 Aberdeen Ave. SE
Kirtland AFB NM 87117-5776
(505) 846-2614

Dean Marvin and Carole Hill
The Aerospace Corporation
P.O. Box 9045
Albuquerque, NM 87119-9045
(505) 846-1454

Abstract

Recent trends in the development of high efficiency, light-weight, compact, reliable and cost-effective space power technologies needed to support the development of next-generation military and commercial satellites will be discussed. Development of new light-weight and reduced volume electrical power system (EPS) technologies are required to enable the design of future "smallsats" with power requirements less than 1500W, to "monstersats" having projected power levels ranging from 10-50kW for commercial communication and military space based radar type satellites. In support of these projected requirements a complement of power generation, power management and distribution, and energy storage technologies are under development at the Air Force Research Laboratory's Space Vehicles Directorate. The technologies presented in this paper include high efficiency multijunction solar cells, alkali metal thermal electric converters (AMTEC), high-voltage (70-130V)/high-efficiency/high-density power management and distribution (PMAD) electronics, and high energy density electrochemical and mechanical energy storage systems (sodium sulfur, lithium-ion, and flywheels). Development issues and impacts of individual technologies will be discussed in context with global presence satellite mission requirements.

INTRODUCTION

In order to meet the demand for increased satellite payload mass and power, and reduce launch vehicle size and cost, an increasing amount of attention is being given to the satellite electric power system (EPS) performance in terms of specific power (W/kg), size, stowed volume, and cost. The EPS is responsible for providing uninterrupted, fault-tolerant electrical power to satellite payload and housekeeping equipment throughout the lifetime of the mission. EPS power requirements are dictated by payload requirements (e.g. channel capacity), antenna characteristics, data rate, and satellite orbit. Today's smallsats and conventional largesats vary in power level and mass from about 1000 watts and 225 kg or less, up to as much as 10 kW and 4,550 kg, respectively. In the case of smallsats, recent trends in shrinking space budgets have pushed mission planners towards the use of cheaper smallsat designs capable of launch on smaller, cheaper, and more easily deployed vehicles. The US intelligence community recently evaluated requirements for future surveillance missions and acknowledged the advantages of lower cost smallsats to address tomorrow's warfighter needs, which include increased flexibility, improved performance, and the ability to launch them easily when needed. In contrast to smallsat applications, mission planners have also acknowledged the need for significantly larger 10-50 kW monstersats to enable next-generation communications, radar, and weapons platform functions. Prime contractors Hughes, Loral, and TRW have projected communications satellite power needs reaching 15-20 kW in the next five years, and military platform power needs are projected to exceed 50 kW over the next decade.

In order to meet projected smallsat and monstersat design requirements, revolutionary advancements in EPS component technology are required over today's conventional technologies. A typical satellite EPS (solar array and support mechanisms, batteries, PMAD electronics, and PMAD cabling/harness) utilizing state-of-practice technology accounts for about 20-30% of the total satellite mass and occupies a significant portion of the satellite volume. A breakout of average EPS component mass for operational DSCSIII, DSP, GPS, and Milstar satellites is shown in Figure 1. For these satellites, the EPS-to-total satellite mass percent of 20-30% corresponds to EPS specific power values of 3-5 W/kg.

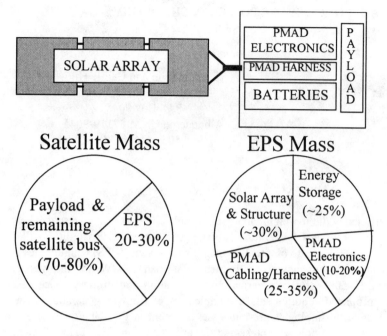

FIGURE 1. Satellite Electric Power System (EPS) Component Mass Fraction Averaged for State-of-Practice DSCSIII, DSP, GPS, and Milstar Satellites.

The goal is to increase total EPS specific power through advancements in component technology to enable a reduction in EPS mass percentage to approximately 10%. A complement of on-going EPS technology development programs at the Air Force Research Laboratory (AFRL) have the potential to increase EPS specific power to 10 W/kg by FY2000 and 13 W/kg by FY2005. Examples of these technologies include multijunction solar cells with up to 1/3 higher efficiency of state-of-the-art (SOA) silicon and GaAs cells, energy storage batteries and flywheels with three times the energy density of SOA batteries, high efficiency power electronics that reduce thermal control loads, and a solar thermal system with integrated (non-photovoltaic) energy generation and (non-electrochemical) energy storage.

SOLAR PHOTOVOLTAICS

For many years, spacecraft have employed photovoltaic technology for energy generation. Silicon solar cells have been used since the 1960s, and gallium arsenide (GaAs) solar cells at 18.5% efficiency are the current state-of-the-art. However, multijunction solar cells represent the next generation of photovoltaic technology, offering substantially improved efficiency and cost-effectiveness.

Multijunction solar cells combine two or more types of semiconductor material, each containing one p/n junction, in a monolithic structure. The improved conversion efficiency of these devices results from the optimized choice of semiconductor bandgaps, which allows each junction to respond to a specific range of light wavelengths. This selective conversion of the incident light spectrum is shown schematically in Figure 2. Concerted development of these cells for space application began in 1991, and in September of 1995 the technology was transitioned to a Multijunction Solar Cell Manufacturing Technology (ManTech) Program. Both dual and triple junction cells are within the scope of the ManTech Program, which is jointly funded by the Air Force Research Laboratory and NASA Lewis Research Center.

The objectives of the ManTech program are to optimize the production process, improve cell efficiency to 24-26%, and improve yield such that the manufacturing cost is limited to 1.15X GaAs/Ge cells, all by 1999. Two US vendors, TECSTAR/Applied Solar Energy Division and Spectrolab Inc., made excellent progress in Phase I toward achieving the program objectives and are building on those successes in Phase II. Both vendors have selected the $GaInP_2$/GaAs/Ge semiconductor combination. Spectrolab has demonstrated a best triple junction cell efficiency of 25.76% with a lot average of 24.2%. Large area triple junction cells, 13.8cm^2, have also been demonstrated at 24.3% with a lot average of 23.3%, showing the feasibility of large area cells with high efficiency. Spectrolab has also assembled a coupon of three triple junction solar cells with an efficiency of 23.5%. The starting efficiencies of the three cells were 24.3%, 24%, and 23.5%, which demonstrates that strings of triple junction cells can be assembled with a small string assembly loss of efficiency. TECSTAR has demonstrated multijunction cells with a best efficiency of 24.7% and a lot average efficiency of 23.8% for 200 cells. Also, large area dual junction cells, 4x4.8cm, have been demonstrated with a best efficiency of 22.1% with an average efficiency of 20.9%.

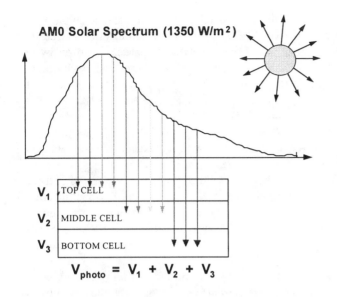

FIGURE 2. Each Layer of a Triple Junction Array Solar Cell Responds to a Band of Incident Light Wavelengths.

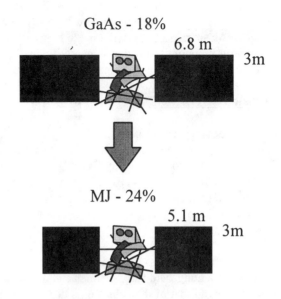

FIGURE 3. Reduction in 10kW Solar Array Area When Multijunction Solar Cells are Used.

Efficiency improvement in solar cells translates directly into reductions in solar array area required to produce a given amount of power. For instance, GaAs arrays can generate up to 225W/m^2, so a 10kW array occupies a substantial 50m^2. However, the 24% conversion efficiency of GaInP/GaAs/Ge multijunction solar cells represents a 33% increase in efficiency and per cell power output over GaAs cells, which translates directly into a 25% decrease in solar array size. Figure 3 illustrates this decrease for a 10kW solar array. For a very high power application, which could require as much as 100kW of generated power, the solar array size could be reduced by more than 100m^2 by employing multijunction solar cells. The reduction in solar array area also leads to a proportional reduction in array mass, and a further reduction in the mass of other subsystems including the attitude control system (ACS) and the array deployment/support structure. Multijunction solar cells therefore represent major improvements in power system specific power (W/kg), and substantial reductions in absolute spacecraft mass.

Even greater benefits can be realized by continuing to increase solar cell efficiency. The Air Force Research Lab has recently begun to investigate concepts in > 30%, ultra-high efficiency energy conversion. The multijunction solar cell concept is being extended with two innovative four-junction solar cell concepts.

First, a mechanically stacked four-junction device consisting of two dual-junction cells connected in series with a metal bond interconnect has been modeled to have 30-35% conversion efficiency. This concept is under investigation by Research Triangle Institute for AFRL. The top tandem cell is leveraged from existing GaInP/GaAs/Ge technology, and the bottom cell is a GaInAsP/InGaAs tandem, which is a new development. The key development areas are the growth of high quality GaInAsP material, which has been demonstrated due to an improved In gas source, and the metal bond interconnect, which is using a recently demonstrated temperature and pressure technique to bond GaAs to In with the necessary optical, electrical, and mechanical properties. Each of the components have been demonstrated with performance approaching that needed for 35% efficiency. Current tasks include optimization of the metal bond interconnect process, development of the GaInAsP/InGaAs tandem cell, and integration of all components into a single structure. Overall, this is a two year effort, which is scheduled to be complete in June 1998.

A second concept under investigation in a joint Sandia National Laboratories/AFRL new start program could push conversion efficiency to 40% or more. This concept employs much improved AlGaInP material and an innovative nitride-based quaternary semiconductor material to form a monolithic four junction solar cell. Historically, large amounts of aluminum have not been used in solar cells because of oxygen contamination in the source and growth chambers, but Sandia has begun to use greatly improved Al sources, which show oxygen contamination no greater than GaAs sources, which will enable high quality AlGaInP material. The nitride-based quaternary is the key to making a monolithic cell that is nearly optimally current matched. Basic materials studies are underway on the nitride at Sandia National Laboratories, and promising results have been produced.

ENERGY STORAGE

All modern spacecraft use batteries to provide electrical power during eclipse, with the exception of a few deep space missions which use radioisotope thermal generators (RTG). The key performance parameters of space batteries are the specific energy (Whr/kg) at the useable depth of discharge (DOD) and the cycle life. Both of these are critical in high power applications - the former because the high power levels lead to large masses, and the latter because large, expensive spacecraft must have long life to be cost-effective. Today, virtually all low earth orbiting (LEO) spacecraft and high power GEO spacecraft use nickel hydrogen (NiH_2) batteries. Present day NiH_2 batteries can provide 7-9 years of service in LEO and 10 years in GEO.

It is difficult to assess the benefit of a particular battery technology based solely on the cell-level Whr/kg since wiring, mounting hardware, thermal control equipment, and charging electronics can contribute additional mass comparable to that of the cells alone, depending on the particular battery technology. The mass of these batteries and advanced energy storage technologies are compared in Figure 4.

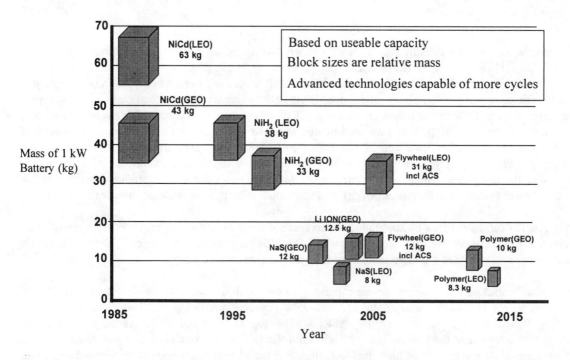

FIGURE 4. Energy Storage Comparisons at the Cell Level for a 1 kW System.

Sodium Sulfur Batteries

The Air Force Research Laboratory has several on-going programs to develop high performance energy storage devices. The sodium sulfur (NaS) cell (Figure 5) is the most mature new electrochemical storage technology. With proven cell-level performance of 150 Whr/kg (40 Ahr, 2V), and a 2-3X improvement over state-of-the-art nickel hydrogen (NiH_2) on the battery level, NaS batteries offer substantial advantages for energy storage. The technology is unique in its high (523K) operating temperature and use of molten electrodes. An extensive test program sponsored by Phillips Lab (now part of AFRL) and conducted at Sandia National Laboratories has shown that NaS

FIGURE 5. Eagle Picher Industries 40 Ahr NaS Cell.

that NaS cells are very robust, having survived multiple freeze-thaw cycles, overtemperature, simulated launch vibration, dead shorts, overdischarge, and discharge at a 4.5C rate without performance degradation. (A "C" rate corresponds to complete discharge in one hour). During this aggressive ground safety test program, the solid ceramic β″alumina electrolyte in several Eagle-Pitcher Industries (EPI) cells were fractured deliberately. Although some cells did breach, no catastrophic failures were observed, which indicates that the cells can be launched while operating, avoiding the need for a primary battery (Unkelhaeuser 1996).

Although NaS must operate at 523K, an adequate, lightweight thermal enclosure will isolate the battery module from the rest of the spacecraft. The cells are endothermic on charge and exothermic on discharge, so with adequate insulation, the module would be virtually self-sustaining. The heat generated by the module could even be used to warm electronics at a different location on the spacecraft. On a battery level, Hughes has designed and built a 16 cell (with switching), 28V, thermally enclosed battery module (See Figure 6) at 117 Whr/kg and 50 Whr/L volumetric energy density (Taenaka, 1990). This work is scheduled to be continued in-house at AFRL with simulated cells (heaters) to further develop the battery module concept.

NaS technology is anticipated to be half the cost (per cell) of NiH_2, and has already demonstrated longer cycle life in both LEO and GEO than competing developmental technologies such as lithium-ion. In addition to completing safety and abuse testing, Phillips Lab had NaS cells in life cycle test at the Naval Surface Warfare Center in both LEO (5000 cycles at 40% DOD) and GEO (300 cycles at 60% DOD) until limited funding forced the test to terminate (Brown 1996). An AFRL-sponsored flight test of a NaS battery will launch on November 19, 1997, to validate operation in zero-g. Figure 7 shows the completed battery experiment (designed and built by the Naval Research Laboratory) that will be flown on the upcoming STS-87 shuttle mission which allows for experiment recovery and cell destructive physical analysis (DPA). If further battery and cell optimization were pursued, NaS technology could provide a viable energy storage system for use by 1999, but lack of industry support and funding commitment by users has shelved further major development efforts.

FIGURE 6. The Hughes High Energy Density Rechargeable Battery (HEDRB).

Lithium-Ion Batteries

The fastest-maturing electrochemical energy storage technology is that of lithium-ion. This technology has come to the fore because of the commercial need for high energy density, rechargeable batteries for consumer electronics (cellular phones, laptop computers, camcorders, etc.). Lithium-ion has already proven itself in that market, but has not yet been demonstrated to be capable of supporting GEO and LEO satellite missions. It is widely believed, however, that lithium-ion technology will mature to the point that it will be capable of sustaining a GEO mission by the year 2002.

Lithium-ion technology promises the same specific energy improvements as sodium-sulfur, but at much lower operating temperatures. (Optimal performance with conventional electrolytes occurs in the 263 to 303K range.) Current generation cells have a proven performance of 120 Whr/kg (20 Ahr, 4.1V), with an energy density of ~250 Whr/L; extrapolating these numbers to the battery level indicates that lithium-ion batteries should have specific energies and energy densities that are three to four larger than nickel hydrogen. Current efforts in the aerospace arena are focusing on increasing the capacity of these cells to 50 Ahr, increasing the cycle life, and optimizing the materials and packaging to reach 150 Whr/kg and 300Whr/L at the cell level. One distinct advantage that lithium-ion technology has when compared to other high energy density cells is its relatively high operating voltage (averaging 3.5V). Because of this feature, fewer cells need to be connected in series to reach the desired battery voltage—assuming a 28V bus, nickel hydrogen requires twenty-four cells; sodium sulfur, sixteen; and lithium-ion, eight. Fewer cells corresponds also to fewer electronics and significantly higher battery reliability (assuming that the cell reliability is high).

FIGURE 7. The Sodium Sulfur Battery Experiment (NaSBE).

Cycle life is perhaps the largest hurdle still facing this technology. Lithium-ion cells have shown more than 600 cycles to 100% DOD at moderate rates (0.5C, where C is the capacity of the cell). However, to be useful for GEO missions, the technology will need to be able to perform 1500 cycles to a maximum DOD of 75% with a maximum discharge rate of 1.0C; the charge rate would be no more than 0.1C. Because of the nature of the eclipses in GEO, most of the cycles would be to much lower DODs. At the rate that the technology is improving, it is most likely that within 5 years, one will be able to incorporate lithium-ion in GEO missions. LEO missions, however, will be much harder to support. These missions will require more than 40,000 cycles with a constant DOD (between 40 and 60%), charging at approximately 0.5C and discharging at approximately a 1.0C rate. There is still considerable speculation about whether or not lithium-ion will ever be able to support LEO missions. LEO missions are therefore looking towards possibly using an electromechanical energy storage system—flywheels.

Flywheels

Flywheels eliminate the need for electrochemical batteries by storing kinetic energy in the flywheel's rotor while also providing large amounts of stored momentum for attitude control. Several programs funded by NASA and the Air Force have produced designs that are very promising for GEO and LEO applications.

Flywheel technology is similar to momentum wheel technology, in that the speeds of the wheels are fluctuating while the units are in operation (as the flywheels are being charged and discharged), but dissimilar in that the speeds are an order of magnitude higher (40,000-60,000 rpm), and energy is being stored, managed, and distributed instead of just exchanged from unit to unit. Recent advancements in carbon fiber composites, magnetic bearings, and high speed control electronics have enabled flywheels for use in space, providing lightweight, long life, efficient operation. Just like momentum wheels, several flywheel units must be configured in a spacecraft to provide attitude control and stability. Although spacecraft attitude control laws are fairly well characterized, the combination of energy storage and attitude control functions has never been implemented simultaneously in a single spacecraft subsystem. Work in integrated power and attitude control systems (IPACS) was started in the mid 70's by NASA, but prototype efforts for spacecraft flywheel technology did not begin until a few years ago, and have focused on the design of a flywheel unit.

Figure 8 shows a cutaway view of a typical space flywheel unit. It is simply a spinning wheel, or rotor, made of high strength carbon fiber composite material. Because it is spinning, it has the capability to generate electricity for a spacecraft. This is done with a motor/generator that supplies a current (by spinning the rotor down) when the spacecraft is in eclipse. Electricity from the solar arrays is stored as mechanical energy in the rotor by spinning it up with the motor/generator. The rotor, made up of a rim, hub, and shaft, is supported in the unit by high magnetic flux bearings. There are usually two sets of radial bearings that support the shaft and prevent it from touching the sides of the unit. One set of thrust, or axial, bearings prevent the shaft from touching the top and bottom of the inside of the unit.

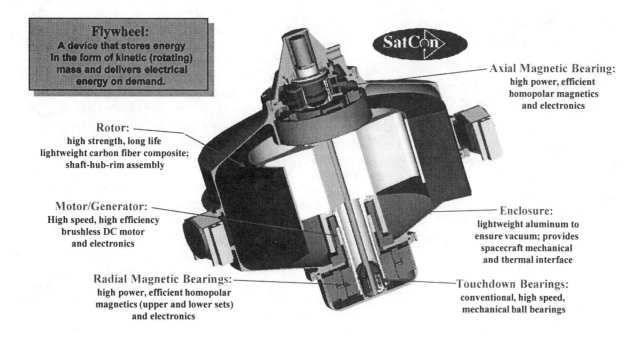

FIGURE 8. Flywheel Unit Design by SatCon Technology Inc.

These bearings will allow the rotor to achieve speeds between 40,000 and 60,000 rpm. Conventional high-speed mechanical bearings are used at lower rpm when the flywheel is spinning on the launch pad, during launch, and through deployment of the spacecraft. Also, if there were an external disturbance to the spacecraft during flight operations, the touchdown bearings would provide rotor stability if the magnetic bearings were not able to instantaneously overcome the disturbance force. For maximum life and power efficiency, a vacuum must exist in the flywheel unit so there is no friction while the rotor is spinning.

Much like a gyroscope, the flywheel unit, while spinning, will oppose rotational changes in its orientation in the direction perpendicular to the plane of the spinning rotor. This is caused by the forces of inertia stored in the wheel. By slightly changing the wheel orientation or the wheel speed, a torque is developed by the wheel, which is in turn applied to the spacecraft, affecting its attitude. These forces have been used in reaction wheels and control moment

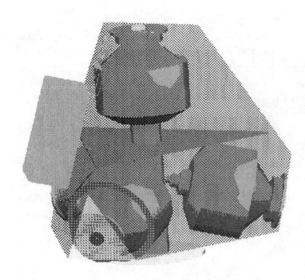

FIGURE 9. Typical Flywheel Tetrahedron Configuration.

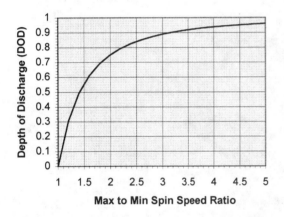

FIGURE 10. DOD Related to Flywheel Speed.

gyros in the past to stabilize or change a spacecraft's orientation, or attitude. Never has this task been integrated with the simultaneous task of supplying power the spacecraft.

Only a few configurations of flywheel units will adequately control the attitude of a spacecraft. A common configuration, shown in Figure 9, consists of four wheels, which will provide attitude control in all three degrees of freedom.

Flywheel energy storage offers many improvements over current electrochemical systems. On the unit level, 100 Whr/kg (nameplate) is achievable (a 4X improvement over NiH_2), and for a configuration of flywheels on a LEO platform, initial prototype systems would offer 30 Whr/kg (useable, including the attitude control system). The useable amount of energy in a flywheel is related to the DOD by the change in wheel speed during discharge (Figure 10), such that most of the stored energy may be extracted without operating the wheel over too large a speed range. High performance charge, discharge, and throughput characteristics are obtained from an optimized, brushless type motor/generator with magnetic bearings. The power losses in actively controlled homopolar magnetic bearings are minimal compared to traditional electrochemical losses from resistance of the battery, charge/discharge equipment, switching, etc. The potential for very high charge and discharge rates exists, and because so much energy can be stored in a flywheel for use during eclipse, large amounts of 'flexible' energy can be supplied at any time. This can become extremely useful for orbit transfer, where electric propulsion would use very high power for continuous thrusting through the entire orbit to perform an optimum transfer. Flywheel systems exhibit very high cycle life because there is no mechanical wear, and there is very little capacity degradation over its lifetime, in contrast to electrochemical energy storage. These advantages will eventually allow for lifetimes of 20-30 years in any orbit. The advantages truly display themselves in LEO, where a flywheel energy storage system could last 100,000 to 150,000 cycles.

When the functions of energy storage and attitude control are coupled, the system mass savings far outpaces any projected electrochemical technology. Costs associated with flywheel technology development are significant, as with any new emerging technology, but life manufacturing costs are projected to be comparable to traditional electrochemical energy storage technologies. When the cost savings associated with greatly reduced mass are taken into account, the life cycle costs are very attractive when compared to a combined electrochemical energy storage system with traditional momentum wheels for attitude control. Once a manufacturing base is established, the life cycle cost should decrease with the price of composite materials and as less expensive methods of manufacturing are developed. There are some remaining issues that need to be addressed by developers, including life cycle testing for composite rotors, rotor and system level reliability/redundancy, and adequate system design configurations for different mission scenarios.

Flywheels are being developed in the Air Force jointly with NASA. The Air Force Research Laboratory and NASA Lewis Research Center (LeRC) are the lead agencies and comprise the only national flywheel program

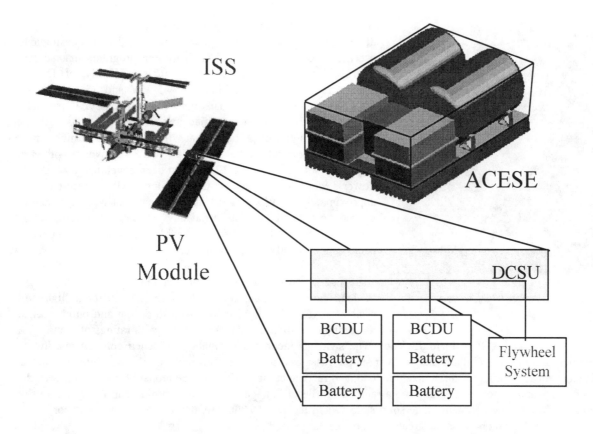

FIGURE 11. Conceptual Design of a 1 Degree Of Freedom Flywheel Attitude Control and Energy Storage Experiment (ACESE) to be flown on ISS in 2001.

focusing strictly on aerospace applications. Other organizations including the Air Force Space and Missile Command, the Office of the Secretary of the Air Force and the other NASA centers, as well as other government agencies such as DARPA, DOE, and the Department of Commerce are contributing to the aerospace flywheel development effort. The program is focused on the development of three basic flywheel sizes: large (>1kWhr), for >3kW loads; medium (250-500Whr), for 500W-3kW loads; and small (<250Whr), for small <500W loads. At AFRL, a multi-disciplinary team approach is being taken to develop the best technology possible. The initial result will be a prototype IPACS delivered by SatCon Technology, Inc. (See Figure 8) in late FY98 that will be tested in three degrees of freedom. Each wheel in this design will have 2.7 kWhr of usable stored energy (8.1 kWhr for the three-wheel system). NASA and AFRL are also working with Boeing on a parallel flight experiment for International Space Station (ISS). The experiment, called the Attitude Control/Energy Storage Experiment (ACESE) will fly on the Space Shuttle in 2001 (Figure 11). Additionally, a Flywheel Test Program will be established soon. Starting in 1999, it will operate in parallel with flywheel development. Phase I will build a composite materials database by testing coupons and sub-elements used in rotor design. Phase II will spin test rotors, concentrating on cycle life and determining the failure mechanisms in the composite structure. Finally, Phase III will consist of qualification testing for particular flywheel units, and will validate flywheels for use in future spacecraft. The Air Force and NASA, in cooperation with DARPA, plan to fund the testing and the construction of the facilities, while the flywheel community supplies hardware that will be used in test units.

ENERGY MANAGEMENT

As the absolute power level of spacecraft increases, the inefficiencies associated with bus voltage regulation and battery charging become important not only because they require oversizing of the energy generation and storage subsystems, but also because of the thermal load that they generate. Regulation architectures with 70-80% efficiency are common in sub-five kilowatt spacecraft, but are unacceptable in a 25 kW spacecraft. High power

FIGURE 12. GaAs DC/DC Converter.

spacecraft also have massive power harnesses when the same 28V distribution from low power systems is used, and so it is desirable to use higher bus voltages. AFRL has three programs that address improvements in this area. The first is a family of DC/DC converters from TRW that interface 45-75V and 75-130V buses with conventional 28V payloads with 87% efficiency. These allow the use of existing flight qualified low voltage payloads with the unregulated high voltage buses that will be required in high power spacecraft. For regulated buses, the 70V to 3.3V converter with 90% efficiency shown in Figure 12 has also been developed by Virginia Power Technologies using GaAs technology. All of these units are approximately 400W/kg. A 70V to 5.5V regulator with high density interconnects and similar efficiency has been developed by Lockheed-Martin East Windsor, which offers very compact packaging and 600W/kg performance.

The use of one-time fuses to protect the power system against payload failure or transients is hard to rationalize in high cost systems. AFRL has funded Rockwell to design and build a set of three flight-qualified solid state switches with ratings of 1 and 10 A at 75 to 130 Vdc and 3 A at 45 to 72 Vdc. The switches incorporate radiation hardened components, hybrid technologies, and offer several advantages over simple fuses as a form of circuit protection. The switch can detect and diagnose power line faults, disconnect the load, and reconnect the load at the operator's discretion. Further, the switch connects and disconnects loads with a ramped profile of voltage versus time, minimizing large current rise and fall rates in the power system. The solid state switch technology incorporates I^2t and thermal memory, thus allowing trip coordination between upstream and downstream switches.

SOLAR THERMAL ENERGY STORAGE AND CONVERSION SYSTEMS

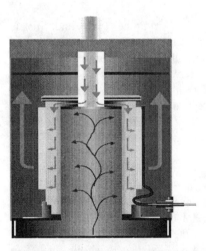

FIGURE 13. Illustration of AMTEC Cell Process.

A revolutionary technology, the alkali metal thermal to electric converter (AMTEC), is being developed at AFRL for incorporation into solar thermal storage and conversion systems. The concept (shown in Figure 13) is a thermally regenerative electrochemical cell in which high enthalpy (pressure) sodium (the alkali metal) is brought into contact with a preferentially conductive β″alumina solid electrolyte (BASE), which conducts Na^+, but not Na^0 or electrons. Electrons are drawn off through an external load and recombined with the Na^+ at a porous electrode on the low enthalpy (pressure) side of the BASE. The neutralized Na atoms evaporate from the BASE, transit a vapor space, and condense on the radiator. The condenser wick collects the Na liquid and pumps it back to the heat source to complete the thermodynamic cycle.

The storage and conversion system (shown in Figure 14) uses a mirror or lens to concentrate sunlight into a receiver cavity. AMTEC devices surrounding the cavity convert the heat to electricity, while phase change materials (e.g. LiF) provide energy storage in place of electrochemical batteries for eclipse power.

This technology has the potential for high energy density, high radiation resistance, and high subsystem specific power, particularly at high power levels, such as those required by space-based radar and the integrated solar upper stage (ISUS) program where the use of AMTEC cells may enhance overall system efficiency when used as a supplementary energy converter. Using AMTEC technology instead of solar cells eliminates sensitivity to environmental radiation, and the phase change material eliminates battery life cycle and temperature control limitations.

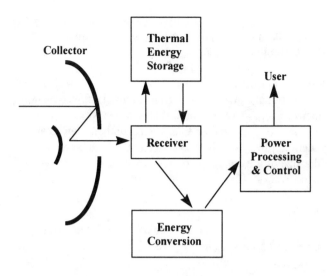

FIGURE 14. Solar Thermal Energy Generation System.

Currently, AMTEC cells manufactured by Advanced Modular Power Systems are in the component demonstration phase and have achieved 18-19% efficiencies. Concentration schemes have been demonstrated on Earth, and inflatable concentrator deployment technology that may be applicable has been flight demonstrated. Advanced solar thermal system designs using this technology and advanced components at the conceptual design level indicate system-level specific powers in excess of 20 W/kg are possible (Underwood 1991). AFRL is supporting an advanced radioisotope power system program being run by DOE and JPL, in which AMTEC is the leading candidate for the conversion technology. If successful, this program will qualify and fly AMTEC conversion before an AF SBR or ISUS flight, providing life and performance data.

According to a design study accomplished by JPL, AMTEC technology is capable of 35-40% energy conversion with the advancement of component technology (Underwood 1991, Sievers 1990). Some of the advanced component technologies needed for AMTEC to reach its potential include thin electrolytes on micromachined substrates, micromachined condensers, and potassium electrolyte material. Under SBIR funding, AFRL is managing a program with Creare, Inc. to develop micromachined substrates for thin-film electrolyte. Currently, the major loss in AMTEC cells comes from resistance across the electrolyte, which is 1100 μm thick. Resistance in the electrolyte is directly related to its thickness, and this technology will allow the electrolyte thickness to be reduced to 10 - 100μm, increasing the cell energy density by 50%. This technology is the major enabler for AMTEC technology to reach 35% efficiency. In a related program, Advanced Modular Power Systems, Inc. parametrically modeled advanced solar AMTEC cells including the advanced components discussed above and several cell geometries including flat plate, multi-tube, and donut shaped configurations (particularly suited for ISUS application). The results of this study have indicated that the radial or donut shaped configuration incorporating advanced AMTEC component technologies could achieve an efficiency of 33%, 975-1200 W/Cell, and 400-500 W/L.

The results of programs and studies conducted by AFRL has shown that solar thermal storage and conversions systems incorporating the AMTEC technology offer clear benefits for several types of missions including ISUS and SBR. Follow-on development programs will integrate advanced AMTEC component technologies to demonstrate AMTEC cell efficiencies up to 35-40%.

COST CONSIDERATIONS

Power systems typically represent 20-30% of the cost of a spacecraft, but in high power, single payload applications the fraction can be higher. There are several ways that advanced power technology can reduce costs. The first is through more efficient components, i.e., higher efficiency solar cells, batteries and power electronics. More efficient solar cells reduce the number of cells needed to produce a given amount of power. Even though the cost per cell is generally higher for more efficient cells, the cost per watt is lower. Mature Si cells are the exception, producing the lowest cost per watt of any commercially available, space qualified technology. However, high power arrays are unmanageably large when constructed using cells with 12% efficiency that is typical of Si. The 24% efficient multijunction ManTech cells described above are expected to be 15% lower in cost per watt than the 18% GaAs cells that they replace. Comparing the cost benefit of GaAs cells with Si cells requires system level considerations. The smaller array that can be obtained with the former leads to not only fewer cells required, but also less labor to assemble them into an array. Further, the lighter weight of finished assembly, due to fewer cells and smaller supporting structure, reduces the weight of other spacecraft subsystems such as attitude control (propellant, momentum wheels, etc). Higher energy density batteries also save weight inherently, as well as requiring less thermal control system mass (e.g., smaller radiators) while improved power electronics efficiency also reduces the thermal control requirement. Flywheel systems will likely require very minimal thermal control, whereas all battery systems now in use have at most a 30°C operating range.

The second approach to addressing high power system costs is to consider revolutionary power systems such as the solar thermal system described above. While the cost of conventional power system components is approximately linear with output over a significant range because the number of components required scales in this way, the cost of such revolutionary systems may be sublinear because the size of parts rather than their number may change. Analyses of this possibility are in progress. The same argument applies to high ratio concentrator arrays, where the cost is not determined by the number of solar cells, but by the size of the collection optics.

CONCLUSIONS

Future satellite missions place special requirements on spacecraft power systems, including reduced mass and volume, high power output, and cost effectiveness. Development programs are in progress at the Air Force Research Laboratory that address these needs in energy generation, storage, and control.

Acknowledgments

The authors of this paper would like to thank Mr. Ralph James and Mr. Mike Brasher for their input, advice, and support.

References

Brown, H. and J. David (1996), "Sodium-Sulfur (NaS) Project Review," Naval Surface Warfare Center Internal Report.

Sievers, R.K. and M.H. Cooper (1990) "Advanced AMTEC Design Options with Composite Electrolyte Membranes", *Proc. 7th Symposium on Space Nuclear Power Systems*, CONF-900109, M.S. El-Genk and M.D. Hoover, eds., American Institute of Physics, New York, 1:239-244.

Taenaka, R.K., et. al. (1990) "High Energy Density Rechargeable Battery," U.S. Air Force Wright Research and Development Center Technical Report, WRDC-TR-89-2131.

Underwood, M.L., et. al. (1991) "Performance Projections of Alternative AMTEC Systems and Devices," *Proc. 8th Symposium on Space Nuclear Power Systems*, CONF-910116, M.S. El-Genk and M.D. Hoover, eds., American Institute of Physics, New York, AIP Conf. Proc. No. 217, 1:472-481.

Unkelhaeuser, T.M. (1996) "Phillips Laboratory NaSTEC Ground Test Results Report," Sandia National Laboratories Technical Report, SAND96-0702.

KINEMATIC PATH PLANNING FOR SPACE-BASED ROBOTICS

Sanjeev Seereeram
Scientific Systems Company
500 West Cummings Park, Suite 3000
Woburn, MA 01801
(617) 933-5355

John T. Wen
Rensselaer Polytechnic Institute
110 8th Street, CII 8123
Troy, NY 12180
(518) 276-8744

Abstract

Future space robotics tasks require manipulators of significant dexterity, achievable through kinematic redundancy and modular reconfigurability, but with a corresponding complexity of motion planning. Existing research aims for full autonomy and completeness, at the expense of efficiency, generality or even user friendliness. Commercial simulators require user-taught joint paths – a significant burden for assembly tasks subject to collision avoidance, kinematic and dynamic constraints. Our research has developed a Kinematic Path Planning (KPP) algorithm which bridges the gap between research and industry to produce a powerful and useful product. KPP consists of three key components: path-space iterative search, probabilistic refinement, and an operator guidance interface. The KPP algorithm has been successfully applied to the SSRMS for PMA relocation and dual-arm truss assembly tasks. Other KPP capabilities include Cartesian path following, hybrid Cartesian endpoint/intermediate via-point planning, redundancy resolution and path optimization. KPP incorporates supervisory (operator) input at any detail to influence the solution, yielding desirable/predictable paths for multi-jointed arms, avoiding obstacles and obeying manipulator limits. This software will eventually form a marketable robotic planner suitable for commercialization in conjunction with existing robotic CAD/CAM packages.

INTRODUCTION

Intelligent robotic systems for space missions, including assembly and maintenance of the International Space Station will rely heavily on autonomous and semi-autonomous robotic path and task planning. Areas such as space assembly and maintenance, undersea robotics, and hazardous environment tasks have been identified as important robotic applications, especially for reduction of human EVA risk. However, these scenarios are unstructured environments for which techniques used in industrial robotics are not suitable. They require autonomous, semi-autonomous and tele-operated robotic systems of significantly greater capabilities than previously used. Recently, emphasis has been placed on achieving greater dexterity through kinematic redundancy, but with a corresponding difficulty in task and motion planning/control.

A recently developed Kinematic Path Planning (KPP) algorithm (Seereeram et al. 1993) has been proposed for path and motion planning of kinematically redundant robots. It can be effectively used for space-based robotic operations (including the shuttle and space station remote manipulator systems, and SPDM arms). The technique is generically applicable to both holonomic and non-holonomic systems (such as mobile robots, free-flying platforms, etc.). The basic KPP technique was designed to alleviate some of the major deficiencies of current autonomous planning algorithms, while retaining the ability to incorporate supervisory input. This latter feature enables the KPP to be used in a variety of planning scenarios, including semi-autonomous and fully autonomous operation.

The path planning problem is posed as a finite time nonlinear root-finding problem. A task error term written as a function of the entire path sequence is iteratively reduced using a modified Newton-Raphson algorithm. This technique is developed to handle various goal task definitions such as Cartesian goal end-point planning, redundancy resolution along a path, or planning with specified intermediate via points. In contrast to potential field techniques, joint and task space inequality constraints are incorporated into the Newton-Raphson procedure by the use of exterior penalty functions. This method iteratively warps infeasible path sequences until all the constraints are met. Additional joint or task space equality constraints can be utilized directly to reduce the search space by projecting the iterative path sequences into the constraint subspace. This yields cyclic joint paths for cyclic Cartesian tip paths, or can be used to meet the kinematic closure requirements for cooperative arm planning. Compared to local model based approaches, this algorithm is less prone to problems such as arm singularities and local minima of potential fields. The algorithm has shown promising results in planning joint path sequences for planar and spatial robots. Previous experimental results have been obtained for applications to spatial seven and nine degree-of-freedom arms.

At Scientific Systems, we have established the feasibility of applying the KPP algorithm to robotic tasks of interest to NASA (Seereeram 1995). We demonstrated the capability for pick-and-place type planning for the SSRMS arm on typical assembly tasks. Currently, the KPP algorithm prototype software exists as a standalone program, with manipulator kinematics being input from geometric specifications, and planning tasks specified in Cartesian or joint space. Obstacle avoidance, joint and task space collision checking and path optimization are all built-in features of the KPP planner. A GUI user-interface is provided which allows the planning engineer to view KPP solutions to manipulator tasks, verifying their feasibility, or altering the solutions interactively via the interface controls.

In previous work, the KPP algorithm has been demonstrated on tasks such as path following for redundant (kinematics) manipulators, hybrid Cartesian endpoint and intermediate via point planning, and redundancy resolution. The KPP is able to incorporate supervisory or operator input, in order to influence the solution and yield desirable paths. This paper summarizes the results obtained using the KPP algorithm for robotic planning/control, and indicates areas of applicability to space-based robotics for the ISS program and beyond.

KINEMATIC PATH PLANNER ALGORITHM

Robotic path planning has traditionally fallen into two categories: geometric approaches and model-based methods (Latombe 1991). Geometric methods typically use the *configuration space (C-space)* representation, in which the path of an n-link robot is reduced to the motion of a single point in an n-dimensional space. Path planning requires construction of the *configuration space obstacles (C-obstacles)* for this particular robot/workspace, and searching for a collision-free path between start and goal configurations. Various tools from geometry, topology and algebra have been utilized to provide the theoretical basis for C-space path planning. These techniques can be divided into three classes: *road-map, cell decomposition* and *potential field* methods. Later researchers have also proposed hybrid schemes which exhibit conceptual and implementation advantages over the basic methods (Hwang and Ahuja 1992).

Kinematic model-based methods combine path planning with redundancy resolution by solving the analytic relationship between joint and task space motion. Given the forward kinematics of the manipulator

$$x(t) = f(\theta(t)) \quad \text{and} \quad \dot{x}(t) = J(\theta(t))\dot{\theta}(t) , \tag{1}$$

where $x(t) \in R^m$ is the task (Cartesian) coordinate and $\theta(t) \in R^n$ is the joint coordinate, the Jacobian (a nonlinear function of joint angles) operator $J(\theta(t)) : R^n \to R^m$ relates the joint velocities to the task-space velocities. Kinematic redundancy means that $n > m$, and Equation (1) possesses an infinite number of joint solutions for a given $\dot{x}(t)$. Recently, there has been significant interest in kinematically redundant mechanisms because of their ability to enhance manipulator dexterity and better avoid joint limits, arm singularities and workspace obstacles.

Basic Approach

In our Kinematic Path Planner (KPP) algorithm (Seereeram and Wen 1995), the main goal is to find a least-effort *feasible* path satisfying joint and task space equality and inequality constraints. Other optimality criteria such as singularity avoidance can also be incorporated – minimum control effort is almost always desirable. The path planning problem is converted into a static nonlinear root-finding problem with a large search space and a comparatively small constraint space. An iterative Newton-Raphson algorithm is applied which guarantees convergence under fairly mild assumptions. Inequality constraints $g(x,\theta) \leq 0$, both in joint space and task space, are handled by a *path-space penalty function* (PSPF) method. With the emphasis placed on finding a feasible path the convergence problem of many global optimization approaches is largely avoided. It has been recently pointed out (Gilbert and Ong 1994) that this approach differs from conventional global optimization problems in that the solution set is relatively large and open – there exists infinitely many solutions to the typical path planning problem. Furthermore, we know *a priori* that if a solution exists, the minimum of the constraint penalty function is zero. By constructing an *initial-value problem* instead of a *two-point boundary value problem* associated with calculus-of-variations solution approaches, computational requirements are moderated. Additionally, our PSPF approach is significantly less susceptible to the singularity and local minima problems inherent in several potential fields based algorithms (Khatib 1986).

Iterative Solution

Given the forward kinematics of a manipulator Equation (1), we wish to find a continuous joint path $\theta(t)$ which follows a desired $x(t)$. By treating $\dot{\theta}(t)$ as a control variable (set $u(t) = \dot{\theta}(t)$), the problem can be stated as follows:

Given the system described by Equation (1) find $\underline{u} \triangleq \{u(t), t \in [0,1]\}$ such that $\underline{x} = \underline{x}_d$, subject to the joint and task space constraints $h(\theta(t)) \leq 0$, $g(x(t)) \leq 0$.

This description allows the task to be defined on any or all of the discretization points over which $x(t)$ is computed, e.g., $\underline{x}_d = x(1)$ is used for Cartesian goal end-point planning, whereas $\underline{x}_d = [x^T(\Delta t) \cdots x^T(1)]^T$ can be used for Cartesian path following (redundancy resolution along a specified path). Equation (1) is written as a nonlinear algebraic equation:

$$\underline{x} = F(\underline{u}) . \qquad (2)$$

The analytic form of $F(\cdot)$ is in general difficult to find, and will not be explicitly required. Equation (2) is cast as a nonlinear zero crossing problem:

$$y \triangleq F(\underline{u}) - \underline{x}_d = 0 , \qquad (3)$$

where y represents the error between the nominal and desired paths. Consider the differentiation of Equation (3) with respect to an *iteration* variable τ:

$$\frac{dy}{d\tau} = \nabla_{\underline{u}} F(\underline{u}) \frac{d\underline{u}}{d\tau} . \qquad (4)$$

To converge to the desired path, a Newton-Raphson iteration in τ is applied. Assume, for each iteration, the system is locally controllable around $\underline{u}$, ie. $\nabla_{\underline{u}} F(\underline{u})$ is full rank. Then choose $\underline{u}$ to satisfy

$$\frac{d\underline{u}}{d\tau} = -\alpha \nabla_{\underline{u}}^{\dagger} F(\underline{u})(F(\underline{u}) - \underline{x}_d) , \qquad (5)$$

where $\nabla_{\underline{u}}^{\dagger} F(\underline{u})$ denotes the generalized inverse of $\nabla_{\underline{u}} F(\underline{u})$. Then $\frac{dy}{d\tau} = -\alpha y$, which implies that the norm of the error, $\|y\|$, decreases monotonically in τ. For practical implementation, (5) is discretized in τ, and $\underline{u}$ is iteratively updated:

$$\underline{u}^{(k+1)} = \underline{u}^{(k)} - \beta_k \nabla_{\underline{u}}^{+} F(\underline{u}^{(k)})(F(\underline{u}^{(k)}) - \underline{x}_d) , \qquad (6)$$

where $\Delta \tau$ is the discretization interval and $\beta_k = \alpha_k \Delta \tau$. Since for $\Delta \tau$ sufficiently small, the approximation is arbitrarily close to (5): $\|y\| = \|F(\underline{u}^{(k+1)}) - \underline{x}_d\|$ as a function of β_k must be strictly decreasing for β_k sufficiently small. Therefore, a line search can be used to determine the best β_k to use.

Constraints

The procedure described by Equation (6) converges to a joint path sequence meeting the task specification $\underline{x}_d$. Practical robotic planning scenarios impose additional equality and inequality constraints, both in joint and task space. Inequality constraints arise naturally out of manipulator joint limits and task space obstacles. Equality constraints may be imposed through fixed joint configurations at specified points along the path – joint path cyclicity is the most common example – or through kinematic constraints such as the closure requirement of dual-arm planning for manipulation of a common payload.

Equality constraints are used directly to reduce the dimensionality of $\underline{u}$. Inequality constraints are handled by a path-space penalty function (PSPF) method. The PSPF is defined as a continuously differentiable function equal to zero if the constraint is met, and strictly positive, monotonically increasing if violated. By incorporating inequality constraint violations into the task error term y in Equation (3), the Newton-Raphson iteration converges to the *constrained* solution. Typically, inequality constraints are enforced at a discrete set of joint configurations throughout the path. While the PSPF method can be applied to the general constrained planning problem in the continuous function space, for robotic planning inequality constraints are usually tested at a finite number of sampling instances along the path.

SAMPLE KPP RESULTS

Some examples are presented which incorporate both joint and task space constraints via the path space penalty functions. Manipulators and the workspace are modeled as polyhedral objects for collision detection and distance calculations. The existing KPP algorithm was implemented and simulations were performed using the MATLAB programming language.

ISS Assembly Tasks

The following results were obtained as part of a NASA SBIR Phase I research effort by Scientific Systems Company (1/95 – 6/95). During this project, SSC applied the existing KPP algorithm to representative tasks of the ISS assembly using the SSRMS arm. Figure 1 illustrates the type of solution found for a PMA module relocation task, while Figure 2 shows the current KPP graphic viewer, illustrating a typical solution for a dual-arm cooperative manipulation of a truss end-module. While this prior SBIR effort established the

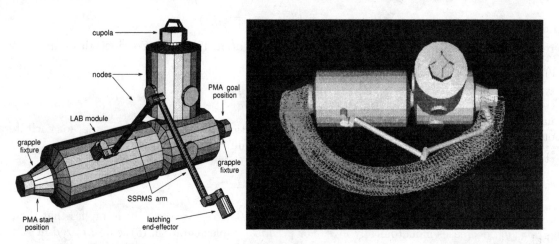

FIGURE 1. PMA Module Relocation Task: (a) Module/Arm Layout, (b) Computed path.

feasibility of using the KPP algorithm for sample ISS tasks, due to limited time, only a few candidate tasks were attempted. It is noteworthy, however, that the ISS tasks attempted represented relatively easy planning problems for the KPP algorithm. All solutions were found within 5~15 iterations of the main KPP loop, with 3~5 iterations needed for local refinement stages. User interaction was demonstrated by using the KPP graphic viewer controls to select desirable configurations for the goal endpoint. Note, however, that the user *was not* required to solve for the joint paths, but simply given the opportunity choose arm configurations – a qualitative task for which the planning engineer generally has a good intuitive feel. The following section

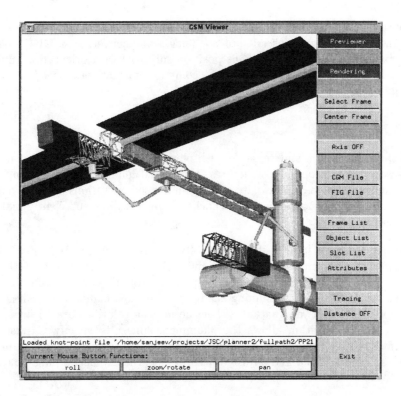

FIGURE 2. KPP Graphic Viewer – Dual-arm Manipulation of Truss End-Module.

briefly describes some diverse robotic planning tasks previously solved using the KPP algorithm.

Other Robotic Planning Case Studies

During the initial development of the KPP algorithm at RPI, the testbed for planning scenarios consisted of dual cooperative nine degree-of-freedom manipulators. An extensive range of robotic planning problems were successfully solved using the KPP algorithm. Key features of the KPP approach include diverse goal task specifications (from end-point only to entire path specification), joint path cyclicity, and robustness to manipulator singularities. Case studies were performed which verified the following desirable aspects for robotic path planning:

Cartesian goal end-point planning: This is perhaps the most common task specification – for instance, in pick-and-place type operations. The KPP incorporates this task as a standard six-DOF coordinate specification for the goal endpoint. Case studies have been performed which involved maneuvering the payload through an opening into a goal position on the other side of an obstacle – a recognized difficulty for planners based on potential field formulations.

Global redundancy resolution along a specified Cartesian path: For a specified trajectory following task (such as in welding or insertion), the manipulator is required to control the end-effector along a pre-specified (position and orientation) trajectory, while resolving kinematic redundancy. The KPP algorithm transparently acommodates this task, providing for joint path solutions which avoid constraints while following the specified trajectory to a desired level of accuracy.

Joint path cyclicity for cyclic Cartesian paths: This has been a difficult problem for kinematically redundant manipulator planning which is acommodated easily by the KPP by incorporating a periodic joint vector constraint (such as the starting arm configuration) into the task. The computed joint path is then cyclic in both joint and task spaces.

Goal task definition variability: The general task and constraint specification of the KPP allows for ad hoc mixtures of the above features – yielding cinsiderable latitude for the planning engineer to specify manipulator tasks and constraints. For instance, case studied have been shown which included free planning segments blended into insertion phases (such as maneuvering an ORU from its pallet location to insertion fixture), and free planning ins position subspace with orientation (pointing) constraints. In all cases, the unmodified KPP was used to compute efficient, practical joint paths satisfying the desired specifications.

ANTICIPATED BENEFITS

The KPP algorithm is a general and versatile path planner, applicable to a wide range of robotic tasks, such as welding/assembly of automobile structures, automated manufacturing systems, material handling in hazardous environments such as nuclear site cleanup, undersea operations and space exploration. Future effort will result in an extensively tested software package, which will be sufficiently mature for commercial testing and acceptance.

For application to ISS manipulator tasks, KPP software can be integrated into NASA's Manipulator Analysis Graphic Interactive Kinematic (MAGIK) simulation environment. The MAGIK program is used by engineers at Johnson Space Center in order to develop plans for robotic assembly of the ISS modules in orbit. At present, however, manipulator joint planning has to be performed manually, which is labor-intensive and error-prone. Additionally, there is no guarantee that the manual solutions are optimal, or even practically implementable. The KPP algorithm allows the planning engineer to specify Cartesian goal end-points, together with desirable configurations and/or via points, and computes joint-space optimal paths to execute the motion. These can be previewed and adjusted according to the engineer's judgment.

Acknowledgment

This research was supported through an SBIR award NAS9–19262 from NASA Johnson Space Center. The support of Mr. Scott Hankins (Automation, Robotics and Simulation Division, JSC) is gratefully acknowledged.

References

Gilbert, E.G., and C.J. Ong (1994) "Robot Path Planning with Penetration Growth Distance," in *Proc. 1994 IEEE Robotics and Automation Conference*, pp 2146–2152, San Diego, CA, May 1994.

Hwang, Y.K., and N. Ahuja (1992) "Gross Motion Planning – A Survey," in *ACM Computing Surveys*, 24(3):219–291, 1992.

Khatib, O. (1986) "Real-time Obstacle Avoidance for Manipulators and Mobile Robots," in *International Journal of Robotics Research*, 5(1):90–98, Spring 1986.

Latombe, J.C. (1991) *Robot Motion Planning*, Kluwer International Series in Engineering and Computer Science: Robotics: Vision, Manipulation and Sensors. Kluwer Academic Publishers, 1991.

Seereeram, S. (1995) "Kinematic Path Planning for Redundant Manipulators Using a Path Space Approach," *NASA SBIR Phase I Final Report*, Scientific Systems Company, Inc., Woburn, MA, 1995.

Seereeram, S., A. Divelbiss and J.T. Wen (1993) "A Global Approach to Kinematic Path Planning for Robots with Holonomic and Non-holonomic Constraints," in *Proceedings of the 1993 IMA Workshop on Robotics*, Minneapolis, MN, January 1993.

Seereeram, S., and J.T. Wen (1995) "A Global Approach to Path Planning for Redundant Manipulators," in *IEEE Transactions on Robotic and Automation*, 11(2), February 1995.

THE GLOBALSTAR SATELLITE CELLULAR COMMUNICATION SYSTEM DESIGN AND STATUS

Fred J. Dietrich
Globalstar L.P.
3200 Zanker Road, San Jose, CA 95164-0670
(408) 473-7188 FAX: (408) 473-7809
e-mail: fred.dietrich@globalstar.com

Abstract

The Globalstar cellular communication satellite system is described, including its use of Code Division Multiple Access (CDMA) as the basic modulation scheme. Use of diversity for signal quality, as well as power control, is described. Complex phased arrays on the satellite are also described.

INTRODUCTION

Globalstar is a satellite-based cellular telephone system that allows users to communicate from anyplace in the world between 70° North and South latitudes. It provides clear communication thanks to Code Division Multiple Access (CDMA) transmission, and avoids outages caused by blockage of signals by using diversity signals from two satellites (Dietrich, 1996).

Globalstar is a partnership of a number of companies. Loral and Qualcomm, Inc. are general partners, and 10 other serve as limited partners, including AirTouch, Alcatel, Alenia, Dacom, Daimler Benz, Elsag Bailey, France Telecom, Hyundai, Space Systems/Loral and Vodafone.

OVERALL DESCRIPTION

The Globalstar system consists of a Walker 48-8-1 constellation, that is, 48 low-orbiting (1400km high) satellites in 8 orbits, inclined 52° with respect to the equator, with six satellites in each orbital plane. They contact users on the 1.6 Gigahertz (GHz) L-band and 2.5 GHz S-band via the hand-held radiotelephone (user terminal, UT) and communicate with the large Gateway ground antennas on the 5 and 7 GHz C-bands. Ground stations (Gateways) use 5.5 meter antennas and connect to the public switched telephone network (PSTN). The user terminals are the same size as a conventional cellular telephone, but with a thicker and longer antenna. Figure 1 shows the elements of the system.

CDMA OPERATION

Code division multiple access (CDMA) is a spread-spectrum technique which was developed for cellular applications and has now been standardized by the Telecommunications Industry Association (TIA) as IS-95 (Monte, 1994).

Globalstar contains a forward link, which consists of an uplink from a Gateway ground station and a down-link to a UT, and a return link, which consists of an uplink from the UT and a downlink to the Gateway.

Globalstar makes extensive use of diversity, where a particular call circuit is put through two or more satellites simultaneously. This allows a mobile user, for example, to drive past a building or row of trees and lose one signal completely, but maintain contact through the second and/or third signals, preventing signal dropouts. A RAKE (multi-channel digital) receiver is used to combine these signals at both ends of the link.

The Globalstar Air Interface (the specification for the link operation) specifies a forward link CDMA waveform that uses a combination of frequency division, pseudo-random code division (Walsh code) and orthogonal signal multiple access techniques.

Frequency division is employed by dividing the available spectrum into nominal 1.23 MHz bandwidth channels. Normally, an MSS Gateway would be implemented in a beam service area with a single radio channel until demand requires additional radio channels. One Walsh circuit has a maximum rate of 4.8 kb/s. The Globalstar Return CDMA Channel also employs PN spreading using a quadrature spreading code of length 2^{15}. Here, however, a

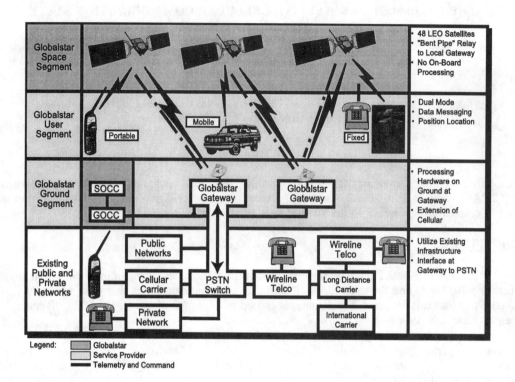

FIGURE 1. The Globalstar System.

fixed code time offset is used. Signals from different handsets (UTs) are distinguished by the use of a very long ($2^{42}-1$) PN sequence whose time offset is determined by the user address. Because every possible time offset is a valid address, an extremely large address space is provided. This also provides a high level of privacy.

To maximize capacity, CDMA requires that the Eb/No received from all UTs is at a similar level as the Gateway. UTs transmitting more power than normal create more interference which reduces the system capacity. To resolve this problem, Globalstar uses dynamic power control on the Return Traffic Channel.

Closed loop power control is also used. For closed loop control, the Gateway compares the signal-to-noise ratio from the UT to a threshold and sends out a command to have it increase or decrease it's transmitted power until the rat ratio corresponds to the threshold in the Gateway. Its time period is typically 50 milliseconds, meaning it does not fix fast fades.

HARDWARE DESCRIPTION

Descriptions of the satellite, ground station and user terminal are described in the following paragraphs.

Satellite Design

The satellite is 3-axis stabilized with the earth- facing panel always parallel to the orbit tangent. A global positioning system (GPS) receiver is used to accurately determine the orbit parameters, and also to supply accurate time and frequency to the satellite systems. Solar panels and a large nickel-hydrogen battery provide power for all phases of the mission. Battery recharge takes place over the oceans, where there is less traffic. The attitude control system uses small (one Newton) thrusters for attitude control. Yaw steering is employed to provide sufficient solar array power during all phases of the mission.

Payload Design

Figure 2 shows a block diagram of the communication subsystem payload, which is a "bent-pipe" transponder. Uplink signals from the Gateway at 5 GHz are downconverted to 2.5 GHz and transmitted to the UT. The UT responds by transmitting at 1.6 GHz, which is upconverted in the satellite to a 7 GHz downlink. Switchable gain control and filtering are provided in each of the 16 channels for circuit control.

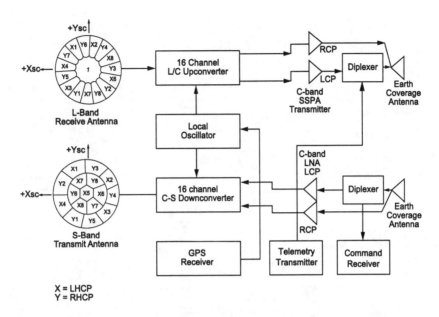

FIGURE 2. Globalstar Satellite Payload.

L- and S Band Antennas

The objective of the L- and S-band array antennas for Globalstar is to form 16 independent, fixed beams each for receive and transmit that cover the visible earth. They provide frequency reuse on the beams to allow reusing the narrow spectrum to a separate transponder in the satellite. The approach taken for implementing these beams is with a separate power divider for each beam, which shapes and steers the beam, and combiners/dividers that share these beam dividers among the 16 channels. The arrays contain active circuits-power amplifiers (PAs) and low noise amplifiers-(LNAs) (Metzen, 1996).

Figure 3 shows a block diagram of the S-band array. It shows single signal paths to two elements. The operation of the entire array can be deduced from this diagram. The transmit array forms 16 simultaneous circularly-polarized beams, as shown in Figure 4a (predicted) and 4b (measured). It can be seen that the desired pattern shape was achieved quite accurately. Figure 5 shows a photograph of this array. The hexagonal shape of the antenna results from the use of an equally-spaced triangle lattice array of 91 radiating elements. The array uses phase only, not amplitude, to form the beams.

The block diagram for the L-band receive array is identical to that for the S-band array except that the amplifier is turned around; that is, it receives a signal from the radiating element and sends it to the power divider network, and it is a low noise amplifier (LNA). Also, there are only 61 instead of 91 elements/amplifiers. The beamshapes are different from those of the S-band array, consisting of a central beam and one ring of 15 outer beams.

C-Band Isoflux Horn Antenna

The C-band antennas are designed to have a pattern that compensates for the space loss variation that occurs on paths from the edge of the earth (10° elevation, maximum path length/loss) to straight down, which is the minimum length/loss. Figure 6 shows measured patterns for both transmit and receive.

User Terminal (UT)

The UT will typically be a hand-held dual-mode unit, although a variety of single-, dual-, and tri-mode units will be available, operating on both the Globalstar system and one or more of several terrestrial cellular systems. In Globalstar operation the UT will transmit an average EIRP of about -10 dBW (maximum -4 dBW), and contains a 3-channel RAKE receiver so that it can receive signals from more than one satellite simultaneously. Globalstar will

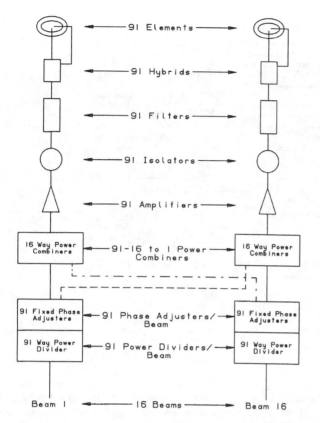

FIGURE 3. Diagram of S-Band Phased Array.

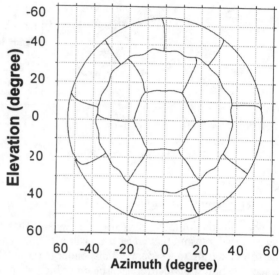

FIGURE 4a. Predicted S-Band Array Pattern.

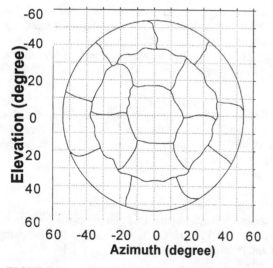

FIGURE 4b. Measured S-Band Array Pattern.

FIGURE 5. Photograph of S-Band Array.

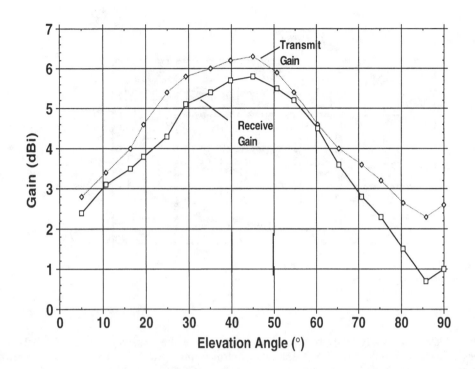

FIGURE 6. Measured Pattern of C-Band Horn Patterns.

FIGURE 7. Photograph of Globalstar User Terminals.

also produce fixed user terminals, which will typically be a solar-powered phone booth in a village. User terminals will be manufactured by Qualcomm, Telital, and Ericsson. A photograph of a UT can be seen in Figure 7.

Gateway Design

A block diagram of the Gateway can be seen in Figure 8. The antenna is approximately 6 meters in diameter. The Gateway contains all the electronics necessary to perform the CDMA communication, including RAKE receivers (multi- digital receivers). It also connects to the PSTN through a switch, and also provides a GSM interface (GSM, or Global System for Mobile Communications, is the European cellular standard).

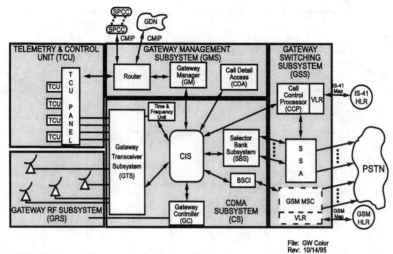

FIGURE 8. Gateway Block Diagram.

Worldwide Network

The Globalstar Data Network (GDN) provides interconnection of the Gateways to the Globalstar Operations Control Center (GOCC), and to the Satellite Operations Control Center (SOCC). This will be a network of links connecting the various sites, including all the Gateways. The GOCC and SOCC are located in the Globalstar facility in San Jose, with an alternate site near Sacramento.

PROGRAM STATUS

Globalstar is on track and on time to deliver service by late 1998. At press time, Globalstar qualification satellite model payloads are undergoing testing at Alcatel, and have passed all of the communication tests imposed on them. Globalstar's four pre-production gateways in Aussaguel, France, Yeo-Ju, Korea, Dubbo, Australia and Clifton, Texas, United States are nearing completion, and contracts have been signed with Qualcomm, Ericsson and Telital to manufacture user terminals. Globalstar's first launch is scheduled for December of this year from Cape Canaveral, Florida, where a Boeing Delta rocket will carry four Globalstar satellites into orbit.

SUMMARY

Globalstar is a technologically superior mobile satellite communication system that is well on its way to becoming reality. It has agreements with service providers in 105 countries.

Acknowledgments

The author would like to acknowledge the contributions of Paul Monte and Phil Metzen to this paper. Their material on CDMA operation and antenna array design/performance have greatly contributed to its content.

References

Dietrich, F. (1996) "Globalstar Update," in *WESCON IC Expo/1996 Conference Record*, Anaheim, CA, 1:200-205.

Metzen, P. (1996) "Satellite Communications Antennas for Globalstar," in *Proceedings of the JINA International Symposium on Antennas*, Nice, France, 1:102-111.

Monte, P. and S. Carter (1994) "The Globalstar Air Interface: Modulation and Access," in *AIAA 15th International Communications Satellite Systems Conference*, San Diego, CA, 1:1614-1621.

TELEPRESENCE BY HIGH-DATA RATE SATELLITE COMMUNICATIONS

Douglas J. Hoder, Michael J. Zernic, Paul G. Mallasch, and Kul Bhasin National Aeronautics and Space Administration Lewis Research Center 21000 Brookpark Road Cleveland, Ohio 44135 (216) 433-3438 and 433-5286	David E. Brooks Sterling Software National Aeronautics and Space Administration Lewis Research Center 21000 Brookpark Road Cleveland, Ohio 44135 (216) 433-5115	David R. Beering American Petroleum Institute 1220 L Street, NW Washington, DC 20005 (312) 856-7323

Abstract

Telepresence allows a person to experience as if he or she is actually present in a different place or time. For remotely located sites, satellite communications are being utilized to develop the medium of telepresence. In this presentation we summarize several experiments conducted using NASA's Advanced Communications Technology Satellite (ACTS) and fiber-optic networks to demonstrate telepresence utilizing data rates ranging from 155 to 622 megabytes per second. In addition, we identify future challenges and directions for telepresence applications over advanced high data-rate hybrid (satellite and terrestrial) networks.

SAMPLE ACTS EXPERIMENTS

ACTS Satellite Hybrid Demonstration Using A Virtual Reality Experiment

Interactive, multi-server virtual reality applications have been used to let people experience the robustness of a broadband digital satellite (ACTS) with fiber-optic networks and enable people to literally "test drive" the technology. As people maneuvered through Terravision (virtual terrain), they could not tell that the data sets that made up the virtual images were being retrieved from various multiple geographically dispersed locations.

Keck Observatory Remote Science

Utilize advanced communications for telescience in remote/hostile conditions to prototype data acquisition, visualization, and control task methods. Caltech astronomers remotely manipulated the Keck Telescope in Hawaii from Los Angeles and computer enhanced images yielding potential impact in future space missions and terrestrial uses.

Supercomputer Global Climate Model

Perform an atmosphere-oceans model simulation in a coupled parallel process between two geographically separated Cray supercomputers. Demonstrated interactive manipulation and collaboration between scientists as if they were co-located.

Redefining Oil Exploration

Use of broadband digital satellites like ACTS allows the process of seismic data collection to become interactive, allowing for dispersed resources to process high volumes of data and personnel to interactively collaborate; thereby actually steering the seismic data acquisition vessel at sea during the acquisition of terabytes of information, dramatically reducing the process time and enabling quicker business decisions.

References

Hoder, Doug, ACTS High Data-Rate Experiments Program, http://www.cgrg.ohio-state.edu/other/actsgsn/expprog.html, visited October 24, 1997.

Zernic, M.J. (1996) "Another ACTS First Draws National Attention", NASA Lewis Research Center, November 1996.

NASA's WIRELESS AUGMENTED REALITY PROTOTYPE (WARP)

Martin Agan, LeeAnn Voisinet, and Ann Devereaux
Jet Propulsion Laboratory
4800 Oak Grove Drive
Pasadena, CA 91109
818-354-3426

Abstract

The objective of Wireless Augmented Reality Prototype (WARP) effort is to develop and integrate advanced technologies for real-time personal display of information relevant to the health and safety of space station/shuttle personnel. The WARP effort will develop and demonstrate technologies that will ultimately be incorporated into operational Space Station systems and that have potential earth applications such as aircraft pilot alertness monitoring and in various medical and consumer environments where augmented reality is required. To this end a two phase effort will be undertaken to rapidly develop a prototype (Phase I) and an advanced prototype (Phase II) to demonstrate the following key technology features that could be applied to astronaut internal vehicle activity (IVA) and potentially external vehicle activity (EVA) as well: 1) mobile visualization, and 2) distributed information system access. Specifically, Phase I will integrate a low power, miniature wireless communication link and a commercial biosensor with a head mounted display. The Phase I design will emphasize the development of a relatively small, lightweight, and unobtrusive body worn prototype system. Phase II will put increased effort on miniaturization, power consumption reduction, increased throughput, higher resolution, and "wire removal" of the subsystems developed in Phase I.

INTRODUCTION

The Wireless Augmented Reality Prototype was conceived in late 1996 as part of NASA's Office of Life and Microgravity Sciences and Applications (OLMSA) ongoing effort to develop technology capabilities to support humanities quest to use and explore space. Initiated in January 1997, WARP is a means to leverage recent advances in communication, display, imaging sensor, biosensor, voice recognition, and microelectronic technologies to develop a prototype system capable of real-time personal display of information relevant to the health and safety of Space Station personnel. The original concept for the system is depicted in Figure 1.

FIGURE 1. Original WARP Concept.

WARP will allow an unteathered astronaut floating in the Space Station, and wearing a lightweight head mounted display (HMD) and outfitted with a suite of miniature biosensors to communicate through a two way wireless communications link to the Space Station communication infrastructure. On the miniature head mounted display (HMD) the astronaut will be able to view his or the other astronauts' biosensor data (e.g., heart rate or oxygen saturation), imagery (e.g., wiring diagrams), or text (e.g., instructions for how to perform CPR on another astronaut) transmitted from the Space Station or Mission Control on earth. A miniature camera worn on the astronaut will allow the viewing of his environment back on earth (e.g., a patient being administered to), and real-time video teleconferencing. Control of these various capabilities will be via voice commands and speech recognition software thus allowing hands free operation.

The Jet Propulsion Laboratory (JPL) is working in conjunction with its partners at the University of California at Los Angeles (UCLA), and McDonnell Douglas Aerospace in the development of WARP. An initial prototype including a subset of the ultimate capabilities will be completed within one year to demonstrate the concept and technologies. This prototype is to be followed by a refined prototype with additional capabilities, reduced size and power requirements (resulting in smaller batteries), and incorporating emerging technologies (e.g., DARPA digital wireless camera, miniature color display technology, battery advances, and solar cell technology).

SYSTEMS ARCHITECTURE

In defining the system architecture, several approaches were investigated that could lead to a Phase I demonstration system sooner than the planned twelve month effort, while at the same time provide sufficient capability for a representative demonstration, and lead naturally into an advanced Phase II system. The requirements for this demonstration system are listed in Table 1 and a conceptual drawing is shown in Figure 2. The initial WARP system architecture will provide for transmission of compressed video and audio from the base station to the astronaut. At the same time it will allow compressed audio and biosensor data to be transmitted from the astronaut to the base station. The transmission of video will allow the head mounted display on the astronaut to present the required biosensor plots as well as other basic PC functions operating at the base station, all via voice control. Essentially the astronaut will be controlling a PC at the base station remotely via the voice link with the display remoted to his local head mounted display. This allows the astronaut to make use of the computational, storage, and connectivity capabilities available at the base station.

TABLE 1. Requirements for Demonstration System.

High Level Requirements
- build ASAP in order to demonstrate technologies and their capabilities
- miniature head mounted display (HMD) on IVA astronaut that resembles the original concept of Figure 1
- commercial biosensor data transmitted from IVA to base station
- voice link to/from IVA astronaut
- voice control of HMD display
- HMD must display biodata, imagery (graphics & text)
- must operate reliably in space station "metal box" environment
- must operate reliably while IVA astronaut is moving about space station
- size, weight, and power (battery powered) must be minimized on the IVA astronaut
Derived Requirements
- transmit compressed video of base station PC VGA display (bio-data/graphics/text) to IVA astronaut HMD
- communications link maximum range = 10 meters
- camera on IVA will be incorporated in Phase II

In an effort to build the WARP demonstration system rapidly, commercial-off-the-shelf (COTS) hardware and software are being utilized to the extent possible. The base station utilizes an industrial passive backplane type PC equipped with commercial boards and voice recognition software. The communication link requires some development for the demonstration system, but to the extent possible commercial based chip sets are being utilized in the design. The video/audio compression is based on a standards based compression product. The biosensor is being selected in collaboration with the NASA Ames Advanced Sensor Systems group and will be commercially available device. The head mounted display is a new design by the industry leading manufacturer of small lightweight HMD's which concentrates on wearability. Integration of all these technologies into a fully functional reliable system is one of the key challenges of WARP.

FIGURE 2. Demonstration System Concept.

Based on the basic system architecture a series of demonstrations have been devised that will allow the rapid validation of the system concept and an incremental building up of capabilities and improved functionality while experimenting and adding/eliminating options in order to optimize the system configuration. These demonstration systems are briefly discussed below.

Demo A

The first demonstration consisted of entirely commercial off-the-shelf (COTS) belt-worn PC/HMD system running a software audio/video codec on the CPU. The belt-worn PC system has a wireless LAN connection to the base station PC. This 100% commercially available system allowed us to evaluate a partial WARP capability implemented in a miniature commercial PC and at the same time provide a limited operational demonstration in the very short term. This demonstration showed the limitations of software codecs (low resolution, not motion friendly, require full processor), and the bulkiness of current HMD's.

Demo B

The second demonstration was intended to be more representative of the ultimate WARP system in that it integrated a hardware audio/video codec together with a breadboard version of the primary candidate for the miniature communication system. Additionally the base station computer was configured as it will be in the ultimate demonstration with a voice recognition operating system that was controlled via the wireless link. This demonstration, while quite large (e.g., rack mounted codecs) evaluated the feasibility of the video codec based approach. The throughput performance and reliability of the communication system was tested (and determined that improvements are needed), and voice recognition software was shown to operate over a compressed audio link.

Demo C

Currently in the process of being fully specified, this demonstration system will have a size on the order of a sub-laptop computer and have most of the desired WARP functionality as shown in the Figure 3 block diagram, but will not be fully customized to minimize power and size. Additionally, Demo C will have the capably to transmit compressed video both to and from the IVA astronaut.

Demo D

This demonstration is being developed concurrently with Demo C, and will result in a miniaturized belt worn version of the system shown in Figure 3 and will have improved communication system performance and improved video image quality. Demo D is the early initiation of the Phase II system.

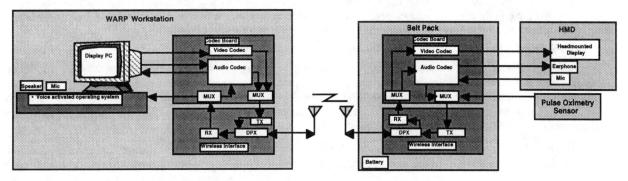

FIGURE 3. Demonstration System Block Diagram.

COMMUNICATION SYSTEM

One of the greatest challenges in developing the WARP is designing a wireless system for indoor communications that will perform reliably in the harsh environment of the Space Station. The first priority was to verify that in fact it is possible to communicate inside a "metal box" that resembles the space station environment. Through extensive testing of the communications system in a metal trailer "space station simulator" such a connection has been shown to be possible and the results of these tests have been incorporated into the design of the demonstration system. The Space Station radio frequency communication channel is equivalent to operating within a large metal box, which has the effect of the transmitted signal reflecting off of the metal surfaces and arriving at the receiver at different times, thus creating destructive self interference. Methods of counteracting this interference that are realizable in a small, low power implementation are being investigated.

The communication system design that must operate in the very high multipath space station environment must necessarily utilize a non-coherent demodulation scheme because of the difficulty of tracking the phase in this environment. At the present time Differential Phase Shifted Keying (DPSK) and Frequency Shift Keying (FSK) have been tested and are candidates for the modulation. Spread Spectrum, antenna diversity techniques, and retransmission schemes are also being investigated as means of improving the reliability of the communications link. Through a combination of these techniques it is expected that channel throughputs greater than 384 kbps can be achieved. A variety of RF frequencies have been tested, but the determining factor in frequency selection will be the availability of a relatively large portion of spectrum in the space station that will allow the transmission of compressed video. It is believed the system stands a good chance of avoiding interference and of being certified in the 2.5 to 2.7 GHz band.

The ability of the WARP project team to build a miniature communication system in a relatively short period of time will necessarily leverage heavily off of the wireless LAN technology that is currently available. The greatest potential for innovation is in the arena of internal space station high bandwidth wireless communications that would allow higher quality video to be transmitted to and from the astronaut. The demonstration system will necessarily utilize commercial components and should be able of achieving data rates approaching 1 Mbps in the space station environment. With an eye toward the future a collaboration has been initiated with UCLA under the WARP effort in which communication system designs and implementations that can reliably achieve 10+ Mbps data rates in the space station are being investigated.

Preliminary Communication System Tests

Extensive, one-way, testing of breadboard 280 MHz and 2.4 GHz radio systems has been done both in the laboratory environment as well as a simulated Space Station module, consisting of a 40 foot fully enclosed metal trailer with various metallic obstructions. The basis for this breadboard system was the Harris Prism wireless LAN chip-set, using simple monopole antennas optimized for 280 MHz and 2.4 GHz, respectively. This chip set allows communication at either differential BPSK or QPSK and implements a pseudo-noise (PN) code spread-spectrum radio with data rates up to 2 Mbps. The Prism system does not provide explicit multipath correction; however, the chip set does have the capability to support antenna diversity, and can support receive-acknowledge protocols to retransmit faded data.

Another chip set, by Broadcom, was used in initial testing because of the potential it offered for achieving multiple megabit/sec data rates. Unlike the Prism, this system did allow for adaptive equalization within the digital functions of the baseband processor. The system did not perform well in the space station simulation, however. It is likely that the equalization technique used, developed specifically for performance in the fairly benign and steady-state conditions of a cable modem application (e.g., not Doppler frequency offset), could not adjust rapidly enough to the dynamic conditions of the simulator with moving human occupants.

The frequency-dependent aspects of signal scattering in the simulator trailer could be easily observed by looking at the received signal on a spectrum analyzer. The received spread signal exhibited varying numbers of 5-10 MHz bands which would dynamically peak and decay versus the overall pattern. The pattern of bands would change drastically from even small motions of people in the simulator, illustrating the large impact of having mobile RF absorbers (i.e., people) in the otherwise static environment.

The simulator test setup used a bit error rate tester (BERT) to produce simulated data at the transmit end and to display channel characteristics at the receive end, including individual bit errors detected. During a test, the receiver portion of the set-up (representing the astronaut) would be moved around in the simulator. It was possible to spatially locate areas where maximum interference was occurring by finding locations that produced the greatest amount of errors.

Various configurations were tested, including: low (72 kbps) through high (1 Mbps) rate data, different transmit power levels, QPSK vs BPSK, and different spatial distributions of obstacles and human observers. In all configurations it was easily possible to arrange the two ends of the test set up such that error-free performance could be obtained running over the course of hours. The distinguishing performance criteria between configurations were strictly the number, size, and depth of peak interference locations which could be found in three dimension space. With the best configurations, finding locations to produce any errors at all could be extremely difficult, and not repeatable. In poorly performing configurations, it was easy to find areas which could continuously produce errors.

Clearly, BER performance of the tests were hard to quantify. Depending on location of the transmitter and receiver, the test set up could perform at better than 10^{-8} or worse than 10^{-4}. The BER experienced in a real-world situation would obviously be some average of the error rates achievable over all the spatial locations traversed by the astronaut during a WARP session. Reducing the number and physical size of the interference regions by choice of radio configuration would make it more likely that the astronaut belt-pack receiver would only cross a fading area momentarily before small movements of the astronaut's body would move it back into a clear zone.

The main result of the testing was to show that by far the most significant factor in reducing the number and size of interference locations was data rate. By example, one test compared the benefit of reducing the power amplifier output by 27 dB (@280 MHz) with the performance improvement obtained by reducing the data rate from 575 Kbps to 287 Kbps, an E_b/N_o difference of about 3 dB. Reducing the transmitter power output had little impact on the system performance. Reducing the data rate by half, however, produced an obvious improvement, going from a reasonably stable system with some interference points to a link in which almost no errors could be induced. The primary reason for such behavior is that when the data rate is reduced the bit period is lengthened and as a result the RMS delay spread of the received multipath signal becomes a smaller fraction of the bit period resulting in less inter-symbol interference.

HEAD MOUNTED DISPLAY

The design of the head mounted display (HMD) began with the artist's concept image of a display and functional requirements. One of these requirements is that it be wearable for the performance of "normal" duties and for extended periods of time. For the headset, it was determined that COTS designs could not provide the wearability and functionality that WARP requires. After an extensive industry survey a manufacturer was chosen that could provide rapid prototyping and in-house design and fabrication; the design leverages heavily on the optics and circuitry of some of their former products produced for other government agencies and industry.

Design efforts focused on producing the most lightweight and wearable HMD possible and resulted in the HMD shown in Figure 4. It will provide its wearer with video display, stereo audio, and a noise canceling microphone. Accommodation has also been made to allow for the addition of a miniature video camera for future integration as communications bandwidth is available. The headset will weigh less than 3.5 ounces when completed, this is approximately 10 ounces less than similarly equipped headsets available that are "state of the industry." The Phase I headset will provide NTSC video through a single bounce optical prism.

FIGURE 4. Mockup of Phase I Demo Head Mounted Display.

The HMD can be worn over prescription eyewear and with bi-focal prescription eyewear. The optic can be "flipped-up" out of the line of sight and the video display will go into "sleep" mode to conserve power until it is brought back down into the field of view; the audio and bio-sensor systems continue operating during "sleep" mode. This headset will not block peripheral vision which is an advantage for a device that is to be worn consistently. The headset also can be adjusted and "formed" to the individual's head shape as needed. Furthermore, the optic and display circuitry can be positioned for either right or left eye dominance. The Phase II HMD is planned to be VGA rather than NTSC in an effort to improve image quality.

AUDIO/VIDEO CODEC

The audio/video codec will be a single board, standards based codec that will accept analog audio and analog video (i.e., NTSC or VGA) and digitize and compress these inputs and multiplex the compressed data with an external data stream. The standards are: ITU H.261 based video, ITU G.722 based audio, and ITU H.221 based multiplexing. The base station A/V codec takes the base station PC display as the video source. The IVA codec outputs the received and decompressed base station display video on the HMD. The codec interface to the communications system is planned to operate at 384 kbps; of this 384 kbps the audio will utilize 32 kbps and the digital bio-sensor data will utilize 8 kbps with the remainder allocated to compressed video. The quality of the video at these data rates is equivalent to video teleconferencing quality video. The codec will be unique to the WARP effort in that it will be an autonomous stand alone board (i.e., no PC required for operation) and be reduced to a size commensurate with the belt worn system.

BIOSENSOR

The serial type data interface on WARP provides reserved bandwidth for bio-sensor data. The WARP Phase I system will incorporate a commercially available Pulse Oximeter sensor. This sensor will allow the measurement of pulse rate and oxygen saturation of the blood. Data from the sensor will be returned to the base station, recorded and can be displayed for the headset user or at the base station for observation by other astronauts. Phase II is planned to include an "on body" wireless human performance and fatigue monitoring system that permits candidate biosensors to communicate wirelessly with a belt-mounted transceiver/control module.

Acknowledgment

The research described in this paper was carried out by the Jet Propulsion Laboratory, California Institute of Technology, under contract with the National Aeronautics and Space Administration.

HIGH-RATE OPTICAL COMMUNICATIONS LINKS FOR VIRTUAL PRESENCE IN SPACE

James Lesh, Keith Wilson, John Sandusky, Muthu Jeganathan, Hamid Hemmati,
Steve Monacos, Norm Page, and Abi Biswas
Jet Propulsion Laboratory
California Institute of Technology
Pasadena. CA 91109
(818) 354-2766

Abstract

As man continues the exploration of space, either from platforms in Earth-orbit or from missions to the bodies of our solar system, there will be an on-going need to keep the public engaged in the process so that the general populace can share in the excitement of discovery and the inquisitive (professional or student) can participate in the data mining. To do this will require the development of high-data-rate communications links for data return and dissemination. However, the cost constraints on future missions will require that these high-rate links be smaller and lighter-weight than even the lower-capacity systems that have been used in the past. NASA has been developing free-space optical communications technology for just such missions. This paper will discuss the technology developments and system demonstrations that have been accomplished to date, and will describe preparations for the space demonstrations that will validate the performance of this new technology.

INTRODUCTION

Past space missions have typically been limited in their data collection abilities by the available sensor technologies. As greater data collection needs were met, the sizes of the spacecraft continued to expand. This was particularly true for the communications systems required to transfer that data to the final user. Today, the technologies for sensors can enable enormously large data collection volumes. However, the economics of space flight have forced significant reductions in the sizes of the platforms on which such sensors must fly. In today's environment, there is a need for much more capability to support those sensors, while, at the same time, requiring less impact (in terms of mass, size, power) on the host spacecraft.

To satisfy this "more for less" requirement, NASA has been developing optical (laser) communications technology. This technology has the ability to enable enormous gains in the available link capacity with systems that are only a fraction of the sizes of the conventional rf systems. This enhancement comes from the narrow "pencil" beams associated with laser beam propagation. This gain does come, however, at a cost of requiring a more precise control of the transmitted optical beam. Such systems in the past were both bulky and required substantial amounts of electrical power. Recently (Chen 1994 and US Patent 1997) a very simple implementation of the beam control system was developed at JPL. This system, called the Optical Communications Demonstrator or OCD has become the basis of a number of planned flight demonstrations of laser communications technology, and for the development of much more capable systems for extending this technology to the far reaches of the solar system.

THE OPTICAL COMMUNICATIONS DEMONSTRATOR

The primary reason for the size, complexity and power consumption of past beam control systems was the number of detectors and steering elements required to track a beacon signal from the target receiver and to point the data-modulated beam back in that direction. The OCD architecture is based on a simplified structure for that process utilizing only one detector array and one steering mirror (past designs used as many as four of each). Figure 1 shows the concept for this system.

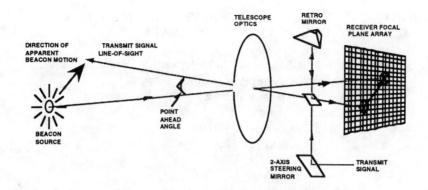

FIGURE 1. Concept Diagram for OCD Beam Tracking System.

The beacon signal is shown in the lower left of the figure. This signal is focused by the telescope onto a single focal plane detector array. The position of that spot identifies the location of the intended receiver. The laser signal to be transmitted is brought in at the lower right of the figure and is deflected off a single 2-axis steering mirror for precision beam pointing. After leaving the steering mirror, the beam reflects off a dichroic element and passes out of the telescope at some angle relative to the beacon-signal axis. To compensate for any cross-velocity of the transmitting system relative to the receiving station, this angle needs to be a specific value called the point-ahead angle. To ensure that the beam is going in the proper direction, a small amount of the transmitted beam is allowed to pass through the dichroic element where it encounters a retro-reflector. The returned beam from that reflector is deflected back toward the detector array and provides a registration of the location of the outgoing beam on that detector. By comparing the offset vector between the locations of the beacon and transmit spots on the detector array a reference vector calculated beforehand to provide the desired point-ahead angle, an error signal can be generated to control the direction of the steering mirror to the proper direction.

The architecture described above has been developed and implemented in an OCD engineering model (see Figure 2).

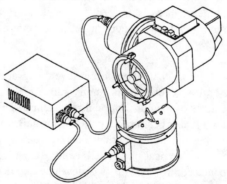

FIGURE 2. OCD Hardware Implementation.

This system was developed as an engineering model for both high-data-rate Earth-orbital flight terminals, and for lower-data-rate (and longer distance) planetary missions. The telescope aperture size is 10 cm in diameter and the total system mass (including coarse-pointing gimbal) is under 15 kg. Although developed as a "laboratory-qualified" engineering model, design choices were made that would not knowingly preclude the flight of the unit in space for demonstration purposes. "High-data-rates" in this context means rates in the 100's of Mbps to multiple Gbps from Earth-orbit to ground. "Lower-data-rates" means below 10's of Mbps.

FUTURE SYSTEM FLIGHT TERMINALS

As stated above, the OCD architecture is being used as the basis for a number of future terminal designs. One such program is the development of a flight-qualified terminal for outer-planet missions under the NASA X2000 Program (Hemmati 1998). This system is currently under development and will be delivered to the JPL Flight System Testbed (FST) in early 2001. A diagram of the system is shown in Figure 3.

FIGURE 3. Diagram of the X2000 Flight Terminal.

The terminal uses a 30 cm diameter telescope and has built-in redundancy in the transmit and receive beam control functions. It will also provide several additional functions to the future flight missions. In addition to the usual uplink and downlink communications, the system will also provide for precise narrow-field science imaging, for 2-way optical ranging, and for reception of possible laser altimeter return signals.

The X2000 terminal will have application to missions traveling to such places as Mars Europa, and Pluto. Additionally, as time goes on, further refinements and technology developments will allow the capabilities of such systems to continue to rise. Figure 4 depicts a roadmap projection for how this capability is expected to expand with time for the Mars, Jupiter, and Neptune programs (Deutsch 1996).

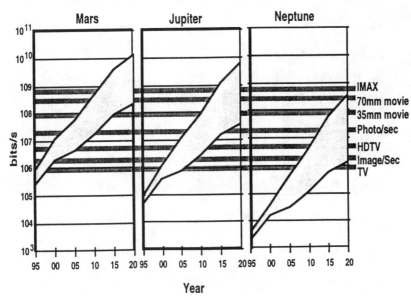

FIGURE 4. Roadmap for Future Optical Communications Capabilities.

The bands represent the bounds of capabilities resulting from an aggressive, and a conservative, investment in the technology. To the right of the figure are indicators of the kinds of service provided to the missions based on reasonable assumptions on coding and data compression. Clearly, IMAX-quality video could be returned from Mars by the 2020 time period.

NEAR-TERM DEMONSTRATIONS

The importance of the OCD terminal architecture cannot be underestimated. It is the basic starting point for both near-earth terminal designs, as well as those for deep-space. It is therefore important that an early demonstration of the OCD principles and performance predictions be carried out in space. Two such demonstrations are under study and preliminary design; one on the Space Shuttle and the other on the International Space Station.

Space Shuttle Demonstration

The first demonstration is the Free-space Optical Communications Assessment Link (FOCAL). It would use the OCD engineering model as the flight terminal which would fly on the Space Shuttle. As the Shuttle flies over an appropriately-equipped ground telescope, a beacon laser signal would be transmitted from the ground to the Shuttle. The flight terminal would use that signal as a pointing beacon against which to direct its downlink beam. The downlink data rate could be as high as several Gbps and still provide 10's of dB of demonstration link margin. While such a link margin is excessive for operational systems, it is felt prudent to keep a large margin for the first several space link demonstrations to cover any uncertainties in the atmospheric beam propagation. Figure 5 depicts the FOCAL demonstration. The FOCAL flight could take place in early-mid 2000.

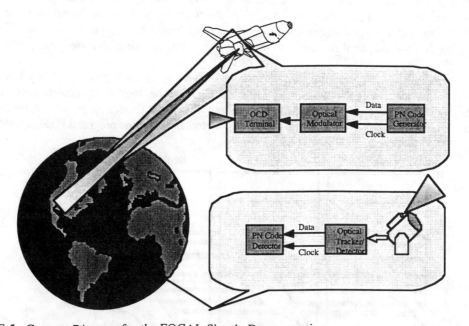

FIGURE 5. Concept Diagram for the FOCAL Shuttle Demonstration.

Space Station Flight Demonstration and Facility

The Space Shuttle provides a convenient opportunity to test out equipment in space for a modest cost. However, the durations of most Shuttle flights do not permit extensive testing of the equipment, especially if such tests are restricted to overflights of a ground station. For more in-depth testing, a longer presence on orbit is needed. One such opportunity was recently realized by an award from the International Space Station (ISS) experiments program office. The ISS demonstration is depicted in Figure 6.

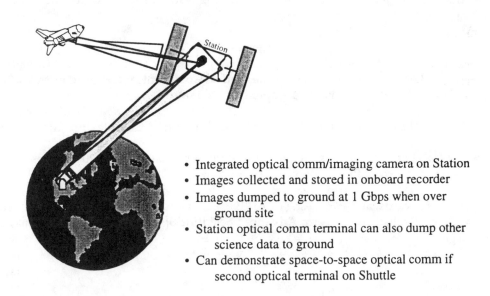

- Integrated optical comm/imaging camera on Station
- Images collected and stored in onboard recorder
- Images dumped to ground at 1 Gbps when over ground site
- Station optical comm terminal can also dump other science data to ground
- Can demonstrate space-to-space optical comm if second optical terminal on Shuttle

FIGURE 6. Space Station Demonstration (With Future Extension to Space-Space Demo).

In this program, the OCD engineering model will again be flown, this time on the ISS. Like the FOCAL demonstration, the flight terminal will utilize a beacon laser signal from the ground for pointing and tracking. Because the terminal will be on orbit for an extended period of time, a more thorough set of experimental measurements can be made. Additionally, the longer time on orbit will also allow reliability data to be collected on the flight terminal.

After the experimental testing has been completed, the flight terminal will be left in place on the station. At this point, it will become a station Facility Instrument and will be made available for use by other station experimenters. Flight demonstrations of instrument (like high-resolution multispectral imagers or synthetic aperture radars) that produce larger amounts of data than can be handled by the normal ISS communications infrastructure will be able to use the optical communications link to download their data.

A follow-on demonstration is also being considered. Given that there is an operational OCD terminal on the Station, it would only be necessary to mount a second unit on the Space Shuttle in order to conduct a space-space optical communications demonstration. At this point, the space-space demonstration is only at the conceptual discussion stage.

CONCLUSIONS

The technology and subsequent demonstrations described above will have a major impact on the architecture and the performance capabilities of future space communications networks. The results of these programs will be useful to both government and commercial space communications needs. The bandwidths afforded by this technology will be important enablers for man's virtual presence in space in the next millennium.

Acknowledgments

The research described in this paper was carried out by the Jet Propulsion Laboratory, California Institute of Technology under contract with the National Aeronautics and Space Administration.

References

Chen, C-C. and J. R. Lesh (1994) "Overview of the Optical Communications Demonstrator," in Proc. of SPIE OE-LASE 94, January 1994, Paper No. 2123-09.

U.S. Patent 5,517,016 "Lasercom System Architecture with Reduced Complexity," May 14, 1996.

H. Hemmati and J. R. Lesh (1998) "Laser-Communication Terminal for the X2000 Series of Planetary Missions," in Proc. SPIE Photonics West 98, San Jose, January 1998.

Deutsch, L. J., C. D. Edwards, and J. R. Lesh (1996) "Extreme Deep Space Communications," 1st IAA Symposium on Realistic Near-Term Advanced Scientific Space Missions, Turin, Italy, June 25- 27, 1996 pages 279-288.

KEY ISSUES IN CONSTELLATION DESIGN OPTIMIZATION FOR NGSO SATELLITE SYSTEMS

Arthur W. Wang
Regulatory Affairs and Spectrum Management, Hughes Communications, Inc.
1500 Hughes Way, Long Beach, CA 90810
Tel:310-525-5034, Fax:310-525-5031, awwang@ccgate.hac.com

Abstract

This paper presents various constellation design criteria for satellite systems in non-geostationary orbits (NGSO). Key design parameters, constraints, and tradeoffs are discussed for two classes of orbits: circular and non-circular. Circular orbits, such as the low earth orbits (LEO), the medium earth orbits (MEO), and the highly inclined geosynchronous orbits (IGSO), have equal coverage period for both north and south hemispheres while non-circular orbits such as the various type of elliptical orbits provide more focused coverage period at certain specific geographic locations. Different services require various constraints including delay, power economics, coverage region, frequency sharing, total capacity, satellite and launch-vehicle numbers. Detailed discussion of the relationship between these constraints and constellations are provided. A comparison between a proposed benchmark MEO system with other proposed broadband NGSO satellite systems is presented to demonstrate the importance of constellation design to enhance frequency-sharing capability. A potential "satellite highway" accommodating families of elliptical geosynchronous satellites is also presented. This is a novel approach to regulatory NGSO constellations which will facilitate sharing valuable resources of spectrum and useful spatial areas.

INTRODUCTION

Currently, the geostationary earth orbit (GSO) belt is congested or fully occupied in certain frequency bands. New NGSO constellations have attracted a great deal of research interest and the results indicate a higher degree feasibility than in the past based on: (1) the explosive demand of multimedia communication for Internet/Intranet traffic; (2) the technical solution for radio/optical inter-satellite links which further strengthens the capability of global satellite network; (3) the progress of low altitude launch vehicles which reduce the launching costs by providing multiple payloads per launch; (4) the use of phase array techniques to track a moving satellite from a user terminal; and (5) the use of active radiated array antenna to steer a narrow beam to anywhere within the satellite field of coverage.

Previous constellation designs (Christopher 1997, Lo 1997, and Rider 1986) usually focused on optimizing constellation parameters to minimize the number of required satellites for system cost reduction. Since the spectrum is becoming a resource which is as limited as the geostationary orbit slots, the focus of this paper is to provide information to facilitate spectrum sharing by constellation design in order to optimize the resource (spectrum or space) usage efficiency. The figure of merit is measured by the how many satellites/systems could co-exist and provide service simultaneously sharing the same frequency band. This constraint for constellation design is more important than before because the radio bandwidth demand for the wireless communication service is larger than the supply of available radio frequencies.

In order to optimize the spectrum usage by proper selection of satellite constellation, we first explore various NGSO constellation design criteria to meet different service requirements. Detailed discussion are presented in section 2 (constellation design constraints). Then a proposed Hughes MEO system designed to facilitate frequency sharing is demonstrated. The results of satellite coverage statistics for future frequency sharing and the comparison of this system with several proposed LEO, MEO, or HEO systems are presented in section 3 (Hughes NGSO satellite system). We also discuss a novel approach to regulate NGSO constellation design. By setting significant parameters, a standardized "satellite highway" is proposed to accommodate several compatible NGSO constellations for spectrum sharing. This concept is proved by demonstrating an example using inclined, highly elliptical geosynchronous orbits (EGSO's). Detailed discussions are presented in section 4 (geosynchronous orbit design constraints).

CONSTELLATION DESIGN CONSTRAINTS

Recently proposed military and commercial global broadband satellite networks using inter-connected GSO satellites or NGSO satellites have brought the worldwide attention. The selection of constellation of each system is made to address the constraints of the service it provides. In this section, we itemize these constraints and services for discussion.

Constellation with propagation delay minimization oriented – For voice traffic with echo-free requirement, most of the designers tend to achieve delay-mitigation by the choice of LEO constellation, since the propagation delays from a GSO and a MEO satellites are about 20~50 and 10~30 times more than from a LEO satellite to an earth terminal. As shown in FIGURE 1 this advantage of LEO systems is technique independent and it explains why so many voice oriented systems have chosen LEO constellations. For examples, GlobalStar^{TM} and Iriduim^{TM} are currently under construction. However, the echo-free constraint does not exclude the choice of a GSO or a MEO alternative because the noticeable echo, voice delay longer than 50 ms, could happen and be the same annoying for systems in any constellation considering delays from other sources such as buffers, processors, and network routing. The propagation delay should not be the single reason to explain echo interferences. The choice of low earth orbit (LEO) constellation is mainly due to the power constraint rather than the propagation delay. Other delay-critical applications like interactive gaming, real-time video conferencing, distance learning, and remote medical operation have the same time-driven constraint for constellation selection. So far, there is no protocol, service, or tectonics which survives only in LEO/MEO environment and excludes the choice of GSO. Conversely, more techniques fail in NGSO environment due to the unavoidable Doppler effect or pointing/tracking errors. Therefore, GSO is still the most reliable constellation for most services.

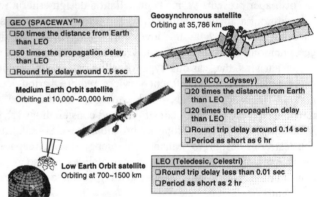

FIGURE 1. LEO, MEO, GEO General Comparison.

Constellation with regional coverage oriented – Due to the difference between the earth self-rotation period (one sidereal day=86,164.1 sec) and the satellite period, the ground footprint of a satellite may not stay in the same area and can migrate through all longitudes. A geosynchronous satellite means the satellite has same period as the sidereal day and thus has footprints either as a single point (geostationary) or a fixed pattern around the center longitude (inclined). Among these type of constellations, the geostationary orbit satellite with 0° inclination angle and time-invariant altitude provides the simplest system design due to the fixed ground coverage and, thus, has been prevalently used. Geosynchronous satellites with the same period as GSO satellite can form various shapes of ground tracks from a "figure 8", a "Shamus", a "Cusp" to a "Pear (or egg)" as shown in FIGURE 2. No matter circular (geostationary) or elliptical, the geosynchronous satellites provide regional coverage only to areas near the center longitudes. These constellations would continuously acquire the most interest from both the regulatory and business points of view.

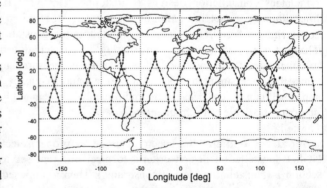

FIGURE 2. Ground Tracks of Geosynchronous Satellites

Constellation with satellite diversity criterion oriented – Since the current GSO belt can not provide enough spatial separation to accommodate all proposed satellite systems at the same band and the capability can no longer meet the demand of global/regional communication, several LEO/MEO systems have been proposed to compete with traditional GSO systems at Ku, Ka, or higher bands to provide global multimedia service. An NGSO communication link would create/receive harmful interference to/from a co-frequency GSO link when the in-line situation of these two links occurs. Various interference mitigation criteria including satellite diversity, site diversity, polar pointing, and high gain antenna have been proposed and studied (ITU WG4A 1996). Among them, the satellite diversity technique -- switching from the interfering NGSO satellite to another -- seems to be the most promising solution because frequent switching is a built-in feature of a NGSO system to perform satellite hand-over. Thus, satellite diversity would not cause extra hardware/software costs.

As shown in FIGURE 3, one typical satellite diversity methodology is the GSO Belt Avoidance. The NGSO satellite terminates service when it flies into the GSO belt within a specified range (0~±θ degrees latitude). The success of this methodology and other similar techniques is heavily affected by the constellation design. Unfortunately most of the currently proposed NGSO systems did not accommodate the spectrum sharing criterion in their constellation design which reduces the opportunity to share spectrum use. In order to provide viable links for all time by applying satellite diversity, the constellation needs to support high dual or more satellites visibility at most latitudes. It is required to maintain an continuous link to another visible satellite while the first satellite creates/suffers interference. The high-dual-satellite-visible constraint could be achieved by either or a combination of increasing satellite number or lowering the elevation angle threshold. The frequency sharing constraint should be imposed to future NGSO constellation designs in order to utilize the valuable spectrum more efficiently. In following sections, we further explore the relationship between constellation design and frequency sharing by introducing spectrum sharing oriented constellation designs. Two examples, a proposed Hughes MEO system and the IGSO satellite highway are presented in section 3 and section 4, respectively.

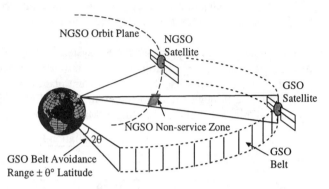

FIGURE 3. Satellite Diversity by Avoiding GSO Belt.

HUGHES NGSO SYSTEM

In this section we present an MEO system which is designed to facilitate spectrum sharing by using methodologies of high dual satellite visibility and repeatable ground tracks. The coverage is optimized to the world population for business advantage. Various results from the comparisons between this MEO system and other proposed NGSO systems are also presented.

High dual satellite visibility – The constellation of the proposed MEO system provides continuous coverage of the populated portions of the world with a minimum elevation angle of 30°. An orbital altitude corresponding to a 6 hour orbital places the satellites between the peak radiation points of the Van Allen radiation belts. This 10,352 km equatorial altitude is high enough to provide excellent coverage using only 20 satellites. Satellites are inclined 55°. The best overall coverage, elevation angle, and diversity results are achieved with a constellation of 4 planes with 5 satellites per plane and with no phase offsets between satellites in adjacent planes. The coverage of the MEO constellation is uniform with respect to longitude and symmetric about the equator. FIGURE 4 shows the diversity coverage characteristics of the MEO constellation with respect to latitude. Continuous single coverage is provided for all latitudes up to 80° N and S. Double coverage exceeds 92% of the day for latitudes within 75° N and S. Triple coverage exceeds 40% of the day for latitudes between 15° and 55°. These latitudes includes most of the world population distribution. With this MEO constellation, some land mass is within the field of view (FOV) of the satellites 90% of the time. So the system capability is not wasted to cover ocean as happened in the LEO constellations.

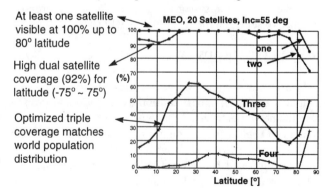

FIGURE 4. Visibility Statistics for MEO Constellation

Repeatable ground tracks – The altitude of MEO satellites is carefully selected in order to provides repeatable ground tracks on the earth surface. Each satellite migrates back to same ground sub-satellite points after 4 revolutions within each sidereal day. The perturbation effects from Sun, Moon, and non-spherical earth are considered in satellite positioning to achieve this feature. Repeatable ground track provides predictable arcs of NGSO satellite routes in the fish eye view from an earth terminal. As a result, it not only reduces the complexity of tracking system design for ground user terminals but also greater sharing feasibility with other NGSO systems.

Comparison with other proposed NGSO systems – Four different NGSO systems are compared with the proposed MEO system. Information regarding altitude, number of planes, number of satellites per plane, total satellite number, inclination angle, phase offset between adjacent planes ,and elevation threshold are listed in TABLE 1. From left to right, column 2 through column 5 represent a low LEO, a high LEO, an MEO, and an IGSO systems respectively. The last column is the 20 satellite Hughes MEO system.

TABLE 1. Constellation Parameters for Various NGSO Systems

Parameters	LEO_A	LEO_B	MEO_A	IGSO_A	Hughes MEO
altitude [km]	700±5	1350	10355	35787	10352
planes/satellites/total	21/40/840	12/6/72	3/5/15	4/3/12	4/5/20
Inclination/Phase [°]	98/0	47/25	50/24	50/90	55/0
elevation threshold [°]	40	22	30	30	30
Corresponding Figure	FIGURE 5(a)	FIGURE 5(b)	FIGURE 5(c)	FIGURE 5(d)	FIGURE 4

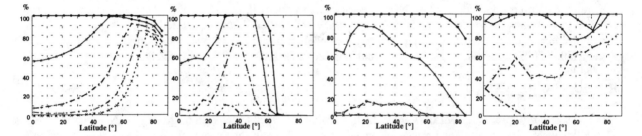

FIGURE 5. Visibility (%) vs. Latitude (combining °N and °S) with Frequency Sharing,(a)LEO_A (b)LEO_B (c)MEO_A (d)IGSO_A.

The visibility statistics of each system are shown in FIGURE 5(a)~(d). None of these systems provide high dual satellite visibility in the latitude ranges needed to share. LEO_A with vertical (south-north) moving satellites designed to simplify intersatellite link structure provides high visibility to high latitude areas but low visibility to low latitude areas where visibility is needed most for GSO arc avoidance. LEO_B with perfect satellite visibility covering only the continental US (CONUS) provides low sharing capability with GSO systems in areas near the equator. MEO_A provides insufficient dual satellite coverage in all areas. IGSO_A provides better dual satellite coverage than the other three systems and is similar to the proposed MEO system. With only 12 satellites, it could be an alternative selection for the proposed MEO system (with 20 satellites) for global coverage.

SATELLITE HIGHWAY

In previous sections, we have reviewed the constellation design by circular orbits. Here we present studies on elliptical geosynchronous earth orbits (EGSO) and propose a potential "satellite highway" which enables multiple satellite networks to share spectrum by forming a family of non-overlapping EGSO contours. This concept is similar to the GSO belt, which facilitates spectrum sharing by forming slots of 2° longitude separation.

Motivation and background – Demand for communication capacity – Despite the advantage of GSO systems which avoid the technical complexities of NGSO systems, the GSO belt can not provide all the communications capacity desired because the demand for satellite communication is growing faster than the technique progress on high frequency hardware. This shortage in desired capacity encourages our study regarding how many satellites/systems the earth surface could accommodate without generating harmful interference.

Maximum number of co-existing satellites – We start from the idealized situation, assume the earth surface could be uniformly filled with bubbles. Each bubble stands for a 3° diameter earth-centered service region away from the adjacent satellites. The extra one degree above typical 2° spacing accounts for the pointing error of a tracking terminal. Though the chosen rectangular packing is not the optimized packing criterion, it is useful for a baseline comparison. Seamless hexagon packing which is used in most wireless cell design feats poorly for a large curved surface such as the satellite field of view. The number of each ring at the same latitude follows the trend of the function of $COS(\psi)$ where ψ is the corresponding latitude. An regional example at North America is shown in FIGURE 6. The number of active non-interfering satellites could increase from 120 (in GSO belt=360/3) to 4582 in this packing manner. Total of 4462 (4582-120) NGSO satellites are equivalent to 37 times more capacity than the GSO belt could provide. Practically, the useful NGSO satellites can be deployed only on certain constellations/orbits, thus we can't reach the maximum satellite number as discussed.

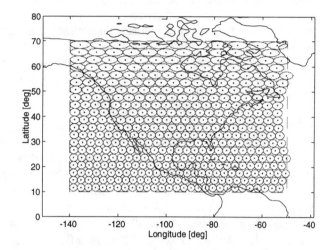

FIGURE 6. Bubble Packing Example at North America

WHY EGSO? – Most of recent proposed broadband systems providing global voice, Internet, and multimedia services are NGSO systems. These systems at LEO/MEO altitudes create regulatory difficulties to fairly allocate the orbital resource among multiple systems at the same band. Geosynchronous orbits could be divided into two groups: the inclined geosynchronous circular orbits (IGSO), and the elliptical geosynchronous orbits (EGSO). The IGSO constellations with zero eccentricity or low eccentricity form balanced or unbalanced figure-8 shaped ground tracks. These ground tracks of different inclination angles converging to a single point may violate minimum distance requirements presented in previous discussion. As a result, only EGSO's with higher eccentricity and non-intersected ground tracks are the appropriate candidates for forming a satellite highway.

Example highway – It is possible to pack satellites into the same region by "cooperative" constellations. From the discussion of previous section, we choose families of EGSO satellites with pear shaped ground tracks to form groups of satellite clusters. As shown in

FIGURE 7, we first select proper eccentricity and cluster separation distance to form many non-overlapping ground tracks. Each ground track is then packed by certain number of satellites with the same size protection region to avoid interference to satellites located in either the same ground tracks or in an adjacent track/cluster. The number of clusters and ground tracks are chosen to be 4 (separated by 90° longitude) and 8 respectively. The maximum number of satellites per cluster is limited by the achievable separation space at the perigee points where the distance between adjacent satellites is the smallest. The pear shape ground tracks provide non-overlapped ground tracks which also simplify the implementation of "satellite highway".

Since there are multiple parameters involved for a satellite highway, a computer program by applying genetic theory may be helpful to achieve the ultimate optimization and maximize the number of satellites could coexist under the same specific protection range. In this example, we only present the best case after several iteration of trials assuming the protection region equals to 3° (earth-centered) between the sub-

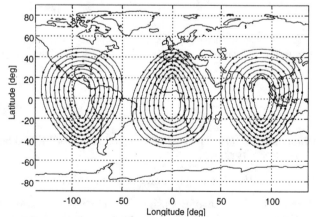

FIGURE 7. Example of Packing the Pear-shaped Ground Tracks with Satellite by 3° Separation

satellite points of two adjacent satellites.

Performance evaluation – The study with multiple parameters including (1) number of ground track groups, (2) inclination angle, (3) eccentricity array, and (4) minimum separation region size could form a large number of combinations for satellite highway candidates. In order to quantify the performance of packing as many satellites as possible to various constellations, a figure of merit is defined by calculating the percentage of feasible NGSO satellites number to the theoretical maximum. The formula is given by

$$\eta(\theta) = N_s(\theta)/N_{total}(\theta) * 100\%,$$

where $N_s(\theta)$ is the number of satellites that could practically fit into certain constellation solution and provide service simultaneously under a protection region with diameter θ degree. $N_{total}(\theta)$ is the number of satellites that could be packed on earth surface under the same protection region. Given the minimum satellite separation of $\theta=3°$, the maximum number of satellites in the i-th track is denoted as $N_i(\theta)$ where θ is the corresponding protection angle. Thus total number of satellites in each group is calculated as $N_s(\theta) = \text{SUM}(N_i(\theta))$. FIGURE 8 shows a example of 118 IGSO satellites packed into nine pear-shaped ground tracks within latitude region [-90° ~ -10°, 10° ~ 90°] which provides an total of 472, (=118*4), satellites globally. A figure of merit of 10.3%, (=472/4582*100%), is certainly higher than 2.61%, (=120/4582*100%), the figure of merit that the geostationary arc could obtain.

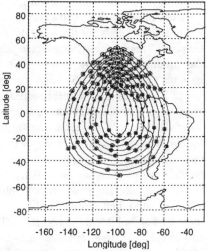

FIGURE 8. 118 IGSO Satellites within [±(10° ~ 90°)] Latitude

CONCLUSIONS

Constellation selection/design plays an important role in recent NGSO broadband satellite systems. In this paper, we present the key issues for spectrum sharing optimization. A study on the spectrum-sharing merits of different constellations is presented. A novel factor in regulating NGSO systems is introduced. The main contribution of this paper includes two different directions: the constellation design and a potential "satellite highway". Regarding constellation design, we discuss the criteria to evaluate the performance of a NGSO satellite network. Several proposed NGSO systems are compared with an MEO constellation optimized from an interference reduction point of view. Regarding "satellite highway", we provide numerical analysis and computer simulation to demonstrate that many satellites could corporately provide service at the same frequency band. The study of the relationship between the interference mitigation methodologies (satellite diversity and site diversity) and the constellation designs (dual satellite visibility and repeatable ground track) provides a new approach to the old problem in spectrum sharing. Promising results have also been obtained.

Acknowledgments

The author wishes to thank the many NI team members at Hughes Communications, Inc. for their technical discussion on various materials presented. Special thanks goes to Patricia Metoyer for her assistance in producing this paper.

References

Christopher, P. (1997) "Satellite Constellations for Millimeter Wave Communication A Comparison," Proceeding of IMSC97, JPL97-11:533-540.

Lo, M. W. (1997) "Constellation Coverage Analysis," Proceeding of IMSC97, JPL97-11:541-542.

Rider, L. (1986) "Analytic Design of Satellite Constellations for Zonal Earth Coverage Using Inclined Circular Orbits," Journal of the Astronautical Sciences, 34(1): 31-64.

ITU Radio Communication Study Group 4A (1996) "Chairman Report," Document 4A/161E.

FIRST CONFERENCE
ON ORBITAL TRANSFER VEHICLES

ROTV AS PART OF A REUSABLE SPACE ARCHITECTURE

Terry H. Philips and Robert O'Leary
Schafer Corporation
2000 Randolph Road, SE, Suite 205
Albuquerque, NM 87102-4267
(505)242-2992, ext. 30 & 242-2992, ext. 31

Frederick Widman
Boeing Defense and Space Group
6633 Canoga Avenue, MS IB-47
P. O. Box 7922
Canoga Park, CA 91309-7922
(818) 586-0975

Extended Summary

REUSABLE ORBIT TRANSFER VEHICLE (ROTVs)

Reusable Orbit Transfer Vehicles (ROTVs) are an integral part of a future reusable space architecture. This architecture will also include reusable launch vehicles (RLVs) such as Military Spaceplane and Commercial RLV, Space Maneuver Vehicles (formerly Mini-Spaceplanes), reusable repair and refuel modules (R3M), and satellites designed for on-orbit repair and routine servicing. A close look at the ΔV requirements to move a notional Military Spaceplane from a 100 nautical mile (nm) by 100 nm orbit to higher orbits demonstrates the lack of feasibility of employing RLVs for missions much above 300 nm. The table below shows it requires nearly a quarter million lbs. of propellant for a one-way trip to GEO for a 100 Klb. Military Spaceplane.

TABLE 1. Propellant Requirements One Way from 100 nmx100 nm to Higher Orbit

Vehicle	300 nm	500 nm	1000 nm	11,000 nm	19,200 nm
100 Klb MSP	5000 lbs	9800 lbs	21,000 lbs	170,000 lbs	240,000 lbs

To access higher orbits, smaller, more efficient vehicles are required. A small ROTV using a high Isp (specific impulse) engine can cut these propellant requirements significantly. Phillips Lab's Integrated Solar Upper Stage (ISUS), as an example, with 900 seconds Isp, can make a trip to GEO using less than half the propellant of an equivalent chemical upper stage. ISUS based ROTVs can offer significant operational capability to move satellites from orbit to orbit. Transit time increases, but the 15-30 day transit time from LEO to GEO represents no operational constraint for routine, non-critical satellite movements.

As a true reusable architecture evolves, mixes of vehicles which operate together will ensure maximum cost effectiveness and operational utility for all vehicles. Military Spaceplane can launch via pop-up delivery SMVs, ROTVs, R3Ms, and satellites. These satellites will be designed for rapid launch, efficient transfer to mission orbit by ROTVs, and on-orbit service of space replaceable units (SRUs) by R3Ms or SMVs. SRUs, the space equivalent of aircraft line reusable units (LRUs) can include batteries, solar panels, processors, and sensors. Rapid transit ROTVs and R3Ms can be designed with hybrid chemical and high Isp powerplants for rapid transit to a higher orbit and mission accomplishment followed by a leisurely and efficient return to LEO for refueling or retasking.

As reusability and reparability of satellites and space vehicles are studies and employed in earness, a true picture will emerge showing the break points between reusable systems and cheap expandable systems. A place for both exists.

The contents of this document have been initially prepared for publication as "Characterization of the Dielectric Properties of the Propellants MON and MMH" in *Space Technology and Applications International Forum - 1998*, January 1998 by The American Institute of Physics.

CHARACTERISATION OF THE DIELECTRIC PROPERTIES OF THE PROPELLANTS MON & MMH

A.A.M. Delil
National Aerospace Laboratory NLR, Space Division
P.O. Box 153, 8300 AD Emmeloord, Netherlands
phone +31 527 24 8229, fax +31 527 24 8210, E-Mail adelil@nlr.nl

Abstract

Future (commercial) satellites will require accurate propellant gauging systems, in order to meet the end-of-life re-orbiting requirement and the need for replacement planning. A capacitive Gauging Sensor Unit is currently being developed for the Meteosat Second Generation spacecraft. Its measurement principle is based on the difference of the dielectric properties of the propellant liquid and vapour. To optimise the sensor accuracy, the dielectric properties of propellants need to be accurately known as a function of the temperature (and pressure). Therefore the dielectric properties of MON (Mixed Oxides of Nitrides) and MMH (Mono Methyl Hydrazine) were to be measured. The test setup and the test results are described in detail.

INTRODUCTION

Future (commercial) satellites must incorporate accurate propellant gauging systems to meet the end-of-life de-orbiting requirement and the need for replacement planning. Earlier discussions (Hufenbach et al. 1997) led to the conclusion that gauging system requirements depend on: Type of Mission (commercial/scientific), Orbit Type (LEO, MEO, HEO, GEO, interplanetary), Overall Delta Velocity Requirement, Propellant Mass Fraction, Mission Profile, Stand Alone/Constellation Spacecraft, Spacecraft Design (operational lifetime & complexity), and Costs of Spacecraft, of Launch (absolute cost, spacecraft mass margin) and of Operation. The degree of accuracy for the propellant mass determination usually is the major driver for the design and implementation of gauging systems. The required accuracy level follows from the weights of the above issues. Other requirements, taken into account for spacecraft design & development, are: Mass Budget (mass of the gauging system and of necessary spacecraft modifications/adaptations), Power Budget, Lifetime of the Gauging System, Aging (of electronics, propellant properties), Amount of the Data (to be transferred on the TTC channels), Costs (recurring/non-recurring, for integration/calibration, for spacecraft interfaces adaptation, operation cost) and Availability of Technology.

For Meteosat Second Generation (MSG) spacecraft series, Eumetsat requires the determination of remaining propellant mass in the tanks (during the last three mission years) to an accuracy of ≈ 4 kg, meaning three months of satellite lifetime. This is an accuracy of ≈ 1.25 kg per oxidizer tank, ≈ 0.75 kg per fuel tank. The Gauging Sensor Unit (GSU), developed for MSG spacecraft, measures the capacitance of the medium between two electrodes to determine the propellant level. The measurement principle is depicted in Figure 1, for a configuration with two electrodes coated with an electrically insulating layer. The GSU (Figure 2) consists of a 125 mm long Platinum profile, the first electrode, embedded in a solid glass tube. The glass tube is within a Titanium frame, the second electrode, providing mechanical protection for the glass tube. The open construction of the Titanium housing allows

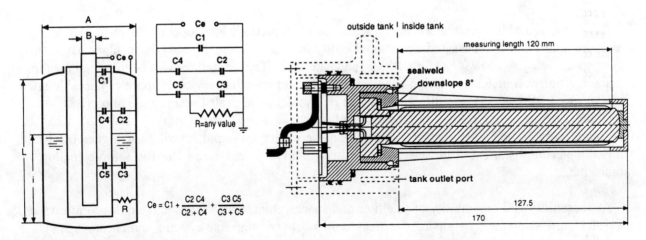

FIGURE 1. Coated Capacitance Probe (Schematic).

FIGURE 2. Propellant GSU for MSG.

the liquid to enter the space between the two electrodes. The liquid level determines the dielectric permittivity of the medium between the electrodes, hence the capacitance (Delil et al. 1997 and Hufenbach et al. 1997). The whole assembly is inserted via a propellant port in the propellant tank of a spin-stabilised spacecraft. The sealing of the sensor consists of a pre-tension type PTFE jacket with an Inconel spring first seal and a Kalrez second seal, placed radially on the glass container. Two PTFE O-ring seals are present as backup. Electronics are mounted on the bottom of the instrument, outside the tank. Wires coming from the Platinum electrode are melted inside a glass feedthrough. The whole assembly mass is 376 grammes, excluding harness. It can withstand a peak load of 200g. It is 200 mm long and has a maximum diameter of 58 mm. The sensor is capable to withstand the environment inside the propellant tank: it can handle temperatures between 273 and 343 K, pressures up to 2.3 MPa. The electronics unit needs 0.3 W peak power, 0.15 W average.

The platinum profile on the glass tube is divided into segments (Hufenbach et al. 1997). Three of them are used for active gauging. The lower one is used for reference purposes only, as it is always emerged in the liquid during the operational mission. When the propellant level reaches the lowest segment, the de-orbit manoeuvre has to start. The different segments are used in an active guarding technique. This means that whenever a certain segment is not wetted, its potential is equal to the wetted (liquid measuring) segment potential. This technique, combined with a smart set of electronics, eliminates secondary effects: field bending and variable parasitic capacities. Over each segment the level has to be measured with an accuracy better than one millimeter. To attain this, just measuring the capacitance is not sufficient as this varies between 30 and 33 pF only. Therefore the loading time of the capacitor segment is measured and converted to a frequency for easier processing. The loading times of each of the four segments are measured in individual sessions of one minute. The measurement of the lowest segment will be used to determine the parasitic capacitance behaviour of the sensor. Due to on ground calibration and knowing that the lowest part is fully emerged in the liquid, the offset signal is eliminated from the level measurement. The liquid level can be determined from the comparison of this reference signal with the signal coming from segments partly emerged in the liquid. The total amount of propellant left in the tank can be determined from this level. The frequency signals will be sent to the ground where the processing is done.

Several errors are to be corrected for. Due to the reference segment at the bottom of the sensor most errors can be eliminated. A disturbing effect is field bending, occurring when two adjacent capacitors have different potentials. Analysis of this effect is hardly possible and accuracy and reliability figures can only be achieved by means of tests. Active guarding proved to reduce the field bending effect to an acceptable level (Mastenbroek et al. 1997) during tests with a GSU Bread Board Model, using MON-1, expected to be non-conductive. Figures 3 and 4 depict measured frequency versus level curves for the segments, resp. the measured GSU resolution. It is clear that, though the test item is only a BBM, the test results for MON-1 are very promising.

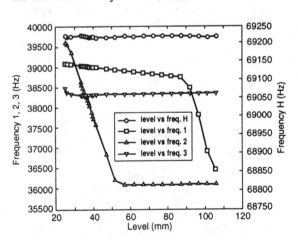 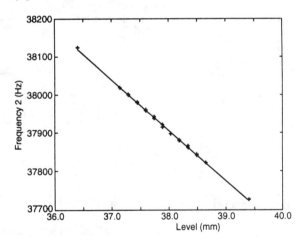

FIGURE 3. MON-1 Test Data for Segmented GSU.   FIGURE 4. GSU Resolution Test Data.

To summarise the conclusions and recommendations (Mastenbroek et al. 1997), it can be said that: the span is within the same range for the three segments, the linearity and the reproducibility is very good, active guarding has led towards a very stable sensor, the required resolution (0.5 mm) is obtained even without any temperature calibration/compensation, a level variation of 0.1 mm can be detected reproducibly under constant temperature conditions, better stability can be obtained by dedicated optimised electronics (small PCB, short wires, efficient routing), optimal accuracy can be obtained by calibrating each individual segment, a proper "level protocol" will considerably reduce parasitic effects and eliminate effects of liquid conductivity, problems will occur when liquids contain dedicated additives influencing the fluid characteristics (e.g. conducting surface layers cause unpredictable errors. Such layers have been found in e.g. petrol). As there is no indication that this effect occurs in MON/MMH, it is assumed that the liquids are homogeneous. As the GSU Titanium electrode is non-coated, the major error source might be the electric conductivity of the liquid. To account for this effect in a naked electrode configuration, the dielectric properties (permittivity & specific conductivity) of MON and MMH had to be determined. This characterisation of electric properties (carried out under contracts for ESA and the Netherlands Agency for Aerospace Programmes) is described in the following chapters.

RESULTS OF MON-1

Figure 5 depicts the test cell and test setup. Table 1 lists the tests performed to characterise MON-1.

TABLE 1. Test Performed to Characterise MON-1.

Test	Name	Purpose and description
1	Pre Empty cell	To check the characteristics of the empty test cell and test setup. The temperature is cycled in the range of 283 to 323 K. The capacitance is measured as a function of the temperature.
2	Calibration	To check the relation between capacitance of test cell and permittivity of the fluid in the test cell (filled with water), cycled between 283 and 323 K. Results are compared to values from literature. At ambient conditions this is also done for 4 other liquids with permittivities in the range 2 to 80.
3	1st batch	To characterise MON-1. The test cell is filled with MON-1 and passivated for 10 hours. Then temperature cycling between 283 and 323 K is done, at a rate of 2 K per hour.
4	2nd batch	To characterise MON-1. Test: same as test 3, but 2 cycles are performed.
5	Post Empty cell	To check the characteristics of test cell and test setup after the MON-1 characterisation. The (cleaned) test cell is empty. Test: same as test 1.

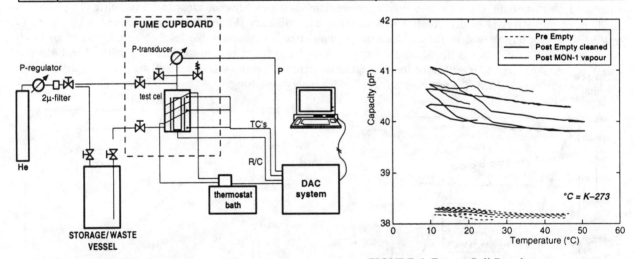

FIGURE 5. Test Cell & Test Setup.   FIGURE 6. Empty Cell Results.

Empty test cell

Figure 6 shows the capacitance of the empty test cell, as a function of the temperature, before (Pre) and after (Post) the MON-1 characterisation tests. The figure shows clearly that after the MON-1 test the capacitance of the test cell is higher than before the test. This is due to the penetration of MON-1 into the Teflon insulator of the test cell. This is visually confirmed (brown colour of Teflon). The permittivity of Teflon is 2.1, the permittivity of MON is 2.7. This explains the slight increase in capacitance. The parasitic test cell capacitance C_p changed from 5.1 to 7.1 pF. As the permittivity (or relative dielectric constant) ε is calculated according to $\varepsilon = (C_{total} - C_p)/C_0$.

The error in the permittivity, due to parasitic capacitance changing, turns out to be 2 % (for permittivity 2.6). During the (Pre & Post) tests with the empty test cell, the resistance of the test cell turned out to be higher than 1 GΩ. The measured MON-1 resistances were in the range of 200 MΩ down to 4 MΩ. This means that the measured resistance of MON-1 is not affected by the resistance of the test cell itself or the measurement method.

Calibration results

Figure 7 shows the permittivity (derived from the capacitance) for water, in the temperature range 283 to 323 K. Values found in the literature are given also. The graph shows that the test setup measures the permittivity of water with an accuracy better than the required 2 %. The "waves" in the measured values are due to the temperature cycling. In bending points the temperature was kept constant for 2 hours, which leads to the more accurate values of these points. To verify the test setup for other permittivities, several liquids were used (Table 2). The literature values are compared with the measured ones.

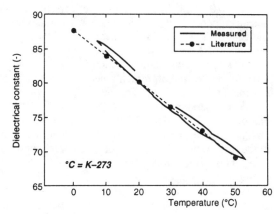

FIGURE 7. Calibration of Test Cell, Using Water.

TABLE 2. Measured and Literature Values of Permittivities of Liquids.

Fluid	ε measured	ε literature	relative error [%]
n-Octane	1.94	1.95	0.3
Ethylacetate	6.04	6.02	0.3
Methanol	33.44	32.63	2.5
Acetone	21.26	20.70	2.7

The measured permittivity is in good agreement (< 3 %) with the values from literature. It can be concluded that in this test setup the permittivity of a liquid can be measured (from 283 to 323 K) with accuracy better than 2 %.

Test results evaluation

The MON-1 characterisation has been performed for two batches. The temperature cycling was from 283 to 323 K. The average pressure during the measurements was 1.18 ± 0.04 MPa (Helium pressurised). The figures 8 and 9 show the temperature cycle, the capacitance and resistance of the first and second batch. For the second batch some extra cycles have been done. These figures show that the capacitance is clearly temperature sensitive and that the resistance is high (> 3M Ω), but not clearly temperature sensitive.

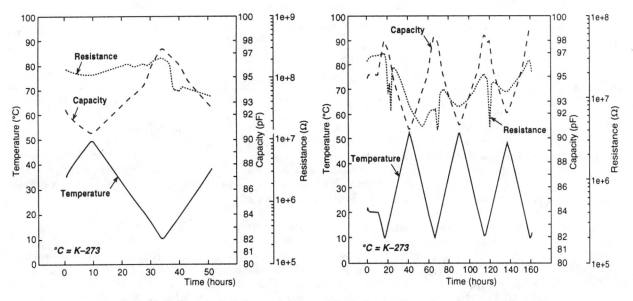

FIGURE 8. Capacitance & Resistance versus Temperature (1st MON-1 Batch).

FIGURE 9. Capacitance & Resistance versus Temperature (2nd MON-1 Batch).

Figure 10 depicts the, from the capacitance derived, dielectric permittivity as a function of the temperature. The figure shows an almost linear temperature dependence (there is some hysteresis below a temperature of 298 K). Also the capacitance slightly increases with time. This is probably due to the aforementioned penetration of MON-1 into the Teflon insulator. The permittivity ε_r as a function of the temperature can be approximated by the linear relation: $\varepsilon_r = 2.833 - 0.0052 * (T-273)$, (T = Temperature in K). The error in the permittivity is the sum of inaccuracies of the test setup (2 %) and of the curve approximation (2 %). This gives a total error of 4 %.

The minimum resistance, measured during the tests of the two batches, turned out to be 3.8 MΩ. Conversion of the lowest measured value of 3.8 MΩ into a specific resistance, gives a minimum value of 14 MΩ m. The fluctuations, at lower temperatures, are not fully understood. As the tests with a empty and cleaned test cell after the MON-1 test show very high resistance (2 GΩ), it is concluded that the test setup can not cause such low resistance values during tests. Also it can be remarked that the test cell interior did not show visible traces of corrosion. Together with the fact that the decreases of the resistance during the MON-1 tests are reproducible and occur after passing the lowest temperature, it is concluded that the resistance decrease is caused by solving/dissolving of some MON-1.

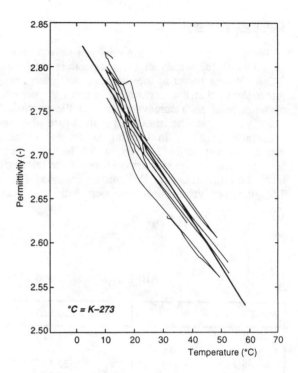

FIGURE 10. Temperature Dependent Permittivity of MON-1.

RESULTS OF MMH TESTS

The MMH characterisation has been done by NLR at DASA-Lampoldshausen. Table 3 lists the tests performed.

TABLE 3. Test Performed to Characterise MMH.

Test	Name	Purpose and description
1	Pre Empty cell	To check the characteristics of the empty test cell and test setup. The temperature is cycled in the range 283 to 323 K. The capacitance is measured as a function of the temperature.
2	1st batch	To characterise MMH. The MMH filled test cell is passivated for 10 hours. The temperature cycling between 283 and 323 K is done at a rate of 2 K per hour.
3	2nd batch	To characterise MMH for the second time. Test: same as test 2.
4	3rd batch	To characterise the MMH for a third time. Test: same as test 2.

The MMH characterisation has been done for three batches. The temperature cycling was from 283 to 323 K. The average pressure during the measurements was 1.2 ± 0.2 MPa (Helium pressurised). During the measurement of the first batch, the resistance and capacitance drifted. The second and third batch results do not show drift at all. An explanation for this is the fact that the test cell was passivated during the testing of the first batch. Therefore the data of the first batch will not be used in the evaluation. Figure 11 shows the from the capacitance derived permittivity as a function of the temperature. The temperature dependence is almost linear and can be approximated by: $\varepsilon_r = 23.1 - 0.11 * (T - 273)$, (T in K). The error in the permittivity is the sum of the test setup inaccuracy (2 %) and the curve approximation of the curve (5 %), yielding a total error of 7 %.

The, from the measured test cell resistance derived, specific resistance of MMH, is depicted in figure 12. The figure shows that the specific resistance is about 400 Ω m at 293 K. The difference between batches 2 & 3 is $\approx$ 25 %, is probably due to small differences in water concentrations. The specific resistance can be approximated by: $\rho = 8.2 * \exp[1135.5/(T - 273)]$, ($\Omega$ m, T in K). During the measurements, the resistance of the test cell was about 100 Ω. The (capacitive) impedance turned out to be minimal 30 kΩ. As the resistance is much lower than the impedance, MMH can be considered to be an electrically conducting fluid, for frequencies up to 10 kHz.

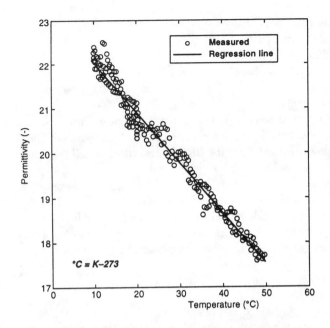

FIGURE 11. Temperature Dependence of Permittivity of MMH.

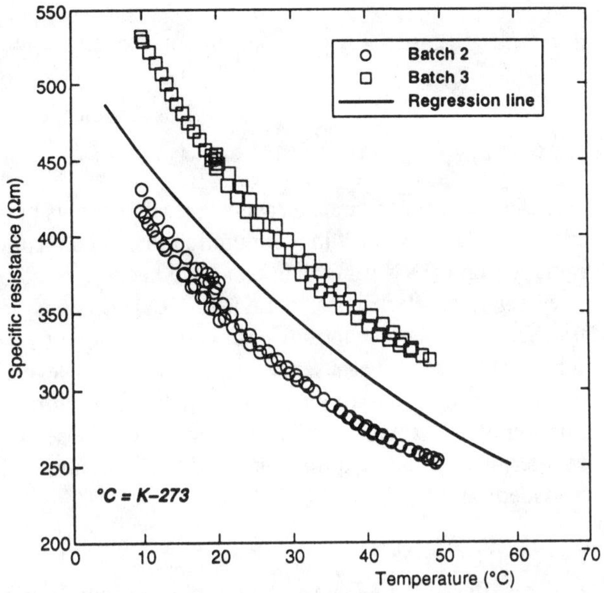

FIGURE 12. Temperature Dependence of Specific Resistance of MMH.

CONCLUDING REMARKS

A test setup to measure the dielectric properties (permittivity and electric conductivity) of propellants has been designed, manufactured and calibrated (using water, n-Octane, Ethylacetate, Methanol, Acetone). In this test setup the above dielectric properties have been experimentally determined for MON-1 and MMH, as a function of the temperature, under MSG propellant tank pressure conditions (Helium at $\approx$ 1.2 MPa). The data will be used to optimise Gauging Sensor Units developed to measure the level in propellant tanks of MSG spacecraft.

MON-1 confirmed to be a non-conductive liquid. MMH turned out to be a conducting liquid. As the electric conductivity is an error source for a propellant gauge with (a) non-coated electrode(s), it is obvious that electrodes covered with an electrically insulating coating are to be preferred for conducting propellants as MMH.

Acknowledgement

The efforts of the NLR colleagues Mr. O. Mastenbroek (test definition, interpretation of test results) and Messrs. A. Pauw and G. van Donk (execution of tests) are highly appreciated. The contributions of Mr. P. van Put (Bradford Engineering) and the efficient support and hospitality offered to the NLR test team during the MMH tests at DASA Lampoldshausen (especially by Mr. W. Müller) are highly appreciated also.

References

Delil, A.A.M. et al. (1997) "Test Cells and Components for Aerospace Thermal Control and Propellant Systems", *NLR TP 97282 U, SAE 972478, 27th International Conference on Environmental Systems*, Lake Tahoe, Nevada, USA, July 1997, and *6th European Symposium on Space Environmental Systems*, Noordwijk, Netherlands, ESA SP-400, I: 289-299.

Hufenbach, B. et al. (1997) "Comparative Assessment of Gauging Systems and Description of a Liquid Gauging Concept for a Spin-Stabilised Spacecraft", *2nd European Spacecraft Propulsion Conference*, Noordwijk, Netherlands, ESA SP-398, 561-570.

Mastenbroek, O. (1997) "Evaluation of Accuracy Verification Test Results for the Segmented GSU BBM, Using MON-1", *GSU-NL-TN-005*.

RADIATION EFFECTS IN LOW-THRUST ORBIT TRANSFERS

James E. Pollard
The Aerospace Corporation
P.O. Box 92957, M5-754
Los Angeles, CA 90009, (310)336-4023

Abstract

A low-thrust orbit transfer vehicle (OTV) and its payload must be designed to survive in the near-Earth radiation environment for a much longer duration than a conventional upper stage. This paper examines the effects of natural radiation on OTV's using data that have become available since 1991 from the CRRES and APEX satellites. Dose rates for microelectronics in LEO-to-GEO missions are calculated for spiral orbit raising and for multi-impulse transfers. Semiconductor devices that are shielded by less than 2.5 mm of aluminum (0.69 g/cm^2) are inappropriate for spiral transfers, because they require hardness levels >100 krad (Si). Shield thicknesses of 6-12 mm reduce this requirement to about 10 krad (Si), which is still an order of magnitude higher than the radiation dose in a 10-year mission at GEO with similar shielding. The dose for a multi-impulse LEO-to-GEO transfer is about 10 times smaller than for a spiral transfer. Estimates of single event upset rates and photovoltaic array degradation are also provided.

INTRODUCTION

The effects of ionizing radiation play a significant role in any spacecraft design, and this is especially true for orbit transfer with low-thrust propulsion. In geocentric orbits, the radiation environment is dominated by toroidal belts of high-energy protons and electrons that gyrate around magnetic field lines while bouncing back and forth between the northern and southern hemispheres (Stassinopoulos 1988). Spacecraft solar cells and microelectronics will degrade or fail at a rate that depends on the level of solar activity, the orbital altitude and inclination, the intrinsic component hardness, and the shielding provided by cell cover glasses or spacecraft structure. Fortunately, geosynchronous equatorial orbit (GEO) is beyond the most damaging portion of the radiation belts at 1500 to 25,000 km altitude, but an orbit transfer vehicle and its GEO payload must survive for 1-6 months in this hostile region.

Progress in designing OTV's has been hindered by lack of data on the radiation environment and its effect on space systems. However, this situation has improved as a result of the Combined Release and Radiation Effects Satellite (CRRES) launched in 1990 (Gussenhoven 1992, Gussenhoven 1993a, Mullen 1993, Gussenhoven 1993b, and Ray 1993). The satellite operated for 15 months in an elliptical orbit (348 km × 33,582 km, $i = 18.2°$) and carried a comprehensive set of instruments and diagnostic packages to quantify the particle fluxes and energies, and the response of microelectronics and solar cells. Estimates of OTV environments and power system properties can also take advantage of the Photovoltaic Array Space Power Plus Diagnostics (PASP Plus) and other experiments launched in 1994 on the APEX satellite (Gussenhoven 1995, Guidice 1996, and Dyer 1996). The spacecraft returned data for 90 days while sampling the low-altitude edge of the inner proton belt (362 km × 2544 km, $i = 70°$). PASP Plus measured space plasma interactions and radiation effects for planar and concentrator arrays at positive and negative bias voltages, along with diagnostic data for correlating the environmental conditions with observed array performance.

In this paper we use CRRES dose models (Kerns 1992) to assess radiation effects on spacecraft microelectronics for low-thrust spiral and multi-impulse transfers from LEO to GEO. It is anticipated that refined models incorporating the PASP Plus database will soon be available to improve the accuracy of our results at lower altitudes. We also consider the impact of CRRES and APEX data on designing OTV's to cope with single event upsets (SEU's) and with degradation of photovoltaic arrays.

DOSE MODEL RESULTS

The standard treatment of low-thrust maneuvers for near-circular orbits with plane-changing provides a suboptimal steering program that defines the relation between altitude and inclination at any point in the mission (Edelbaum 1961 and Kechichian 1992). Figure 1 shows altitude and inclination as functions of elapsed time in a typical LEO-to-GEO transfer, under the assumption of continuous acceleration of constant magnitude with orbital perturbations neglected. The spiral transfer in Fig. 1 applies to a solar electric OTV that produces a very small change in the orbit during a single revolution (acceleration $< 10^{-3}$ m/s^2). This may be compared with Fig. 2, which shows inclination, apogee altitude, and perigee altitude for a multi-impulse transfer that would be typical of a solar thermal OTV having an acceleration $>10^{-2}$ m/s^2 (Frye 1996). The first part of the mission consists of apogee raising with minimal inclination change, and the second part involves perigee raising with simultaneous plane rotation.

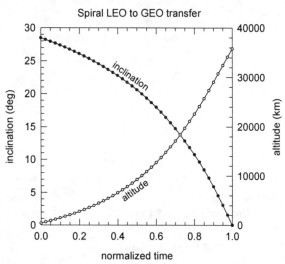

FIGURE 1. Inclination and Altitude vs. Time for a Spiral LEO-to-GEO Transfer. Points correspond to orbits for which radiation dose rates were calculated.

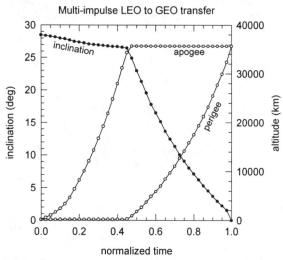

FIGURE 2. Inclination and Altitude vs. Time for a Multi-Impulse LEO-to-GEO Transfer. Points correspond to orbits for which radiation dose rates were calculated.

Using the spiral and multi-impulse trajectories the CRRES models can predict the payload radiation dose rate vs. time, as shown in Figs. 3 and 4 with four different aluminum shield thicknesses. The "quiet" model is based upon data acquired during the first eight months of the CRRES mission, up to 24 March 1991, when geomagnetic conditions were relatively inactive. The "active" model is based upon data acquired during the remaining seven months, when the geomagnetic conditions were highly disturbed. These two models can be considered as bounding values for the radiation environment. For quiet conditions in Fig. 3 there are two main belts of energetic particles, with protons predominating at low altitude and electrons predominating at high altitude. Consequently, an increase in shield thickness gives much greater attenuation of the dose rate at high altitude than at low altitude. During active conditions the slot region between the belts is filled by new populations of high-energy protons and electrons. The spiral transfer in Fig. 3 is analogous to a map of the radiation environment versus altitude, while the multi-impulse mission in Fig. 4 shows a more complex dependence of dose rate on elapsed time, as the apogee and perigee successively pass through the inner and outer belts.

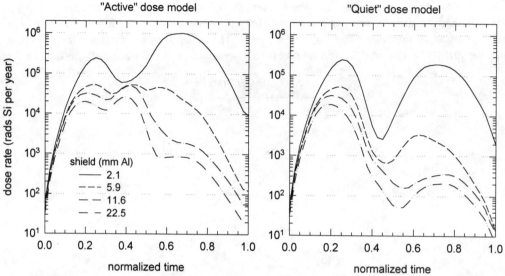

FIGURE 3. Radiation Dose Rate vs. Time for a Spiral LEO-to-GEO Transfer with $\Delta i = 28.5°$. The results are derived from CRRES measurements during active and quiet solar conditions, using four different thicknesses of aluminum shielding for the payload.

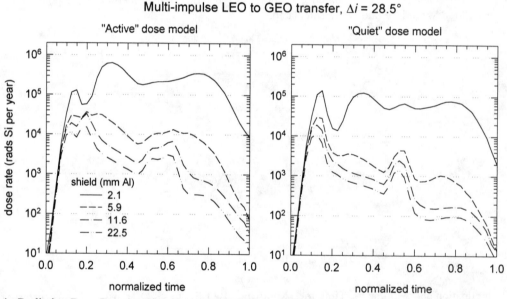

FIGURE 4. Radiation Dose Rate vs. Time for a Multi-Impulse LEO-to-GEO transfer with $\Delta i = 28.5°$.

Integrating the dose rates in Figs. 3 and 4 over the orbit transfer mission gives the total dose vs. outbound trip time shown in Figs. 5 and 6. These results allow the radiation hardness of payload and OTV bus components to be specified for representative shield thicknesses. Payloads that are shielded by less than 2.54 mm (0.686 g/cm² Al) are inappropriate for a spiral transfer, because they require semiconductor hardness levels >100 krad that would lead to costly or impractical spacecraft designs. Figure 5 shows that shield thicknesses of 6-12 mm reduce this requirement to about 10 krad, which allows access to a greater variety of electronic components (Griffin 1991), but is still an order of magnitude higher than the radiation dose in a 10-year mission at GEO with similar shielding. A reusable OTV "space tug" that makes 10-20 spiral passages through the radiation belts during both active and quiet conditions would need a shield thickness of 12-24 mm and a component hardness of 30-100 krad (similar to parts used on the Galileo spacecraft). Secondary shielding using a high-Z layer sandwiched between two low-Z materials is very effective in reducing the radiation dose for components that cannot otherwise be hardened (Mullen 1993 and Fan 1996), although not all susceptible devices can be shielded in this way. Figure 6 shows that a multi-impulse OTV mission with a trip time of 30-60 days will receive a payload radiation dose about 10 times smaller than that of a spiral OTV mission with a 180-240 day trip time, assuming the same payload shielding in both cases.

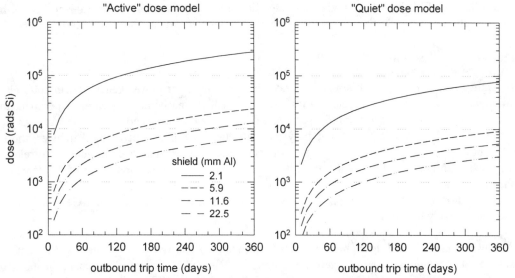

FIGURE 5. Radiation Dose vs. Outbound Trip Time for a Spiral LEO-to-GEO Transfer with $\Delta i = 28.5°$.

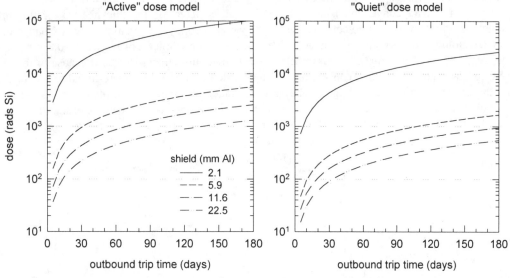

FIGURE 6. Radiation Dose vs. Outbound Trip Time for a Multi-Impulse LEO-to-GEO Transfer with $\Delta i = 28.5°$.

SINGLE EVENT EFFECTS

Susceptibility to single event effects tends to increase as microelectronic device features shrink below 1 μm and critical charge levels go below 1 pC. An SEU or "soft error" is a temporary change in the logic state of a digital circuit that results from the impact of an energetic particle (Peterson 1981 and Griffin 1991). Protons with energies of 100-500 MeV residing in the inner radiation belt (typically 1500-7000 km altitude) produce SEU's as a result of nuclear spallation reactions with circuit materials. At higher altitudes, SEU's are generated mainly via direct ionization by cosmic ray particles, for which there is no effective method of shielding. Energetic particle impacts also produce permanent damage to microelectronics in the form of single event latch-ups and burn-outs.

Measurements of soft error rates on CRRES were made for 60 different types of solid-state devices with shielding that ranged from minimal to heavy (Campbell 1992 and Mullen 1993). CRRES data also afforded an opportunity to test the predictions of SEU rate models (Weatherford 1993 and Reed 1994). The SEU rate induced by protons trapped in the inner belt was observed to be 100-500 times greater than the cosmic ray-induced rate. During the solar flare event of 23 Mar 1991 the upset rate at higher altitudes increased by a factor of 10-100 over the baseline cosmic ray-induced rate, while the inner belt rate was not much changed. On a low-thrust orbit transfer mission, random-access memories will experience 5-50 SEU/kbit/day during the passage through the inner belt, dropping to 0.02-0.5 SEU/kbit/day at higher altitudes during quiet conditions. Several approaches have been developed for dealing with SEU's through device hardening, fault detection, and masking (Kerns 1988).

PHOTOVOLTAIC ARRAY DEGRADATION

Light-weight, radiation-tolerant photovoltaic arrays are candidates for power generation in solar electric orbit transfer missions. In addition to microelectronics tests, the CRRES spacecraft carried an experiment to evaluate radiation effects for solar cells made from Si, GaAs, and GaAs/Ge with various cover glass thicknesses (Ray 1993). Although the elliptical CRRES orbit did not fully replicate the environment of a spiral transfer, the results are certainly the best available for judging cell performance under representative conditions. Correlation of cell degradation rates with the time-dependent flux of high-energy particles showed that most of the damage is caused by protons rather than electrons, and that a dynamic radiation environment can invalidate predictions of cell performance based on a constant flux of particles. Measurements also verified that GaAs/Ge has higher efficiency and degrades less rapidly than Si.

Output power was measured throughout the CRRES mission for GaAs/Ge cells with six different cover glass thicknesses. Solar conditions were relatively quiet until 23 Mar 1991, when a disturbance in the geomagnetic environment increased the flux of high-energy protons. Prior to this event, the cell degradation rate had slowed to almost zero following the initial loss, but thereafter the degradation rate increased. A cover glass thickness of 0.3 mm was the best option for conventional planar arrays, because increasing the thickness to 0.5-0.8 mm did not provide enough additional protection to compensate for the weight penalty. As an alternative to planar arrays, solar concentrator technology allows a much smaller active cell area with a thick cover glass (>0.8 mm), so that the degradation for a 180-day LEO-to-GEO orbit transfer would be no more than 4%-6%. Because of the opportunity for better encapsulation of conductors, concentrator designs tested on PASP Plus were found to reduce the leakage current and the arc rate at high bias voltages. They also degraded less rapidly than planar arrays during radiation exposure. A photovoltaic system capable of operating at high bias voltage is very beneficial for "direct drive" solar electric propulsion, in that it avoids the need for a power converter to step up a low input voltage, thereby reducing the mass and increasing the efficiency of the power processing unit.

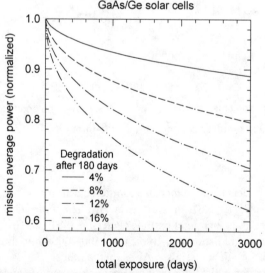

FIGURE 7. Mission Average Power for GaAs/Ge Cells Versus Exposure Duration. The curves are extrapolated from 4%, 8%, 12%, and 16% degradation of the instantaneous power after 180 days of exposure.

To design a reusable solar electric OTV that can make many trips through the radiation belts, we need to extrapolate from the 180-day degradation to much longer exposure times. The results in Fig. 7 are based on published data for output power versus 1-MeV electron fluence with GaAs/Ge cells (Anspaugh 1989), using the fluence values at degradations of 4%, 8%, 12%, and 16% to determine the exposure time that causes a given amount of degradation. For example, an array that produces 1000 W at beginning-of-life and 960 W after 180 days (4% degradation) will generate 840 W after an exposure of 2300 days for a mission average power of 900 W. Hence, a degradation of 4% after a 180-day LEO-to-GEO mission could serve as a target value for a photovoltaic power system appropriate to a reusable solar electric OTV.

Acknowledgments

This work is supported by the Internal Research and Development Program at the Aerospace Corporation. The author is grateful to Bernard Blake for reviewing the manuscript.

References

Anspaugh, B.E. (1989) "Solar Cell Radiation Handbook, Addendum 1: 1982-1988," *JPL Publication 82-69*, Jet Propulsion Laboratory (15 Feb 1989), 37.

Campbell, A.B., P.T. McDonald, and K.P. Ray (1992) "Single Event Upset Rates in Space," *IEEE Trans. Nucl. Sci.*, 39: 1828.

Dyer C.S. et al. (1996) "Measurements of the Radiation Environment From CREDO-II on STRV and APEX," *IEEE Trans. Nucl. Sci.*, 43:2751.

Edelbaum, T.N. (1961) "Propulsion Requirements for Controllable Satellites," *ARS Journal*, 31:1079.

Fan, W.C. et al. (1996) "Shielding Considerations for Satellite Microelectronics," *IEEE Trans. Nucl. Sci.*, 43:2790.

Frye, P.E. (1996) Personal Communication, Boeing North American, Canoga Park, CA.

Griffin, M.D. and J.R. French (1991) *Space Vehicle Design*, American Institute of Aeronautics and Astronautics, Washington, DC.

Guidice, D.A., and K.P. Ray (1996) "PASP Plus Measurements of Space Plasma and Radiation Interactions on Solar Arrays," Paper AIAA-96-0926, *34th Aerospace Sciences Meeting*, 15-18 Jan 1996, Reno, NV.

Gussenhoven, M.S. and E.G. Mullen, (1993a) "Space Radiation Effects Program: An Overview," *IEEE Trans. Nucl. Sci.*, 40: 221.

Gussenhoven, M.S. et al. (1992) "The Effect of a March 1991 Storm on Accumulated Dose for Selected Satellite Orbits: CRRES Dose Models," *IEEE Trans. Nucl. Sci.*, 39: 1765.

Gussenhoven, M.S. et al. (1993b) "CRRES High Energy Proton Flux Maps," *IEEE Trans. Nucl. Sci.*, 40: 1450.

Gussenhoven, M.S. et al. (1995) "Low Altitude Edge of the Inner Radiation Belt: Dose Models from the APEX Satellite," *IEEE Trans. Nucl. Sci.*, 42: 2035.

Kechichian, J.A. (1992) "The Reformulation of Edelbaum's Low-Thrust Transfer Problem Using Optimal Control Theory," Paper AIAA-92-4576-CP, *Astrodynamics Conference*, 9-12 Aug 1992, Hilton Head, SC.

Kerns, K.J. and M.S. Gussenhoven (1992) "CRRESRAD Documentation," PL-TR-92-2201, Phillips Laboratory, Hanscom AFB, MA.

Kerns, S.E. and K.F. Galloway, eds. (1988) "Space Radiation Effects on Microelectronics," (special section), *Proc. IEEE*, 76: 1403.

Mullen, E.G. and K.P. Ray (1993) "CRRES Microelectronics Test Package," *IEEE Trans. Nucl. Sci.*, 40: 228.

Peterson, E.L. (1981) "Soft Errors Due to Protons in the Radiation Belt," *IEEE Trans. Nucl. Sci.*, NS-28: 3981.

Ray, K.P., E.G. Mullen, and T.M. Trumble (1993) "Results from the High Efficiency Solar Panel Experiment Flown on CRRES," *IEEE Trans. Nucl. Sci.*, 40: 1505.

Reed, R.A., et al. (1994) "A Simple Algorithm for Predicting Proton SEU Rates in Space Compared to the Rates Measured in the CRRES Satellite," *IEEE Trans. Nucl. Sci.*, 41: 2389.

Stassinopoulos, E.G. and J.P. Raymond (1988) "The Space Radiation Environment for Electronics," *Proc. IEEE*, 76: 1423.

Weatherford, T.R., et al. (1993) "SEU Rate Prediction and Measurement of GaAs SRAMs Onboard the CRRES Satellite," *IEEE Trans. Nucl. Sci.*, 40: 1463.

NON-TOXIC PROPULSION FOR SPACEPLANE "POP-UP" UPPER STAGES

James B. Eckmann, Robert L. Wiswell and Eugene G. Haberman
Sparta, Inc.
Edwards AFB, California 93524
(805) 275-5444, 5-5724

Abstract

Military spaceplane operations scenarios envision using the "Pop-Up" employment profile to significantly increase the payload to orbit capability of the vehicle. Previous studies have investigated a range of propulsion system and stage design options for a pop-up upper stage (Cotta 1996). Operationally it is desirable to have the upper stage and payload stored as a wooden round that is quickly loaded on the spaceplane when needed. The current study therefore focuses on non-toxic (less-toxic), storable propellant options. These are compared to the use of conventional (toxic) storable bi-propellant, Nitrogen Tetroxide / Monomethyl Hydrazine (N_2O_4/MMH), and cryogenic oxidizer bi-propellant, (LO_2/RP1), options. The non-toxic oxidizers investigated include Hydrogen Peroxide (H_2O_2) and Hydroxyammonium Nitroformate (HANF). The non-toxic fuels include hydrocarbon jet fuel (JP-4), Quadricyclane (C_7H_8), and Methylcubane (C_9H_{10}). The impact of H_2O_2 purity (90% to 100%) and various fuel blends are also evaluated. The comparison includes payload delivery performance, propellant handling issues and technology development needs. The results show that there are propellant combinations that are less toxic than N_2O_4/MMH and yet deliver comparable payload delivery performance. However, there are propellant handling issues and technology development needs that must be addressed. These are discussed.

INTRODUCTION

As currently envisioned, Military spaceplane (MSP) is a rocket propelled reusable launch vehicle (RLV) designed to perform military missions. To accomplish these various missions three employment profiles are possible; orbital, once around, and pop-up. This paper focuses on the pop-up employment profile for the spacelaunch mission. Pop-up profiles are suborbital profiles that use an upper stage to push a payload into orbit. The MSP is capable of placing a small payload (2200 - 4500 kg, 185 km due east) directly into Low Earth Orbit (LEO). However, the use of the pop-up employment profile significantly increases the payload to orbit capability of the spaceplane (≈8500 - 11,300 kg). Several propulsion system and stage design options for a pop-up upper stage to accomplish this mission have been considered (Cotta 1996). This prior work showed that high chamber pressure, pump-fed, bi-propellant propulsion systems gave the best performance of the options considered. It also showed that a stacked toroidal tank stage configuration provided the best performance, followed by a four-cylindrical tank stage configuration.

MSP operations concepts envision the upper stage and payload being stored as a wooden round that is quickly loaded on the spaceplane when needed. A conventional storable bi-propellant Nitrogen Tetroxide / Monomethyl Hydrazine (N_2O_4/MMH) upper stage would be able to meet this mission requirement. However, these propellants are highly toxic and increasingly difficult to manufacture, transport, and handle under todays increased environmental and personnel safety regulations. The current study therefore investigates non-toxic (less-toxic), storable propellant options. These are compared to the use of N_2O_4/MMH and to cryogenic oxidizer (LO_2) propellant options. The comparison includes payload delivery performance, propellant handling issues (toxicity, storability, availability, manufacturability, transportability, safety, etc.), and technology development needs. The non-toxic oxidizers investigated include Hydrogen Peroxide (H_2O_2) and Hydroxyammonium Nitroformate (HANF). The non-toxic fuels include hydrocarbon jet fuel (JP-4), Quadricyclane (C_7H_8), and Methylcubane (C_9H_{10}). The impact of H_2O_2 purity (90% to 100%) and various fuel blends are also evaluated.

PROPELLANT PROPERTIES

The initial screening of candidate propellants considered their present state-of-the-art, availability, and knowledge of their physical properties. The propellants selected are expected to be ready for integrated system demonstration at time periods from 1999 to 2010. Desirable propellant properties include: 1) high energy release per unit mass of propellant, 2) high density or high density impulse to minimize the size and weight of propellant tanks and feed systems, 3) ease of ignition, 4) low vapor pressure for low tank weight and low net positive pump suction head, 5) low freezing point, 6) good storability with high boiling point and resistance to deterioration and/or decomposition,

7) low toxicity of raw propellants, their vapors, and their combustion products, 8) low cost, and 9) available in sufficient quantities. Some of the physical properties of the baseline propellants and the selected less-toxic propellants for this study are given in Table 1. The vapor pressures are stated for 25°C except for liquid oxigen which is at -183°C.

TABLE 1. Propellant Properties.

	Formula	Heat of Formation (kcal/mol)	Specific Gravity (g/cc)	Freezing Point (°C)	Boiling Point (°C)	Vapor Pressure (mmHg)	Threshold Limit Value (ppm)
Oxidizers							
Liquid Oxygen (LOX)	O2	-3.102	1.149	-219	-183	760	NONE
Nitrogen Tetroxide (NTO)	N2O4	-4.680	1.431	-12.2	21.2	925	
Hydrogen Peroxide (H2O2)	H2O2	-44.901	1.444	-1	150	2.1	1
Hydroxyammonium Niroformate (HANF)	CH4N4O7	-45.900	1.500	~15	~150	~0.5 to 5	UNK
Fuels							
Monomethyl Hydrazine (MMH)	CH3N2H3	12.950	0.874	-52.5	87.5	49.6	0.01
Rocket Propellant 1 (RP-1)	CH1.9532	-5.760	0.800	-4.8	217	46	200
Jet Fuel 4 (JP-4)	CH1.954	-5.670	0.770	-60	274	103.4	
Quadricyclane (Quad)	C7H8	72.200	0.985	-44	108		
Methylcubane (MeCu)	C9H10	142.000	1.200	-34			UNK

PERFORMANCE ANALYSIS

The Automated Stage Sizing Program (ASSP) was developed to investigate the upper stage technologies that could be used in a pop-up employment profile. As modeled in ASSP, the MSP flies a ballistic trajectory with its apogee at 185 km. At some velocity shortfall from that required for a circular orbit, the upper stage and payload are "popped out" of the MSP payload bay. The upper stage engine, after a minimum two minute separation time, fires to insert the upper stage and payload into a circular parking orbit. If this parking orbit is not the final orbit destination, subsequent engine firings would be used to provide the velocity required to change orbits.

For the current study, ASSP was used to calculate payload performance of pop-up upper stages for two different space launch missions. One is the GEO mission (35,786 km circular at 0 deg inclination) from 1219 m/s velocity shortfall, as defined in the "Upper Stage Options for Reusable Launch Vehicles" study (Cotta 1996). The other is a direct LEO insertion (185 km circular at 28.5 deg inclination) from 1524 m/s velocity shortfall. For the GEO mission, the upper stage is required to provide 3702 m/s of ΔV with a 0.80 initial stage (including payload) thrust-to-weight ratio. For the LEO mission, the upper stage must provide 1524 m/s ΔV with a 0.98 initial stage thrust-to-weight ratio.

The MSP design imposes volume and total weight constraints that also must be considered. The space plane design used in this study was provided by the Aerospace Corp. during the original development of the ASSP. This vehicle has a 433,635 kg gross mass and is capable of placing a 2722 kg payload into a 185 km circular, 28.5 deg inclination orbit when launched from Cape Canaveral. The payload bay is 3.66 m wide by 7.62 m long and has a maximum mass limit of 18,144 kg. Both the payload and the upper stage, and the associated airborne support equipment (ASE), must be packaged within these constraints.

The payload length for the GEO mission was scaled as a function of the payload mass. The reference payload is a 1361 kg, 3.20 m dia by 3.35 m, HS-601 satellite. The scaling relationship was based on historical satellite data. For the LEO mission the payload length was fixed at 3.35 m. A 3.35 m long payload, after accounting for a 30.48 cm separation clearance on both ends of the payload bay, the 45.72 cm long upper stage to payload adapter ring, and 22.86 cm on either side for the ASE and attach points, leaves a 3.20 m dia by 3.20 m volume for packaging the upper stage.

Two stage configurations were considered: 1) a 4-cylindrical tank configuration, and 2) a stacked toroidal tank configuration. The propellant tanks were sized for the required propellant load assuming 3% ullage, a 3.45 bar tank pressure, and aluminum construction.

A 103.4 bar chamber pressure, pump-fed engine, based on the XLR-132, was assumed for this study. A one-dimensional equilibrium (ODE) code, the ISP Code, was used to predict theoretical engine performance for the various propellant combinations. A curve fit of vacuum Isp as a function of nozzle expansion ratio was then calculated for each propellant combination. These curve fits, and other required propellant data, were added to ASSP. ASSP calculates delivered engine performance by applying a C* efficiency correction to the theoretical engine performance curve fits.

ASSP uses the Excel built-in Solver to maximize the payload mass to orbit, subject to the vehicle and mission constraints. For this study ASSP also optimized the nozzle expansion ratio. The optimum expansion ratio for the GEO mission was the minimum allowed, 30. For the LEO mission the optimum varied from 30 to 110 depending on the propellants and stage configuration. The ASSP results are shown in Figure 1 for the GEO mission and in Figure 2 for the LEO mission.

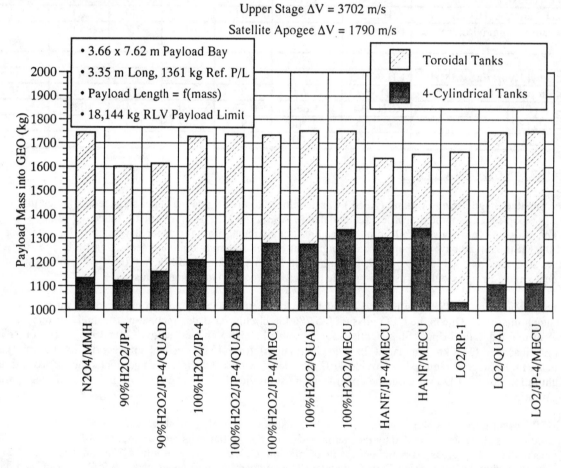

FIGURE 1. Non-Toxic Propellant Effect on Pop-Up Upper Stage Payload Mass to GEO.

Using a 4-cylindrical tanks configuration stage design, the results show a very small (less than 1%) performance decrease by going from the baseline toxic storable N_2O_4/MMH to the near-term availability, less toxic storable 90% H_2O_2/JP-4. A significant performance increase is obtained by using Quadricyclane or Methylcubane fuel additives and/or increasing the H_2O_2 oxidizer purity. 100% H_2O_2/Methylcubane shows an 8% increase in performance over the baseline propellant for the LEO mission, and an 18% increase for the GEO mission. The Hydroxyammonium Nitroformate (HANF) oxidizer increases performance even further for the GEO mission. However, For the LEO mission the HANF oxidizer does not give as high performance as the pure H_2O_2. This is a result of the 18,144 kg payload mass limit. None of the GEO mission cases encounter this limit. All of the pure H_2O_2 and HANF oxidizer cases for the LEO mission do encounter this limit. The HANF oxidizer cases have higher density and operate at a higher O/F ratio than the H_2O_2 cases, and are therefore more severely impacted by the mass limit. The cryogenic LO_2 oxidizer cases do not appear competitive in the 4-cylindrical tanks configuration.

However, using a stacked toroidal tanks configuration stage design changes the results significantly. Not only do the toroidal tanks increase the stage performance, they also change the relative ranking of the propellants considered. In this case, the cryogenic LO_2 oxidizer shows the highest performance for the LEO mission and, with Quadricyclane or Methylcubane fuel, is competitive with H_2O_2 for the GEO mission. The 90% H_2O_2/JP-4 case shows a noticeable (4% to 8%) performance loss compared to the N_2O_4/MMH, and the 100% H_2O_2 / hydrocarbon fuel + additive cases simply buy back that performance loss. The HANF cases show better performance than 90% H_2O_2, but less performance than pure H_2O_2. The 18,144 kg payload mass limit affects the performance potential of all of the toroidal tank cases for the LEO mission, but none of the GEO mission cases.

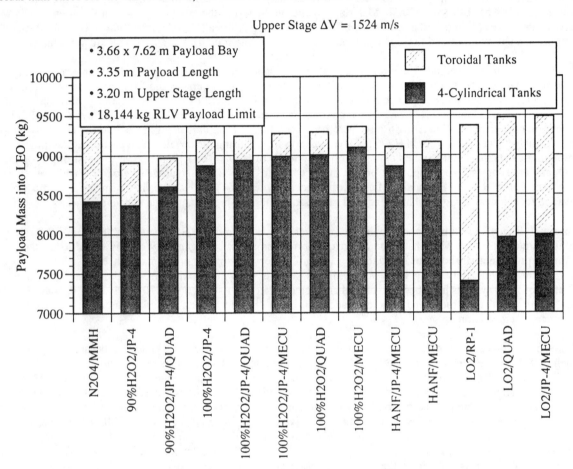

FIGURE 2. Non-Toxic Propellant Effect on Pop-Up Upper Stage Payload Mass to LEO.

These results show that there are potential propellant combinations that are less toxic than N_2O_4/MMH and yet deliver comparable system performance. In fact, for the 4-cylindrical tanks configuration, the higher purity H_2O_2 and advanced fuels increase the payload performance.

PROPELLANT HANDLING CHARACTERIZATION

The expense and time involved in using the current generation of toxic storable propellants includes the direct costs of handling and loading propellants, and storing the loaded stage or payload, as well as the indirect costs of the exclusion of personnel from doing other work during loading and possibly storage operations. An attempt was made to define and quantify these cost and time factors. Operations people from McDonnell Douglas, Lockheed Martin, and the Aerospace Corporation were contacted. None of them was able to define the additional costs and time involved in the use of toxic propellants. They all agreed that there were costs and time loss involved but none could really define or quantify the problem in any detail.

Although quantitative data was not obtained, some cost and time factors were identified. These included the need for double crews suited up for toxic propellant loading and transfer operations, and the time and costs involved in disposing of the contaminated propellants and other contaminated items. In some instances special areas need to be built and set aside for storing and cleaning items contaminated by the toxic propellants or for conducting toxic operations remotely. There are examples of delays in the Shuttle operations where the Shuttle has to be carefully checked for contamination and leaks when it lands, resulting in several hours where the crew must remain on board and people involved in the loading operations are on hold while the checks are done. The NASA Space Propulsion Synergy Team report titled "A Guide for the Design of Highly Reusable Space Transportation" looks at the design features of a reusable space transportation system and prioritizes characteristics (both desirable (+) and undesirable (-)). The top priority item (Top of Top 20 Design Features List) is # of Toxic Fluids (-). Also appearing third on the Middle 22 List is # of Pollutive or Toxic Materials (-) and on the top of the Bottom 22 Design Features List is # of Hazardous Processes (-), and further down is # of Keepout Zones. All in all people agree that it is more difficult to accomplish the short turnaround times that are required for a reusable military system using toxic propellants.

The handling of the propellants recommended by this study for further technology development was investigated. There is little data on the HAN and HANF propellants. Data on Methylcubane and Quadricyclane is also limited. There is a lot of data on 90% H_2O_2 since it was used on the first US Satellites for ACS, for the Scout vehicle ACS, for the turbine drive on the X-1 and X-15 engines, and as the oxidizer for the Rocketdyne AR-2 aircraft superperformance engines. The British also have had a number of programs using 85% H_2O_2. The only British Launch Vehicle "Black Arrow" used H_2O_2 and Hydrocarbon fuel. During the development portion of the program the Royal Aircraft Establishment handled almost 1 million pounds of H_2O_2. The British appear to be strong proponents of H_2O_2 and are sponsoring a symposium on the subject in mid 1998. The Hydrogen Peroxide Handbook prepared for the Air Force Rocket Propulsion Laboratory (Rocketdyne 1967) includes data on material compatibility, passivation techniques, storage and handling, and safety, for a range of propellant grade concentrations.

In informal discussions with several people who worked with H_2O_2 in the U.S. "years ago" there is a notable lack of enthusiasm for its use. Most have some sort of "horror" stories which generally can be traced to contaminated hardware or propellant. It should be noted that "years ago" there were similar problems with N_2H_4 which is now in wide use. The key to safely using H_2O_2 is cleanliness, as it is with most oxidizers. Techniques for cleaning and passivation of H_2O_2 systems have been developed and demonstrated. In strengths of 70%, H_2O_2 is regularly handled in this country in tanker trucks and tank cars and is used in the electronics and paper industries as well as for environmental remediation.

H_2O_2 is often compared with Hydrazine in terms of handling and toxicity. This may be because they are both monopropellants and Hydrazine displaced H_2O_2 on spacecraft systems. Hydrazine is over 10 times as toxic as H_2O_2 in terms of allowable concentrations in air. Both are very corrosive to skin and lungs and other tissues. Hydrazine, however, is a known carcinogen and its manufacturing process includes hazardous processes and intermediates. H_2O_2 is not considered a carcinogen and can be handled differently. Disposal of contaminated material and contaminated propellant is much more complex and costly with Hydrazine. With H_2O_2 simple dilution is an effective decontaminant and decomposition products are water and oxygen. When comparing H_2O_2 with the storable oxidizer that it would replace, N_2O_4, there is an even greater difference in allowable concentrations in air. The short term exposure limit for N_2O_4 is 100 times that of H_2O_2. N_2O_4 decomposes to toxic acids. A NASA/JSC report (Cort 1995) considers H_2O_2 as corrosive and an irritant but non-toxic and recommends it for further consideration as a substitute for N_2O_4 and MMH. In fact overall the toxicity of H_2O_2 appears to be very similar to hydrocarbon fuels. The net result of all of this is that H_2O_2 can be handled using impervious clothing to protect eyes and skin from contact with H_2O_2, but no breathing apparatus is required. It is our understanding that 70% H_2O_2 is transferred in truck and tank car lots without the use of breathing apparatus. TRW is handing significant quantities for the Airborne Laser Program in a similar manner.

Other significant issues impacting the use of H_2O_2 include long term storability and availability. A lot of work was done on H_2O_2 storability in the 1960's with the result that decomposition rates of .02 to .1% per year are possible. This level of decomposition is acceptable for a pop-up upper stage. Availability of H_2O_2 in concentrations of 90% and over is very limited. There are no current "off the shelf" commercial sources. 70% H_2O_2 is readily obtainable in the United States. 85% is available in Europe (and can be shipped to the US). Several of the companies manufacturing the 85% product in Europe have US plants that are making the 70% product and would be interested in making a higher concentration product if there was a market for it. One or two may be interested in making specialty chemical size lots, but this hasn't been done yet. H_2O_2 Inc., a small US company which made 90% H_2O_2, destroyed their plant and will not go back into production until they complete their development of a new method for concentrating the product.

TECHNOLOGY DEVELOPMENT NEEDS

While hydrogen peroxide and several of the fuels are known entities and have been used in the past, additional technology development is needed to provide the necessary design data required for a user to use these propellants in a flight system or experiment. This is especially the situation with the new advanced oxidizers and fuels. Technology questions to be addressed include: scalability and performance at higher thrust levels, long term storability and decomposition rate, ignition and hypergolicity, toxicity, availability and producability, and cost of manufacture and operations. A recommended technology program plan was developed during the study for the purpose of guiding future investments to provide timely transition to flight type experimental demonstrations.

The overall philosophy and approach to the plan is to investigate existing and new non-toxic, high energy fuels with hydrogen peroxide before investigating the new fuels with new oxidizers. The existing fuel of considerable interest is Jet Fuel (JP); JP-8 and JP-10. These fuels are used by the Air Force and Navy, respectively, and are readily available. The crews are trained in handling these fuels. Examples of advanced fuels of interest, include but are not limited to, quadricyclane, methylcubane, bicyclopropylidene, and 1-5 hexadiyne.

The plan uses a building block approach to integrate the new fuels with the existing hydrocarbon JP fuel for the initial tests with hydrogen peroxide. Performance testing will move from using 100% JP, mixtures of JP and advanced fuels, to 100% concentrations of the new advanced fuels. The test matrix was developed to include thrust sizes of 444, 4,448 and 44,482 N thrust to cover the technology requirements for several different potential applications, one of which is the upper stage for MSP. The typical parameters of injector performance, thruster cooling, stable combustion, and reliable ignition will all be investigated at the various application sizes.

New oxidizers will be developed as monopropellants and then will be incorporated into the bipropellant effort. The advanced monopropellant effort will initially investigate the general area of using Nitrate salts. Examples of the propellants of interest are Nitrate/Alkyne mixtures and Nitroformate based propellants. These advanced oxidizers will be integrated with the new advanced fuels and tested using the same test matrix described above.

The plan includes a propellant storability effort. The propellants will be put into containers constructed of various materials to investigate their ability to remain in storage 5-10 years without decomposition or excessive corrosion.

SUMMARY & CONCLUSIONS

The payload delivery performance of propellants that are significantly less toxic than today's storable propellants has been quantified for a MSP pop-up upper stage mission. The near term propellant candidates (H_2O_2/JP) show a slight performance degredation, but the more advanced propellants show comparable and even improved performance. Propellant handling, storability, and availablity issues for these propellants have been discussed and a technology development plan has been proposed. It appears that the technologies may be in hand to overcome the handling and storability issues associated with H_2O_2 that in the past have led to today's toxic storables being used instead. However, these technologies need to be demonstrated before being implemented in flight systems. The cost and operability characteristics of the advanced fuels and Nitrate salts based oxidizers are not well quantified at this time. However, they hold the potential for improved performance, which warrents their continued development.

Acknowledgments

This work was performed for the USAF, Air Force Research Laboratory, Propulsion Directorate, at Edwards AFB, CA.

References

Cotta R., J. Eckmann, and L. Matuszak (1996) "Upper Stage Options For Reusable Launch Vehicles," AIAA 96-3015, American Institute of Aeronautics and Astronautics, Washington DC.

Cort R., E. Hurlbert, J. Riccio, and J. Sanders (1995) "Non-Toxic On-Orbit Propulsion for Advanced Space Vehicle Applications," AIAA 95-2974, American Institute of Aeronautics and Astronautics, Washington DC.

Rocketdyne Division of North American Aviation (1967) "Hydrogen Peroxide Handbook," Technical Report AFRPL-TR-67-144, Air Force Rocket Propulsion Laboratory, Edwards AFB, CA.

STATUS AND DESIGN CONCEPTS FOR THE HYDROGEN ON-ORBIT STORAGE AND SUPPLY EXPERIMENT

David J. Chato
NASA Lewis Research Center
Mail Stop 60-4
21000 Brookpark Road
Cleveland, OH, 44135
(216)977-7488

Melissa Van Dyke
NASA Marshall Space Flight Center
Mail Code EP25
Marshall Space Flight Center, AL, 35812
(205)544-5720

J. Clair Batty and Scott Schick
Space Dynamics Laboratory, Utah State University
1695 North Research Park Way
UMC 9700
North Logan, UT 84341
(435)797-2866 and 797-4426

Abstract

This paper studies concepts for the Hydrogen On-Orbit Storage And Supply Experiment (HOSS). HOSS is a space flight experiment whose objectives are: Show stable gas supply for storage and direct gain solar-thermal thruster designs; and evaluate and compare low-g performance of active and passive pressure control via a thermodynamic vent system (TVS) suitable for solar-thermal upper stages. This paper shows that the necessary experimental equipment for HOSS can be accommodated in a small hydrogen dewar of 36 to 80 liter. Thermal designs for these dewars which meet the on-orbit storage requirements can be achieved. Furthermore ground hold insulation and shielding concepts are achieved which enable storing initially subcooled liquid hydrogen in these small dewars without venting in excess of 144 hours.

INTRODUCTION

Recently, there has been significant interest in Solar-Thermal rocketry (using concentrated sunlight instead of combustion to heat and expand gases for rocket propulsion). There are on-going programs for Solar Thermal rockets in industry (Cady 1996a), Department of Defense (DoD)(Cady 1996b and Jacox 1996), and the National Aeronautics and Space Administration (NASA). Hydrogen is the propellant of choice for solar-thermal rockets due to its high specific impulse. Hydrogen is stored most efficiently as a cryogenic liquid.

NASA has been studying technologies for the management of cryogenic fluids in low gravity for many years. Looking at the cryogenic fluid management technologies for solar thermal stages, the key issues are controlled propellant acquisition and long term storage (Chato 1997). Recent solar-thermal design concepts (Cady 1996a, Cady 1996b, Jacox 1996) have combined these functions into a single system which uses a thermodynamic vent system to remove energy from the liquid storage and provide propellant to the solar thermal collector. Although TVS systems have been studied extensively, they have not been proven in space and it is felt that the change in fluid configuration (see Bentz 1993 for a discussion of zero-g mixing flow patterns) could have a significant effect on their performance. Ground testing with feed system components show that for this coupled system start transients may affect predicted performance. The solar-thermal rocket is sensitive to transient issues because firing occurs in many short bursts and because the vent and feed systems are coupled. Space testing is necessary because of TVS sensitivity to low gravity and the difficulty in simulating the hard vacuum of space while outflowing hydrogen.

EXPERIMENT CONCEPT

The authors have undertaken the design of a small scale experimental spacecraft to investigate these issues. This spacecraft has been named Hydrogen On-Orbit Storage And Supply Experiment (HOSS). Details of the design concept are given in Chato (1997). Key objectives and design features are summarized below.

Objectives

The objectives of this experiment will be:

- Show stable gas supply (steady flow rate, minimal liquid, over an extended period) for storage and direct gain solar-thermal thruster designs.

- Evaluate and compare low-g performance of active and passive pressure control via thermodynamic vent system (TVS) suitable for solar-thermal upper stages.

Approach

The experiment design concept will:

- Launch a small (36-80 liter) hydrogen dewar with liquid acquisition device (LAD), mixer and TVS.

- Operate dewar through several regimes (including short burst transient operation) of active and passive pressure control while monitoring hydrogen flow rates, quality and quantities.

- Scale outflow to match solar-thermal flow rate requirements.

- Verify that liquid free gas is supplied to the overboard vent (the vent simulates the solar-thermal thruster)

DESIGN STUDIES

Chato (1997) established the baseline requirements for HOSS. It became evident, from these requirements that the key element of the spacecraft is the liquid hydrogen dewar. Its weight dominates the spacecraft weight. Its thermal performance determines ground hold capability and the complexity of ground operations. To start the design effort, hardware similar to existing pieces was baselined, including the radiometer dewar built for the Shuttle Pallet Satellite III (SPAS) experiment (unpublished), the LAD from Bentz (1993), and the heat exchanger-mixer assembly from Seigneur (1994). It was clear from our previous work that a 36 liter dewar similar to SPAS would not be capable of storing hydrogen for 30 days (the typical solar-thermal mission) so a larger 80 liter design study was also conducted. The effort was concentrated on two key areas. The first area is mounting existing NASA hardware in the 36 and 80 liter vessel, and defining support structures, valving, plumbing, and mass. The second area considered thermal performance of the dewar system and what ground handling constraints this imposed.

Dewar Design

The first step was to look at a 80 liter dewar design. This design, as shown in figure 1, has an inner vessel 40.6 cm diameter by 72.2 cm long and overall dewar size of 54.9 cm diameter by 94.8 cm long. This design will hold 5.7 kg of hydrogen. The vacuum jacket is made of aluminum. The hydrogen tank, liquid nitrogen guard tank, valves, plumbing and fittings are made of stainless steel. There is a removable lid on the inner vessel to allow access to the mixer, LAD, and heat exchanger. The lid is sealed with a bolted metal seal. Another bolt flange on the vacuum jacket allows access to this lid as well as the internal plumbing. Valves for the panel layout were chosen because of prior use on other programs. Valves include tank fill and vent, liquid nitrogen fill and vent, helium fill and vent, and a TVS valve assembly (described below) as well as an evacuation port for the vacuum jacket. The TVS assembly includes a pyro valve to seal the TVS during ground operation, a solenoid valve to control the TVS operation, a relief valve to protect against dewar overpressure on-orbit, and a tee to vent the TVS flow without thrusting. The NASA heat exchanger has been modified to allow a mixer motor to sit within it thereby greatly reducing the overall length. The design also substitutes an available shorter NASA mixer for the mixer used by Seigner(1994). The dry weight of the 80 liter dewar is estimated at 91.6 kg

The hydrogen tank is insulated using evacuated multi-layer insulation (MLI) and supported using three nested G10 fiberglass tubes. A toroidal tank is attached at the intersection of the outer two tubes. This tank will be filled with liquid nitrogen on the ground to limit heat entering the hydrogen. This technique allows the dewar to remain filled with liquid hydrogen for long periods of time without venting. The liquid nitrogen will be vented prior to launch. A coil of tubing is attached to the inner vessel to allow the liquid hydrogen to be subcooled by flowing liquid helium around the hydrogen

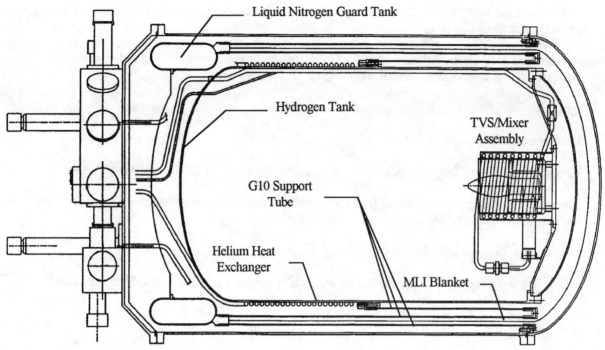

a) Side View

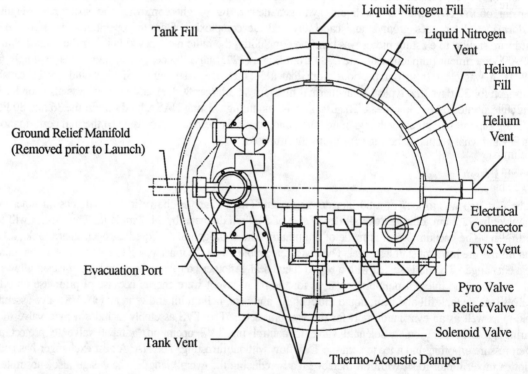

b) Top View

FIGURE 1. 80 Liter Dewar Layout.

tank. This again extends the ground hold capabilities of the tank and provides a means of reducing tank pressure quickly without venting.

A revised 36 liter dewar design, incorporating the design features of the 80 liter dewar was also completed. It was felt that a smaller dewar size and mass might decrease launch costs and this might be more important than storage time. This dewar holds 2.52 kg of hydrogen and has a dry mass of 58.6kg (only 35% less than the 80 liter dewar). Overall size is 38.4 cm diameter 71.6 cm long. One reason that the masses are close is that the external plumbing is identical. The smaller space available on the 36 liter dewar necessitates the use of a stepped vacuum shell lid to achieve the required mounting space.

Results of the design study indicate both dewars are feasible. There is a preference for the 80 liter dewar since the greater volume will make achievement of the experimental objectives easier, and the weight difference between the two designs is not that great. Both designs will be carried forward until spacecraft designs and launch vehicle selection is complete. Volume constraints may yet dictate use of the smaller dewar.

Insulation Designs

Heat load estimates were by computer (SINDA) using a simple lumped element thermal model for each dewar. Heat loads to the hydrogen tank are shown in table 1.

Table 1 Heat Loads on Hydrogen Tank (Watts)

	80 Liter Dewar	36 Liter Dewar
Supports	0.458	0.443
MLI	0.752	0.539
Plumbing	0.299	0.296
Wires	0.0007	0.0007
Pressure Sensor Lines	0.144	0.144
Total	1.66	1.42

The nominal heat load on the 80 liter dewar with this design is 1.66 W (at a 300K shell temperature) resulting in an on-orbit storage time of 18 days. If the shell temperature can be reduced to 250K the heat load is reduced to 0.985 w and the on-orbit storage time will the meet 30 day storage objectives. A 250 K shell temperature could be achieved by the common techniques of using the spacecraft solar arrays to shadow the dewar or coating the tank with specially reflective surfaces (such as silver-Teflon film). Nominal heat load on the 36 liter dewar is 1.42 W resulting in a 9 day orbit life, although this can be increased to 15 days by lowering the shell temperature as described previously. Another source of heat that can be eliminated is the 0.144 watts due to pressure sensor lines. The baseline design specified pressure sensors external to the vacuum space with tubes leading from the inner vessel. These may be replaced by temperature compensated sensors which can be mounted on the inner vessel wall, therefore eliminating the need for long tubes and their associated heat leak.

Most launch vehicles are capable of being launched again after 48 hours so hold times without servicing greater than this are desirable. Ground hold for these insulations without the liquid nitrogen guard was less than desired. For the 36 liter tank the time to rise from the loading temperature of 20K to 25K (saturation pressure of 2,466 torr) is 21 hours. For the 80 liter dewar this time is 42 hours.

The first approach to improving ground hold was the addition of a liquid helium subcooling coil. This coil serves several purposes. Prior to loading hydrogen this coil will be used to cool the tank down to 20K thus eliminating the hydrogen needed for tank chilldown. Periodically after fill liquid helium will be circulated to reduce the liquid hydrogen temperature to 14K pulling the tank pressure down correspondingly to 52 torr. From 14K the hold time to 25K increased to 48 hours for the 36 l dewar and 96 hours for the 80 liter dewar.

With the previously described liquid nitrogen guard tank added to the system hold times are substantially increased. The toriodal tank is filled with enough liquid nitrogen to hold for 48 hours. During this time period heat leak into the tank is absorbed by boiling nitrogen and the net heat leak into the liquid hydrogen is greatly reduced. Assuming the guard tank is allowed to boil dry the hold time for the 36 l dewar is over 144 hours and over 168 hours for the 80 liter dewar. If the guard tank is periodically refilled the hold time is extended greatly.

Tank temperature rise rates with all systems in place are shown in figures 2 and 3. The curves show a shallow rise in temperature until the liquid nitrogen tank is depleted at 48 hours followed by a steeper increase afterwards. Also shown are the liquid nitrogen tank temperature which is constant until the liquid nitrogen is depleted then rises rapidly, and the temperature of the joint of the folded tubes closest to the tank which is constant when the liquid nitrogen is present but afterwards slowly rises in temperature.

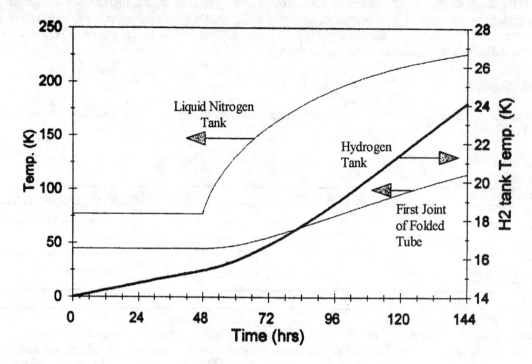

FIGURE 2. 36 Liter HOSS Ground Hold with LN_2 Tank

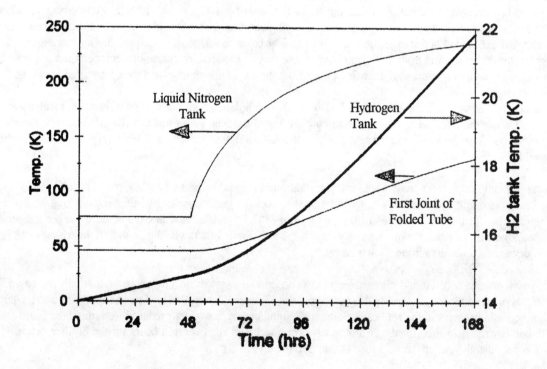

FIGURE 3. 80 Liter HOSS Ground Hold with LN_2 Tank.

Other Design Studies

Launch vehicle and spacecraft bus selection studies are on-going. Results of these efforts will be reported in future. Other design trades were conducted with regard to ground handling and spacecraft mission. Since these trades are closely tied to launch vehicle selection and spacecraft bus design, results are very preliminary. Key results of the ground handling trade indicate that Government Owned equipment from the Wide Field Infrared Explorer (WIRE) satellite (unpublished) may be very useful. Steps are being taken to borrow this equipment for HOSS. Vandenburg Air Force was tentatively selected as the baseline launch site since facilities for handling small hydrogen payloads exist from the Midcourse Space Experiment (MSX; Smola 1996) and WIRE (to be launched shortly on a Pegasus rocket) programs. Spacecraft mission designs confirmed the viability of the sun-synchronous orbit, and one-a-day data download approaches suggested in Chato 1997, but indicate that other approaches may offer benefits also. Discussions with small satellite vendors indicated that HOSS is heavier than most small spacecraft, but also indicated a scale up of their existing design was very feasible.

CONCLUSIONS

This design effort shows that the necessary experimental equipment for HOSS can be accommodated in a small 36 to 80 liter hydrogen dewar and meet all the experiment objectives. Thermal designs for these dewars which meet the on-orbit storage requirements can be achieved. Furthermore, the study has shown that there are insulation and guard techniques for ground processing capable of storing liquid hydrogen in these small dewars without venting in excess of 144 hours. Other areas can now be detailed. Design studies underway include selection of suitable low cost launch vehicles and integration of the dewar design into a satellite bus. These efforts bring the flight of the HOSS experiment inexorably closer.

Acknowledgments

The authors wish to thank: The Utah State Team for their hard work and effort. NASA Marshall's' Preliminary Design Office for their spacecraft study. Orbital Sciences Corporation and Spectrum Astro Incorporated for their patience and assistance.

References

Bentz, M. D., R. H. Knoll, M. M. Hasan, and C. S. Lin (1993) "Low-G Fluid Mixing: Further Results from the Tank Pressure Control Experiment," American Institute of Aeronautics and Astronautics, Reston, VA, AIAA Paper 93-2423, June 1993.

Cady, E. and A. Olsen Jr. (1996a) "Solar Thermal Upper Stage Technology Demonstrator Program," American Institute of Aeronautics and Astronautics, Reston, VA, AIAA Paper 96-3011, July 1996.

Cady, E. and A. Olsen Jr. (1996b) "Cryogen Storage and Propellant Feed System for the Integrated Solar Upper Stage (ISUS) Program," American Institute of Aeronautics and Astronautics, Reston, VA, AIAA Paper 96-3044, July 1996.

Chato, D. J. and , L. Hastings (1997) "An Experiment to Demonstrate Key Cryogenic Technologies for Solar Thermal Rockets," *Space Technology and Applications International Forum (STAIF 97)*, American Institute of Physics, CONF 970115, M. S. El-Genk, ed, AIP Conf. Proc. No. 387, 1:517-522.

Jacox, M. G., F. G. Kennedy, J. Malloy, C. Merk ,and T. M. Miller (1996) "Integrated Solar Upper Stage (ISUS) Space Demonstration System Definition Study" Phillips Laboratory, Kirtland AFB, NM January, 1996. PL-TR-96-1006.

Seigneur, A. (1994) "Design, Analysis, Fabrication and Testing of an Active Heat Exchanger for Use in Cryogenic Fluids," Masters Thesis, Cleveland State University, March 1994.

Smola, James F. et al (1996) "MSX Ground Operations," *John Hopkins APL Technical Digest*, Vol 17, No 2, April 1996.

NAVAL RESEARCH LABORATORY SOLAR CONCENTRATOR PROGRAM

Ilene Sokolsky
AlliedSignal Technical Services Corp.
Naval Research Laboratory, Code 8220
4555 Overlook Ave., SW
Washington, DC 20375
202-767-1154

Michael A. Brown
Naval Research Laboratory, Code 8220
4555 Overlook Ave., SW
Washington, DC 20375
202-767-2851

Abstract

NRL has developed a trough-configuration solar concentrator with ultra-light thin film reflectors that are stretched flat using an edge-tensioning support system. With a geometric concentration ratio of 2.5 to 1, the specific power of the photovoltaic array is approximately doubled by the NRL system. A 300 watt experimental demonstration model of the array/reflector system has been successfully built and deployed, and predicted power levels have been reached in solar simulator testing. Work currently underway includes design, construction, and deployment of a 4 kilowatt array, optimization of reflector border shape and reflector attachment method, testing and modeling of thermal cycling and temperature effects, and assessment of space degradation issues.

INTRODUCTION

NRL has developed a solar concentrator as part of an ongoing program in spacecraft bus technology improvement. This program has three main goals for the concentrator. First, specific power increases of 100% are sought for any given solar cell/coverglass/substrate combination, thereby increasing payload mass fraction and providing for increased payload power requirements. Second, cost savings are sought by reducing the number of solar cells required. Solar arrays can account for as much as 20% of the total spacecraft cost, and the NRL concentrator design would reduce the number of solar cells by 55%, thereby reducing the total solar array cost by about 50%. Finally, NRL's design was required to have a photovoltaic panel exposed during orbit transfer to provide power during this phase of the mission without a secondary power source, and in fact it is now the only solar concentrator system which provides power in the stowed position.

The solar concentrator array is a trough-configuration design, so-called because the reflectors and photovoltaic panel system form a trough, as shown in Figure 1. The two flat reflectors focus sunlight onto the photovoltaic panel, increasing the amount of sunlight on the panel from one sun to as much as 2.5 suns, which is the practical limit for the trough-configuration system. The reflectors are made of thin film, such as 0.5 mil kapton® coated with aluminum or silver to provide the necessary reflective surface. Successfully using thin film is the key to this design, because the film adds very little mass, and it can be rolled up and stowed without blocking the photovoltaic panel, giving the system its unique ability to provide power during orbit transfer.

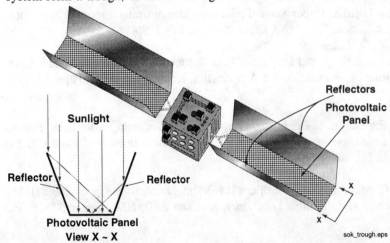

FIGURE 1. Trough-Configuration Solar Concentrators Deployed on a Spacecraft.

The following sections will cover the research and development activities to date, with emphasis on work that is currently underway and planned for the near future.

THEORY

The trough-concentrator concept requires that a light ray hitting the top of the concentrator reflects to the far edge of the solar panel, as shown in Figure 1. Using this constraint, equations (1) and (2) are derived to fully characterize the geometry of the concentrator (Rex 1978).

$$\frac{L}{S} = \left(\frac{A}{S} - 1\right)\sqrt{\frac{1}{2 - \left(\frac{A}{S} - 1\right)}}, \qquad (1)$$

and

$$\sin\theta = \frac{1}{2\sqrt{\frac{1}{2 - \left(\frac{A}{S} - 1\right)}}}. \qquad (2)$$

In these equations, L is the reflector height, S is the photovoltaic array width, A is the overall width of the array/reflector envelope, and θ is the reflector angle, as shown in Figure 2. The relationships of importance here are

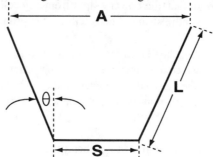

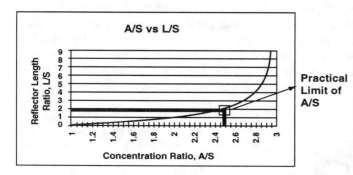
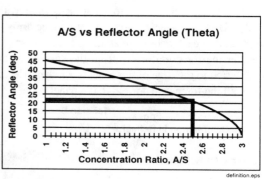

FIGURE 2. Geometric Relationships in Trough Configuration Concentrator.

between A/S, known as the geometric concentration ratio, which indicates the factor by which sunlight on the photovoltaic array is increased, and L/S, which is known as the length ratio, and between A/S and θ. Figure 2 shows these relationships graphically. These graphs illustrate the theory behind the trough-concentrator design, and they also illustrate its limitations. As the concentration ratio A/S increases beyond about 2.5, the length ratio L/S increases rapidly. As a result, the practical concentration ratio for this type of system is about 2.5. The NRL trough-concentrator is designed to provide the maximum concentration ratio of 2.5 to get the best possible performance available from this technology.

KEY ACCOMPLISHMENTS

Much of NRL's design, experimental, and analytical work has been concentrated in the areas of (1) developing a reflector tensioning system that will keep the reflector flat, (2) developing an operational deployment system, (3) testing various parts of system, such as panel reflectance, actual power output, and response to thermal cycling, and (4) analyzing all aspects of the system. This section will focus on NRL's efforts to develop a reflector tensioning system that will keep the reflector flat, and will touch briefly on the other areas.

Flat Reflector

The trough-concentrator concept has been around for many years and was used in a terrestrial application as early as 1911 (Meinel 1976). In terrestrial applications, the mass of the reflector is not critical, while in space applications, low mass is a necessity, and *specific* power output is as important as overall power output. To use the system in space, therefore, it must be very light. Looking at the drawing in Figure 1, it is clear that the reflector area is quite large compared with the cell area, especially when maximizing the concentration ratio. NRL solved the weight problem by constructing the reflector panels out of metallized thin film, such as Mylar® or Kapton®. But using these materials presents a significant challenge in itself, because the system will work only if the reflector is kept flat. Since these materials develop wrinkles easily, developing an effective system to keep them flat is critical.

When the reflector is not flat, reflections onto the photovoltaic panel are not uniform, resulting in less solar input to one or more cells. As a result, power output is reduced. The power reduction, however, effects more cells than just those that receive less sunlight. In fact, the power output from any cell in a string cannot exceed that of the cell that produces the lowest power, so an entire string of cells will be compromised.

To achieve a flat reflector, the film must be placed in tension such that the entire panel is under tensile stress. If this condition is met, and no compressive stress is present, then the panel will be flat. Achieving the desired stress state requires uniform force to be applied around the border of the entire panel. In practice, obtaining the required forces is difficult, because relatively small non-uniformities at the panel's edge can result in large wrinkles that propagate across the entire reflector.

NRL has experimented with a variety of methods to suspend the panel and attain the required tensile stress state, and success has been achieved by using a parabolic or constant radius curve shaped edge with a shear-compliant border. The panels are tensioned at the corners with springs attached to booms. Two booms per reflector -- one on each side -- are connected to the photovoltaic array and are used to suspend the panel. The springs, which are fastened at the end of each boom, are attached to cords (or equivalent) around the entire border of the panel; the edge cords help transfer the force uniformly from the springs to the panel. This system is illustrated in Figure 3. The forces transferred to the edge of the panel when the reflector has constant radius curve edges are shown in Figure 4. In this case, all the forces are normal to the panel edge.

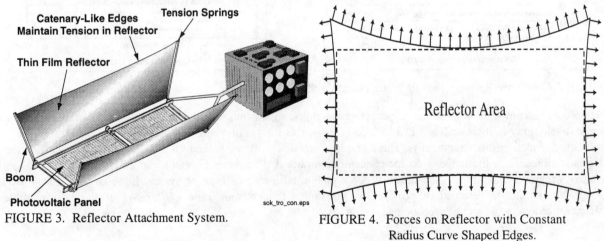

FIGURE 3. Reflector Attachment System.

FIGURE 4. Forces on Reflector with Constant Radius Curve Shaped Edges.

In addition to edge shape, edge cords, and corner tensioning, one more element is needed to provide a flat reflector, and that element is a compliant border. This type of border provides shear compliance, so any shear stresses developed will be taken up by the compliant border without being transferred to the reflector. Figure 5 shows a detail of the compliant border used on the 300 watt demonstration model. This silicon border has been extremely successful in keeping the reflector flat. Features include punched holes that provide the compliance and an edge cord that is encased in a tube and glued to the edge of the border. This border was developed and built by SRS Technologies, Inc. (Huntsville, Alabama), which is under contract to NRL to build several more demonstration reflectors in small and large sizes.

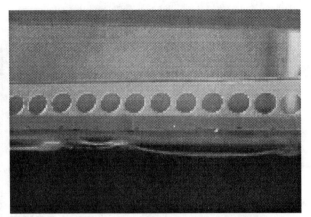

FIGURE 5. Silicon Compliant Border.

Figure 6 shows a detail of a second type of compliant border under consideration. In this case, a wire "paper clip" attached on one end to the reflector panel and on the other to the edge cord provides the compliance needed to keep the reflector flat. A third type of compliant border is also under consideration. This border would use a piece of thicker film around the edge to replace the edge cord and would have slots cut into the film itself to provide compliance. An experimental reflector with this type of border is now being built by SRS Technologies.

Compliant border experimentation will yield a design that meets a number of requirements besides providing a flat reflector, including ease of manufacture, stowing, and deployment, and low mass. Once a final compliant border design is selected, NRL will build a large-scale reflector and move on to a deployment test.

FIGURE 6. "Paper Clip" Compliant Border.

Deployment System

NRL built a 300 watt proof-of-concept model and successfully deployed it in 1996. Figure 7 shows a drawing of the reflector and array during mid-deployment, and Figure 8 shows the system when stowed. The drawing of the stowed system illustrates how the reflector panels are rolled up alongside the photovoltaic array. With this stowing method, the outer panel of the photovoltaic array is exposed during orbit transfer, which provides the NRL solar concentrator system with its unique ability among solar concentrators to provide power during orbit transfer.

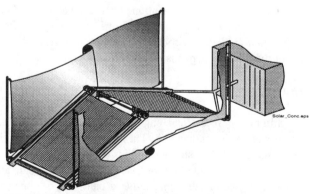

FIGURE 7. Solar Concentrator System at Mid-Deployment.

Design is currently underway for a 4 kilowatt array, which will be similar in concept to the smaller array but will incorporate a variety of structural improvements and boom attachment system improvements. Deployment of this array will take place in late 1997.

Testing

Highlights of NRL's testing program include reflectance, solar simulator, and thermal cycling tests. Reflectance measurements were taken by NRL's Optical Science Department during two different tests, a bi-directional scatter distribution function (BSDF) test and a spectral reflection test. The BSDF is used to determine the angle of incidence and angle of reflectance of laser light from the film surface. The spectral reflectance test shows the wavelengths of light reflected off the metallized thin film. The test results verify that the film reflects the visible spectrum necessary for solar cells with negligible angular distortion.

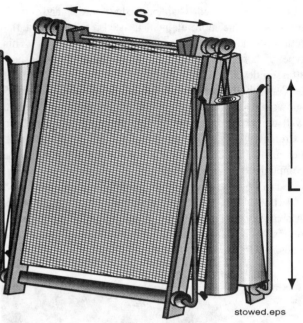

FIGURE 8. Stowed Solar Concentrator System.

Solar simulator testing was carried out at the NASA Goddard Space Flight Center in Greenbelt, Maryland, using its large area pulsed solar simulator (LAPSS) testing facilities. The test showed that the NRL solar concentrator could increase the power output from the photovoltaic panel nearly to predicted levels when flashed by the solar simulator. Power output is predicted to increase by a factor of 2.5 at 301K (test temperature), which is equal to the geometric concentration ratio. (In space, when the panel reaches higher temperatures and experiences reduced efficiency as a result, power output will increase somewhat less, by a factor of 2 rather than 2.5.) Actual power generated during the LAPSS test was about 8% less than expected, due to slight wrinkling in the reflector, lower than expected corner tensioning at the bottom of the reflector, and small deformations on the reflectors that occurred during handling and shipping. These problems are easily corrected, and the system is expected to reach predicted power levels in future tests.

NRL is about to begin the first thermal cycling tests of the reflector, using a test rig with a large copper heat plate instrumented with film heaters. The purpose of this test is to visually inspect the reflector as the temperature increases and decreases to see whether large wrinkles form. Large wrinkles could result in non-uniform solar flux on the photovoltaic panel, as mentioned earlier, which could degrade the power output of the system. In addition, the effect of a temperature gradient within the reflector will be examined. Thermal modeling has indicated that a temperature gradient of about 3K to about 12K from the top to the bottom of the reflector could occur, where the size of the gradient depends on the reflector coating and on whether it is coated on one or both sides.

Analysis

Highlights of recent and ongoing analytical work include boundary shape, stress, deformation, and edge cord stiffness analysis conducted by Dr. Martin Mikulas of the University of Colorado at Boulder, finite element analysis conducted by Dr. Christopher Jenkins of the South Dakota School of Mines and Technology, heat transfer analysis, and analysis of the potential for space degradation. Static analysis of the boundary shape results in equations for the corner loads and angles using either a parabolic or constant radius edge curve. In this analysis, it was initially assumed that the force in the reflector would be transferred from the corners to the reflector panel through edge cords attached directly to the reflector around its perimeter, without a compliant border. Analysis of the edge cord stiffness required for this concept to be successful reveals that the cord would have to be prohibitively stiff for it to keep the reflector flat. Follow-up analysis of the reflector with a compliant border shows that, as long as the border provides shear compliance, reflector flatness is no longer sensitive to edge cord stiffness. The results of this analysis bolstered conclusions based on experiment that the compliant border is the best edge treatment for the solar concentrator application. These conclusions are also supported by finite element analysis.

Two-dimensional heat transfer analysis has concentrated on predicting the temperatures in the reflector and the array, primarily to ensure that the reflector will not undergo unacceptable temperature variations and that the array will run cool enough to prevent the substrate materials from delaminating. Temperature variation within the reflector itself is a cause for concern because hot spots or major temperature gradients could result in differential thermal strain, potentially causing wrinkles to develop. Analysis conducted thus far indicates that the temperature gradient within the reflector does not exceed 12K for a 6-foot reflector, and that the array will maintain an acceptable temperature that does not exceed 408K.

A space degradation study of the reflector system is currently underway to determine the likely effect of micrometeorite damage on the reflector panels, to characterize the likely effect of atomic oxygen and ultraviolet damage to the panels from prolonged exposure to the sun, and to determine the effect on the photovoltaic array of increased temperature due to the concentration effect.

SUMMARY

Research and development efforts on the NRL trough-configuration solar concentrator have thus far proven the concept to be workable. Major milestones accomplished include the deployment demonstration of a 300 watt array and achievement of projected power levels in solar simulator tests at NASA's Goddard Space Flight Center. Confidence in the system is such that NRL is looking for a flight demonstration and is forming a cooperative research and development agreement (CRADA) with TECSTAR Inc. (City of Industry, CA), a major manufacturer of solar cells and panels for the aerospace industry, to develop a commercial product.

Acknowledgments

The authors wish to give special thanks to NRL's Brian Whalen and John Vasquez. Among many other contributions to the solar concentrator program, they spent many hours experimenting with different reflector designs and building the first successful deployment model. Commendation also goes to Paul Gierow and his group at SRS Technologies, who conceptualized and built the first reflector with a compliant border. They continue to build reflectors and solve complex manufacturing problems. Dr. Martin Mikulas of the University of Colorado at Boulder is much appreciated for his invaluable analytical support, excellent ideas, and overall enthusiasm. Thanks also go to Dr. Christopher Jenkins of the South Dakota School of Mines, who provides a high level of sophisticated analytical support.

References

Meinel, A.B., and M.P. Meinel (1976) "Applied Solar Energy, An Introduction," Addison-Wesley Publishing Company, Inc., Reading, Massachusetts:1-35.

Rex, D., and W.D. Ebeling (1978) "Photovoltaic Generators in Space with Concentrating Reflectors," in *Proc. European Symp. on Photovoltaic Generators in Space*, ESA SP-140:221-229.

ULTRALIGHT INFLATABLE FRESNEL LENS SOLAR CONCENTRATORS

Mark J. O'Neill
ENTECH, Inc.
1077 Chisolm Trail
Keller, TX 76248
817-379-0100

Michael F. Piszczor
NASA Lewis Research Center
21000 Brookpark Road
Cleveland, OH 44135
216-433-2237

Abstract

Since 1986, ENTECH and NASA Lewis have been developing refractive solar concentrators for space applications. These Fresnel lens concentrators can be configured as either point-focus dome lenses or line-focus cylindrical lenses. Small point-focus or line-focus lenses can be used to concentrate sunlight onto solar cells in space photovoltaic (PV) arrays. Large point-focus lenses can be used for high solar flux applications. In March 1997, a NASA Phase I SBIR program was initiated to develop ultralight inflatable lenses of both the line-focus and point-focus types. Special program emphasis is being placed on large point-focus lenses for various high-concentration applications, including solar dynamic (SD) power, alkali metal thermal energy conversion (AMTEC), thermophotovoltaics (TPV), and solar thermal propulsion (STP). Key outputs of the Phase I program include conceptual designs, optical performance predictions, micrometeoroid puncture analyses, manufacturing process identification, and functional prototype hardware. This paper summarizes the key results of the Phase I program, leading to the conclusion that inflatable dome lenses will provide excellent high-concentration optical performance, unequaled shape error tolerance, extremely low mass/aperture area ratio, proven manufacturability with space qualified materials, and small make-up gas requirements to maintain inflation on-orbit.

INTRODUCTION

Small, rigid (not inflatable) point-focus and line-focus lenses have recently been developed for space photovoltaic (PV) arrays, including:

- the mini-dome lens Boeing/NASA/ENTECH array, successfully flown in the 1994-95 PASP+ mission, verifying excellent on-orbit performance and durability for the refractive solar concentrator approach (Piszczor & Curtis 1995).
- the arched line-focus BMDO/NASA/ABLE/ENTECH SCARLET 1 array, developed and space-qualified in only six months, but unfortunately lost in the 1995 Conestoga launch failure (Allen et al. 1996)
- the 2.6 kW line-focus BMDO/NASA/ABLE/ENTECH SCARLET 2 array, the power source for both electric propulsion and satellite power for JPL's New Millennium Deep Space One mission to be launched in 1998 (Allen et al. 1996).

Large point-focus lenses can be used for high solar flux applications, including:

- solar dynamic (SD) power
- alkali metal thermal energy conversion (AMTEC)
- thermophotovoltaics (TPV)
- solar thermal propulsion (STP).

To date, such large point-focus lenses have not been fully developed and manufactured, primarily because large rigid (not inflatable) lenses are difficult to implement in a thin light-weight configuration. However, inflatable dome lenses may provide an ideal optical and structural combination for large-area solar concentrators in space. Such inflatable lenses offer a host of advantages over more conventional reflective solar concentrators (parabolic dishes), including shape error tolerance, optical design flexibility, proven manufacturability from space-qualified materials, and a spherical shape which matches the most natural inflatable structure geometry. The following paragraphs further describe the inflatable dome concentrator concept.

CONCENTRATOR DESCRIPTION

Figure 1 shows the inflatable dome lens concentrator concept, while Figure 2 shows a close-up view of a few of the prisms making up the lens. The dome Fresnel lens has a smooth outer (convex) surface and a prismatic inner surface. The shape is spherical. The focal plane is not at the center of the sphere. Instead, the focal plane is

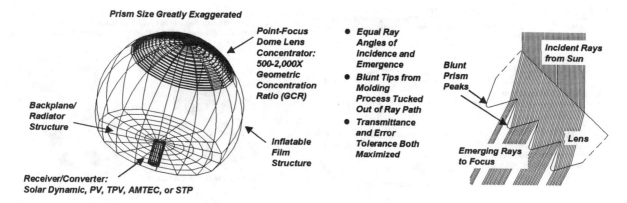

FIGURE 1. Self-Supporting Inflatable Dome Lens   FIGURE 2. Magnified View of a Few Prisms in the Dome Lens

determined by the refractive index of the lens material (e.g., 1.41 for clear space qualified silicone rubber) and the symmetrical refraction condition used to design each prism in the lens. As shown in Figure 2, the angle of incidence of the solar rays striking the smooth outer lens surface is equal to the angle of emergence of the same rays leaving each prism on their way to the focal plane. This symmetrical refraction condition provides many well known optical benefits (O'Neill 1995), including:

- Combined reflection losses at the two lens surfaces are minimized, thereby maximizing transmittance.
- Image spreading due to the finite size of the solar disk, chromatic aberration, sun-pointing errors, prism angle errors, diffraction, and slope errors are minimized, thereby maximizing achievable flux concentration.
- Blunt prism tips resulting from imperfect lens molding are tucked out of the way of the rays, thereby eliminating optical losses due to this common Fresnel lens defect.

The lens can be made from space-qualified silicone (e.g., Dow Corning DC 93-500), with appropriate coatings to provide atomic oxygen (AO) protection for low earth orbit (LEO) and solar ultraviolet (UV) protection for all orbits (e.g., OCLI's metamode coating provides both functions). (This lens material system has been used successfully for the photovoltaic concentrator arrays discussed in a previous paragraph.) The film structure beyond the lens portion of the dome can be made of a lightweight space-qualified material such as aluminized Mylar®. The backplane radiator structure closing out the inflated volume can contain waste heat rejection area for the receiver/converter, which is located at the focal point of the dome lens.

OPTICAL ANALYSIS

The anticipated space applications of the dome lens concentrator include solar dynamic (SD) power, thermophotovoltaics (TPV), alkali metal thermal energy conversion (AMTEC), and solar thermal propulsion (STP). All of these applications will require excellent optical efficiency at relatively high flux concentration levels (i.e., above 500 suns). In the past, reflective parabolic dish solar concentrators have been baselined for such applications. However, a parabolic reflector must be made and maintained at an extremely tight shape accuracy to provide good optical efficiency at high concentration. Small angular deviations from ideal values of the local surface normal (commonly called slope errors) will cause very large image blurring effects for a reflective concentrator. In stark contrast to a reflective concentrator, the symmetrical refraction dome lens is virtually unaffected by slope errors (O'Neill 1995). Figure 3 shows a direct comparison of the effects of slope errors on a parabolic reflector versus the dome lens concentrator. For this apples-to-apples comparison, both concentrators are identical in size (100 unit aperture diameter) and rim angle (45 degrees from the optical axis to the concentrator rim). A slope error of ± 1 degree has been applied to the outer edge of both concentrators, and the resultant image displacements in the focal plane have been calculated. The image spreading is 7.0 units for the reflector and 0.04 units for the lens. Thus the lens provides about *200 times better slope error tolerance* than the reflective concentrator. This insensitivity to shape errors is a unique attribute of the inflatable dome lens concentrator.

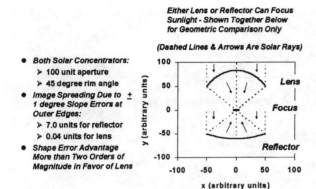
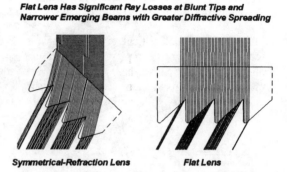

FIGURE 3. Optimal Dome Lens Versus Parabolic Reflector **FIGURE 4.** Optimal Dome Lens Versus Flat Lens

While not shown in Figure 3, the dome lens has a similar slope error advantage (*more than 100 times better*) compared to a conventional flat Fresnel lens (O'Neill 1995). Figure 4 shows a few prisms in the dome lens and in a conventional flat lens. Note that all of the ray deviation is all done at the inner prismatic surface for the flat lens, compared to equal ray deviation at both surfaces for the dome lens. This symmetrical refraction condition is responsible for the maximized transmittance, unequaled slope error tolerance, and other superior optical features of the dome lens.

An optical analysis has been performed to select the proper rim angle (or F/Number) for the lens, and to predict the optical performance of the inflatable dome lens solar concentrator for various applications. Trade studies regarding achievable concentration and sun-pointing error tolerance led to the selection of a 30 degree rim angle for the inflatable dome lens. A ray-trace optical model was then developed to predict net optical efficiency versus geometric concentration ratio for the dome lens. The optical model treats chromatic aberration (refractive index variation over the solar spectrum), finite solar disk size (0.53 degree diameter for Earth orbit applications), and reflection losses at both lens surfaces. The present ray-trace model does not include the effects of diffraction and various errors (slope, prism, and sun-pointing), but these effects have been separately analyzed. Figure 5 shows the results of the optical analysis, including comparative results for a flat Fresnel lens. Net optical efficiency is defined as the concentrated radiant power entering the receiver aperture divided by the solar power incident on the lens aperture. Geometric concentration ratio (GCR) is lens aperture area divided by receiver aperture area.

Note from Figure 5 that the dome lens will provide the following optical performance without any antireflection coatings:

- 89% at 500X GCR
- 87% at 1,000X GCR
- 82% at 1,500X GCR
- 75% at 2,000X GCR.

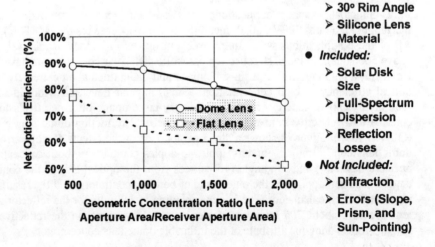

FIGURE 5. Optical Performance Analysis.

290

Note that the flat lens efficiency curve is dramatically lower than the dome lens curve, confirming the optical design optimization of the symmetrical refraction lens approach.

Another potential optical advantage of the dome lens concept relates to flux profile tailoring at the energy receiver/converter. Each prism angle can be slightly varied in design to place the refracted rays at desired locations on the receiver. Since there are thousands of individual prisms which can be so tweaked, virtually any desired radiant energy flux pattern can be produced over the energy receiver. This flux tailoring approach might be useful to smooth out flux spikes over heat transfer surfaces in STP receivers, or to level out electrical current generation distributions in TPV receivers.

MICROMETEOROID PUNCTURE ANALYSIS

One of the key concerns for any inflatable space structure relates to micrometeoroid punctures. Figure 6 is the NASA-estimated micrometeoroid flux versus particle size for LEO (Kessler et al. 1989). When the curve of Figure 6 is integrated over all particle sizes, the total cross-sectional area of all particles encountered in LEO over a full year amounts to 1.5×10^{-7} square meters of particles per square meter of projected target area. This ratio is essentially a porosity factor, which can be applied to any size structure on orbit, to estimate the total puncture area, assuming that the puncture area is equal to the particle cross-sectional area. These results imply that a very large dome lens concentrator (15 meters in diameter) would suffer about 0.26 sq.cm. (0.04 sq.in.) of total puncture area after one full year in LEO. If the particles went all the way through the lens and the backside enclosure, the puncture area would be double, as assumed in the following discussion.

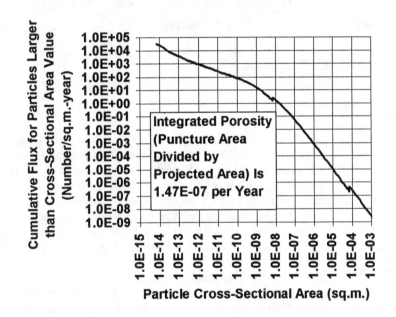

FIGURE 6. Micrometeoroid Puncture Analysis.

The gas which is used to inflate the dome lens will obviously escape to the vacuum of space through the punctures caused by micrometeoroids. The rate of escape can be calculated based on a choked (sonic) flow assumption for each puncture. The gas loss rate then depends on the gas molecular weight, temperature, and pressure. The required inflation pressure to support the dome lens against the low accelerations experienced by most satellites is extremely small (e.g., less than 0.05 Pascal for 0.02 g), due to the small thickness and mass of the lens material. To be conservative, the design pressure (0.7 Pascal [10^{-4} psi]) for the 1996 JPL Inflatable Antenna Experiment (Freeland 1997) was used to estimate gas loss rates in Figure 7. The gas temperature can be estimated from a radiation heat balance on the dome lens surfaces, with the resultant temperature range shown in Figure 7. At the middle of this range, the gas loss rate for hydrogen is less than 2 kg per sq.cm. of puncture area per year. Combining this result with the porosity from Figure 6, the expected loss rate for a 15 meter diameter dome lens concentrator can be estimated at 1 kg of hydrogen per year, after a full year of LEO micrometeoroid punctures. Of course, after two years on orbit, this rate would double, and so on. However, the amount of makeup gas that would be needed to keep the dome lens inflated for many years is fairly small, indicating that a purely inflatable lens would be practical for shorter duration missions. For longer duration missions, some form of rigidization would be desirable sometime after the initial inflatable lens deployment on orbit.

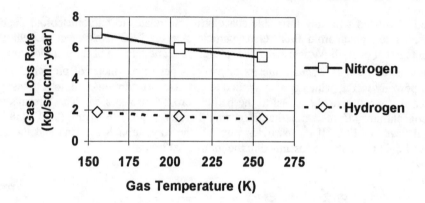

FIGURE 7. Gas Loss Rates.

MANUFACTURING METHOD

Over the past 5 years, a continuous casting process has been developed by 3M to produce high-quality, space-qualified, thin (200 micron) silicone lenses to ENTECH specifications. This process produces a continuous web of lens material 30-50 cm wide by tens of meters in length. Virtually any repeating prismatic pattern can be produced, with geometry variations both cross web (across the 30-50 cm width) and down web for about 1 m, after which the prismatic pattern repeats itself again and again, like a newspaper page on a continuous printing line. This process should be directly adaptable for making segments of the dome lens. These segments can be complete triangular gore sections for smaller domes (less than 2 meters in diameter) or quasi-trapezoidal panel sections for larger domes (perhaps 20-30 different tools for 20-30 different panels will be needed for a 15 meter diameter dome lens).

While the tooling for this process is relatively expensive, high production rates at relatively low costs can be achieved after the tooling has been fabricated. Thus, this process would be ideal for making dozens of identical dome lens concentrators, perhaps for a fleet of solar thermal propulsion (STP) orbit transfer vehicles (OTV's). This manufacturing method has been used successfully over the past 3 years to produce more than 1,000 small photovoltaic concentrator lenses for the SCARLET series of space solar arrays.

PROTOTYPE INFLATABLE CONCENTRATORS

To verify the basic concept of an inflatable silicone Fresnel lens solar concentrator, several functional prototypes have been constructed using available line-focus lens material. Figure 8 shows one of these prototype models, both deflated and inflated. Each of the models produced to date has provided excellent focussing of direct sunlight in outdoor observations. Using a gallium arsenide solar cell, the optical efficiency of one model was measured outdoors at 92% at 7X geometric concentration ratio (GCR). The line-focus models have also verified low concentrator mass per unit aperture area. The lightest model built to date weighs less than 0.8 kg/sq.m. including an aluminum backplane radiator, and less than 0.6 kg/sq.m. excluding the radiator weight, but including the lens, film structure, adhesive, valve, and everything else. These models have also verified the shape error tolerance of the symmetrical refraction lens approach, having good focussing performance even when slightly deflated.

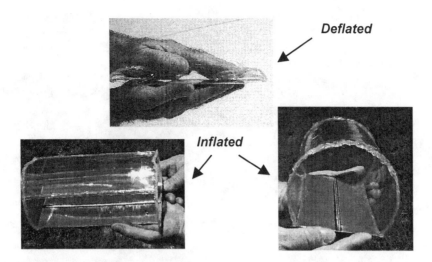

FIGURE 8. Self-Supporting Inflatable Line-Focus Lens.

CONCLUSIONS

Based on the results of this project to date, the large inflatable dome lens solar concentrator appears to offer several key advantages over other solar concentrator approaches, including:

- The spherical shape of the lens is ideal for inflatable deployment, in contrast to parabolic reflectors with their mandatory two-dimensional variation in surface curvature.
- Space-qualified and flight-proven lens material is available (coated silicone rubber).
- A proven high-volume cast-and-cure process is available for manufacturing the lens.
- Shape error tolerance is 200 times better for the lens than for any reflective concentrator.
- The flux profile over the energy receiver can be tailored by adjusting the design of individual prisms.
- Optical analysis predicts excellent high-concentration optical performance for inflatable dome lenses.
- Based on functional prototype line-focus hardware evaluation, a self-supporting inflatable dome lens will provide excellent focussing, with a total mass/aperture ratio between 0.5 and 1.0 kg/sq.m.

Acknowledgments

This paper describes the key results of NASA Phase I SBIR Contract No. NAS3-97073, which was completed in September 1997. The authors would like to recognize and thank their colleagues at NASA Lewis (especially Carol Tolbert, the contract technical monitor), ENTECH, 3M, and ABLE Engineering, who have also contributed to the ongoing development of the inflatable lens concentrator.

References

Allen, D.M., P.A. Jones, D.M. Murphy, and M.F. Piszczor (1996), "The SCARLET Light Concentrating Solar Array," proc. 25th IEEE-PVSC, 1:353-356.

Freeland, R. (1997), Personal Communication, Jet Propulsion Laboratory, Pasadena, CA.

Kessler, D. J., R. C. Reynolds and P. D. Anz-Meador (1989), "Orbital Debris Environment for Spacecraft Designed to Operate in Low Earth Orbit," NASA TM 100471.

O'Neill, M.J. (1986) "Dome Fresnel Lens Concentrator for Space Solar Dynamic Power," proc. 21st IECEC, 3:2039-2044.

O'Neill, M.J. (1995) "Chapter 10 - Silicon Low-Concentration, Line-Focus, Terrestrial Concentrator Modules," in the textbook, *Solar Cells and Their Applications*, Edited by L.D. Partain, Published by John Wiley.

Piszczor, M.F. & H.B. Curtis (1995) "An Update on the Results from the PASP Plus Flight Experiment," proc. 30th IECEC, 1:315-320.

DEVELOPMENT OF AN ADVANCED PULSED PLASMA THRUSTER (PPT) AND PLUME DIAGNOSTIC EXPERIMENT (PDE) FOR DEMONSTRATION ON THE MIGHTYSAT II.1 SPACECRAFT

T. Peterson, E. Pencil,
and L. Arrington
NASA Lewis Research Center
21000 Brookpark Rd.
Cleveland, OH 44135
(216)433-5350

J.R. LeDuc, D.R. Bromaghim,
and J. Malak
Phillips Laboratory
4 Draco Dr.
Edwards AFB, CA 93524
(805)275-5908

J.J. Blandino
Jet Propulsion Laboratory
4800 Oak Grove Dr.
Pasadena, CA 91109
(818)354-2696

W.A. Hoskins and N.J. Meckel
Primex Aerospace Company
P.O. Box 97009
Redmond, WA 98073
(206)885-5010

B. Moore
Trisys, Inc.
5320 W. Alameda Rd.
Glendale, AZ 85310
(602)581-7414

Abstract

The Pulsed Plasma Thruster (PPT) Space Demonstration Program will characterize the performance of an experimental PPT as an orbit raising propulsion system with a 25% reduction in dry mass and twice the power of the LES 8/9 PPT. The project is a partnership of the USAF Phillips Laboratory and NASA Lewis Research Center. The PPT is manifested on the MightySat II.1 satellite scheduled for launch in January, 2000. An extensive breadboard PPT testing program is underway and data from those tests will be used to finalize the design of the flight version PPT. A plume diagnostic experiment (PDE) will be used to characterize contamination rates and effects on surface thermo-optical properties due to PPT plume contamination. These experiments will then undergo full space qualification testing prior to integration with the MightySat II.1 spacecraft.

INTRODUCTION

The Pulsed Plasma Thruster (PPT) Space Demonstration will investigate the performance of an orbit-raising PPT unit as part of the MightySat II.1 (MightySat 1996) spacecraft (Figure 1). A sister payload termed the Plume Diagnostic Experiment (PDE) will determine the effects of plume contamination from the thruster on typical optical surfaces. The PPT Space Demonstration is a cooperative effort between the US Air Force Phillips Laboratory, NASA-Lewis Research Center (NASA-LeRC), and Primex Aerospace Company (PAC). MightySat II.1, known as Sindri, is scheduled for launch in January, 2000. The MightySat program is managed from the Space Experiments directorate of Phillips Laboratory at Kirtland AFB, NM for the purpose of flying Phillips Laboratory technology in a timely and cost-effective manner. Spectrum Astro of Gilbert, AZ is designing and manufacturing the Sindri spacecraft and has completed a preliminary design review in March, 1997.

The MightySat PPT is mounted on the –X face of the MightySat structure, which is not visible in Figure 1. This PPT has been designed to perform a critical orbit raising maneuver to extend the MightySat mission life from 37 days to 1 year when released from the Space Shuttle at an altitude of 370 - 398 km. Although originally designed for a space shuttle insertion orbit, the primary launch vehicle for Sindri is now the Orbital/Sub-Orbital Program (OSP), with an insertion altitude of 509 km. With OSP no orbit raising is necessary to maintain a one-year life, however, these design requirements are maintained because Sindri or future MightySats may be launched from the shuttle. The January, 2000 launch date places the orbit raising maneuver right at the beginning of the solar maximum cycle, which will dramatically increase spacecraft orbit decay rates. This in turn will increase the demands on the MightySat PPT design.

The PPT is being developed at PAC under a NASA-LeRC program. (Meckel 1996) The Primex PPT is based on the Lincoln Experimental Satellite (LES) 8/9 (Vondra 1974) design that was space qualified, but never flown. The MightySat PPT will be a significant improvement over the design of the LES 8/9 thruster due to the incorporation of drastic improvements in technology. The PPT developed in this program will reduce the dry mass of the LES 8/9 system by 25% and will operate at twice the power level. These design factors present a significant challenge for designing the spacecraft interface, such as thermal loading, etc.

The breadboard unit of the PPT shown in Figure 2 is undergoing ground testing at NASA-LeRC in Cleveland, OH. Currently, NASA is conducting measurements of performance and lifetime. Modeling and testing of the conducted and radiated electromagnetic interference (EMI) produced by the thruster and plume will also be performed. Flight qualification and acceptance tests of the PPT will be conducted at Primex Aerospace and at NASA-LeRC, and will include vibration and thermal tests, an electromagnetic conduction (EMC)/EMI evaluation, PPT performance and life evaluations, and verification of other spacecraft interface specifications.

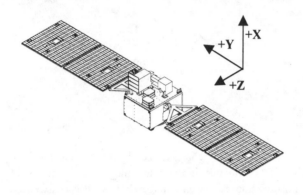

FIGURE 1. MightySat II.1 Spacecraft.   FIGURE 2. LeRC-Primex Breadboard PPT.

MISSION ANALYSIS, PLANNING, AND OPERATIONS

Mission analysis planning was done using an orbit raising code written in-house at Phillips Laboratory. (Tilley 1996) The current concept of operations for the PPT consists of a 30 to 60 day firing upon completion of spacecraft initialization and checkout. This firing will include performance measurements, contamination studies, thermal design verification, and UHF communications compatibility testing. After the initial firing period is complete, other experiments on the spacecraft will perform their operations without risk of contamination from PPT firings. A second 30 to 60 day burn will be accomplished near the end of the MightySat II.1 mission in order to gather contamination data which can be correlated with data from other payloads. A timeline of the PPT operations is shown in Table 1. For a full description of the mission analysis, planning, and operations see LeDuc et. al. (LeDuc 1997)

FLIGHT PPT DESIGN

Due to the critical nature of the orbit-raising maneuver the PPT was intended to perform, very challenging design requirements were set. These requirements are outlined in Table 2.

Physical Dimensions

The PPT system has a mass of 5.83 kg, which is less than half the wet mass of an equivalent system (15,000 N-s) composed of state-of-the-art LES 8/9 components. The PPT envelope is approximately 34.3 cm by 22.9 cm by 12.7 cm. Table 3 shows the system mass breakdown.

TABLE 1. PPT Operations Timeline.

Operation	Time	Duration
PPT/PDE Initialization (w/o PPT firing) PDE Operation Begins	L to L+2wk	2 days
PPT System Check-Out (w/PPT Firing) 1. PPT First-Firing/Spacecraft Compatibility Test 2. PPT Functional Test 3. PPT/ADACS Compatibility Test 4. PPT Performance Determination From ADACS 5. PPT Thermal Compatibility Verification Test	L+2 wk	<1 wk
PPT Orbit Raising Maneuver I 1. PPT Performance 2. PPT Life Determination 3. PPT Thrust Vector/Impulse Bit Variation Characterization 4. PDE Sensor Measurements	L+3 wk to L+7 wk	1 month
PPT Thermal Design Verification	L + 7 wk	1-2 d
PPT/UHF Communication System Compatibility Test	L + 8 wk	1 wk
PPT Orbit Raising Maneuver II 1. PPT System Check-out 2. PPT Performance 3. PPT Life Determination 4. PPT Thrust Vector/Impulse Bit Variation Characterization 5. PDE Sensor Measurements 6. HSI Performance Measurements	L + 6-8 months	3 months

Electrical Design

The main functions of the PPT electronics are to provide the charging voltage for the energy storage capacitor and to provide the command and telemetry functions required to operate the PPT. The PPT electronics are designed to be easily integrated with the host spacecraft. The main capacitor charging circuit is isolated from the chassis to prevent current return through the spacecraft structure. The summary design requirements of the PPT electronics are shown in Table 2.

Thermal Design

The MightySat mission presents a challenge for the thermal design of the PPT system. The thrust requirements to overcome drag and perform the orbit-raising mission drive the power processed by the PPT to more than twice that previously required for spaceflight qualification. The key drivers for the MightySat PPT thermal design are: 100W maximum power handling, PPT non-operating temperature range of $-62°$ C to $+75°$ C, PPT operating temperature range of $-30°$ C to $+85°$ C, and heat transfer across the interface less than 65 W during worst case conditions. Appropriate solar, earth albedo, and background space source and sink values were used in the thermal model for a space shuttle insertion orbit altitude of 398 km. The model was a simple, 15-node conduction and radiation model. The results were then correlated with a more detailed SDRC I-DEAS model. The model was run for three different orbit scenarios to simulate the range of minimum to maximum solar heat inputs. Results are shown in LeDuc et. al. (LeDuc 1997), where it can be seen that the PPT will be able to operate within acceptable temperature limits in the sun and also meet survival limits in eclipse.

Telemetry, Command, and Control

The PPT telemetry comprises seven channels. These include the main capacitor voltage and the discharge initiation voltage for the left and right sparkplugs. Each of the voltage channels is a 0-5 V buffered output. Two resistance temperature detectors (RTDs) provide temperature data on the main power transformer and the surface of the energy storage capacitor. The spacecraft provides a constant current source for the RTDs on these two channels. The final two channels provide the readout from the left and right fuel bar potentiometers, which serve as fuel consumption measurements. A constant current source is also supplied by the spacecraft to these two potentiometers.

PPT command and control is accomplished via two separate 5 V transistor-transistor logic (TTL) pulses. The charging command initiates capacitor charging when it goes to 5 V and terminates charging when it drops back to 0 V. This pulse is typically 500 ms in duration for the 43J MightySat capacitor energy level. The discharge is initiated by a second 5 V TTL pulse on either the left or right sparkplug line. This pulse is 10 µs in duration and is synchronized to occur just after the charging of the capacitor is completed.

TABLE 2. MightySat PPT Design Requirements.

Characteristic	Requirement
Power, main operational	>100 W for > 62% duty cycle
Power, main standby	<0.68 W
Power, command	< 30 mW total for 3 channels
Power, telemetry	< 15 mW total for 4 of 6 channels
Voltage	22 to 34 V
Voltage, command	3.5 to 10 V high; -0.7 to 2 V low
In rush current	< 0.15 A for < 0.1 s
Peak Current	< 7.0 A
Commands	3 channels
Telemetry	7 channels
EMI	
Conducted	MIL-STD-461
Radiated	Tailored MIL-STD-461
Susceptibility	MIL-STD-461

TABLE 3. MightySat PPT Mass Properties.

Component	Mass (kg)
Electronics:	0.855 kg
Capacitor:	1.507 kg
Fuel Bars:	1.04 kg
Structure:	
Horn Assembly:	0.14 kg
Stripline Assemblies:	0.445 kg
Bulkhead Assemblies:	1.527 kg
Other:	0.239 kg
Fasteners:	0.067 kg
Total:	**5.82 kg**

Electromagnetic Interference Management

The EMI management plan for the PPT is accomplished through a combination of shielding and transformer isolation. Because the PPT discharge is a fast pulse, it is necessary to consider both DC and AC isolation. Also, the plasma generated in the discharge can conduct current to the spacecraft structure and potentially cause interference. The EMI/EMC design includes the complete enclosure of the power electronics in an aluminum housing and isolation of the energy storage capacitor from the chassis. Additional aluminum shielding is employed on the propellant feed bar housings and on the electrode housing and horn assemblies. The power electronics circuit incorporates an EMI filter on the input to block conducted interference on the 28 V bus. The high voltage and discharge initiation circuits also incorporate transformer isolation. To complete the shielding, the cabling between the PPT and the spacecraft harness incorporates a shield which is connected to the PPT chassis and the spacecraft ground.

Breadboard PPT Test Program

A number of tests have already been performed or are currently ongoing on the breadboard PPT (Figure 2) and on various components in the development efforts towards a flight PPT design. Previous tests performed include performance, spark plug characterization, capacitor thermal expansion, and capacitor efficiency. The performance tests were conducted on a variety of electrode configurations at various energy levels. This information was used to determine the electrode configuration and energy level that would be most representative of the MightySat mission: 3.81 cm long electrodes with a 20 degree flare at 43 joules.

A spark plug characterization was also performed on a number of new, unused spark plugs to determine the lowest applied voltage at which the capacitor would break down. Results showed that the spark plugs consistently fired at 800 V, but not necessarily below that voltage. Also, there is interest in determining if the spark initiation voltage changes after successive breakdowns over the life of the spark plug. Any voltage change will be identified as part the breadboard life test discussed later. Further testing was conducted at NASA-LeRC to investigate any difference in the mass loss per pulse between a short duration continuous test of 2000 pulses and a longer duration of 10,000 pulses at the energy level proposed for the mission. The difference was less than 1% between the two cases at this energy level. Additional testing was performed at Primex to determine the efficiency and thermal expansion of the capacitor. The thermal tests showed that when the temperature cycled between -20° C and 40° C there was less than a 0.005 cm difference in the diameter of the capacitor, aiding in the design efforts of the flight hardware. Bench top testing was also conducted to determine the efficiency of the engineering model capacitor. Tests showed the efficiency to be 91.6% but as the capacitor skin temperature rose to 34° C, the efficiency dropped to 88.4%.

Testing currently underway at NASA-LeRC includes capacitor life, breadboard PPT life/thermal/contamination, and conducted/radiated EMI testing. The capacitor life test is on going with an engineering model capacitor in a laboratory PPT at the mission power requirement of 100 watts. The test is currently at 12.3 million pulses. Initial conducted/radiated EMI measurements are being made with the breadboard PPT in the glass bell jar facility and will be reported by the end of the calendar year. Thermal testing on the breadboard has been conducted at various power levels including a 7-hour test at full power to demonstrate operation at that power level. Further thermal testing is expected as part of the breadboard PPT life tests. These tests will be used for thermal modeling efforts of the flight PPT.

Testing will continue through the fall with further radiated and conducted EMI tests on the breadboard, followed by a life test of the breadboard and its electronic components. During the life test, contamination measurements will be made using samples of optics from other MightySat II.1 payloads and components. Samples of the Hyper Spectral Imager (HSI) mirror and downwelling sensor, as well as a sample of quartz window material from the satellite star tracker, will be placed in the chamber at the same relative location to the PPT as they will be in on the MightySat spacecraft to determine what effects the PPT plume will have on these instruments. The manufacturer of the flight capacitor has conducted an accelerated bench life test of a number of engineering model capacitors to determine the expected life of the capacitor. The test report is expected by October 15, 1997.

Flight Qualification Testing

Qualification tests will be performed on a qualification PPT unit to demonstrate adequate margins in the flight PPT design and to ensure that design and operational requirements are met. This PPT will be operated in vacuum at temperatures of 48° C on the high end and –24° C on the low end to meet thermal flight qualification levels in the second and third fiscal quarters of 1998 at Primex Aerospace Co. and at NASA-LeRC. The qualification test program will also include performance measurements, vibration testing, conducted and radiated EMI tests, and contamination and life tests.

Flight Acceptance Testing

Flight PPT unit acceptance tests will occur at Primex and at NASA LeRC in the fourth fiscal quarter 1998 prior to delivery of the unit to the MightySat Program Office for spacecraft integration and test activities. A typical acceptance test program will be performed to measure performance parameters and to reveal any manufacturing inadequacies such as workmanship or material defects. Like the qualification unit, the flight unit will undergo a flight acceptance thermal vacuum test where the PPT is operated at temperatures of 48° C and –24° C. In addition, performance tests will be run over the full range of mission requirements.

PLUME DIAGNOSTIC EXPERIMENT

The plume diagnostics experiment (PDE) is a stand alone experiment to be flown on MightySat II.1 to gather data on plume contamination. The potential contamination of sensitive spacecraft surfaces by the plume is a recurring issue among potential users of PPT technology. The solid Teflon fuel of the PPT ablates into a variety of products including high velocity ionic species (C, F, CF, etc.) as well as slower high molecular weight species (CF_4, C_2F_4, C_3F_6, etc.). Five objectives of the PDE are 1) to demonstrate the compatibility of PPTs from a contamination standpoint with current and future DOD, NASA, and commercial small satellite missions, 2) to provide unambiguous assessment of PPT plume deposition in the backflow region, 3) to provide flight data for correlation with ground-based PPT plume contamination measurements, 4) to provide flight data for validation of numerical simulations currently under development, and 5) to develop a low-cost, easily integrated contamination monitoring package design that can be used on other missions.

The PDE consists of two sensor packages each consisting of a quartz crystal microbalance (QCM), calorimeter pair. These two sensors will collectively provide valuable information regarding the contamination effects of the PPT plume on spacecraft surfaces. The information gathered will be compared with that from ground tests to be conducted at NASA LeRC evaluating the effects of plume contamination. Each sensor package will contain a QCM and calorimeter sensor on an aluminum plate mounted on an external surface of the spacecraft bus. An avionics package to support signal conditioning, temperature control, analog to digital conversion and serial communication with the spacecraft will be located on a single card mounted in the card cage within the bus. The avionics will provide the capability for limited data storage and control of sensor operating modes. After analysis, the combined output of the QCM and calorimeter data will be a quantitative measure of contamination in terms of material deposited per unit time as well as the cumulative effect of the deposited material on surface absorptivity and emissivity.

The QCM is used to correlate mass accumulation as a function of time with any specific events of interest in the mission timeline such as thruster firings. In addition, the sense crystal can be heated to bake off material and clean its surface. This is useful in the unlikely event the crystal is saturated, but can also be used to perform thermogravimetric analysis (TGA). In a TGA the crystal is heated in a carefully controlled manner in such a way that discrete changes in frequency can be identified with specific constituents with known heats of vaporization.

For the PDE, the QCMs will be Mark 16 devices made by QCM Research in Irvine, California. These sensors have a optically polished, gold coated sense crystal with a frequency of 15 MHz. The maximum mass sensitivity at this frequency is on the order of 10^{-9} g/cm^2-Hz (at 10K). A 2.4 W heater is capable of raising the crystal temperature to 340 K. Each unit has a mass of approximately 29 grams.

The calorimeters used on the PDE will be fabricated by JPL using a design with a proven flight history and a mass of approximately 40 grams. The calorimeters provide data which can be used to infer changes in the emissivity and absorptivity of the sensor surface due to contamination. During operation, the temperature of, and the heater power applied to the sensor surface is carefully measured. The active sensor surface is an optical solar reflector consisting of a second surface silver coating on a quartz crystal. The calorimeters do not reach thermal equilibrium during an orbit, therefore a thermal radiation model is necessary to relate the measured temperatures to radiant sources as functions of the unknown absorptivity and emissivity and the known physical properties of the crystal. In order to determine the changes in absorptivity and emissivity of the original coating, it is necessary to have a knowledge of the insolation history throughout the mission. A knowledge of angle with respect to the sun or other warm bodies (such as the earth) within one degree or less is desirable.

The calorimeter is equipped with a single resistance heater which serves multiple functions. The primary function of the heater is to maintain the sensor above a predetermined constant temperature throughout the orbit. By relating the heater power required to maintain the setpoint temperature to the overall heat balance, changes in the thermal properties of the coating can be ascertained. In addition the heater can be used to bake out the disk and remove deposited material if necessary as well as maintain the sensor above a predetermined survival temperature of roughly -65 C. For more information on the operation of the calorimeter or QCM, the reader is referred to Reference 4. (Tilley 1996)

Each sensor panel assembly consists of a sensor pair on an aluminum plate coated in silvered Teflon which is mounted to the spacecraft structure with four cylindrical G-10 standoffs. The plate serves as a radiating surface and the entire assembly is thermally insulated from the bus via the fiberglass standoffs. Other components of the panel assembly include a 15 pin micro-d connector, used to connect the sensors to the serial RS-422 cable and a blanket of multi-layer-insulation (MLI) which encloses the area below sensors.

The plume diagnostics experiment support electronics (PDE-SE) package is being developed by TRISYS Inc., of Glendale Arizona and represents an innovative approach to achieving reliable, fault-tolerant operation with a simple low cost design. The PDE-SE uses individual RISC based micro-controllers to control the QCM and calorimeter sensor operation. This will be achieved through a library of commands for specific operating modes, sequences of which are uploaded to the spacecraft. The electronics will be located on a single VME 6U form factor (the card itself is not VME) card located in the card cage. Communication with the two sensor panels will be via a RS-422 command, data, and power interface cable. One of the requirements of the PDE-SE is the ability to store 24 hrs of data for later downloading to the spacecraft computer for downlink. This is achieved with a 128K, 8 bit SRAM data recorder.

CONCLUSION

The PPT Space Demo on MightySat II.1 will be the first spaceflight demonstration of orbit raising propulsion for small satellites using a PPT. Mission analysis has shown that it is feasible to use a PPT to raise a SmallSat from the Space Shuttle payload bay to a one-year orbit. The PPT to be used to demonstrate this mission is currently in the final stages of design. Once the PPT units are manufactured they will undergo rigorous qualification and acceptance testing prior to integration and test with the MightySat II.1 spacecraft and subsequent launch from OSP into a one-year mission orbit.

Acknowledgments

The authors would like to acknowledge assistance from the following people: Mr. Randy Kahn, Capt Bob Costa, Capt Mike Rice, and Lt Brandt Miller from the MightySat II.1 program office; Ron Spores, Greg Spanjers and Jamie Malak, the other members of the PL/RK PPT team; Dave Winner and the PPT engineering group at Primex Aerospace Company; Chris Clark, Scott Fallek, Scott Bussinger, and the rest of the MightySat II.1 spacecraft development team at Spectrum Astro; Dennis Tilley, for performing the initial mission analysis and getting the project off the ground; and Jeff Pobst for transitioning the project to it's current incarnation.

References

LeDuc, J.R., et. al. (1997) "Mission Planning, Hardware Development, and Ground Testing for the Pulsed Plasma Thruster (PPT) Flight Demonstration on MightySat II.1," 33rd Joint Propulsion Conference, Seattle, WA. AIAA 97-2779

Meckel, N.J., et. al. (1996) "Improved Pulsed Plasma Thruster Systems for Satellite Propulsion," 32nd Joint Propulsion Conference, Lake Buena Vista, FL. AIAA 96-2735

MightySat Satellite Program (1996) Point Paper, Phillips Laboratory Space Experiments Directorate.

Tilley, D.L., et. al. (1996) "Advanced Pulsed Plasma Thruster Demonstration on MightySat Flight II.1," Proceedings of 10th AIAA/Utah State Univ. Conference on Small Satellites, Logan, UT, Sept., 1996.

Vondra, R.J., Thomassen, K.I. (1974), "Flight Qualified Pulsed Electric Thruster for Satellite Control, Journal of Spacecraft," 11(9): 613-617.

THE ELECTRIC PROPULSION SPACE EXPERIMENT (ESEX) - A DEMONSTRATION OF HIGH POWER ARCJETS FOR ORBIT TRANSFER APPLICATIONS

D. R. Bromaghim, R. M. Salasovich, J. R. LeDuc
Phillips Laboratory
Edwards AFB, CA
(805) 275-5473, 275-5904, and 275-5908

L. K. Johnson
The Aerospace Corporation
El Segundo, CA
(310) 336-1998

Abstract

The Electric Propulsion Space Experiment (ESEX) is a high power (30 kW) ammonia arcjet space demonstration sponsored by the Propulsion Directorate of the Phillips Laboratory with TRW as the prime contractor. ESEX is one of nine experiments being launched in early 1998 on board the Advanced Research and Global Observation Satellite (ARGOS). ESEX will demonstrate the feasibility of using a high power arcjet for orbit transfer. ESEX is instrumented with various sensors to address all of the expected interactions with ARGOS including electromagnetic interference, contamination, and radiated thermal loading. The performance of the arcjet will also be measured using ground tracking, an on-board GPS receiver, and on-board accelerometer. In addition to the performance and spacecraft interaction studies, ground-based spectroscopic and radiometric measurements will be performed to observe plume species as well as determine the effect of the arcjet firing on the space environment. ESEX is currently undergoing integrated testing with the spacecraft bus and the eight other experiments to verify the full operability of ARGOS while on-orbit. These tests include basic functionality of the system in addition to the normal suite of environmental tests including electromagnetic interference and compatibility, acoustic and pyro-shock testing, and thermal vacuum tests.

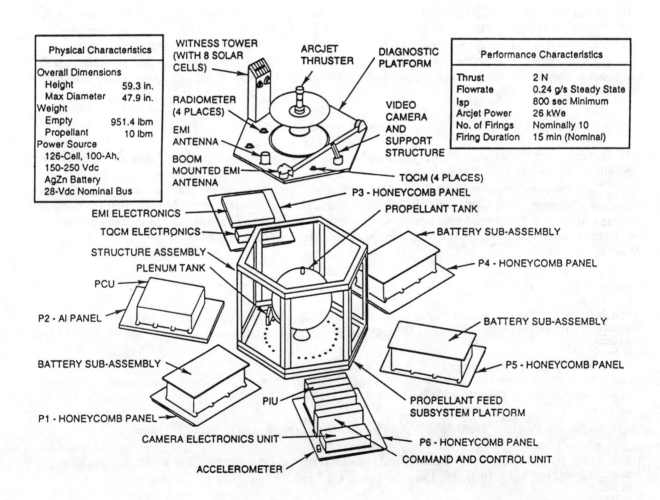

FIGURE 1. Exploded View of the ESEX Flight Unit.

INTRODUCTION

The Electric Propulsion Space Experiment (ESEX) is a space demonstration of a 30 kW ammonia arcjet sponsored by the Phillips Laboratory with TRW as the prime contractor. The experiment will demonstrate the feasibility of a high power arcjet system, as well as measure and record flight data for subsequent comparison to ground results (Kriebel 1992, Sutton 1995, and LeDuc 1996). The flight diagnostic suite includes four thermo-electrically-cooled quartz crystal microbalance (TQCM) sensors, four radiometers, near- and far-field electromagnetic interference (EMI) antennas, a section of eight gallium-arsenide (Ga-As) solar array cells, a video camera, and an accelerometer. ESEX is one of nine experiments being launched in early 1998 on the Advanced Research and Global Observation Satellite (ARGOS). ARGOS is managed by the Space Test Program Office of the Space and Missile Test and Evaluation Directorate at Kirtland AFB, NM. The ARGOS satellite will be launched on a Delta II into a 460 nautical mile, 98.7° inclination orbit.

The ESEX flight system, Figure 1, includes a propellant feed system (Vaughan 1993), power subsystem (Biess 1994), commanding and telemetry modules, on-board diagnostics (Kriebel 1993), and the arcjet assembly (Vaughan 1993). ESEX is a self-contained, hexagonal structure which is thermally isolated from ARGOS. This design allows ESEX to function autonomously, requiring support only for attitude control, communications, radiation-hardened data storage, and 28 Vdc power for housekeeping functions such as battery charging and thermal control.

ESEX completed the flight qualification phase (Sutton 1995) of the development program in July, 1995, and was delivered to the ARGOS prime contractor, Boeing North American (BNA), in early March, 1996. ESEX and the remaining eight experiments have since been, and continue to be, a part of the qualification testing of the ARGOS satellite. Testing to date has included a series of functional verifications, an electromagnetic compatibility (EMC) test, and the acoustic and pyro-shock environment verification. The thermal vacuum/balance test is scheduled to begin in mid-August. Since the ESEX flight unit has already been flight qualified, the integrated space vehicle (ISV) testing primarily serves to verify the interface between ESEX and ARGOS. However, the ISV testing also serves to verify one of the major objectives for the ESEX program - to demonstrate the feasibility of integrating high power electric propulsion systems with operational satellites.

This paper summarizes the ARGOS system-level testing accomplished to date. It also presents the status of the flight planning effort. The paper will also summarize the progress of the experiment development efforts in each of the four science objective areas - optical observations, contamination, electromagnetic compatibility, and arcjet system performance.

ISV TESTING

The ESEX flight unit is unique on the ARGOS spacecraft in that this experiment is mated and de-mated several times throughout the ISV test flow in order to accommodate test requirements and restrictions. This provides a convenient time for ESEX to perform several ground operations including battery installation and propellant loading (Bromaghim 1996). Recent papers have summarized all but the most recent test accomplishments (Salasovich 1997) and give a description of the remaining tests (Bromaghim 1996). The results of the ARGOS acoustic test are summarized below.

Acoustic Testing

For the acoustic testing, the ESEX flight unit was mechanically mated to ARGOS and installed in the acoustic chamber (Figure 2) as soon as the remaining functional tests were completed. As described elsewhere (Salasovich 1997), the first time ESEX was mated, there was an unanticipated gap ranging from 0.5 to 1.0 mm between the six ESEX feet and the ARGOS bulkhead. Originally it was thought that this gap was due to twisting of the ESEX structure induced from changing the mechanical configuration of the flight unit. For this second integration then, the ESEX flight unit was configured the same as it was for the original ESEX measurement (Sutton 1995) and as it will be for the final flight installation. Because of this change, it was anticipated that the gaps would be much smaller. When the integration occurred, however, the gaps were relatively unchanged. Subsequent analysis indicated that the ESEX structure was too rigid to be subject to deformations from opening and closing the access panels. No further measurements of the flatness of either side of the interface are planned, and shims will remain installed for flight.

The acoustic test was conducted in early July, 1997 at the BNA test facilities in Seal Beach, CA. The acoustic test was divided into three runs – one at 140 dB (3 dB below the protoflight level) for 20 seconds, one at the protoflight level of 143 dB for 20 seconds, and the final run at the protoflight level for 40 seconds. Each time, the accelerometer output was evaluated after the run to ensure the vehicle response was as predicted. The ESEX flight unit had two sets of three orthogonal accelerometers installed – one on the interface bulkhead between ESEX and ARGOS, and one on the witness tower behind TQCM sensor #4. During the ESEX flight unit random vibration testing, this TQCM sensor failed due to a known problem with the mounting scheme for the internal crystal.

 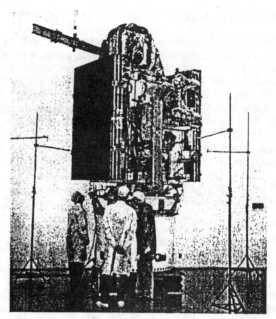

FIGURE 2. Mating the ARGOS and ESEX Flight Units and Installing the Integrated Space Vehicle in the Acoustic Chamber.

Following the 140 dB run, the data was evaluated to ensure proper channelization of the accelerometers, and to perform a preliminary evaluation of the data to assess if any components were in danger of being damaged by the full protoqual run. When looking at the ESEX interface accelerometers, there were significant outages at frequencies near 60 Hz in one the lateral axes (z-axis). Although these levels were significantly higher than the specification, the test continued with the protoqual level run since there was no apparent recourse. Once the full protoqual runs were conducted, a review of the data showed that the response on the z-axis accelerometer was below the limit. Initially, the strong responses were thought to be structural modes of the interface bulkhead as predicted by engineers from BNA and The Aerospace Corporation. The subsequent diminishing of the signals as observed during the protoqual run was attributed to the loosening of the interface fasteners. Subsequent investigation conducted after all of the testing revealed that these signals were actually 60 Hz noise on the analog channels. This also explains the "disappearance" of the signals in the protoqual run when there was more filtering on the analog channels.

The response from the accelerometers on the witness towers was also above the limit in the 140 dB run, but these signals did not decrease at the protoqual levels. The outages occurred at approximately 80 and 110 Hz, but were not a restriction to proceeding with the next runs. The TQCM sensor, and all of the sensors, on the witness tower were tested during the combined systems functional test (CSFT) following the acoustic test and were demonstrated to be fully operational. The CSFT verified the functionality of the remainder of the ESEX flight unit as well as all of the major subsystems on ARGOS and the other eight experiments. This was the same test done on the vehicle before the acoustic test, and is used as a baseline test for all of the environmental testing.

Thermal Vacuum Testing

Following the completion of the post-acoustic CSFT, preparations for the thermal vacuum/balance were initiated. This included the remainder of the thermal blanket installation, but also included some battery operations, installing final hardware to put the vehicle in the flight configuration, and other experiment activities. Once the preparations were complete, the vehicle was installed into the test fixture and moved into the thermal vacuum chamber.

All of the experiments conducted thermal vacuum tests at the component level, so this test served as a verification of the interfaces between all of the experiments and the ARGOS bus. The test also verified the functionality of the ARGOS vehicle components at the temperature extremes including the flight computer, data recorder, command and telemetry system, and all of the other components of the ARGOS space vehicle.

In addition to the system-level verification, this thermal vacuum test allowed several of the experiments, including ESEX, to perform tests which were impossible to accomplish in ambient conditions. For ESEX these included a heater test to verify the thermal control system functioned according to design, and a test to drive the TQCM sensors to the temperature extremes. This is discussed below in the section on the contamination experiments.

Future Ground Testing and Operations

Once the thermal vacuum testing is complete, ESEX will be de-mated to perform a series of ground operations. These have been described elsewhere, and include the installation of the final flight software, loading the ammonia propellant, and installing the activated flight battery (Bromaghim 1996). These activities are divided up as follows: while ESEX is installing the flight software, the remainder of the space vehicle (SV) undergoes a verification of the weight and CG, and a solar array drive test. Once the ESEX software is installed and verified and the SV tests are complete, the ESEX flight unit will be rolled over to ARGOS and connected to the bus via extender cables. At this point, the final factory functional (FFF) test will be conducted to verify all of the ARGOS systems (including all of the experiments) are functional before the test system is dismantled and shipped to the launch site. Due to the time-limited requirement for the activated flight battery, the final closeout of the ESEX flight unit will occur approximately 60 days prior to launch. ESEX will then be re-mated to ARGOS and the ISV will be shipped to the launch site for an early 1998 launch.

FLIGHT OPERATIONS PLANNING

The planning function for flight operations is continuing. An outline of the ESEX firings has been defined which will accomplish all of the mission objectives including the optical observations and the communication tests as described below. This firing plan has been prioritized based on importance to the acceptance of high power arcjet technology. A plan is in work to overlay these firings with the planned releases for the Critical Ionization Velocity (CIV) experiment in order to obtain an integrated Phase II timeline that maximizes the overall scientific return.

Science Team Progress

Development of the science data effort has progressed significantly. The team is divided into four groups based on the type of data that will be returned by the ESEX diagnostics (LeDuc 1996). A brief description, as well as a status, of each of these areas is summarized below.

Optical Observations

The optical experiments include observations from the on-board video camera as well as ground observations of the luminescent thruster plume and radiating thruster body from the AF optical telescopes on Maui. Both scientific and operational objectives are thereby accomplished. DoD system operations planners require plume signature and activity observability data to develop an assessment of the robustness and survivability of arcjet systems. These observations also yield detailed information about excited state populations in the plume, which are in turn directly related to the developmental goals of assessing thruster loss mechanisms.

At this power level, dissociation and ionization are likely to be more important loss mechanisms than for the low power arcjet. The situation is not well understood for high power devices, since neither the best experimental (LIF, absorption, mass spectroscopy, plasma probes) or modeling (DSMC, PIC, etc.) techniques have been applied with the same diligence as accomplished for low power devices. A few tests have been conducted which provide the ESEX science team with spectral data of the high power arcjet in ground facilities. In addition, some plume modeling by Primex Aerospace Company (PAC) has been conducted under contract to Phillips Laboratory.

The experimental objectives will be satisfied by integrating a spectrometer and calibrated CCD detector in series with an atmospheric tracker and compensator onto the 1.6 meter diameter telescope at the Maui Space Surveillance Site (MSSS, formerly AMOS). Since the arc itself is a bright (>~ 100W) UV source, and since the arcjet plume radiates primarily in the UV-visible spectrum (arc plasma temperatures allow electronic state excitation to dominate the plume emission), much of the observational effort will be devoted to the visible and near UV. Specifically, the NH(A-X) transition, centered near 335 nm, and the hydrogen Balmer series will be targeted for Boltzman analysis to determine population distributions of the excited state populations and comparison to equilibrium temperatures.

Secondary objectives include gathering data on atomic, molecular, and ionic nitrogen emissions, blackbody radiation from the thruster body, and a survey spectrum to gather data on emitters not seen in ground tests (such as those arising from metastable states or from interactions between the plume and ambient background on orbit). Observations will also be compared to the PAC observations and plume models. The space flight observations constitute a collaboration between the AF Space Command and Phillips Laboratory telescope operators, University of Maryland observational astronomer Ken Kissell, and the ESEX science team.

Contamination

The contamination experiment includes the TQCM sensors, the radiometers, and the Ga-As solar array witness sample. The primary goal of this effort is to assess the contamination of critical spacecraft surfaces as a result of the arcjet firing

The TQCM is capable of detecting a number of different materials – especially as the crystal temperature decreases. There are four TQCMs onboard ESEX located at different positions with respect to the arcjet (Figure 1). Modeling conducted under contract with TRW indicates that the sensor closest to the arcjet will potentially see the greatest flux of anode material, however it will get too hot to measure propellant constituents during a 15-minute firing. Current plans dictate a cold gas release of GN_2 and ammonia, as well as a four-minute hot firing prior to the start of nominal operations (Bromaghim 1996). These operations (which result in a lower thermal loading) in conjunction with the flexibility to alter the duration of all of the ESEX firings may allow this sensor to detect some constituents. The other three sensors are located farther away from the arcjet so that they can be maintained at cooler temperatures. Unfortunately, this also means the sensors will be subject to less mass flow - making detection difficult.

In order to measure the lowest temperature these sensors can attain, a test was performed in the ARGOS thermal vacuum/balance test. This test attempted to cool the sensors to -100°C while the heat sink (i.e. the ESEX flight unit) is at a nominal orbital temperature. If these lower temperatures can be attained, then there is a potential for measuring deposition on-orbit of the CIV releases of CO_2 and Xe, as well as the ESEX cold gas release and the short duration firing as described above.

Electromagnetic Compatibility

The electromagnetic (EM) compatibility experiments will utilize the data from the near- and far-field EMI antennas, as well as ground communications tests to determine the effect of the arcjet on standard spacecraft communications. This includes radiated EMI transmit frequency measurements at 2, 4, 8, and 12 GHz, as well as quantified testing of the bit error rate on Space-Ground Link Subsystem (SGLS) transmissions. This area also collaborates with another experimenter, as described below, to observe the effect of the arcjet firing on UHF transmissions.

In order to demonstrate the one of the three communications test concepts, a series of bit error rate tests (BERT) was conducted over the Camp Parks Test Facility in Pleasanton, CA using the MSTI-3 and MSX spacecraft. The purpose of the test was to prove the concept and gather preliminary BERT data with an on-orbit vehicle. The test made use of the spacecraft signal that accommodates a ranging channel and associated pseudo-random noise (PRN). For this test the PRN uplink was replaced with a BERT code (2047 pattern @ 1.024 Mbps) to accurately characterize generic spacecraft communications. The space vehicle transponder received the signal and remodulated the detected information onto the downlink carrier. The ground system received the mirrored BERT signal and determined a bit error rate based on a comparison with the original transmission. This test verified the feasibility of the technique and allowed for an assessment of the effect of transmit power and modulation index.

A synergistic experiment is planned with the Coherent Electromagnetic Radio Tomography Experiment (CERTO). CERTO is a UHF beacon operating at 150.012 and 400.032 MHz with full ground illumination along most of the orbit. Receivers on the ground will use differential phase techniques to derive the electron density of the ionosphere. CERTO will be operated before, during, and after arcjet operation to quantify the effect of arcjet operation on electron density in the upper atmosphere.

Performance

ESEX is using a variety of methods for measuring the performance of the arcjet including the on-board accelerometer, remote tracking, and spacecraft GPS telemetry. This area will also determine the effectiveness of the other arcjet hardware including the power conditioning unit (PCU) and the propellant feed system.

Several issues arise when using Global Positioning System (GPS) data or SGLS tracking for a performance measurement of a low thrust device like the ESEX arcjet. The resolution of the GPS measurement, for example, is highly dependent on what type of receiver is located on-board the vehicle, ARGOS in this case. For the ESEX measurements, the raw data from the GPS constellation is most important, but is not available. To compensate for this unfortunate configuration problem, ESEX will be performing some limited post-processing of the data output from the on-board receiver. For the tracking measurements, the resolution is highly dependent on the number of contacts attainable over a relatively short (i.e. 24-hour) period. Both of these issues are being evaluated and will be resolved before the flight.

In order to support the thrust vector analysis on-orbit, the alignment of the arcjet relative to the diagnostic deck was measured to form a baseline for the final alignment measurement. This measurement was made because the arcjet exit plane cannot be conveniently measured while ESEX and ARGOS are mated. The next time the alignment will be measured is after the final mating, following all of the ESEX ground operations. The final measurement will determine the alignment between the reference point on the diagnostic deck (established by this most recent measurement) and the center of gravity of the ARGOS vehicle.

SUMMARY AND CONCLUSIONS

The ESEX flight unit and the ARGOS satellite are proceeding through the flight qualification testing. Most of the pre-environmental functional testing has been completed and the environmental testing, starting with the acoustic tests, will proceed in mid-July. The planning for the flight operations and the science data reduction continues and emphasis will continue to be placed on maximizing the science return from the ESEX flight.

Acknowledgments

The authors wish to thank the ESEX science team at the Phillips Laboratory including Greg Spanjers, Keith McFall, and Jamie Malak; and the TRW program manager, Mary Kriebel, for her expertise on the flight hardware.

References

Biess, J. J. and A. M. Sutton (1994), "Integration and Verification of a 30 kW Arcjet Spacecraft System," AIAA Paper 94-3143.

Bromaghim, D. R. and A. M. Sutton (1996), "Electric Propulsion Space Experiment Integration and Test Activities," AIAA Paper 96-2726.

Kriebel, M. M. and N. J. Stevens (1992), "30-kW Class Arcjet Advanced Technology Transition Demonstration (ATTD) Flight Experiment Diagnostic Package," AIAA Paper 92-3561.

LeDuc, J. R., et. al. (1996), "Performance, Contamination, Electromagnetic, and Optical Flight Measurement Development for the Electric Propulsion Space Experiment," AIAA Paper 96-2727.

Salasovich R. M., Bromaghim D. R., and Johnson L. K. (1997), "Diagnostics and Flight Planning for the US Air Force Phillips Laboratory Electric Propulsion Space Experiment (ESEX)," AIAA Paper 97-2777.

Sutton, A. M., D. R. Bromaghim, and L. K. Johnson (1995), "Electric Propulsion Space Experiment (ESEX) Flight Qualification and Operations," AIAA Paper 95-2503. Presented as a JANNAF Paper, December, 1995.

Vaughan, C. E. and J. P. Morris (1993), "Propellant Feed Subsystem for a 26 kW Flight Arcjet Propulsion System," AIAA Paper 93-2400.

Vaughan, C. E., R. J. Cassady, and J. R. Fisher (1993), "Design, Fabrication, and Test of a 26 kW Arcjet and Power Conditioning Unit," IEPC-93-048.

PERFORMANCE EVALUATION OF A THERMIONIC CONVERTER WITH A MACRO-GROOVED EMITTER AND A SMOOTH COLLECTOR

Mohamed S. El-Genk and Yoichi Momozaki
Institute for Space and Nuclear Power Studies/Department of Chemical and Nuclear Engineering
The University of New Mexico, Albuquerque, NM 87131
(505) 277-5442, FAX: -2814, email: mgenk@unm.edu

Abstract

The performance of a thermionic converter with a planar, macro-grooved, Molybdenum emitter and a smooth Molybdenum collector was investigated experimentally. The emitter's concentric macro-grooves were 0.5 mm deep, 0.5 mm wide, and 1.0 mm apart and the inter-electrode gap was 0.5 mm wide. The emitter temperature was varied from 1573 to 1773 K, the collector temperature was 1073 and 1173 K, and the cesium pressure, P_{Cs}, in the inter-electrode gap was varied from 50 - 500 Pa. A peak electric power density of 1.2 W/cm^2 and a peak conversion efficiency of ~8.3%, were achieved at a relatively low emitter temperature of 1773 K and a high collector temperature of 1173 K. When the emitter and collector temperatures were 1673 and 1073 K, the maximum peak conversion efficiency at $P_{Cs} = 300$ Pa was 8.1%, and the peak electric power density was ~1.0 W/cm^2. At emitter and collector temperatures of 1573 and 1073 K, respectively, the maximum peak power density (0.46 W/cm^2) and the maximum peak conversion efficiency (5.4%) were ~20% higher and 2% lower, respectively, than for a larger gap of 1.5 mm (0.4 W/cm^2 and 5.5%, respectively).

INTRODUCTION

Thermionics (TI) technology has been under development for more than four decades. These static converters had been developed mostly for space power applications. In such applications, typical emitter (cathode) and collector temperatures are 1850-1900 K and 800-900 K, respectively, and inter-electrode gap is typically 0.1 to 0.5 mm wide. The gap is filled with cesium vapor at a pressure that could be as high as 3 torr (400 Pa). For space applications, the electrodes are usually made of high temperature refractory metals, such as tungsten, molybdenum, or rhenium. These materials have relatively high bare work functions (4-5 eV), but their cesiated work functions could be as low as 1.2 eV.

When a TI converter is operated in the ignited mode, increasing the cesium pressure increases the electric power output up to a point, beyond which scattering losses causes the electric power output to decrease. The optimum cesium pressure, corresponding to the peak electric power, depends on the type of the electrode material and the electrode temperatures. For a given cesium pressure, raising the emitter temperature or lowering the collector temperature increases emission current and the electric potential across the external load. The temperature of the collector, however, should not be lower than that corresponding to the minimum cesiated work function of its material.

On the other hand, increasing the emitter temperature increases material loss from emitter surface by sublimation. Emitter material loss can also occur due to the formation of volatile oxides, if oxygen partial pressure in the inter-electrode gap is in excess of 10^{-8} torr (Gun'ko and Smirnova 1983; Korykin and Oberzumov 1990, and Paramonov and El-Genk 1996a, 1996b and 1997). High emitter temperatures also impose a technical challenge to the integrity of the insulation material and the metal-ceramic brazes in the converter, hence shortening its performance lifetime.

Recently, there has been a growing interest to develop low temperature TI converters for terrestrial applications, such as topping cycles for natural gas and oil fired electric power plants and for diesels fired engines. In these applications, the source temperature would be several hundred degrees lower than in space applications. A practical approach to enhance the performance of low temperature thermionic converters is to use grooved electrodes. When these converters operate in the so called "hybrid mode", the breakdown or discharge ignition in the emitter's grooves occurs at low cesium pressure. Such a breakdown provides: (a) electrons for cesium ionization in non-grooved portion of the emitter surface, which operates in the diffusion mode, and (b) cesium ions, that diffuse to the non-grooved portion of the emitter and neutralize the space charge, hence, lowering the optimum cesium pressure (El-Genk and Luke 1997).

Several experiments have been performed using electrodes with macro- and micro-grooved electrodes, for a wide range of electrodes temperatures, cesium pressures, inter-electrode gap sizes, with mixed results (Shimada 1977 and 1979, Tskhakaya et al. 1983, Atamasov et al. 1984, and Lee et al. 1990). In general, TI converters with grooved electrodes have performed better than converters with smooth electrodes. The best performance has been reported for emitter temperatures of 1700-1750 K, with an increase of 25-50% in the conversion efficiency and the electric power density. However, due to the lack of systematic studies that use same experimental setup and procedures, reported results were inconclusive and sometimes contradictory (Paramonov and El-Genk 1996c).

Shimada (1977 and 1979) tested a thermionic converter with molybdenum (Mo) emitter that had 0.5 mm square longitudinal grooves, with a 1.0 mm separation. The non-grooved portion of the emitter surface was coated with rhenium, which has a high bare work function, and hence lower cesiated work function, than Mo. He reported a 300% increase in the electric power output, compared to a smooth electrodes converter. Tskhakaya et al. (1983) tested a converter with a smooth zirconium oxide collector and a Mo emitter that had grooves similar to that of Shimada's, but the tops were coated with platinum. They compared the electric power output with that for a converter that had smooth platinum emitter and zirconium oxide collector. Although the performance of the former was better than the latter, Tskhakaya et al. (1983) were unable to duplicate the large increase in electric power output reported by Shimada (1977 and 1979). Their optical observations of plasma discharge confirmed that, while the grooves supplied ions to the rest of the gap, the converter did not operate in the hybrid mode described by Shimada. Tskhakaya et al. (1983) results also showed that the optimum gap size for a grooved emitter converter (1.6 mm) was higher than that for a smooth electrode converter (1.0 mm). Tskhakaya et al. (1983) also tested a converter with a grooved emitter, similar to that of Tskhakaya and Yarygin (1983), and a zirconium oxide collector that had the same grooves as the emitter. The collector's grooves were laid in the perpendicular direction to the emitter's grooves. The electric power output of this converter was better that of Tskhakaya et al. (1983).

Atamasov et al. (1984) tested a converter consisting of a single crystal, smooth tungsten emitter and five micro-grooved collectors in the same test device. The inter-electrode gap was 0.35 mm wide. They applied an axial magnetic field to eliminate edge effects and electron scattering among various collectors. The longitudinal grooves in all collectors were 0.1 mm wide and 0.15 mm apart, but had different depths from 0.1 to 0.5 mm. The measured peak electric power was 20-60% higher than with a smooth collector, and occurred at 0.1-0.2 V higher terminal voltage. Recently, Lee et al. (1990) performed a series of experiments with four electrode combinations, namely: (a) smooth electrodes, (b) micro-grooved emitter and smooth collector, (c) smooth emitter and micro-grooved collector, and (d) micro-grooved electrodes. The emitter grooves were 0.075 mm wide, 0.25 mm deep, and 0.075 mm apart, and the collector grooves were 0.1 mm wide, 0.25 mm deep, and 0.125 mm apart. The emitter was made of tungsten chloride, the collector material was niobium, and the size of the inter-electrode gap varied from 0.125 to 0.5 mm. Their results showed a 0.1 V increase in the output voltage of the converter (b), a small increase in the electric power output and a 50% increase in the emission current density of converter (c), and no discernible difference in the performance of converter (d), compared to converter (a).

The objective of this work was to experimentally investigate the performance of a TI converter with a planar, macro-grooved, Molybdenum emitter and a smooth Molybdenum collector. The concentric macro-grooves of the emitter were of the same dimensions as those of Shimada (1977): 0.5 mm wide, 0.5 mm deep, and 1.0 mm apart 0.5 mm deep, 0.5 mm wide, and 1.0 mm apart. This work and that of El-Genk and Luke (1997) were the first to use concentric grooves, in order to reduce edge losses of ions from the inter-electrode gap. The inter-electrode gap for the present converter was 0.5 mm wide. In the experiments, the emitter temperature was varied from 1573 to 1773 K, the collector temperature was 1073 and 1173 K, and P_{Cs} was varied from 50-500 Pa. The results of the present converter, for emitter and collector temperatures of 1573 and 1073 K, respectively, were compared with those for an identical converter that had a larger gap size of 1.5 mm, and was tested at the same condition in the same test facility (El-Genk and Luke 1997).

EXPERIMENT SETUP

In the present converter, the emitter and the collector are parallel flat ends, hollow molybdenum cylinders, 16 mm in outside diameter (emission surface area of 200 mm^2). The assembled converter was tested in vacuum (10^{-8} torr) at well controlled conditions in the Thermionic Test facility at the University of New Mexico's Institute for Space and Nuclear Power Studies. The cesium vapor was supplied to the inter-electrode gap from the cesium reservoir, via a connecting tube. The cesium pressure in the gap was adjusted by varying the temperature of its reservoir and was determined from the measured cesium reservoir temperature, using the saturation pressure correlation of Alcock et al. (1984). The range the Cs reservoir temperature in the experiments was 473-603 K, and the corresponding P_{Cs} in the inter-electrode gap was 10-500 Pa. The temperature of the Cs reservoir was controlled to within ±1 K.

The emitter and collector were heated using tungsten wires that were wound near the inside surface of the electrodes. Electron bombardment was also used to heat the electrodes at high temperatures. The temperatures of the electrodes were measured using tungsten-rhenium thermocouples. The emitter temperatures in the experiments were varied from 1573 to 1773 K, and the collector temperature was 1073 K and 1173 K. The emitter temperature was controlled to within ±5 K and the collector temperature was controlled within ±1 K.

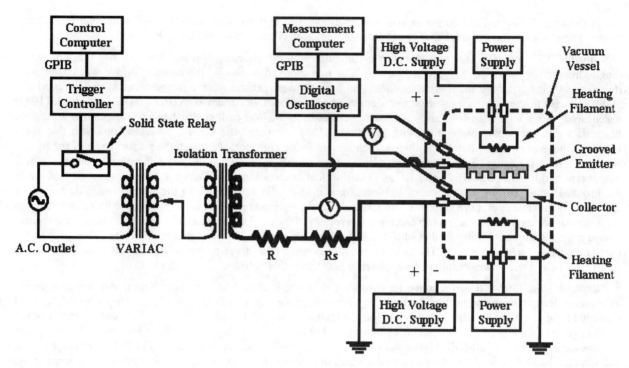

FIGURE 1. A Line Diagram of the Experimental Setup.

Before each experiment, the emitter and the collector temperatures were kept at 1100 K and 900 K, respectively, to stimulate cesium absorption onto the electrodes. The electrodes were kept at these temperatures for 3 hours, during which the cesium pressure in the interelectrode gap was kept constant at 500 Pa. After the cesium sorption onto the electrodes reached equilibrium, the temperature of the cesium reservoir was changed to the desired value, prior to increasing the emitter and the collector temperatures. After reaching equilibrium at the new electrodes and cesium reservoir temperatures, the current-voltage (I-V) characteristics of the device were measured by applying a 60 Hz sinusoidal voltage pulse (Fig. 1). The duration of the pulse was typically 33 ms, which corresponds to two full sinusoidal waves. However, the measurements were recorded only during the first full wave, to avoid possible change in the initial conditions due to the discharge. The voltage sweep was from zero (short circuit current) to the open circuit voltage (zero current). A load resister was connected to the external circuit of the converter (Fig. 1) to prevent ignition, except when the sinusoidal waves were applied to the electrodes, in order to maintain the cesium pressure in the inter-electrode gap almost the same in subsequent voltage sweeps. The terminal voltage was measured directly across the electrodes (Fig. 1), while the electric current was determined from the measured voltage across the shunt resister, R_s, in the external circuit. The experimental data were recorded, at intervals of 4 μs, by a digitizing oscilloscope and stored in a PC for subsequent analysis.

RESULTS AND DISCUSSION

The experimental measurements for the present converter with a grooved emitter and a smooth collector were compared, as functions of cesium pressure, at different emitter and collector temperatures. Best estimate values of the converter's efficiency were obtained and also compared at different operating conditions. The peak conversion efficiency and electric power density, at emitter and collector temperatures of 1573 and 1073 K, respectively, were compared with those for an identical converter that had a larger gap of 1.5 mm (El-Genk and Luke 1997), to assess the effect on the converter performance.

I-V Characteristics

Figures 2a and 2b show the measured I-V characteristics for emitter temperatures of 1573-1773K, collector temperatures of 1173 K and 1073 K, respectively, and different P_{Cs}. As delineated in these figures, the short circuit current increased and the I-V characteristics shifted to the right, to higher voltage, as P_{Cs} was increased. A current density in excess of 5.5 A/cm^2 was measured at an emitter temperature, T_{EM}, of 1673 K, collector temperature, T_{co} of 1073 K, and P_{Cs}=312 Pa (Fig. 2b). For the same T_{EM} and P_{Cs}, raising the collector temperature decreased the converter's short current density. The effects of the changing the collector temperature on the I-V characteristics,

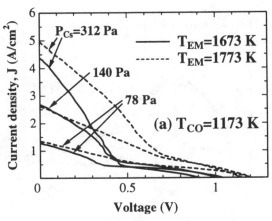
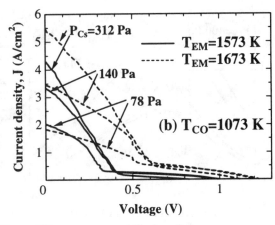

FIGURE 2. I-V Characteristics for Different Emitter and Collector Temperatures and Cesium Pressure.

for an emitter temperature of 1673 K are delineated in Figure 3, at P_{Cs} of 140 and 312 Pa. As shown in this figure, lowering the collector temperature by 100 K, from 1173 to 1073 K, particularly at the high P_{Cs}, significantly increased the current density of the converter. For example, at a cesium pressure of 312 Pa, lowering the collector temperature from 1173 to 1073 K, increased the short circuit current density of the converter by 25% (from 4.35 to 5.45 A/cm^2). A further decrease in the collector temperature was not possible in the present test facility, because the collector assembly did not have an independent cooling system. Such capability will be incorporated in future tests. The present results suggest that operating the collector at a typical temperature value of 800 K, could more than double the current density of the present converter.

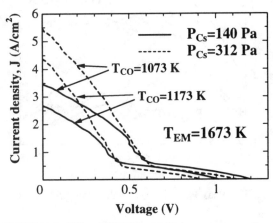

FIGURE 3. Effect of Collector Temperature on the I-V Characteristics.

Load Following Characteristics

Figures 4 and 5 present the measured load-following characteristics and the conversion efficiency curves of the present TI converter. In the load following characteristics, the measured electric power density of the converter was plotted versus the load resistance. In the portion of the characteristics to the right of the peak power density is of practical interest, since in it the TI converter is load following. Therefor, in this portion of the characteristics as the load resistance increases (i.e. less demand for electric power), the electric power density of the converter decreases. Conversely, as the demand for electric power increases (i.e., load resistance decreases), the electric power density of the TI converter increases.

Figures 4a and 4b show that for the same collector temperature of 1173 K, increasing the emitter temperature and or the cesium vapor pressure, increased both the peak electric power density and peak efficiency of the converter, and shifted them to lower load resistance (or low voltage). Figures 4 and 5 also show that increasing the cesium pressure decreased the load resistance (or load voltage) corresponding to the peak conversion efficiency and peak electric power density. The load resistance corresponding to the peak conversion efficiency, however, was always larger that that for the peak electric power density. Increasing the cesium vapor pressure also shifted the peak electric power density to a lower load resistance (lower voltage). Figure 4a indicates that a peak electric power density of about 1.2 W/cm^2 and a peak conversion efficiency of 8.1% were measured at P_{Cs} of 400 Pa. This peak electric power density was almost halved and the peak conversion efficiency decreased to 5.6%, as the emitter temperature decreased by only 100 K, to 1673 K. Similar results are demonstrated in Figures 5a and 5b, at a lower collector temperature of 1073 K and emitter temperatures of 1673 K and 1573 K, respectively. When the collector temperature was reduced from 1173 (Fig. 4b) to 1073 K (Fig. 5a), a peak power density of more than 1.1 W/cm^2 and a peak conversion efficiency of 8.3% were measured at the same emitter temperature as Fig. 4b (1673 K). Again, when the emitter temperature was reduced from 1673 (Fig. 5a) to 1573 K (Fig. 6b), the peak electric power density

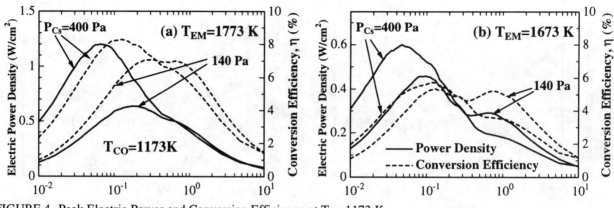

FIGURE 4. Peak Electric Power and Conversion Efficiency at T_{CO}=1173 K.

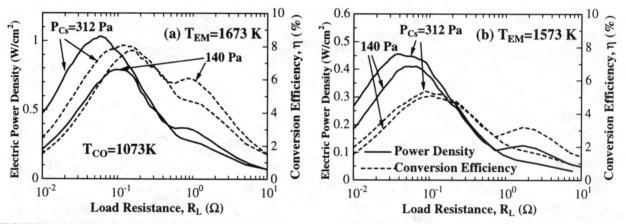

FIGURE 5. Peak Electric Power and Conversion Efficiency at T_{co}=1073 K.

at P_{Cs} = 312 Pa, was more than halved, from 1.1 to 0.45 W/cm^2, and the peak conversion efficiency decreased to ~5%.

Peak Electric Power Density

Figure 6 plots the measured peak electric power density for the present converter, at different combinations of emitter and collector temperatures and for cesium vapor pressure from 50 to 500 Pa. This figure shows that for the same collector temperature of 1173 K, the peak electric power density at an emitter temperature of 1173 K was much higher than that at an emitter temperature of 1673 K, and the difference increased with increased cesium pressure. For example, at a cesium pressure of 200 Pa, the peak electric power at emitter temperatures of 1773 K and 1673 K was 0.9 and 0.5 W/cm^2, respectively. At a higher cesium pressure of 400 Pa, the electric power density at an emitter temperature of 1773 K increased to 1.2 W/cm^2, while that for the lower emitter temperature of 1673 K, remained almost unchanged at 0.5 W/cm^2. The highest peak electric power density for the former emitter temperature (1.2 W/cm^2), however, occurred at a higher cesium pressure (400 Pa) than the latter (~250 Pa).

Figure 7 also show that decreasing the collector temperature significantly increases the peak power density of the grooved emitter converter. The cesium pressure corresponding to the maximum peak electric power density, however, increased as the collector temperature decreased. For example, at an emitter temperature of 1673 K and cesium pressure of 300 Pa, decreasing the collector temperature form 1173 to 1073 K, increased the maximum peak electric power density of the present converter from 0.6 W/cm^2 to as much as 1.05 W/cm^2, a 75% increase. Figure 6 shows that the maximum peak electric power for the higher collector temperature (1173 K) occurred at ~200 Pa, while that for the lower collector temperature (1073 K) occurred at about 300 Pa.

Effect of Inter-Electrode Gap Width

Figure 7 compares the measured peak electric power density and conversion efficiencies of two identical converters with grooved emitters and inter-electrode gaps of 1.5 and 0.5 mm. The results presented in Figure 7 are

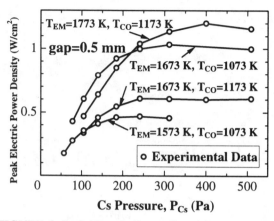
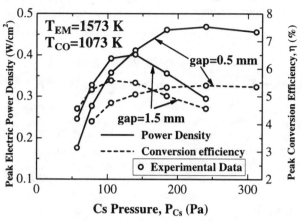

FIGURE 6. Peak Electric Power Densities.

FIGURE 7. Effect of Gap size on Peak Power Density and Conversion Efficiency.

for a low emitter temperature of only 1573 K and a collector temperature of 1073 K. As this figure shows, the maximum peak electric power for the 0.5 mm gap converter was 0.46 w/cm², which was 15% higher than that for the 1.5 mm gap converter. The maximum peak conversion efficiency for the former (5.4%), however, was about 2% lower than for the latter (5.5%). The results delineated in Fig. 7 clearly show the strong effect of the inter-electrode gap size on the performance of the grooved emitter converters.

Conversion Efficiency

The present test facility does not allow for direct measurements of the heat input to the emitter or of the heat rejection from the collector. Best estimates of the conversion efficiency of the present converters were calculated, based on the measured emission current and electrode temperatures, as:

$$\eta = P_e / [P_e + Q_{rad} + Q_{ec} + Q_{con}]. \tag{1}$$

In this equation, the thermal power transfer in the inter-electrode gap by conduction, Q_{con}, is usually small, compared to that due to electron cooling, Q_{ec}, and by thermal radiation, Q_{rad}. These thermal power transfer contributions can be given as:

$$Q_{rad} = \sigma \varepsilon_{eff} F A_s \left(T_{EM}^4 - T_{CO}^4 \right), \tag{2a}$$

$$Q_{ec} = \left(\frac{2kT_{EM}}{q} + \phi_{EM} \right) I, \text{ and} \tag{2b}$$

$$Q_{con} = k_{Cs} (T_{EM} - T_{CO}) / (g + \delta), \tag{2c}$$

where g is the width of the inter-electrode gap and δ is the temperature jump distance, which can be given (Kitrilakis and Meeker 1963) as:

$$\delta = 1.53 \times 10^{-5} [(T_{EM} + T_{CO}) / P_{Cs}]. \tag{3a}$$

The thermal conductivity of the cesium vapor in the inter-electrode gap is given as (Kitrilakis and Meeker 1963):

$$k_{Cs} = 1.65 \times 10^{-4} \sqrt{T_M}, \tag{3b}$$

where,

$$T_M = (2/3) [(T_{EM}^{1.5} - T_{CO}^{1.5}) / (T_{EM} - T_{CO})]^2. \tag{3c}$$

In equation (2a), the effective emissivity, ε_{eff}, of the Molybdenum electrodes was taken equal to 0.11 (Paramonov and El-Genk 1996a), the calculated radiation view factor for the inter-electrode, F, using the relationship of two parallel disks was 0.94. The work function of the emitter, ϕ_{EM}, was determined at the voltage corresponding to the peak electric power output. The peak electric power corresponds to the transition from the obstructed ignited mode to the unobstructed ignited mode on the I-V characteristics. This emitter work function was determined from the Richardson - Dushman equation (Hatsopoulos and Gyftopoulos 1973), using the measured load current and emitter temperatures as:

$$\phi_{EM} = (T_{EM}/11{,}606)\ln\left(A_s \kappa T_{EM}^2/I\right) \quad (4)$$

In equation (4), the projected surface area of the emitter, A_s, is 2.0×10^{-4} m^2 and the emission constant, κ, for the emitter material (Mo) was taken equal to 55×10^4 A/m^2 K^2 (Culp 1979).

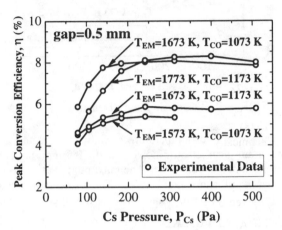

FIGURE 8. Peak Conversion Efficiency.

Figure 8 shows that for $P_{Cs} > 250$ Pa, the highest peak conversion efficiencies were those for emitter and collector temperatures of 1773 and 1073 K, respectively. At these temperatures, the maximum peak efficiency achieved was as much as 8.3%, at a cesium pressure of 300Pa. Increasing the emitter or lowering the collector temperature by 100K each, caused the maximum peak conversion efficiency of the converter to decrease slightly to 8.1% and shifted the corresponding cesium vapor pressure to 400 Pa. For an emitter temperature of 1673 K, increasing the collector temperature to 1173 K, dropped the maximum peak conversion efficiency to 5.9%. For an emitter and collector temperatures of 1573 and 1073 K, the maximum conversion efficiency was even lower at 5.3%. The results in Fig. 8 clearly demonstrate the excellent potential for achieving significantly higher peak conversion efficiencies (12-15%) with present converter, at a relatively low emitter temperature of 1773 K, when the collector temperature is lowered to 800 K.

SUMMARY AND CONCLUSIONS

Experiments were conducted to investigate the performance of a thermionic converter with a planar, micro-grooved emitter and a smooth collector. In the experiments, the emitter temperature was varied from 1573 to 1773 K, the Cesium vapor pressure in the inter-electrode gap was varied from 10 to 500 Pa, and the collector temperature was 1073 and 1173 K. The inter-electrode gap size for the present converter was 0.5 mm and the electrodes were made of polycrystalline.

Results showed that the grooved emitter converter gave a peak electric power density of 1.2 W/cm^2 and a peak conversion efficiency of ~ 8.3%, when tested at a relatively low emitter temperature of 1773K and a high collector temperature of 1173K. The cesium pressure corresponding to the peak power (400 Pa) was lower than that corresponding to the peak efficiency (300 Pa). When the collector temperature was decreased to 1073 K, the electric power density and conversion efficiency increased. When the emitter and collector temperatures were 1673 and 1073K, the peak conversion efficiency was ~ 8.1%. For emitter and collector temperatures of 1573 and 1073 K, respectively, the maximum peak electric power density for the present converter (0.46 W/cm^2) was about 15% higher than that for an identical converter that had a larger gap of 1.5 mm (0.4 W/cm^2). The maximum peak conversion efficiency for the present converter (5.4%), however, was about 2% lower than that for 1.5 mm gap converter, with a grooved emitter (5.5%).

The reported results are very promising, because a simple physical modification of the emitter surface resulted in a significant increases in the electric power output and the conversion efficiency of the converter, when operated at relatively low emitter temperature and a collector temperature that as high as 1073 K. The latter is about 270 K higher that that for conventional thermionic converters for space applications, for which achieving a conversion efficiency > 8.3%, would require operating at an emitter temperature >1800 K. The structural development of macro-grooves into the emitter surface is much less costly and more reliable than modifying the emitter work function through the introduction of oxygen traces in the inter-electrode gap, increasing the emitter temperature, or using a cesium-barium vapor mixture in the inter-electrode gap. However, further studies are needed to investigate the effect of using a grooved collector as well as of decreasing the collector temperature to 800 K, on the performance of the present thermionic converter with a grooved emitter.

The ability to achieve a peak conversion efficiency > 8.3% and a peak electric power density of 1.2 W/cm^2 for the present converter design, is very promising for low temperature terrestrial application, such as in topping cycles for oil, coal, and natural gas fired electric power plants. Additional research to demonstrate a further increase in the conversion efficiency to the 15% range, while keeping the emitter temperature below 1800 K and collector temperature at 800 K, would certainly expand the prospect of using TI converters in many terrestrial thermal-to-electric energy conversion applications.

Acknowledgment

This research was sponsored by the University of New Mexico's Institute for Space and Nuclear Power Studies.

References

Alcock, C. B., V. P. Itkin and M. K. Horrigan (1984) "Vapor Pressure of the Metallic Elements," *Canadian Metallurgical Quarterly*, 23: 309-313.

Atamasov, V. D., A. N. Evdokimov, V. A. Zhirkov, N. A. Ivanova, S. A. Skrebkov, S. V. Timashev, and I. A. Urtmintsev (1984) "Experimental Study of Cesium Arc Thermionic Converters with Multicavity Anodes," *Sov. Phys. Tech. Phys.*, 29: 42-45.

Baksht, F. G., G. A. Dyuzhev, A. M. Martsinovskiy, B. Ya. Moyzhes, G. Ye. Pikus, E. B. Sonin, and V. G. Yurev (1978) *Thermionic Converters and Low-Temperature Plasma*, Department of Energy Report No. DOE-tr-1, Technical Information Center/U.S. Department of Energy.

Culp, D. W. (1979) *Principle of Energy Conversion*, McGraw-Hill, New York, NY.

El-Genk, M. S. and J. Luke (1997) "Performance Comparison of Thermionic Converters with Smooth and Micro-Grooved Electrodes," *J. Energy Conversion and Management* (in Review).

Ender, A. Ya., V. I. Kuznetsov, B. G. Ogloblin, A. N. Luppov, and A. V. Klimov (1994) "Ultra-High Temperature Thermionic System for Space Solar Power Applications," in *Proc. 11th Symposium on Space Nuclear Power and Propulsion*, M. S. El-Genk, ed., CONF-94101, American Institute of Physics, AIP Conference Proceedings 301, 2: 861 - 867.

Geller, C. B., C. S. Murray, D. R. Riley, J.-L. Desplat, L. K. Hansen, G. L. Hatch, J. B. McVey, and N. S. Rasor (1996) *Final Report of the High Efficiency Thermionics (HET-IV) and Converter Advancement (CAP) Programs*, Report No. WAPD-T-3106, Bettis Atomic Power Laboratory, Westinghouse Electric Corp., West Mifflin, PA.

Gun'ko, V. M. and P. V. Smirnova (1983) "High Temperature Oxidation of Tungsten in Presence of Cesium Vapor," J. Phys. Chem., 57(10): 2581-2583.

Hotsopoulous, G. N. and E. P. Gyftopoulos (1973) *Thermionic Energy Conversion - Vol. I: Processes and Devices*, MIT Press, Cambridge, Mass.

Kitrilakis, S. and M. Meeker (1963) "Experimental Determination of the Heat Conduction of Cesium Gas," *Adv. Energy Conversion*, 3: 59 - 68.

Korykin V. A. and V. P. Oberzumov (1990) "Dynamics of Changes in Thermionic Energy Converter Collector Properties Under Conditions of Mass Transfer in the Interelectrode Gap," *Proc. 25th IECEC*, 2: 332 - 339.

Lee, C., D. Lieb and G. Miskolczy (1990) "Performance of Thermionic Converters with Structured Emitters and Collectors," *Proc. 25th IECEC*, 2: 294-299.

Paramonov D. V. and M. S. El-Genk (1996a) "Effect of Oxygen on Performance and Mass Transport in a Single Cell Thermionic Fuel Element," in *Proc. 31st IECEC*, IEEE, Washington. D.C., paper #96200.

Paramonov D. V. and M. S. El-Genk (1996b) "An Analysis of Ya-21 Thermionic Fuel Elements Test Results," in *Proc. 13th Symp. Space Nuclear Power and Propulsion,* M. S. El-Genk, ed., CONF-960109, American Institute of Physics, AIP Conference Proceedings 361, 3: 1395-1400.

Paramonov D. V. and M. S. El-Genk (1996c) "A Review of Cesium Thermionic Converters With Developed Emitter Surfaces," *J. Energy Conversion and management*, 38(6): 533 - 549.

Paramonov, D. V. and M. S. El-Genk (1997) "Effect of Oxygen on the Operation of a Planar Thermionic Converter for Isothermal and Isoflux Conditions," *J. Energy Convers. Mgmt.*, 38: (in print).

Shimada, K. (1977) "Low Arc Drop Hybrid Mode Thermionic Converter," in *Proc. 12th Intersociety Energy Conversion Engineering Conference*, Society of Automotive Engineers, 2: 1568-1574.

Shimada, K. (1979) "Recent Progress in Hybrid Mode Thermionic Converter Development," *in Proc. 14th Intersociety Energy Conversion Engineering Conference*, 2: 1890-1893.

Tskhakaya, V. K., L. P. Chechelashvili, and V. I. Yarygin (1983) "Hybrid Operating Mode of a Thermionic Converter with a Grooved Collector," *Sov. Phys. Tech. Phys.*, 28: 869-870.

---**Nomenclature**---

English

- A_s: Surface Area (m²)
- F: Radiation View Factor
- g: Width of the Inter-Electrode Gap (m)
- I: Current (A)
- J: Current Density (A/cm²)
- k: Boltzman's Constant (1.38×10⁻²³ J/K)
- k_{Cs}: Thermal Conductivity of Cesium (W/mK)
- P_{Cs}: Cesium Pressure (Pa)
- P_e: Electric Power (W)
- q: Electronic Charge (Coulomb)
- Q_{con}: Thermal Power Transfer by Conduction (W)
- Q_{ec}: Electron Cooling (W)
- Q_{rad}: Thermal Radiation (W)
- R: Resister (Ω)
- R_L: Load Resistance (Ω)
- T_{CO}: Collector Temperature (K)
- T_{EM}: Emitter Temperature (K)
- T_M: Inter-Electrode Gap Mean Temperature (K)

Greek

- δ: Temperature Jump Distance (m)
- η: Conversion Efficiency
- ε_{eff}: Effective Emissivity
- ϕ_{EM}: Work Function of Emitter (V)
- κ: Emission Constant (A/m²K²)
- σ: Stephan-Boltzman Constant (W/m²K⁴)

Subscript

- CO: Collector
- con: Conduction
- Cs: Cesium
- e: Electric
- ec: Electron Cooling
- eff: Effective
- EM: Emitter
- L: Load
- M: Mean
- rad: Radiation
- s: Surface

A RIGOROUS APPROACH FOR PREDICTING THERMIONIC POWER CONVERSION PERFORMANCE

Albert C. Marshall
Defense special Weapons Agency
1680 Texas Street
Albuquerque, NM 87185-0744
(505) 853-0946

Thermionic power conversion devices have been considered for use as a source of electrical power for electric propulsion of space craft. Methods for improving the performance of thermionic converters have been studied and successfully applied. However, puzzling anomalies have prevented the achievement of the full potential, as predicted by the standard approach for calculating net thermionic currents. Recently, electron reflection at the electrode surfaces has been identified as a possible cause of anomalous thermionic performance. Unfortunately, the standard methods used to predict net thermionic currents do not correctly account for electron reflection. The net current density for a vacuum-type conversion device is typically predicted by subtracting the collector current density from the current density computed for the emitter. This method is invalid, however, when electron reflection effects are included in the emission constant *(A)*. A reformulation of the Richardson-Dushman equation is recommended to prevent improper usage of the emission constant, and a revised approach for computing net currents has been developed to correctly account for electron reflection. Energy and angle dependence of electron emission and reflection are included in the net current density formulation. The reformulated equation predicts net currents identical to the original equation when reflection is insignificant, but predicts very different net currents when reflection is important. The ability to predict and understand thermionic behavior may lead to design modifications for thermionic converters, resulting in performance improvements. The revised method has been developed for vacuum-type diodes. An extension of the basic approach to include plasma-type converters is planned.

CONDUCTIVELY COUPLED MULTI-CELL TFE WITH ELECTRIC HEATING PRETEST ABILITY

Yuri V. Nikolaev, Rafail Ya. Kucherov,
Stanislav A. Eryomin, Oleg L. Izhvanov,
Vladimir U. Korolev, Nikolai V.
Lapochkin, and David L. Tsetshladze,
Research Institute of SIA LUTCH
Podolsk, Moscow Region, 142100,
Russian Federation
(095) 137-98-76

Thomas A. Lechtenberg, Lester L. Begg
GENERAL ATOMICS
PO BOX 85608 San Diego, Ca, USA
(619) 455-2482

Abstract

Problems associated with the development of a multi-cell thermionic fuel element (TFE) with ability of electric heating test are discussed. A conceptual design of such TFE with trilayer emitter stack is proposed. Trilayer emitter stack consists of a strong emitter fuel clad coated with a high temperature oxide ceramic. Emitter tungsten coatings applied to a ceramic and they separated one after another by insulated gaps. Modern materials that should be base to build this trilayer emitter are presented. Results of calculational investigations of TFE output parameters are included. Results of the preliminary test of TFE and it's components are presented. It is shown that proposed TFE conceptual design from one side allows to provide high output parameters inherent to multi-cell design, and from other side to gain advantages of single cell TFE, such as TFE and reactor nuclear safety, reliability, work cost savings.

INTRODUCTION

Currently it is important and pressing problem of thermionic power conversion to develop a thermionic fuel element of high output power and with a free access to emitter fuel cavity, that will provide an opportunity of TFE test with electric heaters. Thermionic nuclear power system (NPS) based upon such TFEs will possess high power to mass parameters together with advantages of non nuclear testing during the whole cycle of NPS development and fabrication. Such a NPS will be loaded with a nuclear fuel only at the last phase of its preparation to launch, that essentially increases system reliability, reduces cost and term of the experimental development and ensure its nuclear safety during fabrication and prelaunch testing.

Single cell TFE (fig.1A) with a single emitter of an entire reactor core length (Nikolaev et al. 1993) is the most known of TFEs designs with an open emitter fuel cavity. The benefit of this TFE design is its simplicity and associated with its reliability. But the same time single cell TFE has only limited possibilities of output parameters improvement due to the electrodes ohmic losses. From other side mult-icell TFEs with an open fuel cavity provide essentially higher output parameters, but their design is more complicated than a single cell one. Different multi-cell, TFEs designs and an analysis of their output parameters and reliability are presented in the report (Lapochkin et al. 1997). It is shown, that a problem to ensure integral leak tightness of metal ceramic unit (MCU) with strict size limitations in a reactor operating conditions is a main one for a TFE design with traditional intercells metal ceramic units. Increasing of a number of thermionic power cells (NPS of on average and high power) makes this problem more complicated. Trilayer emitter stack with a common fuel clad of an entire reactor core length is used in the other TFE design with an open emitter cavity (fig.1B). Such a design ensure high TFE's leak tightness reliability as for a single cell TFE. The main problem of such TFE development is a problem to ensure workability and electric insulation characteristics of the trilayer emitter stack. It should be noted, that low value of electric voltage of stack insulation, that does not exceed a voltage of one TFE (several volts), and application of the perspective materials promise a principle possibility to ensure the trilayer stack reliability.

This work is a preliminary consideration of the main tasks associated with a development of the TFE with trilayer emitter stack. This design envisages a free access to emitter fuel cavity and consequently a possibility of TFEs test with electric heaters.

TFE DESIGN AND MATERIALS

Fig.1b shows TFE design schematic. TFE main structural and functional units are:
emitter stack, collector stack, metal ceramic units, spacers and intercel's commutation adapters.

Emitter stack consists of a fuel clad, coated with a ceramic. Emitters of several power cells are applied onto ceramic surface. The last emitter is shorted to a fuel clad which serves as a TFE emitter terminal. Collector stack consists of collector niobium clad coated from both sides with a ceramic. Insulated one from another collectors of several power cells are applied onto the internal ceramic layer. The last collector is shorted to a collector clad that serves as a TFE collector electric terminal. External ceramic layer separates TFE from a reactor coolant. Metal ceramic units that insulate TFE electrodes one from another and the electrodes from reactor body, and that confine cesium plenum consist of ceramic insulators connected with metal adapters and of the bellows that compensate difference in thermal expansion of different TFE components. A ceramic insert with a spiral cesium channel is introduced into a high voltage metal ceramic unit to increase its electric strength in cerium vapors.

Ceramic spacers fix mutual disposition of electrodes and insulators in axial and diametrical directions ensuring required value of an interelectrode gap. Thermionic power cells are connected in seria by an intercell commutation, which has enough flexibility to compensate the difference in thermal expansion of an emitter and collector.

Due to similarity of design schematics, identity of a number of structural and functional TFEs components and conditions of their operating modes the high effective technologies and materials, that have been developed earlier for single cell TFEs and multi-cell TFEs of different design, may be used for multicell TFE with trilayer emitter stack. These components, in particular are: metal ceramic composition $Nb-Al_2O_3$ (collector stack); scandium oxide (spacers); single crystal alumina (MCU insulators) special emissive coatings of the electrodes (emissive pair tungsten-tungsten); special insulation coatings on the collector stack (two layer external insulation ceramic-enamel).

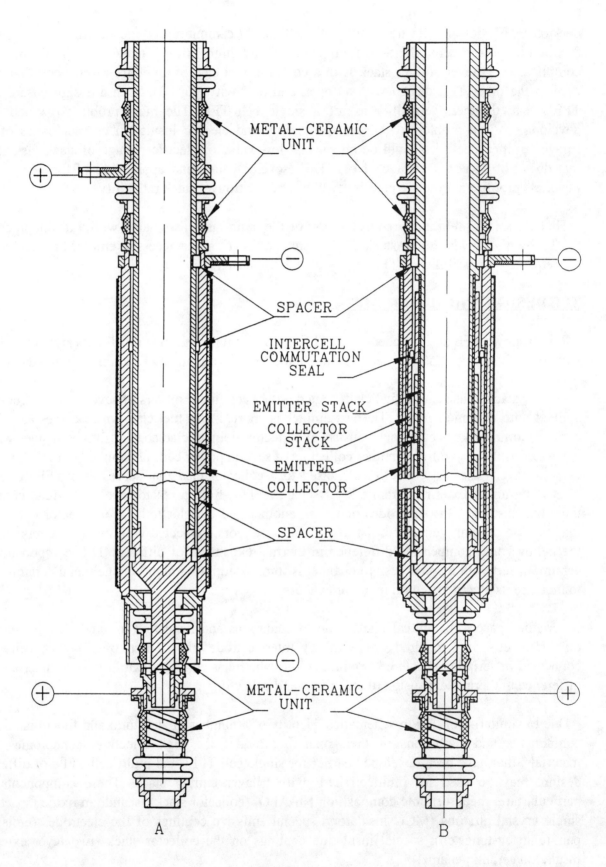

FIGURE 1. Design Schematics of a Single-Cell TFE (A) and Multi-Cell TFE (B)

Emitter stack is the most high temperature TFE unit and it is expedient to develop it on the base of single crystal molybdenum or tungsten and alumina that incorporates special transient layers. Single crystal alloy molybdenum-niobium or tungsten-niobium that possesses good high temperature strength, low creep, good compatibility with different fuel materials may be recommended as an emitter fuel clad. It is expedient to apply additional scandia coating onto the open surface between emitters to improve alumina coating stability in cesium plasma.

TFE OUTPUT PARAMETERS AND WORKABILITY

The calculations of output parameters of the described above TFE were performed. The length of an active heating zone was accepted to be ~ 600 mm. This value is typical for a number of known reactors-converters based upon multicell TFEs. Under complete alignment of heat release along the heating length and optimal value of electric resistance of the intercell commutation adapters the number of thermionic power cells will be equal to 10-12.

Emitter stack internal diameter was taken as 17.3 mm, fuel clad thickness 1 mm, emitter 0.8 mm, collector 1 mm, interelectrode gap - 0.4 mm. Dependences of specific output electric power and TFE efficiency versus output voltage of separate power cells for the different emitter temperatures are presented in fig.2. From the figure it can be seen that the TFE has high power effectiveness. An example, even at moderate level of an emitter temperature 1870 K the value of TFE specific output power is equal approximately to 5.5 W/cm^2, and efficiency to 12%, that 2-3 times more than corresponding values for a single cell TFE.

Calculations of the emitter stack deformation due to fuel swelling and estimation of the strain state of the emitter stack separate layers under thermocycling were performed. Calculations were applied to an emitter stack with fuel clad made of strengthened single crystal alloy Mo-Nb, W-Nb 1 mm thickness and ~ 30% fuel element void volume. It is shown, that with admitted emitter deformation value TFE service life at temperature 1870 K exceed 7 years. Under these conditions there should not be an emitter stack fractionating, and possible cracks formation in a ceramic layer will not impact thermal and insulation emitter stack properties.

Workability of a trilayer emitter multi-cell TFE and its main structural and functional components is confirmed by the results of experimental investigations of multi-cell and single cell TFE of similar design (Begg et al. 1993, Lapochkin et al. 1995, Nikolaev et al. 1996). Metal ceramic units and spacers workability have been investigated for several years with the use of vibration and thermal rigs and nuclear channels. These test demonstrated positive results for service life of three years and more. There are also the results of the previous investigations of experimental trilayer emitter stack and experimental prototypes of trilayer emitter TFE consisting of 3 and 5 power cells with electric and nuclear heating for the service life more than 1000 years. TFEs experimental prototypes have been manufactured on the base of the earlier existing technology (emitter stack was made of trilayer metal ceramic molybdenum-alumina-molybdenum). Emitter temperature was equal to 1870 K. No significant decreasing of TFEs output power was observed. The values of output power were corresponded to the calculated ones, and electric strength of an emitter insulation exceeded TFE output voltage (several volts), that proved the stable TFE and its components operation.

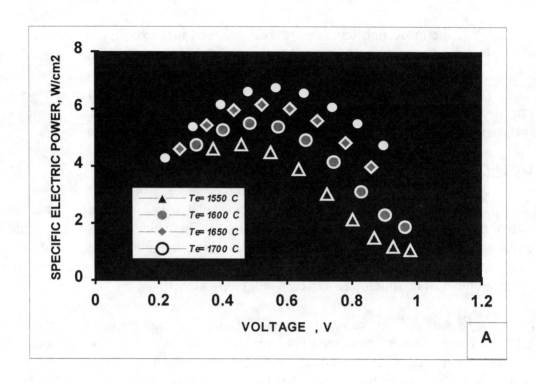

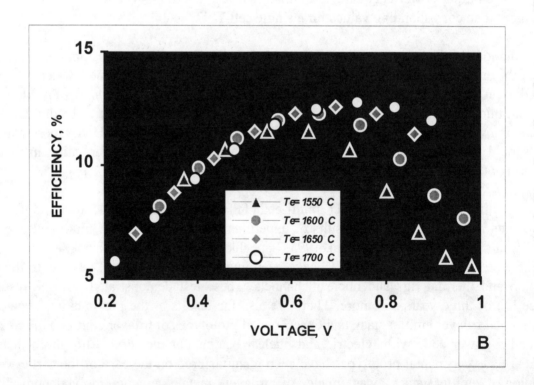

FIGURE 2. Dependence of Specific Electric Power (A) and Efficiency (B) of a Multi-Cell TFE Versus Output Voltage of its Single Cell.

SUMMARY

The results of the analysis of the conceptual design of a multi-cell TFE with trilayer emitter stack show a principle possibility to develop high effective, long life conductively coupled multi-cell TFE with test ability without nuclear fuel. It makes possible financial and time savings during TFE and reactor development and ensure NPS nuclear safety.

Acknowledgments

This work was done by joint efforts of the authors of RI of SIA "LUTCH" (Russia) and GENERAL ATOMICS (USA).

References

Begg L.L. et al. (1993) "Review of Thermionic Technology:1983 to 1992" in A Critical Review of Space Nuclear Power and Propulsion 1984-1993, El-Genk, M.S. ed., AIP Press. New York, USA, 121-146

Lapochkin N.V. et al. (1995) "High-Voltage Metal-Ceramic Assembly of Thermionic NPP", in *Proc. of 12th Symposium on Space Nuclear Power and Propulsion*, M. S. El-Genk ed., Albuquerque, NM, USA, 1:289-294.

Lapochkin N.V. et al. (1997) "The Conceptual Design Analysis", in *Proc. of Space Technology and Applications International Forum (STAIF-97)*, M. S. El-Genk ed., Albuquerque, NM, USA, 3:1553-1558.

Nikolaev Yu. V. et al. (1993). "Comparative Analysis of Concepts of Single-cell and Multi-cell TFE of Thermionic NPS". in *Proc. of 10th Symposium on Space Nuclear Power and Propulsion*, M. S. El-Genk ed., Albuquerque, NM, USA, 3:1347-1354.

Nikolaev Yu.V. et al. (1995) "Multi-Cell TFE's of Thermionic NPP's in *Proc. of 12th Symposium on Space Nuclear Power and Propulsion*, M. S. El-Genk ed., Albuquerque, NM, USA, 2:699-704.

Nikolaev Yu.V.. et al. (1996) "Investigation of TFE Collector Insulation of Thermionic NPP", in *Proc. of Space Technology and Applications International Forum (STAIF-96)*, M. S. El-Genk ed., Albuquerque, NM, USA, 3:1221-1226.

RE-START: THE SECOND OPERATIONAL TEST OF THE STRING THERMIONIC ASSEMBLY RESEARCH TESTBED

Francis J. Wyant	David Luchau	T. D. McCarson
Sandia National Laboratories	TEAM Specialty Services,	New Mexico Engineering Research
MS 0744	Incorporated	Institute
P.O. Box 5800	11030 Cochiti SE	901 University Blvd. SE
Albuquerque, NM 87185-0744	Albuquerque, NM 87123	Albuquerque, NM 87106
(505) 846-8889	(505) 291-0182	(505) 272-7403

Abstract

The second operational test of the String Thermionic Assembly Research Testbed—Re-START—occurred over the period from June 9 to June 14, 1997. This test series was designed to help qualify and validate the designs and test methods proposed for the Integrated Solar Upper Stage (ISUS) power converters for use during critical evaluations of the complete ISUS bi-modal system during the Engine Ground Demonstration (EGD). The test article consisted of eight ISUS prototype thermionic converter diodes electrically connected in series. Results demonstrated the high temperature structural performance of the re-engineered diode mounting assembly, measurable degradation in electrical performance of seven of the test diodes, and the susceptibility of the diode array to load conditions during fast heat up ramps.

INTRODUCTION

An initial test of the prototype ISUS diode array was conducted in the period of March 10 through March 29, 1997 (Boonstra 1997). The intent of this test, "START", was to measure the performance and conversion efficiency of the 8-diode array thermally coupled to an electrically heated graphite block (receiver), simulate the sun-eclipse cycle and monitor the array's electrical performance during the "shadowed" periods, and to characterize the electrode and plasma properties of the individual diodes under nominal operating conditions. Insofar as multiple problems, principally electrical shorting to ground and between adjacent diodes, plagued the performance of START there was limited value placed on its outcome. The problems uncovered did, however, lead to the redesign of the diode support structure and several recommendations regarding other changes to the test setup.

Driven by the need to evaluate the effectiveness of the new diode support structure prior to the system level EGD, Phillips Laboratory and the Defense Special Weapons Agency jointly decided to perform a second test on the ISUS bi-modal system prototype diode array. The objectives of this second test, called "Re-START", were to evaluate the structural integrity of the re-engineered diode mounting assembly at high temperatures, measure the output power and current-voltage (I-V) response of the diodes under a variety of thermal and electrical transient conditions, and to provide an operational data base for use in the ISUS power management and distribution (PMAD) system development effort (Baez 1997).

Both the START and Re-START tests took place in the Baikal test stand located at the New Mexico Engineering Research Institute (NMERI) OTV Laboratory in Albuquerque, New Mexico (Wold 1993).

TEST ARTICLE

The Re-START test article consisted of seven prototype diodes (James 1997) previously tested during the START test and one diode provided by Lockheed Martin Electro-Optical Systems (LM EOS) to replace the eighth diode, which had failed during START. The replacement diode was one that had previously undergone ~1200 hours of life testing at LM EOS. The array was electrically connected in the same order as was employed during the START test (i.e., the number 1 diode for START was the number 1 diode for Re-START, etc.). The eight diode array was tied to ground between the fourth and fifth diodes, thus limiting the potential difference above and below the structure at the output electrodes to that generated by four diodes in series.

Changes made for the purpose of conducting Re-START were:

1. The graphite receiver assembly was not used; instead the array diode emitters were heated by direct radiation from the electric heater (this is the operational configuration for START test run #3).
2. A new diode mounting assembly was installed. This structural member was reengineered to prevent movement of the diodes at operational temperatures.
3. The zirconium-oxide felt insulation material was replaced with hard-fired ceramic insulators.
4. The array diode thermocouples were replaced by sheathed, ungrounded thermocouple probes as previously recommended, and additional thermocouples were installed to measure temperatures at three locations on the new diode mounting assembly.
5. Diode electrical interconnections were routed external to the Baikal Stand vacuum chamber for ease in reconfiguring the diode array circuit. The vacuum chamber integrity did not need to be compromised in order to bypass one or more of the diodes in the array.
6. A different diode of the same design was employed as a replacement for #8 in the original START array.

The new diode mounting assembly (Fig. 1) consists of eight saddle blocks mounted on a single mounting ring segment. One saddle block supports one diode with alumina insulator segments (four per diode) arranged within the grooves machined in the saddle support and the clamp block upper section. The insulator segments fit within the grooved segment around the diode collector-heat pipe interface flange. Additional alumina insulating material was inserted between the diodes and multifoil insulation to prevent shorting to ground. Alumina ceramic insulation was also inserted between adjacent diode hot shoes to prevent inter-diode shorting during operation.

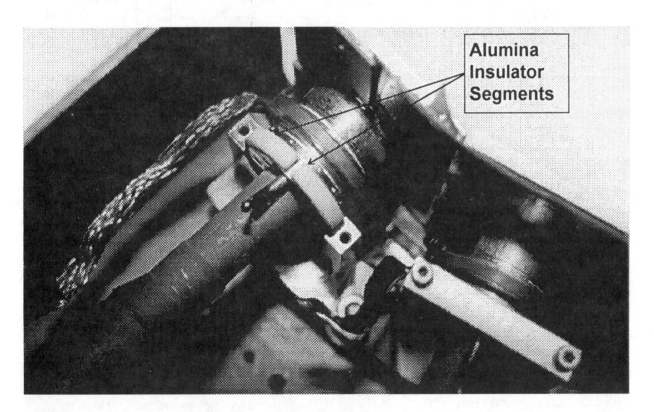

FIGURE 1. Diode Mounting Assembly Ceramic Insulator Segments.

TEST APPROACH

Testing of the START diode array was conducted on the basis of average array emitter temperature. The overall Re-START test emitter temperature profile is shown in figure 2. As indicated in the figure, the test was divided into four phases. Each phase was defined primarily by the rate at which the average array emitter temperature was changed.

Diode operating temperatures ranged between 1600 K and 2100 K during the course of the test. Thermal transients included heat up rates of 100, 300, and 500 K/h. Cool down rates included -300, -500, and -2000 K/h. Electrical transients were induced by the changing thermal conditions and by manual adjustments of the external resistive load on the array. Dynamic I-V sweep measurements were made on the array and individual diodes at each temperature plateau.

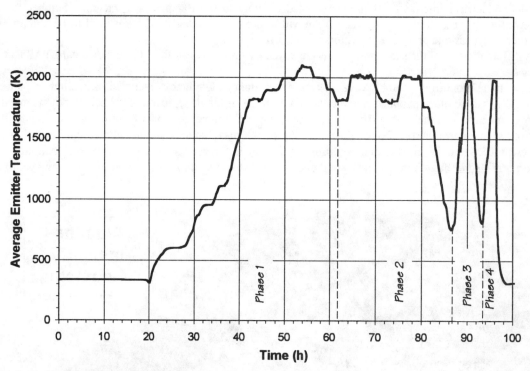

FIGURE 2. Re-START Test Profile.

Phase 1: Initial Heat Up and Baseline Data

The nominal array average emitter temperature transients occurred at ~100 K/h. Beginning at 1800 K, the emitter temperatures were progressively increased in 100 K steps from 1800 K to 2100 K, and back down to 1800 K with a hold plateau at each step. At each plateau a dynamic I-V sweep was conducted and the peak array power level was found and maintained during the remainder of the hold time by adjusting the external load resistance.

Phase 2: Thermal and Electrical Transient Data

The array average emitter temperature transients nominally occurred at 300 K/h during Phase 2 operations. A number of emitter temperature cycles were performed to obtain data on the array's response to thermal transients within the limited operating range of the diodes (1800 - 2000 K). Current-voltage sweeps and load bank transients were conducted at each of the steady state plateaus in this phase, and the peak array power level was achieved and maintained during the remainder of the hold time by adjusting the load resistance. At the end of the Phase 2 cycles, the load bank resistance was set to its minimum value (~44 mΩ), and the average array emitter temperature was decreased to 800 K at a nominal rate of -300 K/h.

Phase 3: Fast Heat Up With Minimum Load

The initial conditions at the beginning of Phase 3 were that the average array emitter temperature was ~800 K and the load bank was set to 44 mΩ (minimum value). The initial nominal heat up rate was slightly greater than 500 K/h. Power to the electric heater was manually decreased when diode #4 indicated a low voltage condition (less than -2 V) at about 1500 K average emitter temperature (T_E). Heater power was decreased until the problem cleared; diode #4 was electrically bypassed and the heat up ramp continued. At T_E ~1600 K, the bypass shunt was removed from diode #4 and the heat up transient was completed at T_E ~2000 K. Another series of I-V sweeps were completed at the high temperature plateau and the peak array power level was set and maintained during the remainder of the hold time by adjusting the load resistance accordingly. The load was again set at 44 mΩ prior to the start of the Phase 3 cool down ramp.

Phase 4: Fast Heat Up With Maximum Load

The conduct of Phase 4 operations was very similar to that described for Phase 3, above, with the exception of setting the load to its maximum value (~900 mΩ) prior to initiating the 500 K/h heat up ramp and prior to the final rapid cool down. No alarms were activated during the Phase 4 heatup transient. However, diode #4 voltage did go as low as -1.5 V at one point during the ramp up. In addition, I-V sweeps and peak array power settings were accomplished at the temperature plateau, as usual. The final cool down ramp was initiated by reducing power to the electric heater as quickly as possible. The resulting cool down rate on the diodes approached -2000 K/h over the range 2000 - 500 K.

RESULTS

A post test evaluation of the structural performance of the redesign diode support structure was performed, and the operational data was analyzed to assess the electrical performance of the START array and individual diodes.

Support Structure Performance

Overall, the redesigned diode mounting assembly performed very well throughout the Re-START test: All of the thermocouples attached to the diode mounting assembly structure stayed within specification (i.e., less than 1000°C) and there was no discernible distortion of any of the mounting hardware found during post-test disassembly and inspection. Upon disassembly, the top of the diode mounting ring was found to be discolored at locations between adjacent saddle blocks. The additional alumina pieces placed between the emitter flange and the multifoil insulation were also found to be discolored in the area near where the insulator was compressed between the diode emitter flange and multifoil insulation. The blue/black discoloration was on the side facing the multifoil insulation.

The alumina ceramic located between adjacent hot shoes and the alumina tube insulators placed between the hot shoe multilayer insulation and the multifoil thermal insulation did not show any discoloration. All discoloration of the mounting assembly components and insulators were on the heat pipe side of the multifoil thermal barriers. It was also found that all 32 of the diode mounting assembly alumina insulator segments survived the test intact and without any discernible damage. No breaks, cracks, or chipping of these pieces were noted.

Diode Electrical Performance

In summary, the peak electrical performance of the START array was about half of its design value: 113 watts measured during Re-START versus 250 watts. Seven of the eight diodes in the string were determined to be degraded. Figures 3 and 4 show the DC I-V characteristics of the diodes at T_E ~1820 K and 2000 K, respectively. The reduction of electrical performance for three of the diodes (2, 3, and 4) appears to be due to a loss in the dielectric strength of the interelectrode insulators. The performance degradation of three other diodes (1, 6, and 7) is due to the interelectrode gap being too small (ionization rate in the interelectrode gap being small). The last diode that degraded (5) has indications that its loss of output is due to a decrease in dielectric strength and to a leakage of oxygen into the interelectrode gap. Finally, even though diode 8 does not show signs of having a degradation in its electrical output, it has indications that its dielectric strength decreases as T_E is increased. This decrease in the dielectric strength of diode 8 as T_E is increased is possibly due to the extended length of operation of this diode. Therefore, there may be a lifetime issue in the dielectric strength of these diodes.

CONCLUSIONS

Results indicate that the diode mounting assembly maintained its structural integrity over the test temperature range. No indications of thermal stress or distortions were found upon disassembly and inspection following the test. Seven of the eight diodes in the array exhibited some degree of operational degradation in electrical performance possibly due to internal shorting pathways between the emitter and collector. The eighth diode was one received from Lockheed Martin Electro-Optical Systems just prior to installing the array in the test chamber. Its output exceeded all of the other diodes by a significant amount, yet it did show some degradation in dielectric strength as T_E increased.

Additional ceramic insulators placed between the grounded structure and the diode electrode components eliminated the electrical shorting problems encountered during the original START test. Use of sheathed ungrounded thermocouples also eliminated many of the diagnostic problems that plagued the START test; however, at least three of the new tungsten-rhenium high temperature thermocouples failed during Re-START.

The transient electrical response of one of the ISUS prototype diodes indicates they may be susceptible to load conditions during fast heat ups. Consequently, it is recommended that a resistive load commensurate with an output current of 1 - 5 amperes at design temperature be placed across the array prior to startup. It is also recommended that hot shoe heat up rates and cool down rates be limited to no more than 300 K/h.

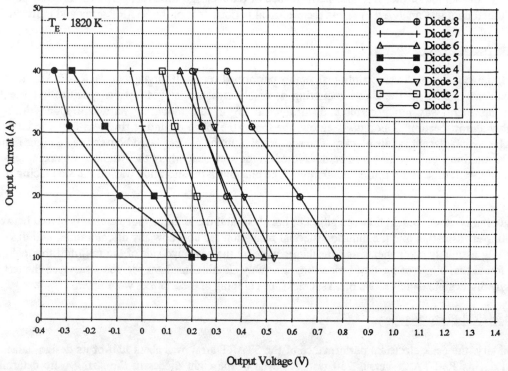

FIGURE 3. DC I-V Characteristics of the Diodes in the Array at $T_E \sim 1820$ K.

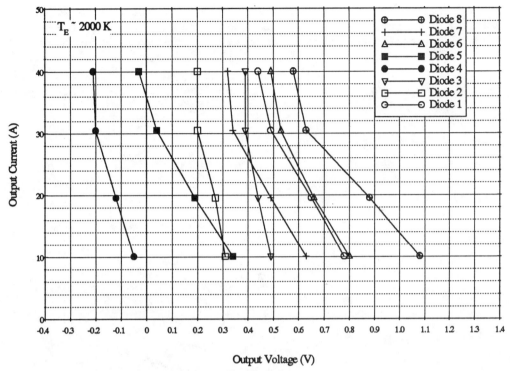

FIGURE 4. DC I-V Characteristics of the Diodes in the Array at $T_E \sim 2000$ K.

Acknowledgments

The work presented in this paper was conducted under the joint sponsorship of the U.S. Air Force Phillips Laboratory (PL) and the U. S. Defense Special Weapons Agency (DSWA). The authors would like to thank Dr. Bernard Wernsman, formerly with the New Mexico Engineering Research Institute, for his analysis of diode performance.

References

Baez, A. N. and G. L. Kimnach (1997) "The Integrated Solar Upper Stage Engine Ground Demonstration Power Management and Distribution Subsystem Design," in Proceedings of the Space Technology and Applications International Forum (STAIF-97), CONF-970115, M. S. El-Genk, ed., American Institute of Physics, New York, AIP Conf. Proc. No. 387, 1:251-256.

Boonstra, R. and J. L. Desplat (1997) "ISUS Program START Test Report", available from the ISUS Program Office, USAF Phillips Laboratory, Kirtland AFB, NM 87117.

James, E. L., W. D. Ramsey, and G. J. Talbot (1997) "Thermionic Converters for ISUS," in Proceedings of the Space Technology and Applications International Forum (STAIF-97), CONF-970115, M. S. El-Genk, ed., American Institute of Physics, New York, AIP Conf. Proc. No. 387, 1:479-484.

Wold, S. K. (1993) "Thermionic System Evaluation Test (TSET) Facility Construction: A United States and Russian Effort," in Proceedings of the 10th Symposium on Space Nuclear Power and Propulsion, CONF-930103, M. S. El-Genk and M. D. Hoover, eds., American Institute of Physics, New York, AIP Conference Proc. No. 271, 2:731-735.

EXPERIMENTAL TESTING OF A FOAM/MULTILAYER INSULATION (FMLI) THERMAL CONTROL SYSTEM (TCS) FOR USE ON A CRYOGENIC UPPER STAGE

Leon J. Hastings
Propulsion Laboratory/EP63
National Aeronautics and Space Administration
George C. Marshall Space Flight Center
Huntsville, Alabama 35812
leon.hastings@msfc.nasa.gov
205-544-5434

James J. Martin
Propulsion Laboratory/EP63
National Aeronautics and Space Administration
George C. Marshall Space Flight Center
Huntsville, Alabama 35812
jim.martin@msfc.nasa.gov
205-544-6054

Abstract

An 18-m^3 system-level test bed termed the Multipurpose Hydrogen Test Bed (MHTB has been used to evaluate a foam/multilayer combination insulation concept. The foam element (Isofoam SS-1171) protects against ground hold/ascent flight environments, and allows the use of dry nitrogen purge as opposed to a more complex/heavy helium purge subsystem. The MLI (45 layers of Double Aluminized Mylar with Dacron spacers) is designed for an on-orbit storage period of 45 days. Unique MLI features included; a variable layer density (reduces weight and radiation losses), larger but fewer DAM vent perforations (reduces radiation losses), and a roll wrap installation which resulted in a very robust MLI and reduced both assembly man-hours and seam heat leak. Ground hold testing resulted in an average heat leak of 63 W/m^2 and purge gas liquefaction was successfully prevented. The orbit hold simulation produced a heat leak of 0.22 W/m^2 with 305 K boundary which, compared to historical data, represents a 50-percent heat leak reduction:

INTRODUCTION

The development of high-energy cryogenic upper stages is essential for the efficient delivery of large payloads to various destinations envisioned in future programs. A key element in such upper stages is cryogenic fluid management (CFM) advanced development/technology. Reliable thermal protection for a cryogenic vehicle or space platform throughout its entire mission is critical to establishing overall success and is a necessity for future missions. Due to the cost of, and limited opportunities for, orbital experiments, ground testing must be employed to the fullest extent possible. Therefore, a system-level test bed, termed the Multipurpose Hydrogen Test Bed (MHTB), which is representative in size of a fully integrated space transportation vehicle liquid hydrogen propellant tank, has been established at the Marshall Space Flight Center (MSFC). Although upper stage studies have often baselined the foam/multilayer insulation (FMLI) combination concept, hardware experience with the concept is minimal, and it was therefore selected for MHTB testing. The foam/MLI combination, along with several unique MLI installation and design features, was evaluated during the subject test program. The test article, facility, approach, and results are discussed in subsequent sections.

TEST ARTICLE ELEMENTS

The major test article elements consist of the test tank and environmental shroud with supporting equipment, cryogenic insulation subsystem, and test article instrumentation. Technical descriptions of each of these elements are presented in the following sections.

Test Tank and Supporting Equipment

The MHTB aluminum tank is cylindrical in shape with a height of 3.05 m, a diameter of 3.05 m, and 2:1 elliptical domes as shown in Figure 1. It has an internal volume of 18.09 m^3 and a surface area of 34.75 m^2 with a resultant surface area to volume ratio of 1.92 1/m. The tank is ASME pressure-vessel coded for a maximum operational pressure of 344 kPa and was designed to accommodate various CFM concepts as updated versions become available. The low heat leak composite legs and other tank penetrations are equipped with LH$_2$ heat guards so that more accurate measurement of the tank insulation performance can be made.

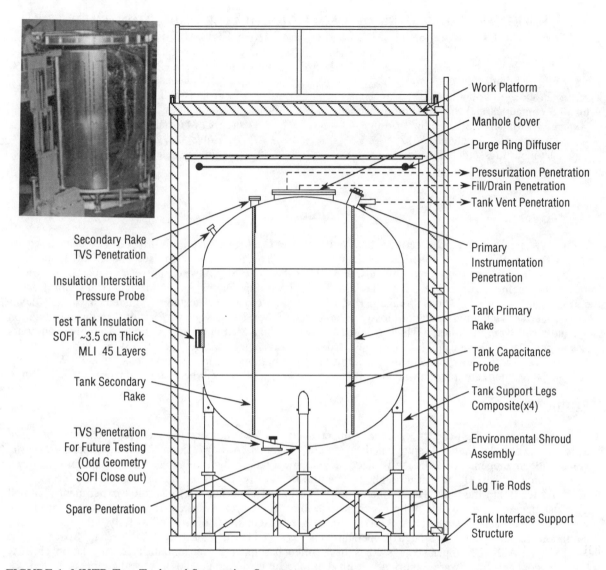

FIGURE 1. MHTB Test Tank and Supporting Structure.

The tank is enclosed within an environmental shroud which contains a ground-hold conditioning purge (similar to a payload bay) and imposes a range of uniform temperatures on the insulation external surfaces during orbit hold simulations. The shroud is 4.57 m high by 3.56 m in diameter, and contains a purge ring for distributing dry nitrogen.

Cryogenic Insulation Subsystem

The MHTB insulation concept consists of a foam/multilayer combination. The foam element enables the use of a payload bay type purge as opposed to the complex helium purge bag subsystem normally required with MLI on cryogenic tankage during ground-hold periods. That is, the foam assures surface temperatures adjacent to the MLI inner layer at or above 117 K to preclude gaseous nitrogen liquefaction. Additionally, the foam reduces the heat leak during the ground-hold and ascent flight periods. The spray-on foam insulation (SOFI), termed Isofoam SS-1171, was applied directly to the tank surface, by a robotic process, with an average thickness of 3.53 cm, the minimum that could be applied with available equipment and procedures.

A 45-layer MLI blanket, placed over the SOFI, provides thermal protection while at vacuum conditions. The blanket is composed of double-aluminized Mylar (DAM) radiation shielding and separated by a combination of B4A Dacron netting and B2A bumper strips. Unique, innovative features of the MLI concept include utilization of a variable-density (layers per unit thickness) concept for the radiation shields to provide a more weight-efficient insulation system and the use of fewer, but larger, perforations for venting during ascent to orbit. As illustrated in Figure 2, the variable density was accomplished using bumper strips (of variable thickness) to provide more layers in warmer regions, and fewer layers in the colder regions (where radiation blockage is less important). The variable density provides a maximum theoretical heat leak reduction of 50 percent compared with uniform-density MLI. The vent hole perforation pattern, which provides a 2-percent open area, is unusual in that the perforation size is large (1.27 cm diameter) and the holes are more widely spaced (7.6 cm). The larger holes reduce the radiation view-factor (hence, the radiation exchange) between layers thereby enabling a maximum theoretical heat leak reduction of 35 percent. Additionally, the virtually seamless insulation enabled by the MLI roll-wrap installation technique further reduces heat leak. However, the lack of seams, together with the vent hole distribution, does decrease the vent-down rate during ascent flight and orbital injection.

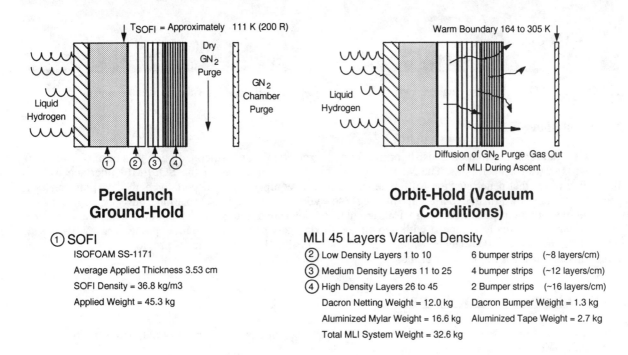

FIGURE 2. MHTB Cryogenic Insulation Concept.

MLI Installation

The MLI installation technique was unique relative to the usual aerospace approach. A commercial roll wrapping technique was utilized for the barrel section application (Fig. 3) wherein the DAM and Dacron netting consisted of continuous sheets or strips. The dome insulation was prefabricated on a flat table and installed layer by layer, interleaving each dome layer with the corresponding barrel blanket layer (Fig. 4). Each of the tank penetrations were closed out with both pour-foam and MLI. The MLI performance around the penetrations was enhanced utilizing a temperature matching technique wherein the MLI layers along the penetration longitudinal axis were attached at penetration positions predicted to have the same temperature as that particular MLI layer. This approach minimized temperature gradients and, therefore, heat transfer parallel to the insulation layers. Using the preceding MLI installation techniques on a 3-m diameter tank set (hydrogen and oxygen) would result in an estimated savings of 2400 man-hours. The insulation component weights resulted in SOFI and MLI weight totals of 45.3 kg, and 32.6 kg, respectively.

FIGURE 3. MHTB Roll-Wrapping Process.   FIGURE 4. MHTB Dome Blanket Installation.

Instrumentation

The test article instrumentation consists primarily of thermocouple and silicon diodes to measure insulation, fluid, and tank wall temperatures. The MLI interstitial pressure is measured at the SOFI/MLI interface using a thin-walled probe that penetrates the MLI. The probe is also equipped with a port for both dew point and gas species sampling. Two of the four composite legs, the vent, fill/drain, pressurization, pressure sensor probe, and manhole pump-out penetrations are instrumented to determine the solid conduction component of heat leak. The tank is internally equipped with two silicon diode rakes and a capacitance probe. The rakes provide temperature gradient measurements within both ullage and liquid, in addition to providing a backup to the continuous liquid level capacitance probe.

TEST FACILITY AND PROCEDURES

The facility and procedures used for test preparations and the simulation of ground-hold, ascent flight, and orbit-hold conditions are summarized below.

Facility Description

Testing was performed at the MSFC East Test Area thermal vacuum facility, Test Stand 300. The vacuum chamber is cylindrical in shape and has usable internal dimensions of 5.5 m in diameter and 7.9 m in height. The chamber-pumping train consists of a single stage gaseous nitrogen (GN_2) ejector, three mechanical roughing pumps with blowers, and two 1.2-m diameter oil diffusion pumps. Liquid nitrogen cold walls surround the usable chamber volume, providing cryopumping and thermal conditioning. The facility and test article shroud systems in combination enabled simulation of orbital conditions (vacuum levels of 10^{-8} torr and temperatures of 80 to 300 K). The GN_2 ejector system enables a rapid pump-down capability to simulate the ascent flight portion of a mission. Two solid-state video cameras were mounted inside the environmental shroud to view the test article lower dome and sidewall during the ground-hold and ascent flight test phases

A key facility capability was the pressure control subsystem used to maintain steady-state MHTB ullage pressure. The subsystem was composed of several flow control valves (located in the vent line), each of which was regulated through a closed-loop control system. This control loop manipulated the valve positions based on a comparison between the measured ullage pressure and the desired set point. An MKS "0–133 kPa" absolute and MKS "0–133 Pa" differential-pressure transducers positioned outside the chamber were used to measure ullage pressure. Boiloff was measured by a series of four flow meters which covered the expected range of flowrates.

Test Procedures

Prior to testing, the vacuum chamber, tank, fill, drain, and vent lines were conditioned with purge and/or vacuum cycles to expel all trapped contaminates. After the vacuum-purge cycles were completed, the chamber was continuously purged with dry GN_2. Several hours were required to saturate and equilibrate the LH_2 at the set point pressure after tanking. During ground-hold testing, a payload bay environment was simulated with an inert GN_2 purge at an average flowrate and temperature of 5 kg/min and 273 K, respectively. The ullage pressure was maintained in a range of 110.316 to 124.106 kPa with a tolerance of ± 0.0689 kPa. The pressure-rise rate within the tolerance band was held to within .689 kPa/hr to control venting oscillations. The ground-hold test duration, after tanking and ullage-pressure control was established, was typically one to two hours.

The ascent flight portion of a heavy lift launch vehicle or Space Shuttle mission was simulated as far as possible using the vacuum chamber GN_2 ejector system, i.e., the first 120 seconds of flight (covering ambient pressure to 30 torr). The two video cameras were activated for the first three minutes of pump-down to observe any MLI "billowing." Upon completion of the ascent flight simulation, establishment of steady-state vacuum and thermal conditions (within both the chamber and MLI) had to be achieved for the on-orbit test phase. Two key criteria must be met:

1) Insulation equilibrium is assumed to exist once transients of no more than 0.55 K in six hours are measured in any section of the insulation, and the outer MLI shield is at the prescribed set point temperature imposed by the environmental shroud.

2) Thermal equilibrium of the stored LH_2 must be maintained through precise ullage-pressure control throughout. Ullage pressures were maintained at set points in the range of 110.316 to 124.106 kPa with a tolerance of ± 0.00689 kPa.

Additionally, the interstitial MLI pressures must be 10^{-5} torr or less to preclude convective heat transfer, and the tank dome mass must be in thermal equilibrium, i.e., not adding to the vented gas enthalpy.

DATA REDUCTION AND EVALUATION APPROACH

The general methodology used for thermal performance data reduction and the specific approaches utilized to compare the ground-hold and orbit-hold simulation data with predictions are described below.

Data Reduction

Digital data were recorded at sample rates ranging from 1.0 to 0.017 Hz. These raw data were then time-averaged over the steady-state period of interest to obtain measurement values required to calculate thermal performance. The heat input was expressed as an energy balance across the tank boundary by equating the total measured boiloff with the sum of heat flow through the insulation, the penetrations, and the rate of energy storage (if any):

$$\dot{Q}_{boiloff} = \dot{Q}_{insulation} + \dot{Q}_{conduction} + \frac{\Delta U_{system}}{\Delta t}.$$

The terms $\dot{Q}_{boiloff}$ and $\dot{Q}_{conduction}$, were defined using the test data. The thermal storage term $\frac{\Delta U_{system}}{\Delta t}$ (energy flow into or out of the test tank wall, insulation and fluid mass) is driven by the fluid saturation temperature which varies as ullage pressure fluctuates. The storage term was eliminated since the ullage pressure was maintained within a tight control band about the set point. The insulation performance term, $\dot{Q}_{insulation}$, could then be determined from the other measured quantities.

The $\dot{Q}_{boiloff}$ term represents the total energy vented as boiloff and includes both the evaporated fluid latent heat and the sensible heat absorbed while the vented gas passes through the ullage space (also known as ullage superheat):

$$\dot{Q}_{boiloff} = \dot{m}h_{fg}\left(\frac{\rho_{satliq}}{\rho_{satliq} - \rho_{satvap}}\right) + \dot{m}(h_{vent} - h_{satvap}).$$

The latent heat term of the above equation contains a density ratio, which accounts for the increased volume of gas (and hence remaining energy) resulting from the decrease in liquid volume due to boiloff losses.

The solid conduction term $\dot{Q}_{conduction}$ represents the heat flow through the tank support legs, vent assembly, and other fluid lines. Solid conduction was evaluated by using the Fourier heat transfer equation (Ozisik 1985) with known structural geometry, material properties, and a measured temperature difference as follows:

$$\dot{Q}_{conduction} = \left(\frac{A}{L}\right)\int_{T_{cold}}^{T_{hot}} K(T)dT.$$

The heat input through the TCS, $\dot{Q}_{insulation}$, was then assessed using experimental data, fluid properties, and the assumption that energy storage in tank material and fluid are, and can be written, as follows:

$$\dot{Q}_{insulation} = \dot{Q}_{boiloff} - \dot{Q}_{conduction}.$$

Ground-Hold Simulation

Tank heat leak during the ground-hold phase was estimated using the Systems Improved Numerical Differencing Analyzer and Fluid Integrator (SINDA'85/FLUINT) program (Cullimore 1988). The primary heat leak source was the insulation acreage which was modeled assuming solid conduction through the SOFI and gaseous conduction through the stagnant GN_2 filling the MLI layers. The boundary condition on the outer MLI was imposed as convective heat transfer to the surroundings.

Orbit-Hold Simulation

For comparison, the performance of a standard constant density MLI blanket was calculated based on the following assumptions: 1) the blanket consists of one Dacron B4A layer for every DAM layer, resulting in an approximate packing density of 27 layers/cm; 2) the DAM is perforated with the standard small closely spaced holes; and 3) the number of layers in the standard blanket was adjusted to match the variable-density blanket applied weight per square meter. This resulted in a standard blanket with 55 layers and an approximate weight of 0.94 kg/m².

A one-dimensional model was set up to evaluate the MLI heat leak, starting with a cold boundary condition at the SOFI surface and a warm boundary at the environmental shroud. The MLI was modeled with the semiempirical equation commonly referred to as the *Lockheed equation* (Keller 1974) while the heat transfer from the MLI to its surroundings (SOFI and shroud) were treated as pure radiation. The *Lockheed equation*, which was developed from flat plate calorimeter test data (and hence provides ideal performance), contains three terms representing the modes of heat transfer through a blanket: radiation, solid conduction, and gaseous conduction.

$$\dot{Q}_{mli} = \frac{7.07 \times 10^{-10} \varepsilon_{TH}(T_{hot}^{4.67} - T_{cold}^{4.67})}{N_{shield} - 1} + \frac{7.30 \times 10^{-8} \overline{N}_{layers}^{2.63}(T_{hot} - T_{cold})(T_{hot} + T_{cold})}{2(N_{shield} - 1)} + \frac{1.46 \times 10^4 P(T_{hot}^{.52} - T_{cold}^{.527})}{N_{shield} - 1}.$$

The model was built using a Microsoft Excel spreadsheet and MLI surface temperature, iterated until the solution reached convergence for the heat leak. The output of this model will be discussed along with the experimental results in the next section.

RESULTS AND DISCUSSION

TCS performance was recorded during three tests. A total of 4 vacuum and tank fill/drain cycles were performed over a 10-month period totaling 1,525 hours at vacuum conditions and 1,335 hours at LH$_2$ temperatures. The vacuum chamber was maintained at ambient pressure with a dry GN$_2$ purge (dew point of -54 °C) for the simulated ground hold phase and at a vacuum level in the low 10^{-6} torr range (with LN$_2$ coldwalls engaged). During orbit-hold simulation the warm boundary (controlled by the environmental heater shroud) was maintained constant at selected temperatures.

Ground-Hold Simulation

Table 1 reports the test environmental conditions and test insulation thermal performance for the simulated ground-hold phase. Three tests were performed, producing average insulation heating rates ($\dot{Q}_{insulation}$) ranging from 2111 to 2225 W (60.7 to 64.0 W/m^2). Penetration heating contributed a very small fraction of the total heating. The environmental shroud GN$_2$ purge was maintained at a flow rate of approximately 5 kg/min at nearly constant temperature during each test (ranged from 270 to 279 K). The average heat leak was numerically predicted to be 2172 W (62.5 W/m^2) which corresponds favorably with that measured. The temperature distribution through the insulation was averaged based on measured data and is presented in Figure 5. The SOFI successfully prevented purge gas liquefaction except for a small localized area near an odd-shaped penetration peculiar to the MHTB on the lower dome.

TABLE 1. TCS Steady State Ground-Hold Performance

Test #	Test Conditions							Measured TCS Performance (W)			TCS Heat Flux W/m^2
	Chamber		MLI Purge			Chamber Press (torr)	Ullage Range (%)	$\dot{Q}_{boiloff}$ (W)	$\dot{Q}_{penetrations}^{c}$ (W)	$\dot{Q}_{insulation}^{d}$ (W)	$\dot{Q}_{insulation}/A_{tank}^{e}$
	Gas	Temp (K)	Gas	Rate (kg/min)	Temp (K)						
P9502	GN$_2$	290	GN$_2$	5.7	279	753	4 - 10^b	2116	4.95	2111	60.7
P9601	GN$_2$	279	GN$_2$	4.9	270	743	4 - 12^b	2213	4.64	2208	63.6
P9602A^a	GN$_2$	279	GN$_2$	4.7	270	752	1 - 23	2227	2.47	2225	64.0
Numerical Prediction	GN$_2$	300	GN$_2$	5.0	300	760	2	2177^f	5.38	2172	62.5

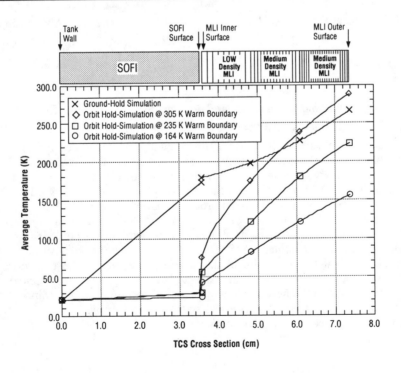

FIGURE 5. Average Measured TCS Temperature Profiles.

Ascent Simulation

Key items of interest observed during the ascent simulation included both TCS structural integrity under broadside venting loads, and the time required to achieve TCS thermal equilibrium. The two video cameras positioned within the environmental shroud provided views of the entire lower dome, and a portion of the tank sidewall enabled MLI observation during ascent. During the rapid evacuation (transition from 760 to 35 torr in approximately 120 seconds) the roll-wrapped MLI was observed to expand only slightly, which is attributed to its seamless robust construction.

The transient heat leak and chamber pressure experienced during ascent is illustrated in Figures 6 and 7 for Tests P9502 and P9601, respectively. The ground heat leak at the beginning of the ascent (at approximately 1000 minutes) is in the 2000-W range. The heating rate fell much faster during the second test (P9601) and reached the 10-W range approximately 1000 minutes sooner than in the first test (P9502), supporting the idea that increased vacuum exposure is beneficial. The large pressure and heat leak excursions, at approximately 4500 minutes on Test P9502, were attributed to outgassing with subsequent smaller heating rate spikes attributed to minor outgassing. Such outgassing effects were not experienced during the second test (P9601). Another observation is the temporary heating rate drop due to MLI layer cooling below steady-state values during evacuation. This is clearly visible in Figure 7 at 3,000 minutes where the heating rate is approximately 30 percent below the steady-state value. An additional 5,000 minutes past this point were required to obtain steady-state boiloff.

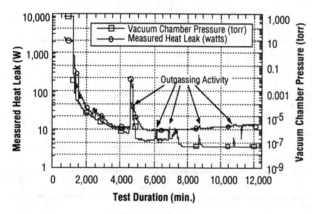

FIGURE 6. Test P90502 Ascent Simulation Heating Response.

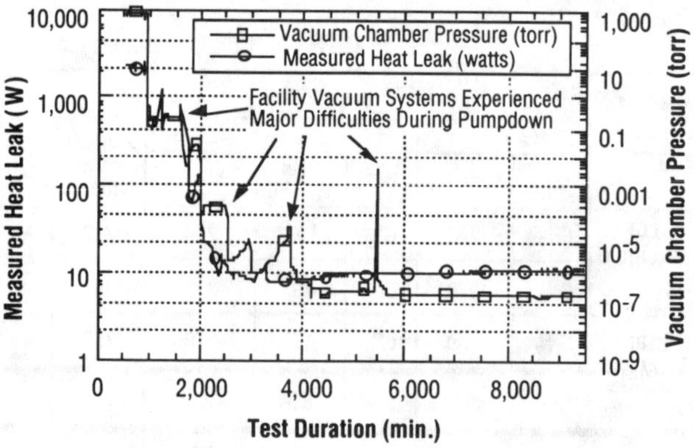

FIGURE 7. Test P9601 Ascent Simulation Heating Response.

Orbit-Hold Simulation

The results of the three orbit-hold simulations are documented in Table 2. The insulation heating rates ($\dot{Q}_{insulation}$) ranged from 10.93 to 2.98 W (0.31 to 0.085 W/m^2) for warm boundaries ranging from 305 to 164 K. These performance values are illustrated in Figure 8 along with the calculated performance of a standard constant-density MLI blanket. The variable-density MLI outperformed the standard MLI by approximately 50 percent at the highest boundary temperature and both systems performed equally at the lowest boundary temperature. The first test (P9502) yielded higher heating than the second test (P9601), which was expected due to decreased outgassing. The third test (P9602B) produced boiloff rates nearly as high as that recorded on the first test due to significant MLI damage which occurred at the conclusion of Test P9601. During a pressurized liquid expulsion test the tank seal developed a leak which over-pressured and tore the manhole cover MLI (approximately 1.5 m^2). The outer DAM layer was also torn loose, exposing approximately 11m^2 of Dacron (resulting in 44 DAM shields on a third of the tank surface area). The two video cameras located within the environmental shroud failed to reveal the damage. However, the incident accidentally demonstrated TCS robustness. In comparison with the best previously measured performance of traditional MLI, a 50-percent heat leak reduction was achieved (Fig. 9). Boiloff losses of 0.117 and 0.16 percent per day were measured with and without heat guards, respectively, at the warm boundary condition of 305 K.

TABLE 2. TCS Steady-State Orbit-Hold Performance.

Test #	Test Conditions						Measured TCS Performance (W)			TCS Heat Flux W/m^2
	Initial Conditions	Chamber Press (torr)	Interstitial[a] Press (torr)	Heat Guards	Heater Shroud Temp (K)	Ullage[b] Range (%)	$\dot{Q}_{boiloff}$ (W)	$\dot{Q}_{penetrations}$[c] (W)	$\dot{Q}_{insulation}$[d] (W)	$\dot{Q}_{insulation}/A_{tank}$[e]
P9502	Vacuum Chamber Rapid Evacuation to Orbit Conditions After Completion of Ground Hold Test	6 x 10^{-8}	---	Off	305	12 - 17	13.10	2.39	10.71	0.31
		9 x 10^{-8}	---	Off	164	17 - 21	5.34	0.96	4.38	0.13
P9601	Vacuum Chamber Rapid Evacuation to Orbit Conditions After Completion of Ground Hold Test	2 x 10^{-7}	---	Off	305	25 - 30	11.07	2.41	8.66	0.25
		6 x 10^{-8}	---	On	305	25 - 30	7.89	0.25	7.64	0.22
		2 x 10^{-7}	---	Off	305	25 - 30	10.90	2.39	8.51	0.24
		9 x 10^{-8}	---	Off	164	30 - 35	3.90	0.92	2.98	0.085
P9602B[f]	Vacuum Chamber Evacuated to 10^{-5} torr and Test Article Vacuum Conditioned Prior to Tanking of LH$_2$	5 x 10^{-8}	8 x 10^{-6}	Off	235	5 - 8	8.41	1.59	6.82	0.20
		4 x 10^{-8}	4 x 10^{-6}	On Legs Only	235	5 - 8	7.28	0.76	6.52	0.19
		4 x 10^{-8}	1 x 10^{-7}	Off	305	8 - 12	12.87	2.40	10.47	0.30
		4 x 10^{-8}	1 x 10^{-7}	On Legs Only	305	8 - 12	12.11	1.18	10.93	0.31

a For tests P9502 and P9601 the MLI interstitial pressure measurement system failed to operate.
b Liquid level estimate continuous silicon diode temperature rake (continuous liquid level sensor not operational).
c Includes the sum of all penetration solid conduction (legs, fill/drain, vent etc.)
d Q insulation calculated as Q boiloff - Q penetrations.
e Tank surface area taken as 34.75 square meters.
f MLI on tank upper dome damaged prior to test P9602A.

Averaged temperatures measured at various layer locations and on the SOFI are presented in Figure 5. This figure also illustrates that the SOFI layer provides essentially no thermal protection at vacuum conditions. In fact, the SOFI surface temperature is only slightly warmer than the LH$_2$. The majority of the TCS temperature gradient occurs across the MLI which, for the 305 K boundary set point, is radiation dominated as illustrated by the 4th-order temperature profile shape. The MLI temperature profile is nearly linear with a 164 K boundary, indicating that solid conduction is the dominating mode. Figure 5 also illustrates the excess thermal energy stored in both the SOFI and MLI during ground-hold. This excess TCS energy must be rejected both into the LH$_2$ and to the surroundings during the transition to the orbit-hold, requiring a substantial amount of test time. For all tests the vacuum chamber pressure was maintained below the 10^{-6} torr range. The interstitial pressure probe results were questionable, but typically indicated a pressure level in the mid 10^{-6} torr range or less.

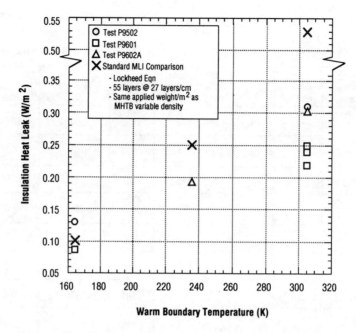

FIGURE 8. TCS Steady-State Orbit-Hold Performance.

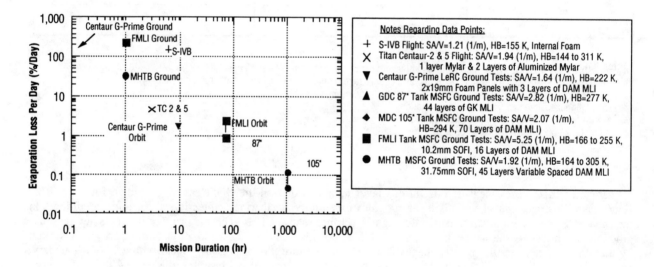

FIGURE 9. Historical Comparison of Flight and Test Data.

CONCLUSIONS

An 18-m^3 hydrogen test article, termed the MHTB, was utilized to experimentally evaluate the performance of a foam/multilayer thermal control concept. The Isofoam SS-1171 spray-on foam insulation (SOFI) thickness averaged 3.5 cm and was designed to protect against ground-hold/ascent flight environments, and it enabled the use of a dry nitrogen purge as opposed to the more complex/heavy helium purge subsystem normally required. The MLI (45 layers of double-aluminized Mylar with Dacron spacers) was designed for an on-orbit storage period of 45 days and included several unique features: a variable-density MLI lay-up; larger but fewer DAM perforations for ascent venting; and roll-wrap installation of the MLI with a commercially established process. The MLI roll-wrap installation process resulted in a very robust, virtually seamless insulation with an estimated man-hour savings of 2400 hours per liquid hydrogen and oxygen tank set (3 m diameter). Further, the installation concept enables a more repeatable, consistent product as compared with individually constructed MLI blankets.

Insulation performance was measured during three test series conducted between September 1995 and May 1996. Ground-hold testing produced the expected average heat leak of 63 W/m² at a foam surface temperature of 170 K. The SOFI successfully prevented purge gas liquefaction except for a small localized area near an odd-shaped penetration peculiar to the MHTB. It is concluded that SOFI-type insulation is a feasible means for eliminating the need for helium purge-bag subsystems. The simulated ascent test successfully demonstrated MLI-venting with the larger (1.27 cm diameter), more widely spaced (7.6 cm apart) perforations.

The simulated orbit-hold periods produced heat leaks ranging from 0.085 to 0.22 W/m² at warm boundary temperatures of 164 K and 305 K, respectively. This orbit-hold-measured insulation performance, when compared to the best previously measured performance of a traditional MLI system, achieved a 50-percent heat leak reduction. This substantial performance improvement is attributed to the variable-density lay-up, the vent perforation pattern, and the almost seamless MLI roll-wrap installation. It should also be noted that additional time will be required to attain steady-state orbital heat leak on "first flights" due to TCS outgassing. Overall, the MHTB insulation demonstrated excellent performance for all mission phases representative of a cryogenic upper stage.

Acknowledgments

The authors extend their appreciation to Glen McIntosh, Cryogenic Technical Services, for the innovative MLI design, to the Lockheed-Martin Corporation for the test tank and raw materials, and to the NASA-MSFC team responsible for originating, assembling, and testing the MHTB.

References

Cullimore, B. A. et al. (1988) Systems Improved Numerical Differencing Analyzer and Fluid Integrator Version 2.2, Martin Marietta Document, (NASA Contract NAS9-17449) MCR 86-594.

Keller, C. W., G. R. Cunnington, and A. P. Glassford (1974) "Thermal Performance of Multilayer Insulations," Final Report, Lockeed Missiles & Space Company, Contract NAS3-14377.

Ozisik, M. N. (1985) "Heat Transfer - A Basic Approach," McGraw-Hill Book Company, New York, 83.

Nomenclature

A Area
ε_{TH} Total hemispherical emissivity
h_{fg} Heat of vaporization
h Enthalpy
K Thermal conductivity
L Length
$\dot{m}$ Mass flow rate
N Number of MLI shields
$\overline{N}$ Average MLI layer density
P MLI interstitial pressure
ρ Density
$\dot{Q}$ Heat leak rate
T Temperature
t Time
U Internal Energy

STUSTD LH2 STORAGE AND FEED SYSTEM TEST PROGRAM

A. D. Olsen
The Boeing Company
Huntington Beach, CA 92647-2099
(714) 896-4852

Abstract

This paper describes the proposed ground testing of a liquid hydrogen storage and feed system (LHSFS) which is part of the Solar Thermal Upper Stage Technology Demonstrator (STUSTD). Some background information is provided to highlight the benefits of solar thermal upper stages and the type of development work that is required. A functional description of the LHSFS is provided along with detailed descriptions of individual components. A test program is defined which points out the objective of the various tests and describes how the tests will be conducted. Completion of this test program will ready the Cryogen Storage and Propellant Feed System (CSPFS) for incorporation into integrated CSPFS/engine ground-based tests that will provide the data required to design a flight experiment with acceptable risk levels.

INTRODUCTION

This paper describes the proposed ground testing of a liquid hydrogen storage and feed system (LHSFS) which is a part of the Solar Thermal Upper Stage Technology Demonstrator (STUSTD). Solar thermal upper stage propulsion represents a high performance means of transporting space payloads from Low Earth Orbit (LEO) to Geosynchronous Equatorial Orbit (GEO). The solar thermal propulsion concept investigated in the STUSTD program consists of stored cryogenic hydrogen propellant, lightweight deployable solar concentrators, and a direct gain engine. The concentrators direct solar energy to the engine where hydrogen is heated to high temperatures (in excess of 2500 K) and then expanded through a nozzle to generate thrust. While engine performance for this system is quite high (specific impulses of over 800 s), thrust levels are small (approximately 2.2 N for the test design). As a result, LEO to GEO orbit transfer requires a longer time than achieved with conventional chemical propulsion systems. The design to be tested here utilizes a series of 140 "burn" and coast cycles over a 30 day period to achieve the desired payload transfer.

A critical aspect of the solar propulsion system is the reliable and efficient storage of cryogenic hydrogen both during launch, as well as during the LEO to GEO orbit transfer mission when the solar engine is operational. The STUSTD LHSFS test program is intended to demonstrate this technology. The LHSFS consists of a 2 m^3 cryogenic fluid storage tank equipped internally with liquid acquisition devices (LAD's), active and passive thermodynamic vent systems (TVS), instrumentation, and fluid lines. System features external to the tank are helium (He) purged multi-layer insulation (MLI), heaters, instrumentation, structural supports, and fluid lines and components.

OBJECTIVES

The technical objectives for the STUSTD LHSFS testing are:

(1) Demonstrate MLI purge evacuation and performance,
(2) Deliver near-saturated hydrogen vapor at 158.6-241.3 kPa at 0.91 kg/hr for 140 cycles over 30 days,
(3) Demonstrate venting/lockup for 140 burns over 30 days (with no additional overboard venting),
(4) Demonstrate use of MLI and heaters to assist control of tank pressure at 310.3 ± 27.6 kPa,
(5) Demonstrate subcooled LAD/TVS operation in one-g using internal LH_2 tank temperature profile, and
(6) Demonstrate LH_2 loading/unloading and LAD filling for ground operations.

Testing of the LHSFS will be performed at the NASA Marshall Space Flight Center (MSFC) in Huntsville, Alabama. The purpose of this ground test program is to demonstrate the capability of the LHSFS to provide reliable, efficient storage and supply of cryogenic hydrogen while exposed to the space mission environment. This environment will be provided by the MSFC 20-foot diameter vacuum chamber at Test Stand 300, Position 302, in the East Test Area.

An overall test schedule is shown in Figure 1. Installation of the LHSFS test article into the vacuum chamber is planned for the Feb '98 time frame, followed by checkout and test activities during March and April '98.

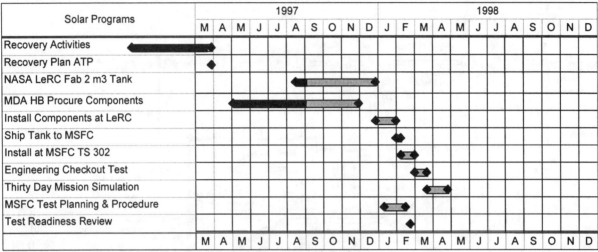

FIGURE 1. Liquid Hydrogen Storage and Feed System Test Schedule.

System Functional Description

A schematic of the LHSFS is shown in Figure 2. The system is designed to store LH_2 and deliver GH_2 to the engine interface at 206.9 kPa and 0.91 kg/hr. The exterior of the tank is fitted with electrical heaters and He purged multi-layer insulation. Tank penetrations and lines are shown for hydrogen (LH_2) fill and drain, He pressurization and purge, venting, and instrumentation/power connections. Remote Operated Valves (ROV's) are indicated in the various fluid lines. A micrometering valve is placed upstream of the simulated absorber engine feed point for system adjustments (pressure and flow) during initial characterization tests. This GH_2 feed system flow then passes through a flowmeter and orifice and is vented to atmosphere.

The interior of the tank contains four LAD channels. The LAD channels are triangular in cross-section, with screened fluid passages facing the tank wall. The screens acquire LH_2 from the tank by using phase separation through surface tension forces. During initial tank fill, boil-off hydrogen and residual helium purge gas in the LAD channels will travel through vent lines that exit the top of the LAD's (indicated by dashed lines in the schematic). These lines are manifolded together and exit the tank via a single vent line.

Tank pressure during mission simulation testing will be maintained at 310.0 ± 27.6 kPa. During an engine burn simulation (tank vent), LH_2 in the LAD channels flow into lines at the base of each channel. This LH_2 then passes through Joule-Thomson expanders, where the flow experiences a 69.0 kPa pressure drop which results in a temperature drop of approximately 1.1 K. The cooled hydrogen then travels back through the LAD channels inside tubing bonded to the interior apex of each channel. The cooled fluid and channel walls serve as a heat sink to remove heat from the tanked hydrogen through condensation and convection. This constitutes the passive TVS. The fluid then exits the top of each LAD channel, is combined in a manifold, and is fed into the active TVS system.

The active TVS consists of a heat exchanger and pump. Hydrogen from the passive TVS travels through a heat exchanger coil and exits the tank. The active TVS pump draws LH_2 from the tank bulk fluid at a flow rate of approximately 0.11 kg/s and routes it through the warm side of the heat exchanger. This flow is then discharged inside the tank at an exit velocity of approximately 0.76 m/s. The active TVS pump and heat exchanger thus provide LH_2 cooling and bulk tank fluid circulation. The heat exchanger is rated for a heat extraction rate of 75 W.

Hydrogen leaves the tank at approximately 22.9 K. This flow then passes through a heat exchanger external to the tank, where it is warmed by an aluminum plate that has been heated to a temperature of 200 K by a 0.9 kW electrical heater. The increase in fluid temperature from the heat exchanger is required for the non-cryogenic valve system downstream. Downstream controls and instrumentation include on/off valves, the micrometering valve, and temperature and pressure sensors.

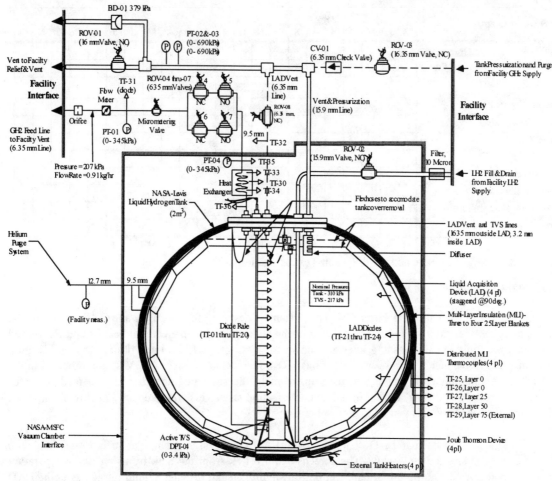

FIGURE 2. Liquid Hydrogen Storage and Feed System (LHSFS) Schematic.

Tank

The tank is an ellipsoidal 2 m^3 pressure vessel made from 2219 aluminum and has a dry weight of 1085 N. The tank has a wall thickness of 1.524 mm increasing to three times that in the weld areas. The tank major axis is approximately 1.78 m; its minor axis (height) is approximately 1.25 m. It has a maximum operating pressure of 344.75 kPa, has been proofed at 517 kPa, and has a design burst pressure of 689.5 kPa. The top and bottom of the tank contain 0.61 m and 0.0762 m diameter access openings respectively. Both openings are closed with cover plates and Conoseal gaskets. The tank is mounted in a hexagonal metal frame by 12 low heat conduction, fiberglass composite struts.

Liquid Acquisition Device and Passive Thermodynamic Vent System

The LAD and passive thermodynamic vent system consists of four curved channels of triangular cross-section, Joule Thomson devices, and associated fluid lines. The four LAD channels are spaced 90° apart in the tank. Metal screens are located along the wall facing sides of the channels for LH$_2$ acquisition. The channels are mounted to the tank wall via formed fiberglass clip assemblies bonded to the inner tank wall and stand off approximately 6.35 mm from the tank wall. The bottoms of the LAD channels are equipped with Joule-Thomson devices (Lee brand Visco Jets). The Visco Jet connects to 3.175 mm tubing formed into a 180° bend that reenters the LAD channel, continuing the entire length of the channel and passing out the top. The four lines are manifolded into a single 6.35 mm line that supplies the active TVS. The LAD channels are also fitted with 6.35 mm vent lines at the top, which are manifolded together in the top of the tank and provide an exit for venting excess gas formed during tank filling.

Active Thermodynamic Vent System

The active TVS is located at the bottom of the tank. It consists of an 0.218 m high by 0.375 m OD heat exchanger and a centrally located 1.2 W (input power) pump. The cold side of the heat exchanger is supplied by the 6.35 mm passive TVS line and consists of a 23 wrap coil of 3.56 mm ID tubing which forms the interior of the assembly. LH_2 passing through this coil exits the tank via a 6.35 mm line. The warm side of the heat exchanger is supplied with LH_2 suctioned from the base of the tank by the vertically oriented pump. The warmer LH_2 is routed through an exterior coil of 44.45 mm square flow channels containing 3 turns and is discharged back into the tank through a vertical 53.85 mm ID pipe.

Diode Temperature Probe

The temperature probe consists of a Teflon bar attached to the tank fill tube by stainless steel clamps, screws, and nuts. Twenty silicon diodes are mounted to the bar to perform tank fluid temperature and level sensing. The output wires are held in place by additional clamps.

Tank Cover

The top cover of the tank contains five smaller diameter cover plates mounted to it. All penetrations in and out of the tank for fluid flow and instrumentation/power connections are made through these feed-through cover plates. Two of the feed-through cover plates have an OD of 0.117 m, and the remaining three have an OD of 0.08 m. The cover plates are fastened with six bolts each and are sealed with Conoseal gaskets.

Engine Feed Line Heat Exchanger

The absorber engine feed line heat exchanger consists of a 9.525 mm aluminum tube welded in a serpentine pattern to an electrically heated aluminum plate. The assembly will be mounted to the tank support structure and covered with MLI. Nine 100 W Kapton heaters are bonded under the plate.

Multi-Layer Insulation, Helium Purge, and Tank Heaters

The multi-layer insulation (MLI) consists of layered blankets made of double aluminized Kapton film (DAK) and alternating layers of Dacron mesh (B4A). The top and bottom areas of the tank will contain 4 sets of 25 layer blankets (100 layers total), while the middle sections will contain 3 sets of 25 layer blankets (75 layers total). The tank support struts and penetrations will be individually spiral wrapped. The MLI blankets attach to each other with sewn Velcro strips. Exterior web straps are also used to secure the blankets to the tank. These straps attach to the struts and permit the lower insulation to be locked in place while the upper insulation is removed to enable access to the tank via the manhole cover. This insulation will be electrically bonded and grounded to the tank support structure. The MLI contains a pattern of 1.45 mm holes covering approximately 3% of the surface area to enable gas evacuation. Underneath the first layer of MLI and attached to the tank, will be three loops of lightweight, low conductivity, 9.525 mm perforated Teflon tubing to supply helium purge gas. This purge system will be turned on prior to introduction of cryogenic fluids to displace all condensable gases in the MLI that could potentially freeze. The purge system enables tank filling in a non-vacuum environment such as that encountered immediately prior to launch. Four 10 W electrical resistance Kapton strip heaters are attached directly to the bottom half of the tank external surface with epoxy. Power is supplied individually to the heaters via a wiring harness routed to the facility power source.

TEST DESCRIPTIONS AND OPERATIONS SEQUENCE

A description of the planned test operations, including objectives and chronological sequence are presented below. The actual test operations sequence will be subject to the continual stream of data and information obtained during test execution. At the start of the operations the tank will contain He and residual air.

Tank Safing and Inerting

The objective of this operation will be to remove all condensable gases from inside the LHSFS. Prior to this operation the chamber will have been evacuated to 0.133 kPa or less and maintained at that value for 4 hours to fully remove all condensable gases from the test article. If at any time after initial chamber evacuation the chamber requires re-pressurization and opening, the MLI He purge will be activated. This will inhibit the entrance of moisture into the test article MLI.

Tank safing and inerting will be accomplished by three cycles of tank evacuation and helium fill (purge) of the entire system. At the conclusion of this operation, the tank will be at approximately ambient temperature and filled with He at a pressure of 48.27 ± 6.895 kPa.

LH_2 Fill

The objective of this operation is to perform tank LH_2 fill and demonstrate the MLI GHe purge. This test will be conducted with the MLI purge on and the vacuum chamber filled with one atmosphere of nitrogen (dew point of 21.94 K or lower). The cold wall nitrogen flow will be off and the wall temperature will be at steady state conditions. Data will be obtained at a frequency of 1 Hz (low rate). The MLI purge is maintained for the simulated launch that follows.

Simulated Launch

The objective of this test is to demonstrate the LHSFS exposure to a launch depressurization. This test will immediately follow the LH_2 Fill operation. There is no requirement for tank pressure stabilization prior to commencement of the simulated launch. The MLI GHe purge will be turned off at the same moment that the chamber vacuum pumps are activated to simulate a launch depressurization. A Space Shuttle launch profile (approximately 7 minutes) will be simulated. After on orbit conditions (i.e. vacuum at 1.33×10^{-4} Pa) are achieved, the tank will be monitored for pressure stabilization via PT-02 and PT-03. Stabilization will be marked by pressure fluctuations in the tank of less than ±0.6895 kPa/hr for four consecutive hours.

Baseline Boil-off (without cold wall)

The purpose of this test is to obtain data to determine the heat leak into the tank system. This test will initiate any time after stabilization has been achieved. The cold wall nitrogen flow will be off and the wall at steady state temperature. The boil-off rate will be determined by making use of the facility vent flow rate measurement system. The facility Tank Pressure Control System (TPCS) will be used in the vent line downstream of ROV-01 to maintain a constant tank pressure of 124.1 ± 0.0069 kPa. The suggested test duration is 10 hours. With initial tank fill level above 92%, ROV-01 will be opened and data will be acquired at the low (1 Hz) sampling rate while boil-off occurs. Predicted boil-off rate for this test is 0.0817 kg/hr.

Baseline Boil-off (with cold wall)

The purpose of this test is to obtain data to determine the heat leak into the tank system under a cold wall boundary condition. After a steady state cold wall temperature of 94.44 K has been attained, the test will begin. This cold wall boundary condition will be maintained for all subsequent testing until the end of the thirty day mission simulation described later. This test will make use of the facility vent flow measurement system and will use the facility TPCS to maintain the tank pressure at 124.1 ± 0.0069 kPa. The tank initial fill level will be noted from temperature probe readings. Predicted boil-off rate for the cold wall boil-off test is 0.0247 kg/hr.

Engineering Tests

An engineering test sequence is to be implemented at this point to develop a database for the LHSFS before the 30-day mission simulation. The expected duration of these tests is to be about ten days. Data obtained during these tests will help characterize and verify expected performance of the passive and active TVS, the tank heaters, and the valve system. The data acquisition system will start at the low rate for the tests and transition to the high rate (10 Hz) when required. The tank will be brought to a pressure of 310.275 ± 27.58 kPa during this test operation.

Top Off and Stabilization

The objective of this operation is to top off and stabilize the tank in preparation for the mission simulation. ROV-02 will be activated to fill the tank to 92%. The system will then be monitored for stabilization. Stabilization will have been achieved when pressure fluctuations are less than or equal to 0.6895 kPa/hr for four consecutive hours. This operation is expected to take approximately 12 hr.

Pressurize to 310.275 kPa

Targeted tank pressure for the entire mission simulation is 310.275 ± 27.58 kPa. Two of the four 10 W tank heaters will be activated while monitoring tank pressure. When tank pressure reaches 310.275 ± 27.58 kPa, the heaters will be turned off. Pressurization rates will be monitored. The active TVS will be operated (pump on, no venting) and the heaters will be turned on as required to reach a saturation pressure of 310.275 ± 6.895 kPa. The testing will then immediately proceed to the 30 day mission simulation.

Thirty Day Mission Simulation

The objective of this test is to simulate a LEO to GEO mission. Specifically, the system will demonstrate tank pressure control at 310.275 ± 27.58 kPa in the external vacuum provided by the chamber, and the corresponding engine feed at 0.9072 kg/hr. The tank heaters, feed line heat exchanger, absorber engine feed line valves, and the active TVS pump will be operated at various times during the mission simulation. The 30 day mission simulation will consist of 140 alternating burn and coast periods as would occur in an actual LEO to GEO transfer. The tank heaters are only scheduled for activation during coast periods and the Active TVS only during burn periods. A dedicated flow meter (MSFC supplied) will be used to measure the feed line flow rate.

CONCLUSIONS

Solar thermal upper stages offer performance advantages over chemical stages. Payload delivered to GEO can be doubled or a fixed payload can be placed in orbit with a smaller launch vehicle. A key system is the LHSFS, which is required to store cryogen for up to 30 days and deliver liquid to the engine at the proper interface pressure and flowrate. Boeing has designed a LHSFS that will be used to validate performance in a relevant environment. Tests have been defined which will fully characterize performance of the system at 1-g. Several issues cannot be resolved in a 1-g environment including propellant acquisition. Follow on tests will be used to integrate the LHSFS with the engine system. Eventual flight tests of a complete solar thermal upper stage are expected in the 2000 to 2001 timeframe.

Acknowledgments

The author would like to thank Doug Richards for his help in preparing the LHSFS test plan.

References

Richards, D. R., Cady, E. C (1996) "Test Plan For The Solar Thermal Upper Stage Technology Demonstrator Liquid Hydrogen Storage and Feed System," McDonnell Douglas Aerospace, September 1996.

INTEGRATED SOLAR UPPER STAGE (ISUS) ENGINE GROUND DEMONSTRATION (EGD)

Charles T. Kudija and Patrick E. Frye
The Boeing Company
Rocketdyne Propulsion and Power
P.O. Box 7922
Canoga Park, CA 91309-7922
(818) 586-3611

Abstract

The Integrated Solar Upper Stage (ISUS) Engine Ground Demonstration (EGD) Program sponsored by the Air Force Phillips Laboratory (PL) conducted a full-up ground demonstration of a solar thermal power and propulsion system at NASA Lewis Research Center in mid-1997. This test validated system capability in a relevant environment, bringing ISUS to a Technology Readiness Level (TRL) of 6, and paving the way for a flight demonstration by the turn of the century. The ISUS technology offers high specific impulse propulsion at moderate thrust levels and high power, radiation-tolerant electrical power generation. This bimodal system capability offers savings in launch vehicle costs and/or substantial increases in payload power and mass over present day satellite systems. The ISUS EGD consisted of the solar receiver/absorber/converter (RAC), power generation, management, and distribution subsystems, solar concentrator, and cryogen storage/feed subsystems. Simulation of a low Earth orbit (LEO)-to-Molniya orbit transfer (30-day trip time) as well as characterization of on-orbit power production was planned for this ground test. This paper describes the EGD test integration, setup and checkout, system acceptance tests, performance mapping, and exercise of the system through a mission-like series of operations. Key test data collected during the test series is reported along with a summary of technical insights achieved as a result of the experiment. Test data includes propulsion performance as derived from flowrate, temperature, and pressure measurements and the total number of thermal cycles.

INTRODUCTION

The Engine Ground Demonstration (EGD) test series was conducted at the NASA LeRC Tank 6 vacuum facility from July 8th through September 19th, 1997. This test series successfully validated system level feasibility of the solar thermal propulsion mode for the ISUS hardware. A total of 117 simulated hydrogen "burns" or blowdowns were performed. This represents 54% of the planned 216 burns from the Reference Flight System (RFS) mission profile. A peak temperature of 2200 K was achieved in the receiver during bakeout. The peak gas temperature achieved was in excess of the 2012 K reading taken by thermocouples attached to the exterior of the gas outlet tube beneath a multi-layer insulation (MLI) package. The calculated specific impulse for the hydrogen bulk temperature (at an estimated ΔT of 100 K above this measured wall temperature) is 758 sec. This calculation assumes the full 5 g/s flowrate and 100:1 nozzle of the RFS. Table 1 summarizes the accomplishments of the EGD test series.

TABLE 1. Summary of EGD Test Series.

EGD Test Property	Value
Peak RAC body temperature	> 2200 K
Peak gas temperature	> 2012 K
Correlated specific impulse	758 sec
Heat exchanger effectiveness	> 90 %
Total number of burns	117
Total time RAC @ temperature	320 hrs
Fraction of RFS flowrate	~ 33 %
MLI Performance (non-aperture losses @ 2300 K)	~ 4700 W
Cooling rate	312 K/min

EGD TEST SETUP

The EGD test series represents the culmination of a two-year team effort by Boeing (Rocketdyne and Huntington Beach), Babcock & Wilcox (B&W), Harris Corporation, and NASA LeRC. The principle hardware components of the EGD test article include the receiver cavity, multi-layer insulation package, secondary concentrator, support structure, advanced facets, and a hydrogen pre-heater. Figure 1 presents the layout of the EGD test article within the Tank 6 test facility.

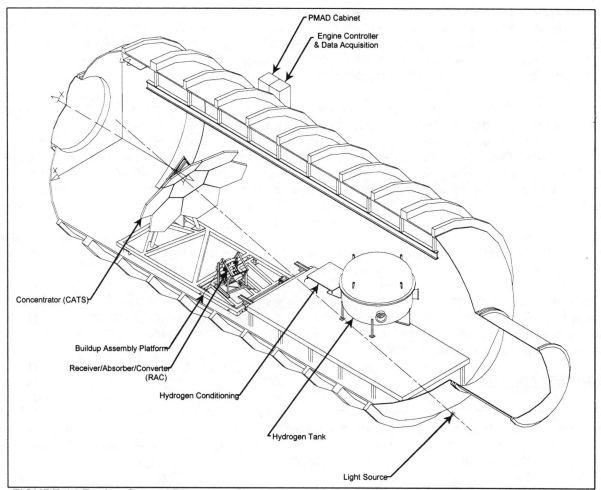

FIGURE 1. Engine Ground Demonstration Layout at NASA-LeRC Tank 6 Facility.

The Tank 6 vacuum facility consists of a 4.6-m diameter by 22-m long main chamber that can be pumped down to 3×10^{-6} torr base pressure. Twenty 0.8-m diameter oil diffusion pumps (ODPs) are used to pump down Tank 6 and are backed by four Roots Blowers and three Kinney roughing pumps. A three-section nitrogen coldwall used to simulate the space thermal environment was operated at partial capacity during the test.

Optical power is provided by a solar simulator located at one end of Tank 6. The solar simulator consists of nine 30-kWe Xenon arc lamps that together provide up to 1.8 kW/m² of solar energy at a subtended angle of < 1° over a 4.5 meter diameter test area. The input solar flux from the Xenon arc lamps passes through an optical train consisting of a collector, lens, turning mirror, and quartz window before illuminating the concentrator located in Tank 6.

The concentrator array consists of seven panels arranged edge-to-edge with six triangular facets mounted to each panel. The array is attached to a frame that is mounted to a tripod support structure. The concentrator array and support structure were fabricated and installed by Harris Corporation during the previous Solar Dynamic Ground Test Demonstrator (SDGTD) program.

Ten advanced facets were fabricated by Harris subcontractor Composite Optics Incorporated (COI) and installed by Harris personnel as part of the EGD program. These facets use a thin composite face sheet backed by a triangular-grid structure with a replicated epoxy front surface. Silver is deposited on the front surface to achieve a reflectivity greater than 90 percent. Silicon dioxide is deposited as a protective coating over the silver layer. These facets represent a 40 percent reduction in area density and a reflectivity increase of 4 to 6 percent. Also, the COI facets were fabricated on a toroidal mold improving the concentration ratio of each facet, as compared to the SDGTD facets that were fabricated on spherical molds. Clear Teflon film was placed over the concentrator facets to provide protection against contaminants outgassed from the high temperature RAC components.

The receiver/absorber/converter (RAC) hardware including the rhenium coated graphite cavity, secondary concentrator, MLI package, and support framework was located at the focal point of the EGD test. The receiver cavity consists of a graphite cylindrical main body and a pair of endcaps. All surfaces of the receiver (inside and outside) are coated with CVD rhenium. The endcaps are penetrated by insulated rhenium, propellant inlet and exit lines. The exit tube contains a sonic throat ensuring choked H_2 flow during propulsive operations. The MLI package consists of several layers of tungsten and molybdenum foils. Two secondary concentrators, one each with an iridium specular surface and a rhodium specular surface, were prepared and used, each for a portion of the EGD system test. These were non-imaging compound parabolic concentrators used at the primary focal point to increase flux entering the RAC cavity, thereby reducing aperture losses. For each, the body consisted of a rhenium-coated graphite blank, and a specular reflective coating(s) was applied to the optic surfaces.

Gaseous H_2 supplied from a truck at the Tank 6 facility was fed to the H_2 preheater. The preheater used waste heat from the receiver body to preheat the H_2 flow prior to entering the receiver. The liquid H_2 supply equipment originally planned for the EGD test was not available at the time of the test and the test proceeded with the gaseous H_2 supply.

EGD TEST ACTIVITY

The original test activity sequence at Tank 6 consisted of subsystem checkout tests followed by propulsion system testing using in turn both gaseous and liquid hydrogen, remote reconfiguration to support electric power production, and electric power system testing. In reality, a subset of this plan was carried out, and is described here with summary results.

The concentrator was checked out using optical survey techniques developed by Harris. Testing in vacuum with light levels at about one sun and a water-cooled black body cavity calorimeter confirmed the concentrator performance analysis and optical survey predictions within 5 percent. The achieved concentration ratio was 5% lower than specified, due wholly to substitution of 10 original facets for 10 advanced facets that proved unusable. On the other hand, of the 20 planned advanced facets, the 10 advanced facets from COI performed at a level consistent with the overall CATS performance specification had all new facets come from that supplier. An additional calorimetry test, with a water-cooled, silvered secondary concentrator at the primary focus, confirmed the accuracy of the compound parabolic concentrator optical design. These tests confirming the overall performance of the EGD optical train linking the Harris optical analysis and optical survey tools to empirical power data in vacuum, were a major result of the checkout tests that lead up to EGD.

After the concentrator performance test, the RAC was installed at the primary focus with the passively cooled secondary concentrator (iridium surface) in place. Following instrumentation checks, the RAC was heated via the optical train intermittently over a period of days to a series of temperature plateaus baking out absorbed gasses, manufacturing contaminants, oxide layers, and other impurities. While the tank was open on July 18th, 1997, the other secondary concentrator (rhodium surface) was installed in anticipation of transition into propulsion testing. The bakeout was then completed after about 20 hours of operation above 1500 K. It was terminated by closing the shutter at noon on July 21st, 1997 when the RAC body temperature was above 2200 K. The RAC cooled passively over a period of two hours with the last hour interrupted occasionally by shutter openings to keep the CATS facets from cooling too much. This transient provided the first empirical record of heat loss from the RAC through the aperture and MLI package. This passive cooldown showed greater than expected thermal losses at a rate consistent with an estimated 4700 watts through the MLI at a cavity temperature of 2300 K.

Heating was resumed and the first of a series of burns was simulated by passing hydrogen gas through the RAC at about 1.7 grams per second. The flowrate, and the amount of hydrogen discharged for each of these burns was planned in advance to simulate a representative sample of the propulsive maneuvers expected for the baseline Reference Flight System (RFS). However the temperatures of the RAC at the beginnings of the burns were in the 2120 K to 2150 K range rather than the planned 2200 K to 2300 K range. A decision was made on the evening of July 21st, 1997 to collect the detailed profile of an extended burn to benchmark the existing system model with empirical data, and a 27-minute blowdown was initiated. On-site analysis of the data confirmed that the capability of the RAC to sustain high outlet gas temperature over time as it is discharged is greater than previously accounted for in the system modeling. This empirical profile data, when plugged back into the system model, showed that RAC burn initiation at 2150 K was sufficient to demonstrate the RFS mission performance. The RFS orbit transfer schedule was adjusted for this result and the sample of RFS propulsion testing was continued with adjustment of the RAC temperature target to 2150 K at the start of each burn. The sequence was completed late on July 23rd, 1997.

Figure 2 shows a picture of the EGD test in-progress during a heat-up cycle with shutters open and lamps at full power. The picture was taken from behind the receiver and gives a view of the reflection of the solar image in the receiver aperture. Figure 3 shows another picture of the EGD test in-progress moments later with shutters closed and full hydrogen flow. The reflection of re-radiated energy out of the cavity aperture and additional losses near the gas exit tube can be seen. These pictures were extracted from a video taken during initial sequence of propulsion testing.

FIGURE 2. Shutters Open and Full Lamp Power.   FIGURE 3. Shutters Closed and Full H$_2$ Flow.

The propulsion testing then focused on achieving additional thermal cycles (burns) using a typical (mid-transfer) amount of propellant to discharge during each burn for the remainder of the week. Seventeen cycles were completed over the next two days with anomalous heatup duration noticed the morning of July 25th, 1997. Further review of the data revealed a discrepancy in flowmeter readings, indicating a probable leak, and a change in facet thermal response indicating possible contamination. The test sequence was terminated Friday evening and the tank was opened for inspection the following week.

Hardware examination confirmed the audible presence of a leak near the RAC inlet and the visible presence of contamination on the Teflon film over the concentrator facets along with other effects. Furthermore, an examination of contaminant deposited on the front face of the RAC MLI package revealed a pattern consistent with a localized contaminant source in the vicinity of the inlet tube where it enters the MLI wrap. There was no apparent crack or corrosion of that portion of the RAC body that was visible from the cavity interior.

A decision to proceed with more thermal cycle testing was made. The used Teflon film was changed out for new, and the hydrogen supply setting was increased to provide sufficient flow through the RAC in spite of the leak which had by now been localized to the RAC inlet region. On August 4th, 1997 the testing was continued. Eleven more cycles were obtained as the differential between inlet and outlet flowmeter readings increased. Testing was stopped when it became apparent that the RAC could no longer be driven to temperatures that would be useful for thermal cycle accumulation. Upon opening the tank and removing the MLI from the inlet line, the leak source became apparent. A portion of the inlet tube appeared corroded. A month passed while the RAC was dismounted and partially disassembled, the inlet line repaired, and a source of ultra pure hydrogen obtained. During this time the solar simulator lamps were changed, a decision was made to concentrate on propulsive thermal cycle accumulation. The thermionic diode array testing was deferred until the RAC could be fitted with an electric heat source in the Baikal facility located at the New Mexico Engineering Research Institute (NMERI).

Two sessions of propulsion testing each lasting a work week were conducted in September, 1997. These tests conducted from the 8th through the 12th and the 15th through the 19th accumulated an additional 67 thermal cycles. These cycles added to the 50 carried out in July and August, 1997 sum to a total of 117 propulsion burns. This represents more than half of the 216 burns expected in the rebaselined RFS orbit transfer profile, albeit in some cases at somewhat lower initial RAC temperatures.

EGD TEST RESULTS

Temperature profiles are presented in Figure 4 for three heat-up cycles and a pair of hydrogen blowdowns conducted early in the test series. The temperature measurements are made with thermocouples and an optical pyrometer. The thermocouples were mounted to the receiver body and H_2 gas exit tube. The optical pyrometer was mounted between two concentrator facets with a view into the receiver cavity. As shown, heating is characterized by a gradual increase in RAC temperature, and blowdowns associate with relatively steep thermal transients.

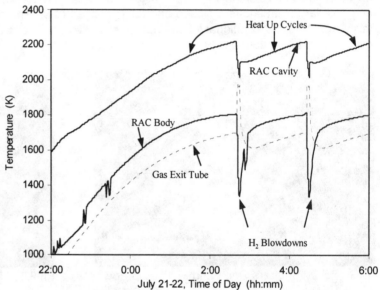

FIGURE 4. EGD Heatup and Blowdown Cycles.

Figure 5 presents the temperature profiles for the receiver during the first H_2 blowdown on July 22nd, 1997. The thermocouple reading on the propellant exit tube is seen to decrease very slightly over the course of the 3.57-minute blowdown. The temperature reading for the thermocouple mounted near the receiver inlet tube shows a rapid decrease during the blowdown. These data give an indication of excellent exhaust temperature sustainment, high heat exchanger effectiveness, and a substantial axial temperature gradient across the receiver.

Figures 6 and 7 show the H_2 flowrate and pressure traces for the same blowdown. Data from these plots in addition to the gas exit tube temperature from Figure 5 were used to predict a value for burn average specific impulse. A ΔT of 100 K was assumed between the bulk gas and exhaust tube temperatures. Temperature and pressure data were then compared and flowrate data scaled to calculations of delivered specific impulse for the RFS design. The methodology used to predict performance was the simplified JANNAF methodology of CPIA publication 246. The burn average specific impulse was determined to be approximately 746 seconds for the first burn conducted on July 22nd, 1997. The peak specific impulse for the corresponding exhaust tube temperature measured during the test series was estimated to be 758 seconds using the same method.

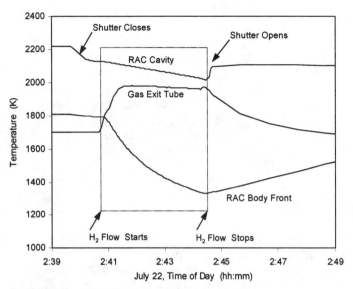

FIGURE 5. EGD RAC Temperature Profiles.

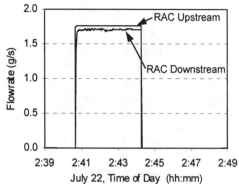

FIGURE 6. EGD Flowrate Profile.

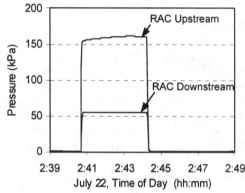

FIGURE 7. EGD Pressure Profile.

The relative stability of operation during the last two weeks of the EGD test can be attributed in part to the very stable and predictable performance of the test facility in producing the test boundary conditions. In balance, the EGD test stability can be attributed to the apparent robustness of the EGD test article itself once the initial problems of inlet tube leakage and contamination had been resolved. A post-test inspection conducted at NASA-LeRC found the test article to be in excellent condition.

The performance data collected on the concentrator during checkout testing and on the RAC during the first stages of the EGD test provide a substantial empirical confirmation of the basic soundness of the ISUS system approach. The ability to sustain H_2 exhaust temperatures over a wide range of burn durations (2 to 27 minutes) was validated, substantiating the high performance propulsion operation of the ISUS technology. EGD is an example of a propulsion system that is compatible with later use as an electric power production system. Further confirmation of the electric power function awaits the completion of RAC testing in the near future at the NMERI facility.

Acknowledgments

The work presented in this paper was conducted under the sponsorship of the Air Force Phillips Laboratory (PL) as part of Contract/F29601-95-C-0226. The PL program manger for the Integrated Solar Upper Stage Engine Ground Demonstration and Technology Development Program was Mike Jacox. The EGD test director was Duane Beach of NASA-LeRC. The authors would like to express their gratitude to the members of the ISUS team around the country at Harris Corporation, Babcock & Wilcox and Boeing Huntington Beach. Most particularly they wish to thank and acknowledge the exceptional support that they received from NASA LeRC personnel during the test activities.

References

Kudija, C., P. Frye, (1997), "Integrated Solar Upper Stage (ISUS) Technology Development Program and Engine Ground Demonstration (EGD) Test Plan", RD97-127, Boeing, Rocketdyne Division, Canoga Park, CA, May 1997.

Chemical Propulsion Information Agency (CPIA), "JANNAF Rocket Engine Performance Prediction and Evaluation Manual", CPIA Pub 246, The Johns Hopkins University, APL, Silver Spring, MD, April 1975.

PROPULSIVE SMALL EXPENDABLE DEPLOYER SYSTEM (ProSEDS) SPACE DEMONSTRATION

Les Johnson
Program Development/PS02
National Aeronautics and Space Administration
George C. Marshall Space Flight Center
Huntsville, Alabama 35812
les.johnson@msfc.nasa.gov
205-544-0614

Judy Ballance
Chief Engineers Office/EE61
National Aeronautics and Space Administration
George C. Marshall Space Flight Center
Huntsville, Alabama 35812
judy.ballance@msfc.nasa.gov
205-544-6042

Abstract

The Propulsive Small Expendable Deployer System (ProSEDS) space experiment will demonstrate the use of an electrodynamic tether propulsion system. The flight experiment is a precursor to the more ambitious electrodynamic tether upper stage demonstration mission which will be capable of orbit raising, lowering and inclination changing—all using electrodynamic thrust. ProSEDS, which is planned to fly in 2000, will use the flight-proven Small Expendable Deployer System (SEDS) to deploy a tether (5-km bare wire plus 15-km spectra) from a Delta II upper stage to achieve ~0.4N drag thrust, thus deorbiting the stage. The experiment will use a predominantly 'bare' tether for current collection in lieu of the endmass collector and insulated tether approach used on previous missions. ProSEDS will utilize tether-generated current to provide limited spacecraft power. In addition to the use of this technology to provide orbit transfer of payloads and upper stages from low-Earth orbit (LEO) to higher orbits, it may also be an attractive option for future missions to Jupiter and any other planetary body with a magnetosphere.

INTRODUCTION

Since the 1960's there have been at least 16 tether missions. In the 1990's, several important milestones were reached, including the retrieval of a tether in space, successful deployment of a 20-km-long tether, and operation of an electrodynamic tether with tether current driven in both directions—power and thrust modes (Johnson 1997). A list of known tether missions is shown in Table 1. The ProSEDS mission, to be flown in 2000, is sponsored by NASA's Advanced Space Transportation Program Office at The George C. Marshall Space Flight Center (MSFC).

TABLE 1. Known Tether Flights.

NAME	DATE	ORBIT	LENGTH	COMMENTS
Gemini 11	1967	LEO	30 m	spin stable 0.15 rpm
Gemini 12	1967	LEO	30 m	local vertical, stable swing
H-9M-69	1980	suborbital	500 m	partial deployment
S-520-2	1981	suborbital	500 m	partial deployment
Charge-1	1983	suborbital	500 m	full deployment
Charge-2	1984	suborbital	500 m	full deployment
ECHO-7	1988	suborbital	?	magnetic field aligned
Oedipus-A	1989	suborbital	958 m	spin stable 0.7 rpm
Charge-2B	1992	suborbital	500 m	full deployment
TSS-1	1992	LEO	<1 km	electrodynamic, partial deploy, retrieved
SEDS-1	1993	LEO	20 km	downward deploy, swing & cut
PMG	1993	LEO	500 m	electrodynamic, upward deploy
SEDS-2	1994	LEO	20 km	local vertical stable, downward deploy
Oedipus-C	1995	suborbital	1 km	spin stable 0.7 rpm
TSS-1R	1996	LEO	19.6 km	electrodynamic, severed
TiPS	1996	LEO	4 km	long life tether

ProSEDS FLIGHT EXPERIMENT OVERVIEW

A flight experiment to validate the performance of the bare electrodynamic tether in space and demonstrate its capability to produce thrust is planned by NASA for the year 2000. The ProSEDS (Propulsive Small Expendable Deployer System) experiment will be placed into a 400-km circular orbit as a secondary payload from a Delta II launch vehicle (Fig. 1). Once on orbit, the flight-proven SEDS will deploy 15 km of insulating Spectra tether attached to an endmass, followed by 5 km of predominantly bare wire tether (Fig. 2). Upward deployment will set the system to operate in the generator mode, thus producing drag thrust and electrical power. The drag thrust provided by the tether, with an average current of 0.5A, will deorbit the Delta II upper stage in approximately 17 days, versus its nominal ≥6-month lifetime in a 400-km circular orbit (Fig. 3).

FIGURE 1. Artist Concept of ProSEDS on a Delta II.

Approximately 100 W of electrical power will be extracted from the tether to recharge mission batteries and to allow extended measurements of the system's performance. A plasma contactor will be attached to the Delta II to complete the circuit and emit electrons back into space. Performance and diagnostic instruments mounted on the Delta II will be used to correlate the propulsive forces generated by the electrodynamic tether and the existing plasma conditions. These instruments will measure plasma density, temperature, energy, and potential. ProSEDS will be the first tether mission to produce electrodynamic thrust, use a bare wire tether, and recharge mission batteries using tether-generated power.

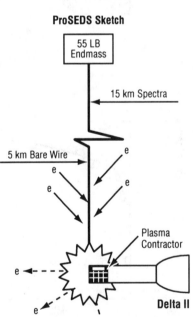

FIGURE 2. ProSEDS Sketch.

Electrodynamic Tethers

The ProSEDS flight experiment will demonstrate electrodynamic propulsion (through drag thrust) in space. From theoretical analyses and preliminary plasma chamber tests, bare tethers appear to be very effective anodes for collecting electrons from the ionosphere and, consequently, attaining high currents with relatively short tether lengths (Colombo 1981). A predominantly uninsulated (bare wire) conducting tether, terminated at one end by a plasma contactor, will be used as an electromagnetic thruster. A propulsive force of $\mathbf{F} = I\mathbf{L} \times \mathbf{B}$ is generated on a spacecraft/tether system when a current, I, from electrons collected in space plasma, flows down a tether of length, $\mathbf{L}$, due to the electromotive force (emf) induced in it by the geomagnetic field, $\mathbf{B}$. Preliminary tests indicate that a thin uninsulated wire could be 40 times more efficient as a collector than previous systems (Fig. 4).

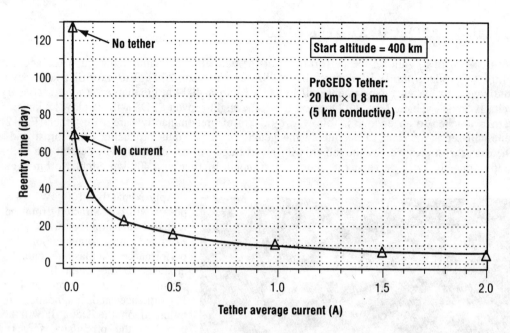

FIGURE 3. Predicted Demonstration of ProSEDS Propulsive Drag Thrust. The Delta II Reentry Time is Shown as a Function of Tether Average Current. Data Provided by Enrico Lorenzini/SAO.

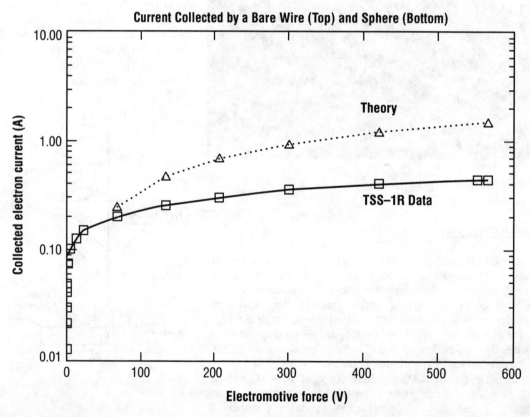

FIGURE 4. Current Collected by a Bare Wire Versus a Sphere. Data Provided by Jim Sorensen and Nobie Stone/NASA.

FUTURE APPLICATIONS FOR ELECTRODYNAMIC TETHERS

The main advantage of electrodynamic tethers is that they can be used as propellantless (no resupply required) space propulsion systems. Tethers take advantage of the natural plasma environment and sunlight to provide thrust and power. For example, if solar arrays and an external power supply are used, an emf can be generated in the tether such that current collected from the ionosphere produces thrust rather than drag. This thrust can then be used to raise the orbit of the system or change its inclination—all without propellant or rocket engines. It is envisioned that this type of propulsion could be used on a reusable upper stage to provide a low-recurring-cost alternative to chemical stages.

FIGURE 5. Electrodynamic Tether Upper Stage.

The electrodynamic tether upper stage could be used as an orbital tug to move payloads within low-Earth orbit (LEO) after insertion. The tug would rendezvous with the payload and launch vehicle, dock/grapple the payload and maneuver it to a new orbital altitude or inclination within LEO without the use of boost propellant. The tug could then lower its orbit to rendezvous with the next payload and repeat the process. Such a system could conceivably perform several orbital maneuvering assignments without resupply, making it a low-recurring-cost space asset. The ProSEDS itself could be used operationally to extend the capability of existing launch systems by providing a propellantless system for deorbiting spent stages. The launch service provider need not carry additional fuel for the soon-to-be-required deorbit maneuver, thus allowing all the onboard fuel to be used for increasing the vehicle's performance. Similarly, satellites thus equipped could safely deorbit at their end of life without using precious onboard propellant. Both of these applications would help reduce the increasing threat posed by orbital debris.

An electrodynamic tether system (Fig. 6) could be used on the *International Space Station* (*ISS*) to supply a reboost thrust of 0.5-0.8 N, thus saving up to 6,000 kg of propellant per year (Johnson 1996). The reduction of propellant needed to reboost the *ISS* equates to a $2B savings over its 10-year lifetime (Johnson 1996). Other advantages of using the electrodynamic tether on *ISS* are that the microgravity environment is maintained and external contaminants are reduced. Yet another use for electrodynamic tethers is the exploration of any planet with a magnetosphere, such as Jupiter. Jupiter's rapid rotation produces a condition where a tether can produce power and raise its orbit passively and simultaneously. MSFC is working with the Jet Propulsion Laboratory (JPL) to determine the use of electrodynamic tethers for future Jovian missions such as the Europa Orbiter and Jupiter Polar Orbiter (Fig. 7).

FIGURE 6. *ISS* with Electrodynamic Tether System for Reboost.

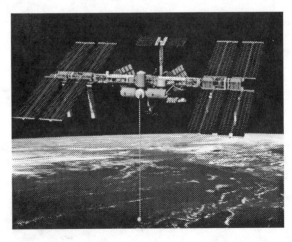

FIGURE 7. Jovian Electrodynamic Tether Concept.

CONCLUSIONS

Tether technology has advanced significantly since its inception over 30 years ago. The recent successes of the SEDS system show that tethers are ready to move from experiment and demonstration to application. One of the most promising applications for tethers is space propulsion and transportation. The use of electrodynamic tether propulsion for reusable upper stages, planetary missions, space station, and launch vehicle deorbit applications will soon be demonstrated with the ProSEDS mission. The ProSEDS mission will demonstrate and validate the production of power in space using a bare wire tether which produces drag thrust propulsion.

Acknowledgments

This paper was prepared for the ProSEDS project, which is sponsored by the Advanced Space Transportation Program Office of the National Aeronautics and Space Adminstration (NASA) located at Marshall Space Flight Center (MSFC). Data contributions to this paper were provided by Nobie Stone and Jim Sorensen of MSFC and Enrico Lorenzini of Harvard-Smithsonian Center for Astrophysics.

References

Colombo, G. (1981) "Study of Certain Launching Techniques Using Long Tethers," NASA-CR-164040.

Johnson, L., R. Estes, and E. Lorenzini (1997) "Space Transportation Systems Using Tethers," in Sixth Alumni Conference of the International Space University, Houston, TX, NASA Conference Publication 3355, 89-96.

Johnson, L., J. Caroll, R. Estes, B. Gilchrist, E. Lorenzini, M. Martinez-Sanchez, J. Sanmartin, and E. Vas (1996) "Electrodynamic Tethers for Reboost of the International Space Station and Spacecraft Propulsion," in 1996 AIAA Space Programs and Technology Conference, Huntsville, AL, AIAA 96-4250.

THERMIONIC CONVERTERS FOR GROUND DEMONSTRATION TESTING

Gregory J. Talbot, William D. Ramsey and Edmund L. James
Lockheed Martin
600 E. Bonita
Pomona, CA 91767-2737
(909) 624-8021

Abstract

Sixteen thermionic converters were fabricated for the Engine Ground Demonstration (EGD) tests of the Integrated Solar Upper Stage (ISUS). EGD power generation tests mapped the array performance under simulated semi-synchronous orbit conditions. Results from these tests have been reported elsewhere. This paper describes converter development and the manufacturing techniques used to ensure good performance, reliability and minimum unit-to-unit variations.

INTRODUCTION

The EGD converters were operated in series for ease of regulation and power conversion. Serial converter operation relied on uniform converter performance for optimum efficiency. Variations in performance would have reduced array efficiency if any converter operated outside its current density-voltage (J-V) envelope. Under the worst case temperature and current conditions, low output converters reduce array voltage by functioning as an in-series impedance with a forward voltage drop. As both thermal inputs and loads varied during the EGD tests, consistent outputs were a high priority.

DEVELOPMENT

Performance uniformity required stringent controls over dimensional tolerances and assembly procedures. Controls minimized unit-to-unit variations by establishing criteria for cleanliness, cesium handling and dimensional tolerance. The EGD converter development effort focused on operational gap predictability and lifetime performance expectations.

Gap Control

The emitter-collector gap was produced by a complex interplay of thermal conductivity and radiation involving the emitter, its support, and the collector-heat pipe assembly. Conductivity was regulated through materials properties while radiation losses were functions of emissivity, heat shielding and the thermal environment. Converter design controlled 3 of the 4 radiation factors.

In order to fit ISUS solar receiver geometry, dimensions of the prototype shoe were increased from 3.8×10^{-5} m^2 to 5.0×10^{-5} m^2. The converter thermal model predicted the larger area would increase the shoe thermal gradient and cause emitter distortion. Tests confirmed the prediction. A non-uniform gap reduced performance and, in the worst case, produced an emitter-to-collector electrical short.

To limit the gradient to $<1.0 \times 10^4$ K/m, the multi-layer insulation (MLI) over shoe was increased 50% and the test facility was modified. In order to ensure uniform heating, the shoe was placed in a fully enclosed oven for all converter tests. Tests found these changes reduced the shoe thermal gradient to $\sim 7.0 \times 10^3$ K/m.

The emitter support minimized thermal conductivity between the emitter and heat pipe assembly. For EGD, the ISUS system MLI surrounded these supports for a significant fraction of length to minimize ISUS solar receiver losses. This design effected gap dimensions as it allowed non-uniform radiation losses from the support. Tests simulating the ISUS environment found that wrapping the full length of the support MLI integrated radiation losses and avoided gap distortion.

Twenty-three percent of the collector-heat pipe assembly length was within the thermionic cavity. Hence thermal expansion due to insufficient cooling would drive the collector toward the emitter. The gap was increased ~2.5 x 10^{-5}m and hafnium carbide plasma was sprayed on the condenser exterior surface increasing its emissivity from .25-.28 to .6-.7. Tests in a "hot" radiating environment found no occurrence of emitter-collector shorting.

Lifetest

The passive cesium pressure control was unique to ISUS converters. During converter prototype development, an optimum cesium volume was identified empirically. It yielded the maximum power output for emitter temperatures of 1900K-2200K. Use of a precise volume eliminated the need for an active control loop to regulate cesium pressure in the thermionic cavity and reduced system complexity but raised questions regarding loss of cesium over time. A 1200-hour lifetest found there was no deterioration in converter performance, but longer tests are needed.

EGD Heat Pipe

Heat pipe design was modified to accommodate EGD operation. Early in the program heat pipe groove widths were selected for an orbital, micro-gravity environment. However, during EGD, the converter plane was near-vertical and the heat pipe major axes orientation ranged from horizontal to $\pm$vertical, i.e. gravity forced wicking to wicking against gravity (anti-g). Narrowing the grooves and covering them with 100-mesh tantalum screen produced the lift for needed anti-g operation.

MANUFACTURING

A sign-off sheet or traveler was generated for each serialized converter. The traveler provided a convenient means to track fabrication through each step of assembly, to identify the operator completing the step, show the date of the completion and the source of parts or materials used in the assembly. Traceability allowed any problems to be quickly evaluated.

Components and Subassemblies

The thermionic converter consisted of two major sub-assembles: the emitter-emitter support and heat pipe-collector. The two were joined only after completing an extensive series of checks that verified integrity and dimensional tolerance. Vacuum leak checks were performed after each weld or braze to confirm joint integrity before continuing. Five leak checks were performed during heat pipe assembly and three during the emitter assembly.

Heat pipe functional tests were conducted once the assembly was completed and before it was joined to the emitter assembly. Functional tests verified operation in any orientation, anti-g and with g, and ranked pipe wicking efficiency. Functional tests used a rf coil to heat the collector (evaporator) and adiabatic sections nominally within the thermionic cavity. Tests consisted of maintaining the evaporator temperature at the minimum operating level until equilibrium was established. After recording equilibrium data, the evaporator temperature was increased to the operational maximum and equilibrium established. Each heat pipe was then rotated π radians in regards to gravity and the tests were repeated. A 40K temperature difference between the evaporator and condenser was acceptable.

Fit checks were performed before every weld and cross-checked before the final welds. Cross-check prior to the final weld ensured electrode gap and electrode parallelism were within the <1.3 x 10^{-5}m tolerance. After the weld the measurements were repeated to verify weld shrinkage dimensions and the assembly was leak checked before conditioning the converter for cesium loading.

Performance Tests

Converter performance mapping tests were conducted prior to delivery. Converters not meeting the 30 watts acceptance criterion were reprocessed and re-tested. As mentioned above, these single converter tests were conducted with the shoe fully within an oven to minimize the thermal gradient. Data were recorded manually at thermal equilibrium.

During the performance tests, 5 or more I-V data sets were recorded every 100K from 1900K-2200K. Each data set included temperatures of: the emitter; emitter support; the electrical isolator; and two points on the heat pipe. Data sets were recorded every 10 amperes by adjusting a constant current supply and/or a resistive load bank.

Supply currents were verified with a calibrated 2.0×10^{-1} ohms shunt in series with the emitter. Calibrated DDMs monitored output voltage through dedicated leads on the emitter and collector. Periodic isolation checks were performed to confirm electrical isolation of the leads. Converter output voltage read from a battery powered DDM eliminated any possible ground loops.

Table 1 shows the individual EGD converter output voltages with a 30 amperes load at nominal operational temperatures. The data were taken from performance test data mentioned above. The TID number represents placement in the EGD array.

TABLE 1. Individual Converter Output Voltage at 30 Amperes Load.

Emitter Temp. (K)	TID #1 (V)	TID #2 (V)	TID #3 (V)	TID #4 (V)	TID #5 (V)	TID #6 (V)	TID #7 (V)	TID #8 (V)
1900	0.71	1.0	0.74	0.79	0.75	0.71	0.96	0.85
2000	0.85	1.19	0.88	0.93	0.98	0.96	0.97	1.02
2100	1.19	1.10	1.10	1.08	1.18	1.26	1.24	1.23
2200	1.39	1.40	1.31	1.20	1.33	1.38	1.46	1.42

Emitter Temp. (K)	TID #9 (V)	TID #10 (V)	TID #11 (V)	TID #12 (V)	TID #13 (V)	TID #14 (V)	TID #15 (V)	TID #16 (V)
1900	0.93	0.67	0.83	0.65	0.82	0.80	0.68	1.0
2000	1.1	0.91	0.98	1.19	1.19	1.0	0.91	1.32
2100	1.26	1.04	1.20	1.12	1.31	1.30	1.20	1.44
2200	1.46	1.27	1.32	1.40	1.50	1.40	1.29	1.51

Table 2 uses the values listed in Table 1 to show the average, mean, standard deviation (root-mean-squared) and percentage of deviation from average individual output voltage. The percentage of deviation from average ranged from 14% to 6% as a function of emitter temperature which reflects converter optimization at higher emitter temperatures.

TABLE 2. Array Performance Uniformity Comparison

Emitter Temp (K)	Converter (Ave.) Voltage (V)	Mean Converter Voltage (V)	Standard Deviation (σ) (V)	Ratio of Std Deviation to Ave. Voltage (%)
1900	0.806	0.825	0.112	14
2000	1.024	1.09	0.128	12
2100	1.203	1.27	0.099	8
2200	1.378	1.30	0.083	6

Figure 1 shows the estimated I-V curves for the array of 16 converters. The estimate included voltage losses due to interconnect impedance from measurements taken with interconnect hardware during the performance tests. Interconnect impedance between adjacent converters ranged from 1.1 to 1.5×10^{-3} ohms as a function of emitter temperature.

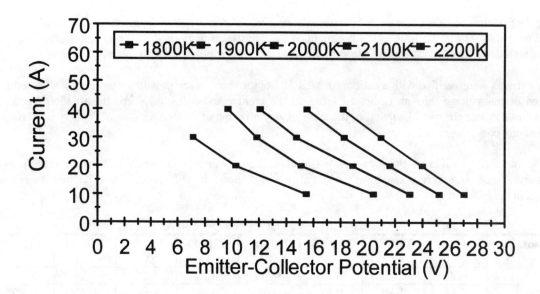

FIGURE 1. Estimated 18 Converter Array Current vs. Voltage.

Figure 2 plots the estimated output voltage versus power as a function of emitter temperature from the data in Chart 1. As power was current squared, interconnect losses are more significant at higher current levels with a maximum loss of 40 watts drop at 2200K.

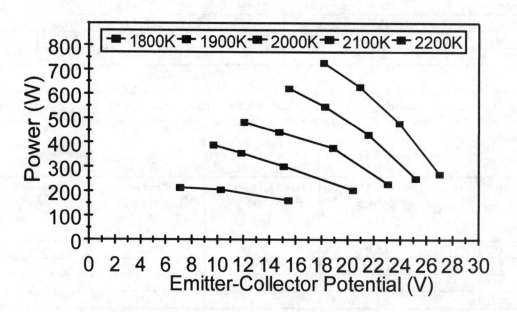

FIGURE 2. Estimated 16 Converter Array Output Power vs. Voltage

Figure 3 is a photograph of 16-converters mounted on the support, arranged in a circle approximating the EGD test configuration. Figure 4 is a close up of the parallel interconnect geometry used to minimize impedance.

Acknowledgment

The work described was done with funds provided by Babcock & Wilcox of Lynchburg, VA.

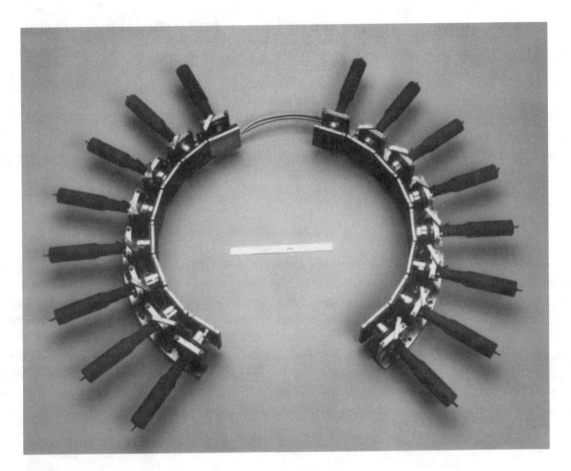

FIGURE 3. Overview of Converters Mounted on the EGD Support.

FIGURE 4. Close-up of the Converter Interconnects Assembled on the EGD Support.

JSUS SOLAR THERMAL THRUSTER AND ITS INTEGRATION WITH THERMIONIC POWER CONVERTER

Morio Shimizu, Kunihisa Eguchi,
Katsuya Itoh
and Hitoshi Sato
National Aerospace Lab. (NAL)
Jindaiji-Higashi, Chofu, Tokyo,
182, Japan
+81(422)47-5911

Tadayuki Fujii
National Research Institute for
Metals (NRIM)
Sengen, Tsukuba, Ibaragi,
305, Japan
+81(298)53-1151

Ken-ichi Okamoto
and Tadashi Igarashi
Tokyo Tungsten Co., Ltd.
(TOTAN)
Iwasekoshi, Toyama, Toyama,
931, Japan
+81(764)37-7181

Abstract

This paper describes solar heating test results of a single crystal Mo thruster of solar thermal propulsion (STP) with super high-temperature brazing of Mo/Ru for hydrogen-gas sealing, using the paraboloidal concentrator of 1.6 m diameter newly installed in NAL in the Japan Solar Upper Stage (JSUS) research program. The designed thruster has a target Isp about 800 sec for 2,250 K or higher temperatures of hydrogen propellant. Additionally, tungsten CVD-coating was applied to a outer surface of the thruster in order to prevent vaporization of the wall material and Mo/Ru under the condition of high temperature over 2,500K and high vacuum. Also addressed in our paper is solar thermionic power module design for the integration with the STP receiver. The thermionic converter (TIC) module is of a planar type in a Knudsen-mode operation and provides a high conversion efficiency of 23% at the TIC emitter temperature of nearly 1,850 K for a heat input flux of 24 W/cm^2.

INTRODUCTION

The solar thermal propulsion (STP) is considered as a promising high performance upper stage capable of transferring payloads from LEO to higher orbits such as GEO very effectively. One of the most important investigation items with respect to STP is a thruster made of refractory metals or ceramics because of the high operating temperature up to about 2,500 K involved. This paper describes design, fabrication, and preliminary experimental results of a 20 mm dia. of thruster as part of the Japan Solar Upper Stage (JSUS) research program in NAL. In our previous test, carbon sheet gasket and the 1.6 m dia. cut-in-half parabodiodal concentrator were used for 700 sec Isp (1850 K) of single crystal Mo thruster assessment (Shimizu 1996). In this test, a precise solar concentrator of 1.6 m dia. full paraboloid with a focal length of 0.65 m is employed for higher temperature, and Mo/Ru brazing and tungsten CVD-coating are applied to the thruster for 800 sec of Isp at 2,250 K.

Additionally in the JSUS research program, a design work has been performed on integration of a thermionic power generator with solar cavity receiver for the STP bimodal system. It has also been a joint work between NAL and SIA-LUTCH in Russia (Eguchi 1996a and 1996b). To achieve the high efficiency TIC power generation at the operating temperatures below 2,000 K, a Knudsen-mode thermionic converter with a small interelectrode gap is conceptually designed. Its configuration and output performance are specified in our paper.

THRUSTER MATERIAL SELECTION

Currently, two independent solar thermal propulsion R&D programs have been in progress, called the Solar Thermal Upper Stage (STUS) and the Integrated Solar Upper Stage (ISUS), respectively, in NASA (Cady 1996) and USAF (Kennedy 1995). The USAF Phillips Laboratory (PL) has been in a primary position for the STP study over 15 years STP investigation (Shoji 1992). So far STUS program has adopted pure tungsten or tungsten alloy for STP thrusters, but these materials become brittle due to recrystallization at high temperatures.

A STP basic study initiated in NAL a few years ago. Actually, manufacturing technology of rhenium, which is considered as one of the most suitable materials for solar thrusters, has been not advanced in Japan, so it is difficult to choose suitable materials for STP thruster. Fortunately, the National Research Institute of Metals (NRIM) of the Science and Technology Agency (STA) has patented and developed single crystal molybdenum and tungsten. These are ideal materials from a viewpoint of high material strength, non-possibility of the recrystallization embrittlement (Fujii 1995), and non-reaction to H_2, N_2 and He at high temperatures.

PROPERTIES OF SINGLE CRYSTAL MOLYBDENUM

Both of molybdenum and tungsten have many advantages such as a high melting temperature, useful elevated temperature strength, high thermal conductivity, low thermal expansion and good resistance to liquid metal corrosion. Thus, molybdenum, tungsten and their alloys are widely utilized in various industrial fields. However, these materials undergo a severe loss of ductility after recrystallization or welding. It is generally accepted that this problem is due to intergranular embrittlement, and the greatest weak point of these metals.

Recently, NRIM et al. has succeeded in establishing a new technology to develop commercial scale single crystal molybdenum and tungsten from hot-rolled sheet doped with a certain amount of CaO and/or MgO by means of the intentional secondary recrystallization (Fujii 1995). FIGURE 1 shows the effects of the dopants on the grain growth behavior in hot-rolled molybdenum sheets. The grain growth in doped molybdenum was initially restricted: however, the subsequent abnormal grain growth behavior at high temperatures was promoted. Consequently, a single crystal sheet could easily be produced.

FIGURE 2 depicts the stress and elongation properties of the produced single crystal and polycrystal molybdenum specimens at various temperatures. The single crystal ultimate stresses (σ_u) and 0.2% proof stresses ($\sigma_{0.2}$) are about a half of the polycrystal ones, and single crystal breaking elongations are about double of polycrystal ones, namely single crystal molybdenum is much more ductile than polycrystal one.

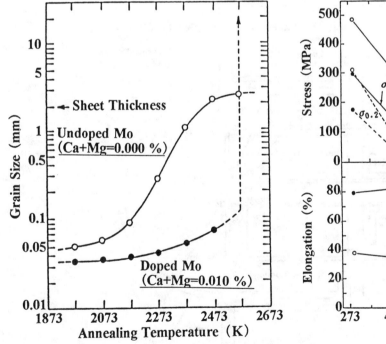

FIGURE 1. Effects of Dopants on the Secondary Grain Growth of Mo.

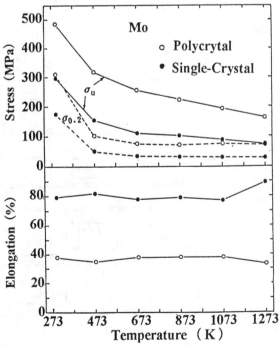

FIGURE 2. Comparison of Tensile Properties between Single and Polycrystal Mo.

THRUSTER CONFIGURATION AND EXPERIMENTS

A cavity type of solar thruster was made of single crystal molybdenum, as illustrated in FIGURE 3. The diameters of the outer chamber and the cavity of the inner chamber are 20 mm and 10 mm, respectively. The propellant gas is supplied through two pipes fit on the opposite side of the outer chamber with each other. These elements are jointed with screw for high mechanical strength and sealed with very high temperature brazing with molybdenum-ruthenium (Mo/Ru, 2,320 K of the nominal melting point) for sealing (Hiraoka 1989), instead of carbon sheet gasket in the previous test (Shimizu 1996). As a result, the operating temperature is expected to reach 2,300 K (namely approximately 800 sec of Isp for H_2). In order to achieve higher reliability and performance, the thruster is CVD-coated with tungsten so that the tungsten layer prevents not only the Mo/Ru but also the Mo chamber wall from spilling out and vaporizing away under high temperature and vacuum condition. The temperatures of outer chamber surface and propellant gas were measured by Ir/Ir-40%Rh (up to about 2,300 K) thermocouples of 0.25 mm in diameter. The thruster was surrounded with carbon felt to insulate thermally.

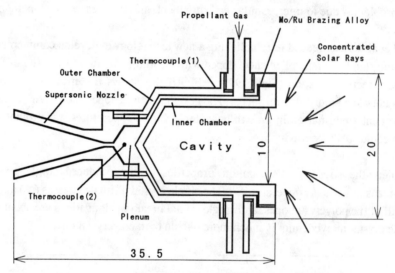

FIGURE 3. Structure of Solar Thruster (20 mm in Diameter) Made of Single Crystal Mo.

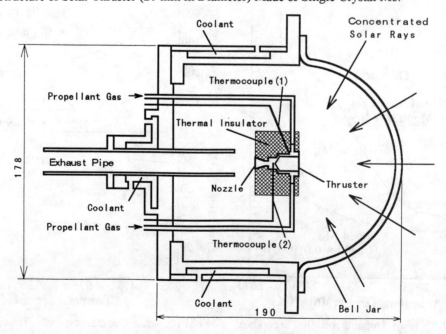

FIGURE 4. Solar Thruster Mounted in the Vacuum Chamber with Bell Jar Window.

The thruster was installed in the small vacuum chamber with a bell jar window (Pyrex glass) through which concentrated solar rays were applied, as shown in FIGURE 4. The chamber was vacuumed with a rotary pump, and the vacuum chamber pressure was maintained at about 0.3 kPa, while the propellant gas (N_2 or He) was injected from the thruster nozzle at 1 to 10 SLM of propellant flow rate and 0.4 MPa of max. chamber pressure.

The vacuum chamber was fixed on the solar concentrator tracking the sun so that the focal point of the solar concentrator corresponded to the cavity aperture of the thruster. The 1.6 m dia. of on-axis full paraboloid concentrator was precisely made of 15 mm thickness of glass back-coated with aluminum, and its focal length and the ideal solar image diameter are 0.65 m and 7 mm (really about 10 mm), respectively. Because of cutting the solar concentrator in half in the previous test, internal stresses near the cutting line were relieved and small amount of change from the original paraboloidal shape was generated. Then, the solar image at the focal point was deformed from the original disk to ellipse-like shape with over twice the area, namely the concentration rate decreased. This time the full paraboloidal mirror was adopted, doubling the mirror area and gaining higher concentration ratio due to decreasing the above mentioned mirror deformation. The test was performed successfully, and the cavity aperture of 10 mm in dia. is enough for catching almost energy in the solar image.

TEST RESULTS AND DISCUSSION

Typical experimental results using N_2 and He gas propellant for safety are shown in FIGURE 5. Thses data indicate that the highest value of the chamber gas temperature T_2 reached about 2,300 K (approximately 800 sec of Isp for H_2), at 0.4 MPa of the maximum chamber pressure, much higher than the 1,850K and 700 sec in the previous test using the cut-in-half concentrator (Shimizu 1996). Furthermore, under the much higher vacuum condition in space than the experimental vacuum chamber, T_2 could be kept against higher gas flow rate for much higher thrust with the multilayer insulation (MLI) instead of the carbon felt. It is difficult to fabricate very light weight and precise solar concentrator of space use for higher temperature cavity, from which the radiation loss increases a lot through the cavity aperture. Then, we supposed that the propellant temperature of 2,300 K (800 sec of Isp) was appropriate to the cavity type of STP. Then, the cavity type of solar thruster technologies are considered to be basically established, including the single crystal Mo, Mo/Ru brazing and tungsten CVD coating. The next step of the cavity type of STP thruster may be large scale thruster development, to be tested by the 10 m diameter of precise and segmented solar concentrator (4 cm of solar image dia.) of Tohoku Univ. in Japan.

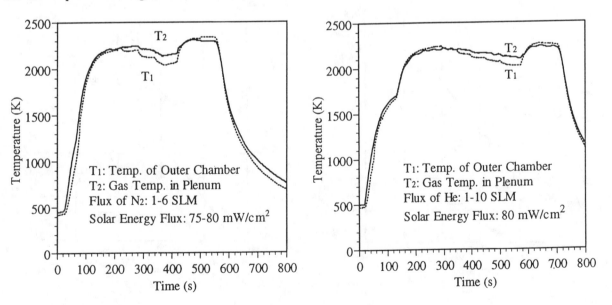

FIGURE 5. Typical Results of Solar Thruster Test with N_2 and He propellant.

DESIGN OF JSUS THERMIONIC POWER CONVERTER

To minimize heat losses for the TIC and STP-receiver integration, the thermionic power modules are arranged circumferencially on an outer wall surface of the cavity receiver. The module configuration is schematically illustrated in FIGURE 6. Heat is transported to the TIC emitter by thermal radiation. The emitter heating surface is coated by a layer of titanium carbide. The TIC collector is cooled with a sodium heat pipe to keep its temperature of 1,100 K constant. The heat resistance between TIC and HP is estimated about 0.3 W/K. A small-gap TIC design in Knudsen-mode with planar electrodes is applied, and the emitter and collector materials are single crystal tungsten and molybdenum single crystal alloy, respectively. The gap spacing is kept 20 μm with three spacer pieces of scandium oxides (Si_2O_3). The more-detailed structure is specified in the reference (Eguchi 1996a).

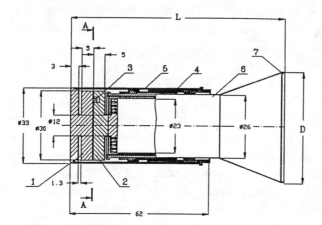

	Units	Materials
1	Emitter	Single crystal alloy W-Ta-Re
2	Collector	Single crystal alloy Mo-Re
3	Spacer	Sc_2O_3
4	Ceramic seal	Nb-Al_2O_3-Nb
5	Holder	W
6	Heat pipe	Shell and wick – Nb Working fluid - Na
7	Ceramic seal	Nb-Al_2O_3-Nb

FIGURE 6. Configuration and Specifications of Knudsen-Mode TIC Design.

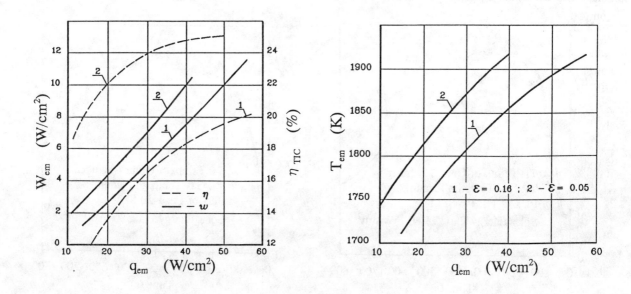

FIGURE 7. Calculated Performance of Knudsen-Mode TIC Designs.

Shown in FIGURE 7 is calculated performance of electric power density and efficiency as a function of heat input flux to the emitter. The calculations are also compared for two different electrode emissivities (ε) of 0.16 and 0.05, where the TIC and collector are both 30 mm in diameter. The lower value is given for the coated collector to reduce TIC radiation loss, as compared to the conventional type of the higher emissivity. Cesium vapor pressures range from 133 to 266 Pa as optimum values. It is found from the calculated data that for the lower-emissivity collector type, the emitter heat flux of q_{em} is 24.4 W/cm^2 at a specified temperature, T_{em}, of 1,850 K, and as a result the TIC power density w_{em}, and efficiency η_{TIC} reach 5.7 W/cm^2 and 23.2%, respectively.

CONCLUSIONS

A 20 mm dia. of single crystal Mo (NRIM has patented) STP thruster was fabricated and tested successfully. The propellant gas (N_2 and He) temperature within the thruster chamber reached about 2,300 K (about 800 sec of Isp for H_2) at 0.4 MPa of the maximum chamber pressure, using the 1.6 m diameter of full paraboloidal concentrator and adopting Mo/Ru high temperature brazing, and tungsten CVD-coating on the thruster prevents Mo/Ru and Mo from vaporizing away at high temperature and vacuum operations. Therefore, the cavity type of solar thruster technologies of 800-sec-Isp class are considered to be basically established.

As a power generator candidate for the JSUS-STP bimodal system, the Knudsen-mode thermionic converter is conceptually specified, and its conversion efficiency is estimated 23% at the TIC emitter temperature of nearly 1,850 K for a heat input flux of 24 W/cm^2.

Acknowledgments

The authors are pleased to acknowledge the continuing guidance and encouragement of Dr. T. Toda.

References

Shimizu, M., et al. (1996) "Solar Thermal Thruster Made of Single Crystal Molybdenum," IAF-96-S.4.01.

Eguchi, K. (1996a) "Design Review on Space Thermionic Power Experimental Systems with Solar Heat," NAL Contract Report AB-3397.

Eguchi, K., et al. (1996b) "Design Analysis of Solar Thermionic/AMTEC Cascade Converter for Future Space Power Technology," in *Proc. 20th International Symposium on Space Technology and Science* (20th ISTS-GIFU), Gifu, Japan, 96-i-o4, pp. 1093 - 1098.

Cady, E., et al.(1996) " Solar Thermal Upper Stage Technology Demonstration Program," AIAA 96-3011.

Kennedy, C. F. and M. Jacox (1995) " The Integrated Solar Upper Stage (ISUS) Program," AIAA 95-3628.

Shoji, J. M., et al. (1992) "Solar Thermal Propulsion Status and Future," AIAA 92-1719.

Fujii, T. (1995) "A New Technique for Preparation of Large-Scaled Mo and W Single Crystals and Their Multilayer Crystals for Industrial Applications," in *Proc. Japan-Russia-Ukraine International Workshop on Energy Conversion Materials* (ENECOM 95), Sendai, Japan.

Hiraoka, Y., and T. Fujii (1989) " Welding and Joining of Single Crystals of BCC Refractory Metals," in *Proc. of 12th Plansee Seminar '89 in Austria*, 265 - 279.

MULTI-FOIL© INSULATION FOR HIGH TEMPERATURE APPLICATIONS; SUMMARY OF RACCET AND EGD EXPERIENCE FOR THE ISUS PROGRAM

Gabor Miskolczy and Joe Burchfield
Thermo Energy Conversion Laboratory
85 First Avenue
Waltham MA 02254-9046
(781) 622 1357, Fax. (781) 622 1026

John Malloy
NovaTech
10108 Timberlake Road
Lynchburg VA
(804) 239 3787, Fax. (804) 239 6232

Abstract

This paper summarizes the recent experience with high temperature MULTI-FOIL thermal insulation and compares the results with those obtained in the 1960s. In contrast to other energy conversion systems such as Brayton, thermoelectric, AMTEC, or thermo photovoltaic, the thermionic ISUS (Integrated Solar Upper Stage) operates at high temperatures, up to 2500 K. The ISUS uses a solar receiver, a concentrator heating a graphite absorber, which is part of the RAC (Receiver Absorber Converter) module, further described by Rochow (1996) and Westerman (1995) The MULTI-FOIL surrounds the absorber and minimizes the heat losses from it. During the propulsion phase of the mission the thermionic converters, the third part of the RAC, are thermally shielded from the absorber by a movable Multi-Foil assembly. For the power phase the insulation at the converter is lowered, exposing the converters to the hot absorber.

INTRODUCTION

The Multi-Foil assembly consists of 95 layers of 0.025 mm thick foils, 35 of which are tungsten the balance molybdenum. The innermost, hottest foil is 0.25 mm thick tungsten as is the outermost molybdenum foil. The general assembly, shown in Figure 1, consist of two cylindrical sections one fixed and one movable. Each assembly is about 6 mm thick interlocked using 0.13 mm thick molybdenum foils. The assembly is completed with top and bottom planar circular foil assemblies. The top assembly has a central hole for accepting solar energy. Additional holes are provided for the hydrogen gas inlet and outlet tubes, which are also insulated with interlocking foil assemblies.

The RACCET assembly, a subscale version of the RAC, was tested in the solar test facility at Edwards Air Force Base, CA in 1996. The insulation heat loss results were reported by Miskolczy (1997). The heat losses were much higher than originally predicted using the methods of Huffman (1976 and 1979) and Dunlay (1966), during the design phase, up to 6000 W versus the predicted 1000 W. The most likely reason for the discrepancy was the distortion of the foils at the joints, resulting in a gap through which radiation could stream. Design effort was directed to reduce the likelihood of such distortion in the EGD testing.

These tests were performed in the simulated solar test facility, Tank 6 at NASA Lewis Research Center in Cleveland, OH. This facility was described by Shaltens (1995).

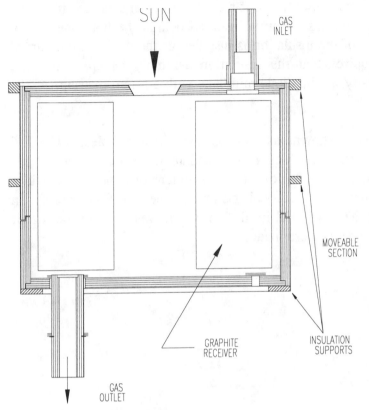

FIGURE 1. Insulation Assembly.

These tests were primarily aimed at testing the propulsive mode of operation in the ISUS system. Unlike the test of RACCET at Edwards (Miskolczy 1997) this test was not equipped with water cooled calorimeters, the heat loss though the insulation system could not be measured directly. The heat losses were estimated by the rate of cooldown. These losses also exceeded the original predictions.

HEAT LOSS MEASUREMENTS

The heat losses were estimated from the cooldown curves for the EGD testing in the absence of cooling flow. Malloy (1997) determined that at a RAC temperature of 2300 K, as measured with an optical pyrometer the total losses were 7,350 W of which 2610 W were the radiation loss through the 45.7 mm black body aperture. The remaining losses of 4740 W were ascribed to the insulation system. Further examination of the cooldown curve showed that 88% of the loss was fourth power dependent radiation and 12% was conduction dependent. These losses are very similar to those obtained during the RACCET testing.

There are two heat loss paths not included in the above: streaming losses at the propellant outlet, inlet, and additional radiation loss where the secondary concentrator mates to the cone-shaped opening of the insulation package. Videotapes taken during the EGD testing revealed a significant hot spot at the propellant outlet tube, where there was a poor fit between the bottom planar foils and the cylindrical wrap of foils around the outlet tube. This could represent an additional loss of 1000 W. The second area of poor fit occurred where the secondary concentrator joined the conical opening in the top planar foils. The overheating and distortion of the foil support did not leave enough space for a conical shield, which was therefore omitted. The secondary was at 1500 K when the aperture in the foils was at 2300 K. This could account for and additional loss of 1100 W.

THERMIONIC DIODE INSULATION

The array of thermionic diodes which surround the heat receiver assembly, is also insulated with Multi-Foil insulation. A portion of this insulation is installed behind the hot shoe during the fabrication of the diode. The balance of the insulation is installed as the diodes are mounted on their strongback support. This requires that the insulation assembly be split. For the RACCET the assembly was split vertically, at the diode midplane with a 5 mm overlap. At the split, as shown in Figure 2.

The locations of the hot shoes is also shown, as well as the stitching holes. The foil assembly is mounted on a right angle bracket which is fastened to the water cooled diode support. This arrangement allows for slight variation of the diode dimensions and still provides insulation. The diodes were tested with an electrical heater at the NMERI facility in Albuquerque NM, in conjunction with the RACCET insulation which was severely oxidized, thus no meaningful thermal loss measurements were made.

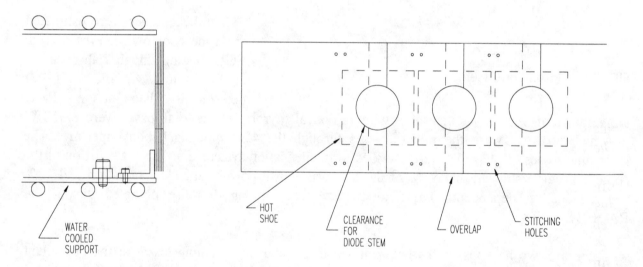

FIGURE 2. Diode Insulation for RACCET.

A different approach was used for the diode insulation for EGD. Here, the insulation was split horizontally, as shown in Figure 3.

The overlap and the stitching holes are also shown with the outline of the diode hot shoes. The assembly is supported by brackets which are attached to the intermediate support plate to which the diode strongback is attached. In this method the insulation is held more rigidly with

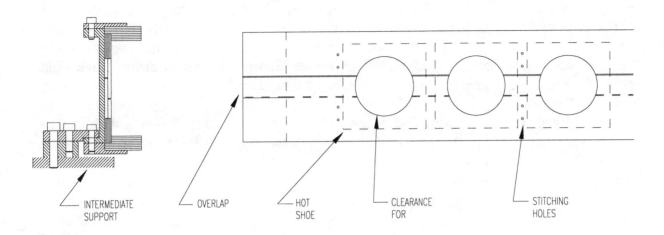

FIGURE 3. Diode Insulation for EGD.

little adjustment for diode variation possible. This insulation and the diodes will be attached to the EGD and tested with an electrical heater at the NMERI facility.

Acknowledgment

This work was performed under contract with BWXT in Lynchburg VA.

References

Dunlay, J.B. et al. (1966) "Vacuum Foil-Type Insulation for Radioisotope Power Systems," *Thermo Electron Report, AOO 3634,* Waltham MA.

Huffman, Fred N. (1976) "Application of Multi-Foil Insulation to the Brayton Isotope Power System and Conceptual Design of Multi-Foil for the Flight System," *Thermo Electron Report TE3209-100-76,* Waltham MA.

Huffman, Fred N., S., Matsuda, G. Miskolczy and C.C. Wang (1979) "Meltdown Characteristics of Multi-foil Thermal Insulation," in *Proceedings of 14th IECEC,* 1390-1395.

Malloy, John, (1997), "Projected Heat Loss from EGD Experiments and their Impact on Predicted Performance in Thermionic Converter Operation." *NovaTech Document 8/14/97.*

Miskolczy, Gabor, et al (1997) "A Multi-Foil insulation System for a Solar Bimodal Power and Propulsion System: Design Fabrication and Testing." *Proceedings of 32nd, IECEC,* Honolulu HI, 431-433.

Shaltens, Richard K. and Robert V. Boyle, 1995 "Initial Results from the Solar Dynamic (SD) Ground test Demonstration (GTD) Project at NASA Lewis," in *Proceedings of the 30th IECEC*, Orlando FL, 363-368.

Rochow, R. F. and Barry J. Miles (1996) "Power Generation Considerations in a Solar Bimodal Receiver," in *Proceedings 31st IECEC*, Washington DC, 345-350.

Westerman, Kurt O. and Richard F. Rochow (1995) "Solar Bi-Modal Power and Propulsion System for Satellite Applications," in *Proceedings of International Solar Energy Conference,* Maui HI.

TESTING OF A RECEIVER-ABSORBER-CONVERTER (RAC) FOR THE INTEGRATED SOLAR UPPER STAGE (ISUS) PROGRAM

Kurt O. Westerman and Barry J. Miles
BWX Technologies, Inc.
P.O. Box 785
Lynchburg, VA 24551
Tel. (804) 522-6758/6764
Fax (804) 522-6999

Abstract

The Integrated Solar Upper Stage (ISUS) is a solar bi-modal system based on a concept developed by Babcock & Wilcox in 1992. ISUS will provide advanced power and propulsion capabilities that will enable spacecraft designers to either increase the mass to orbit or decrease the cost to orbit for their satellites. In contrast to the current practice of using chemical propulsion for orbit transfer and photovoltaic conversion/battery storage for electrical power, ISUS uses a single collection, storage, and conversion system for both the power and propulsion functions. The ISUS system is currently being developed by the Air Force's Phillips Laboratory. The ISUS program consists of a systems analysis, design, and integration (SADI) effort, and three major sub-system development efforts: the Concentrator Array and Tracking (CATS) sub-system which tracks the sun and collects/focuses the energy; the Receiver-Absorber-Converter (RAC) sub-system which receives and stores the solar energy, transfers the stored energy to the propellant during propulsion operations, and converts the stored energy to electricity during power operations; and the Cryogenic Storage and Propellant Feed Sub-system (CSPFS) which stores the liquid hydrogen propellant and provides it to the RAC during propulsion operations. This paper discuses the evolution of the RAC sub-system as a result of the component level testing, and provides the initial results of systems level ground testing. A total of 5 RACs were manufactured as part of the Phillips Laboratory ISUS Technology Development program. The first series of component tests were carried out at the Solar Rocket Propulsion Laboratory at Edwards AFB, California. These tests provided key information on the propulsion mode of operations. The second series of RAC tests were performed at the Thermionic Evaluation Facility (TEF) in Albuquerque, New Mexico and provided information on the electrical performance of the RAC. The systems level testing was performed at the NASA Lewis Research Center Solar Simulator Facility (Tank 6) in Cleveland, OH.

INTRODUCTION

In the early 1990's, there was a great deal of interest in space nuclear power and propulsion systems. A number of space nuclear technology development programs were being conducted at that time; however, they were all terminated due to the lack of a well defined user, high development costs, or anti-nuclear political pressure. At that time, engineers at BWX Technologies began investigating alternate non-nuclear concepts that could take advantage of the technologies developed in these nuclear programs. In 1992, a solar bi-modal concept was developed and patented. In 1994, the United States Air Force commissioned a study to assess the benefits of a solar bi-modal concept. The next year, the Integrated Solar Upper Stage technology demonstration program was initiated to develop and demonstrate the technologies necessary to prove the feasibility of a bi-modal solar upper stage. The program included separate contracts for each of the major sub-systems: the Receiver-Absorber-Converter (RAC), the Concentrator Array and Tracking Sub-system (CATS), and the Cryogenic Storage and Propellant Feed Sub-system (CSPFS). This paper will discuss the development and testing of the RAC sub-system by BWX Technologies and its partners. Two series of RAC tests were performed. The first test, the Receiver-Absorber-Converter Configuration and Electrical Test (RACCET), was a component level test designed to benchmark the design and provide feedback for design of the system level test article. The second test, the Engine Ground Demonstration (EGD) was designed to demonstrate the performance of the RAC in a systems level test.

FIGURE 1. Integrated Solar Upper Stage (ISUS).   FIGURE 2. Receiver-Absorber-Converter (RAC).

Integrated Solar Upper Stage (ISUS) Description

The Integrated Solar Upper Stage, Figure 1, provides both orbit transfer propulsion and on-orbit electrical power for spacecraft. The solar energy is collected by the primary concentrator array and focused into the receiver-absorber-converter through a secondary concentrator. The thermal energy storage material in the receiver cavity absorbs the solar energy. In the orbit transfer phase of the mission, a series of apogee and/or perigee burns are executed to propel the spacecraft to its mission orbit. During each burn, hydrogen propellant flows from the cryogenic storage tank to the RAC. The propellant is heated as it passes through flow channels in the energy storage material. The hot propellant is then expelled through a nozzle to provide thrust for the spacecraft. After each burn, solar energy is again used to reheat the RAC. The heat/thrust cycle is repeated for each subsequent orbit. Once the spacecraft reaches its final orbit, thermionic diodes convert the thermal energy directly into electricity, providing mission power for the spacecraft. The thermal mass of the receiver cavity provides enough stored energy to produce continuous power through periods of eclipse, without the need for large storage batteries.

Receiver-Absorber-Converter (RAC) Description

The RAC consists of the following components: receiver cavity/thermal energy storage, thermionic energy converters, secondary concentrator, insulation package, propellant pre-heater, and support structure. The receiver cavity consists of a graphite center body section and two endcaps. When assembled, these components form an annular ring with an outer diameter of 30cm, a cavity diameter of 14cm, and a length of 21cm (see Figure 2). The center body section is fabricated with 195 flow passages, each with a diameter of 3mm. The flow passages were sleeved with rhenium tubes with a wall thickness of 7mm. The faces of the center body were then coated with rhenium using the CVD process. The rhenium serves to protect the graphite from erosion by the hot hydrogen propellant. The inlet and exit tubes were placed in the appropriate endcap and the inside surfaces were then coated with rhenium. Next, the endcaps were pressed onto the center body to form the cavity. For the RACCET test article, holes were drilled through the side of the endcaps and through the center body. Graphite pins were pressed into the holes to secure the endcaps and carry the pressure load during hydrogen flow. This process cracked the graphite cavity in several places. Due to the cost and time to manufacture a replacement test article, the team decided to proceed with the damaged cavity for the RACCET test. The outer surface of the assembled cavity was then coated with about 70 mils of rhenium. Concurrently, the design engineers modified the RAC design for the EGD test article. The endcaps were redesigned with an elliptical head allowing the rhenium coating to carry the pressure load and eliminating the need for the pinning operation.

The RAC cavity is surrounded by multi-layer insulation (MLI) to minimize heat loss. The MLI is fabricated from a series of ninety 1mil tungsten and molybdenum foils, each separated by a thin layer of oxide powder. The MLI insulation package is split into two sections, a fixed upper section and a moveable lower section. The lower section remains in close contact with the upper section during propulsion operations. During power operations, the lower section of MLI is moved about 5cm to expose the thermionic converters to the cavity. The sixteen thermionic converters are placed around the circumference of the MLI package. The secondary concentrator is affixed to the entrance of the receiver cavity. For the RACCET test, a water-cooled silver secondary concentrator was used. For the EGD test article, an uncooled secondary concentrator was used. This concentrator had a high temperature reflective surface, which enabled it to operate at temperatures of about 1500K.

RACCET Test Setup and Objectives

The RACCET test was designed to demonstrate the feasibility of the RAC design in both the power and propulsion modes, and to provide design and operational feedback to the team for the EGD test. The Edwards Air Force Base Solar Rocket Propulsion Laboratory (SRPL) was selected as the primary test site to take advantage of the existing solar collector. However, the BWXT team had to design, procure, and install special test equipment to perform the desired tests. This equipment included: a 36 inch diameter by 36 inch high stainless steel vacuum chamber with a multi-port instrumentation feedthrough collar, a 50 kVA DC power supply/controller, a 25 kWt tungsten mesh heater, a 1000 l/s turbomolecular vacuum pump, propellant mass flow measurement/control devices, a custom "clamshell" cooling enclosure, and temperature measurement instrumentation. Figure 3 shows the RACCET test article in its water-cooled "clamshell" at the SRPL. The RACCET test consisted of three phases. The first was on optical performance test which was designed to demonstrate that the RAC could be heated to the necessary operating temperatures (2500K) using a primary and secondary collector. The second phase was an electrically heat propulsion test which was designed to demonstrate that sufficient heat could be transferred to the propellant to provide high impulse thrust. An electric heater was used in place of the solar collection system to enable testing during non-sun conditions. The third phase of the test was an electrically heated power test to demonstrate the feasibility of operating a string of thermionic converters to produce electricity. Due to time and funding constraints, the third phase of testing was moved to the Thermionic Evaluation Facility (TEF) in Albuquerque, NM.

RACCET Operations and Test Results

Prior to the start of the RACCET, a series of calorimetry tests were performed to determine the power delivered to the entrance of the secondary concentrator by the SRPL collector and to calibrate the normal incidence pyroheliometer (NIP). A water-cooled copper blackbody cavity was provided by BWXT for this purpose. The team decided to perform the electrically heated propulsion test prior to the solar heated test because results were needed to support the EGD RAC design effort. As discussed above, the RACCET test article had been cracked during fabrication and the coating process was unable to seal all of the leak paths. Therefore, both the test plan and objectives had to be modified to account for this condition. The RAC heatup could be performed as intended; however, the hydrogen leak could shut down the vacuum system enabling air to reach the test article. Since oxidation is a tremendous concern at these temperatures, the team decided to run the test in a helium overpressure, rather than vacuum, to ensure that air could not reach the test article. However, the helium reduced the effectiveness of the MLI insulation and therefore the test article cooled much faster than desired. This rapid cool down made it impossible to characterize a complete burn cycle, but it did allow the measurement of propellant temperature at the beginning of a burn for the prototypic flow rate.

A summary of the RACCET test results is provided in Table 1. A peak propellant temperature of 2200K was obtained when the RAC starting temperature was 2300K. These results validated the thermal efficiency of the RAC design. However, other results indicated the need for design modifications prior to EGD. During flow testing, the pressure drop through the RAC was an order of magnitude higher than predicted. Rhenium foam inserts had been placed in the inlet and exit tube to reduce heat loss due to radiation. Post test analysis showed that these inserts had become plugged, most likely due to contaminants in the hydrogen gas supply. The team decided to remove these inserts for the EGD test and accept the heat loss associated with the open tubes. The heat loss through the MLI was greater than expected and the EGD package was redesigned to provide better insulation around the joints. The MLI was also severely discolored, indicating contamination (oxidation).

The second phase of testing was conducted using the optical heat input from the SRPL collector. The RAC test article was rotated onto its side to allow the secondary concentrator to face the fixed SRPL collector. A total of four solar heat up tests were conducted. The peak temperature reached during these tests was 2175K. The major factor limiting the temperature was contamination of the secondary concentrator. The secondary was absorbing nearly 6.8 kWt when the peak temperature was reached, an increase of 5.5 kWt over its starting condition. If the absorption of the secondary concentrator had not increased, calculations show that a peak temperature of 2500K would have been reached. This information led the team to design a high temperature concentrator for the EGD test. The high temperature would greatly reduce the likelihood that contaminants would condense on the secondary concentrator.

The third phase of the RACCET test was conducted at the TEF facility in Albuquerque, New Mexico. The test article was placed in the Baikol test stand with the string of eight thermionic converters installed. The RAC was heated using the tungsten mesh heater to temperatures of about 2300K. The output power of the string was well below expectations. The individual converter j-v curves indicated a large variation from

FIGURE 3. RACCET Test Article at SRPL Facility.

converter to converter and also showed that all were operating well below the levels measured during their individual acceptance tests. Much of the problem appeared to stem from the multi-layer insulation on the back of the converters. The 1 mil foils were easily bent and would cause shorts in the string if they touched an adjacent converter. It also appeared that the converter hot shoe spacing might not have been adequate. While the spacing was designed with thermal growth in mind, the hot shoes had to be perfectly parallel or they could touch one another and short. Post test verification showed that all but one of the converters was operating as initially tested. The niobium mounting structure appeared to have sagged under the weight of the converters during testing. This may have led to converter shorting as well. This information was fed back to the design team and a new support structure/insulation scheme was designed for the EGD test.

TABLE 1. RACCET Test Summary.

Peak RAC Temperature (Electric Heater)	2550K
Peak RAC Temperature (Solar)	2175K
Total Time at Temperature > 2100K	11 Hours
Total # of Thermal Cycles	14
Total Number of Propulsive Burns	10
Calculated Peak Specific Impulse	816 seconds
Calculated Average Impulse over 30 Min. Burn	696 seconds

Engine Ground Demonstration Test Setup and Objectives

The Engine Ground Demonstration (EGD) test was performed at the NASA Lewis Research Center solar simulator facility "Tank 6". The facility is comprised of a large vacuum chamber (25' diameter x 50' length), a nine lamp solar simulator, and a primary concentrator. The test was designed to demonstrate the operation of the ISUS system in a near prototypic environment (zero g could not be simulated). The goal was to perform as much of the orbit transfer and electrical power operations as possible given the time and fiscal constraints of the program. The primary factors that limited test time were the degradation of the simulator's lamps and turning mirror, and funding for NASA and/or contractor personnel.

The solar simulator was capable of providing a broad range of power to the primary concentrator; however, the power level directly affected the life of the lamps. The program was only able to afford two sets of lamps, which was insufficient to reach full performance levels for the entire duration of the propulsion mission. Since the

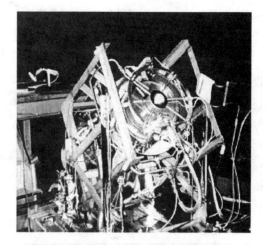

FIGURE 4. EGD RAC on BAP. FIGURE 5. EGD RAC Test Article.

characteristics (temperature and duration) of each of the 265 burns vary greatly from low earth orbit to geostationary orbit, the team decided to run a representative number of burns over the range of orbital distances. For example, the team would start with burns 1, 11, 21, 31, and so on. Once burn 261 was reached, the cycle would be repeated with 2, 12, 22, 32 ...

The RAC test article was assembled and instrumented at the NASA Lewis facility. Harris, the CATS contractor, modified the primary concentrator with facets obtained from Composite Optics. The RAC test article was installed onto the build-up assembly platform (BAP) as shown in Figure 4. The BAP was then placed in the Tank 6 facility for testing. Figure 5 shows the fully instrumented test article in the tank, just prior to start of the test.

EGD Test Operation and Results

The results of the Engine Ground Demonstration were very promising. A significant portion of the actual propulsion mission profile was demonstrated and peak exhaust temperature goals were achieved. Table 1 provides a summary of the test data from the EGD. Figure 6 shows the RAC temperature data for a few representative burns in the propulsion test. The total test time was limited by the life of the solar simulator lamps. Two sets of lamps were procured for the testing and the test was stopped when the available power in the second set of lamps could no longer get the RAC to a temperature where data would be useful (>1500K).

TABLE 2. EGD Test Summary.

Peak Temperature	~2250K
Total # of Propulsive Burns	117
# Hours Above 1500K	320
Percent of Total Propulsion Time Completed	30

In general the testing went very well. However, as with all tests, several changes to the test plan were required due to unanticipated circumstances. After about 40 thermal cycles, the test loop developed a significant hydrogen leak, which resulted in suspension of the test. The test article was taken out, disassembled, and inspected. The leak turned out to be in the inlet tube of the RAC, where the rhenium tube wall near the weld joint had noticeably eroded. This could not be caused by hydrogen because hydrogen doesn't react with rhenium. A sample of the hydrogen gas supply was analyzed and significant amounts of oxygen were found. We concluded that the tube had oxidized and the hydrogen carried the rhenium oxide away. A new hydrogen tanker was ordered with ultra pure hydrogen, (<5 ppm oxygen). The rhenium inlet tube was cut out and replaced with a new one. The RAC was borescoped to see if it had been damaged internally by the oxygen impurity. The inside looked pristine, leading us to conclude that the hot rhenium inlet tube had served as a getter for the oxygen and thus the damage was limited to that area. Testing was resumed following the inlet tube replacement and the test was run until lamp power diminished. We had originally intended to perform some electrical power testing; however, due to the time constraints at the test site, the decision was made to perform the power testing at the TEF facility in Albuquerque later in the year. These results will be presented at a later date.

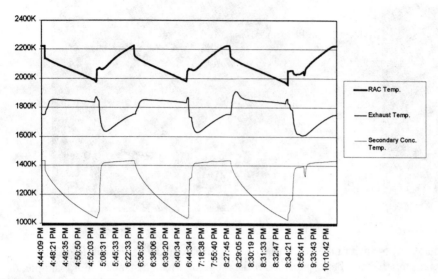

FIGURE 6. EGD Test Data.

The secondary concentrator was covered with a thin white power, but the performance of the concentrator did not seem to be affected. The temperature of the secondary remained nearly constant throughout the test, and no drop in RAC temperature was observed. No dark discoloration of the surface was evident, as we saw in the RACCET test. The white residue came off of the surface fairly easily. The residue is being examined to determine its nature and possible source.

CONCLUSIONS

The Integrated Solar Upper Stage (ISUS) has clearly demonstrated its high specific impulse propulsive capabilities. Target exhaust gas temperatures were achieved in the Engine Ground Demonstration. The exchange of heat from the RAC to the propellant was slightly higher than calculated, resulting in a flatter exhaust gas temperature over the duration of the burn. This means that the average specific impulse was greater than predicted and that the mission requirements can be met with a lower peak RAC temperature. The test brought to light the need for contaminant free propellant (hydrogen) to minimize system degradation. The test verified the performance of the high temperature secondary concentrator, which was developed by BWXT. There was very little degradation in the performance of the secondary concentrator during EGD, in contrast to the tremendous degradation seen in the low temperature secondary concentrator during RACCET. Further testing is necessary to demonstrate the electrical power capabilities of the RAC and to verify full lifetime performance.

ACKNOWLEDGMENTS

This testing was conducted under contract to the Air Force Phillips Laboratory. The solar bi-modal concept is the property of BWX Technologies, Inc.

INTERDIFFUSION OF RHENIUM AND POCO GRAPHITE

Jinglong Li and Ralph H. Zee
Materials Research & Education Center
Department of Mechanical Engineering
201 Ross Hall
Auburn University
Auburn, AL 36849-5341
(334)844-3320

Abstract

Graphite is being considered as the prime candidate as the energy storage material for space propulsion. One of the anticipated problems is the loss of graphite at high temperature. Rhenium has been proposed as a diffusion barrier for graphite to prevent excessive loss of the material. However, the mutual diffusion of graphite and rhenium has not been measured, and much of the data is based on extrapolation with high degree of error. A technique to study the interdiffusion of rhenium and carbon without a marker was developed based on Rutherford Backscattering Spectroscopy (RBS) technique. The graphite samples used in this experiment were Poco graphite purchased from Unocal. The specimens used for rhenium deposition were in the form of round discs cut from a 25.4 mm diameter bar. The round discs were mechanically polished down to a thickness of 2 mm. A thin layer of rhenium was then sputter-deposited onto the polished Poco graphite surface using a RF sputtering system. The deposition rate was determined to be 10 nm/min at a RF power of 20 Watts and an Ar gas pressure of 2.67 Pa. The samples coated with rhenium were given different diffusion anneals at temperatures between 1373 K and 2273 K in a vacuum electron beam furnace. The diffusion constants were obtained by fitting the RBS profiles with appropriate diffusion equations. The diffusion constants for rhenium in Poco graphite range from 9×10^{-18} m^2/s at 1373 K to 7×10^{-17} m^2/s at 2273 K, whereas the diffusion constants for carbon in rhenium films range from 6×10^{-17} m^2/s at 1373 K to 9×10^{-16} m^2/s at 2273 K. Both Arrhenius plots of rhenium in graphite and carbon in rhenium clearly show a change in the slope with temperature. This feature indicates that there are two diffusion mechanisms involved. At lower temperatures, rhenium atoms diffuse along the open pores in Poco graphite, whereas at temperatures above 2073 K diffusion of rhenium through graphite lattice becomes dominant. Similarly, the carbon atoms diffuse along grain boundaries of rhenium at lower temperature. Diffusion of carbon atoms through rhenium lattice becomes important at temperatures above 1773 K.

INTRODUCTION

Graphite is frequently considered for ultrahigh temperature applications. Graphite possesses significant attributes for aerospace applications where weight is a critical factor. However, graphite evaporates when it is operating at high temperatures. Rhenium, a transition element with extreme scarcity, possesses a unique combination of properties which draw a special attention for many applications demanding high-temperature strength, wear resistance and corrosion resistance. The superior performance of rhenium can be attributed to its chemical inertness, excellent high temperature strength and room temperature ductility, and resistance to carburizing. Rhenium does not form a stable carbide (Isobe 1989). Rhenium can be used in contact with graphite to form a protection layer. As a coating on graphite materials used in low-oxygen environments, rhenium has a significant solubility for carbon with no carbide formation, ensuring an excellent bond between the two materials.

The coating of rhenium on graphite has been proposed as a diffusion barrier of carbon for a thermal absorber for space thermionic power systems. In this concept, parabolic mirrors are used to focus solar energy onto the graphite cavity. Much concern exists over excessive evaporation of carbon at the high operation temperature of 2400 K. Excessive evaporation may coat the parabolic reflectors, thereby reducing their efficiency. However, the resistance of the evaporation of the coating depends strongly on the interdiffusion of rhenium and carbon. Therefore, research on the diffusion processes in rhenium thin films is of special importance. Unfortunately, the studies on diffusion in graphite are poorly documented. Essentially no previous systematic work exists on diffusion of rhenium in graphite and carbon in rhenium. Traditionally, diffusion coefficients are measured by radioactive tracers. This technique is not feasible because of environmental concern. Interdiffusion is usually measured by the introduction of an inert marker (Kirkendall 1942). This technique is not practical for the studies of diffusion on thin films. In this paper, we report that a technique based on Rutherford Backscattering Spectroscopy (RBS) (Chu 1978) to measure the mutual diffusion behavior of rhenium and carbon has been successfully developed at Auburn University.

EXPERIMENTAL

Poco AXF 5Q coupons of graphite were used in this research. Poco graphite is made by a non-conventional proprietary method, which is assumed to be isostatic hot-pressing without the use of any binder. With respect to its physical properties, it is macroscopically isotropic. Poco graphite may be regarded as an agglomerate of structurally nearly perfect polycrystalline graphite particles. Coating of rhenium onto graphite was achieved by a vacuum sputtering system. Diffusion annealing was conducted by a vacuum electron beam furnace. RBS studies were conducted in the Leach Nuclear Science Center at Auburn University to determine the diffusion profiles.

Sample Preparation

Graphite round disc samples were cut from a Poco graphite bar (diameter 25.4 mm). The Poco graphite was purchased from Unocal. The round discs were then mechanically polished down to a thickness of 2 mm. The final polish was performed with a TXTMET® 1000 polish cloth purchased from Buehler®. A polished surface of graphite sample is necessary to prevent misinterpretation of the profiles obtained from RBS. A rhenium target for sputtering deposition was made by cutting a rhenium square foil 0.1 mm thick, 99.98% (purchased from Aldrich®). A polished graphite sample was placed in the front of rhenium target at a distance of approximately 30 mm in a sputtering system. The sputtering chamber was evacuated to a vacuum of 1.22×10^{-4} Pa by a cryopump. A grade 5.0 Ar gas was used for sputtering. A thin film of rhenium was sputtered onto the graphite surface by RF sputtering at an Ar gas pressure of 2.67 Pa. For all the samples reported in this paper, an RF power setting of 20 Watts was used. A sputtering rate of 10 nm/min was obtained by fitting the RBS spectra.

Diffusion Annealing

All the samples for diffusion annealing were cut into squares measuring 5 mm×5 mm from the as-deposited round wafer. The samples were heated in a contamination free environment to minimize oxidation of the graphite (graphite is highly susceptible to react with oxygen to form CO and CO_2). An ultra-high vacuum electron beam heating furnace was used for this purpose. A base pressure of 1.22×10^{-6} Pa was achieved by an ion pump and a titanium sublimation pump. Samples were annealed in a vacuum better than 10^{-4} Pa during the entire experiment. Due to the wide range of temperature needed in this study (1373 K to 2273 K), two different means of temperature measurement were employed. The low temperature (1373 K to 1773 K) part was measured by a type C thermocouple and the high temperature by a two-color pyrometer (1773 to 2273 K).

RBS Measurement

All the RBS measurements were performed using a 2 MeV α particle beam produced at the Auburn University Dynamitron accelerator. Diffusion constants were obtained by fitting RBS spectra with the program RUMP (Doolittle 1985). The fitting was done by selecting available diffusion equations which were most suitable to boundary conditions of the rhenium/graphite diffusion couples.

RESULTS

Different annealing temperatures were employed to obtain the temperature dependence of diffusion. For each annealing condition, the RBS spectrum was analyzed to determine the diffusion coefficient at that temperature. Figure 1 shows a RBS spectrum obtained for a specimen prior to annealing (in the as-deposited condition) whereas figure 2 shows a similar spectrum but after annealing. The spreading of the rhenium peak is a clear indication that interdiffusion has occurred. The RUMP program was used to translate these RBS spectra into concentration profiles from which diffusion constants were obtained. The experimental results obtained to date are summarized in table 1. The as-deposited thicknesses indicated in table 1 are for the rhenium thin films. The Arrhenius plots (Fig. 3 and Fig. 4) for the diffusion of rhenium in Poco graphite and carbon in rhenium were obtained by plotting the logarithm of diffusion constants against the reciprocal of temperatures.

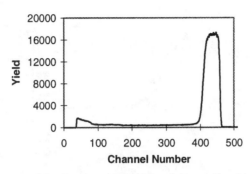

FIGURE 1. A Typical RBS Spectrum of As-Sputtered Rhenium on Poco Graphite. Sputtering Time is 1500 s.

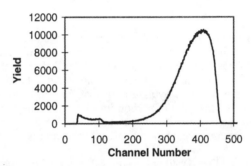

FIGURE 2. A RBS Spectrum of Rhenium on Poco Graphite after Annealing at 2273 K for 360 s.

TABLE 1. Summary of the Experimental Results.

Sample number	As-deposited rhenium thickness (nm)	Annealing temperature (K)	Annealing time (s)	Diffusion constant, Re in C (m^2/s)	Diffusion constant, C in Re (m^2/s)
Re/C-04	425	1373	120	9×10^{-18}	6×10^{-17}
Re/C-18b	170	1473	1800	1.2×10^{-17}	6.4×10^{-17}
Re/C-15	290	1773	3000	1.3×10^{-17}	6.5×10^{-17}
Re/C-10	475	1873	600	1.4×10^{-17}	9×10^{-17}
Re/C-11	320	1973	900	1.48×10^{-17}	2×10^{-16}
Re/C-13	380	2073	1200	1.55×10^{-17}	4×10^{-16}
Re/C-18	170	2173	1200	5×10^{-17}	6×10^{-16}
Re/C-07	130	2273	360	7×10^{-17}	9×10^{-16}

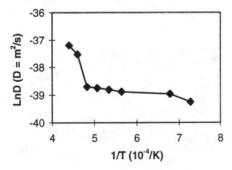

FIGURE 3. Arrhenius Plot for Diffusion of Rhenium in Poco Graphite.

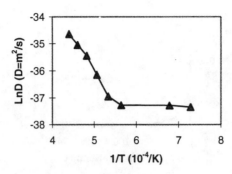

FIGURE 4. Arrhenius Plot for Diffusion of Carbon in Rhenium.

DISCUSSION

In this study, two diffusion constants i.e. rhenium in graphite and carbon in rhenium were obtained on the assumption that the thickness of rhenium is larger than the diffusion length of carbon in rhenium. Due to lack of a suitable computer program for this study and limitation of RUMP (RUMP is limited on rhenium film thickness of approximately 300 nm), the diffusion constants obtained to date should be treated as relative values and not absolute. The diffusion constants were obtained by assuming that the diffusion of rhenium into graphite satisfies the equation for thin films, whereas the diffusion of carbon into rhenium obeys the distribution of error function. The latter assumption is not true when the thickness of rhenium thin film is less than the diffusion length of carbon in rhenium. A computer program is being developed to overcome this dilemma.

From the Arrhenius plots (Fig. 3 and Fig. 4), there are two slopes for both the diffusion of rhenium in graphite and carbon in rhenium, indicting that there is a change in the diffusion mechanism. There are two different explanations for the curvature observed in a number of the diffusion plots. The first one is due to the coexistence of two distinct diffusion mechanisms with different entropies and enthalpies. The second one is based on a single mechanism but with an activation energy which is a function of temperature. The presence of a sharp transition (Fig. 3 and Fig. 4) suggests that the former explanation is more likely.

For the diffusion of rhenium in graphite, two mechanisms are speculated to be responsible for the curved Arrhenius plots. One possible mechanism is diffusion along open pores in the Poco graphite structure. A RBS spectrum of as-deposited rhenium on graphite (Fig. 1) exhibits a long tail on lower energy side of rhenium peak, which indicates that rhenium atoms migrate into graphite even without diffusion annealing. The only possible path for this migration is along open pores in the Poco graphite. Microscopy analysis shows porosity of the Poco graphite as shown in figure 5. The as-deposited rhenium already coats the open pore surface resulting in apparent rhenium diffusion even prior to annealing. The other mechanism is diffusion through lattice (bulk diffusion). As temperature increases, rhenium atoms have sufficient energy to overcome the barriers and migrate into the graphite lattice. The Arrhenius plot for diffusion of rhenium in graphite indicates clearly that there are two mechanisms responsible for the change in slope, i.e., rhenium atoms migrate along pores at lower temperatures and through lattice at higher temperatures. The slope of the Arrhenius plot is related to the activation energy of diffusion for the particular mechanism. According to figure 3, the activation energy for diffusion of rhenium in carbon at low temperature (surface diffusion through pores) is 18 kJ/mole whereas the energy for bulk diffusion at high temperature is 295 kJ/mole.

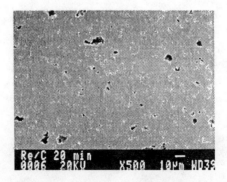

FIGURE 5. SEM Micrograph of Porosity Structure of Sputtered Rhenium on Graphite Surface.

A diffusion behavior similar to that of rhenium in carbon was observed for the diffusion of carbon in rhenium (see figure 4). Again a change in slope in the Arrhenius plot was observed indicative of a change in mechanism. However, the two mechanisms in this case are most likely to be grain boundary diffusion at low temperature and lattice diffusion when the temperature is above 1773 K. The activation energy for grain boundary diffusion is 4 kJ/mole whereas that of lattice diffusion is 176 kJ/mole. Results show that the bulk diffusion of rhenium in graphite (295 kJ/mole) is higher than that of carbon in rhenium. This is due to the fact that carbon atoms occupy the interstitial sites in rhenium and therefore migrate with less distortion resulting in a lower activation energy. On the contrary, rhenium atoms are most likely substitutional in nature and migrate via a vacancy mechanism requiring a higher energy.

CONCLUSIONS

A thin layer of rhenium has been successfully sputter-deposited onto Poco graphite for diffusion studies. A technique without using a marker rhenium has been developed to measure the mutual diffusion constants of rhenium and carbon. Activation energies for the various processes were obtained from the data. The Arrhenius plots clearly show that there are two diffusion mechanisms involved in the diffusion process. It appears that activation energy for diffusion of rhenium on graphite surface is 18 kJ/mole whereas diffusion of carbon in rhenium grain boundary requires a very low energy of 4 kJ/mole. At high temperature diffusion is controlled by the lattice process with carbon in rhenium (interstitial mechanism) and migrating with an energy of 176 kJ/mole and rhenium in carbon requiring a higher energy of 295 kJ/mole (vacancy mechanism).

Acknowledgment

This work is supported by the Air Force Phillips Laboratory in Albuquerque, NM which the authors gratefully acknowledge.

References

Chu, W. K., J.W. Mayer, and M.-A. Nicolet (1978) *Backscattering Spectrometry*, Academic Press, Inc. New York, 54-88.

Doolittle, L.R. (1985) "Algorithms for the Rapid Simulation of Rutherford Backscattering Spectra," *Nucl. Instr. And Meth.*, B9: 344-351.

Isobe, Y., M. Tanaka, S. Yamanaka, and M. Miyake (1989) "Chemical Vapour Deposition of Rhenium on Graphite," *J. Less-Common Met.*, 152:177-184.

Kirkendall, E.O. (1942) "Diffusion of Zinc in Alpha Brass," *Trans. AIME*, 147:104-110.

SECOND CONFERENCE ON APPLICATIONS OF THERMOPHYSICS IN MICROGRAVITY

REAL-TIME X-RAY TRANSMISSION MICROSCOPY FOR FUNDAMENTAL STUDIES SOLIDIFICATION: AL-AL$_2$AU EUTECTIC

Peter A. Curreri
Space Sciences Laboratory NASA
Marshall Space Flight Center, AL
35812
(205) 544-7763

William F. Kaukler
Center for Microgravity and
Materials Research, The University
of Alabama in Huntsville, AL
35899

Subhayu Sen
Universities Space Research
Association
MSFC, Huntsville, AL 35812

Abstract

High resolution real-time X-ray Transmission Microscopy, XTM, has been applied to obtain information fundamental to solidification of optically opaque metallic systems. We have previously reported the measurement of the solute profile in the liquid, phase growth, and detailed solid-liquid interfacial morphology of aluminum based alloys with exposure times less than 2 seconds. Recent advances in XTM furnace design have provided an increase in real-time magnification (during solidification) for the XTM from 40X to 160X. The increased magnification has enabled for the first time the XTM imaging of real-time growth of fibers and particles with diameters of 5 μm. We have previously applied this system to study the kinetics of formation and morphological evolution of secondary fibers and particles in Al-Bi monotectic alloys. In this paper we present the preliminary results of the first real-time observations of fiber morphology evolution in optically opaque bulk metal sample of Aluminum-Gold eutectic alloy. These studies show that the XTM can be applied to study the fundamentals of eutectic and monotectic solidification. We are currently attempting to apply this technology in the fundamentals of solidification in microgravity.

INTRODUCTION

Eutectics are polyphase alloys that (especially when grown in a one dimensional thermal gradient) can possess uniform rod, plate, or more complex regular structures. Eutectic alloys have applications for specialty alloys such as magnets and superconductors, and high volume structural alloys. There has been much interest in the study of solidification processes under the reduced sedimentation and convection environment available in low-gravity (Curreri and Stefanescu 1988). One very interesting, still unresolved, finding (Curreri and Larson 1988) is that directional solidification of on-eutectic alloy in low gravity often dramatically changes the eutectic interphase spacing, as well as eutectic grain size in equiaxed eutectic growth. Many eutectics with considerable commercial importance (Fe-C, Al-Si) solidify with a unfaceted primary phase and a faceted secondary phase. Since the faceted phase has anisotropic growth kinetics it is very difficult to control phase alignment by thermal gradient during solidification. Thus these alloys are usually referred to as irregular eutectics.

Current high resolution X-ray sources and high contrast X-ray detectors have advanced to allow systematic study of solidification dynamics and the resulting microstructure. We have employed a state-of-the-art sub-micron source with acceleration voltages of 10-100 kV to image solidification of metal alloys. In-situ X-ray imaging of Al-Pb and Al-In monotectic alloys (Curreri and Kaukler 1996 and Curreri and Kaukler 1995) showed for the first time that the isoconcentration lines of the solute boundary layer in metals are not necessarily parallel to the growth interface as has been assumed in some theories. Further, it was observed that striations in the solidified crystal may not decorate the interface position and shape. It was also shown that metal monotectic alloys at the monotectic composition do not necessary grow in a coupled manner. The ability of the XTM to integrate cross sectional microstructural features was utilized to study the details previously unreported formation of striations in Al-Pb (Kaukler 1996 and Kaukler 1997). The process of morphological instability and cellular growth was imaged for Al-Cu and Al-Ag alloys (Curreri and Kaukler 1996). The dynamics of solid/liquid interface shape evolution near an insoluble particle was studied with XTM (Sen 1997) showing the limits to the applicability of some analytical models. The results discussed above indicated that limitations of classical quench methods could be overcome providing more precise study of the dynamics.

The application of XTM for the study of solidification fundamentals was limited by a demonstrated resolution limit of 25 μm for real-time imaging. The study of the dynamics of formation of secondary eutectic and monotectic droplets and fibers requires a resolution on the order of 5 μm. It was suggested (Kaukler and Curreri 1996) that improvement of furnace and detector design could enable the XTM to achieve this required resolution.

The latest furnace configuration has allowed us to achieve resolutions of the order of 5 μm in real-time (using an existing intensifier and video camera imaging arrangement) for Al-Bi alloy monotectic alloy (Curreri, Kaukler and Sen 1997). This enabled the study of the bulk solidification dynamics of metal alloy monotectic fibers and droplets for the first time.

In this paper we present the preliminary results of the first real-time observations of solidification for optically the eutectic alloy Au-Al_2Au. For this study we selected the on-eutectic composition of Al-2 atom % Au. The alloy solidifies (Piatti and Pellegrini 1976) with faceted Al_2Au fibers or lamellar plates in an unfaceted Al matrix. The high X-ray absorption difference between Au and Al provide excellent contrast for XTM. Single phase Al-Au was previously (Beech 1984) studied, utilizing radiographic with about 50 μm resolution, to determine the dimensions of the solute boundary layer in the liquid and to determine the kinetics of solute depletion band formation.

EXPERIMENTAL METHODS

Projection radiography (see FIGURE 1) using a micro-focus x-ray source offers magnification, adequate resolution and with suitable detector technology, adequate contrast to observe solidification in metals. Resolution, which is limited by the x-ray spot size can approach micrometer values. Radiography by projection permits placing the specimen in a furnace between the x-ray camera and the source.

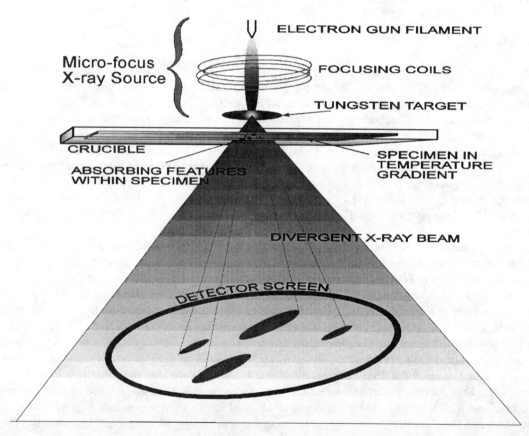

FIGURE 1. Schematic Diagram of the Working Principle and Component Arrangement of the X-ray Transmission Microscope. Microfocus X-Ray Source Configuration is at the Top; the Sample and Crucible in the Middle; the Magnification of the X-Ray Image is Shown at the Bottom.

Simply, the furnace is a modified horizontal Bridgman-Stockbarger design that uses a rectangular, sintered boron nitride crucible. There were no contact interactions observed even after prolonged heating in air. For the work presented here, the specimen sits 3×10^{-3} m from the furnace housing exterior and had dimensions of about $(3 \times 70 \times 1) \times 10^{-3}$ m thick. For these solidification experiments, magnifications in the range of 150 were obtained on the video monitor. The images shown here were digitized from the video camera/x-ray intensifier.

The apparatus has been described in detail elsewhere (Curreri and Kaukler 1996 and Kaukler and Curreri 1996). Typical x-ray image intensifiers or converters degrade detectability as an inverse function to feature size. This limit of contrast by feature size is known as the modulation transfer function, MTF, of the camera/converter system. A simple interpretation of the MTF is that it represents the ratio of the outgoing signal to the incoming one. Tiny features, with 100% contrast, when imaged through the system in question may produce only a 5% contrast on the output. As a result, spatial resolution alone is not satisfactory to describe the capability of the system. Poor image quality is compounded by the fact that most specimen features do not produce a high contrast input signal to start with.

The Al-Au alloy was prepared from 99.999 % pure metals, by vacuum induction melting in a graphite crucible. The alloy was chill cast in a copper mold and then rolled to 1mm thickness sheet from which the samples were cut. Chemical analysis of the alloy indicated 7 wt % Au with an experimental error of ±1 %.

RESULTS AND DISCUSION

Most of the results we furnish in this paper are in the form of morphological observations and will be exhibited as radiomicrographs. The discussion contrasts these observations with some findings from the literature so the material is combined.

Aluminum Gold Eutectic

We selected the Al-Au eutectic system for a number of reasons. It offers excellent radiographic contrast due to the high X-ray absorptivity of the Al_2Au minority phase and the relatively low X-ray absorptivity of the majority Al phase. The XTM contrast was predicted utilizing a model (Curreri and Kaukler 1996 and Kaukler and Curreri 1996) described previously. Al-Au alloy has also been reported (Piatti and Pellegrini 1976) to grow in a eutectic-like morphology over an extended range of thermal gradient, G, growth rate, R and ratio G/R. Al-Au eutectic is one of the irregular faceted nonfaceted systems which are less well understood theoretically than the regular nonfaceted-nonfaceted eutectic systems. It has also been shown (Piatti and Pellegrini 1976) to exhibit both lamellar and rod structures.

The eutectic composition for the $Al-Al_2Au$ eutectic has been reported to be 5 (Piatti and Pellegrini 1976) and 7.5 (Massalski 1986) wt. % Au. The more accepted value is 7.5 wt. % Au. The extended range of composition in which $Al-Al_2Au$ eutectic solidifies with eutectic-like structure has been reported (Piatti and Pellegrini 1976) as a function of composition and G/R. Eutectic-like morphologies were reported within the range of 6 to 11 wt. % Au for G/R of 1×10^8 K sec m^{-2}. The range widens with increasing G/R to about 3 - 12 wt. % Au at G/R of about 1×10^{10} K sec m^{-2}. The thermal gradient for our XTM furnace for the experiments discussed here was 5200 K/m and the growth rates were between 5 and 0.5 μm/sec. Thus within our range of G/R values of 1×10^9 to 1×10^{10} K sec m^{-2}, and the nominal composition of 7 wt. % Au, it is predicted that the alloy will solidify with eutectic like microstructure for the entire experimental range we investigated.

First Real Time In-situ Observations of Eutectic Metal Fiber and Plate Solidification

To our knowledge, these data represent the first real-time in-situ observations of eutectic solidification of optically opaque metal with the resolution necessary to image secondary fibers, flakes and lamellae. FIGURE 2 gives an X-ray radiograph showing the real time solid/liquid interface for steady state solidification at R = 2 μm/sec. The right side is the solid where the Al_2Au phase takes the form of an aligned array of irregular fibers. Some fibers are extending into the liquid to a distance of about 20 μm. The growth direction is to the left in a thermal gradient of 5200 K/m as it is for the FIGURES 2-5. The structure at 5 μm/sec is similar but with finer fiber spacings. There is no corresponding figure shown. FIGURE 3 shows steady state solidification at R = 1.0 μm/sec. The Al_2Au phase now manifests more irregular plate or flake like structure with coarser spacing. Note the large Al_2Au phase fiber or plate that is extending into the liquid about 50 μm. FIGURE 4 shows the structure for R = 0.5 μm/sec. The solid on the right side is growing with large coarse plates of the Al_2Au phase most of which are greater than 100 μm wide. Note the considerable extension of the intermetallic phase into the liquid. The spacing is larger and more irregular at this very low R. Changing the angle of the X-ray beam revealed the structure to consist of complex irregular 2-dimensional structures or arrays. These data indicates that a fiber to lamella or plate transition (Piatti and Pellegrini 1976) occurs between R = 1 and 0.05 μm/sec at this G.

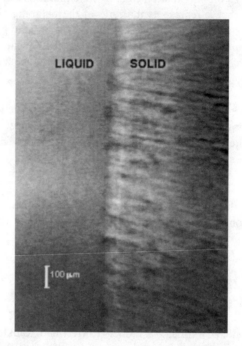

FIGURE 2. Radiomicrograph of Al-Au Eutectic During Solidification at a Rate of About 2 μm/sec.

Comparison with Previous Experiments

The Al-Au system has been studied previously by Beech et. al. (Beech 1984 and Miller 1975) utilizing x-ray transmission radiography with resolution of about 50 μm. Beech employed a graphite vacuum furnace that was transparent to x-rays for melting and solidifying the metal. They also used an x-ray sensitive Vidicon camera. The experiments studied the solute field in the melt ahead of the solid/liquid interface. Periodic film exposures were analyzed by a densitometer calibrated to give gold concentration in the liquid. The solute boundary layer 1/e value thus determined gave a value of Au diffusion in liquid Al (assuming no convection) of 7×10^{-12} $m^2 sec^{-1}$. Beech also observed and measured the kinetics of solute depletion in the solid that occurs behind the solid/liquid interface at zero velocity. (He determined that these kinetics were consistent with that expected by the diffusion of Au from the solid back into the melt.) Beech studied the single phase (solid solution) Al-2 wt % Au composition. Although the resolution of Beech's instrument could identify cellular structure, it would have been insufficient to examine the Al_2Au intermetallic fibers or lamellae.

During our experiments many of the images also show a solute boundary layer in the liquid at solid liquid interface. We also observed the *initial transient* portion of growth where the gold solute built up ahead of the moving solid interface. During solidification from the parent metal at R = 1 μm/sec, and G = 5200 K/m the solute buildup caused a constitutional undercooling of the interface and subsequent formation of eutectic cells. Very fine (>5 μm) intercellular fibers were found using a real-time x-ray film camera which can provide higher resolution then the real time CCD. A few millimeters later this cellular structure spontaneously changes to planar solidification with thicker fibers (>5 μm diameter). We also reproduced the observation of solute depletion in the solid that occurs behind the solid liquid interface at zero velocity. FIGURE 5 shows a radiomicrograph of the solid liquid interface at zero velocity for about 2 minutes. The solute depletion band in the solid at the S/L interface is clearly shown. Thus we confirmed that the solution depletion mechanism also occurs in the poly-phased Au-Al$_2$Au eutectic.

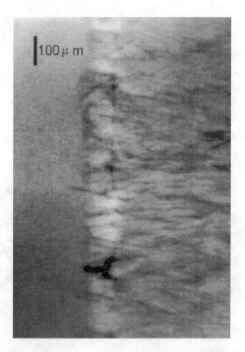

FIGURE 3. Real-time Image of the Solid-liquid Interface with Al-Al$_2$Au Eutectic Solidifying to the Left With the Rate of 1 μm/sec.

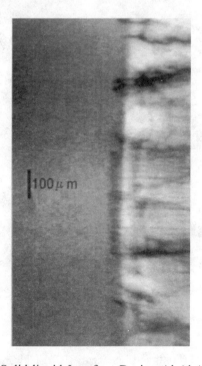

FIGURE 4. Radiomicrograph of the Solid-liquid Interface During Al-Al$_2$Au Eutectic Solidification at the Slow Rate of About 0.5 μm/sec.

Fiber Thickening in the Melt

The tendency of the intermetallic phase to form ahead of the isotherm leads to an interesting observation that could only have been made by in-situ x-ray microscopy. FIGURE 5 also shows the Al_2Au phase fibers extending into liquid with the solid liquid interface at zero velocity for a couple of minutes. The solid on the right side was grown with a rate of 2 μm/sec and formed fibrous Al_2Au. Clearly visible are the coarsening intermetallic phases that extended from the interface into the melt during growth. The solid behind the interface shows the diffusive depletion of the Au content (white band) developing during the stagnation of the interface. The larger intermetallic phases in the melt eventually break free from the matrix where they originally formed and fall to the bottom. Note the thickening of the Al_2Au phase fibers extending into liquid and that one fiber or plate is about 100 μm in length and about 25 μm thick. The existence of such a process and subsequently its kinetics would be difficult to study with solidification and quench techniques conventionally used to study the solid liquid interface morphology. We are not aware that this phenomena has been previously reported.

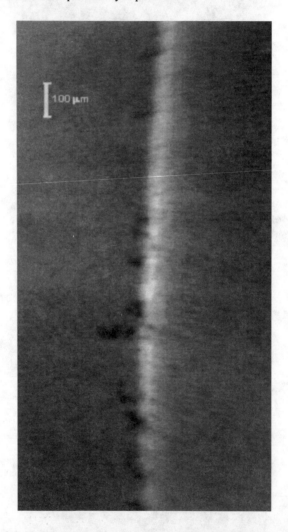

FIGURE 5. Real-time Image of the Solid-liquid Interface Taken After About Two Minutes of Zero Velocity.

Irregular Eutectic Structure

The Al-Al_2Au eutectic alloy solidifies (Piatti and Pellegrini 1976) with faceted Al_2Au fibers or lamellar plates in an unfaceted Al matrix. The faceted minority phase is highly anisotropic and thus grow in an irregular manner which is more difficult to align by thermal control during solidification. The solidification of regular eutectics is fairly well understood (Jackson and Hunt 1966 and Magnin and Trivedi 1991) where the interfacial undercooling, ΔT, is a function of the solidification rate, R, and the phase spacing, λ, but the actual spacing selected during solidification is (thought to be) selected to lie on the extremum (minimum undercooling), point on that functional curve. The theory of irregular eutectic growth was not advanced as quickly. Flemings realized that a theory of faceted nonfaceted eutectics could not assume equilibrium at the solid liquid interface and for usual experimental conditions the solid liquid interface could no longer be assumed isothermal (Flemings 1974). Theories for irregular eutectics have since been developed and applied to Fe-C (Magnin and Kurtz 1987) and Al-Si (Magnin, Mason and Trivedi 1991) systems. This initial XTM study of Al-Al$_2$Au eutectic does not attempt to differentiate between current theories but some pertinent observations can be made. The Al$_2$Au fibers or plates extend into the liquid with distances from on the order of 50 µm. These extended fibers are not coupled with the primary aluminum matrix. This can be inferred directly by XTM observation and indirectly since fiber thickening and detachment occurred at zero velocity.

CONCLUSIONS

The high X-ray absorption difference between Au and Al provide excellent contrast for XTM. To our knowledge these data represent the first real-time in-situ observations of eutectic solidification of an optically opaque metal with the resolution necessary to image secondary fibers, plates and lamellae. The Al$_2$Au phase can extend well into the melt (as much as 10X the diameter) from the solid/liquid interface. The data indicates that a fiber to lamella or plate transition occurs between R = 1 and 0.05 µm/sec. It was observed over time (several minutes), that the coarsened, extended portions of the intermetallic phase loosened from the matrix and settled to the bottom of the crucible. We are not aware that this phenomena has been previously reported.

Acknowledgments

Thanks are extended to Pat Salvil in the foundry at Materials & Processes Lab. for casting the alloys and having the chemical analysis performed. We wish to thank NASA's Microgravity Science Division NRA Program for funding this work.

References

Beech, J. (1984) "The Formation of Solute Depleted Bands in Aluminum-Gold Alloys," *Journal of Crystal Growth*, 67:385-390.

Curreri, P. A. and D. M. Stefanescu (1988)"Low-Gravity Effects During Solidification," *Metals Handbook*, American Society of Metals International: Metals Park, Ohio,15(9):147-158.

Curreri, P. A., Kaukler, W. F., Sen, S., and Peters, P. N. (1997) "Application of Real-Time X-ray Transmission Microscopy to Fundamental Studies of Metal Alloy Solidification," to be published in *Proceedings of the 9th International Symposium on Experimental Methods for Microgravity Materials Science*, Spacebound'97, Montreal, Quebec, The Metallurgical Society, Inc., Warrendale, Penn.

Curreri, P.A. and W. F. Kaukler (1996)"Real-Time X-Ray Transmission Microscopy of Solidifying Al-In Alloys," *Metallurgical Transactions 27A* (3):801-808,.

Curreri, P.A. and W. F. Kaukler (1995) "Real-Time X-Ray Transmission Microscopy of Solidifying Al-In Alloys," presented at The Metallurgical and Materials Society Annual Meeting, Las Vegas, NV, Feb. 12-16, 1995) published in *Proceedings of 7th International Symposium on Experimental Methods for Microgravity Materials Science*, Robert Schiffman, Ed., The Minerals, Metals and Materials Society, 93-101.

Curreri, P.A. and W. Kaukler (1996) "X-Ray Transmission Microscopy Study of the Dynamics of Solid/Liquid Interfacial Breakdown During Metal Alloy Solidification," *Presented at 8th International Symposium on Experimental Methods for Microgravity Materials Science*, Feb 4-8, 1996, Anaheim, CA, 125 TMS Annual Meeting, to be published in proceedings.

Curreri, P.A., D.J. Larson, and D.M. Stefanescu (1988) "Influence of Convection on Eutectic Morphology," in *Solidification Processing of Eutectics*, D. M. Stefanescu, G.J. Abbaschian and R.J. Bayuzick, eds., The Metallurgical Society, 47-64.

Flemings, M. (1974) *Solidification Processing*, Mcgraw-Hill Inc., New York, 107.

Jackson, K. A., and Hunt, J. D. (1966) "Lamellar and Rod Eutectic Growth," *Transactions of the Metallurgical Society of AIME*, 236:1129-1142.

Kaukler, W. F. and P. A. Curreri (1996)"Advancement of X-ray Microscopy Technology and it's Application to Metal Solidification Studies," presented at the *1996 SPIE Technical Conference in Space Processing of Materials*, Aug. 4, 1996, Ed. N. Ramachandran, 2809:34-44.

Kaukler, W. F. and P.A. Curreri, (1996) "X-Ray Transmission Microscopy of Al-Pb Monotectic Alloys During Directional Solidification," Presented at *8th International Symposium on Experimental Methods for Microgravity Materials Science*, Feb 4-8, 1996, Anaheim, CA, 125 TMS Annual Meeting, in print.

Kaukler, W., F. Rosenberger, and P. A. Curreri (1997) "In-situ Studies of Precipitate Formation in Al-Pb Monotectic Solidification by X-ray Transmission Microscopy," *Metallurgical Transactions*, 28A:1705-1710.

Magnin, P. and Kurz, W. (1987)"An Analytical Model of Irregular Eutectic Growth and it's Applications to Fe-C," *Acta Metall*. 35(5):1119-1128.

Magnin, P. and Trivedi, R. (1991) " Eutectic Growth: A Modification of the Jackson and Hunt Theory," *Acta Metall. Mater*, 39(4):453-467.

Magnin, P., Mason, J. T., and Trivedi, R. (1991) "Growth of Irregular Eutectics and the Al-Si System," *Acta Metall. and Mater*, 39(4):469-480.

Massalski, T. B. (1986) *Binary Alloys Phase Diagrams,* The American Society for Metals, Metals Park, Ohio, 88-90.

Miller, W. J., Stephenson, M. P., and Beech, J. (1975) "A Technique for the Direct Observation of Alloy Solidification," *J. Phy*. 8E:33-37.

Piatti, G. and Pellegrini, G. (1976) "The Structure of the Unidirectionally Solidified Al-Al2 Au Eutectic," *J. Mat. Sci,* 11:913-924.

S. Sen, D.M. Stefanescu, B.K.Dhindaw, W.F. Kaukler and P.A. Curreri (1997) AIAA paper 97-0451, 35th *Aerospace Sciences Meeting Proceedings*, Reno NV, Jan. 6-10, 1997.

HIGH TEMPERATURE ELECTROSTATIC SAMPLE LEVITATOR AS A FUTURE CONTAINERLESS MATERIALS PROCESSING FACILITY IN SPACE

Won-Kyu Rhim
Jet Propulsion Laboratory
California Institute of Technology
4800 Oak Grove Drive
Pasadena, CA 91109

Extended Abstract

Investigation of deeply undercooled melts will open up the possibilities of studying basic phenomena of thermodynamics, nucleation and solidification processes. Such research work is particularly important to understand the processes of metastable solids which are formed from the non-equilibrium state of undercooled melt. There are a wide variety of metastable states, ranging from metastable crystalline phases (supersaturated and grain-refined alloys) to amorphous metals. A detailed understanding of the thermodynamics, the nucleation and crystal growth conditions can lead to comprehensive understanding of the criteria for the formation of such metastable states. However, at the present time, only limited information is available about the thermophysical parameters as a function of undercooling. The demands for accurate thermophysical property values have also been strong in the electronics industry which constantly demands high quality semiconductor materials for high density integrated circuit devices. In order to simulate the crystal growth for optimization of the growth process, the accurate thermophysical properties of molten semiconductors are essential input parameters.

Thermophysical properties of high temperature molten materials are difficult to determine accurately because of the experimental problems associated with taking measurements at high temperatures in the presence of gravity. In the presence of gravity, convective flows are generated if there exist density gradients in a melt. In the high temperature materials processing, for reasons of maintaining purity of sample materials and attaining deeply undercooled states of melts, the sample has to be isolated from the container walls using some kind of levitators. However, the levitation of a high density melt against the gravity requires strong levitation forces which in turn induce undesirable flows in the melt. Such flows in melts would make measurements of certain thermophysical properties either impossible or at best erroneous. In microgravity environment of space, these disturbing effects are greatly reduced, therefore, more accurate measurement results are expected.

Different kinds of sample positioning (levitating) devices have been investigated in the past, some of which have even gained a few space experiences. However, the only serious sample positioner which was designed for high temperature materials processing experiments in space is the TEMPUS, the German electromagnetic sample positioner. In this presentation I would like to introduce a new sample positioning device and discuss about its capabilities of measuring various thermophysical properties and studying non-equilibrium solidification process. At Jet Propulsion Laboratory, we have developed a high temperature electrostatic levitation system for the ground based applications (Rhim et al. 1993, and Rhim 1997). The system is operated in a high vacuum level ($\sim 10^{-8}$ torr) and can heat up a sample of 2000 K. A typical sample diameter is 2~3 mm. Advantages of the electrostatic levitation technique over other levitation techniques, especially the electromagnetic levitation is decoupling of levitation and heating elements and a wide selection of samples to be levitated. The sample can be levitated at any temperature between room and maximum temperatures. Both conductive and non-conductive (including semi-conductive) materials can be levitated. For any thermophysical property measurements, diagnostic devices must be incorporated with the levitation technique. The devices must be based on non-invasive techniques and at the same time compatible with the levitation mechanism. We have developed the diagnostic techniques such as a high speed pyrometer, static as well as oscillating sample imaging and analysis. These techniques allow us to measure the thermophysical properties which include the true temperature, the emissivities, the density (specific volume) (Chung et al. 1996, and Ohsaka et al. 1997), total hemispherical emissivity (Rulison et al. 1995, and Rhim et al. 1997), specific heat (Rulison et al. 1995, and Rhim et al. 1997), surface tension, and viscosity (Rhim et al., submitted for publication, and Ohsaka et al., submitted for publication). We also envision adding the capabilities which will allow us to measure the thermal and electrical conductivities and to determine the liquid structures in the near future.

This work represents one phase of research carried out at the Jet Propulsion Laboratory, California Institute of Technology, under contract with the National Aeronautics and Space Administration.

References

Chung, S. K., D. Thiessen and W. K. Rhim (1996), "A Non-Contact Measurement Technique for the Density and Thermal Expansion of Molten Materials," Rev. of Sci. Instrum. 67(9): 3175-3181.

Ohsaka, K., S. K. Chung and W. K. Rhim, "The Specific Volumes and Viscosities of the Ni-Zr Liquid Allows and Their Correlation with the Glass Formability of the Allows," Applied Phys. (submitted).

Ohsaka, K., S. K. Chung, W. K. Rhim, A. Peker, D. Scruggs and W. L. Johnson (1997), "Specific Volumes of the $Zr_{41.2}Ti_{13.8}Cu_{12.5}Ni_{10.0}Be_{22.5}$ Allow in the Liquid, Glass and Crystalline Phases," Appl. Phys. Lett. 70(6): 726.

Rhim, W. K. (1997), "Present Status of High Temperature Electrostatic Levitator Technology," Microgravity Quarterly 6: 79-83.

Rhim, W. K., K. Ohsaka and R. E. Spjut, "A Transient Technique for Simultaneous Measurements of Surface Tension and Viscosity of Undercooled Melts," Rev. Sci. Instru. (submitted).

Rhim, W. K., S. K. Chung, A. J. Rulison and R. E. Spjut (1997), "Measurements of Thermophysical Properties of Molten Silicon by a High Temperature Electrostatic Levitator," Int. J. Thermophysics 18(2): 459-469.

Rhim, W. K., S. K. Chung, D. Barber, K. F. Man, Gary Gutt, A. Rulison and R. E. Spjut (1993), "An Electrostatic Levitator for High Temperature Containerless materials Processing in 1-g," Rev. Sci. Instrum. 63: 2961.

Rulison, A. and W. K. Rhim (1995), "Constant Pressure Specific Heat to Hemispherical Total Emissivity Ratio for Undercooled Liquid Nickel, Zirconium and Silicon," Metall. & Mat. Trans. 26B: 503-508.

THE MICROGRAVITY APPLICATIONS PROMOTION PROGRAMME OF THE EUROPEAN SPACE AGENCY, ESA

H. U. Walter, R. Binot, E. Kufner, and O. Minster
ESA/ESTEC, Postbus 299 - 2200 AG Noordwijk
The Netherlands +31 (71) 565 5311, Fax. +31 (71) 565 3661

Abstract

The overall strategic objective of ESA'S Microgravity Applications Promotion Programme is to generate a European activity using the International Space Station as a facility for industrial R & D. "Applications" of microgravity may be understood as the exploitations of the ISS for applied research and as a testbed for the development of technology and processes useful on Earth and for long-duration space flight. The objective is therefore to develop projects in order to : optimise applied ground-based physical and biological processes; generate data and materials samples relevant for industrial R & D; and investigate space-related physiological changes relevant for long-duration space flight, which may also be relevant for clinical applications on Earth. The Microgravity Applications Promotion Programme has the objective to foster and to develop a first generation of projects with involvement from industry, which should demonstrate that microgravity is indeed a useful tool for industrial R & D. These projects will be prepared by precursor experiments using traditional carriers and opportunities in the early utilisation of the ISS.

PROGRAMME STRATEGY AND OBJECTIVES

The promotion of the Applications of Microgravity has the objective of preparing the utilisation of the ISS for application-oriented research. The long-term objective is to develop projects in order to:

- optimise ground-based physical processes
- generate data and materials samples relevant for industrial R & D
- investigate space-related physiological changes occurring for long-duration space flight, which may also be relevant for clinical applications on Earth

In <u>Microgravity Physical Sciences and Applications</u>, the aim is to exploit the space environment in terms of the learning potential and the generation of data useful for ground-based process optimisation and the understanding of phenomena critical for ground-based production. This includes the precision measurement of certain thermophysical properties needed for the modelling of earth-based industrial processes. It may also be feasible to produce benchmark samples to probe the performance and the potential of an advanced material, and furthermore to produce new metastable alloys and glasses by containerless processing.

In <u>Medicine</u>, the amplification of symptoms and the acceleration due to the space environment of certain physiological changes motivate the investigation of the underlying mechanisms, and one may exploit this condition to develop and to test treatments for such ailments in an accelerated fashion. The latter will be necessary to protect astronauts during extended missions. Furthermore, such countermeasures could also be applicable to the treatment of diseases on Earth, for example those affecting the cardiovascular system and the regulation of body fluids, the muscular atrophy and bone demineralisation, the immune system and the physiological response to radiation exposure. Portable devices using non-invasive techniques with remote operation developed to monitor various physiological functions of astronauts could be of interest to industry for terrestrial applications (Telemedicine). The medical instruments industry interested in developing and marketing such devices should be involved in the development of the flight instruments whenever possible, which could in turn be a stepping stone for additional Earth applications.

In <u>Cell Biology</u>, the related mechanisms can be studied at cellular level, complementary to the holistic approach with humans, animals and plants. Examples are cell structure, cell motion, cell growth, differentiation, proliferation and signal transduction.

<u>Biotechnology</u> deals with the practical application of biological organisms to productive activities and to environmental management, which includes agriculture, biomedicine and the management of the environment, this last being of particular interest for longer duration space flight and more generally for humans living in a confined habitat. Practical applications of the microgravity condition thus far were the electrofusion of cells and electrophoretic separation.

The Microgravity Applications Promotion Programme has the objective of fostering and developing a first generation of application-oriented projects, as far as possible with active involvement from industry. These projects should demonstrate that microgravity is indeed a useful tool for application-oriented research and ultimately for industrial R & D. These projects will be prepared by precursor experiments using traditional carriers and opportunities in the early utilisation of the ISS.

The overall approach to develop applications of microgravity can be summarized as follows:

- Identifying ongoing R & D activities where microgravity could be useful as a tool.
- Identifying and defining application-oriented projects based on an assessment of ongoing activities and results obtained worldwide.
- Fostering the dialogue between the present traditionally scientific users and potential new users from non-aerospace industry.
- Soliciting proposals (Announcements of Opportunity/peer and expert evaluation).
- Exploiting precursor flight opportunities such as the traditional Drop Tower/Drop Tube, Aircraft, Sounding Rockets, FOTON, Spacehab, and early ISS opportunities, whereby maximum use will be made of already existing and newly developed flight experiment hardware in the EMIR and MFC programmes.
- Coordinating with European national space agencies the development of flight experiment hardware and ground-based research financing.
- Fostering international cooperation with the U.S.A. and with Japan, Russia and Canada.

The guiding principle for setting up such a programme should be to foster an organic evolution by building as much as possible on the expertise and know-how in Europe and by expanding ongoing basic research projects towards applications whenever possible as a first step. Sufficiently large European project teams should be created, since they are essential to the efficient development, operation and exploitation of facilities and to carrying out long term and coherent research and development programmes.

SPECIFIC PROJECTS

Several projects have been initiated. They are described briefly in the following.

Osteoporosis

Osteoporosis represents a very major health problem in modern societies with an increasing percentage of elderly. Osteoporosis does not only affect postmenopause women, some 15 % of those affected are male.

It has been observed since the early days of manned spaceflight that there is substantial bone demineralization and bone loss with perfectly healthy astronauts even after relatively short space flight. This observation has been confirmed and various countermeasures such as exercise and also pharmaceutical treatment are being employed.

While the underlying biochemical mechanisms for this accelerated process are yet to be elucidated space flight conditions may be employed as an accelerated testbed to investigate the process itself and to evaluate the efficiency of various countermeasures and treatments. One major element in the present initiative is the development of an instrument for the quantitative, three-dimensional and high resolution characterization of bone density and bone architecture.

Since in vivo observation of the bone microstructure with high resolution is today not clinically feasible, this newly developed instrument will be a major contribution towards the better understanding of bone evolution and a more accurate prediction of risks. There is therefore the clinical application of this new instrument on Earth. Secondly, this investment will be used to investigate quantitatively the efficiency of various treatments such as exercise, mechanical constraints, diet, drugs during space missions.

The lead centre for this activity is MEDES, Toulouse, France.

Crystal Growth of Cadmium Telluride (CdTe) and related Compounds

CdTe X-ray and γ-ray detectors have high potential in real time dental imaging or mammography including 3-D tomography. This promises quicker and more reliable medical diagnostics, while exposing the patient to lower radiation doses.
CdTe infrared detectors enable high resolution thermal imaging and high data rate optical telecommunication. In addition, the photorefractive properties can be employed for high performance devices in optical ultrasonics nondestructive testing.

Today, the commercialisation of advanced CdTe detectors is impeded by the difficulty to grow large CdTe single crystals with the required quality. The melt growth method usually employed for the production of CdTe crystals is the vertical Bridgman technique. Due to the contact with the crucible impurities are incorporated which are inhomogeneously distributed in the crystal. Crucible contact contributes also to extensive twinning. In addition, the crucible-induced mechanical stress in a material with poor mechanical properties leads to residual stress and very high dislocation densities.

One technique to be used in microgravity is to provide for the progressive detachment of the melt from the crucible during directional solidification.

When the material solidifies from a detached melt its impurity content is lower and no mechanical stress is induced by the differential contraction of the crucible and the crystal during cooling. As a result, the density of twins and dislocations in the crystal is decreased by several orders of magnitude. Laminar convective flow can be imposed in the melt by applying a rotating magnetic field in order to minimize concentration fluctuations at the solid-liquid interface and, thereby, enhance the homogeneity of the crystal.

The primary objective of the current project is to understand and to control the detachment process. This can only be investigated quantitatively in microgravity, where hydrostatic pressure is virtually eliminated and free surfaces are exclusively shaped by capillary forces. Several experimental demonstration of this new approach have already been made.

To develop this new process accurate measurements of a number of physico-chemical parameters such as surface and interface tensions, vapour composition and pressure and their dependency on temperature are necessary first. Second, theoretical models have to be developed as a basic for the numerical modelling of the phenomena involved. Finally optimisation and validation of the process under microgravity conditions will be the most important element of the project.

Relevant tests are already planned in the Advanced Gradient Heating Facility (AGHF) in October 1998.

Another technique is the growth from the vapour phase. This process takes place at significantly lower temperatures than growth from the melt, thus reducing thermal stress and dislocations. In addition -with semi-closed configurations- contact to the walls of the growth ampoule can be avoided. This has the same benefits to the crystal quality as explained above. There is compelling evidence that gravity-driven convective flows in the vapour phase have a detrimental effect on the compositional homogeneity of the crystals, particularly with large dimensions.

A thorough understanding of the convective flows and of their coupling with the growth process would be a major contribution in the optimisation of the production of CdTe crystals from the vapour phase on Earth.

The lead Centre for this project is the Institute for Crystallography, Freiburg, Germany.

Precision Measurement of Diffusion Coefficients Related to Oil Recovery

Modern techniques allow the exploration of oil reservoirs in depths up to 7000 m. Since the exploitation costs of resources increase with depth, Oil Companies are interested in methods which allow an accurate prediction of the capacity of a given reservoir.

The projects Diffusion Coefficients in Crude Oil (DCCO) and Soret Coefficients in Crude Oil (SCCO) are aiming at the development of a numerical code, which shall provide accurate Phase Diagrams (see Figure) and which shall provide the basis for a more reliable prediction of the vertical and lateral extent of an oil reservoir and thus its volume. The code will be based on laws describing diffusion processes, determining the concentration of constituents in hydrocarbon mixtures. These diffusion processes are driven by concentration differences (mass diffusion) and by temperature gradients (Soret diffusion).

Consequently the numerical model to be developed requires accurate coefficients of the mass diffusion and the Soret diffusion, unaffected by convection and buoyancy. Therefore, microgravity provides the ideal environment for the precise measurement of these coefficients.

In the Project DCCO the mass diffusion coefficients of realistic hydrocarbon mixtures will be measured with high precision in a Getaway Special experiment (see Figure), which is scheduled to fly on the Space Shuttle in March 1998 (STS-91).

The precise measurement of Soret diffusion coefficients will be performed in the Getaway Special experiment SCCO, planned for autumn 1999.

The lead Centre for this project is the Microgravity Research Centre, Université Libre de Bruxelles, Belgium

Ultraprecise Measurement of Time

This project has been proposed for flight on External Payload Adapters of the ISS. It is a proposal by an international team of scientists from Europe, the USA and Australia. Presently the feasibility of this project and the financing aspects are being evaluated. A decision is expected in early 1998.

ACES is a proposal to fly on the ISS a newly developed cesium atomic clock that is exploiting the microgravity condition to measure time with unprecedented precision. By frequency comparison with other onboard and ground-based ultrastable clocks such as H-masers, trapped ion clocks or cryogenic oscillators its superior performance as theoretically predicted will be demonstrated.

This will first allow to study the physics of cold atom clocks in a new range of frequency stability (10^{-16}-10^{-17}) and accuracy. Second, ACES will provide a high quality universal time reference for users all around the Earth with unprecedented precision. It opens opportunities in various application fields such as Earth observation and geodesy, navigation, advanced telecommunication systems and time and frequency transfer. Third, ACES will perform relativistic measurements and tests compatible with the circularity of the ISS orbit and its dynamical position accuracy. Finally ACES will validate this ensemble of clocks for future space missions in fundamental physics.

OTHER INITIATIVES TO SOLICIT NEW PROJECTS

The Microgravity Application Programme will try to capitalize on the very substantial investments made in Europe during the past 15 years in basic microgravity research. Building and expanding from this solid scientific basis and expertise appears to be the most promising perspective for the development of application - oriented microgravity research projects. With this philosophy in mind seven Topical Teams have been created about one year ago. They address the following topics :

- The Influence of steady and alternating Magnetic Fields on Crystal Growth and Alloy Solidification.

- Convection and Pattern Formation in Morphological Instability during Directional Solidification.

- Metastable States and Phases.

- Equilibrium and Dynamic Properties of Adsorbed Layers.

- Dust Aggregation and Related Subjects.

- Double Diffusive Instabilities with Soret Effect.

- Thermophysical Properties of Fluids.

The assignment of these teams is to form European Networks and to include researchers from industry in their discussion on procuring applications projects in their particular field. Once projects have been identified, proposals will be submitted to ESA. Selection of projects will be done on the basis of the advise by external peers and experts.

STABILITY OF A CAPILLARY SURFACE IN A RECTANGULAR CONTAINER

Mark M. Weislogel
NASA Lewis Research Center
M.S. 500/102
Cleveland, OH 44135
(216) 433-2877

K.C. Hsieh, NYMA
NASA Lewis Research Center
M.S. 500/102
Cleveland, OH 44135
(216) 433-8106

Abstract

The linearized governing equations for an ideal fluid are solved numerically for the stability of free capillary surfaces in rectangular containers against unfavorable disturbances (accelerations, *i.e.* Rayleigh-Taylor instability). The preliminary results are expressed graphically in terms of a critical Bond number as a function of system contact angle. A critical wetting phenomena in the corners is shown to significantly alter the region of stability for such containers when contrast to simpler geometries such as the circular cylinder or the infinite rectangular slot. Such computational results provide additional constraints for the design of fluids systems for space-based applications.

INTRODUCTION

Particularly since the inception of space flight a number of studies have been conducted to identify the stability limits of capillary surfaces to unfavorable disturbances (accelerations). The motivation for such investigations is generally to obtain design characteristics and performance limitations for in-space fluids management systems. For example, it is essential to understand the potentially destabilizing effect of a thruster firing on the liquid fuel in a partially-filled tank. In this paper a brief review of interfacial stability of the Rayleigh-Taylor-type will be provided which focuses on the restricted set of container geometries for which solutions are offered in the literature. In light of the growing need for design specific solutions for an ever increasing number of fluids systems applications, *i.e.* fluids experiments in space (Singh 1996), a new problem is outlined and solved for the stability of capillary surfaces in containers of rectangular cross-section. The results of this investigation, presented in terms of a critical Bond number as a function of container aspect ratio and contact angle, may be readily added to the repertoire of the space systems designer.

Review

Surface tension forces dominate fluid interface behavior for low Bond number systems, $B \ll 1$, where

$$B \equiv \frac{\rho g R^2}{\sigma},$$

where ρ is the density difference across the interface, σ is the surface tension, R is a characteristic dimension of the container, and g is the acceleration field strength taken positive in the direction of the density gradient. As B approaches $O(1)$, however, a critical balance is reached and, depending on the orientation of the acceleration field, further increases in g can cause destabilization of the interface and reorientation of the fluid to a perhaps undesirable location within the container. The precise value of B at which such a "reorientation" might occur ($\equiv B_{cr}$) is an important design parameter for any capillarity-controlled fluid system and is particularly significant for fluids management processes in space.

TABLE 1. Correlation Constants for Eq. 3.

R_i/R_o	C_1	C_2
0	0.81	2.59
0.1	1.30	1.99
0.25	1.83	0.83
0.5	0.22	1.39
0.75	0.28	1.01
1.0	0.74	0.41

Critical Bond number analyses for confined geometries were performed indirectly as early as Duprez (1854) and Maxwell (1890) for the stability of pinned interfaces in circular and rectangular containers. These investigations are restricted to predominately flat surfaces originating out of an assumption of either a 90° contact angle condition or a pinned contact line. Treating the contact angle as a variable and thus allowing significant curvature of the interface, as is most common in capillary systems, solutions were obtained by Concus (1968) for the circular cylinder, Concus (1963) for the infinite slot, and Seebold et al. (1967) for the annulus. The experimental works of Masica et al. (1964) and Masica and Petrash (1965) concerning the cylinder and Labus (1969) concerning the annulus are also noteworthy. Solutions for spherical containers are presented by Reynolds and Satterlee (1966) and a number of solutions are reported for semi-bounded surfaces such as wall bounded drops and bubbles by Reynolds and Satterlee (1966), pedant drops by Wente (1980), and liquid bridges by Coriell et al. (1977). Unbounded liquid layers are treated by Yiantsios and Higgins (1989) with pertinent references contained therein.

Correlations for B_{cr} as a function of the contact angle may be derived from the numerical results of Concus (1968 and 1963) and Seebold et al. (1967), respectively. For the infinite rectangular slot one finds

$$B_{cr} = 0.71 + 1.74 \sin\theta, \tag{1}$$

where B_{cr} is defined using the slot half-width, and for the cylinder

$$B_{cr} = 0.81 + 2.59 \sin\theta, \tag{2}$$

where B_{cr} is defined using the cylinder radius. Larger values $B > B_{cr}$ lead to instability and breakup of the interface. Similar results for the annulus may be obtained, however, these depend heavily of the radius ratio R_i/R_o, where R_i and R_o are the inner and outer radii, respectively. In Fig. 1, correlations for B_{cr} for free annular surfaces are plotted versus contact angle for a variety of radius ratios. The curves are determined using the form

$$B_{cr} = C_1 + C_2 \sin\theta, \tag{3}$$

and the correlation constants C_1 and C_2 as listed in Table 1. Note that B_{cr} is again defined on the outer radius, R_o, and that $R_i/R_o = 0$ recovers the correct form of eq. 3 for the circular cylinder, eq. 2. Observation of eqs. 1–3 and Fig. 1 reveals that B_{cr} is nonzero and positive for all values of the contact angle θ. In addition, the solutions are found to be symmetric about $\theta = 90°$.

The commonality between circular cylinders, slots, and annular containers are the smooth continuous boundaries within which the fluids are confined. For the case of a container with an interior corner, the situation is altered significantly. As mathematically demonstrated by Concus and Finn (1969), when $\theta < 90° - \alpha$ (or $\theta > 180° - \alpha$), hereafter referred to as the Concus-Finn condition, a critical wetting condition is established resulting in complete wetting of the corner by the fluid. α is the corner half-angle. Such surfaces are unconditionally unstable to all adverse accelerations, thus, for fluid-container systems satisfying the Concus-Finn condition, $B_{cr} = 0$. Therefore, for problems such as the stability of a capillary surface in a rectangular container where $\alpha = 45°$, nonzero values for B_{cr} may only be obtained for contact angles in the range $45° < \theta < 135°$. A sketch of the different interfacial regimes is provided in Fig. 2 for a container of square cross-section. The condition of Fig. 2b is investigated here since the cases of Fig. 2a and 2c are unconditionally unstable for $B > 0$.

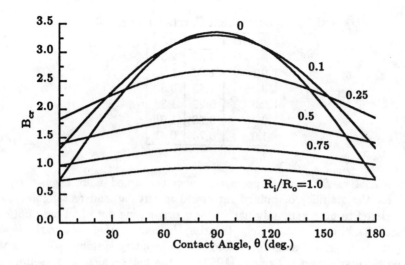

FIGURE 1. Correlations of B_{cr} with θ for Annular Interfaces for a Variety of Radius Ratios, R_i/R_o (Numerical Data of Seebold *et al.* (1967)).

ANALYTICAL SOLUTION

Maxwell (1890) predicted the stability of an inverted capillary surface in a rectangular container of half-length a and half-width b. Fig. 3 depicts the geometry under discussion, where the more dense fluid is below the interface and **g** acts positive in the positive z-direction. Maxwell's solution is derived by minimizing the surface-plus-gravitational energy and assumes a pinned, predominately flat interface ($\theta \approx 90°$). His result may be cast in terms of B_{cr} such that

$$B_{cr} \equiv \frac{\rho g b^2}{\sigma} = \frac{\pi^2}{4}\left(1 + 4\left(\frac{b}{a}\right)^2\right), \tag{4}$$

where $0 \leq b/a \leq 1$. This solution approach may be extended to the case of perfect slip at the contact line where θ is fixed at 90°. The result is

$$B_{cr} = \frac{\pi^2}{4}\left(1 + \left(\frac{b}{a}\right)^2\right). \tag{5}$$

As is commonly observed in practice, comparison of eqs. 4 and 5 shows that stability is significantly enhanced by the pinned condition. Note also from eq. 5 that for $a/b \to 0$, $B_{cr} \to \pi^2/4$ which is equivalent to Concus' (1963) solution for the infinite slot for $\theta = 90°$ (see eq. 1), as well as to the solution for unbounded liquid layers where the disturbance wavelength $\lambda = 4b$, Yiantsios and Higgins (1989). No further analytical solutions are possible which allow appreciable variation in θ.

NUMERICAL SOLUTION

The numerical solution to the idealized equations of fluid motion are overviewed below in a like manner to Concus (1963), the dimensions of the problem being extended to analyze the surface $h = h(x, y, t)$. Conservation of mass leads to the 3-dimensional Laplace equation for the velocity potential, ϕ. The kinematic condition is applied at the free surface and the pressure jump condition across the interface due to capillary forces is then incorporated into Bernoulli's law for a transient, inviscid, incompressible, and irrotational fluid. The resulting second order nonlinear partial differential equation is subject to the contact angle condition along the container walls. At this point the equations are linearized and normal modes are introduced for the velocity potential and for a small perturbation $\eta(x, y, t)$ to the leading order static interface shape $H(x, y; B)$. The numerical solution to the resulting governing equation requires solution to the eigenvalue

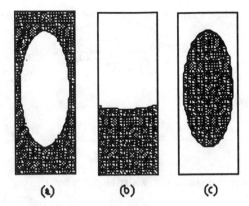

FIGURE 2. Sketch of Wetting Regimes in a Square Cross-Sectioned Container (Note: Container is Bisected Across Diagonal): a. $\theta \leq 45°$, b. $45° < \theta < 135°$, c. $\theta \geq 135°$.

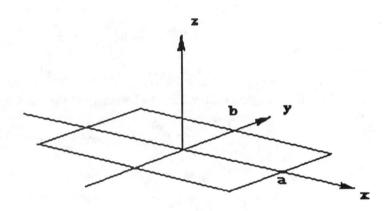

FIGURE 3. Rectangular Solution Domain, $a \times b$.

problem via the evaluation of the determinant of the solution matrix for the disturbance η. A negative (positive) determinant implies growth of the disturbance and thus instability (stability). B_{cr} is determined by the zero value.

The mathematical solution detail is provided below in dimensionless form. Lengths are scaled such that $x \sim a$, $y \sim b$, $z \sim b$ and the surface height as measured from the x-y plane is $h \sim b$. Note that $\epsilon = b/a$ and that $0 \leq \epsilon \leq 1$. Time is scaled by $(b/g)^{1/2}$, pressure by σ/b, and the velocity potential by $(gb^3)^{1/2}$. Subscript notation for differentiation is employed throughout. The equations are posed for the full domain ($|x| \leq 1$, $|y| \leq 1$).

Generalized Equations

Continuity of mass leads to
$$\nabla^2 \phi = 0, \tag{6}$$
subject to $\phi_n = 0$ on the container walls, where n is the normal to the wall. The kinematic condition on the free surface is
$$\epsilon^2 \phi_x h_x + \phi_z h_z - \phi_y + h_t = 0 \tag{7}$$

on $z = h(x,y,t)$. The pressure jump condition due to curvature of the free surface is given by

$$-P = \nabla \cdot \frac{\nabla h}{\left(1 + |\nabla h|^2\right)^{1/2}} \equiv \mathcal{L}h, \tag{8}$$

where $\mathcal{L}$ is an operator on h given by

$$\mathcal{L}h \equiv \frac{\epsilon^2 h_{xx}\left(1 + h_y^2\right) + h_{yy}\left(1 + \epsilon^2 h_x^2\right) - 2\epsilon^2 h_x h_y h_{xy}}{\left(1 + \epsilon^2 h_x^2 + h_y^2\right)^{3/2}}. \tag{9}$$

The contact angle condition along the container walls is given by

$$\mathbf{n} \cdot \mathbf{k} = \cos\theta,$$

where $\mathbf{k}$ is the inward normal to the walls and $\mathbf{n}$ is the outward normal to the free surface given by

$$\mathbf{k} = (\mp 1, 0, 0) \quad \text{along} \quad x = \pm 1,$$
$$\mathbf{k} = (0, \mp 1, 0) \quad \text{along} \quad y = \pm 1,$$

and

$$\mathbf{n} = \left(1 + \epsilon^2 h_x^2 + h_y^2\right)^{-1/2} (-\epsilon h_x, -h_y, 1),$$

respectively. Thus, the contact angle conditions at the boundary of the surface are

$$\pm \frac{\epsilon h_x}{\left(1 + \epsilon^2 h_x^2 + h_y^2\right)^{1/2}} = \cos\theta \quad \text{along} \quad x = \pm 1 \tag{10}$$

and

$$\pm \frac{h_y}{\left(1 + \epsilon^2 h_x^2 + h_y^2\right)^{1/2}} = \cos\theta \quad \text{along} \quad y = \pm 1. \tag{11}$$

Incorporating eq. 8 into Bernoulli's equation for a transient, ideal fluid yields

$$B\left(\phi_t + \frac{1}{2}\left(\epsilon^2 \phi_x^2 + \phi_y^2 + \phi_z^2\right) + h\right) + \mathcal{L}h = C, \tag{12}$$

which is applicable on $z = h(x,y,t)$. C in the above equation is a constant, and in the most general sense $C = C(t)$ and is determined by the volume of fluid present in the container, which is here assumed steady in time.

Linearized Governing Equations

Introducing the perturbation

$$h = H(x,y) + \eta(x,y,t), \tag{13}$$

normal modes are selected for ϕ and η such that

$$\phi = \phi'(x,y,z)\cos(\omega_i t), \tag{14}$$

$$\eta = \eta'(x,y)\sin(\omega_i t). \tag{15}$$

Substituting eqs. 13–15 into eq. 12, neglecting nonlinear terms, and noting $\omega_i = 0$ for neutral stability, yields the simplified Bernoulli equation

$$B(H + \eta) + \mathcal{L}(H + \eta) = C, \tag{16}$$

where the prime notation for η has been dropped for clarity. Assuming $\eta/H \ll 1$, the zeroeth order solution for the interface shape H may be determined from eq. 16 to be

$$BH + \mathcal{L}H = C, \qquad (17)$$

where $\mathcal{L}H$ is the operation of eq. 9 on H. Eq. 17 is subject to

$$\pm \frac{\epsilon H_x}{\left(1 + \epsilon^2 H_x^2 + H_y^2\right)^{1/2}} = \cos\theta \qquad \text{along} \qquad x = \pm 1$$

and

$$\pm \frac{H_x}{\left(1 + \epsilon^2 H_x^2 + H_y^2\right)^{1/2}} = \cos\theta \qquad \text{along} \qquad y = \pm 1.$$

The first order solution for the perturbation η is given by

$$B\eta + \tilde{\mathcal{L}}\eta = 0, \qquad (18)$$

subject to

$$\epsilon\eta_x \left(1 + H_y^2\right) - \epsilon\eta_y H_x H_y = 0 \qquad \text{along} \qquad x = \pm 1$$

and

$$\eta_y \left(1 + \epsilon^2 H_x^2\right) - \epsilon^2 \eta_x H_x H_y = 0 \qquad \text{along} \qquad y = \pm 1.$$

The operation $\tilde{\mathcal{L}}\eta$ is defined by

$$\tilde{\mathcal{L}}\eta = \frac{1}{\left(1 + \epsilon^2 H_x^2 + H_y^2\right)^{3/2}} \left[\epsilon^2 \eta_{xx}\left(1 + H_y^2\right) + \eta_{yy}\left(1 + \epsilon^2 H_x^2\right) - 2\epsilon^2 \eta_{xy} H_x H_y \right.$$

$$\left. + 2\epsilon^2 \left(\eta_y H_{xx} H_y + \eta_x H_x H_{yy} - \eta_y H_x H_{xy} - \eta_x H_y H_{xy}\right)\right]$$

$$- \frac{3\mathcal{L}H}{\left(1 + \epsilon^2 H_x^2 + H_y^2\right)} \left(\epsilon^2 \eta_x H_x + \eta_y H_y\right), \qquad (19)$$

and is determined by expanding $\mathcal{L}(H + \eta)$ in powers of η retaining only terms of $O(\eta)$. In the limit $\epsilon \ll 1$,

$$\tilde{\mathcal{L}}\eta = \frac{\eta_{yy}}{\left(1 + H_y^2\right)^{3/2}} - \frac{3\eta_y H_y H_{yy}}{\left(1 + H_y^2\right)^{5/2}},$$

which recovers the result of Concus (1963) for the infinite slot.

Numerical Solution Detail

In the numerical solution procedure eq. 17 is discretized based on a fourth-order central-differencing scheme. For the calculation of the static shape of the free surface $H(x, y; B)$, the Newton iteration method with successive under-relaxation is used to address the nonlinearity of the governing equation. In determining the eigenvalue, $B = B_{cr}$, the same discretization as that used for solving the static interface shape is used and can be expressed in the following form

$$[\mathbf{A} + B\mathbf{I}]\eta = 0. \qquad (20)$$

The determinant of the coefficient matrix in eq. 20 is zero at $B = B_{cr}$. Since the coefficient matrix $[\mathbf{A}]$ is dependent on the static shape and thus B, the overall solution procedure involves an iteration between eq. 20 via eq. 18 and the calculation of the static shape, eq. 17. A bisection method is used to determine B_{cr} in the iteration process. It should be noted that the determination of the critical Bond number has to be based on the solution from the full domain (*i.e.* all four solid boundaries included). This is due to the fact that an asymmetric disturbance leads to the fundamental subharmonic mode instability.

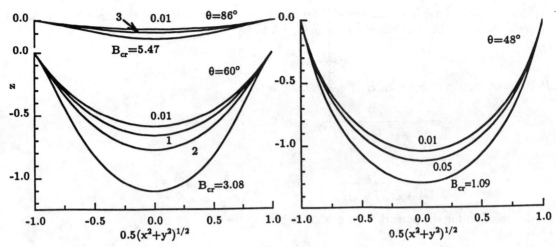

FIGURE 4. Surfaces Profiles Across Container Diagonal of a Square Container for a Variety of Contact Angles and Bond Numbers.

As $\theta \searrow 45°$, the nonlinearity of the static eq. 17 increases dramatically, requiring a smaller relaxation factor for the Newton iterations. Owing to this fact, the run time per case increases as the contact angle decreases. For a 60×60 grid system, the typical run time per case on an SGI Indigo II with a single-processor is one hour for contact angles in the vicinity of $90°$. However, the run time becomes significantly longer, reaching 24 hours when the contact angle is close to $45°$. Expectedly, solutions for B_{cr} are found to be symmetric about $\theta = 90°$.

RESULTS

In Fig. 4, surfaces profiles across the diagonal of a square cross-section container ($\epsilon = 1$) are compared for a variety of Bond numbers with contact angle fixed. It is apparent that the effect of θ on the interface is indeed significant.

The base-state interface shape $H(x, y)$ for the case $\epsilon = 1$, $\theta = 60°$ is presented 3-dimensionally in Fig. 5 for $B_{cr} = 3.08$. Slight inflections of the interface near the corners are observed which can also be discerned in Fig. 4, $\theta = 60°$, $B_{cr} = 3.08$. These decrease with decreasing θ, see Fig. 5, $\theta = 48°$.

The numerical results for B_{cr} are presented in Fig. 6 as a function of θ for the aspect ratio $\epsilon = 1$. The numerical results of Concus (1963) for the infinite slot are also provided via eq. 1. The region of stability denoted by the area below the curves is obviously altered for the rectangular section of this study when compared to the infinite slot. This is attributable to the restricted range of contact angles allowing for stable interfaces which cover the solution domain. The rapid decrease of B_{cr} towards zero for $\theta \lesssim 48°$ reveals the sensitivity of the surface to the critical contact angle condition which is satisfied for $\theta \leq 45°$. It is useful to note that for $\theta \gtrsim 48°$, B takes normally anticipated values, $O(1)$. The effect of contact angle hysteresis on such stability results is likely to delay the instability while equilibrium conditions are established at the contact line. If the disturbance has temporal periodicity, the effect of hysteresis could be to significantly increase the stability the interface, particularly for contact angles near $45°$. However, as found in recent space experiments, extended periods of thermal and mechanical disturbances in the presence of a steady background acceleration such that $B \gtrsim B_{cr}$ will ultimately bring about the predicted instability.

It is of value to note that the rectangular geometry of this investigation is fundamentally different from the infinite rectangular slot as seen by the limiting case of $\epsilon \to 0$. One might expect the solution to agree in this limit, however, the presence of the corners dramatically alters the base state surface profile for $\theta \searrow 45°$, or $\theta \nearrow 135°$. In other words, the problem differs significantly as the Concus-Finn condition is approached.

These numerical results are termed "preliminary" for several reasons. The fact that an inflection point appears in the static interface shape raises questions concerning the impact of the additional space dimension

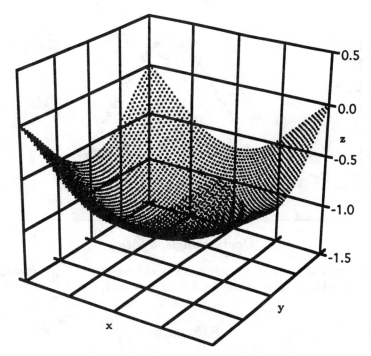

FIGURE 5. Surface Shape $H(x, y; B_{cr})$ for $\theta = 60°$ in a Square Cross-Sectioned Container, $\epsilon = 1$.

in the solution of the problem in x and y. Such inflections are not found in the cases of the slot and the cylinder which are 1-dimensional problems. The fact that in the limit $\theta \to 90°$, B_{cr} determined numerically does not agree with eq. 5 ($\triangledown$ on Fig. 6) is also disconcerting and may even suggest an error in the computations. Calculations were performed in the rectangular domain with $\theta = 90°$ on $x = \pm 1$. These conditions mimic the slot problem studied by Concus (1963). Comparisons of the computed interface shapes to the results for the slot are favorable for all $B \leq B_{cr}$, but at B_{cr} from eq. 1, the numerical results determine a dynamically *stable* interface. This suggests a deficiency in the numerical approach when calculating the determinant of the coefficient matrix given by eqs. 18–20. The discrepancy is currently under investigation. Once these concerns are resolved, calculations for B_{cr} can be completed for a range of aspect ratios, ϵ. The qualitative nature of the computational results, however, will remain unchanged as those presented in Fig. 6.

CONCLUSION

The governing equations and boundary conditions for the determination of the dynamic stability of capillary surfaces in rectangular containers are presented for numerical solution. Preliminary calculations for a square cross-sectioned container are performed. The results reveal that, though stability is enhanced for large contact angles near 90° when compared to the circular cylinder, the range of contact angle yielding positive values for the critical Bond number is significantly reduced due to a corner wetting phenomena governed by the Concus-Finn condition. B_{cr} is determined to be $O(1)$ for $48° \lesssim \theta \lesssim 132°$, but diminishes rapidly to zero as θ approaches 45° from above, or 135° from below. These conclusions are at least qualitatively correct. Efforts are continuing to resolve suspected numerical difficulties in the solution approach.

Acknowledgments

This work was supported by the Fluid Physics Branch of the Microgravity Science Division at NASA's Lewis Research Center. The authors wish to thank summer high school students C. Rogers, D. Swanson, and S. Harasim for assistance with the computations.

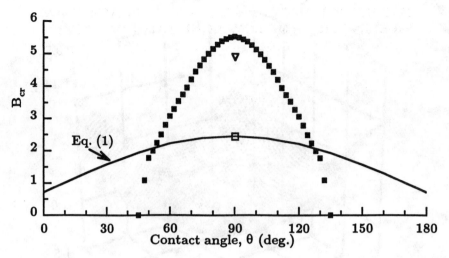

FIGURE 6. B_{cr} versus θ for $\epsilon = 1$; ■ Preliminary Numerical Results, —— Eq. 1 for the Infinite Rectangular Slot, and ▽ and □ are Eqs. 4 and 5 for the Case $\epsilon = 1$, $\theta = 90°$.

References

Concus, P. (1968) "Static Menisci in a Vertical Right Circular Cylinder," *J. Fluid Mech.*, 34:481-495.

Concus, P. (1963) "Capillary Stability in an Inverted Rectangular Tank," *Advances in the Astronautical Sciences*, Western Periodicals Co., North Hollywood, CA, 14:21-37.

Concus, P. and R. Finn (1969) "On the Behavior of a Capillary Free Surface in a Wedge," *Proc. Nat. Acad. Sci. U.S.A.*, 63: 292-299.

Coriell, S.R., S.C. Hardy, and M.R. Cordes (1977) "Stability of Liquid Zones, *J. Colloid and Int. Sci.*, No. 1, 60:126–136.

Duprez, M. (1854) "Sur un cas particularier l equilibre des liquides," par F. Duprez, *Nouveaux Mem. de l'Acad. de Belgique*, 1854.

Labus, T.L. (1969) "Natural Frequency of Liquids in Annular Cylinders under Low Gravity Conditions," NASA TN D-5412.

Masica, W.J. and D.A. Petrash (1965) "Hydrostatic Stability of the Liquid-Vapor Interface in a Gravitational Field," NASA TN D-2267.

Masica, W.J., J.D. Derdul, and D.A. Petrash (1964) "Hydrostatic Stability of the Liquid-Vapor Interface in a Low-Acceleration Field," NASA TN D-2444.

Maxwell, J.C. (1890) "Capillary Action, in the Scientific Papers of James Clerk Maxwell," Cambridge University Press, London.

Reynolds, W.C. and H.M. Satterlee (1966) "Liquid Propellant Behavior at Low and Zero g," in NASA SP-106, ed. H.N. Abramson, chpt. 11.

Seebold, J.G., M.P. Hollister, and H.M. Satterlee (1967) "Capillary Hydrostatics in Annular Tanks," *J. Spacecraft*, No. 1, 4:101-105.

Singh, B.S. (1996) "Third Microgravity Fluid Physics Conference," NASA Conference Publication 3338, Workshop Proceedings, Cleveland, Ohio, July 13–15.

Wente, H.C. (1980) "The Stability of the Axially Symmetric Pedant Drop," *Pacific J. of Math.*, 88:421–470.

Yiantsios, S.G., B.G. Higgins (1989) "Hydrodynamic Stability of an Inverted Liquid Film," *Phys. Fluids A*, 1:1484-1501.

HYDRODYNAMIC DRYOUT IN TWO-PHASE FLOWS: OBSERVATIONS OF LOW BOND NUMBER SYSTEMS

Mark M. Weislogel and John B. McQuillen
NASA Lewis Research Center
M.S. 500/102
Cleveland, OH 44135
(216) 433-2877 and 433-2876

Abstract

Dryout occurs readily in certain slug and annular two-phase flows for systems that exhibit partial wetting. The mechanism for the ultimate rupture of the film is attributed to van der Waals forces, but the pace towards rupture is quickened by the surface tension instability (Rayleigh-type) of the annular film left by the advancing slug and by the many perturbations of the free surface present in the $Re_g \sim O(10^3)$, $Re_l \sim O(10^4)$, and $Ca \sim O(10^{-1})$ flows. Results from low-gravity experiments using three different test fluids are presented and discussed. For the range of tests conducted, the effect of increasing viscosity is shown to eliminate the film rupture while the decrease of surface tension via a surfactant additive is shown to dramatically enhance it. Laboratory measurements using capillary tubes are presented which reveal the sensitivity of the dryout phenomena to particulate and surfactant contamination. From such observations, dryout due to the hydrodynamic–van der Waals instability can be expected in a certain range of flow parameters in the *absence* of heat transfer. The addition of heat transfer may only exacerbate the problem by producing thermal transport lines replete with "hot spots." A caution to this effect is issued to future space systems designers concerning the use of partially wetting working fluids.

INTRODUCTION

The rising ambitions of aerospace industries and space administrations worldwide will require parallel increases in space systems performance and reliability, Carpenter (1996). In terms of fluid systems design, a greater understanding and application of low-gravity (low-g) two-phase flows will be necessary to meet the new requirements, Salzman (1996). Several example applications which convey the potential importance of two-phase systems are efficient thermal transport/rejection systems, tankage and fluids management, and recycling of fluid wastes. Thus, one of NASA's primary science objectives is to develop the fundamentals, experience, and confidence necessary to exploit the benefits of two-phase flows for in-space applications, Singh (1996).

Of the myriad published works concerning two-phase flows, relatively few have been devoted to the effects of low-g. Of this subset, most have centered on observations of isothermal systems, the mapping of flow regimes, and the predictions of transitions between flow regimes, Bousman and Dukler (1993), Colin *et al.* (1995), and Jayawardena and Balakotaiah (1997). Unfortunately, the availability of long duration, decidedly steady, low-g, two-phase flow data is at best extremely limited if not nonexistent. As a result, very few tools are available for the designer with which to predict system performance.

In this paper, observations of low-g, slug and annular two-phase flow are discussed that highlight the role of wetting in particular fluid-conduit systems. An interfacial-van der Waals instability is shown to cause rupture of the annular film resulting in local dryout between rewetting events as the liquid slug passes over the expanding "dry spot." For the idealized problem of laminar flow, a brief review of the excellent literature relating to the subject is provided and a collection of the pertinent results of other investigators is offered that includes important design parameters such as the characteristic rupture time. This characteristic time in turn suggests a critical slugging frequency necessary to prevent the dryout. Because the dryout phenomenon is caused by wetting characteristics and encouraged by hydrodynamics and system contamination, concerns are raised for potential systems where continuous annular films are desired for optimum performance. The addition of heat transfer may only worsen the situation, enhancing the instability via local thermocapillary flows and inhibiting or even preventing rewetting by superheating the "dry spots" and thus reducing the heat exchange properties of the flow.

REVIEW

An informative introduction to the effect of partially wetting fluids on flow regimes in capillary tubes is given by Barajas and Panton (1993). In the brief review to follow, the fluid flows are treated as continuous phases without entrained bubbles or drops of the adjacent phase.

Molecular, Planar Film Rupture Instability

Thin planar liquid films are susceptible to a well known rupture mechanism caused by van der Waals attractions which are resisted only by surface tension forces. If the film is sufficiently thin and the wetting conditions of the fluid(s) and the solid are favorable (or unfavorable depending on the point of view), perturbations to the free surface of the film can grow leading to a rupture event where the liquid spontaneously recedes from the rupture location leaving a "dry spot." An early development of the linear theory describing the instability is reported by Ruckenstein and Jain (1974), and a full nonlinear analysis is reported by Williams and Davis (1982). The latter investigation reveals that rupture times decrease up to an order of magnitude compared to those determined using the linear theory as the initial amplitude of the perturbation is increased. The analysis is carried further to include the effects of evaporation, condensation, and thermocapillarity by Burelbach et al. (1988), who also provide an excellent background and introduction to the processes leading to dryout.

A brief outline of the linear dynamic stability theory applied to the planar geometry serves to introduce the key variables for later discussion. Employing the augmented Young-Laplace equation for the pressure in the liquid film, the lubrication approximation leads to an evolution equation for the thickness of the film, h. This equation may be solved to show that a small, normal modes disturbance of wavenumber k can destabilize the film if $0 < k < k_c$, where

$$k_c = \left(\frac{A}{2\pi\sigma h_o^4}\right)^{1/2} \qquad (1)$$

is the critical wave number, h_o is the unperturbed film thickness, σ is the surface tension, and A is the Hamaker constant. The coefficient A, often $\sim O(10^{-20}\text{J})$, is related to the molecularly attractive van der Waals forces between the fluid(s) and the solid. Thus, A governs the wetting properties of the system, i.e. the static contact angle, θ_s. $A < 0$ implies a wetting condition (stability) while $A > 0$ implies nonwetting (instability). Viscous forces only affect the rate of the rupture process which takes place for very thin films $h_o \sim O(100\text{-}1000\text{Å})$. For thicker films, other mechanisms are necessary in order to locally thin the film to the point where van der Waals attractions are significant enough to rupture the film.

Hydrodynamic Core-Annular Flow Instability

Annular liquid films in circular tubes have been demonstrated to be unstable to a Rayleigh breakup-type, surface tension instability. Long wavelength axial perturbations to an initially uniform annular film lead to the formation of periodically-spaced lobes which coalesce to form liquid slugs provided the film is thick enough. This phenomena is thoroughly addressed in the literature and the works of Hu and Joseph (1989) and Aul and Olbricht (1990) and references contained therein serve well as an introduction. The salient points of several studies are collected here that are applicable to annular flows where the more viscous fluid

1 is in contact with the tube of radius R. The core fluid 2 is used to define the Reynolds number Re.

1. There are two mechanisms which can lead to instability in such core-annular flows: a surface tension instability dominates at low Re while a Reynolds stress instability dominates at high Re. The former instability may result in large drops of fluid 1 being entrained in the core fluid 2, while the latter instability may result in the production of an emulsion of fluid 1 in the core fluid 2, Hu and Joseph (1989).

2. For films where $h_o/R \lesssim 0.3$, shear stresses on the interface act to stabilize the flow for moderate Re. For films where $h_o/R \gtrsim 0.3$, no stabilization of the interface is possible by interfacial shear stress and the flow is destabilized by surface tension (Rayleigh instability), Hu and Joseph (1989).

3. In the case of no core-phase flow, films where $h_o/R \gtrsim 0.09$ destabilize to form a train of liquid slugs, Gauglitz and Radke (1988).

4. The characteristic time t_σ for lobe formation marks the onset of capillary instability. It is shown by Hammond (1983) for a fluid film of viscosity μ to be

$$t_\sigma \sim \frac{\mu R^4}{\sigma h_o^3}. \tag{2}$$

5. The condition $Bo \ll 1$ given by Frenkel et al. (1987) is required to neglect the effects of gravity assumed perpendicular to the tube axis. In this case,

$$Bo = \frac{\Delta \rho g h_o R}{\sigma}, \tag{3}$$

with $\Delta \rho$ being the density difference across the interface and g the acceleration field strength, i.e. gravity.

Film Rupture in Annular and Slug-Annular Flows

The experiments of Aul and Olbricht (1990) employed a wetting fluid ($A < 0$) such that the thin films produced by the annular film instability did not further destabilize leading to rupture by van der Waals attractions. However, the combined effects of surface tension, the annular geometry, and the condition $A > 0$ do lead to rupture. For the critical disturbance wavenumber scaling $k \sim R^{-1}$, characteristic of the annular film instability due to surface tension, the characteristic rupture time can be shown to be approximately

$$t_{\sigma A} \sim \frac{\mu R^4}{\sigma h_o^3} \left(1 + \frac{AR^2}{\sigma h_o^4}\right)^{-1} \tag{4}$$

In the limit $AR^2/\sigma h_o^4 \gg 1$, the scaling for the disturbance wavenumber is given by eq. 1 for the locally planar interface, and this scaling applied to eq. 4 leads to

$$t_A \sim \frac{\mu \sigma h_o^5}{A^2}. \tag{5}$$

As observed in eqs. 1–5, the film thickness h_o is central to predictions of the flow. Fortunately, for slug flows, the mean liquid slug velocity U may be used to determine h_o. In an experimental work, Chen (1986) qualifies a number of applicable relationships for the film thickness left behind by an advancing liquid slug for a variety of capillary numbers $Ca \equiv \mu U/\sigma$ in the range 10^{-7} to 10^{-2}. Two relationships are selected for discussion here:

$$\text{I.} \quad h_o = 0.5 R Ca^{1/2} \quad \text{for} \quad 10^{-3} \lesssim Ca \lesssim 10^{-1}, \tag{6}$$

$$\text{II.} \quad h_o = 1.3 R Ca^{2/3} \quad \text{for} \quad 10^{-6} \lesssim Ca \lesssim 10^{-3}. \tag{7}$$

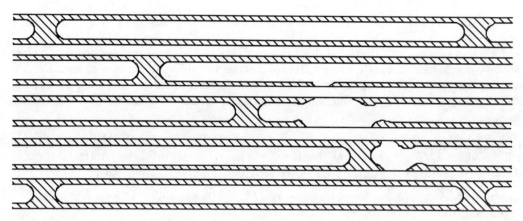

FIGURE 1. Sketch of Rupture Sequence in Low-g Slug Flow. Flow is Left to Right.

TABLE 1. Thermophysical Properties of Test Fluids.

Fluid	ρ (kg/m^3)	μ (kg/m·s)	σ (N/m)	θ_s (deg.)
Water	1000	0.00085	0.07	80
Aq. Glycerol	1127	0.006	0.067	68
Water/Zonyl	1000	0.00085	0.020	67
PDMS 1cs	816	0.000816	0.0174	44

Depending on the value of Ca for the slug flow, the lobe formation time of eq. 2 gives

$$t_{\sigma_I} \sim 8 \left(\frac{\sigma R^2}{\mu U^3} \right)^{1/2} \quad \text{and} \quad t_{\sigma_{II}} \sim 0.45 \frac{\sigma R}{\mu U^2}, \qquad (8)$$

where it is seen in both Ca regimes that $t_\sigma \propto R$, but the dependence on μ/σ has been inverted when compared to eq. 2. Using eq. 7 for h_o, the rupture time scale of eq. 5 shows $t_A \propto \mu^{13/3}$. In addition, it may be possible that eqs. 6 and 7 could be used to approximate h_o for large amplitude roll waves in strictly annular flow. In this case, U would be taken as the wave velocity.

If the time between sequential liquid slugs t_b is greater than t_σ, and if $t_\sigma \gg t_A$ for $h_o \sim 100$–1000Å, then the annular film might be expected to rupture between rewetting events. It should be noted that $t_b = L_b/U$ where L_b is the length of the "bubble" between slugs. Knowledge of the mean liquid slug length L_s may be used directly to compute a critical slugging frequency $\omega_{cr} = U/(L_s + Ut_\sigma)$ below which film rupture may be expected. In the limit $L_b \gg L_s$, $\omega_{cr} \simeq t_\sigma^{-1}$.

LOW-GRAVITY OBSERVATIONS

Isothermal, two-phase flow experiments are conducted on NASA Lewis's Learjet aircraft flying parabolic trajectories. Low-g levels of approximately $\pm 10^{-2} g_o$ are established for up to 20-25s. Details of the experimental apparatus are reported by McQuillen and Neumann (1995); the length to diameter ratio for the test section is 80. For a 12.7 and 25.4mm diameter acrylic tubes, a large number of tests are conducted for a variety of air-liquid Reynolds number ratios $0.1 \lesssim Re_g/Re_l \lesssim 10$ using water, a 50% by mass aqueous glycerol mixture, and water with a surfactant additive–Zonyl at 0.5% by mass. Using water as a control, the aqueous glycerol solution is chosen to isolate the effect of increasing viscosity and the surfactant solution is chosen to isolate the effects of decreasing surface tension. The Reynolds number ratio regime is selected to access the slug-annular flow configuration, Jayawardena and Balakotaiah (1997). The relevant thermophysical properties of the test fluids are presented in Table 1 including the static contact angle θ_s on the acrylic. Hysteresis in θ_s for each fluid is significant and on the order of $\pm 20°$.

A selection of test results for conditions resulting in rupture of the annular films between the slugs (or large roll waves) are listed in Table 2, denoted by *, along with comparative tests in which rupture did not occur. The subscript s on velocities tabulated denotes a superficial value. The large roll wave velocity is given for U_{slug} for the annular flow pattern. A sketch of a typical rupture event for an idealized flow is provided in Fig. 1. In some cases, dry sections several diameters in length result from the rupture process as the liquid film recedes from the initial rupture location. Standard deviations in the measured values of film thickness h_o/R are large and range between 33% and 95%, as tabulated by Bousman (1995). The measured values for h_o/R are compared to predictions using eq. 6 and are listed in the table. The comparisons are poor for pure water and for the aqueous glycerol solution. The former discrepancy may be due to the usage of the roll wave velocity in the evaluation of Ca, while the latter is attributable to Ca being beyond the range of applicability of eq. 6 since for the aqueous glycerol, $Ca \gtrsim 0.1$. The films are thick enough to form new liquid slugs according the static results of Gauglitz and Radke (1988), and on occasion, temporary slug formation due to film rupture and recoil was observed for water in the $R = 12.6$mm tube. Test data was also taken for the aqueous glycerol and water/Zonyl solutions in the $R = 12.6$mm tube. These will be reported in detail at a later date.

Discussion: Slug and Slug-Annular Flows

From one point of view, the slugs lay down the annular films as they proceed down the tube. This will always be the result for perfectly wetting liquids. However, for partially wetting systems, a critical velocity must be exceeded before such films can be drawn. This threshold is often termed the film pulling limit and may be estimated empirically using $Ca \simeq 0.004\theta_s^{2.83}$, combining the works of Weislogel (1996) and Li and Slatterly (1991), where θ_s here is in radians. Slug flow with Ca less than this limit should not be expected to possess connecting annular films between the slugs, and the flow might best be described as a train of gas and liquid slugs. For low liquid flow rates, partially wetting systems can exhibit flows that are not at all rotationally symmetric, and the rivulet flow regime reported by Barajas and Panton (1993) aptly describes such cases.

The annular film deposited by a liquid slug rapidly becomes stationary within $t_\mu \simeq h_o^2/4\nu$ before being overrun by the following slug. This is discerned by observing the rapid deceleration of small bubbles of diameter smaller than h_o which act as tracer particles within the films. (For the water and water–zonyl fluids, the viscous time scale is $0.06 \lesssim t_\mu \lesssim 0.27$s for the tests in Table 2, $R = 6.3$mm.) This feature of the flow implies that surface shear is small and that the surface tension or Reynolds stress instabilities for the film proceed from a predominantly quiescent condition. However, the film surfaces are by no means unperturbed. In fact, though there is little streamwise flow in the films, deflections δ in the approximately static free surface are observed on the order of $\delta/h_o \sim O(1)$. The source of these essentially normal disturbances to the film are in part due to energy production in the transitional core (Reynolds stress) flow where $1050 \lesssim Re_g \lesssim 3200$. For the liquid slug, $9500 \lesssim Re_l \lesssim 60000$, with Re based on tube diameter and liquid slug velocity. The large disturbances also signal nonlinear effects which accelerate both the surface tension instability, as shown by Hammond (1983), as well as the film rupture instability, as shown by Williams and Davis (1982).

Because the liquid slug lengths are often much smaller than the annular film lengths, the residence time t_b of the annular film can be approximated as ω_{slug}^{-1}. Thus, from Table 2 for the Water/Zonyl fluid, $0.17 \lesssim t_b \lesssim 0.7$s. Perhaps coincidentally, these times are certainly the correct order of magnitude for t_σ, which by eq. 2 gives $0.08 \lesssim t_\sigma \lesssim 0.75$s for the same tests!

Film rupture is not observed for the water tests in the $R = 6.3$mm tube ($Re \sim 1600$) but is readily present in the $R = 12.6$mm tube ($Re \sim 3200$). The latter is characterized by the larger Reynolds number of the core flow triggering the Reynolds stress instability. The aqueous glycerol films are not observed to rupture for either size tube. The fact that $t_A \propto \mu^{13/3}$ may explain this finding. Provided A is similar for water and the aqueous glycerol, the increase of μ by a factor of 7 for the latter results in an increase in t_A by a factor of 4600!

The water/Zonyl films rupture readily in every test conducted. The effect of decreased surface tension for this fluid increases t_σ, but decreases t_A, and lowers the Reynolds number necessary to excite the Reynolds

TABLE 2. Sample Low-g Slug, Annular, and Transition Slug-to-Annular (TAS) Test Data.

Fluid	R (mm)	Pattern	U_{gs} (m/s)	U_{ls} (m/s)	U_{slug} (m/s)	ω_{slug} (Hz)	h_o/R meas.	h_o/R Eq. (6)
Water	6.3	TAS	1.12	0.07	1.4	2.5	0.17	0.07
Water	6.3	TAS	2.25	0.07	–	2.6	0.14	–
Water	6.3	TAS	2.0	0.20	2.5	3.5	0.18	0.09
Water	6.3	Annular	4.0	0.20	2.4	5.1	0.14	0.09
Aq. Glycerol	6.3	Slug	1.08	0.08	2.4	1.5	0.18	0.23
Aq. Glycerol	6.3	TAS	2.3	0.09	1.6	2.8	0.17	0.19
Aq. Glycerol	6.3	TAS	2.46	0.20	3.0	4.4	0.18	0.26
Aq. Glycerol	6.3	TAS	4.46	0.20	3.0	6.1	0.16	0.26
Water/Zonyl*	6.3	Slug	0.98	0.07	1.3	1.4	0.15	0.13
Water/Zonyl*	6.3	TAS	2.4	0.07	1.4	1.7	0.12	0.13
Water/Zonyl*	6.3	TAS	2.37	0.20	2.2	3.3	–	0.15
Water/Zonyl*	6.3	Annular	4.2	0.20	1.9	5.0	0.12	0.15
Water/Zonyl*	6.3	Annular	0.07	4.0	1.6	2.6	0.10	0.14
Water/Zonyl*	6.3	Annular	0.07	10.0	1.9	5.9	0.07	0.15
Water*	12.6	Slug	0.5	0.2	0.47	1.5	–	0.04
Water*	12.6	TAS	2.1	0.1	1.9	0.8	–	0.08
Water*	12.6	TAS	2.0	0.26	1.4	3.0	–	0.07
Water*	12.6	Annular	10.8	0.1	1.0	2.3	–	0.06
Water*	12.6	Annular	10.5	0.24	1.3	4.8	–	0.06

stress instability. Surfactant effects have been previously reported to excite flow instabilities, Troian et al. (1990). The film rupture is also enhanced by bubbles, entrained in the annular films, that burst (coalesce with core phase), nucleating rupture sites in the film. Particulates in the flow, or sharp irregularities in the tube wall, are also naturally suspect as nucleation sites for rupture as suggested by repeated rupture events at a given location.

EFFECT OF CONTAMINANTS

Laboratory experiments are conducted using small diameter capillary tubes to observe the impact of particulate and surfactant contaminants on the rupture phenomena for an idealized slug-annular flow. The experimental approach is depicted in Fig. 2 where a capillary tube is employed that is coated with a fluoropolymer over half its length. The test fluid perfectly wets the lefthand uncoated glass portion of the tube, $\theta_1 = 0$, but exhibits partial wetting with the righthand coated portion of the tube, $\theta_2 = \theta_s$ as listed in Table 2 for 1cs PDMS (polydimethylsiloxane).

The simple test procedure is as follows: the liquid index is drawn down the tube by an infusion pump at a known rate. The flow is left to right in Fig. 2b. When the receding meniscus of the slug has traveled as many as 30 diameters past the wetting discontinuity at $z = 0$ the flow is halted, Fig. 2c. The wetting discontinuity has thus served as a means for "stretching" a long annular film across a partially wetting surface. The surface tension instability for the annular film begins to take effect immediately, but not exactly in the same manner as described above for infinite annular films. When the liquid index is halted the pressure drop across the annular film is approximately $-\sigma/R$ while that across the slug is approximately $-2\sigma/R$. Thus, the fluid thins fastest near the junction of the annular film and the slug meniscus as the film drains locally into the slug, Fig. 2d. When the film is sufficiently thin, it ruptures by the van der Waals instability leaving the slug isolated on the righthand side of the tube while the annular film recedes to the lefthand side of the tube, Fig. 2e. The rupture is most frequently initiated at the top of the tube due to the draining effect of gravity.

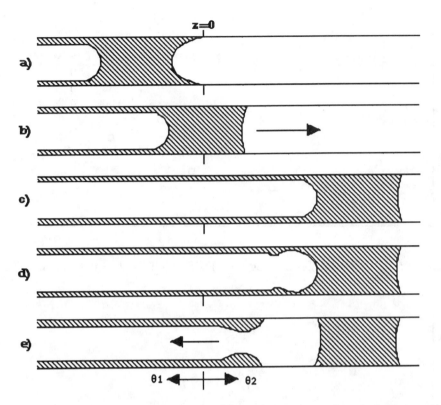

FIGURE 2. Film Rupture in Tube with Wetting Discontinuity at $z = 0$. $\theta_1 = 0$, $\theta_2 > 0$: a. Initial Condition, b. Forced Flow, Left to Right, c. Flow Halted, d. Capillary Instability Fastest at Junction of Slug Meniscus and Annular Film, e. Film Ruptures, Retracts left Isolating Slug on Right.

Film thicknesses are varied by changing the slug velocity and the time necessary for film rupture is recorded from video images. Fluid properties are also readily varied, but only the effects of particulate contaminants and soluble surfactants are reported here.

1-g Experiment Results and Discussion

Results of several tests using 1cs PDMS in a tube 1.64mm in diameter are shown graphically in Figs. 3 and 4. The test fluid is filtered using $2\mu m$ paper. Fig. 3 reveals the significant scatter in the rupture time for tests performed with $Ca = 7.4 \times 10^{-5}$. Eq. 2 yields $t_\sigma \simeq 3200s$ which is much larger than the largest of the rupture times plotted on Fig. 3, where $35 \lesssim t_{rupt} \lesssim 281s$. An analysis of the flow between the annular film and the slug meniscus needs to be performed to modify t_σ of eq. 2 for this test configuration. A preliminary scaling yields a modified $t_\sigma \sim \mu R^3 / \sigma h_o^2 \simeq 7s$ for the condition of Fig. 3. The films did not always rupture in exactly the same location, and, in many cases tests could not be conducted because of repeated, premature rupture of the films in the annular portions of the tubes obviously caused by a particulate $\sim O(1\mu m)$ attached to the wall.

Fig. 4 reveals the rupture time as a function of time in a tube which was blown dry with an aromatic duster containing tetrafluoroethane. The decrease in t_{rupt} over the approximately 2.5 hour period of repeated testing in the single tube illustrates that surfactants can significantly enhance film rupture.

SUMMARY

Film rupture leading to dryout in certain low-g, gas-liquid, slug and annular flows is observed to be enhanced by decreasing liquid viscosity, decreasing surface tension, increasing core flow Reynolds number, and

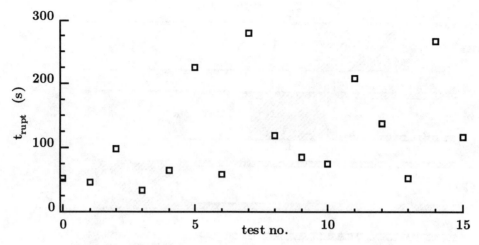

FIGURE 3. Scatter of t_{rupt} for repeat Tests Using 1cs PDMS in a Tube $R = 0.82$mm and $Ca = 7.4 \times 10^{-5}$.

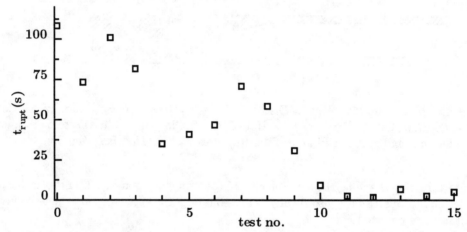

FIGURE 4. General Decrease in t_{rupt} in tube $R = 0.82$mm: Test Conditions of Fig. 3 with Surfactant Effect. Tests Indicated by Test Numbers are Conducted Approximately 10-15 min. Apart.

increasing static contact angle (increasing A). Particulate and surfactant contaminants are also expected to enhance the rupture process as is qualitatively demonstrated herein. Since dryout can result in diminished system performance, special consideration should be made when selecting working fluids for two-phase systems in space. When the selection of a partially wetting fluid cannot be avoided, knowledge of the rupture phenomena is useful to guide system design. An applicable example is noted for the case of room temperature water in stainless steel tubing where, for instance, the poorer heat transfer characteristics of the dryout-ridden slug-annular flows might require longer evaporator sections.

Acknowledgments

This work was supported by the Fluid Physics Branch of the Microgravity Science Division at NASA's Lewis Research Center. The authors wish to thank summer students M. Sekerak and M. Harasim for experimental support.

References

Aul, R.W. and W.L. Olbricht (1990) "Stability of a Thin Annular Film in Pressure-Driven Low-Reynolds-Number Flow Through a Capillary," *J. Fluid Mech.*, 215:585–599.

Barajas, A.M. and R.L. Panton (1993) "The Effects of Contact Angle on Two-Phase Flow in Capillary Tubes," *Int. J. of Multiphase Flow*, No. 2, 19:337–346.

Bousman, W.S. and A.E. Dukler (1993) "Studies of Gas-Liquid Flow in Microgravity: Void Fraction, Pressure Drop and Flow Patterns," *Proc. of the 1993 ASME Winter Ann. Mtg.*, New Orleans, LA, AMD-Vol. 174/FED-Vol. 175, December.

Bousman, W.S. (1995) "Studies of Two-Phase Gas-Liquid Flow in Microgravity," NASA CR-195434, Lewis Research Center, Cleveland, OH.

Burelbach, J.P., S.G. Bankoff, and S.H. Davis (1988) "Nonlinear Stability of Evaporating/Condensing Liquid Films," *J. Fluid Mech.*, 195:463–494.

Carpenter, B.M. (1996) "Process Research in Support of the Human Exploration and Development of Space: Issues in Program Development," *Engineering, Construction, and Operations in Space V*, Proc. of the 5th Int. Conference on Space '96, 1:509–515.

Chen, J.D. (1986) "Measuring the Film Thickness Surrounding a Bubble Inside a Capillary, *J. Colloid Int. Sci.*, No. 2, 109:341–349.

Colin, C., J. Fabre, and J. McQuillen (1996) Bubble and Slug Flow at Microgravity Conditions: State of Knowledge and Open Questions," *Chem. Eng. Comm.*, 141-142:155–173.

Frenkel, A.L., A.J. Babchin, B.G. Levich, T. Shlang, and G.I. Sivashinsky (1987) "Annular Flows Can Keep Unstable Films from Breakup: Nonlinear Saturation of Capillary Instability," *J. Colloid and Int. Sci.*, No. 1, 115:225–233.

Gauglitz, P.A. and C.J. Radke (1988) "The Dynamics of Liquid Film Breakup in Constricted Cylindrical Capillaries," *Chem. Engng Sci.*, 43:1457.

Hammond, P.S. (1983) "Nonlinear Adjustment of a Thin Annular Film of Viscous Fluid Surrounding a Thread of Another Within a Circular Pipe," *J. Fluid Mech.*, 137:363–384.

Hu, H.H. and D.D. Joseph (1989) "Lubricated Pipelining: Stability of Core-Annular Flow. Part 2," *J. Fluid Mech.*, 205:359-396.

Jayawardena, S.S. and V. Balakotaiah (1997) "Flow Pattern Transition Maps for Microgravity Two-Phase Flows," *AIChE J.*, No. 6, 43:1637–1640.

Li, D, and J.C. Slattery (1991) "Analysis of the Moving Apparent Common Line and Dynamic Contact Angle Formed by a Draining Film," *J. Colloid Int. Sci.*, No. 2, 43:382–396.

McQuillen, J.B. and E.S. Neumann (1995) "Two-Phase Flow Research Using the Learjet Apparatus," NASA TM-106814, Lewis Research Center, Cleveland, OH, May.

Ruckenstein, E. and R.K. Jain (1974) "Spontaneous Rupture of Thin Liquid Films," *J. Chem. Soc.*, Faraday Trans. II, 70:132–147.

Salzman, J.A. (1996) "Fluids Management in Space-Based Systems," *Engineering, Construction, and Operations in Space V*, Proc. of the 5th Int. Conference on Space '96, 1:521–526.

Singh, B.S. (1996) "Third Microgravity Fluid Physics Conference," NASA Conference Publication 3338, Work- shop Proceedings, Cleveland, Ohio, July 13–15.

Troian, S.M., E. Herbolzheimer, and S.A. Safran, (1990) "Model for the Fingering Instability of Spreading Surfactant Drops," *Phys. Rev. Letters*, No. 3, 65:333–336.

Weislogel, M.M. (1997) "Steady Capillary Flow in Partially Coated Tubes," *AIChE J.*, 43:3:645–654.

Williams, M.B., S.H. Davis (1982) "Nonlinear Theory of Film Rupture," *J. Col. Int. Sci.*, No. 1, 90:220-228.

APPLICABILITY OF THEORETICAL MODELS TO THE EVALUATION OF THERMAL CONDUCTIVITY IN MICROGRAVITY

G. Passerini and F. Polonara
Dipartimento di Energetica
Università di Ancona
Via Brecce Bianche
60100 Ancona, ITALY
Tel +39 71 220 4432 - Fax +39 71 280 4239

F. Gugliermetti
Dipartimento di Fisica Tecnica
Università di Roma "La Sapienza"
Via Eudossiana
00184 Roma, ITALY
Tel +39 6 4814 334

Abstract

In several problems concerning heat transfer in liquids the thermal conductivity coefficient is required at different values of temperature and pressure, but often reliable experimental data are not available. In these conditions prediction methods are necessary to calculate the thermal conductivity coefficient. Both theoretical equations and empirical or semi-empirical equations were proposed by various authors, the first ones deriving from rigorous physical considerations, the second ones being based both on theoretical considerations and on experimental evidence. Theoretical equations should be obviously preferred, but often their use can be accepted with reasonable accuracy (±5%) for simple molecular structures only. Empirical or semi-empirical equations on the other hand seem to be more useful for engineering purposes because of their ease of use and their larger range of validity. In this work a review of theoretical models is critically presented and explained with respect to their applicability in microgravity conditions. It will be shown that heat transfer phenomena in liquids are likely to be influenced by gravity forces only in a negligible way and how a set of experiments on pure fluids would be exceptionally useful both to enhance theoretical studies and to acquire new experimental thermal conductivity data useful for engineering purposes.

INTRODUCTION

Exact values of thermal conductivity coefficient λ of liquids are often required in several engineering problems. Thus, many sets of experimental data concerning the most important fluids have been published by various authors in order both to cover and to complete the range of temperatures in which data are needed. The latter item forced authors to support the new sets of experimental data with new correlation whose task is usually that of supplying data for all the temperatures within the explored range. In several cases these formulas have a Taylor Series form and they are often the result of a mere interpolation of experimental points. Some of these correlations moreover are proposed for a temperature range larger than the explored one and so they are validated against some other sets of data.

The efforts to complete information about liquid thermal conductivity also led to the development of estimation and prediction methods. Estimation methods are empirical or semi-empirical. They move from experimental data but extend the knowledge to a wider temperature range usually requiring the knowledge of only few data. Prediction methods are usually based on some physical and chemical theories and they should not require the knowledge of any experimental datum.

Collections and reviews of such formulas are largely available in literature (Latini°1978, Latini°1987, Reid 1987, and Passerini 1995) so the central problem is that of their applicability in microgravity conditions. Engineers, to design or adapt new devices for microgravity applications, obviously need reliable data since space applications require maximum reliability and efficiency. Unfortunately no experimental data, acquired in microgravity, seem to be available while the problem of the use of the existing data seems not addressed as well. Thus, we will try to describe how the heat flux processes in liquids could change due to the absence of gravity and whether or not usual prediction formulas could be applied. It will be shown that heat transfer phenomena are likely to be influenced by gravity forces only in a negligible way for what concerns engineering purposes. This results come from the analysis of the theories of the liquid state concerning heat flux and thermal conductivity. At the same time the need of experimental measurement of liquid thermal conductivity coefficient in microgravity will be stressed. Scientific community needs reliable liquid thermal conductivity data evaluated in microgravity to finally assess both the accuracy of experimental data acquired on Earth and the reliability of theories on the liquid state.

LIQUID THERMAL CONDUCTIVITY AND NON-EQUILIBRIUM STATISTICAL MECHANICS

Expressions for momentum and energy fluxes in terms of non equilibrium distribution functions have been derived from the equations of classical hydrodynamics and the non-equilibrium distribution functions by several authors (Born 1947, Eisenschitz 1955, and Irving 1950).

Introduction to Non-Equilibrium Statistical Mechanics

The probability that the coordinates $\mathbf{r}_i$ and moments $\mathbf{p}_i$ of a system of N particles of mass m and no internal degree of freedom have specified values at a given moment is given by distribution function f_N which satisfies the normalisation condition:

$$\int \ldots \int f_N(\mathbf{r}_1, \mathbf{r}_2, \ldots, \mathbf{r}_N, \mathbf{p}_1, \mathbf{p}_2, \ldots, \mathbf{p}_N) \prod_{i=1}^{N} d\mathbf{r}_i d\mathbf{p}_i = N! \qquad (1)$$

The time dependency of the distribution function f_N satisfies a continuity equation in phase space which corresponds to the hydrodynamic equation of continuity for an incompressible fluid. This equation is known as Liouville's equation:

$$\frac{\partial f_N}{\partial t} + \sum_{i=1}^{N} \left(\frac{1}{m} \mathbf{p}_i \cdot \nabla_{\mathbf{r}_i} f_N + \mathbf{F}_i \cdot \nabla_{\mathbf{p}_i} f_N \right) = 0, \qquad (2)$$

where $\mathbf{F}_i$ is the force acting on the i-th molecule due to the sum of the gradients of the separate pair potentials of its (N-1) neighbours. Assuming a spherical symmetry it will be

$$-F_i = \sum_{i=1}^{N} \sum_{j=1}^{N} \frac{\partial \psi_{ij}}{\partial r_i}. \qquad (3)$$

Using the substantial derivative, Liouville's equation may be written in a far more compact form:

$$Df_N / Dt = 0. \qquad (4)$$

Gibbs showed that this form of Liouville's equation is reversible with respect to time inversion, whereas transport processes are irreversible. Kirkwood (Kirkwood 1946) showed that, when lower order distribution functions $f_h (h \ll N)$ are averaged over a certain time interval τ, this procedure, better known as "coarse graining" (Rice 1960), leads to the expected irreversible behaviour. τ, determined on the basis of analogies with Brownian motion, is assumed by Kirkwood to be greater than the time interval (t'-t) for which we can consider the force acting on molecule at time t' statistically independent with the force acting at time t. This time is small compared with a force relaxation time, but is longer than that related to force fluctuations due to microscopic Brownian motion. This assumption plays a similar role to that of molecular chaos assumption in the kinetic theory of gases. Lower order distributions f_h can be derived fro f_N by averaging with respect to the positions and the momenta of the remaining (N-h) particles:

$$f_h(\mathbf{r}_1, \mathbf{r}_2, \ldots, \mathbf{r}_h, \mathbf{p}_1, \mathbf{p}_2, \ldots, \mathbf{p}_h, t) = \int \ldots \int f_N(\mathbf{r}_1, \mathbf{r}_2, \ldots, \mathbf{r}_N, \mathbf{p}_1, \mathbf{p}_2, \ldots, \mathbf{p}_N, t) \prod_{i=h+1}^{N} d\mathbf{r}_i d\mathbf{p}_i, \qquad (5)$$

which, on coarse graining becomes:

$$\overline{f}_h(\mathbf{r}_1, \mathbf{r}_2, \ldots, \mathbf{r}_h, \mathbf{p}_1, \mathbf{p}_2, \ldots, \mathbf{p}_h, t) = \frac{1}{\tau} \int_0^{\tau} f_h(\mathbf{r}_1, \mathbf{r}_2, \ldots, \mathbf{r}_h, \mathbf{p}_1, \mathbf{p}_2, \ldots, \mathbf{p}_h, t+s) ds. \qquad (6)$$

When $h = 1$ equation (6) becomes

$$f_1(\mathbf{r}_1, \mathbf{p}_1, t) = \int \ldots \int f_N(\mathbf{r}_1, \mathbf{r}_2, \ldots, \mathbf{r}_N, \mathbf{p}_1, \mathbf{p}_2, \ldots, \mathbf{p}_N, t) \prod_{i=2}^{N} d\mathbf{r}_i d\mathbf{p}_i, \qquad (7)$$

while, for $h = 2$, it becomes

$$f_2(\mathbf{r}_1, \mathbf{r}_2, \mathbf{p}_1, \mathbf{p}_2, t) = \int \ldots \int f_N(\mathbf{r}_1, \mathbf{r}_2, \ldots, \mathbf{r}_N, \mathbf{p}_1, \mathbf{p}_2, \ldots, \mathbf{p}_N, t) \prod_{i=3}^{N} d\mathbf{r}_i d\mathbf{p}_i. \qquad (8)$$

Equations (7) and (8) are respectively known as singlet and doublet distributions in (momentum configuration) phase space. Integration of $\overline{f_1}$, $\overline{f_2}$ over the momenta respectively gives:

$$n_1(\mathbf{r}_1,t) = \frac{N!}{(N-1)!}\int \overline{f_1}(\mathbf{r}_1,\mathbf{p}_1,t)d\mathbf{p}_1, \tag{9}$$

and

$$n_2(\mathbf{r}_1,\mathbf{r}_2,t) = \frac{N!}{(N-2)!}\iint \overline{f_2}(\mathbf{r}_1,\mathbf{r}_2,\mathbf{p}_1,\mathbf{p}_2,t)d\mathbf{p}_1 d\mathbf{p}_2, \tag{10}$$

which are the singlet and pair densities in configuration space. It can be demonstrated that singlet density is simply equal to the number density $n_1 = n \equiv N/V$ while n_2, which represents the probability that a first molecule is at $\mathbf{r}_1$ and a second is at $\mathbf{r}_2$, can be related to non-equilibrium radial distribution function g_2. Assuming $\mathbf{r}_2 - \mathbf{r}_1 \equiv \mathbf{R}$ we can write

$$n_2(\mathbf{r},\mathbf{r}+\mathbf{R},t) = n(\mathbf{r},t)\cdot n(\mathbf{r}+\mathbf{R},t)\cdot g_2(\mathbf{r},\mathbf{R},t), \tag{11}$$

which is in the same form as the equilibrium relation

$$n_2(\mathbf{r},\mathbf{r}+\mathbf{R}) = n(\mathbf{r})\cdot n(\mathbf{r}+\mathbf{R})\cdot g_2(\mathbf{r},\mathbf{R}), \tag{12}$$

or, when the fluid is isotropic,

$$n_2(\mathbf{R}) = n_2 \cdot g_2(\mathbf{R}), \tag{13}$$

where $g_2(\mathbf{R})$ is the equilibrium radial distribution function.

Theory of Kirkwood et al.

Kirkwood et al. theory on thermal conductivity of liquids follow the work of Kirkwood and his school which derived expressions for transport properties of liquids by means of non-equilibrium statistical mechanics theories (Kirkwood 1946, 1947, 1949, and 1952). In a first stage Irving and Kirkwood (1950) derived an expression for heat flux vector. They assumed the heat flux composed of two terms letting $\mathbf{J}_q = \mathbf{J}_k + \mathbf{J}_\psi$ being $\mathbf{J}_k$ the heat flux due to transport of kinetic energy -thus a convective contribution- and $\mathbf{J}_\psi$ the heat flux due to molecular interaction arising from the Brownian motion. For these two terms they proposed:

$$\mathbf{J}_k = \frac{Nm}{2}\int \left|\frac{\mathbf{p}}{m}-\mathbf{u}\right|^2 \left(\frac{\mathbf{p}}{m}-\mathbf{u}\right)\overline{f_1}(\mathbf{r},\mathbf{p},t)d\mathbf{p}, \tag{14}$$

and

$$\mathbf{J}_\psi = \frac{1}{2}\int\left[\psi(\mathbf{R})\mathbf{1}+\frac{\mathbf{RR}}{R}\frac{d\psi(R)}{dR}\right]\left[\mathbf{J}'_2(\mathbf{r},\mathbf{r}+\mathbf{R})-\mathbf{u}(\mathbf{r},t)\cdot n_2(\mathbf{r},\mathbf{r}+\mathbf{R},t)\right]d\mathbf{R}, \tag{15}$$

where $\mathbf{u}$ is the centre of mass, $\mathbf{1}$ is the unit dyad and $\mathbf{J}'_2$ is given by expression

$$\mathbf{J}'_2(\mathbf{r}_1,\mathbf{r}_2,t) = N^2 \iint \frac{\mathbf{p}_1}{m}\overline{f_2}(\mathbf{r}_1,\mathbf{r}_2,\mathbf{p}_1,\mathbf{p}_2,t)d\mathbf{p}_1 d\mathbf{p}_2. \tag{16}$$

Kirkwood and his colleagues (Zwanzig 1954) combine distribution functions $\overline{f_1}$ e $\overline{f_2}$, in form of Fokker-Plank equations, with expressions (14) and (15) and with the Fourier's law obtaining:

$$\lambda_{conv} = \frac{nk^2T}{2\xi} - \frac{k^2T^2}{6\xi}\left(\frac{\partial n}{\partial T}\right)_p, \tag{17}$$

and

$$\lambda_\psi = \frac{n^2\pi kT}{3\xi}\int_0^\infty R^3\left(R\frac{d\psi}{dR}-\psi\right)\cdot g'_2(R)\cdot \frac{d}{dR}\left(\frac{\partial}{\partial T}\ln g'_2(R)\right)dR + \\ + \frac{n^2\pi kT}{\xi}\int_0^\infty R^2\left(\psi-\frac{R}{3}\frac{d\psi}{dR}\right)\cdot\left(\frac{\partial}{\partial T}\ln g'_2(R)\right)dR \tag{18}$$

where ξ is a friction constant related to the self diffusion coefficient.

Since convective, kinetic, contribution and intermolecular force contribution are assumed independent, the thermal conductivity coefficient can be written as $\lambda = \lambda_{conv} + \lambda_\psi$. It can be seen in equation (17) that the convective term is directly dependent on the number density $n=N/V$ while the intermolecular term depends on n^2. Thus, in a first approximation, we can conclude that the convective term is negligible with respect to the intermolecular one. Kirkwood et al. equations were checked against simple liquids using Lennard-Jones 12:6 potential (Lennard-Jones 1937). The deviations were up to 30% making this model not useful for engineering purposes.

Theory of Rice and Kirkwood

In a second stage, Rice and Kirkwood tried to simplify the expression related to intermolecular contribution (Rice 1959) but their new equation

$$\lambda_\psi = -\frac{kT}{12\xi}\left[\left(\frac{N}{V}\right)^2 \cdot \int_0^\infty R^2 \nabla^2 \psi(R) g_0'(R) d^3R\right] \tag{19}$$

showed consistent weaknesses being deviations between results and experimental data sometimes greater than 50%. The convective component was still judged negligible.

THEORY OF HORROCKS AND MCLAUGHLIN

Horrocks and McLaughlin (1959, and 1960) assume a quasi-crystalline structure for the liquid composed of spherical molecules interacting with a Lennard-Jones 12:6 potential. Molecules are located on a face centred cubic lattice in which there is a certain number of vacancies. When a temperature gradient dT/dx is applied to the liquid, the average difference in thermal energy between successive molecules in adjacent planes at a distance l is ldU/dx. This excess energy is assumed to be transferred down the temperature gradient by means of two distinct mechanisms: a vibrational mechanism and a convective mechanism.

For what concerns vibrational contribution, the difference in energy ldU/dx is transferred down the temperature gradient with a frequency $P\nu$ being P the probability that energy is transferred when two molecules collide and ν the mean vibrational frequency of the quasi-lattice. The rate of flow of heat dQ/dT across the unit area normal to the temperature gradient is

$$dQ/dT = -2n \cdot P\nu \cdot l \cdot (du/dx), \tag{20}$$

where n is the number of molecules per unit area and the factor 2 takes into account that a molecule in a complete cycle of vibration will cross two times a plane perpendicular to its motion direction. Introducing the one-dimensional equation of Fourier

$$dQ/dT = -dT/dx, \tag{21}$$

and letting

$$du/dx = (du/dT) \cdot (dT/dx), \tag{22}$$

they obtain

$$\lambda_{vibr} = 2n \cdot P\nu \cdot l \cdot (du/dT). \tag{23}$$

Horrock and McLaughlin assume that the motion of particles which generate convective contribution is made of particle hoppings from occupied sites to vacancies, "holes", in the quasi-lattice. In absence of temperature gradients the frequency of such hoppings from one layer in the liquid to the next will be

$$\varsigma = \varsigma_0 \cdot (n_h/N_A), \tag{24}$$

where n_h is the number of holes in the liquid and N_A Avogadro's number. The theory of rate processes (Glasstone 1941) gives for ς_0 the expression:

$$\varsigma_0 = \frac{kT}{2\pi m} \cdot \frac{1}{v_f^{1/3}} \cdot e^{-\frac{e_0}{kT}}, \tag{25}$$

where v_f is the free volume of the liquid, e_0 is the height of the boundary potential restricting a molecule to its cell, m is the molecular mass and k is the Boltzmann constant.

The rate of transport of energy across unit area of the reference plane will be

$$(dQ/dT) = -2n \cdot \varsigma \cdot l \cdot (du/dx), \tag{26}$$

where the factor 2 accounts for the equivalence to the net transport of "hot" molecules moving across the reference plane down the temperature gradient and of "cold" molecules moving up.

Introducing equations (21) and (22), the convective contribution can be written as

$$\lambda_{conv} = 2n \cdot \varsigma \cdot l \cdot (du/dT). \tag{27}$$

To obtain useful expressions, Horrocks and McLaughlin introduce expressions to relate both the frequency of vibration and the frequency of convective movement to physical parameters and obtain the following equations:

$$\lambda_{vibr} = P \cdot \left\{ \frac{\varepsilon}{\pi^2 r_0^2 m} \cdot \left[11 C_{14} \cdot \left(\frac{r_0}{a}\right)^{14} - 5 C_8 \cdot \left(\frac{r_0}{a}\right)^8 \right] \right\}^{1/2} \cdot C_v \sqrt{2} \cdot \frac{(1 - n_h/N)}{a}. \tag{28}$$

$$\lambda_{conv} = \frac{n_h}{N} \cdot \left(\frac{kT}{2\pi m}\right)^{1/2} \cdot \frac{C_v}{v_f^{1/3}} \cdot \frac{\sqrt{2}}{a} \cdot \left(1 - \frac{n_h}{N}\right) \cdot e^{-\frac{e_0}{kT}}, \tag{29}$$

where ε is the depth of potential with respect to the equilibrium diameter r_0 which is related to collision diameter as $r_0^3 = \sqrt{2}\sigma^3$, a is the distance between nearest neighbours, C_{14} and C_8 are lattice summation constants (Glasstone 41), C_v is the molecular specific heat, n_h is the total number of holes and N is the total number of molecules.

On applying equations (28) and (29) to simple liquids it appears that the convective contribution to thermal conductivity is negligible, less than 1% for liquid Argon at its normal boiling point. Horrocks and McLaughlin compare experimental thermal conductivity data of six compounds (Argon, Nitrogen, Carbon Monoxide, Methane, Benzene, Carbon Tetrachloride) with the values predicted by means of equation (28). Deviations at different temperatures were meanly about 20%. This means that the method is not completely acceptable for engineering purposes, but is of extraordinary importance from a theoretical point of view since it shows that the heat flux mechanisms, the authors assumed, is coherent with experimental evidence.

HEAT TRANSFER THEORIES AND MICROGRAVITY

It is generally difficult to derive from liquid state theories practical results for engineering purposes in the field of heat transfer. Numerical calculations show qualitative agreement with the actual behaviour of liquids, but they evidence a very poor accuracy. For the aim of this work, the importance of such theories is due to the description of the mechanisms of heat transfer.

Several rigorous models were analysed, based on two different theoretical approaches. All these models show qualitative accordance with experimental evidence so it can be concluded that the assumptions they are based on are, in principle, acceptable. The kind of molecular interaction assumed in both original theories seems unaffected by the presence of gravity forces and this should let to suggest that heat flux itself is unaffected as well. In fact, all the models assume that heat flux is made of two independent terms. The first term (vibrational or Brownian term) takes into account the related, well known, heat flux phenomenon. The second term ("convective" term) is due to a kind of molecular drift and should not be confused, due to its denomination, with any kind of gravity-driven convective phenomena. It is a kind of "micro-convection" active at molecule neighbour level and the related term is considered negligible with respect to the previous one. Neither the convective, nor the vibrational phenomena are explicitly influenced by gravitational forces.

Nevertheless, we could expect a general increase of molecular order due to the absence of an external force acting on the molecules of the liquid. This should allow the molecules to move more efficiently so to slightly increase thermal conductivity. Based on the study of intermolecular and intramolecular forces, we expect that such effect is negligible, but at the moment no experimental proof exist to support this conclusion. Thus, it would be exceptionally useful to perform a series of experiments on liquid thermal conductivity in microgravity. By collecting few sets of experimental data we could reach three advantages. First of all, we could achieve reliable sets of data to be used for design purposes. Then we could validate, by comparison, all the currently available prediction formulas and experimental data sets to assure their applicability in microgravity conditions. A third, not less important, return should arise for what concerns the theoretical study of heat transfer in liquids as soon as the relevance of gravity forces on molecular interaction is assessed.

CONCLUSIONS

Both the theories based on the non-equilibrium statistical mechanics, and those based on Lennard-Jones molecular interaction assumptions, seem to prove that heat flux in liquids is due to molecular mechanisms which are affected by the presence of gravity forces only in negligible ways.

Theoretical studies show weaknesses for what concerns the overall precision of the proposed values of thermal conductivity coefficient. It would be exceptionally useful to acquire new sets of experimental data in microgravity conditions to compare the related thermal conductivity coefficients with those evaluated on Earth. This procedure would let scientists to enhance their theories of heat transfer in liquids and would supply engineers both with sets of new reliable data and with the usual sets of data, acquired on Earth, validated by comparison.

Acknowledgements

This work has been supported in part by the Commission of the European Communities in the framework of the JOULE2 Programme, and by the Ministero dell'Universita' e della Ricerca Scientifica e Tecnologica of Italy.

References

Born, M., H. S. Green (1947)"A General Kinetic Theory of Liquids II.", *Proc. Roy. Soc.*, **A190**:455-463.

Eisenschitz, R. (1955), "Virial Theorem for the Flow of Energy", *Phys. Rev.*, **99**:1059-1961.

Glasstone, S., K. J. Laidler and H. Eyring (1941), *The Theory of Rate Processes*, McGraw-Hill, New York.

Horrocks, J. K. and E. McLaughlin (1960), *Trans. Faraday Soc.*, **56**:206-.

Irving, J. H. and J. G. Kirkwood (1950), "The Statistical Mechanics Theory of Transport Processes IV." *J. Chem. Phys.*, **18**:817-829.

Kirkwood, J.G. (1946), "The Statistical Mechanics Theory of Transport Processes I", *J. Chem. Phys.*, **14**:180-201.

Kirkwood, J.G. (1947), "The Statistical Mechanics Theory of Transport Processes II", *J. Chem. Phys.*, **15**:72-86.

Kirkwood, J.G., F. P. Buff and H. S. Green H. S. (1949), "The Statistical Mechanics Theory of Transport Processes III", *J. Chem. Phys.*, **17**:988-994.

Kirkwood, J.G.,V.A. Lewinson, B.J. Alder (1952), "Radial Distribution Functions and Equations of state of Fluids Composed of Molecules Interacting According to the Lennard-Jones Potential", *J. Chem. Phys.*, **20**:929-938.

Latini, G., C. Baroncini and P. Pierpaoli (1987), "Liquids Under Pressure. An Analysis of Methods for Thermal Conductivity Prediction and a General Correlation", *High Temperatures-High Pressures*, **19**:43-50.

Latini, G. and M. Pacetti (1978), *Thermal Conductivity 15*, Plenum Press, New York, 245-253.

Lennard-Jones, J. E. (1937), "The Equations of State of Gases and Critical Phenomena", *Physica*, **4**:941-956.

McLaughlin, E., (1959), "Viscosity and Self-Diffusion in Liquids", *Trans. Faraday Soc.*, **55**:28-38.

Passerini, G. (1995*), Thermal Conductivity and Dynamic Viscosity of Organic Compounds in Their Liquid State*, PhD Thesis.

Reid, R. C., J. M. Prausnitz and B. E.Poling (1987), *The properties of Gases & Liquids*, McGraw-Hill, New York,

Rice, S. A. and J. G. Kirkwood (1959), "On an Approximate Theory of Transport in Dense Media", *J. Chem. Physics*, **31**:901-908.

Rice, S. A. and H. L. Frish (1960), "Some Aspects of the Statistical Theory of Transport", *Ann. Rev. Phys. Chem.*, **11**:187-272

Zwanzig, R. W., J. G. Kirkwood, I. Oppenheim and B. J. Alder (1954), "The Statistical Mechanics Theory of Transport Processes VII", *J. Chem. Physics*, **22**:783-790.

THERMOCAPILLARY FLOW NEAR A CORNER IN A THIN LIQUID LAYER

R. Balasubramaniam
National Center for Microgravity Research on Fluids and Combustion
NASA Lewis Research Center
Cleveland, OH 44135
(216)433-2878

Abstract

The thermocapillary flow in a thin liquid layer present in a differentially heated cavity is analyzed in the limit of creeping flow. Away from the end walls, it is shown that the flow field is parallel. The turning of the flow is accomplished in regions near the end walls. The method of matched asymptotic expansions is used to analyze the flow fields in these regions. An analytic solution has been obtained to the inner equations using the method of biorthogonal series expansions. The coefficients in the infinite series expansion for the streamfunction have been determined numerically after truncation of the series.

INTRODUCTION

When the free surface of a liquid is subjected to variations in temperature along it, a stress is exerted on the free surface because of the change in surface tension with temperature. This stress is transmitted to the liquid by the action of viscous forces and a flow is induced in the liquid from its hot portion toward the cooler side. Such a flow is called thermocapillary flow and has been studied vigorously in the last two decades owing to the prospect of materials processing in space. An early account of the knowledge in the field is given by Ostrach (1982). The results of a recent shuttle experiment that investigated thermocapillary flows in a liquid layer is described by Kamotani, Ostrach and Pline (1994).

The flow in a thin rectangular liquid layer induced by the thermocapillary stress at its free surface has been analyzed by Sen and Davis (1982) when viscous forces are predominant, *i.e.*, the Reynolds number is small compared to unity. In the limit when the aspect ratio ϵ is small ($\epsilon = H/L$, H is the depth of the pool and L is its width) it was shown that the leading order flow is a recirculating parallel flow. Conditions at the end walls, *viz.*, the no slip and no penetration conditions, cannot be imposed on the parallel flow solution; thus it is valid only in regions away from the end walls. In the language of asymptotic expansions, the parallel flow solution is an outer solution. In order to analyze the velocity field subject to the conditions at the end walls, an inner solution must be determined that approaches the parallel flow away from the end walls. This problem has been analyzed by Sen and Davis (1982) by using the method of matched asymptotic expansions. They included the deformation of the free surface and calculated the solution to leading order and a higher order in ϵ.

In what follows the problem of thermocapillary flow in a thin layer is revisited when the free surface is undeformed. While Sen and Davis compute the leading order inner solution numerically, an analytical solution is provided below for the streamfunction at leading order. It is worthy of note that de Socio (1979) analytically solved the problem of combined thermocapillary and natural convection in a rectangular cavity in the Stokes limit by using the method of biorthogonal series described by Joseph (1977) and Joseph and Sturges (1978). It is not convenient to obtain the structure of the flow as a recirculating parallel flow with turning layers at the end walls from the limit $\epsilon \to 0$ in de Socio's analysis. The analysis provided below uses Joseph's biorthogonal series method in a more natural way for the boundary conditions that obtain in a thin liquid layer.

PROBLEM FORMULATION AND SOLUTION

It is assumed that the Reynolds and Marangoni numbers are negligibly small. The analysis is performed in a cartesian coordinate system. The x and y coordinates are scaled by L and H respectively. The free surface is assumed to be undeformed and located at $y = 1$. The end walls are located at $x = 0$ and 1, while the bottom boundary is present at $y = 0$. Temperature in the liquid layer is scaled by $\Delta T = T_1 - T_2$, where T_1 and T_2 are the temperatures at the walls at $x = 0$ and $x = 1$ respectively. The thermal conditions at the remaining boundaries are assumed to be such that the temperature field is one-dimensional and is given by $T(x,y) = 1 - x$. The thermocapillary stress at the free surface is therefore uniform. The stream function ψ for the flow is scaled by $\frac{(-\sigma_T)\Delta T \epsilon H}{\mu}$, where σ_T is the rate of change of surface tension with temperature that is assumed to be a constant, typically negative and μ is the viscosity of the liquid. The streamfunction satisfies the following equation.

$$\epsilon^4 \frac{\partial^4 \psi}{\partial x^4} + 2\epsilon^2 \frac{\partial^4 \psi}{\partial x^2 \partial y^2} + \frac{\partial^4 \psi}{\partial y^4} = 0. \tag{1}$$

Outer solution

In the limit $\epsilon \to 0$, the leading order outer streamfunction ψ_o is determined as follows:

$$\frac{\partial^4 \psi_o}{\partial y^4} = 0, \tag{2}$$

$$\psi_o(x,0) = \frac{\partial \psi_o}{\partial y}(x,0) = 0, \tag{3}$$

$$\psi_o(x,1) = 0, \tag{4}$$

$$\frac{\partial^2 \psi_o}{\partial y^2}(x,1) = 1. \tag{5}$$

The condition in Eq(5) arises from the thermocapillary stress at the free surface. The solution is:

$$\psi_o(x,y) = \frac{1}{4}y^2(y-1). \tag{6}$$

Thus ψ_o is a function of y alone and represents a recirculating parallel flow. The conditions at the walls $x = 0, 1$ cannot be imposed on ψ_o. Thus ψ_o is an outer solution.

Inner solution

In order to obtain a solution valid near the end walls, an inner streamfunction ψ_i must be considered near each end wall. Due to symmetry, it is necessary to consider only the flow near one end wall; this is chosen as the wall at $x = 0$. Then, the inner variable is $\eta = \frac{x}{\epsilon}$. The inner solution must be matched to the outer streamfunction in a region away from the wall. We restrict to the analysis of the inner solution at leading order, which is governed by the following equations and boundary conditions:

$$\nabla^4 \psi_i(\eta,y) = \frac{\partial^4 \psi_i}{\partial \eta^4} + 2 \frac{\partial^4 \psi_i}{\partial \eta^2 \partial y^2} + \frac{\partial^4 \psi_i}{\partial y^4} = 0, \tag{7}$$

$$\psi_i(0,y) = \frac{\partial \psi_i}{\partial \eta}(0,y) = 0, \tag{8}$$

$$\psi_i(\eta,0) = \frac{\partial \psi_i}{\partial y}(\eta,0) = 0, \tag{9}$$

$$\psi_i(\eta,1) = 0, \tag{10}$$

$$\frac{\partial^2 \psi_i}{\partial y^2}(\eta,1) = 1, \tag{11}$$

$$\psi_i(\eta \to \infty, y) \to \frac{1}{4}y^2(y-1). \tag{12}$$

The last equation above represents the matching condition between the inner and outer solutions. It is convenient to pose the inner problem slightly differently. Let

$$F(\eta, y) = \psi_i(\eta, y) - \frac{1}{4}y^2(y-1), \tag{13}$$

$F(\eta, y)$ satisfies the biharmonic equation (Eq 7) as well. The boundary conditions that are altered are:

$$\frac{\partial^2 F}{\partial y^2}(\eta, 1) = 0, \tag{14}$$

$$F(0, y) = -\frac{1}{4}y^2(y-1), \tag{15}$$

$$F(\eta \to \infty, y) \to 0. \tag{16}$$

The remaining boundary conditions are the same. The biharmonic equation is frequently encountered in the theory of elasticity. The solution for $F(\eta, y)$ may be written as:

$$F(\eta, y) = \sum_{n=1}^{\infty} H_n \frac{e^{-k_n \eta}}{k_n^2} f_n(y), \tag{17}$$

where,

$$f_n(y) = \sin(k_n y) - y \sin k_n \cos[k_n(1-y)], \tag{18}$$

and k_n are the roots of

$$2k_n = \sin(2k_n); \quad k_n \neq 0; \quad Re[k_n] > 0. \tag{19}$$

The function $f_n(y)$ is called the Papkovich-Fadle eigenfunction and satisfies the fourth order differential equation (the reduced biharmonic equation) given below:

$$\frac{d^4 f_n}{dy^4} + 2k_n^2 \frac{d^2 f_n}{dy^2} + k_n^4 f_n = 0, \tag{20}$$

with boundary conditions

$$f_n(0) = f_n'(0) = 0, \tag{21}$$

$$f_n(1) = f_n''(1) = 0. \tag{22}$$

The Papkovich-Fadle eigenfunctions considered by Joseph (1977) and de Socio (1979) are slightly different and satisfy $f_n = f_n' = 0$ at the end points. It is evident that the values of k_n that satisfy Eq(19) must be complex numbers. The only real root $k = 0$ leads to an inadmissible solution that violates Eq(16). Roots of Eq(19) occur in all the four quadrants in the complex plane and are complex conjugates. In order to satisfy Eq(16) we need to restrict to roots in the right half-plane. The eigenvalues k_n are arranged as complex conjugates in the order of increasing magnitudes. For large n, k_n is given by (Joseph and Sturges, 1978):

$$\{k_n, k_{n+1}\} = \left[\left(n + \frac{1}{4}\right)\pi - \frac{\ln[(4n+1)\pi]}{(4n+1)\pi} + o\left(\frac{\ln n}{n}\right)\right] \mp$$
$$i\left[\frac{1}{2}\ln[(4n+1)\pi] + \left(\frac{\ln[(4n+1)\pi]}{(4n+1)\pi}\right)^2 + o\left(\frac{\ln n}{n}\right)^2\right]. \tag{23}$$

In general H_n is a complex number as well whose real and imaginary parts must be determined so that the boundary conditions on $F(0, y)$ and $\frac{\partial F(0,y)}{\partial \eta}$ are satisfied.

Biorthogonality condition

Joseph (1977) has derived a biorthogonality condition satisfied by the eigenfunctions $f_n(y)$ which must be used to determine H_n. The technique is illustrated below. Let

$$p_{n_1}(y) = f_n(y), \tag{24}$$

$$p_{n_2}(y) = \frac{f_n''(y)}{k_n^2} = \frac{p_{n_1}''(y)}{k_n^2}. \tag{25}$$

The adjoint system is defined as:

$$q_{n_2}(y) = p_{n_1}(y); \qquad q_{n_1}''(y) = -k_n^2 q_{n_2}(y). \tag{26}$$

Eq(20) may be written in vector form as:

$$\frac{d^2 \mathbf{p_n}}{dy^2} + k_n^2 \mathbf{A} \cdot \mathbf{p_n} = 0, \tag{27}$$

$$\frac{d^2 \mathbf{q_n}}{dy^2} + k_n^2 \mathbf{A^T} \cdot \mathbf{q_n} = 0, \tag{28}$$

where,

$$\mathbf{p_n} = \begin{pmatrix} p_{n_1}(y) \\ p_{n_2}(y) \end{pmatrix}; \qquad \mathbf{q_n} = \begin{pmatrix} q_{n_1}(y) \\ q_{n_2}(y) \end{pmatrix}; \qquad \mathbf{A} = \begin{pmatrix} 0 & -1 \\ 1 & 2 \end{pmatrix}. \tag{29}$$

In a standard way, the following biorthogonality relation can be derived:

$$\int_0^1 \mathbf{q_m} \cdot \mathbf{A} \cdot \mathbf{p_n} \, dy = 0; \qquad m \neq n. \tag{30}$$

It is worthy of note that the biorthogonality relation is obtained for boundary conditions of the form given in Eqs (21) and (22) as well as those considered by Joseph (1977). When $m = n$, it can be verified by direct integration that:

$$\int_0^1 \mathbf{q_n} \cdot \mathbf{A} \cdot \mathbf{p_n} \, dy = -2 \tan^2 k_n. \tag{31}$$

Determination of the coefficients H_n

The boundary conditions to be satisfied at $\eta = 0$ are:

$$F(0, y) = \sum_{n=1}^{\infty} \frac{H_n}{k_n^2} p_{n_1}(y) = -\frac{1}{4} y^2 (y - 1), \tag{32}$$

$$\frac{\partial F}{\partial \eta}(0, y) = \sum_{n=1}^{\infty} -\frac{H_n}{k_n} p_{n_1}(y) = 0. \tag{33}$$

Differentiating Eq(32) twice with respect to y and using the relation $p_{n_1}'' = k_n^2 p_{n_2}$, the above equations may be written as:

$$\begin{pmatrix} 0 \\ -\frac{1}{2}(3y - 1) \end{pmatrix} = \sum_{n=1}^{\infty} \begin{pmatrix} \frac{H_n}{k_n} p_{n_1}(y) \\ H_n p_{n_2}(y) \end{pmatrix} = \sum_{n=1}^{\infty} H_n \begin{pmatrix} p_{n_1}(y) \\ p_{n_2}(y) \end{pmatrix} + \sum_{n=1}^{\infty} \left(-H_n + \frac{H_n}{k_n} \right) \begin{pmatrix} p_{n_1}(y) \\ 0 \end{pmatrix}. \tag{34}$$

The operator $\int_0^1 \begin{pmatrix} q_{m_1}(y) \\ q_{m_2}(y) \end{pmatrix} \begin{pmatrix} 0 & -1 \\ 1 & 2 \end{pmatrix} \begin{pmatrix} \cdot \\ \cdot \end{pmatrix} dy$ is applied to both sides. This yields:

$$\int_0^1 (3y - 1) \left(\frac{1}{2} q_{m_1}(y) - q_{m_2}(y) \right) dy = -2 H_m \tan^2 k_m + \sum_{n=1}^{\infty} H_n \left(\frac{1}{k_n} - 1 \right) \int_0^1 q_{m_2}(y) p_{n_1}(y) \, dy, \tag{35}$$

which can be simplified to:

$$-\sec k_m \tan k_m = -2H_m \tan^2 k_m + \sum_{n=1}^{\infty} H_n \left(\frac{1}{k_n} - 1\right) B_{nm}, \qquad (36)$$

where,

$$B_{nm} = \frac{1}{2(k_m - k_n)^2}\left[(\sin k_n - \sin k_m)^2 + 2\sin k_m k_n\right] - \frac{1}{2(k_m + k_n)^2}\left[(\sin k_n + \sin k_m)^2 - 2\sin k_m k_n\right]$$

$$-\frac{\sin k_m \sin k_n \sin(k_m + k_n)}{(k_m + k_n)^3} - \frac{\sin k_m \sin k_n \sin(k_m - k_n)}{(k_m - k_n)^3}. \qquad (37)$$

When $m = n$, B_{mm} reduces to:

$$B_{mm} = \frac{1}{2} - \frac{1}{3}\sin^2 k_m - \frac{\sin^2 k_m}{2k_m^2}. \qquad (38)$$

RESULTS AND SUMMARY

The unknown coefficients H_m were determined numerically from Eq(36) after truncation of the infinite series. Sixty terms were retained in the series corresponding to thirty values for k that are complex conjugates which satisfy Eq(19). The convergence of the series solution to the boundary conditions, Eqs (32) and (33) was verified. The deviations are of the order 10^{-6} and 10^{-3} respectively. The streamlines corresponding to the flow near the end wall are displayed in FIGURE 1. It is seen that beyond a distance of 1.5 units of the liquid depth from the end wall, the flow in the layer is predominantly parallel. This is in agreement with the results of Sen and Davis. The shape of the streamlines near the corner is a little different from what Sen and Davis have obtained. The present results are similar to the results of de Socio in this respect. The flow near the wall at $x = 1$ can be obtained immediately by symmetry.

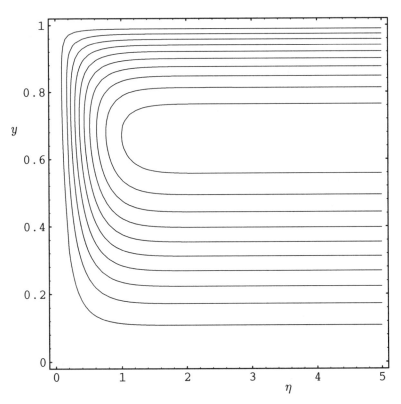

FIGURE 1. Ten Equally Spaced Streamlines for $0 \leq \eta \leq 5$, $0 \leq y \leq 1$. $\psi_{max} = 0$ at the Bottom, Left and Upper Boundaries. $\psi_{min} = -0.037$ at $\eta = 5, y = 2/3$.

Acknowledgments

The author would like to thank Prof. R.S. Subramanian, Clarkson University, for helpful discussions during the course of this investigation.

References

de Socio, L.M. (1979) "Convection Driven by Non-Uniform Surface Tension," *Letters in Heat and Mass Transfer*, **6**:375-383.

Joseph, D.D. (1977) "The Convergence of Biorthogonal Series for Biharmonic and Stokes Flow Edge Problems. Part I," *SIAM J. Appl. Math.*, **33**(2):337-347.

Joseph, D.D., and Sturges, L. (1978) "The Convergence of Biorthogonal Series for Biharmonic and Stokes Flow Edge Problems. Part II," *SIAM J. Appl. Math.*, **34**(1):7-26.

Kamotani,Y., S. Ostrach and A. Pline (1994) "Analysis of velocity data taken in Surface Tension Driven Convection Experiment in microgravity," *Phys. Fluids* **6**(11):3601-3609.

Ostrach, S. (1982) "Low Gravity Fluid Flows," *Ann. Rev. Fluid Mech.*, **14**:313-345.

Sen, A.K., and S.H. Davis (1982) "Steady thermocapillary flows in two-dimensional slots," *J. Fluid Mech.*, **121**:163-186.

Nomenclature

A	Matrix defined in Eq(29)
B_{nm}	Coefficients of a matrix defined by Eqs(36) and (37)
f_n	Papkovich-Fadle eigenfunction with index n defined by Eq(18)
F	Scaled streamfunction defined by Eq(13)
H	Height of the liquid layer, m
H_n	Coefficients (complex) in the expression for F in Eq(17)
k_n	Complex numbers defined by Eq(19)
L	Length of the liquid layer, m
p_n, q_n	Vectors with components $\{p_{n_1}, p_{n_2}\}, \{q_{n_1}, q_{n_2}\}$
T	Scaled temperature in the liquid layer
T_1, T_2	Temperature of the walls at $x=0$ and $x=1$ respectively, K
x, y	Scaled co-ordinates
ϵ	Aspect ratio $\frac{H}{L}$
ΔT	$T_1 - T_2$
η	Inner region co-ordinate $\frac{x}{\epsilon}$
μ	Viscosity of the liquid, $\frac{kg}{ms}$
σ_T	Temperature coefficient of surface tension, $\frac{N}{mK}$
ψ	Scaled streamfunction in the liquid layer
Subscripts	
i	inner region
m, n	indices for the components of a vector
o	outer region

GRAVITATIONAL EFFECTS ON DIAMOND AND CARBON FIBER FABRICATION

Elliot B. Kennel
Applied Sciences Inc
PO Box 579
Cedarville OH 45314-0579
Ekennnel@Apsci.com
937-766-2020

Chi Tang
Space Exploration Associates
PO Box 579
Cedarville OH 45314-0579
Tang@Apsci.com
937-766-2050

Abstract

One of the most interesting materials developments in the past several years has been the fabrication of carbon nanostructures such as C-60 and filaments based on the carbon lattice. Such structures, referred to informally as "buckyballs," "buckytubes" or "fullerenes," are fabricated by dissociating hydrocarbons in a high temperature chemically catalyzed environment. When the constraint of gravity is removed, it is very likely that exciting new chemical forms can be created, perhaps never before seen on earth. This is especially true for fiber forms, which are entrained in the natural thermal convection (i.e., density driven convection which depends upon the force of gravity) of the furnace. In addition, similar gas phase pyrolosis reactions are used to manufacture diamond fibers and diamond films. Gravity constrains the growth of diamond by requiring that they adhere to a nucleation point on a surface, which leads to limitations on the size of individual single crystals which can be grown. Accordingly, in microgravity, it may be possible that diamond of arbitrarily large size might be grown. Such materials can be expected to have a value which can justify their production in space.

INTRODUCTION

The ability to make diamond films from pyrolysis of hydrocarbon products is well known (Field, 1992). The basic technique is to use some method (e.g., hot filament, microwaves or laser radiation) to decompose hydrocarbon molecules in the presence of hydrogen. Although the carbon tends to be created in the form of graphite as well as diamond, the graphitic atoms are scavenged by the excess hydrogen. Thus a diamond crystal can be nucleated and grown on a surface.

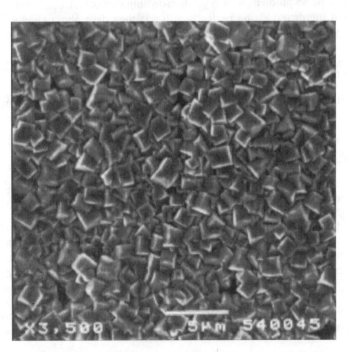

FIGURE 1. Typical Polycrystalline Diamond Film.

However, it has been very difficult to produce single crystals of diamond. The reason is that diamond is nucleated on a substrate (say, on silicon) typically at thousands of places at once. The diamonds then grow in clusters, producing a polycrystalline film as shown below. This type of growth is relatively easy to achieve, as it occurs over a fairly wide range of methane-to-hydrogen ratios (approximately from 1:100 to 1:10) and temperatures (950 to 1100 C). Typically, the diamond growth is much faster at 111 surfaces than for other crystal planes in the diamond lattice.

As can be easily seen in the Figure 1, the individual crystals shown excellent crystallinity, but the size of individual crystals is limited by the nucleation density. In addition, diamond can only be deposited on the exposed portions of the surface.

Diamond films with lower nucleation density result in larger individual crystals. The limiting case, however (e.g., a single nucleation site producing an individual monocrystal of diamond) is difficult to evaluate under terrestrial gravity.

In addition to diamond and graphite, there are several forms of carbon which can be created from a plasma discharge such as C-60 and related molecules (often referred to as "Buckyballs or Fullerenes due to their similarity to structures designed by the late Buckminster Fuller) and nanofibers (graphite single wall or multiwall tubes). The growth dynamics of all of these depend in a significant way upon gravity, as thermal buoyancy in the gas leads to a turbulent condition in the gaseous precursors.

DISCUSSION

The production of diamond as well as carbon nanofibers can be accomplished in a hydrocarbon pyrolysis furnace. Dissociation of hydrocarbon gases can be accomplished by a variety of means including laser, microwave and hot filament techniques. The hot filament method is attractive for space applications because low voltage DC can be used directly, so that high voltage power conditioning equipment is not required, and moreover avoids the complexity and mass associated with items such as microwave generators, lasers and the like.

Diamond from the gas phase

In space it is easily conceivable to grow a diamond in three dimensions simultaneously in the gas phase. It is envisioned that in microgravity, diamond could be seeded with individual single crystals, which would then grow independently. There would be no problem with crystals crowding each other out as is the case on a substrate. It is probably not necessary that the diamonds be perfectly still. If they drift about slowly in the chamber, they will continue to grow when they drift into the growth zone, much like nucleating rain drops.

It is envisioned that diamonds produced in this manner would grow in all directions at once, rather than in one direction as is the case for polycrystalline diamonds on a substrate.

It seems very likely that such diamonds would be much larger and higher quality than diamonds grown on the earth using this method. In fact, the limits to diamond growth using this microgravity technique are not known at this time.

Carbon Nanofibers from Hydrocarbon Gas

Within recent years, the practical production of carbon nanotubes has been enabled, resulting the ability to manufacture microscopic highly-ordered carbon fibers with exceptional properties.

One such fiber, referred to as PYROGRAF^{TM}-III, is produced by a catalytic process using hydrocarbon-containing gas with no substrate (Lake and Ting 1995). The catalyst functions as a nucleation point for carbon fiber growth, bonding securely in a graphitic bond. Rather than forming a plane, however, the graphitic sheet wraps around itself to form a cylindrical sheath, as depicted very schematically in Figure 1 (for the sake of simplicity, only a few atoms are shown).

Thus, the metal catalyst acts as a sieve for pyrolytic carbon atoms, with the result being that carbon fibers are nucleated and grown from the catalyst particle. These fibers can grow up to several mm long.

PYROGRAF^{TM} III is made under a joint venture with Applied Sciences, Inc. (the sister company of Space Exploration Associates) and Delphi Chassis (formerly Delco Chassis) of Dayton, a division of General Motors.

The properties of PYROGRAF^{TM} III are significantly different than those of conventional carbon fiber. When heat treated to achieve a high index of graphitization, the electrical conductivity becomes nearly metallic. It also has an extremely high surface area to volume ratio owing to its submicron diameter. The highly ordered graphitic lattice results in nearly metallic electrical conductivity and extremely small size.

As shown in Figure 4, the fibers can be straight for some distance, but eventually start to become kinked and ultimately tangled. It is very likely that this is due at least in part to thermal buoyancy effects, causing a turbulent gas condition in the presence of gravity.

The growth process can be visualized according to the following sequence, as shown in Table 1. In microgravity, it is likely that thermal buoyancy effects would be eliminated. As a consequence, fiber growth would occur in a more laminar environment. Thus, the fiber morphology can be expected to be entirely different than the chaotic, entangled mass which is currently grown.

Support for this argument is obtained by examining the case for Pyrograf™-I. This fiber is grown using a solid substrate and very slow gas flow. The flow fields are restricted by the substrates which act like baffles, thus resulting in a basically laminar condition. In this case, fibers grown on the top of the substrate grow as a flat mat, comprised of straight fibers of up to a few cm long. Conversely, those fibers growing on the bottom of the substrate (i.e., in negative gravity), grow as a short staple tow. Thus it is clear that the fiber growth is dramatically changed by the orientation of the fibers with respect to gravity.

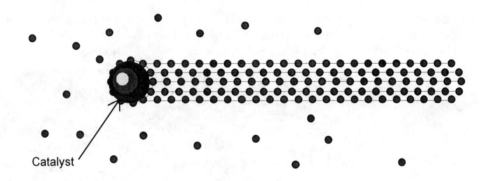

FIGURE 2. Nucleation of Carbon Fiber by Metal Catalyst

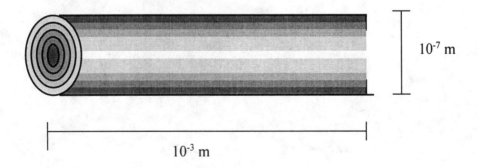

FIGURE 3. Illustration of Fiber Geometry.

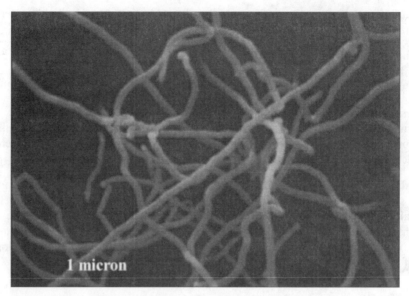

FIGURE 4. Scanning Electron Micrograph of Pyrograf™-III fibers As-Grown.

One of the most interesting properties of Pyrograf™-I is its high thermal conductivity, measured at 1950 W/mK at ambient temperature. By comparison, copper is only 401 W/mK and aluminum is only 237 W/mK. Only diamond, with a thermal conductivity as high as 2320 W/mK, is higher.

FIGURE 5. Fiber Grown on Carbon-Carbon Substrates Produces a Mat in Positive Gravity; but a Short Staple Tow in Negative Gravity. Thus the Effect of Gravity on Fiber Growth is Evident.

TABLE 1. Fiber Growth Sequence and Effect of Gravity.

Step 1. Gas heating	Using microwaves or electrical heaters, the temperature is raised to ~1000 C.
Step 2. Fiber nucleation	The fiber is nucleated, and carbon atoms attach themselves to the growing nanotube.
Step 3. Fiber growth	The fiber lengthens very quickly. Thermal buoyancy causes localized turbulence causing the fiber shape to become kinked and erratic.
Step 4. Termination of Growth	Due to thermal buoyancy and other factors, the fiber leaves the growth zone and growth is terminated

SUMMARY

Based on the above observations, the growth of carbon fiber and diamond from the gas phase is heavily influenced by gravity. In a microgravity environment, it is reasonable to expect that new fiber morphologies can be created. It is also possible that, using different gas mixtures, single crystal diamonds can be grown, possibly of large size.

Acknowledgment

This research was performed using internal research and development funds of Applied Sciences Inc.

References

Field, J. E. (1992), *The Properties of Natural and Synthetic Diamond*, Academic Press, New York.

Ting, J. M., and M. L. Lake (1994) "Passivation of Carbon Fiber by Diamond Deposition," *Diamond and Related Materials* 3:1249-1255.

Ting, J. M., and M. L. Lake (1994) "Diamond-coated carbon fiber," *Diamond and Related Materials* 3:1249-1255.

Ting, J. M., and M. L. Lake (1995) "Vapor Grown Carbon Fiber Reinforced Carbon Composites," *Carbon*, 33(5):663-667, 1995.

EFFECTS OF GRAVITY ON CAPILLARY MOTION OF FLUID

S. H. Chan, T. R. Shen and G. D. Proffitt
Department of Mechanical Engineering
University of Wisconsin-Milwaukee
Milwaukee, WI 53201
(414) 229-5001

B. Singh
NASA Lewis Research Center
Cleveland, OH 44135
(216) 433-5396

Abstract

A new experimental approach provides good visualization of the fluid capillary motion along the rectangular groove corners. A hydrodynamically controlled rewetting model, which considers the capillary motion of both the bulk fluid front as well as the fluid along the groove corner, is presented to predict the transient rewetting distance and the corner motion length at initial plate temperatures lower than the rewetting temperature, and has been found to be in relatively sufficient agreement with the experimental data. It has been found that gravity has considerable effects on capillary motion of fluid along the rectangular corners. It is concluded that for the condition investigated, namely, non-evaporating surface tension driven flow along rectangular grooves under microgravity conditions, the capillary bulk front consideration is sufficient and there is no need to consider fluid capillary corner motion for rectangular groove surfaces.

INTRODUCTION

In our prior investigation (Chan, et al. 1995), the capillary motion of a bulk fluid front, in a rectangular shaped groove, was considered, while fluid motion along the rectangular groove corners was ignored due to difficulties with observation of the corner motion using conventional working fluids, such as isopropanol. Numerous researchers have studied the characteristics of fluid motion along triangular grooves, as well as other groove geometry, but it is to the authors' best knowledge that no prior investigations have been conducted on the capillary transient fluid motion along the groove corners of a rectangular shaped groove.

In a rectangular groove geometry, each groove has two corners, in which fluid may be gathered with much smaller radius, thus higher capillarity, than the bulk fluid front. It is then of interest to study the effects of gravity on the capillary motion of fluid along the rectangular groove corners, in order to better understand the rewetting characteristics of a thin liquid film flowing over a grooved plate. During the course of the research on this subject, both experimental and theoretical investigations have been performed. Because our previous working fluid, isopropanol, and other common working fluids are transparent and effectively colorless on their own, visual data of the corner front motion was difficult to obtain with standard visualization techniques. In order to observe the experimental fluid motion along the rectangular groove corners, a special fluid, Spectroline® SP-21, was used. This fluid is composed of 92.6% ethyl alcohol, 7.4% acetone (by weight) and minute amount of fluorescent "brightener" that makes the fluid glow bright green when observed under ultraviolet light. When the advancing liquid front was observed with the aid of a microscope, this fluid provides excellent visualization of the fluid corner motion. In the theoretical investigations, a hydrodynamically controlled rewetting model, which considers the capillary motion of both the bulk fluid front as well as the fluid along the groove corners, is presented to predict the transient rewetting distance and corner motion length at initial plate temperatures lower than the rewetting temperature.

EXPERIMENTAL INVESTIGATIONS

Figure 1 shows a schematic detail of the experimental apparatus utilized in this investigation. As can be seen from the schematic, a grooved test plate was mounted in the test fixture with a horizontal inclination angle ϕ. The grooved test plate, with groove width equal to 0.4 mm and depth 0.5mm, was fabricated from 101 Oxygen-Free Copper stock so as to inhibit high temperature surface oxidation that could affect the test results. The SP-21 fluid was placed in the experimental reservoir mounted on a vertical travel, pneumatic positioning system, which was utilized to ensure that the fluid reservoir travel, with respect to the test plate, was consistent from run to run and for all ϕ values. An UV light source was mounted in the test fixture so as to illuminate the entire grooved test section.

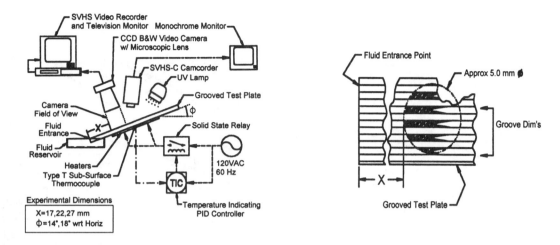

FIGURE 1. Experimental Apparatus and Camera Field of View

Liquid corner motion was observed and recorded with the aid of a black and white CCD video camera and a microscopic lens assembly. The camera field of view, which is also illustrated in Figure 1, was recorded by a Mitsubishi® SVHS VCR and simultaneously viewed on a 19" monitor. The fluid bulk motion was recorded and observed via a JVC® SVHS-C camcorder and monochrome monitor combination. The plate temperature was controlled throughout the experimental operations by an Omega® AutoTune Indicating, PID controller. Prior to beginning the experimental runs, the test plate was thoroughly cleaned to ensure that minimal oxidation and other impurities were present. Both of the test tapes were played back on the Mitsubishi® SVHS VCR in a frame-by-frame fashion (1/30 second per frame) to determine the position of the fluid front and the length of the corner motion versus time. The detailed experimental procedure is given in (Shen 1997).

In the experiment, ϕ was adjusted to 14° and 18°, respectively and the initial plate temperatures were set at 25°C and 80°C, respectively. The CCD camera field of view position (X, Figure 1) were set to be 17mm, 22mm, and 27mm, respectively. Figure 2 shows the experimental results for $\phi=14°$ and $\phi=18°$. It should be pointed out that from the fluid bulk motion recordings, it was very difficult to distinguish whether the observed fluid front was the corner front or the bulk liquid front or some point in between. Since the observed corner motion length was relatively short, the overall error of the liquid front position, estimated to be ± 2.5mm, will be very small.

From Figure 2, we can see that for a larger ϕ, both the liquid bulk and liquid corner motion move forward at a lower speed and corner motion length relative to the liquid front position is larger. This indicates the presence of gravitational effects on the capillary motion of the fluid. When the ϕ is increased, the gravitational component acting on the fluid along the groove axial direction increases accordingly, thus, causing both the liquid bulk and corner motion to progress at a lower speed. Since most of the liquid mass is located in the bulk portion with only a very small mass in the groove corner portion, the gravitational force is expected to have more effect on the bulk flow than on the corner. This is why the corner motion length relative to the liquid front position is increased when the inclination angle increases. If the plate is placed in a horizontal position, the gravitational component along the groove axial direction becomes zero, thus, both bulk front motion and corner motion of the fluid can be expected to progress forward at similar rates, in turn, leading to a much shorter corner motion length. This matter will be discussed further in the next section.

Figure 2 also illustrates the initial plate temperature effect on the rewetting velocity and the corner motion length. As can be seen from the figure, a higher initial plate temperature leads to a higher rewetting velocity and a longer liquid corner motion length. This is because the liquid viscosity decreases much faster than the surface tension, as the temperature is increased. Thus, the liquid viscous force is much smaller at higher temperatures, resulting in a higher rewetting velocity and a larger liquid corner motion length.

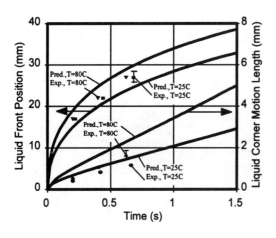 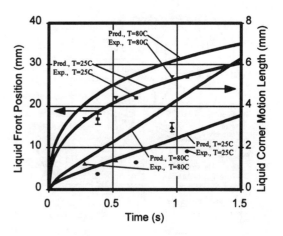

FIGURE 2. Transient Liquid Front Position and Corner Motion Length at $\phi = 14°$ (Left) and $\phi = 18°$

NUMERICAL MODEL

The control volumes V_1 and V_2, and the cross-sections A_1 and A_2, are shown in Figure 3 (Note: the ϕ is not depicted in this figure). In control volume V_1, surface 1 is a stationary surface, through which fluid flows into the control volume at a transient local mean velocity u_{1m}, and has the cross-section area A_1. Unlike the stationary surface 1, surface 2, with the cross-section area A_2, moves in the positive axial direction with the bulk front at velocity u_{s2}. In addition, fluid flows through surface 2, from control volume V_1 into control volume V_2, with a transient local mean velocity of u_{2m}. The fluid front, with a minimum capillary radius, moves in the positive axial direction with the velocity u_f. For control volume V_1, the mass and momentum equations are:

$$\frac{d}{dt}\int_{V_1}\rho dV = \rho u_{1m}A_1 - \rho(u_{2m}-u_{s2})A_2 \tag{1}$$

$$\frac{d}{dt}[\rho u_{V1m}V_1] = \rho u_{1m}^2 A_1 - \rho u_{2m}(u_{2m}-u_{s2})A_2 + F_{p1} - F_{v1} - F_{g1} \tag{2}$$

where ρ is fluid density, u_{V1m} is the fluid mean velocity in control volume V_1; F_{p1}, F_{v1} and F_{g1} are the capillary pressure force, viscous force and the gravitational force, in control volume V_1, respectively. For control volume V_2, the mass and momentum equations are:

$$\frac{d}{dt}\int_{V_2}\rho dv = \rho(u_{2m}-u_{s2})A_2 \tag{3}$$

$$\frac{d}{dt}[\rho u_{V2m}V_2] = \rho u_{2m}(u_{2m}-u_{s2})A_2 + F_{p2} - F_{v2} - F_{g2} \tag{4}$$

where u_{V2m} is the fluid mean velocity in control volume V_2; F_{p2}, F_{v2} and F_{g2}, are the capillary pressure force, viscous force and the gravitational force in control volume V_2, respectively. It should be noted that the purpose of this investigation is to find the transient fluid bulk front velocity (u_{s2}) and fluid corner front velocity (u_f), though the above equations do not contain u_f. All other variables, u_{1m}, u_{2m}, u_{V1m}, and u_{V2m} are expressible in terms of u_{s2} and u_f, while the expressions of F_{p1}, F_{v1}, F_{p2}, F_{v2}, V_1 and V_2 are obtainable (Shen 1997) after some necessary assumptions are made: the liquid flow is laminar and one-dimensional; the liquid is in an isothermal state; the vapor pressure is constant and equal to one atmospheric pressure; the fluid bulk front keeps its shape as shown in cross-section c-c of Figure 3, where the intrinsic meniscus shape is circular with $R_2=0.5l_2$ and the liquid is just in touch with the groove bottom surface; at the liquid entrance $R_1=\infty$ or $1/R_1=0$, and $1/R$ changes linearly from $x=0$ to $x=x_2$; in control volume V_2, the intrinsic meniscus shape is circular, with the radius of curvature decreasing linearly from $R_2=0.5l_2$ at $x=x_2$ to $R_c=R_{min}$ at $x=x_f$; the liquid height l_x in section b-b of Figure 3 is decreased linearly from l_1 at $x=0$ to $0.5l_2$ at

$x=x_2$. The local liquid velocity (u_x), at any time (t) changes linearly from u_{1m} at $x=0$ to u_{2m} at $x=x_2$ in control volume V_1 and from u_{2m} at $x=x_2$ to u_f at $x=x_f$ in control volume V_2.

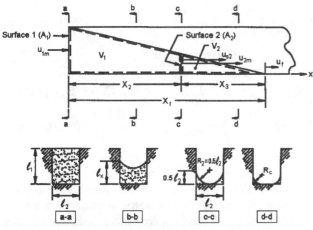

FIGURE 3. Numerical Model Geometric Detail

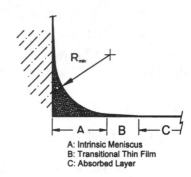

FIGURE 4. Corner Front Detail

When using the Young-Laplace equation to calculate the capillary pressure force, the radius of curvature cannot be zero. The radius of the corner motion front (at $x=x_f$) is assumed to be R_{min}, which is expected to be a very small, non-zero value. In order to determine R_{min}, an extended meniscus concept, similar to the prior study (Ha and Peterson 1996), was used. We assumed that in the cross-section of the corner front (at $x=x_f$), there is a transitional thin film and an adsorbed layer region adjacent to the intrinsic meniscus, as shown in Figure 4. In the intrinsic meniscus region, the capillary pressure is the dominant driving force affecting the liquid volume. Although the capillary pressure may still be present in the transitional thin film, the disjoining pressure effect is assumed dominant in this region. Based on this argument, the following equation can be set up in the junction of the intrinsic meniscus and transitional thin film at the corner front by matching the capillary pressure to the disjoining pressure:

$$P_v - P_l = \sigma / R_{min} = \overline{A} / \delta_j^3 \tag{5}$$

where $\overline{A}$ is the modified Hamaker constant and is equal to $A/6\pi$, σ is the surface tension and δ_j is the thickness of the liquid at the junction. There are many complex analytic formulae for calculating the Hammaker constant. Several $\overline{A}$ values have been reported by different investigators. One approach for obtaining approximate values for the unknown system Hamaker constants is in terms of known A values for individual components (Israelachvili 1991), as follows

$$A_{132} \approx (\sqrt{A_{11}} - \sqrt{A_{33}})(\sqrt{A_{22}} - \sqrt{A_{33}}) \tag{6}$$

where A_{ii} is the Hamaker constant for pure component i, and tabulated by the author. Since the major component of the fluid we used is ethyl alcohol, a copper/ethanol/vapor system is taken as an approximation for our system. From the table we have $A_{11} = 40 \times 10^{-20}$, $A_{22} = 0$ and $A_{33} = 4.2 \times 10^{-20}$. $\overline{A}$ can be evaluated as $\overline{A} = A_{132} / 6\pi = 4.6 \times 10^{-21}$. Due to the microscopic magnitude of this theory, the thickness of the thin film, in this junction region, is extremely small. It was reported (Ha and Peterson 1996) that $\delta_j = 3 \times 10^{-9}$m for an axial distance from the leading edge $x=3.23$mm and $\delta_j = 7.8 \times 10^{-9}$m for $x=7.47$mm. As an approximation, a mean value of their results, $\delta_j = 5.4 \times 10^{-9}$m was used. Substituting $\overline{A}$, δ_j, and a measured value of the surface tension ($\sigma=0.016$ N/m) into Equation (5), we obtain $R_{min} = 1.7 \times 10^{-6}$.

With the Equations (1), (2), (3), (4) and the major assumptions listed previously, the following two equations, for the unknowns u_f and u_{s2}, can be readily derived:

$$\frac{du_f}{dt} = \frac{B_1 C_6 - B_2 C_{11}}{C_6 C_{10} - C_5 C_{11}} \quad (7)$$

$$\frac{du_{s2}}{dt} = \frac{B_2 C_{10} - B_1 C_5}{C_6 C_{10} - C_5 C_{11}} \quad (8)$$

where

$$B_1 = [\rho(A_{m1}x_2 + V_{o1})]^{-1} \{F_{p1} - F_{v1} - F_{g1} - \rho(C_{10}u_f + C_{11}u_{s2})A_{m1}u_{s2} + \rho A_1(C_7 u_f + C_8 u_{s2})^2 - \rho A_2[C_1 u_f + (1-C_1)u_{s2}](C_1 u_f - C_1 u_{s2})\} \quad (9)$$

$$B_2 = [\rho(A_{m2}x_3 + V_{o2})]^{-1} \{F_{p2} - F_{v2} - F_{g2} - \rho A_{m2}(C_5 u_f + C_6 u_{s2})(u_f - u_{s2}) + \rho A_2[C_1 u_f + (1-C_1)u_{s2}](C_1 u_f - C_1 u_{s2})\} \quad (10)$$

and $C_1 = A_{m2}/A_2$, $C_2 = 2(1-\pi/4)/A_{m2}$, $C_3 = (\ell_1^2 + 0.25\ell_2^2 + 0.5\ell_1\ell_2)/3$, $C_4 = (\ell_1^2 + 0.75\ell_2^2 + \ell_1\ell_2)/12$, $C_5 = C_1 C_2(C_3 - C_4) + C_2 C_4$, $C_6 = C_2(1 - C_1)(C_3 - C_4)$, $C_7 = A_{m2}/A_1$, $C_8 = (A_{m1} - A_{m2})/A_1$, $C_9 = \int_0^1 X A_1(X) dX$, $C_{10} = C_7 + [C_9(C_1 - C_7)]/A_{m1}$, $C_{11} = C_8 + [C_9(1 - C_1 - C_8)]/A_{m1}$.

In the above expressions, A_{m1} and A_{m2} are the average cross-section areas of the control volumes V_1 and V_2, respectively. It can be proven that A_{m1} and A_{m2} are constant values in terms of the geometric assumptions we have made. $A_1(X)$ is the cross-section area at $X=x/x_2$ in the control volume V_1. F_{p1} and F_{p2} are calculated in terms of the Young-Laplace equation and the assumptions of the radii of curvature. F_{v1} and F_{v2} are calculated using the approach of laminar flow in noncircular ducts. ρV_{o1} and ρV_{o2} are virtual masses created to account for entrance effects (Levine et al. 1976; Dreyer et al. 1994). $V_{o1} = 0.916 r_{e1} A_1$, $V_{o2} = 0.916 r_{e2} A_2$, where $r_{e1} = 2(0.25\ell_1\ell_2/\pi)^{0.5}$, $r_{e2} = (A_2/\pi)^{0.5}$.

Since the fluid properties, such as surface tension, viscosity and density, of the Spectroline® SP-21, are not available from the manufacturer, we were forced to experimentally determine these properties. A Stormer®, falling weight type viscometer was used to measure the viscosity of the fluid. The measured viscosity, surface tension, and density of the fluid are: $\mu_{T=25°C} = 8.7\times10^{-4}$ Nm/s², $\mu_{T=52°C} = 4.8\times10^{-4}$ Nm/s², $\sigma_{T=25°C} = 0.016$ N/m, $\sigma_{T=52°C} = 0.015$ N/m, $\rho_{T=25°C} = 790$ kg/m³, $\rho_{T=52°C} = 780$ kg/m³. The estimated error is less than 20% for viscosity and surface tension, and less than 5% for density. In the calculations, the reference temperature method was used to determine the fluid properties. For simplicity, an average of the liquid temperature and initial plate temperature was used as the reference temperature.

In order to provide theoretical data to compare the test results with, Equations (7) and (8) were solved simultaneously using a fourth-order Runge-Kutta numerical method. The resulting calculated values are shown in Figures 2 and 5. Comparison of theoretical data to experimental values is shown in Figure 2 for $\phi = 14°$ and $\phi = 18°$, and two initial plate temperatures of 25°C and 80°C. The predicted transient liquid front position and liquid corner motion length were found to be in good agreement with the experimental data. Looking at both the predicted and experimental results, it is obvious that the liquid front velocity decreases with increasing plate inclination angle but increases with increasing initial plate temperatures, and the liquid corner motion length increases with increasing plate inclination angle and initial plate temperature. As discussed in the prior section, increasing the plate inclination angle will increase the gravitational force component acting on the fluid, in the negative axial direction, which seems to have more effect on the bulk flow than on the corner. As the fluid temperature is increased, the liquid viscosity decreases significantly, but with the same temperature increase, the surface tension of the fluid is changed very little. In going through a simple momentum analysis of the groove system, it becomes apparent that this temperature increase will result in a larger positive resultant force acting on the fluid than at lower temperatures.

To further understand the effects of gravity on capillary motion of the fluid along rectangular groove corners, three different inclination angles ($\phi=0°$, 10° and 20°), at the initial plate temperature of 25°C, were input into the model, with the resulting data shown in Figure 5. It can be observed from Figure 5 that in the 20° case, where the gravitational component acting on the liquid in the negative axial direction is quite large, the fluid corner motion length is about 13% of the total liquid length. For the 0° case, where there is no gravitational force acting on the liquid in the negative axial direction, the fluid corner motion length is greatly reduced to less than 3% of the total liquid length. Thus, we can reach the following conclusions :

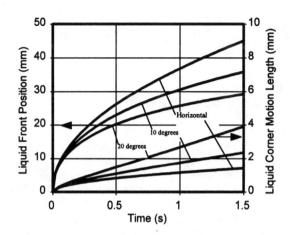

FIGURE 5. Predicted Transient Liquid Front Position and Corner Motion Length (T= 298°K)

CONCLUSION

Gravity has considerable effects on the capillary motion of fluid along rectangular groove corners. Under microgravity conditions, the gravitational force acting on the fluid is near zero and the corner motion length, with respect to the liquid bulk front, is significantly reduced. It is therefore concluded, that for the condition investigated, namely, non-evaporating surface tension driven flow along rectangular grooves under microgravity conditions, the corner motion component of fluid flow in a rectangular groove can be ignored and any further theoretical calculations can be based solely on the bulk motion model.

Acknowledgments

The authors would like to acknowledge the support from the NASA Lewis Research Center regarding this investigation.

References

Carey, V. P.(1992) *Liquid-Vapor Phase- Change Phenomena,* Taylor & Francis, 1: 67-70.

Chan, S. H., J. D. Blake, T. R. Shen and Y. G. Zhao (1995) "Effects of Gravity on Rewetting of Capillary Groove Surface at Elevated Temperatures - Experimental and Theoretical Studies," ASME Journal Heat Transfer, 117(1): 1042-1047.

Dreyer, M., A. Delgado, and H. J. Rath (1994) "Capillary Rise of Liquid Between Parallel Plates Under Microgravity," *J. Colloid and Interface Science*, 163(1): 158-168.

Ha, J. M. and G. P. Peterson (1996) "The Interline Heat Transfer of Evaporating Thin Films Along a Micro Grooved Surface," ASME Journal of Heat Transfer, 118(1): 747-753.

Israelachvili, J. (1991) *Intermolecular and Surface Forces*, 2nd Edition, Academic Press, 1: 200-201.

Shen, T. R. (1997) "Rewetting of Capillary Groove Surface - Experimental and Theoretical Studies," *A Thesis for the Degree of Master of Science,* University of Wisconsin-Milwaukee.

EXPERIMENTAL STUDY AND MODELING OF THE EFFECT OF LOW-LEVEL IMPURITIES ON THE OPERATION OF THE CONSTRAINED VAPOR BUBBLE

J. Huang, M. Karthikeyan, J. Plawsky, and P.C. Wayner, Jr.
The Isermann Department of Chemical Engineering
Rensselaer Polytechnic Institute
Troy, NY 12180-3590
Tel: (518)276-6199, Fax: (518)276-4030, wayner@rpi.edu

Abstract

A Constrained Vapor Bubble (CVB) formed in a 3x3x40 mm fused silica cell of square cross section was studied. This small-scale device is capable of transferring high heat flux using the latent heat of vaporization of liquid. Capillarity is the driving force to recirculate the working fluid. Experiments were conducted at different operating conditions to obtain the temperature profiles of the cell. The effect of low-level impurities was analyzed from a thermodynamic point of view. A two-dimensional model was developed to describe the temperature distribution in the evaporator of the CVB. An area averaged heat transfer coefficient was used to account for the complicated heat transfer mode on the inside wall of the cell. The results demonstrated that small amounts of a second component drastically affects the temperature distribution and heat transfer characteristics of the evaporation process due to distillation.

INTRODUCTION

A generic vapor bubble formed and constrained by underfilling an evacuated enclosure with a liquid has many uses. A specific Constrained Vapor Bubble (CVB) formed in a 3x3x40 mm fused silica cell of square cross-section is a large version of Cotter's (1984) micro heat pipe. Studies on micro heat pipes have been reported (Babin *et al.* 1990 and Khrustalev *et al.* 1994). The Constrained Vapor Bubble is presented in Figure 1. For a complete wetting system, the liquid will adhere to the walls of the chamber. For a finite contact angle system, some of the walls will have only a small amount of adsorbed liquid that changes the surface properties at the solid-vapor interface. Liquid will fill a portion of the corners in both cases. If the temperature at End (2) is higher than at End (1) because of an external heat source, energy flows from End (2) to End (1) by means of conduction in the walls and by a combined evaporation, vapor flow, and condensation mechanism. Heat applied to End (2) of the CVB vaporizes the liquid in this region and the vapor is forced to move to End (1) where it condenses, releasing the latent heat of vaporization in the process. The curvature of the liquid-vapor interface changes continually along the axial length of the cell because of viscous losses, and the vaporization/condensation process. This capillary pressure difference between the evaporator and the condenser regions causes the working fluid to flow back to End (2) from End (1) along the right-angled corner regions. These corner regions act as liquid arteries and hence replace the wicking structure of a conventional heat pipe.

Of particular interest herein is the effect on the performance of Constrained Vapor Bubble of a small amount of a second component. Since distillation naturally occurs in this device under non-isothermal conditions, there is an accumulation of the low vapor pressure component at the hotter end of the device. For example, we found experimentally that a change in the purity of the initial charge from 99.8% to 99% had a dramatic effect on the temperature profile in the CVB. This system is being used to experimentally characterize both the interfacial force field under equilibrium conditions and transport processes under non-equilibrium conditions (DasGupta *et al.* 1995, Karthikeyan *et al.* 1996 and Huang *et al.* 1997). The CVB experimental setup consisted mainly of the silica cell (partially filled with pentane liquid), the Pyrex connecting tubes, a resistance heater, and four miniature (5x5x2.4 mm) coolers. Figure 2 shows the location of the thermocouples and a portion of the experimental setup. The experimental procedure has been reported by Karthikeyan *et al.* (1997). In this paper, we report additional experimental results and theoretical analysis of the temperature profile of the CVB.

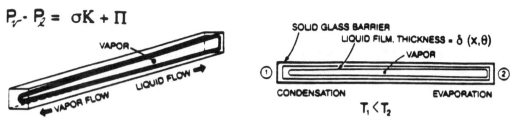

FIGURE 1. Constrained Vapor Bubble Concept.

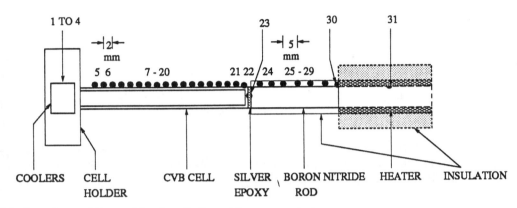

FIGURE 2. Experimental Setup (Numbers Indicate Thermocouples).

THEORETICAL ANALYSIS

In this section, we will derive an equation to demonstrate the effect of low-level impurities on the temperature distribution of the CVB from a thermodynamic point of view. A two-dimensional temperature model is also presented.

Interfacial Thermodynamics

Herein, we derive an equation that relates the concentration of liquid to the temperature. From a thermodynamic point of view, this equilibrium for mass transfer model can help us understand qualitatively the effect on the performance of the CVB of a second component in liquid. Starting from the Gibbs-Duhem equations for both liquid and vapor phases near the interface region, the chemical potentials can be expressed as:

$$x_1 d\mu_{l1} + x_2 d\mu_{l2} = -S_l dT + V_l dP_l ,\qquad(1)$$

$$y_1 d\mu_{v1} + y_2 d\mu_{v2} = -S_v dT + V_v dP_v ,\qquad(2)$$

where x_1 is the mole fraction of component 1 in liquid, y_1 the mole fraction of component 1 in vapor, and $x_2 = 1 - x_1$, $y_2 = 1 - y_1$. At equilibrium, $\mu_{l1} = \mu_{v1} = \mu_1$, $\mu_{l2} = \mu_{v2} = \mu_2$. Subtracting equation (2) from (1), we have:

$$(x_1 - y_1)(d\mu_1 - d\mu_2) = -(S_l - S_v)dT + V_l dP_l - V_v dP_v .\qquad(3)$$

Taking into account the influence of capillary pressure and the disjoining pressure on the liquid-vapor interface, the augmented Young-Laplace equation is represented as

$$P_v - P_l = \sigma K + \Pi ,\qquad(4)$$

where K is the curvature of the interface and Π is the disjoining pressure. Substituting equation (4) into equation (3), and noting that $S_v - S_l = h_{fg}/T$, equation (3) becomes:

$$(x_1 - y_1)(d\mu_1 - d\mu_2) = \frac{h_{fg}}{T} dT + (V_l - V_v) dP_v - V_l d(\sigma K + \Pi). \tag{5}$$

Assuming $V_v - V_l = RT/P_v$ and noting that $d\mu_1 = RTd(\ln x_1 P_{v1}^{sat})$, and $d\mu_2 = RTd(\ln x_2 P_{v2}^{sat})$, $V_l = 1/\rho_l$, also assuming that the ratio $\frac{P_{v1}^{sat}}{P_{v2}^{sat}} = $ constant, and $y_1 \approx 1$, and rearranging equation (5):

$$d(\ln P_v) = d(\ln x_1) + \frac{h_{fg}}{RT^2} dT - \frac{1}{\rho_l RT} d(\sigma K + \Pi). \tag{6}$$

Integrating between the evaporator, x_1, and the condenser, $x_1=1$, and neglecting the pressure difference caused by capillarity and the disjoining pressure, we obtain:

$$\ln x_1 = \frac{h_{fg}}{RT_e T_c}(T_c - T_e), \tag{7}$$

where R is the ideal gas constant, h_{fg} the latent heat of vaporization. T_e is the temperature at the beginning of the evaporator, and T_c the temperature at the beginning of the condenser.

Equation (7) is used later to evaluate the temperature rise that is caused by the liquid concentration gradient. It is obvious from equation (7) that a decrease in concentration of pentane will cause an increase in the temperature difference between the evaporator and the condenser of the CVB.

Two-Dimensional Model of the Evaporator

Huang *et al.* (1997) developed a two-dimensional temperature model for the evaporator of the CVB. The temperature distribution of the silica wall in the evaporator of the CVB was treated as a two dimensional problem. Assuming a symmetric temperature distribution, the silica wall of the CVB was treated as a slab, taking x as the axial distance starting at 1 mm (thickness of the cell wall) after the beginning of the cell, and y the radial distance starting from the inside wall of the cell. The heat transfer condition for the outside wall of the cell is natural convection. The condition for the inside wall of the cell is more complicated. There are thermal resistance associated with the flowing liquid and the evaporation process at the liquid-vapor interface. But since the temperature of the liquid is unknown, we assumed an overall forced convection heat transfer mode between the wall and the vapor phase. An area average overall heat transfer coefficient, h_i, was used to account for the wall-liquid convection, liquid evaporation and liquid-vapor convection. The governing Laplace equation was solved analytically with its boundary conditions. Let $\theta = T - T_\infty$, where T_∞ is the outside surrounding temperature. The solution of θ is combined of two parts, $\theta = \theta_1 + \theta_2$, where:

$$\theta_1 = C_1 y + C_2, \tag{11}$$

$$C_1 = \frac{hh_i(T_\infty - T_v)}{k_s(h + h_i) + hh_i l}, \quad C_2 = -\frac{hlC_1 + h_i(T_\infty - T_v)}{h + h_i}, \tag{12}$$

$$\theta_2 = \sum_{n=1}^{\infty} C_n \cosh \lambda_n (L_e - x)(\cos \lambda_n y + \frac{h_i}{k_s \lambda_n} \sin \lambda_n y), \tag{13}$$

where λ_n is the characteristic solution of the following transcendental equation:

$$k_s \lambda (h + h_i) \cot \lambda l = (k_s \lambda)^2 - hh_i, \tag{14}$$

and,

$$C_n = \frac{Q[\sin \lambda_n l + \frac{h_i}{k_s \lambda_n}(1 - \cos \lambda_n l)]}{k_s A_c \lambda_n^2 [\frac{h_i^2}{k_s^2 \lambda_n^2}(\frac{l}{2} - \frac{\sin 2\lambda_n l}{4\lambda_n}) + \frac{l}{2} + \frac{\sin 2\lambda_n l}{4\lambda_n} + \frac{h_i}{2k_s \lambda_n^2}(1 - \cos 2\lambda_n l)] \sinh \lambda_n L_e}, \tag{15}$$

where Q is the heat input to the CVB, A_c the cross-sectional area of the silica wall, and k_s the thermal conductivity of the wall. T_v is the vapor temperature. h_i is the area average heat transfer coefficient which averages out the heat transfer between the silica wall and liquid, and the liquid and vapor over the inside perimeter of the CVB evaporator. h is the heat transfer coefficient between the outside wall of the cell and the air. The value of h was obtained by fitting the fin equation to the experimental temperature profile of a dry run (dry cell without liquid inside). l is the thickness of the wall of the cell and L_e the length of the evaporator.

RESULTS AND DISCUSSION

Equation (7) was used to evaluate the experimental temperature profiles. Three sets of experimental data were analyzed. The operating conditions of these experiments are listed in Table 1. Note that negative β means that the evaporator is raised higher than the condenser of the CVB. It is also important to notice that Q is the heat input from the beginning of the bubble which does not necessarily start from the beginning of the CVB. In our experiments with low grade pentane (Experiments I & III), we observed that the non-evaporative impurities accumulated at the beginning of the cell (evaporator end). A schematic of the evaporator of the CVB is shown in Figure 3. We found that the bubble started at 4 mm from the end of the cell (x=4 mm) for Experiment I, 5 mm from the end of the cell for Experiment III and x=0 for Experiment II. Since there is non-evaporative liquid accumulated at the beginning of the CVB for Experiments I and III, the real concentration of pentane at the beginning of the bubble is lower than 99% for both Experiment I and III.

Figure 4 shows the temperature profiles of these three experiments. It is obvious that the temperature at the beginning of the CVB for Experiment I and III is higher than that of the Experiment II, although the heat input for Experiment I is lower than that for Experiment II. This is due to the non-evaporative liquid that accumulated at the beginning of the cell for both Experiment I and III. We can see a clear temperature drop in the evaporator due to the low thermal conductivity of the silica wall. Practically, we want to avoid this by using a metal as the wall material. However, it is advantageous for us here to see the meniscus and analyze the temperature difference in order to understand the heat transfer mechanism in this region. In Experiment II, we expect that there is intensive evaporation at the beginning of the evaporator, therefore, the liquid temperature is much lower than that of the wall. This was proven to be true by the analysis using the two-dimensional model. However, due to the existence of non-evaporative liquid, the liquid temperature at the beginning of the evaporator is almost the same as the wall temperature for Experiments I and III. Using equation (7), we found that the concentration of pentane at the beginning of the bubble is about 72% for Experiment I. This caused a 10 K temperature drop between the evaporator and the condenser. In Experiment III, more non-evaporative liquid accumulated at the beginning of the cell. Our analysis shows that the concentration of pentane at the beginning of the bubble in the evaporator section is only 47%, and a temperature drop of more than 20 K occurred between the evaporator and the condenser. We note that the amount of liquid at the hotter end is small relative to the amount of liquid in the complete set-up. One the other hand, we also note that the assumptions regarding Raoult's Law are very approximate for large temperature differences.

The analysis of the temperature profile using the two-dimensional model is demonstrated in Figure 5. The experimental outside wall temperature is compared with the model results. The area average overall heat transfer coefficient h_i for experiment I, II, III were found to be 400 W/m²K, 1000 W/m²K and 150 W/m²K, respectively. The value of h_i is a direct indication of the efficiency of the heat transfer. The higher the value of h_i is, the greater the efficiency of the heat transfer in the evaporator. We can see that the values of h_i strongly depend on the

TABLE 1. Operating Conditions of the Experiments.

Experiment	Q(W)	T_v(K)	T_∞(K)	Inclination Angle β (°)	Purity of pentane (%)
I	0.1388	303.15	299.25	-3.80	>99.0
II	0.1609	301.15	299.15	-1.95	>99.8
III	0.2215	305.55	299.15	-7.50	>99.0

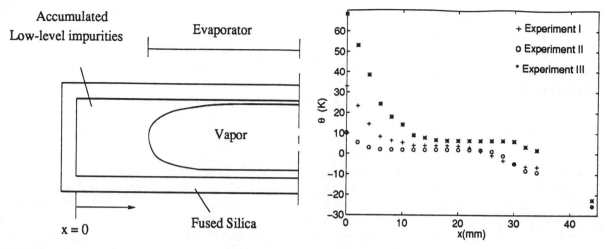

FIGURE 3. Schematic of the Evaporator of the CVB.

FIGURE 4. CVB Temperature Profile.

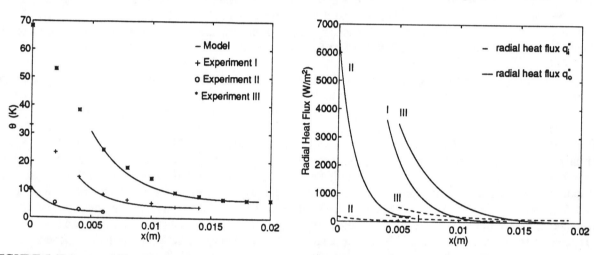

FIGURE 5. Evaporator Temperature Profile.

FIGURE 6. Radial Heat Flux Profile in the Evaporator.

operating conditions of the CVB. In experiments I and III, the non-evaporative substances in pentane accumulated in the beginning of the evaporator, and hence elongated the length of the evaporator and reduced the efficiency of the evaporation. Also, higher inclination angles for experiments I and III made it difficult for pentane to flow back to the evaporator. All these factors made h_i for experiments I and III to be much lower than that of experiment II. Compare experiment I to III, the higher heat input and inclination angle of experiment III made the length of the evaporator longer, and h_i lower (inclination angle effect only) than those of experiment I. For all three experiments, the area average heat transfer coefficients are not very high because a large portion of the inside wall of the cell was covered by a non-evaporative ultra-thin film. While the heat transfer coefficient near the interline region (where most of the evaporation occurs) could be very high, the overall heat transfer coefficient was not so high.

Figure 6 shows the axial distribution of the radial heat flux in the evaporator of the CVB. q_i'' is the radial heat flux toward the liquid vapor phase, and q_o'' toward the surroundings:

$$q_i'' = h_i(T_i - T_v), \qquad q_o'' = h(T_s - T_\infty), \qquad (16)$$

where T_i is the inside wall temperature, and T_s the outside wall temperature. It is clear that the radial heat flow rate due to evaporation (integration of q_i'' over the distance) is always higher than that due to natural convection.

The highest q_o'' occurred in case III for the highest heat input. The highest q_i'' for case I and III are almost identical though the heat input in case III is much higher. This is due to the higher inclination angle and distillation effect in case III which reduces the overall heat transfer coefficient. The highest q_i'' for experiment II demonstrated that for a pure liquid with small inclination angle, the radial heat flux due to evaporation can be very high in spite of a low heat input.

CONCLUSIONS

Based on the experimental results and the theoretical analysis, the following conclusions are reached:

- The effect on the temperature of a second component with a low vapor pressure in the liquid has been demonstrated from a thermodynamic point of view.

- A two dimensional temperature model has been developed for the evaporator of the CVB.

- The model has been used successfully to evaluate the effect of low-level impurities on the operation of the CVB.

- A change in the purity of the initial charge from 99.8% to 99% had a dramatic effect on the temperature profile in the CVB.

Acknowledgments

This material is based on work supported by the National Aeronautics and Space Administration under grant # NAG3-1399. Any opinions, findings, and conclusions or recommendations expressed in this publication are those of the authors and do not necessarily reflect the view of NASA.

References

Babin, B. R., G. P. Peterson and D. Wu (1990) "Steady State Modeling and Testing of a Micro Heat Pipe,", *J. Heat Transfer*, 112: 595-601.

Cotter, T. P. (1984) "Principles and Prospects of Micro Heat Pipes," *Proc 5th Int. Heat Pipe Conf.*, Tsukuba, Japan, 328-335.

DasGupta, S., J. L. Plawsky and P. C. Wayner Jr. (1995) "Interfacial Force Field Characterization in a Constrained Vapor Bubble Thermosyphon," *AIChE Journal*, 41(9): 2140-2149.

Huang, J., M. Karthikeyan, J. L. Plawsky and P. C. Wayner Jr. (1997) "Two-Dimensional Analysis of the Evaporator of a Constrained Vapor Bubble," *Proc. 10th Intl. Heat Pipe Conf.*, Stuttgart, Germany.

Karthikeyan, M., J. Huang, J. L. Plawsky and P. C. Wayner Jr. (1996) "Initial Non Isothermal Experimental Study of the Constrained Vapor Bubble Thermosyphon," ASME, HTD, *1996 National Heat Transfer Conference*, Houston, TX, 327:121-129.

Karthikeyan, M., J. Huang, J. L. Plawsky and P. C. Wayner Jr. (1997) "Experimental Study and Modeling of the 'Adiabatic' Section of the Constrained Vapor Bubble Heat Exchanger," *AIP Conference Proceedings 387, Part 2*, Space Technology and Applications International Forum-1997, Albuquerque, NM, 641-646.

Khrustalev, D. And A. Faghri (1994) "Thermal Analysis of a Micro Heat Pipe," *Proc 7th Int. Heat Pipe Conf.*, Paper No. A-41, Minsk, USSR, May, 21-25.

BUBBLE FLOW IN REDUCED GRAVITY

Timothy L. Brower
Center for Engineering Infrastructure and Sciences in Space (CEISS)
Colorado State University
Fort Collins, Colorado 80523
(970) 491-8573

Abstract

The combination of buoyancy and thermocapillarity can cause a flow that impedes or alters the motion of a bubble in reduced gravity. In analyzing such a flow, it is important to account for the dynamics occurring on the surface of the bubble. A first-order solution to the governing equations of motion and energy is presented for small Reynolds numbers. This solution provides the means to determine the terminal velocity, the direction of motion, the external flow and temperature fields of a bubble. Its features explicitly account for all gravity levels by using a Bond number and consider the negative variation of the surface tension with temperature. Streamline patterns outside the bubble are computed to obtain an overall picture of the flow over the range $0 \leq Re \leq 20$. The results of the analysis compare the homogeneous to the first-order solutions by illustrating differences to five unique flow regimes as the Bond number changes from small to very large values, i.e., from thermocapillary to gravity dominated flow. Examination of the streamline patterns allows for the determination of how the migration of the bubble can be affected.

INTRODUCTION

Immiscible fluids in the form of bubbles suspended in a continuous medium under a temperature gradient comprise an important class of flow situations in reduced gravity. Local changes in surface tension, caused by a temperature gradient along the surface, i.e., thermocapillary flow, induce motion of the bubble interface in the direction of greatest surface tension. This flow produces a net force that acts on the bubble, and induces bubble motion in the direction of the hot temperature region when the manner in which the surface tension changes with temperature is negative, i.e., $(\partial\sigma/\partial T)^* < 0$. A star superscript denotes a dimensional quantity. If the gravity vector and the temperature gradient are collinear and in the same direction, the motion induced by thermocapillarity is opposite to that caused by buoyancy and the resulting motion of the bubble and its flow field is not intuitive. Brower and Sadeh (1997) report the homogeneous or zeroth-order solution to this problem in the limit as $Re \to 0$. In this work, five flow regimes are described as the bubble migrates from thermocapillary to gravity dominated flow.

As in many low-speed flow situations, one can hypothesize whether the inertia can be neglected in order to simplify the solution. Brower and Sadeh (1997) allow Re to become very small, i.e., in the limit as $Re \to 0$, such that the nonlinear inertia terms in the governing equations are neglected with respect to the viscous terms. In this work, Re is defined in terms of the dynamics driving the flow at the bubble surface:

$$Re = \frac{\rho^* R^{*2}}{\mu^{*2}} \left|\frac{\partial \sigma}{\partial T}\right|^* \cdot \left(\frac{dT}{dz}\right)^*_\infty, \tag{1}$$

where ρ^*, μ^*, R^*, $|\partial\sigma/\partial T|^*$ and $(dT/dz)_\infty^*$ are the density and kinematic viscosity of the continuous fluid, the bubble radius, the absolute value of the change in surface tension with the change in temperature, and the temperature gradient far from the bubble, respectively. According to Eq. (1) for a given fluid-fluid combination, i.e., ρ^*, μ^*, and $(\partial\sigma/\partial T)^*$ are known, the only way for $Re \to 0$ is when $(dT/dz)_\infty^* \to 0$. In the most extreme case of $(dT/dz)_\infty^* = 0$, surface tension is static and no thermocapillary flow exists, a condition that is studied elsewhere. It seems reasonable that to better understand this flow situation, one must consider the effects of inertia. Hence, a first-order solution to the governing equations for small Re is proposed.

FLOW SITUATION

The flow situation considered is that of a steady migration of a single bubble through a continuous fluid of infinite extent under the combined action of a constant gravity force and a thermally induced surface-tension force. Both fluids are assumed incompressible with constant thermophysical properties, except the surface tension that varies linearly with temperature. The bubble is assumed to maintain a spherical shape, i.e., its radius R^* = constant.

A sketch of this flow situation is provided in Fig. 1 and shows a bubble moving upward in the negative z-direction at its steady freestream or terminal velocity as denoted by U_T^*, i.e., the velocity attained after equilibrium is reached. Here, the gravity force F_g^*, or buoyancy, induces upward motion due to a lighter bubble. On the other hand, as the local gravity g^* decreases, downward motion of the bubble is induced by the thermocapillary force F_σ^* due to the manner in which the surface tension changes with temperature, i.e., $(\partial\sigma/\partial T)^* < 0$. Thus, the bubble migration is only in the z-direction and is determined by the net effect of the gravity and thermocapillary forces. In Fig. 1, r^* denotes the radial coordinate.

Bond Number

The analysis conducted by Brower and Sadeh (1993) suggests that a generalized definition of the Bond number given by:

$$Bo = \frac{\rho^* g^* R^{*2}}{\mu^* U_R^*}, \quad (2)$$

be employed to assess the relative importance of the gravity and the thermally induced surface-tension forces in flows of immiscible fluids. Here, the reference velocity is:

$$U_R^* = \frac{|\frac{\partial\sigma}{\partial T}|^* (\frac{dT}{dz})_\infty^* R^*}{\mu^*}. \quad (3)$$

In bubble investigations where no temperature gradient exists, it is appropriate to use the terminal velocity for a U_R^*. However, the terminal velocity can be near zero in the flow situation herein, thus, a reference velocity that accounts for the dynamics at the bubble surface is more applicable. A Bond number near unity is then the boundary between flows dominated by the gravity force (Bo >> 1) and thermally induced surface-tension driven flows (Bo << 1).

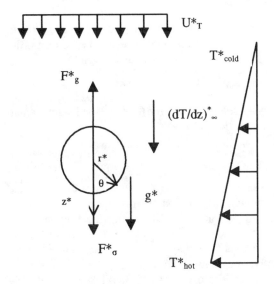

FIGURE 1. Flow Situation.

ANALYSIS

In order to examine the migration of a bubble and the flow induced by its motion under the simultaneous action of gravity and thermocapillary forces when the Reynolds number is slightly greater than zero, it is necessary to include the inertia and the convective terms in the equations of motion and energy, respectively. If Re is sufficiently small, it is likely that the flow departs only slightly from the zeroth-order solution given by Brower and Sadeh (1997). Then, one can estimate the effect of inertia and convection by approximating the dependent flow variables via a power series expansion in terms of Re. In this expansion, the first-order solution is a function of the zeroth-order solution. Both solutions are linearly independent such that their superposition gives the final approximation to the flow. The inherent assumption with this approach is that the subsequent ordered solution contributes little to the overall approximation. Nevertheless, the first-order approximation offers valuable insight into the role that inertia plays and is an important step to undertake in understanding the flow situation.

The results of this analysis, reported in Brower (1997), give the solution for the Stokes stream function and temperature in the continuous fluid flow region as:

$$\psi(r,\theta) = GA\frac{\sin^2\theta}{2}(r^2 - \frac{r}{A}) + C\frac{\sin^2\theta}{2}(r^2 - \frac{1}{r}) + Re[a_0(1-\frac{1}{r^2}) - \frac{E_1}{r^2} + \frac{G^2A}{8}(-r^2+\frac{r}{A}) + \frac{GC}{8}(-r^2-\frac{1}{r})]\sin^2\theta\cos\theta], \quad (4)$$

$$T(r,\theta) = [r + \frac{X}{r^2}]\cos\theta$$

$$+ Ma\,[-\frac{c_0}{Pr\,3r^3} + \frac{c_2}{Pr\,r} - (\frac{3G}{8})r + GX(\frac{3}{8r^2}) + (\frac{c_0}{Pr\,r^3} + \frac{C}{4r} + CX(\frac{1}{2r} + \frac{1}{4r^4}) + GA(\frac{r}{8A}) + GAX(\frac{1}{2r} - \frac{5}{8Ar^2}))\cos^2\theta], \tag{5}$$

where the constants in Eqs. (4) and (5) are given by:

$$A = 1, \quad C = -\frac{1}{2}, \quad G = \frac{Bo}{3}, \quad E_1 = \frac{GA}{8}(-1 + \frac{1}{A}) - \frac{C}{4}, \quad X = \frac{1}{2}, \quad c_0 = \frac{(3H_2 + H_4)}{3} - H_2, \quad c_2 = H_3 + \frac{H_4}{3},$$

$$a_0 = \frac{c_0}{5} + \frac{Pr}{5}[\frac{C}{4} + \frac{3CX}{4} + GA(\frac{1}{8A}) + GAX(\frac{1}{2} - \frac{5}{8A})] - \frac{3GE_1}{5} + \frac{2GE_2}{5} - \frac{G^2 A}{40} - \frac{GC}{20},$$

$$H_2 = Pr\,[\frac{C}{4} + \frac{3CX}{4} + GA(\frac{1}{8A}) + GAX(\frac{1}{2} - \frac{5}{8A}) + \frac{Y}{7\alpha}(GD + \frac{3C}{4})], \quad H_3 = Pr\,[G(-\frac{3}{8} - \frac{3X}{4})],$$

$$H_4 = Pr\,[-\frac{C}{4} - \frac{3CX}{2} + GA(\frac{1}{8A}) + GAX(-\frac{1}{2} + \frac{5}{4A})], \quad Y = \frac{3}{2}, \quad D = \frac{1}{2}, \quad E_2 = -GE_1 - \frac{GA}{16}(-2 + \frac{1}{A}) + \frac{C}{16}.$$

Here, α is the thermal diffusivity and the Marangoni number (Ma) is the product of the Reynolds and Prandtl numbers, i.e., Ma = Re Pr. No first-order correction to the terminal velocity of the bubble results in the first-order analysis, i.e., it remains unchanged from the zeroth-order solution. The terminal velocity is then given by:

$$U_T = \frac{Bo}{3} - \frac{1}{2}. \tag{6}$$

Hence, under the conditions specified herein the terminal velocity is solely a function of the Bond number and agrees with the homogeneous solution offered by Young, Goldstein and Block (1959).

Stream Function Examination

To illustrate the flow patterns about the bubble, Figs. 2 through 6 show streamlines as a function of the Bond and Reynolds numbers. The figures that depict Re = 0 represent the homogeneous solution of Brower and Sadeh (1997). In all the figures, the gravity vector and temperature gradient are downward, consistent with the flow situation shown in Fig. 1. An arrow at the bubble center shows its direction of motion as determined by the terminal velocity Eq. (6). Arrows to indicate the direction of flow mark selected streamlines. Positive and negative magnitudes of the streamlines are denoted by (+) and (-), respectively, at representative locations in the figures. The bubble surface and other boundaries in which flow does not pass are noted by a zero streamline, i.e., $\psi = 0$. In all cases, the Prandtl number and thermal diffusivity are unity.

The streamlines presented in Fig. 2 illustrate the flow field for Bo = 0 over the range $0 \leq Re \leq 20$. Purely thermocapillary flow propels the bubble downward for all Re. At Re = 0, the streamline pattern is symmetrical fore and aft of the bubble. Such symmetry is lost in the presence of inertia, which causes the streamlines to become more asymmetrical with increasing Re. At Re $\approx$ 12 - 14, a closed recirculating wake or standing eddy forms on the downstream side of the bubble. This region of flow, defined by a curve where $\psi = 0$, grows as Re increases. The emergence of this eddy at Bo = 0 was described by Thompson (1979) to occur at Re = 16/3. The condition when the terminal velocity of the bubble is zero as determined by Eq. (6) is illustrated in Fig. 3 for Bo = 1.5. Here, the thermocapillary and gravity forces act simultaneously on the bubble such that it remains motionless for all Re. As Re increases, a region of flow defined by $\psi = 0$ appears below the bubble. This region moves closer to the bubble with increasing Re and may be regarded as a separatrix since it separates one region of flow from another (Shankar and Subramanian, 1988).

In Fig. 4, the streamline pattern outside the bubble is shown for Bo = 2 over the range $0 \leq Re \leq 10$. Here, the net motion of the bubble is upward and one observes an outer toroidal vortex region of flow. As Re increases, the dividing streamline defining the peripheral boundary of the outer toroidal vortex moves closer to the upstream edge of the bubble and the vortex region of flow elongates in the downstream direction. The flow characterized as Re increases is suggestive of the streamline pattern expected for a jet of fluid emanating from the forward surface of the bubble. This jet strikes the freestream, turns, and creates a region of recirculation. The toroidal vortex virtually expands the radius of the bubble and correspondingly increases its influence in the surrounding fluid.

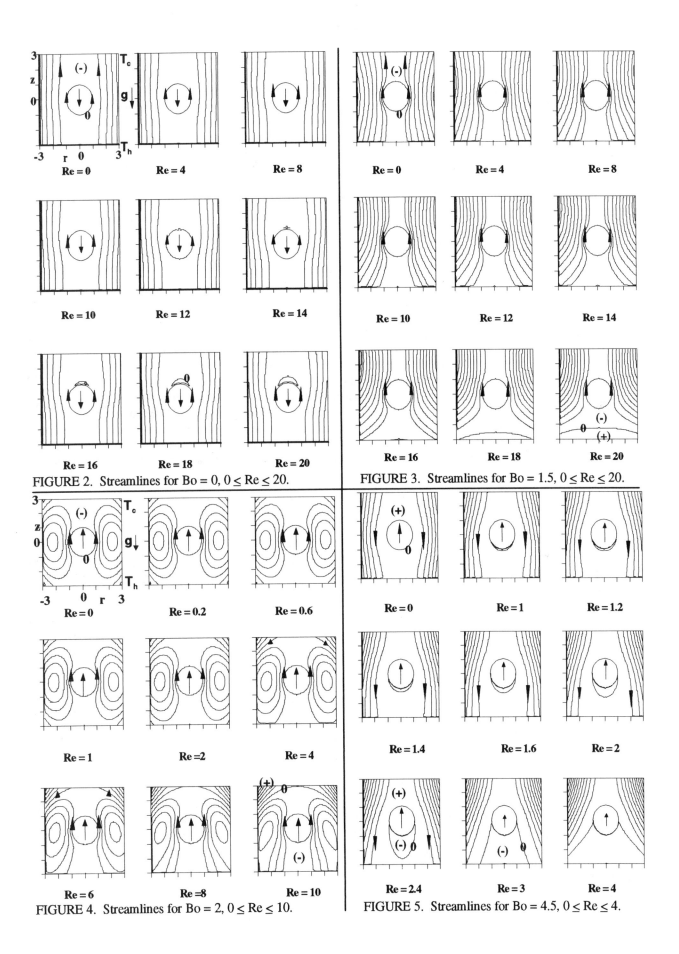

FIGURE 2. Streamlines for Bo = 0, 0 ≤ Re ≤ 20.

FIGURE 3. Streamlines for Bo = 1.5, 0 ≤ Re ≤ 20.

FIGURE 4. Streamlines for Bo = 2, 0 ≤ Re ≤ 10.

FIGURE 5. Streamlines for Bo = 4.5, 0 ≤ Re ≤ 4.

An interesting development in the flow field occurs when the bubble migrates upward at Bo = 4.5 over the range 0 ≤ Re ≤ 4.5 as pictured in Fig. 5. At this flow condition, no surface flow exists and the outer toroidal vortex disappears. At Re ≈ 1, a standing eddy forms on the aft hemisphere of the bubble. As Re increases, the eddy grows and elongates aft of the bubble in the flow direction. This flow resembles the flow about a solid sphere. For Bo = 6, as depicted in Fig. 6, gravity dominates the flow with no indication of thermocapillarity and the lighter bubble rises opposite to the gravity vector. Notice that surface flow is now downward in the direction of the freestream. Apparently, the reversal of surface flow is not a consequence of the presence of inertia, because the same surface flow behavior exists in the homogeneous case. The emergence of a standing eddy on the downstream side of the bubble is delayed slightly when compared with the Bo = 4.5 case, i.e., it appears at Bo ≈ 1.4 vs. Bo ≈ 1. However, inertia hastens its growth and subsequent disappearance. The streamline pattern changes character in the downstream direction for Re ≥ 2, suggesting that higher ordered terms to the solution might be needed.

The apparent divergence of the streamlines in the continuous fluid flow region as Reynolds number increases suggests an upper limit to Re for a valid solution. For Bo = 0 and 1.5, the streamlines given in Figs. 2 and 3 are as expected for all Re ≤ 20. The flow begins to diverge for Re > 10 for the Bo = 2 case, and at Bo = 4.5 and 6, a breakdown for Re > 3 and 2.4 is displayed in Figs. 4 and 5, respectively. The validity of the approximation thus depends not only on the perturbation quantity, i.e., Re, but also on Bo.

Temperature Examination

Isotherms at Bo = 0 over the range 0 ≤ Re ≤ 20 in Fig. 7 illustrate the effect that inertia has on the temperature field. As the bubble moves downward, purely thermocapillary flow exists. The fore and aft symmetry of the temperature field shown in the Re = 0 case is lost when inertia is included and a thermal wake develops on the downstream side of the bubble. Upstream, a temperature "cap" makes its appearance in the continuous fluid at Re ≈ 12 and enlarges with increasing Re. This temperature cap occurs because the circulating gas inside the bubble pulls the colder fluid from the aft bubble surface forward and, in turn, impinges on the forward bubble surface. Hence, the inside forward bubble surface creates a temperature sink that affects the adjacent outer fluid temperature.

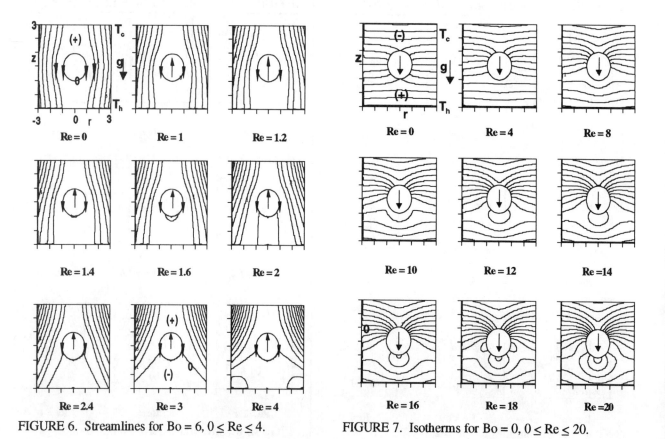

FIGURE 6. Streamlines for Bo = 6, 0 ≤ Re ≤ 4.   FIGURE 7. Isotherms for Bo = 0, 0 ≤ Re ≤ 20.

Examination of isotherms for Bo < 1.5 indicate that a thermal wake develops downstream of the bubble as it migrates upward. However, there is no indication of the temperature cap on its forward hemisphere. Brower (1997) gives a full description of the isotherm patterns for all Bo and Re.

CONCLUDING REMARKS

The solutions to the equations of motion and energy presented here provide the means to find the terminal velocity, the direction of motion, the external flow and temperature fields of a bubble for small Re. When this solution is applied to the case in which the gravity vector and the temperature gradient are collinear, such that the thermocapillary and buoyancy forces are opposite, one can describe five distinct flow regimes as the bubble covers the range $0 \leq Bo \leq 6$. First is purely thermocapillary driven flow, where gravity plays a negligible role, i.e., Bo << 1. Here, the bubble migrates opposite to its surface motion in a direction toward the hot temperature region. Inertia changes the homogeneous solution by the asymmetric bending of the streamlines and causes separation of the flow on the aft hemisphere of the bubble. In the second regime at Bo = 1.5, the bubble becomes stationary in the continuous fluid. Including inertia induces a separatrix in the hot temperature region below the bubble as Re increases. The third flow regime occurs over the range 1.5 < Bo < 4.5. Here, the lighter bubble moves opposite to the gravity vector as buoyancy plays a more important role. In this regime, a toroidal vortex, i.e., a doughnut-shaped region of recirculating flow perpendicular to its direction of motion, emerges outside the bubble. When inertia is present, this vortex structure virtually expands the influence of the bubble in the downstream direction. A unique flow situation characterizes the fourth regime at Bo = 4.5. All surface flow ceases at this Bond number and the bubble continues to move upward opposite to the gravity vector. The presence of inertia causes separation of the flow on the aft hemisphere of the bubble. Finally, the fifth flow regime is dominated by gravity for Bo > 4.5. In this regime, the migration of the lighter bubble is always upward with no evidence of the outer toroidal vortex and surface tension effects can be neglected.

Analysis of the first-order approximation reveals differences in the flow and temperature fields when compared to the zeroth-order results. The flow about the bubble is altered in all regimes, particularly in regimes two and three, from what conventionally might be expected. Hence, further study of these flows is warranted to assess the affect they might have on space fluid flow systems that operate in a confined reduced gravity thermal environment.

Acknowledgments

The author thanks Willy Z. Sadeh (1932-1997), for his assistance and encouragement throughout this research. Additional partial support from the National Aeronautics and Space Administration's (NASA) Graduate Student Researchers Program and the NASA Colorado Space Grant Consortium is also recognized.

References

Brower, T.L. (1997) "Droplet Flow in Reduced Gravity," Ph.D. Dissertation, Colorado State University, Fort Collins, CO.

Brower, T.L. and W.Z. Sadeh (1993) "Bond Number in Low-Gravity Environments," *Fluid Mech. Phenomena in Microgravity*, 1993 ASME WAM, New Orleans, LA, Nov. 28-Dec. 3, AMD/FED, 174/175:145-154.

Brower, T.L. and W.Z. Sadeh (1997) "Thermocapillary Drift of a Droplet in Reduced Gravity," in *Proc 48th International Astronautical Congress*, IAF-97-J.1.03, Future Projects in Microgravity, Turin, Italy, 6-10 October.

Shankar, N. and R.S. Subramanian (1988) "The Stokes Motion of a Gas Bubble Due to Interfacial Tension Gradients at Low to Moderate Marangoni Numbers," *J. Colloid and Interface Sci.*, 123, 2:512-522.

Thompson, R.L. (1979) "Marangoni Bubble Motion in Zero Gravity," Ph.D. Dissertation, University of Toledo.

Young, N.O., J.S. Goldstein and M.J. Block (1959) "The Motion of Bubbles in a Vertical Temperature Gradient," *J. of Fluid Mechanics*, 6:350-356.

A THEORETICAL APPROACH TO THE EVALUATION OF GRAVITY INFLUENCE ON HEAT AND MASS TRANSFER IN LIQUIDS

G. Latini, G. Passerini and F. Polonara
Dipartimento di Energetica
Università di Ancona
Via Brecce Bianche
60100 Ancona, ITALY
Tel +39 71 220 4432 - Fax +39 71 280 4239

G. Galli
Dipartimento di Fisica Tecnica
Università di Roma "La Sapienza
Via Eudossiana
00184 Roma, ITALY
Tel +39 6 4814 332

Abstract

The evaluation of the behaviour of liquids in terms of transport properties has often been carried out trying to adapt well known theories already developed to understand the characteristics of gases and/or solids. Thus, among such theoretical archetypes, a rough distinction can be made introducing the terms "gas-like" and "solid-like" to address models respectively based on an extension of theories originally conceived to study gases and solids. The latter models seem, in general, more reliable and accurate and many efforts have been made to introduce a "convective" term in heat, mass and momentum transfer equations and a new "vibrational" term, as for solids having a strong lattice structure. The presence of such two terms has suggested to study the behaviour of liquids, for what concerns heat and mass transfer, with respect to the presence or not of gravity itself. This paper presents an attempt to assess both the actual contribution of "convective" terms to heat, mass and momentum transfer in liquids and the related effects of gravity.

INTRODUCTION

The study of transport properties of fluids in their liquid state has always been a hard challenge for scientific research. As a result, the liquid state itself is not yet completely defined for what concerns transport properties and, for heat, mass and momentum transfer, the liquid state is somewhat a "limbo". In a liquid we cannot theorise and find a kind of order as for the solid state, while we cannot apply the well known hypothesis of low, or negligible, interaction between molecules as for the gaseous state. However most of the theories on the liquid state move from those of the solid state or those of the gaseous state trying to consider a liquid as a somehow partially disordered solid or a kind of very dense gas.

On those assumptions, several models have been introduced, on a theoretical basis, to explain and quantify heat and mass transfer in liquids. Scientists tried to derive a theory both from those of the gaseous state and from the ones originally conceived for homogeneous and isotropic solids. The former consider liquids from the point of view of dense gases and were named "Gas-Like Models", the latter are based on the lattice theories of solids and were named "Solid-Like Models". Despite the efforts of scientists neither the assumption of low interaction between molecules, typical of models for gaseous state, nor the assumption that a well stated order exists, typical of models for solid state, could be easily applied. However the latter models seem, in general, more reliable and accurate. Thus many scientists concentrated their researches on solid-like models and many efforts have been dedicated to couple a "convective" term in mass and momentum transfer equations to the usual "vibrational" term, typical of solids having a lattice structure.

SOLID-LIKE QUASI-LATTICE MODELS

In a solid, the atoms or molecules occupy 'sites' according to a definite geometric pattern known as a lattice which tends to extend through the whole of the crystal. In a simple cubic lattice such sites are situated at the eight corners of cubes. This regular geometric pattern lets to say that the solid possesses a "long-range order" and, if we choose two any internal molecules, a definite geometrical relation exists which joins the sites of the two molecules. Any molecule will also have a certain number z of nearest neighbours on its adjacent sites and z is said to be the "co-ordination number" of the lattice.

At the absolute zero, atoms or molecules are at rest at their sites. If the solid is heated each particle starts to vibrate about its lattice site and the amplitude of vibrations increases as the temperature rises. However, the centres of vibration keep to maintain their original well defined geometry and the long range order is still preserved. As the temperature rises, the solid comes near to its melting point and, while the ordering influence of molecular forces tends to preserve the geometry of lattice, the increase of the molecular thermal vibrations tries to disrupt the lattice. As the second effect prevails, the solid melts into a liquid and the long range order is destroyed but a good deal of "short range order" remains. In fact, the molecular arrangement near a chosen central particle will be fairly regular, very similar to that of the corresponding solid but, for distances equal or greater to 30 Å, no spatial geometric relation between the central particle and the others can be found. This molecular arrangement is said to be "quasi-lattice" or "quasi-crystalline".

There is an experimental evidence that the increase of volume in melting is not due to an increase of the distance between particles but it is mainly due to a general decrease of the co-ordination number z. This means that the number of sites in the lattice structure with no associated particle increases: we refer to these free position as to "holes" or "vacancies" in the lattice structure. From this point of view, the melting process can be regarded as an injection of holes inside the lattice structure. It must be pointed out that the existence of holes or vacancies in the liquid lattice has been taken as assumption by several author on attempting to theorise the liquid state (Cernushi 1939, Ono 1947, Peek 1950, and Rowlinson 1951).

In mathematical terms, the quantity which describes the molecular arrangement is named "radial distribution function" denoted by $g(r)$. For a simple liquid as Argon with molecular spherical symmetry each molecule can be regarded as a small sphere of diameter σ and the force between two molecules will depend only on the distance between their centres. Now, let σ be the diameter a molecule regarded as a small sphere and $\rho(r)$, "number density", the number of molecules per unit volume at a distance r from a given molecule. Due to the short range order, $\rho(r)$ will be sometimes greater than the mean number density n (total number of molecules divided by the total volume), sometimes smaller. Thus, the radial distribution function $g(r)=\rho(r)/n$ will show a general behaviour as in figure 1.

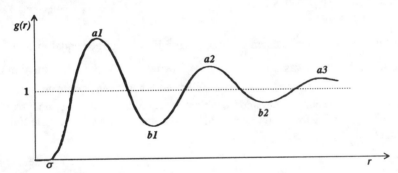

FIGURE 1 - A Typical Radial Distribution Function for a Simple Liquid Plotted as a Function of the Distance r from a Given Molecule Centred in 0.

Figure 1 shows that $g(r)$ is near zero for distances less than molecular diameter, increases to the unit value, then oscillates around the unit. According to the theory, the peaks a1, a2, a3 correspond to shells of respectively first, second, third nearest neighbours while b1 and b2 correspond to somewhat emptier regions between shells. Following investigations performed by X-ray and neutron diffraction there is an experimental evidence that the actual radial distribution function of liquids roughly follows the behaviour shown in figure 1. After such knowledge we can conclude that quasi-lattice models could be used to evaluate properties of substances in their liquid state. The same experiments showed (Eisenstein 1942) that, for liquid Argon just above melting point, the number of nearest neighbours decreases from 12 of the solid (typical of face centred cubic structure) to approximately 10.5. The same value of co-ordination number seems acceptable for most of organic liquids.

LENNARD-JONES AND DEVONSHIRE THEORY OF THE LIQUID STATE

The first lattice theory applied to liquids was due to Lennard-Jones and Devonshire (Lennard-Jones 1937a). They developed equations to evaluate equilibrium properties of liquids in terms of lattice geometry and intermolecular forces. They assumed, for liquids far from critical region, that molecules interact with spherically symmetric fields of force. Thus, any terms in transport equations which depend on the existence of a force field surrounding the molecules could be evaluated from the knowledge of fundamental parameters of the assumed molecular interaction.

To evaluate both equilibrium and transport properties of a real liquid we firstly need some knowledge of the actual force field surrounding the molecules. Lennard-Jones proposed (Lennard-Jones 1937b) a form of intermolecular potential $\psi(r)$ derived from the Mie equation (Mie 1903). The Lennard-Jones potential is:

$$\psi(r) = 4\varepsilon \left[\left(\frac{\sigma'}{r}\right)^{12} - \left(\frac{\sigma'}{r}\right)^{6} \right], \tag{1}$$

where r is the distance between molecular centres and σ' is the value of r for which $\psi(r) = 0$. It is known as molecular cross-section and usually taken equal to molecular diameter σ. ε is a molecular pair potential. It has been shown (Hirschfelder 1954) that it is likely to assume both for ε and for σ experimental values derived from temperature dependency of viscosity coefficient.

Lennard-Jones and Devonshire considered a liquid of volume V divided into spherical cells of volume $a^3/\sqrt{2}$ where a is the distance between nearest neighbours. Then, assuming that molecular interaction could be determined by means of equation (1) and that the quasi-lattice liquid had face centred cubic symmetry they were able to evaluate the partition function for a generic molecule

$$q = \frac{(2\pi mkT)^{3/2}}{h^3} \cdot v_f \cdot q_{int} \cdot e^{-\Phi_0/NkT}. \tag{2}$$

In equation (2) Φ_0 represent the energy of the system when all the molecules are in their equilibrium sites, thus at the centres of cells. Based on the model it could be given by

$$\Phi_0 = Nz\varepsilon \cdot \left[1.2045 \cdot \left(\frac{V_0}{V}\right)^2 - 0.5055 \cdot \left(\frac{V_0}{V}\right)^4 \right], \tag{3}$$

where z, the co-ordination number was taken equal to 12 and V_0 was defined equal to $N \cdot \sigma^3$.

The free volume in equation (2) was taken $v_f = \chi V/N$ being $\chi = 2\pi G\sqrt{2}$ the fraction of volume available to molecules. The term G is function of V/V_0 and kT/ε and has been extensively tabulated for example in (Buehler 1951). By means of equation (2) and statistical mechanics, all the thermodynamic properties could, in principle, be evaluated. Lennard-Jones and Devonshire subsequently tried to evaluate the actual number of holes in the liquid which could be given as

$$n_h = N \cdot \frac{e^{-w/kT}}{1 + e^{-w/kT}}, \tag{4}$$

where w represents the work of formation of a hole and is temperature dependent. The work of formation could be seen as composed by two terms, the first one due to the disruption of local intermolecular forces, the second one related to the work to be done against the external pressure P to create a hole of size δv. Denoting by w' the first term, equation (4) could be written as

$$n_h \approx N \cdot e^{-\frac{w' + P\delta v}{kT}}. \tag{5}$$

Lennard-Jones and Devonshire assumed (Lennard-Jones 1939) that the work of formation was mainly due to repulsive forces thus obtaining:

$$w' = w_0 \cdot (V_0/V)^4. \tag{6}$$

They assumed that the term $P\delta v$ was negligible and that $W_0 = \alpha \cdot \varepsilon$ so that equation (5) becomes:

$$n_h = N \cdot \frac{-\alpha\varepsilon \cdot (V_0/V)^4}{kT}. \tag{7}$$

Lennard-Jones and Devonshire obtained for simple molecules $\alpha \approx 1$. This is in acceptable agreement with experimental results as in (Eisenstein 1942).

MCLAUGHLIN AND HORROCKS THEORIES OF THE LIQUID STATE

Horrocks and McLaughlin developed in early sixties two interesting theoretical approaches to transport properties of liquids. In a first work McLaughlin (McLaughlin 1959) assumed a purely "convective" mechanism of mass and momentum transfer. As a matter of fact, he inferred a somewhat "micro-convection" at a molecular level made of particle hopping from occupied sites to holes in the liquid. Each hopping was to happen in presence of both a molecule holding sufficient energy and a neighbouring hole. In a second work Horrocks and McLaughlin (Horrocks 1960) kept the same model for the liquid state to extract equations for the liquid thermal conductivity coefficient. This time they postulated the presence of both a usual "vibrational" and a "convective" mechanism for energy transport.

In the first work, McLaughlin moves from the previously discussed model of Lennard-Jones and Devonshire and introduces a modified rate theory approach (McLaughlin 1959) to obtain transport coefficients of liquids composed of spherically symmetric molecules. Assuming the hopping mechanism for the movement of molecules in the liquid, he derives a form of the self-diffusion coefficient based on the same hole-diffusion equation in solids:

$$D = a^2 \cdot \frac{kT}{h} \cdot e^{-\frac{(w'+P\delta v)}{kT}} \cdot e^{-\frac{\Delta g^*}{kT}}, \qquad (8)$$

where Δg^* is the difference in Gibbs free energy between a molecule at site centre and at the top of the boundary energy barrier and a is the distance between nearest neighbours, thus the distance between a molecule and an eventual hole. At this point McLaughlin introduces the Stokes-Einstein equation in the form modified by Sutherland (Sutherland 1905):

$$\frac{kT}{D\pi r' \eta} = 4, \qquad (9)$$

where η is the viscosity and $r' = \sigma/2$ is the radius of the Brownian particle. He substitutes equation (8) in equation (9) to obtain the coefficient of dynamic viscosity as

$$\eta = \frac{h}{4\pi r' a^2} \cdot e^{\frac{(w'+P\delta v)}{kT}} \cdot e^{\frac{\Delta g^*}{kT}}. \qquad (10)$$

To eliminate the free energy of activation, McLaughlin makes use of the Eyring method (Eyring 1935, Eyring 1936) and proposes the following equations for the self-diffusion and dynamic viscosity coefficients:

$$D = \left(\frac{kT}{2\pi m}\right)^{1/2} \cdot \frac{a^2}{v_f^{1/3}} \cdot e^{-\frac{\varepsilon \cdot (V_0/V)^4}{kT}} \cdot e^{-\frac{e_0}{kT}}, \qquad (11)$$

and

$$\eta = \frac{(2\pi m kT)^{1/2}}{2\pi \sigma a^2} \cdot v_f^{1/3} \cdot e^{-\frac{\varepsilon \cdot (V_0/V)^4}{kT}} \cdot e^{\frac{e_0}{kT}}, \qquad (12)$$

where e_0 is the height of a potential barrier related to the Eyring free energy of activation theory.

McLaughlin theory shows a quite good agreement with experimental evidence being deviation between predicted values and experimental data less than 10%. In the subsequent work, Horrocks and McLaughlin (Horrocks 1960) apply a similar approach to study the thermal conductivity of simple liquids. Firstly they show that both a "vibrational" and a "convective" mechanism contribute to thermal conductance of liquids and that the two mechanisms are independent. Thus, they write the thermal conductivity coefficient as:

$$\lambda = \lambda_{vibr} + \lambda_{conv}. \qquad (13)$$

To evaluate the vibrational contribution they again assume a quasi-crystalline liquid, being the quasi-lattice face centred cubic and consider spherically symmetric molecules which interact according to Lennard-Jones 12:6 potential. The result is the following equation:

$$\lambda_{vibr} = P \cdot \left\{ \frac{\varepsilon}{\pi^2 r_0^2 m} \cdot \left[11 C_{14} \cdot \left(\frac{r_0}{a}\right)^{14} - 5 C_8 \cdot \left(\frac{r_0}{a}\right)^8 \right] \right\}^{1/2} \cdot C_v \sqrt{2} \cdot \frac{(1 - n_h/N)}{a}, \qquad (14)$$

where P is a probability factor, ε is the depth of potential with respect to the equilibrium diameter r_0 which is related to collision diameter as $r_0^3 = \sqrt{2}\sigma^3$, a is the distance between nearest neighbours, C_{14} and C_8 are lattice summation constants (Glasstone 41), C_v is the molecular specific heat, n_h is the total number of holes and N is the total number of molecules. Then, on the same assumptions, Horrocks and McLaughlin introduce a new equation to evaluate the convective term of liquid thermal conductivity as:

$$\lambda_{conv} = \frac{n_h}{N} \cdot \left(\frac{kT}{2\pi m}\right)^{1/2} \cdot \frac{C_v}{v_f^{1/3}} \cdot \frac{\sqrt{2}}{a} \cdot \left(1 - \frac{n_h}{N}\right) \cdot e^{-\frac{e_0}{kT}}, \qquad (15)$$

where e_0 is an activation energy and v_f is the free volume.

Discussing their achievements Horrocks and McLaughlin conclude that the contribution of convective component to thermal conductivity coefficient is negligible, mainly due to the presence of the term n_h/N. Since this second work of Horrocks and McLaughlin seems less accurate than the previous -deviations between experimental data and predicted values sometimes reach the 30%- a first objection could be that the latter conclusion is not completely true. In particular the term n_h/N seems not really small having experiments shown that the typical increase of volume of about 10% as a solid melts is mainly due to the increase of the number of holes. However, other authors reach the same conclusions on negligibility of convective term in liquid thermal conductivity coefficient (Zwanzig 1954, Rice 1959, Rice 1961a, and Rice 1961b).

QUASI-LATTICE THEORIES AND TRANSPORT PROPERTIES IN MICROGRAVITY

In previous chapters we found that quasi-lattice theories can be used to derive transport coefficients for liquids reaching results in acceptable agreement with experimental evidence. Self-diffusion and dynamic viscosity coefficients were derived assuming a kind of "micro-convection" in the liquid regarded as a quasi-crystal, a kind of drift of particles and/or of holes. For thermal conductivity coefficient a new vibrational term was introduced, this term apparently prevailing on the convective one; this second theory seems to contain some inadequacies, but it appears consistent and shows that the model can be able to explain the behaviour of pure substances in their saturated liquid state when transport properties are investigated. At the same time, such theories seem important both for the applications in microgravity and for the experiments to be performed in microgravity.

As a first approximation and at theoretical level the interaction between gravity and transport phenomena in pure compounds in their liquid state can be neglected. At a molecular level the mechanisms of heat, momentum and mass transfer can be seen as approximately similar to those of solids. This similarity is more evident for heat transfer (thermal conductivity) as the usual vibrational transport phenomenon seems to prevail on the convective term. Now, the absence of gravity could only produce a certain increase of molecular order which, in principle, could ease molecular movements. However we have seen that molecular interaction and hoppings occur at a neighbouring level and there is an experimental evidence that, at such level, a solid-like order is maintained in liquids. Thus we could conclude that effects of gravity on transport phenomena seem negligible.

On the other hand some inadequacies of theoretical models have to be taken into account so it can be concluded that the described kind of micro-convection, made of particle hoppings, can be only accepted as a useful model to have a reliable estimation of the transport coefficients while there is an evidence that purely theoretical models cannot predict transport property coefficients with a precision comparable with that of experimental data. The absence of gravity involves that all gravity-driven convective processes are absent and it should be outlined that such convective motions should not be confused with the micro-convection. This should allow to perform experiments on heat, mass and momentum transfer with an accuracy better than that reachable on Earth. Moreover it must be remembered that new experimental devices have been introduced during past years which completely avoid the use of gravity itself as an experimental constant.

CONCLUSIONS

Quasi-lattice theories appear consistent and in good agreement with experimental data. They show that the mechanisms of heat, mass and momentum transfer in pure liquids should not be influenced by the presence of gravity. This lets to conclude that in microgravity the usual experimental and predicted values of transport properties coefficients can be used.

On the other hand the transport coefficients do not appear to be predicted with an accuracy comparable with that of experimental data so that it is important to achieve some reliable experimental thermal conductivity and dynamic viscosity data by experiments performed in microgravity; these experimental data will allow to suggest a better and more physically grounded theory for the transport phenomena.

Acknowledgements

This work has been supported in part by the Commission of the European Communities in the framework of the JOULE2 Programme, and by the Ministero dell'Universita' e della Ricerca Scientifica e Tecnologica of Italy.

References

Buheler, R., K. Wentorf, J. O. Hirschfelder, and C.F. Curtiss (1951), "The Free Volume for Rigid Sphere Molecules", *J. Chem. Physics*, **19**:61-71.

Cernushi, A., H. Eyring (1939), "Elementary Theory of the Liquids", *J. Chem. Physics*, **7**:547-551.

Eisenstein, R., and R. G. Gingrich (1942), "The Diffraction of X-Rays by Argon in the Liquid, Vapor, and Critical Regions", *Physic. Rew.*, **62**:261-270.

Eyring, H. (1935), "Activated Complex in Chemical Reactions", *J. Chem. Physics*, **3**:107-115.

Eyring, H. (1936), "Viscosity, Plasticity and Diffusion as Examples of Absolute Reaction Rates", *J. Chem. Physics*, **4**:283-291.

Glasstone, S., K. J. Laidler, and H. Eyring (1941), *The Theory of Rate Processes*, McGraw-Hill, New York.

Hirschfelder, J. O., C. F. Curtiss, and R. B. Bird (1954), *Molecular Theory of Gases and Liquids*, Chapman and Hall, London.

Horrocks, J. K., and E. McLaughlin (1960), "Thermal Conductivity of Simple Molecules in the Condensed State", *Trans. Faraday Soc.*, **56**:206-212.

Lennard-Jones, J.E., and H.C. Devonshire (1937a), "Critical Phenomena in Gases", *Proc. Roy. Soc.*, **A163**:53-70.

Lennard-Jones, J. E. (1937b), "The Equation of State of Gases and Critical Phenomena", *Physica*, **4**:941-956.

Lennard-Jones, J.E., and H. C. Devonshire (1939), "Critical and Coöperative Phenomena IV.", *Proc. Roy. Soc.* **A170**:464-484.

McLaughlin, E., (1959), "Viscosity and Self Diffusion in Liquids", *Trans. Faraday Soc.*, **55**:28-38.

Mie, G. (1903), "The Equations of State of Gases and Critical Phenomena", *Amm. Phys. Lpz.*, **11**:657-666.

Ono, Y. (1947), "Lattice Theories of the Liquid States", *Mem. Fac. Eng. Kyushu Univ.*, Japan, **10**:191-197.

Peek, F. M., and J. Hill (1950), "On Lattice Theories of the Liquid State", *J. Chem. Physics*, **18**:1252-1255.

Rice, S. A., and J. G. Kirkwood (1959), "On an Approximate Theory of Transport in Dense Media" in *J. Chem. Physics*, **31**:901-908.

Rice, S. A., and A. R. Alnatt (1961a) "On the Kinetic Theory of Dense Fluids VI. The Singlet Distribution Function for the Rigid Spheres with Anactractive Potential", *J. Chem. Physics*, **34**:2144-2155.

Rice, S. A., and A. R. Alnatt (1961b) "On the Kinetic Theory of Dense Fluids VII. The Dublet Distribution Function for the Rigid Spheres with Anactractive Potential", *J. Chem. Physics*, **34**:2156-2165.

Rowlinson, J.S., and C.F. Curtiss (1951), "Lattice Theories of the Liquid States", *J. Chem. Physics*, **19**:1519-1529.

Sutherland, R. C. (1905), "Some Aspects of the Statistical Theory of Transport", *Phil. Mag.*, **9**:871-882.

Zwanzig, R. W., J. G. Kirkwood, I. Oppenheim and B. J. Alder (1954), "Statistical Mechanics Theory of Transport Processes VII." in *J. Chem. Physics*, **22**:783-790.

EXPERIMENTAL INVESTIGATION OF REDUCING STARTUP TIME ON CAPILLARY PUMPED LOOP WITH EHD ASSISTANCE

Bingjian Mo, Michael M. Ohadi
and Serguei V. Dessiatoun
Mechanical Engineering Department
University of Maryland
College Park, MD 20742
(301) 405-5263

Jeong (Jake) H. Kim and Kwok Cheung
Naval Research Laboratory
Spacecraft Engineering Department
Code 8221
Washington, DC 20375
(202) 767-6996

Abstract

The capillary pump loop (CPL) is the current state-of-the-art space cooling system. It provides higher cooling capacity than most heat pipes, more installation flexibility, and much greater distance of heat transport due to the small diameter of wickless transport lines. Major disadvantages of the CPL include long and complicated startup procedures and the possibility of depriming at high heat input and load variation. The presented work was an experimental study to characterize the startup process for an EHD-assisted CPL system. Startup is achieved by an almost stable differential pressure and average temperature at the evaporator wall. When the electric field is applied, it interacts with the vapor/liquid distribution inside the core and the wick. It also provides an additional pumping effect of liquid to the evaporator surface. As a result, less time is needed to build up the meniscus. Furthermore, the instability-induced EHD pumping at liquid-vapor interface pushes the liquid-vapor interface near the evaporator wall to enhance the phase-change. These EHD-enhanced mechanisms collaborate to reduce the required duration at different regimes and hence realize the EHD-reducing startup time for a CPL system. Experimental data showed that about 50% startup time, reduction was attainable.

INTRODUCTION

Capillary pumped loop (CPL), a wickless condenser heat pipe, is a passively pumped two-phase heat transport device that has demonstrated performance capabilities up to a magnitude greater than that of the conventional state-of-the-art heat pipe. For a CPL system, the most important design parameter other than the steady-state maximum power capacity is the transient startup process. If a CPL can not start up reliably and/or it experiences deprime condition, then CPL failure is likely since repriming the evaporators usually requires a repeat of the startup process. Startup might not only represent the highest stress on the evaporator wick, but it also has major impact on the design of the reservoir and its temperature controller, as well as the suitability of CPLs to fulfill the thermal management. This report investigates experimentally the reduction of CPL startup time with the assistance of electrohydrodynamics.

Electrohydrodynamics (EHD) is a new and promising technique that has demonstrated proven potential for significantly reducing the heat exchanger size/volume while providing on-line/on-demand control for the heat transfer surface. Its applicability for substantial increase of heat transfer coefficients for industrially significant fluids such as refrigerants R-134a and R-123, PolyAlphaolefin (PAO), and certain aviation fuel has already been demonstrated. Unlike the conventional methods, the additional pressure drop penalty is minimal and the electric power consumption is typically negligible. Polarization electrohydrodynamic force (EHD pumping) can also assist or substitute for capillary forces to collect, guide, and pump condensate in regions of high electric field intensity, whereas the saturated vapor of the dielectric fluid is rejected to locations where the electric field is less intense. Due to the dominance of polarization EHD force over surface tension force, the dielectrophoresis phenomenon may significantly improve the heat transfer capability of two-phase thermal devices using poor capillary dielectric fluids. The EHD technique has high payoff potential and promising space applications, including use in liquid-cooled thermal control systems and potential implementation for thermal management of almost all major subsystems. The terrestrial applications of the EHD technique are extensive as they include commercial heat exchanger equipment for refrigeration and air conditioning, electronic cooling, cryogenic and process industry applications, and laser medical/industrial cooling.

Because in the EHD technique the magnitude of heat transfer enhancement is directly proportional to the supply voltage, an on-line/on-demand control feature to the heat transfer surface is provided. This feature is particularly attractive for developing higher heat transfer capacity and vapor-tolerant CPL system. EHD coupling also enhances CPL startup procedure and improves the heat transfer coefficient. All these merits further promote the readiness of CPL technology for space thermal control applications.

For a CPL system, startup is identified by achievement of a steady temperature in the vapor line. Frozen startup refers to the heat input to the "cool" loop. The wick must be fully wetted, and the liquid zone in the evaporator should be cleared of bubbles before starting the CPL loop. This is usually accomplished by heating the reservoir and pressurizing the system to collapse all bubbles. Since the evaporator is the warmest part of the loop, collapsing bubbles in that component requires filling the entire loop with liquid before allowing startup.

Kiper et al. (1990) and Cullimore (1991 and 1993) presented the startup transient analysis in a CPL system. Maidanik et al. (1993) showed that the startup process could be mainly characterized by three different durations. They also predicted the minimum heat capability for the reliable startup. Results of this experimental study identified the different characteristics of the startup process as well as the startup time reduction with EHD assistance for CPL performance enhancement.

EXPERIMENTAL APPARATUS

By employing EHD technique, we investigated enhanced performance of a CPL system. The schematic diagram of the EHD-assisted CPL setup is shown in figure 1. The key component was the EHD-assisted evaporator. The allowable heat transport capacity was designed for up to 1500W. Other main components included the cooling loop, the control (reservoir) loop, and a different depriming conditions control. A brief description of the major components of the experimental apparatus is given in the following sections.

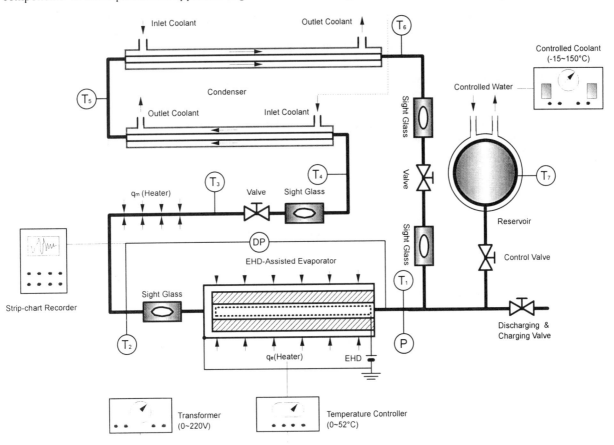

FIGURE 1. Schematic of the EHD-Assisted CPL

EHD-Assisted Evaporator Test Section

An EHD-assisted evaporator was designed and fabricated to simulate conditions in a space CPL system. A 1" CPL pump was modified to accommodate the EHD-assisted test section. A heater was placed at the surface of the evaporator. A total of fourteen thermocouples were mounted along the inside wall of the evaporator to measure the average wall temperature. A differential pressure measurement was used to evaluate the evaporator behavior for the CPL system. Seven additional thermocouples were probed inside of the loop to measure whole loop performance. An electrode was inserted in the liquid side of the wick. The voltage potential came from the high voltage supply. A high voltage electric field was established between the liquid side and vapor side in the evaporator wick. With the electrode inside, we quantified the performance improvement at the EHD-assisted evaporator. Consequently, a better understanding of the EHD-assisted CPL mechanism is expected.

Cooling(condensing) Loop

The stability of the EHD-assisted CPL setup was maintained utilizing a cooling (condensing) loop. Cold water generated by the chiller (750W capacity at T=5~10°C) circulated in the condenser side and removed heat from the phase change process in the evaporator. The amount of heat removal, as well as the wall temperature at the evaporator, were controlled by adjusting the water set-temperature. For this purpose, a temperature control system was installed inside the chiller loop.

Control (Reservoir) Loop

The function of the reservoir in the CPL system is to control the operating temperature and facilitate the startup procedure for the loop. Instead of an electric heater, a small chiller with temperature control is installed for the reservoir control. The temperature was controlled in the range of -10~120°C. By setting the chiller temperature, we defined the reservoir temperature and thus fixed the operating temperature condition for the CPL system.

Different Depriming Condition Control

Two adjustable valves were arranged at the liquid line and vapor line of CPL loop. At different power levels, we adjusted the opening of the two valves to find the depriming condition. Thus, we were able to realize the different depriming condition control for the EHD-assisted CPL setup. For better visualization, four self-designed sight glasses were also arranged in the system to monitor the loop operation.

Experimental Procedure

A typical experimental procedure began by turning on the two-chiller system. Chilled water was then applied to the condenser for about a half hour. At the same time, the controlled temperature of the reservoir was set at the operating temperature typically close to 25°C to push the liquid into the loop and thus wet the wick inside the evaporator. When the reservoir temperature reached the operating temperature, the power to the evaporator was applied to start up the CPL system. The differential pressure transducer reading was monitored and the evaporator wall temperatures were observed to ensure proper loop function. After the loop operation stabilized, high voltage was turned on to realize the EHD assistance.

Afterwards, the system was allowed to reach the steady state conditions where the fluctuations in wall temperatures and the average DP reading was less than 1%. Then, the local wall temperatures along the test section, the loop temperatures, and the high voltage current were measured. Depending on the type of parametric study, the condition settings were changed accordingly for the next data point. Then, the above procedure was repeated.

Data Reduction

The average wall temperature of the evaporator was calculated using the following equation:

$$T_{ave} = \frac{\sum_{i=1}^{N} T_{W,i}}{N},\qquad(1)$$

where $T_{w,i}$ is the local wall temperature and N is the number of thermocouples(14) mounted on the evaporator wall. The heat transfer rate of the test section (Q_h) was evaluated by:

$$Q_h = VI,\qquad(2)$$

where V and I are the heater's applied voltage and current, respectively. The ratio of EHD power consumption to the total heat transferred to the test section was also calculated as,

$$\frac{Q_{EHD}}{Q_{Total}} = \frac{Q_{EHD}}{Q_h + Q_{EHD}},\qquad(3)$$

in which,

$$Q_{EHD} = \phi i,\qquad(4)$$

where ϕ and i are applied high voltage potential and discharge current, respectively. For the different operating temperature and system depriming conditions, we collected data for the DP reading, evaporator wall and loop temperatures, and different heat transport capabilities at a variety of applied high voltage. Analysis of the data clearly suggests that the EHD reduced startup time, improved heat transport capability and prevented the depriming condition.

EXPERIMENTAL RESULTS

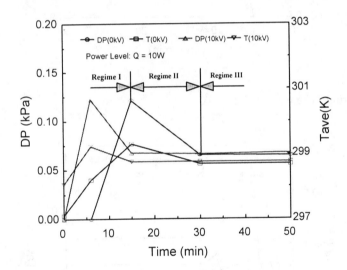

FIGURE 2. Reducing Startup Time with EHD at 10W

Figure 2 shows our typical experimental results to reduce the startup time with EHD assistance. With EHD on or off conditions, we identified the startup procedures by three different regimes. Taking EHD off condition as shown in fig.2, we classified the startup process to the following three regimes.

Regime I: Initial period of startup involved heating of the evaporator, characterized by steady growth of temperature in the evaporator wall. The differential pressure at the evaporator stayed constant (zero) due to the absence of vapor phase.

Regime II: The second stage was the boiling of working fluid in the evaporator where the highest temperature was attained. Boiling was characterized by the sharp jump of differential pressure and a subsequent quick drop.

Regime III: The final startup stage, identified as the successful startup process, consisted of a gradual increase in vapor pressure and temperature in the vapor line. In the case of a successful startup, the temperature of the evaporator wall remained stable and constant. Otherwise its unlimited growth continued until operation failure.

In addition to the reliable startup, reduced startup time is also important to enhancing the CPL startup process. We implemented the startup improvement with EHD assistance as shown in Fig. 2. At a power level of 10W, the startup time reduced from 30 minutes to 15 minutes with an applied voltage of 10kV. Similar performance improvements were achieved at other power levels, such as 20W and 50W illustrated in Figs. 3 and 4.

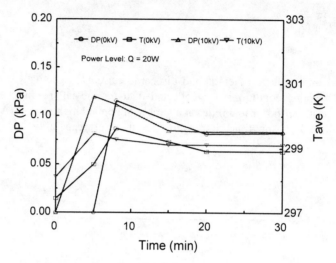

FIGURE 3. Reducing Startup Time with EHD at 20W

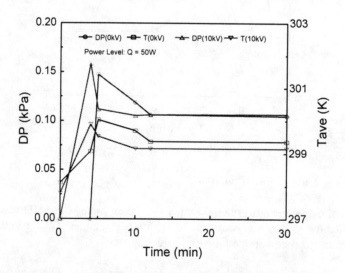

FIGURE 4. Reducing Startup Time with EHD at 50W

MECHANISM OF EHD ENHANCEMENT

Three factors contribute to the EHD-enhanced mechanism of reducing startup time for a CPL system: EHD interacting bubble behavior, thermally-induced pumping, and EHD-induced instability.

For EHD interacting bubble behavior, the fluid with a higher dielectric constant will always move toward the stronger field strength and vice versa (Pohl, 1965). When applying high voltage between liquid phases, the electric field will generate an electric force density to pump the liquid. The expression for thermally-induced pumping is given by Melcher (1981). In phase change processes, the major EHD effect is created due to the liquid-vapor interface instability. The interfacial electric stress either pulls or pushes the interface depending on the gradient of applied electric field strength. The EHD forces have to be much stronger due to the difference in dielectric constants of liquid and vapor phases during a phase change process. At the liquid-vapor interface, the dielectric constant takes a singularity. When an electric field is applied in a process having liquid-vapor interface, Maxwell stresses cause the interface to move due to induced electric pressures.

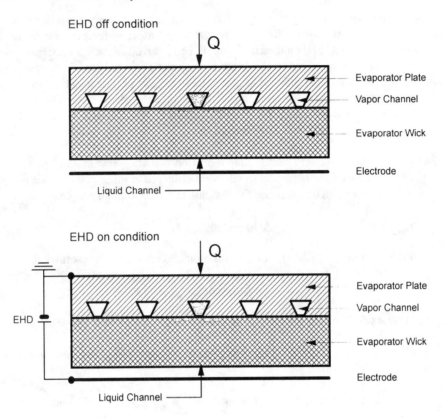

FIGURE 5. EHD-Reducing Startup Time for a CPL System

Figs. 2-4 refers to the non-EHD condition, where the wick inside the evaporator needed time to develop the menisci in order to produce capillary force. The first Regime I was characterized by the gradual temperature increase at the evaporator wall. At lower heat input, more time was required. Regime II was followed by saturated/superheated bubbles first generated at the inter-surface of the wick and plate (area farther from wick surface was easier due to the heat conduction at the evaporator plate). It was characterized by the highest wall temperature and a sharp peak in the differential pressure. After that, DP and wall temperature decreased because of the vapor bubbles expansion and released latent heat from the liquid inside the wick. Finally they were almost stable to finish the successful startup process.

On the EHD condition, the general behavior of the three-regimes appeared the same as that for the non-EHD condition. However, due to electric field coupling, the characteristic performances were somehow different.

In Regime I, when the electric field was applied, bubbles were rejected to a less intense electric-field location due to the EHD-interacting bubble behavior effect. When bubbles first formed, this EHD interaction already took effect and thus formed much larger bubbles to speed up the menisci-established process. At the same time, thermally-induced pumping also combined to enhance the heat transfer and hence reduce the required time for Regime I. After some of the menisci formed, instability-induced EHD pumping put the liquid-vapor interface up, thereby assisting the formation of a complete menisci (~0.01psi at 10kV) to reduce the duration of Regime II.

All these combined EHD effects reduced the duration necessary to finish the startup process and realize the performance improvement. The mechanism is depicted in Fig. 5. At high power level, the EHD effect was less obvious, e.g. 50W shown in Fig. 4. This was because the menisci formed easier at the higher power level.

CONCLUSIONS

The present work was an experimental study characterizing the startup process for a CPL system. Startup was achieved by establishing an almost stable differential pressure and average temperature at the evaporator wall. When the electric field was applied, EHD-interacting bubble behavior effect and the thermally-induced pumping influenced the bubble behavior by speeding up the onset of evaporation or boiling. Thus, less time was needed to develop the menisci before the onset of boiling (at Regime I). From there on, the instability-induced EHD pumping pushed the liquid-vapor interface near the evaporator wall to enhance the phase-change heat transfer. These EHD-enhanced mechanisms collaborated to reduce the required durations at different regimes. Hence EHD-reduced startup time for a CPL system yielded enhanced performance.

Experimental data showed that about 50% startup time reduction was attainable at a low power level of 10W with applied high voltage of 10kV. The same reduction trends were observed at other power input levels. At high power levels, EHD-assisted startup time reduction was less evident due to higher capillary force at this condition.

Acknowledgments

Financial support of this work by a consortium of sponsoring members is greatly acknowledged.

References

Cullimore, B.A.(1991) "Start up Transients in CPLs," in *AIAA 26th Thermophysics Conference*, AIAA91-1374.

Cullimore, B.A. (1993) "Capillary Pumped Loop Application Guide" in *23rd International Conference on Environmental Systems*, Colorado Springs, CO.

Kiper, A., et al.(1990) "Transient Analysis of a CPL Heat Pipe," in *AIAA/ASME 5th Joint Thermophysics and Heat Transfer Conference*, AIAA 90-1658.

Ohadi, M., S. Dessiatoun, B. Mo, J. Kim, K. Cheung and J. Didion(1997) "An Experimental Feasibility Study on EHD-assisted Capillary Pumped Loop," in *Space Technology and Applications International Forum*, AIP Conference Proceedings, Vol: 387, 567-572.

Pohl, H.A.(1965) "Dielectrophoresis - The Behavior of Neutral Matter in Non-uniform Electric Fields," Cambridge University Press, New York.

Maidanik, Y., et al.(1993) "Experimental and Theoretical Investigation of Startup Regimes of Two-phase Capillary Pumped Loops," in *23rd International Conference on Environmental Systems*, Colorado Springs, CO.

Melcher, J. R.(1981) "Continuum Electromechanics," The MIT Press.

A STUDY OF THE FUNDAMENTAL OPERATIONS OF A CAPILLARY DRIVEN HEAT TRANSFER DEVICE IN BOTH NORMAL AND LOW GRAVITY
PART 1. LIQUID SLUG FORMATION IN LOW GRAVITY

Jeffrey S. Allen
University of Dayton
c/o NASA Lewis Research Center
21000 Brookpark Road
Cleveland, Ohio 44135
216-433-3087

Kevin Hallinan
Department of Mechanical
and Aerospace Engineering
University of Dayton
300 College Park
Dayton, Ohio 45469
937-229-2875

Jack Lekan
Mail Stop 500-216
NASA Lewis Research Center
21000 Brookpark Road
Cleveland, Ohio 44135
216-433-3459

Abstract

Research has been conducted to observe the operation of a capillary pumped loop (CPL) in both normal and low gravity environments in order to ascertain the causes of device failure. The failures of capillary pumped heat transport devices in low gravity; specifically; evaporator dryout, are not understood and the available data for analyzing the failures is incomplete. To observe failure in these devices an idealized experimental CPL was configured for testing in both a normal-gravity and a low-gravity environment. The experimental test loop was constructed completely of Pyrex tubing to allow for visualization of system operations. Heat was added to the liquid on the evaporator side of the loop using resistance heaters and removed on the condenser side via forced convection of ambient air. A video camera was used to record the behavior of both the condenser and the evaporator menisci simultaneously. Low-gravity experiments were performed during the Microgravity Science Laboratory (MSL-1) mission performed onboard the Space Shuttle Columbia in July of 1997. During the MSL-1 mission, a failure mechanism, heretofore unreported, was observed. In every experiment performed a slug of liquid would form at the transition from a bend to a straight run in the vapor line. Ultimately, this liquid slug prevents the flow of vapor to the condenser causing the condenser to eventually dryout. After condenser dryout, liquid is no longer fed into the evaporator and it, too, will dry out resulting in device failure. An analysis is presented to illustrate the inevitable formation of such liquid slugs in CPL devices in low gravity.

INTRODUCTION

Capillary pumped loops (CPL's) are used to transfer heat from one location to another using the latent heat of the working fluid. The basic CPL operation is illustrated in Figure 1. Heat is added to a porous wick where the liquid evaporates and the vapor recondenses at the cold end of the loop. The liquid is then driven back to the evaporator by the capillary pressure difference between the condenser meniscus and the menisci in the porous media in the evaporator. The advantages of CPL's are that they are entirely passive, requiring no power, and can transfer heat over distances as large as 10 or more meters. These features are very desirable for spacecraft operations. Unfortunately, CPL's have proven to be unreliable in low-gravity operations.

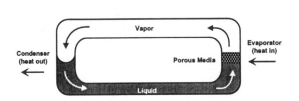

FIGURE 1. Schematic of Operations in a Capillary Pumped Loop (CPL)

Previous research of high power capillary pumped loops has revealed operating characteristics which at times fall well-below design predictions of the heat transport capability. Richter and Gottschlich (1994) have observed that design predictions for low temperatuer CPL's significantly over estimate the experimentally realized heat transport. In the experimental work of Ku, Kroliczek, McCabe, and Benner (1988) pressure oscillations across the evaporators with a magnitude as large as 700 Pa (0.1 psi) were observed. Later, Ku (1995), reported that when these pressure oscillations are present, the maximum heat transport can be as little as 10% of the design value. Finally, the CPL GAS can, CAPL, CAPL-2, View-CPL, and TPF low-gravity technology demonstration experiments, all based out of NASA Goddard Space Flight Center, have reported difficulty in start-up of capillary pumped loops.

The Capillary-driven Heat Transfer (CHT) experiment was conceived to study the fundamental fluid physics phenomena thought to be responsible for the failure of CPL's in low-gravity operations.

EXPERIMENT

The principle component of the Capillary-driven Heat Transfer (CHT) experiment is a glass test loop which is illustrated in Figure 2. Within the loop is a three-way valve; referred to as the control valve. The control valve directs liquid flow from the reservoir into the condenser leg and/or the evaporator leg of the test loop. During the experiment, the control valve allows for liquid flow from the condenser leg to the evaporator leg while isolating the reservoir. The reservoir is constructed from a 10cc gas tight syringe with a screw type plunger which allows for repeatable fills. The liquid leg of the test loop is constructed of Pyrex capillary tubing. The vapor leg of the test loop is constructed from 10mm diameter Pyrex tubing. Two conical transition sections connect the vapor leg to the liquid leg. The conical sections are designed to be capillary traps in low gravity so as to preferentially locate the evaporator meniscus. A capillary pumping potential is established by the difference between the pressure drop across the condenser meniscus in the 10mm tube and the pressure drop across the evaporator meniscus in the capillary tube (1mm or 4mm in diameter). The difference in the condenser and evaporator diameters allows for the capillary pumping potential to be maintained even with a large temperature difference between the evaporator and condenser. For these experiments, the test fluid is spectroscopic grade ethanol which perfectly wetted the test loop.

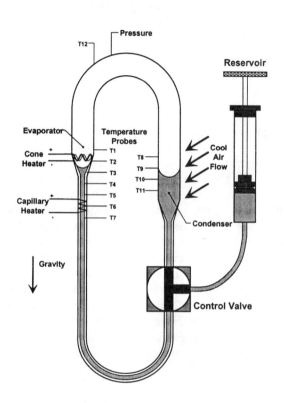

FIGURE 2. Schematic of Capillary-driven Heat Transfer (CHT) Experiment Test Loop

Heat is applied to the evaporator by either of two heaters. The first, referred to as the cone heater, is a serpentine wire attached to the conical transition section. The second, referred to as the capillary heater, is a spiral wound wire located on the capillary tubing approximately 15mm from the cone heater. When using the cone heater the wall temperature gradient along the thin film region of the meniscus is cooler towards the liquid side. The capillary heater reverses this temperature gradient. Cooling of the condenser meniscus is accomplished by forced air convection at ambient temperature. Instrumentation consists of 7 thermocouples (Type T, 40 AWG) along the length of the evaporator leg and 4 thermocouples along the length of the condenser leg. An additional thermocouple and a pressure transducer are connected to the vapor leg. The position and behavior of both the condenser meniscus and the evaporator meniscus are recorded using a single video camera.

Low gravity testing was obtained during the Microgravity Science Laboratory - 1 (MSL-1) flown on board the Space Shuttle Columbia in July, 1997. The experiment was operated in the Middeck Glovebox (MGBX) by the payload crew members. In low gravity, the test loop is filled using capillary pressure so as to allow the evaporator and condenser menisci to find a natural equilibrium position. The low gravity experiment begins with the evaporator and condenser menisci in static equilibrium as left following the filling process. Either the cone heater or the capillary heater is then set to a fixed heat input and turned on. The experiment is now monitored until evaporator dryout occurs. After evaporator dryout, the heater is turned off and the liquid in the test loop is reoriented back to the static equilibrium position. The heat input is then adjusted to the next setting and the experiment repeated.

OBSERVATIONS

One of the first experimental observations, irrespective of the gravitational environment, was the formation of a continuous liquid film over the entire length of the vapor leg. At low heat inputs the liquid film was continuous from the evaporator meniscus to the condenser meniscus. At higher heat inputs the liquid film appeared to begin slightly beyond the evaporator meniscus and extended to the condenser meniscus. This, in and of itself, is not a surprising result. However, soon after formation it became obvious that this liquid film was not stationary. Under normal gravity conditions, the liquid film would drain into the condenser meniscus and into the evaporator conical section. The latter film drainage affected the stability of the evaporating meniscus by introducing cold fluid into localized areas of the meniscus. In low gravity, the liquid film also drained, but the draining mechanism in this instance is capillarity. Variations in the capillary pressure caused the liquid film to drain and collect in the vapor leg of the test loop.

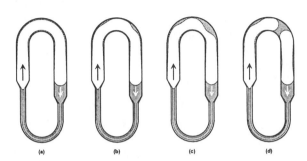

FIGURE 3. Low-Gravity Liquid Accumulation and Slug Formation in the Vapor Leg of the Test Loop

The process of the liquid pooling and eventual formation of a liquid slug in low gravity is illustrated in Figure 3. As each experiment progressed, the liquid began to pool in the outer radius of the bend in the vapor section of the test loop. This pool of liquid would grow until a slug of liquid would form, completely bridging the vapor line. The formation of this liquid slug is independent of the heat inputs. The heat input only affects the rate at which the liquid pools in the outer radius of the bend. After the formation of the liquid slug, system failure is imminent. Figures 4 and 5 show the formation of the liquid slug in the vapor leg

FIGURE 4. Liquid Accumulation and Formation of a Liquid Slug During a Low-Gravity Experiment

FIGURE 5. Recession of the Condenser Meniscus After Formation of a Liquid Slug in the Vapor Leg.

and the subsequent effects on the condenser meniscus during one of the low-gravity experiments. In Figure 4, the liquid film begins to pool in the outer radius of the bend region eventually bridging. Prior to complete bridging of the vapor line, there is no adverse effect on the CPL operation due to the pooling of the liquid in the bend of the vapor leg. However, after the liquid slug has formed, both vapor flow and liquid film flow to the condenser meniscus was eliminated. Subsequently, as liquid is continually fed to the evaporator, the condenser meniscus begins to recede as shown in Figure 5. Eventually, the condenser meniscus recedes into the capillary tube thereby eliminating the pressure feeding liquid into the evaporator. At this point the evaporator dries out and the system fails.

DISCUSSION OF OBSERVATIONS

An annular liquid film inside a straight tube having a vapor core has been shown repeatedly to be fundamentally unstable to any long-wave disturbances by Gauglitz and Radke (1989), Aul and Olbricht (1990), Hu and Patankar (1994), and numerous other researchers. Such a liquid film will always breakup and form periodically spaced annular lobes or liquid slugs. In the Capillary-driven Heat Transfer (CHT) experiment, however, the formation of the liquid slug arises from a pressure gradient within the liquid film and is not due to long wavelength instabilities.

In the absence of gravity, the flow of the liquid film into the bend occurs as a result of capillary pressure differences between various regions within the tube. More appropriately, capillarity results in drainage of the annular liquid film for a low Bond number system where the Bond number is defined as $Bo = (\rho g R^2)/\sigma$. A low gravity environment is naturally a low Bond number system. The following analysis applies to small diameter tubes in normal gravity as well as large diameter tubes in low gravity.

In order to study the dynamics within the annular liquid film, four distinct regions of the film will be examined (see Figure 6). The first region (region 1) is where the liquid accumulates in the bend; that is, the outer portion of the tubing bend. Region 2 is the inside radius of the tubing bend and is located 180° from region 1 at the same centerline location (See Figure 6, section A-A). Region 3 comprises the straight portion of the vapor line. And region 4 is the condenser meniscus. Initially, the condenser meniscus is assumed to be relatively far from the bend. The liquid film will also be studied with the meniscus in the vicinity of the bend; i.e., when there is no region 3. For the purposes of this analysis the liquid film is assumed to be of uniform thickness, t. The inside diameter of the tube is $2R_i$ and the centerline radius of the bend in the tube is R_b.

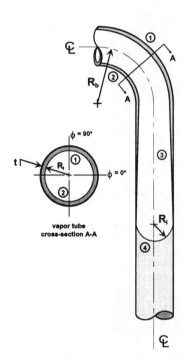

FIGURE 6. Geometry of the Annular Liquid Film

The pressure drop across a liquid-vapor interface is known from the Laplace-Young equation:

$$P_v - P_l = \sigma \left(\frac{1}{R_1} + \frac{1}{R_2} \right); \quad (1)$$

where P_v is the vapor pressure, P_l is the liquid pressure, R_1 and R_2 are the principle radii of curvature, and σ is the surface tension. One of the principle radii of curvature is the same for each of the four regions and is equal to $R_i - t$. The second principle radius curvature varies from region to region. In region 1 that radius of curvature is $R_b + (R_i - t)$. Similarly, $-[R_b - (R_i - t)]$, ∞, and $R_i - t$ are the second principle radius of curvature for regions 2, 3, and 4, respectively. In order to simplify the analysis, a film thickness ratio is defined as $\delta = t/R_i$. Also, the ratio of the radii is defined as $\Gamma = (R_i - t)/R_b$. Substituting the appropriate principle radii of curvature and the two new definitions into the Laplace-Young equation results in a set of expressions for the pressure drop across the liquid surface for each of the four regions of the liquid film.

Region 1: $\quad P_v - P_l|_1 = \dfrac{\sigma}{R_i(1-\delta)} \left[1 + \dfrac{\Gamma}{1+\Gamma} \right] \quad (2)$

Region 2: $\quad P_v - P_l|_2 = \dfrac{\sigma}{R_i(1-\delta)} \left[1 - \dfrac{\Gamma}{1-\Gamma} \right] \quad (3)$

Region 3: $\quad P_v - P_l|_3 = \dfrac{\sigma}{R_i(1-\delta)} \quad (4)$

Region 4: $\quad P_v - P_l|_4 = \dfrac{2\sigma}{R_i(1-\delta)} \quad (5)$

The interest in this study, however, is in the pressure drop within the liquid film between the four regions. The first pressure drop considered is between the inner bend radius and the outer bend radius, or from region 2 to region 1. For an isothermal system without significant vapor flow there is no pressure drop in the vapor between regions 1 and 2. Likewise, the pressure drop in the vapor between the other regions is also assumed to be negligible. In considering a system with a condensing vapor this assumption would have to be revisited.

Now, equations 2 through 5 are rearranged so as to calculate the pressure drop within the annular liquid film between the various regions. Shown below in the left column, equations 6 through 9 describe the pressure difference within the annular liquid film when the condenser meniscus is relatively far from the tubing bend. In the right column, equations 10 through 12 describe the liquid film pressure drops for the case where the meniscus is in the vicinity of the tubing bend. In the latter case, there is no straight section (region 3) of the liquid film.

$$\Delta P_l|_{2-1} = \frac{\sigma}{R_i(1-\delta)} \left[\frac{2\Gamma}{1-\Gamma^2}\right] \quad (6) \qquad \Delta P_l|_{2-1} = \frac{\sigma}{R_i(1-\delta)} \left[\frac{2\Gamma}{1-\Gamma^2}\right] \quad (10)$$

$$\Delta P_l|_{3-1} = \frac{\sigma}{R_i(1-\delta)} \left[\frac{\Gamma}{1+\Gamma}\right] \quad (7) \qquad \Delta P_l|_{1-4} = \frac{\sigma}{R_i(1-\delta)} \left[\frac{1}{1+\Gamma}\right] \quad (11)$$

$$\Delta P_l|_{2-3} = \frac{\sigma}{R_i(1-\delta)} \left[\frac{\Gamma}{1-\Gamma}\right] \quad (8) \qquad \Delta P_l|_{2-4} = \frac{\sigma}{R_i(1-\delta)} \left[\frac{1}{1-\Gamma}\right] \quad (12)$$

$$\Delta P_l|_{3-4} = \frac{\sigma}{R_i(1-\delta)} \quad (9)$$

If flow in the annular liquid film is assumed to be steady and dominated by viscous effects, then the governing equations can be reduced to the lubrication approximation and the velocity in the liquid film scales as:

$$U \sim \frac{\delta R_i^2}{\mu} \frac{\Delta P_l}{L}; \quad (13)$$

where L is the length scale over which the pressure difference, ΔP_l, occurs. The appropriate length scales for the pressure differences are shown in Equations 14 through 20.

$$L_{2-1} = \frac{\pi}{2} R_i(1-\delta)(2) \quad (14) \qquad L_{2-1} = \frac{\pi}{2} R_i(1-\delta)(2) \quad (18)$$

$$L_{3-1} = \frac{\pi}{2} R_i(1-\delta) \left[\frac{1+\Gamma}{\Gamma}\right] \quad (15) \qquad L_{1-4} = \frac{\pi}{2} R_i(1-\delta) \quad (19)$$

$$L_{2-3} = \frac{\pi}{2} R_i(1-\delta) \left[\frac{1-\Gamma}{\Gamma}\right] \quad (16) \qquad L_{2-4} = \frac{\pi}{2} R_i(1-\delta) \quad (20)$$

$$L_{3-4} = \frac{\pi}{2} R_i(1-\delta) \quad (17)$$

The potential for liquid flow can be rewritten in terms of the Capillary number, where $Ca = \mu U \sigma$ by combining the expressions for the pressure drop in the liquid film, ΔP, the length scale associated with the pressure drop, L, and the velocity scale in the liquid film, U. When the condenser meniscus is relatively far from the tubing bend, the resulting expressions for the Capillary numbers between each of the regions are shown in Equations 21 through 24. The Capillary numbers for when the condenser meniscus is near the bend are expressed in Equations 25 through 27.

$$Ca|_{2-1} \sim \frac{2}{\pi} \left(\frac{\delta}{1-\delta}\right)^2 \left[\frac{\Gamma}{1-\Gamma^2}\right] \quad (21) \qquad Ca|_{2-1} \sim \frac{2}{\pi} \left(\frac{\delta}{1-\delta}\right)^2 \left[\frac{\Gamma}{1-\Gamma^2}\right] \quad (25)$$

$$Ca|_{3-1} \sim \frac{2}{\pi} \left(\frac{\delta}{1-\delta}\right)^2 \left[\frac{\Gamma}{1+\Gamma}\right]^2 \quad (22) \qquad Ca|_{1-4} \sim \frac{2}{\pi} \left(\frac{\delta}{1-\delta}\right)^2 \left[\frac{1}{1+\Gamma}\right] \quad (26)$$

$$Ca|_{2-3} \sim \frac{2}{\pi} \left(\frac{\delta}{1-\delta}\right)^2 \left[\frac{\Gamma}{1-\Gamma}\right]^2 \quad (23) \qquad Ca|_{2-4} \sim \frac{2}{\pi} \left(\frac{\delta}{1-\delta}\right)^2 \left[\frac{1}{1-\Gamma}\right] \quad (27)$$

$$Ca|_{3-4} \sim \frac{2}{\pi} \left(\frac{\delta}{1-\delta}\right)^2 \quad (24)$$

During the Capillary-driven Heat Transfer (CHT) experiment, the thickness of the annular liquid film was calculated to be on the order of 250 μm. The inside diameter of the vapor leg of the test loop is

10 mm and the bend radius of the test loop is 20 mm. Therefore, for the CHT experiment the film thickness ratio, δ, is 0.05 and the radius ratio, Γ, is 0.2375. Based upon these parameters, the Capillary numbers for the CHT experiment are calculated and shown in Table 1 for both the case where the condenser meniscus is relatively far from the tubing bend and for the case where the condenser meniscus is near the tubing bend. The Capillary numbers for both positions of the condenser meniscus are

TABLE 1. Capillary Numbers for the Annular Liquid Film in the CHT Experiment

Meniscus far from bend		Meniscus near the bend	
Regions	$Ca \cdot 10^5$	Regions	$Ca \cdot 10^5$
2 - 1	44	2 - 1	44
3 - 1	7	1 - 4	142
2 - 3	17	2 - 4	231
3 - 4	176		

represented pictorially in Figure 7, where the length of the arrow representative of the magnitude of the liquid flows. In the case where the meniscus is relatively far from the tubing bend (Figure 7a), the liquid flow into the bend is much larger than the liquid flow out of the bend. Also, the liquid flow into the condenser meniscus is an order of magnitude higher than any other liquid film flows in the system. This implies that there is always significant drainage into the condenser meniscus even in low gravity systems and that the liquid will accumulate in the outer region of the tubing bend in any low Bond number system.

Similarly, for the case where the condenser meniscus is in the vicinity of the tubing bend (Figure 7b), the liquid flow into the meniscus is an order of magnitude greater than any other liquid flow in the annular film. This has the effect of draining the bend region of the tube and can prevent the formation of the liquid slug *in that location only*. The prevention of the liquid slug formation was observed during one particular run of the CHT experiment where the condenser meniscus was very near the bend. A liquid slug still formed, however, on the other side of the test loop in the evaporator section of the vapor leg.

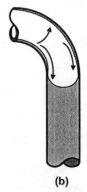

FIGURE 7. Illustration of the Direction and Relative Magnitude of Liquid Film Flows for Two Cases; (a) the Condenser Meniscus Relatively Far from the Tubing Bend and (b) the Condenser Meniscus in the Vicinity of the Tubing Bend

The variation in the Capillary number between the various regions is only a function of the geometry of the system and is not a function of the properties of the liquid. Therefore, the liquid film flows illustrated in Figure 7 are typical of the flows in any annular liquid film within a bend in a low Bond number system.

CONCLUSIONS

During the operation of the CHT experiment both in normal gravity and in low gravity a liquid film formed over the entire vapor leg of the loop. In normal gravity, the draining of this film affected the stability of the evaporating meniscus, but was not directly responsible for evaporator dry out. In low gravity, however, the drainage of the annular liquid film resulted in an accumulation of liquid in the outer radius of the bend in the tubing. The liquid in this region eventually bridged the vapor leg blocking the flow of vapor to the condenser meniscus. Subsequently, both condenser dryout and evaporator dryout ensued and the system failed. The flows in the annular liquid film will always collect in the bend of the tube because of capillary forces. This mode of failure was not appreciated before the on-orbit experiment. It is very probable that previous low-gravity capillary pumped loop experiments exhibited poor performance due to the formation of liquid slugs in the vapor line since all of the designs to date have incorporated a significant number of bends in the vapor leg.

Acknowledgments

The authors would like to acknowledge the support for this research from NASA Microgravity Science under grants NAG3-1391 and NAG3-1919.

References

Aul, R. W. and W. L. Olbricht (1990) "Stability of a thin annular film in pressure-driven, low-Reynolds-number flow through a capillary", in *J. Fluid Mech.*, 215:585-599.

Gauglitz, P. A. and C. J. Radke (1990) "The Dynamics of Liquid Film Breakup in Constricted Cylindrical Capillaries", in *J. Colloid Interface Sci.*, 134(1):14-40.

Hu, H. H. and N. Patankar (1994) "Non-Axisymmetric Instability of Core-Annular Flow", in *Two Fluid Flows - With or Without Phase Change*, AMD-Vol. 184, A. Narain, D. A. Siginer, and K. M. Kelkar, eds, The American Society of Mechanical Engineers, New York, presented at the 1994 International Mechanical Engineering Congress and Exposition, Chicago:33-39.

Ku, J. (1995) "Start-Up Issues of Capillary Pumped Loops", 9th International Heat Pipe Conference, Albuquerque, New Mexico.

Ku, J., E. J. Kroliczek, M. McCabe, and S. M. Benner (1988) "A High Power Spacecraft Thermal Management System", in *AIAA Paper No. 88-2702*.

Richter, R. and J. M. Gottschlicht (1994) "Thermodynamic Aspects of Heat Pipe Operation", in *J. Thermophysics and Heat Transfer*, 8(2):334-340.

Nomenclature

- L - length scale for ΔP
- P - pressure
- R_i - inside radius of tubing
- R_b - bend radius of tubing
- t - thickness of liquid film
- U - liquid velocity scale

- μ - absolute viscosity
- σ - surface tension

- δ - ratio of film thickness to inside radius, t/R_i
- Γ - ratio of radii, $R_i(1-\delta)/R_b$
- Bo - Bond Number, $\rho g R^2/\sigma$
- Ca - Capillary number, $\mu U/\sigma$

subscripts

- 1 - outer portion of the tubing bend
- 2 - inner portion of the tubing bend
- 3 - straight section of the tubing
- 4 - condenser meniscus region

- l - liquid
- v - vapor

TESTING AND EVALUATION OF SMALL CAVITATING VENTURIS WITH WATER AT LOW INLET SUBCOOLING

S. G. Liou, I. Y. Chen and J. S. Sheu
Department of Mechanical Engineering
National Yunlin University of Science and Technology
Yunlin, Taiwan, ROC
(TEL)886-5-5342601,ext.4141 (FAX)886-5-5312062
LIOUSG@FLAME.YUNTECH.EDU.TW

Abstract

Cavitating venturi (CV) has been widely used as a flow control device in many different industries. In 1990, cavitating venturi was selected as the baseline flow control device in the Space Station Freedom's(SSF's) two-phase active thermal control system(ATCS). However, the design and the operation of the CVs used in SSF's ATCS is quite different in many ways from that typically used in the industry, such as low mass flow rate, small size, low pressure difference between inlet and outlet, and low inlet subcooling. During the prototypic ATCS' testing at NASA/Johnson Space Center, a phenomenon called overflow associated with throat superheat was observed. Although data was obtained and analyzed, no useful correlation for the superheat at rechoking was acquired. The objective of this study is to conduct a performance test on small CVs under low inlet subcooling. Water is used as the working fluid. Data acquisition and analysis are carried out under normal choked flow, over flow and recovery conditions. The effects of CV's size, fluid temperature, flow condition and inlet subcooling on CV performance are evaluated. Analysis of the test results showed that the superheat necessary for the onset of nucleation in pool boiling can be applied for the estimation of superheat required at rechoking for the CVs. With this postulated superheat and the predetermined CV loss coefficient, a equation as a function of inlet subcooling is recommended for predicting the pressure ratio at the recovery for the choked flow control in a mechanically pumped system.

INTRODUCTION

Active thermal control systems (ATCSs) are used on space vehicles to collect waste heat at distributed sites and dissipate the heat through radiation to space. The current state-of-the-art uses a single phase coolant loop to perform this function as on the US Space Shuttle and the US-led portion of the International Space Station (ISS). As space vehicles become larger and require higher electrical power levels, two-phase (liquid and vapor) ATCSs become an attractive alternative. Two-phase systems require less operating power than the single-phase systems, and offer the potential for lower overall system mass. Mechanically pumped two-phase ATCSs send liquid to each heat acquisition site where it is partially evaporated by the local heat load. The resulting two-phase mixture travels to a radiator array where the vapor is condensed and subcooled prior to being returned to the pump inlet. Two-phase ATCSs for space applications must use flow control devices (FCDs) upstream of each heat acquisition site to control the flow of liquid coolant. These FCDs can be active flow control valves, as in the two-phase ATCS on the Russian segment of ISS (Ungar and Mai 1996), or can be passive flow control device (PFCDs) as in the US two-phase ATCS designs. PFCDs have the advantages of requiring no electrical power for flow control and not being subject to wear or breakage. Possible PFCDs for use in two-phase ATCSs are cavitating venturis, orifices, flow nozzles, and capillary tubes (Ungar and Chen 1996). Recently cavitating venturis (CVs) had been selected as the baseline flow control device in the ATCS on the US Space Station Freedom (SSF) in 1990 (Raetz and Dominick 1992), and were used in every test of the Ground Test Article (GTA), a prototypic SSF ATCS two-phase thermal loop, at the NASA Johnson Space Center since 1988.

The venturi is an important component in many liquid flow systems, used for flow measurement and flow regulation, as well as for cavitation research. When liquid flow venturi cavitates, it can provide a choked flow regime useful for flow control in thermal-fluid systems. If liquid's inlet pressure and temperature are fixed, a wide range of downstream pressures may be imposed with no effect on flow. Cavitating venturis have been used extensively in a variety of industrial and aerospace applications as flow control devices. For all the CV applications, it is important to know its all-liquid loss coefficient and the cavitation number corresponding to the inception of cavitation (Hammitt et al. 1976). These applications include regulating the flow of propellants to the rocket engines, regulating water flow to the automatic sprinkler systems, and regulating flow to the parallel thermal-hydraulic devices (Fox 1977). In these applications, the CV inlet subcooling (the pressure difference between the inlet pressure and the saturation pressure at the inlet temperature) is typically in the order of 700 to 1400 kPa.

For the SSF prototypic two-phase ATCS testing, CVs were installed at the upstream of each heat acquisition device (evaporator) to provide the desired liquid ammonia flow rate according to its maximum heat load in the GTA system. Each CV would maintain a constant flow rate at giving upstream conditions as long as the downstream pressure remains below a critical value. However, those venturis were intended to be operated under conditions that are quite different from the past experiences, such as small flow rates, very small sizes, low pressure difference across CVs (40 to 125 kPa), and very low inlet subcooling (40 to 140 kPa). Because the GTA's two-phase pump is operated at low head rise to conserve electric power, the inlet subcooling for the CV would have been lower than the cases in normal CV applications. During operation of the prototypic ATCS GTA in 1991, an anomalous CV overflow behavior was observed at these low values of inlet subcooling. Following the discovery of CV overflow, a series of experiments were performed in the GTA tests to gain understanding on the CV's overflow phenomenon and to define its limits (Ungar, Dzenitis and Sifuentes 1994).

Although the unchoking and the overflow phenomena of the CVs found in the GTA ammonia testing were understood, the limits of the throat superheat required to induce cavitation from liquid overflow was not correlated. Recovery from overflow is still a problem for a system of paralleled CVs with low inlet subcooling. Such a system may have many possible operating flow rates since each CV in the loop may be operated in either choked or overflow mode at pressure ratios lower than a critical value. This unstable operation is not a acceptable mode of operation to the thermal systems, especially, for a mechanically pumped spacecraft two-phase ATCS.

The objective of this study is to conduct a test on cavitating venturi at low inlet subcooling to observe the phenomena of normal operation, overflow and recovery from overflow. The results will be correlated and applied not only to the spacecraft's mechanically pumped ATCSs designed at low inlet subcooling, but also to the general applications in industry. The specific technical objectives of this study are (1) to construct and operate a CV test loop at low inlet subcooling, using water as the working fluid to observe the phenomena of normal choked cavitating operation, unchoked flow, overflow, and the cavitation inception for recovery to choked flow from overflow, (2) to analyze the test results and obtain a correlation of the throat superheat at recovery with the effects of fluid properties, flow conditions, and size of CVs used in the test, (3) to predict the mode of operation of a cavitating venturi at low values of inlet subcooling and to give recommendations for maintaining the desired choked flow rate in various CV applications.

CAVITATING VENTURI'S OPERATING CHARACTERISTICS

A typical cavitating venturi consists of a upstream converging section, a short and straight section called the throat, and a downstream diverging section or diffuser. Figure 1 shows the schematic and dimensions of the Fox venturis used in this study. The cavitating venturi is designed to provide cavitation at the throat so that the flow rate is independent of the downstream pressure. With cavitation, the flow at the throat is choked and the throat pressure is equal to the saturation pressure at the inlet temperature. Under this condition, the upstream pressure and the throat pressure are both constant, as well as the mass flow rate. The cavitation at the throat causes the liquid flow to separate with a vapor layer in the diffuser. The point where the liquid flow reattaches in the diffuser is a function of the downstream pressure. As the downstream pressure increases, the reattachment point moves upstream toward the throat. Once the

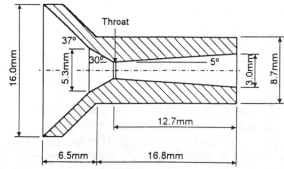

FIGURE 1. Fox's Cavitating Venturi

downstream pressure reaches a critical value such that no cavitation will occur at the throat, then the flow rate will decrease with further increase in the downstream pressure. At this point, the venturi will act like a standard single-phase liquid venturi. The unchoked liquid flow rate is then a function of the pressure drop across the venturi. When the downstream pressure is brought below the critical value, the venturi should then return to its normal operation as a choked venturi.

Flow Equations and Relevant Parameters

For a steady and choked flow condition, the static pressure will be a minimum at the throat, and is assumed to be equal to the saturation pressure at the inlet temperature. The cavitating venturi's mass flow rate (M_c) can then be determined from the Bernoulli's equation as:

$$M_c = A_{th}\sqrt{2\rho[P_{in} - P_{sat}(T_{in})]}. \tag{1}$$

For a non-choked condition, a flow loss coefficient must be applied to account for the friction loss through the venturi. The cavitating venturi's mass flow rate (M_{liq}), the averaged liquid velocity at the throat (V_{th}), and the all-liquid venturi's loss coefficient (K_v) can be expressed as:

$$M_{liq} = A_{th}\sqrt{2\rho(P_{in} - P_{out})/K_v}, \tag{2}$$

$$V_{th} = M_{liq}/(\rho A_{th}), \tag{3}$$

$$K_v = (P_{in} - P_{out})/[(\rho V_{th}^2)/2]. \tag{4}$$

Under either condition, the actual CV's measured mass flow rate, M_{act}, can be expressed in dimensionless form as:

$$M_r = M_{act}/M_c. \tag{5}$$

For the case of choked flow, the mass flow ratio is unity, i.e. $M_r = 1$. The venturi's downstream pressure can also be expressed in dimensionless form as the pressure ratio, P_r:

$$P_r = (P_{out} - P_{sat})/(P_{in} - P_{sat}). \tag{6}$$

Based on Eq. (6) and Eq. (2), the mass flow ratio for the unchoked all-liquid flow can be expressed as:

$$M_{r,liq} = M_{liq}/M_c = \sqrt{(1 - P_r)/K_v}. \tag{7}$$

The critical pressure ratio, $P_{r,crit}$, at the intersection of the choked flow ($M_r=1$) and the all liquid flow, Eq. (7), can be obtained as:

$$P_{r,crit} = 1 - K_v. \tag{8}$$

Overflow Problem Description

In the classic theory of the operation of CV, cavitation is normally assumed to occur whenever the pressure ratio, P_r, is less than the critical value, $P_{r,crit}$. The all-liquid flow is assumed to occur only when the P_r exceeds $P_{r,crit}$ with $M_r<1$. However, the CV's performance characteristics that are not usually observed in typical applications were noticed in the prototypic SSF ATCS GTA tests at low values of inlet subcooling (in the order of 100 kPa). The CV did show a tendency to become unchoked after the pressure ratio exceeds $P_{r,crit}$ with $M_r<1$. However, as the pressure ratio was progressively decreased below this critical value, the flow tend to remain unchoked, causing $M_r>1$. The GTA cavitating venturi test also confirmed that the venturis continued to persist as a all-liquid operation until the pressure ratio is dropped to a value much lower than $P_{r,crit}$. Only then, the CV resumes its choked condition. During the overflow, the pressure at the throat, P_{th}, is lower than the saturation pressure at the inlet temperature, $P_{sat}(T_{in})$. Thus, the liquid at the throat is superheated.

Throat Superheat at Cavitation Inception

The rechoking point is a function of the throat superheat required to incite cavitation at the throat. Once cavitation is induced, CV is back to normal choked operation. During all-liquid flow the throat superheat can be expressed as:

$$dP_{sup} = dP_{sub}[(1-P_r)/K_v - 1], \qquad (9)$$

where the throat superheat is defined as:

$$dP_{sup} = P_{sat}(T_{in}) - P_{th}. \qquad (10)$$

Since the pressure at the throat (P_{th}) is not measured, the throat superheat at the cavitation inception is estimated using Eq. (9), based on the measured values of P_r, dP_{sub} and K_v at the instant. By using the Eq. (9) concerning the dP_{sup} at the cavitation inception, the pressure ratio at the rechoking point can be rewritten as:

$$P_{r,rechoked} = 1 - K_v(dP_{sup}/dP_{sub} + 1). \qquad (11)$$

Hamitt et al. (1976) conducted a venturi's cavitation tests on various CV geometries and flow conditions. They found that the cavitation number corresponding to the cavitation inception is affected by the loss coefficient, throat Reynolds number, liquid temperature, and the presence of gaseous nuclei in the liquid. Their cavitation number "τ" is defined as:

$$\tau = (P_{th} - P_{sat})/(\rho V_{th}^2/2). \qquad (12)$$

Although their τ data was typically positive, a limited number of τ data has a low value of -0.2 which meant liquid was superheated at the throat. Since the superheat at rechoking point varies with the inlet subcooling, the cavitation number is not adequate to describe the cavitation inception for cavitating venturis.

On the other hand, the onset of nucleation in pool boiling had been widely studied (Collier and Thome 1994). The superheat required for vapor nucleation based on the radius of spherical vapor (r^*) can be shown as:

$$dP_{sup} = 2\sigma/r^*. \qquad (13)$$

Hsu (1962) also used Eq. (13) to determine the wall superheat requirement for the onset of nucleation condition in a pool boiling by assuming $r_c = r^*$, where r_c is the cavity mouth radius or the surface roughness.

EXPERIMENT

The CV test loop is made of insulated copper tubes with 0.75 cm ID as shown in Figure 2 (Sheu 1997). In this loop, pure water is stored in a 10 liter stainless steel tank which has a electric heater and refrigerating circulator for temperature control (293K-363K). The container's feed line can also be used for system pressure control, however, only atmospheric data is presented in this study. A stainless steel centrifugal pump with constant speed is used to provide the working fluid to the CV test section. A filter is mounted downstream of the pump to screen out any significant solid impurities and noncondensable gas flowing into CVs to enhance cavitation (Hammitt 1980). There is a heater block located upstream of the test section to allow minor increase of the water temperature at the CV entrance. As shown in Figure 2, three venturis are installed, with one in each test branch. Two of them have fixed throat diameter of 0.91 mm and 1.16 mm. The third one is a adjustable throat area venturi, which has the function to vary its choked flow rate by adjusting the throat area. However, its test data is not presented in this study. A ball valve upstream of each CV is used to adjust the inlet pressure (or CV inlet subcooling), and the CV's back pressure and its pressure ratio are varied through a needle valve located downstream of each CV. The instrumentation for the CV test loop is also shown in Figure 2. The flow meter has a calibrated measurement range from 700-2000 cc/min, with an accuracy of $\pm 0.2\%$. Thermocouple has an accuracy of $\pm 0.5\%$. The pressure transducers had a range from 0-1374 kPa, and an accuracy of $\pm 1\%$ of measured value. The differential pressure transducers had a range from 0-106 kPa, and an

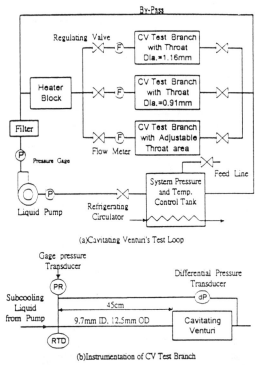

FIGURE 2. Simplified Schematic of CV's Test Loop

accuracy of $\pm 1\%$ of measured value. Data from the tests was collected through a data acquisition system, with a sampling rate of one data point per second. Afterwards, Excel was used to calculated various parameters, i.e., inlet subcooling, loss coefficient, mass flow ratio, pressure ratio, and the superheat at all the overflow points.

Before starting the CV test, water in the container is heated up to 348K for degassing. During each test process, the downstream needle valve was slowly cycled between the extreme positions. Variations of the pressure ratio and mass flow ratio are analyzed in order to determine the loss coefficient and identify the choked, unchoked, overflow and the rechoked operations. All CV mapping data are taken at set points with five water temperatures (298, 308, 330, 343 and 355K), and three inlet pressures (239, 274, 313kPa). The inlet subcooling are ranging from 180 to 310 kPa.

TEST RESULTS AND DISCUSSION

The process of the overflow and recovery are observed for the 0.91 and 1.16mm CVs in this study. A typical result at 343K is shown in Figure 3. Once the pressure ratio exceeded the critical value($1-K_v$), the CV operation changed to all-liquid flow. And this all-liquid overflow persisted even when the pressure ratio was decreased well below the critical value. The dashed line in Figure 3 is the prediction of the overflow path for the 1.16 mm CV based on Eq. (7). It is also observed that the all-liquid data of the 0.91 mm CV has shown some inconsistency for $P_r>0.8$, which may be attributed to transient unsteadiness during testing. Figure 4 shows the loss coefficient vs. pressure ratio for the 1.16mm CV at 343K. In the all-liquid flow regime for $P_r>1-K_v$, the value of K_v

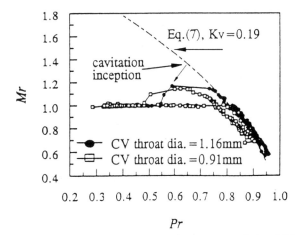

FIGURE 3. Overflow and Recovery at 343K for 0.91 and 1.16mm CVs

is approximately equal to 0.19. Apparently, K_v is still close to the liquid loss coefficient at cavitation inception ($P_r = 0.75$). Similarly, the value of K_v for the 0.91 mm CV was measured to be 0.22. Figure 5 shows that at a higher inlet subcooling of 277 kPa, flow rechoked at lower pressure ratio, but there is no obvious difference between the inlet subcoolings of 208 and 242.5 kPa. However, at the same inlet pressure of 313 kPa but with five different inlet temperatures, Figure 6 shows that the higher the inlet temperature (smaller inlet subcooling), the higher the rechoking pressure ratio.

As shown in Figure 3, 5 and 6, the value of the P_r at the cavitation inception is slightly greater than the rechoked P_r. Therefore, the measured P_r at cavitation inception was used in the calculation of the throat superheat at

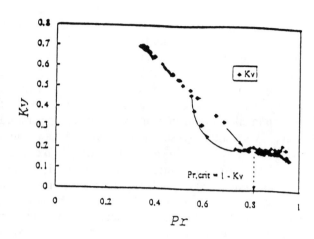

FIGURE 4. Loss Coefficient vs. Pressure Ratio at 343K for 1.16mm CV

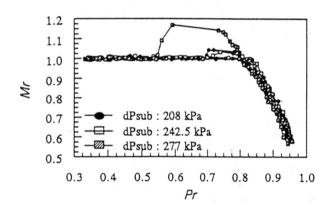

FIGURE 5. Mass Flow Ratio vs. Pressure Ratio at 343K for 1.16mm CV

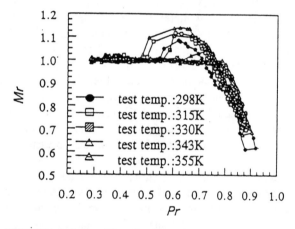

FIGURE 6. Mass Flow Ratio vs. Pressure Ratio at 313 kPa for 0.91 mm CV

rechoking. Figure 7 shows the variation of the superheat at cavitation inception with inlet subcooling for the 0.91 and 1.16 mm CVs. The superheat data for the 0.91 mm CV is linearly increased with the inlet subcooling.

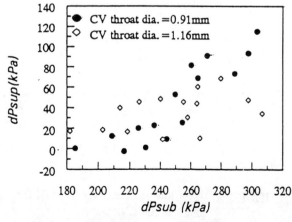

FIGURE 7. Superheat at Rechoking vs. Inlet Subcooling

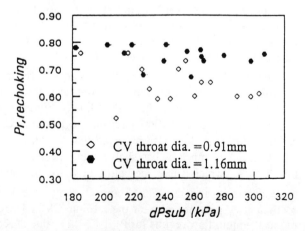

FIGURE 8. Pressure Ratio at Rechoking vs. Inlet Subcooling

However, this trend is not observed for the 1.16 mm CV. Figure 8 shows the pressure ratio at cavitation inception versus inlet subcooling for the 0.91 and 1.16 mm CVs. There is no clear effect of the inlet subcooling to the pressure ratio at rechoking in the inlet subcooling range of 180 to 310 kPa. Nevertheless, the value of P_r at rechoking for the 0.91 mm CV is less than the 1.16 mm CV because of the higher loss coefficient for the 0.91 mm CV. Figure 9 shows the P_r rechoking data versus the inlet subcooling, the curved line of the P_r prediction is based on Eq. (11) and the observed maximum superheat of 70 kPa for the 1.16mm CV. Similar result with the observed maximum superheat of 120 kPa for the 0.91 mm CV is shown in Figure 10. The results of Figures 9 and 10

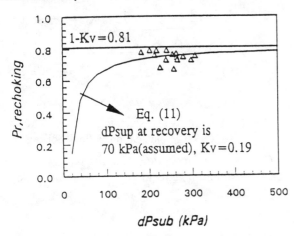

FIGURE 9. Pressure Ratio at Rechoking vs. Inlet Subcooling - 1.16 mm CV

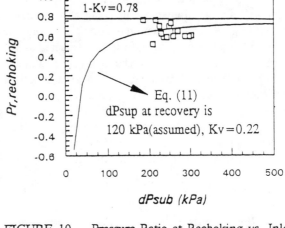

FIGURE 10. Pressure Ratio at Rechoking vs. Inlet Subcooling - 0.91 mm CV

indicate that the P_r rechoking points scatter around the present predictions. The P_r rechoking data for the GTA CV ammonia testing (Chen and Navickas 1993) is also compared using Eq. (11) as shown in Figure 11. Although large data scattering in Figures 9, 10 and 11 is due to the large variation of the throat superheat at cavitation inception, the trend of P_r at rechoking versus inlet subcooling is consistent with the prediction of P_r rechoking using Eq. (11). Figures 9, 10 and 11 also show that the pressure ratio required for rechoking will approach to the critical pressure ratio ($P_{r,crit} = 1-K_v$) as the inlet subcooling increases to the order of 1000 kPa. This is the reason that the overflow problem was not reported in the typical industrial applications.

Apparently, Eq. (11) is very useful in determining the pressure ratio required for rechoking. To use Eq. (11) for

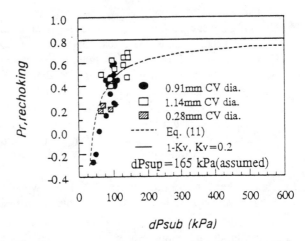

FIGURE 11. Pressure Ratio at Rechoking vs. Inlet Subcooling, GTA Ammonia Data

determining the pressure ratio at rechoking for a given inlet subcooling, the values of the loss coefficient and the throat superheat must be known. The loss coefficient is usually empirically obtained, however, no superheat correlation is available at the cavitation inception for CVs. Since the liquid velocity near the diffuser surface is assumed to be very small due to viscosity for a all-liquid flow CV, it is reasonable to utilize Eq. (13) to determine the superheat required at the cavitation inception for CVs. The surface cavity sizes for a commercially finished drilling surface is typically in the ranges of 1.8 - 6.3 μm for an average application (Avallone and Maumeister III 1987). Assuming the cavity size for the two tested Fox CVs is in the above range, then, superheat required at cavitation inception for the five tested temperatures can be calculated using Eq. (13). Figure 12 shows the results of the observed superheat calculated from the test data using Eq. (9) and that obtained from Eq. (13) with cavity sizes of 1 and 5 μm. The predicted superheat curves with cavity sizes of 1 and 5 μm are in the same order of the large scatter data due to the chaotic nature of the nucleation boiling phenomena. Based on this comparison, the tested CV surface cavity size may be approximately assumed as 2 μm. However, using the same Fox CVs in the GTA ammonia testing, the maximum data was obtained at 230 kPa, which implies the surface cavity size is 0.2 μm (Ungar and Sifuente, 1994). This discrepancy between the water and ammonia data indicates that the CV throat superheat at cavitation inception isn't dominated by the onset of nucleation in pool boiling. Should nucleation dominates, the superheat corresponding to ammonia and water is expected to differ by the ratio of 2.5: 1(due to the difference in surface tension). That means water would give a 650 kPa superheat, if the surface cavity size of Fox CVs was indeed 0.2 μm. Apparently, the superheat for CV at cavitation inception is not only due to nucleation. Since

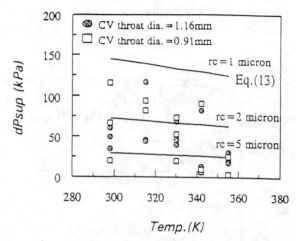

FIGURE 12. Superheat at Rechoking and Predictions of Onset of Nucleation in Pool Boiling

the Reynolds number was observed as an important factor for CV cavitation inception (Hamitt et al 1976), the factors influence the onset of boiling in subcooled convection flow should be considered in the future study.

CONCLUSION

Cavitating venturis with water at low inlet subcooling were tested to observe the cavitation inception for changing the liquid overflow to cavitating flow, the results were analyzed and can be concluded as follows:

- It is important to know the CV loss coefficient prior to the determination of the critical pressure ratio for changing the choked flow to all-liquid flow and the critical pressure ratio of cavitation inception for rechoking.
- At a fixed temperature, the throat superheat at rechoking becomes more obvious at higher inlet subcooling. However, at a fixed pressure, the higher inlet temperature implies the flow rechoke at a higher pressure ratio.
- At fixed inlet temperature and pressure, the CV with bigger throat diameter will have smaller rechoked pressure ratio because of the lower loss coefficient.
- According to the comparison of the test results and the analysis, the pressure ratio at rechoking can be reasonably predicted using Eq.(11) at given loss coefficient and inlet subcooling.
- The superheat based on the onset of nucleation in pool boiling, Eq.(13), can be presumed for the superheat required for CVs in Eq.(11) at a given liquid temperature and surface cavity size.

Cavitating venturis operating under choked flow would be the best choice for a space-based mechanically pumped ATCS. To avoid the dual mode operation, the downstream pressure of the CV must satisfy $P_r < P_{r,crit}$. If overflow induces in CVs, the combination of Equations (11) and (13) would permit a better determination of the pressure ratio at cavitation inception. Hopefully, with more test data in the future, a correlation concerning CV's behavior on throat superheat at rechoking can be established, which will be very useful for predicting the choked flow recovery accurately.

Acknowledgments

The authors wish to acknowledge the support provided by the National Science Committee, Taiwan, R.O.C. Research Grant No. NSC 86-2221-E-224-020 for making this study possible. We do wish to specially acknowledge the invaluable assistance of Dr. E. K. Ungar at the NASA Johnson Space Center.

Reference

Avallone, E.A. and T. Baumeister III (1987) *Mark's Standard Handbook for Mechanical Engineers*, 9th Edition, McGraw-Hill Book Company, New York, NY, 1987.

Chen, L.L. and J. Navickas (1994) "The Behavior of Small Cavitating Venturis in Low Subcooling and Pressure Loss Range", *Advances in Cryogenic Engineering*, B:1059-1064.

Collier, J.H. and J.R. Thome (1994) *Convective Boiling and Condensation*, third edition, Clarendon Press, Oxford, United Kingdom, 1994.

Fox, Z. (1977) "Cavitating Venturis and Sonic Nozzles," National Conference on Fluid Power, 33rd Annual Meeting, Chicago, IL, October 25-27, 1977.

Hamitt, F.G., O.S.M. Ahmed, and J-B Hwang.(1976) "Performance of Cavitating Venturi Depending on Geometry and Flow Parameters," Proc. ASME Cavitation and Polyphase Flow Forum, I: 18-21.

Hammitt, F.G. (1980) *Cavitation and Multiphase Flow Phenomena*, McGraw-Hill, New York, NY, 1980.

Hsu, Y.Y. (1962) "On the Size and Range of Active Nucleation Cavities on a Heating Surface," Trans. ASME J. of Heat Transfer, 84:207-.

Raetz, J. and J. Dominick (1992) "Space Station External Thermal Control System Design and Operational Overview," SAE paper 921106, Proc. 22nd International Conf. on Environmental Systems, Seattle, WA, July 1992.

Sheu, J.S. (1997) "The Study of Cavitating Venturi Performance at Low Inlet Subcooling", M. S. Thesis, Mech. Eng. Dept. of National Yunlin Univ. of Technology, Taiwan, July 1997.

Ungar, E.K., J.M. Dzenitis and R.T. Sifuentes (1994) "Cavitating Venturi Performance at Low Inlet Subcooling: Normal Operation, Overflow and Recovery from Overflow," ASME Symposium on Cavitation and Gas-Liquid Flows in Fluid Machinery and Devices, FED 190:309-318, ASME, 1994.

Ungar, E.K. and Mai, T.D. (1996) "The Russian Two-Phase Thermal Control Systems for the International Space Station: Description and Analysis," AIChE Symposium Series-Heat Transfer, 92(310):19-24.

Ungar, E.K. and I.Y. Chen (1996) "Passive Flow Control Devices for Use on Space-Based Mechanically Pumped Two-Phase Thermal Control Systems," AIAA paper 96-3976, 31st National Heat Transfer Conf., Houston, TX, August, 1996.

Nomenclature

A_{th}	: throat area (mm^2)	$P_{r,rechoking}$	: pressure ratio at the rechoked point
K_v	: all liquid venturi loss coefficient	P_{sat}	: saturation pressure temperature (kPa)
M_{act}	: actual measured mass flow rate (kg/s)	P_{th}	: pressure at the throat (kPa)
M_c	: CV choked mass flow rate (kg/s)	dP_{sub}	: CV inlet subcooling (Pin-Psat(Tin)) (kPa)
M_{liq}	: CV all-liquid mass flow rate (kg/s)	dP_{sup}	: CV throat superheat (kPa)
M_r	: dimensionless mass flow ratio	r_c	: surface roughness (μm)
$M_{r,liq}$	: mass flow ratio for unchoked all liquid flow	r^*	: radius of spherical vapor (μm)
		T_{in}	: CV inlet temperature (K)
P_{in}	: CV inlet pressure (kPa)	σ	: surface tension (N/m)
P_{out}	: CV outlet pressure (kPa)	ρ	: liquid's density (kg/m^3)
P_r	: dimensionless pressure ratio	τ	: cavitation number
$P_{r,crit}$	: pressure ratio under choked condition		

THERMAL DESIGN AND ANALYSIS OF THE SPARTAN LITE SPACECRAFT

J. Warren Tolson and Daniel F. Powers
Swales Aerospace
5050 Power Mill Road
Beltsville, MD 20705
(301) 902-4099 or 4096

Abstract

The Spartan Lite spacecraft is an expendable instrument carrier bus with a generic design capable of accommodating 50 kg. class science instruments. The spacecraft is stowed prior to operations in the Space Shuttle cargo bay in an extended "Hitchhiker" canister. The Orbiter carries the spacecraft into space and ejects it into a free flying, low earth orbit. The spacecraft then assumes either a stellar or solar inertial attitude for a 3 to 12 month science observing mission. The Spartan Lite thermal design is passive, with the exception of a payload powered heater bus. Challenges in the thermal design included working with limited heater power, a wide range of possible on-orbit environments, and a requirement to provide thermal boundary conditions that accommodate a variety of science instruments. This paper will present the thermal design and analysis up to the Critical Design Review status for the Spartan Lite spacecraft. A description of analytical tools used, results from analysis, and thermal design implementation will be included.

INTRODUCTION

The original Spartan (200 series spacecraft) program concept was to provide a spacecraft carrier bus to the science community capable of extending the free flight duration for sounding rocket type experiments. The concept continues now with the development of a series of new generation carriers manifested to fly around the turn of the century. See Table 1. The Spartan program allows payloads to take advantage of the Space Transportation System (STS) or Space Shuttle through usage of the Spartan carriers. The Spartan program philosophy is to have the majority of spacecraft subsystems reusable and generic in design with the carrier to instrument interface tailored to accommodate specific missions. The project goal is to utilize existing designs, equipment, and resources of GSFC and industry to provide a series of high performance, low cost spacecraft to the science community. The approach is to use existing technology to reduce cost and time. However, technology upgrades to the new carriers will be implemented tot he electronics, avionics, and communications to improve and simplify satellite control and ground operations. The new carrier busses will be designed to accommodate a wide range of instrument requirements. Variations in these requirements include size, weight, on-orbit pointing, power levels, and mission durations to name a few. This paper will focus on the thermal design and analysis of the Spartan Lite carrier.

TABLE 1. Spartan Spacecraft Carrier Busses.

Carrier	Mission Duration	Mission Profile	Instrument Class	Current Status
Spartan 200	40 to 50 hours	Shuttle launched/recovered; autonomous	sounding rocket size	4 existing carriers; total of 6 flights; 1 flight pending 11/97
Spartan Lite	3 to 18 months	Shuttle launched/expendable	50 kg	CDR 5/97
Spartan 250	up to 12 days	Shuttle launched/recovered; commandable	Spartan 200 type	pre PDR
Spartan 400	12+ months	Shuttle launched/recovered; commandable	1M+, 1,000 kg	CDR 10/97

GENERAL DESCRIPTION

The Spartan Lite Spacecraft is a three axis, fine pointing spacecraft capable of stellar or solar inertial missions. It is designed to be launched from the Space Shuttle or an expendable launch vehicle and thus is a non recoverable satellite. One of the unique and attractive features of this spacecraft is that the principal investigator can control the spacecraft and receive data at his or her institution during science operations. Baseline science altitude is 400 to 450 kilometers at a 28 to 57 degree inclination. Minimum mission life is three months with a design life of 12 months.

The general mechanical configuration of Spartan Lite is shown in Figure 1. The primary structure is octagonal and made of cast aluminum (A356-T6). The instrument volume is cylindrical (d = .36 m., h=1.02 m.) and fills the inside of the primary structure with the exception of the aft bay where the Attitude Control System (ACS) reaction wheels are housed. The instrument is supported by the primary structure on kinematic mounts. Spacecraft electronics are mounted around the exterior perimeter of the primary structure. Eight solar panels constructed of aluminum honeycomb are configured in a flower petal arrangement as shown. These arrays fold up over the electronics in the stowed configuration. The solar arrays are reversible to accommodate stellar or solar missions. The instrument aperture remains at the solar array end of the carrier for both stellar and solar configurations. Spacecraft total mass including the instrument is approximately 147 kg. which includes an instrument mass no greater than 45 kg.

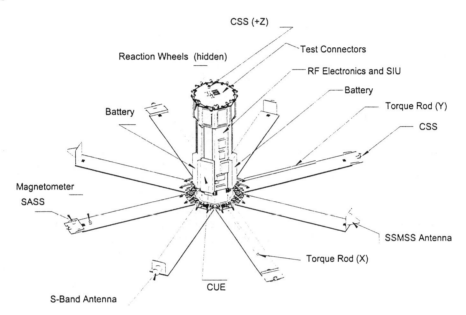

FIGURE 1. Spartan Lite Mechanical Configuration.

Spacecraft attitude determination is accomplished using a Global Positioning System (GPS) receiver, solar sensors and magnetometer for coarse pointing. The science instrument is required to provide the fine error signals. Attitude control is accomplished using reaction wheels and torque rods. The Central Unit Electronics (CUE) performs all attitude determination and control calculations. This processor serves as the interface between the attitude determination and control hardware.

Total orbit average spacecraft power required is approximately 130 watts with 40 watts allocated to the instrument. The power subsystem utilizes Gallium Arsenide solar cell arrays and Nickel Cadmium battery cells. The CUE performs all power subsystem processing.

Command and control is accomplished using a satellite to ground RF transmitter and receiver. In addition, a Space Mobile Satellite Services Transceiver (or cellular phone) is an integral but not required feature of the satellite design.

THERMAL REQUIREMENTS

Thermal requirements for the major Spartan Lite spacecraft components are presented in Table 2. The definition of each set of limits is defined as follows. Non operational limits are the range over which the components must be capable of operation but not necessarily perform within specification. These limits also apply when the spacecraft is powered down and stowed in the cargo bay. Operational limits are defined as the range over which components must be capable of continuous operation within specifications.

TABLE 2. Spartan Lite Thermal Requirements.

Component	T_{non-op} min (K)	$T_{operational}$ min (K)	$T_{operational}$ max (K)	T_{non-op} max (K)
Central Unit Electronics	253	273	313	333
Power Control Electronics	253	273	313	333
Batteries	253	273	293	313
S-Band Transmitter & Receiver	253	273	323	343
SMSS Transceiver	253	273	313	333
RF Combiner & Diplexer	253	273	313	333
GPS Receiver	253	273	313	333
Reaction Wheels	253	263	313	353
Torque Rods	228	238	343	363
Solar Arrays	173	193	363	383
Carrier Structure	233	253	313	333
Carrier Aft Cover	233	253	323	343
Instrument	253	273	313	333

ENGINEERING ASSUMPTIONS

The engineering assumptions presented in this section were made to provide a conservative approach to the Spartan Lite spacecraft thermal analysis. The variation in each of the presented parameters envelops conditions the spacecraft may be exposed to on-orbit for most missions.

While the spacecraft is free flying, a stellar or solar inertial attitude is assumed relative to the longitudinal (pointing) axis of the spacecraft. Off sunline pointing must be no greater than 5° during the daylight portion of each orbit. This requirement is driven by the power system. At night there is no restriction on pointing.

The orbit is assumed to be circular with an altitude between 300 and 450 kilometers. For modeling purposes the orbit plane crosses the equator at 6am and 6pm. The beta angle (defined as the angle between the orbit plane and the solar vector) is unrestricted. In other words, a 100% sunlit orbit is possible. The assumed on orbit environments are presented in Table 3.

Surface optical properties are presented in Table 4. All emittance values are assumed to be hemispherical. End of life properties assume 3 years degradation. Note that the solar array absorptivity for the hot case assumes all energy absorbed remains on the array. The solar array effective absorptivity for the cold case accounts for 18% efficient solar array cells. The effective radiation coupling per unit area from the inner to outer layer of the multi-layer insulation (MLI) blankets range from 0.01 to 0.03 for hot and cold cases, respectively.

Spacecraft component power dissipations are presented in Table 5. All spacecraft and instrument power has been varied +/-10% to account for uncertainty. The average spacecraft battery and heater powers are based on results from a worst cold case. Note that the S-Band transmitter is active for up to 15 minutes once every 12 hours and thus orbit average power dissipation is negligible. Also note that 5 Watts of transmitter power is RF energy.

TABLE 3. Environmental Parameters.

Heat Source	Cold	Nominal	Hot
Solar UV (W/m^2)	1290	1355	1420
Albedo (% solar)	0.25	0.30	0.35
Earth IR (W/m^2)	205	235	270

TABLE 4. Surface Optical Properties.

Property	α	ε
Black Anodize	0.65	0.77
Silver Teflon	0.15 - 0.08	0.75 - 0.78
3 mil Kapton (on VDA)	0.55 - 0.43	0.77 - 0.79
Gallium Arsenide	0.88 - 0.72	0.78 - 0.82
White Paint	0.35 - 0.20	0.84 - 0.87
Aluminum Tape	0.20	0.03

TABLE 5. Transient & Orbit Average Component Power Dissipations (Watts).

Component	Transient	Orbit Average
Central Unit Electronics	10 @ 100%	10
Power Control Electronics	4 @ 100%	4
Battery Charge Electronics	14 @ 60%	8
Main Regulator Electronics	13 @ 60%	7
S-Band Transmitter	45 @ 15 min./12 hrs.	0
S-Band Receiver	6 @ 100%	6
SMSS Transceiver	1 @ 100%	1
GPS Receiver	10 @ 100%	10
Reaction Wheels (3)	(3) x 8 @ 100%	24
Torque Rods (3)	(3) x 1 @ 100%	3
Instrument	40 @ 100%	40
Spacecraft Heaters	12 @ 67%	7
Battery Heaters	(3) x 5 @ 59%	8
Total		128.0
Battery heat dissipation		6

THERMAL DESIGN PHILOSOPHY

The Spartan Lite spacecraft thermal design goal was to isothermalize the spacecraft carrier, electronics and instrument as much as possible. This was accomplished by grouping components with the same temperature requirements into common thermal zones. The primary advantage of this concept is to minimize the number of required heater circuits.

Radiator area on the primary structure is provided to reject heat from the instrument. The instrument to structure thermal design interface is mission specific. Examples could include instrument to structure radiation; instrument radiation directly to space through cutouts in the structure; heat straps and/or heat pipes. The reaction wheels reject heat to the primary structure by conduction and radiation. The structure radiators as well as the aft bulkhead reject this energy from the wheels to space.

The electronics are designed to reject heat to their baseplates. The electronics boxes have the flexibility to be mounted facing outward or inward relative to the spacecraft. Thus each box has the capability to use its baseplate as a radiator to space or as a heat sink to the structure. Electronics boxes utilizing the latter technique depend on the spacecraft radiators to reject their heat.

Since the batteries have tighter temperature requirements, their boxes are thermally isolated from the spacecraft and reject their heat to space. It is also necessary to isolate the solar arrays to avoid significant heat transfer to and from the structure as the arrays temperature cycle throughout each orbit.

THERMAL DESIGN METHODS

The Spartan Lite spacecraft utilizes a passive thermal design with the exception of heaters and thermostats. General thermal design methods are presented in this section with individual hardware specifications presented in the following section.

The carrier primary structure is cast aluminum for good conductivity with structural ribs to minimize circumferential temperature gradients. All spacecraft instrument radiators are an integral part of the cast structure further reducing any gradients. Additional area on the structure is dedicated as a heat sink for the RF components. The solar array hinges are designed such that interface conduction across each hinge does not exceed 0.035 W/K.

All electronics boxes are contiguous five-sided housings including the baseplate. A protective cover closes out the housing and components. Boxes with enough baseplate radiator area to reject their internal dissipation are mounted facing outward, radiating the majority of their heat to space. Additionally, their closeout covers are heat sunk to the structure to aid in isothermalizing the electronics and the carrier structure. The electronics boxes which are too small to dissipate their internal heat to space via radiation have their baseplates heat sunk to the carrier and utilize the spacecraft radiators to reject their heat. An interface material, such as Cho-Therm, will ensure good contact from these boxes to the structure. The RF transmitter is configured in this manner, however, its primary purpose is to utilize the mass of the carrier structure to damp out the transient temperature rise during periodic turn on. The battery boxes are isolated from the carrier and configured with their baseplates facing outward to radiate all of their heat to space.

MLI blankets are used to control heat rejection throughout the spacecraft. All external surfaces are insulated with the exception of radiators and mechanical interfaces or mechanisms. MLI construction consists of 12-15 layers of aluminized mylar separated by Dacron netting between layers. The outer layer facing space has a 3 mil Kapton coating overlaying the vapor deposited aluminum (VDA). The MLI is attached to the spacecraft using mechanical fasteners, G-10 fiberglass buttons with snaprings, velcro and transfer adhesive.

Selective surface coatings are also used for passive thermal control. Internal surfaces are black anodized to optimize radiation heat rejection and to isothermalize the structure. All primary structure, electronics box, and battery box radiators are covered with silver Teflon tape. The structure aft cover is a combination of white paint and aluminum tape for a solar pointed spacecraft and covered with MLI for a stellar pointed spacecraft. Exposed areas on the structure necessary for mechanical interfaces are painted white. The solar array non-active side is painted white to maximize heat rejection during sunlight.

Kapton foil resistance heaters and bimetallic mechanical thermostats are used for active thermal control of the primary structure and the battery boxes.

THERMAL ANALYSIS

All radiation couplings and environmental heat loads were calculated via the Thermal Synthesizer System (TSS) computer program. The geometric math model (GMM) created by TSS had sufficient detail (224 nodes) such that the heat flow of each electronics box, battery pack, radiator panel, etc. could be traced. The specular component of the surface coatings was taken into account where applicable. The spacecraft attitudes were varied with respect to beta angle and pointing axis rotation to determine the worst overall hot and cold spacecraft environments. Nominal case environments were determined based on a typical Space Shuttle orbit. Figure 2 shows a plot of the GMM.

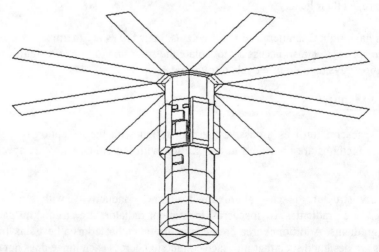

FIGURE 2. Spartan Lite Geometric Math Model.

The Systems Integrated Numerical Differencing Analyzer - 1985 Version (SINDA'85) computer program was utilized to build a 239 node model which calculated orbit average and transient temperature predictions. Orbit average temperatures are presented in Table 6. Temperatures are rounded to the nearest degree. Cold, nominal, and hot case temperature predictions were determined for both solar and stellar pointed spacecraft.

Table 6. Spartan Lite Orbit Average Temperature Predictions.

Component	Solar Spacecraft			Stellar Spacecraft		
	Cold (K)	Nominal (K)	Hot (K)	Cold (K)	Nominal (K)	Hot (K)
Central Unit Electronics	279	288	304	278	288	304
Power Control Electronics	279	289	304	278	288	303
Batteries	276	276	287	276	276	289
S-Band Transmitter	278	282	297	278	280	296
S-Band Receiver	279	288	304	279	286	303
SMSS Transceiver	279	288	303	279	286	302
RF Combiner	277	286	303	276	285	302
RF Diplexer	277	286	303	276	285	302
SIU	277	286	333	276	285	302
GPS Receiver	282	292	308	282	291	307
Reaction Wheels	282	292	306	278	287	302
Torque Rods	259	288	324	259	289	324
Solar Arrays	291	308	354	291	308	354
Carrier Structure	269/279	278/288	291/305	265/279	273/286	287/304
Carrier Aft Cover	262	270	279	270	278	292
Instrument	284	294	309	282	292	308

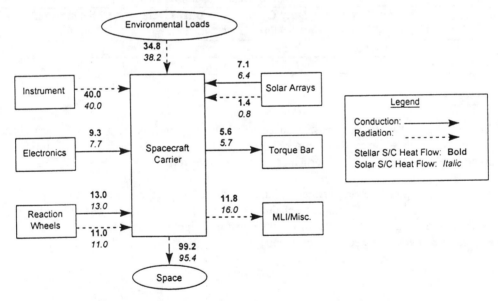

FIGURE 3. Spartan Lite Nominal Case Energy Balance (watts).

Figure 3 presents the nominal case heat flow for both a stellar and solar pointed spacecraft. Powers are in watts.

THERMAL DESIGN SPECIFICATIONS

Thermal design specifications for each spacecraft component is presented in Table 7. Required heater design for spacecraft components is presented in Table 8.

TABLE 7. Spartan Lite Thermal Design Hardware Specifications

Component	Mounting Configuration	$A_{radiators}$ (m^2) Available	$A_{radiators}$ (m^2) Required
Central Unit Electronics	baseplate radiates to space; cover coupled to structure	.0531	.0444
Power Control Electronics	baseplate radiates to space; cover coupled to structure	.1092	.0784
Batteries	baseplate radiates to space; isolated from structure	.0508	.0124
GPS Receiver	baseplate radiates to space; cover coupled to structure	.0355	.0355
S-Band Transmitter	baseplate heat sunk to carrier; cover radiates to space	.0142	.0142
S-Band Receiver	baseplate heat sunk to carrier; cover radiates to space	.0110	.0110
SMSS Transceiver	baseplate heat sunk to carrier; cover radiates to space	.0103	.0103
SIU	baseplate heat sunk to carrier; cover radiates to space	.0077	.0077
Reaction Wheels (3)	Heat sunk to mid bulkhead; radiates to aft structure	n/a	n/a
RF Combiner & Diplexer	baseplate heat sunk to carrier; cover radiates to space	.0026	.0026
Torque Rods (3)	heat sunk to carrier; isolated from solar arrays	.0347	.0347
Solar Arrays (8)	isolated from structure	n/a	n/a
Primary Structure	n/a	stellar; .9860 solar; .9860	stellar; .4164 solar; .2581
Aft Cover	conduction & radiation to structure	stellar; none solar; .1574	stellar; none solar; .1180
Instrument	kinematically mounts to carrier	mission specific	mission specific

TABLE 8. Heater Specifications

Component	Heater Power Available (w)	Duty Cycle Stellar	Duty Cycle Solar	Thermostat on (K)	Thermostat off (K)
Battery Boxes	5 (each box)	59%	65%	273	278
Primary Structure	12	67%	49%	275	281

CONCLUSIONS

The Spartan program has shown consistent success without straying from its original project goals. The Spartan Lite spacecraft continues to expand on this philosophy while providing further opportunities for the science community to take advantage of high performance, low cost spacecraft.

The most challenging engineering element of the Spartan Lite spacecraft will be to provide a generic design which will handle each experiments unique mechanical configuration. Through the use of basic thermal engineering design principles and the growing number of possible heat pipe applications, Spartan Lite promises to continue to provide affordable carriers for the science community.

Acknowledgments

This paper was prepared with the knowledge acquired while working with the Spartan Project Office at NASA GSFC, through a contract rewarded to Swales Aerospace, Inc. Thank you to the Spartan Project, the Thermal Group at Swales, and the Thermal Group at GSFC.

References

Ackerman, Norman (1990) "Thermal Design Manual," NASA GSFC, Greenbelt, Maryland.

Spartan Project (1997) "Spartan Lite Critical Design Review," NASA GSFC, Greenbelt, Maryland, Chapter 4.

IMPROVED COOLER PERFORMANCE USING SPECTRALLY SELECTIVE THERMAL COATINGS

Dave Neuberger, Norm Ackerman, and George Harris
Swales Aerospace
5050 Powder Mill Road
Beltsville, Maryland, USA 20705
Phone: (301) 595-5500; Fax: (301) 595-2871

Abstract

The GOES Imager and Sounder Radiant Coolers are controlled to run at temperatures around 100 K. Future instruments may have added detectors and additional detector heat that will cause the radiant cooler temperatures to rise if design changes are not implemented. Thermal analyses show that lowering the radiant energy from the cooler sun shield (temperatures range between 170 K and 250 K) and/or the Solar Sail Astromast (temperatures range between 270 K and 310 K) adsorbed by the 100 K cooler patch (detector radiator) can significantly lower cooler temperatures if the patch hemispherical emittance is not lowered substantially. The existing cooler patch is an open honeycomb with black paint (Z-307) and had an extremely high emittance even at 100 K. The proposed approach is to replace the open honeycomb with a coating that is spectrally selective with low absorptance out to 10 micrometers and high absorptance beyond 20 micrometers. Several coating formulations were developed and parametric thermal analyses were conducted to select the coating formulation for final coating verification. The coating formulation selected was Ag/Al_2O_3 (14,000 Å)/TiO_2 (6,000 Å)/Al_2O_3 (14,000 Å) vacuum deposited to a highly specular substrate. The thermal radiative properties were: solar absorptance, 0.09, hemispherical emittance at 100 K, 0.80, IR absorptance (200 K blackbody), 0.78, and IR absorptance (300 K BB), 0.65. To take advantage of the low solar absorptance of this cooler patch coating, a change in the Astromast coating was proposed that would keep its solar absorptance/emittance ratio the same (approximately 1.0), but significantly lower the emittance, and thereby lower the IR irradiance on the emitter. The net results reduce the patch temperature by approximately 9 K. The paper will also contain descriptions of the environmental tests and measurements conducted on the coatings and the results of the thermal parametric studies on the cooler patch.

INTRODUCTION

The latest GOES spacecraft series employs synchronous orbit, three-axis stabilized spacecraft with separate Imager and Sounder instruments on-board. Each instrument has a radiant cooler designed to operate at temperatures less than 100 K.

The radiant cooler is designed to keep direct sun off of the radiator surfaces below incidence angles of 25 degrees. The radiator surfaces do have a view to the cooler sunshields as well as the spacecraft Astromast. The goal of this endeavor was to:

a) Modify the Astromast design to reduce the thermal backloading from the Astromast onto the GOES Radiant Coolers, and
b) Modify the Radiant Cooler patch coating to reduce the absorbed thermal energy from the spacecraft Astromast and from the sunshields of the radiant coolers.

Modifications to the Astromast that were investigated included changing the coatings of the Astromast components and changing the shape of the Astromast components. The best of those changes are discussed in this paper. Modifications to the patch (detector radiator surface) involve coating the radiant cooler patch with a spectrally selective coating. Finally the effect of combining the best Astromast configuration with the spectrally selective coating on the patch was investigated.

A solar sail was used to counterweight the large single solar array used on the current GOES I-M spacecraft. The Astromast is shown in Figure 1 with the north panel of the GOES spacecraft and its two radiant coolers. The Astromast is 17.7 meters in length including the solar sail. The Astromast contains 126 bays, each bay is made up of three vertical longerons, 3 horizontal battens, and 6 diagonals. One bay is shown in Figure 2.

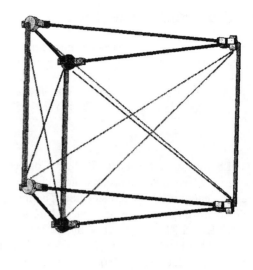

FIGURE 1. Astromast with the North Panel of the GOES Spacecraft and its Two Radiant Coolers.

FIGURE 2. One of the Astromast Bays.

COATING DEVELOPMENT

The goal of the coating development was to reduce solar and IR absorbed energy from the Astromast to the cooler patch by applying a combination of selective coatings on both the cooler patch and the mast structure.

The underlying guidelines were that the mast design structure not be significantly changed and there be no changes in the coating that would significantly change the mast on-orbit temperature range.

DEVELOPMENT APPROACH

Specifically, the approach was to:
a) Reduce the mast thermal emittance, ε, without raising the mast temperature (maintain the solar absorptance/thermal emittance ratio at about one),
b) Reduce the patch solar absorptance, α_s,
c) Lower the patch infrared absorptance of a blackbody source between 200 and 300 K,
d) Keep the patch thermal emittance at 100 K, as high as possible, and
e) Accomplish this using no tapes or coatings that would change the mast mechanical properties.

To lower the emittance while maintaining an α_s/ε of 1, requires lowering the αs also. Thin film coatings with which GSFC has extensive experience, seem to be the coating system of choice because of their ability to provide a wide variety of α_s/ε. These coatings have been shown to be stable in the expected GOES orbital environment, and the IR spectral absorptance can be tailored to be spectrally selective.

THE MAST COATING

The structure of the mast was a composite with a matte finish, so attempts were made to develop coatings for matte (non-specular) and specular surfaces. The matte finish samples were made with G10 roughened with 1500 grid cloth and the specular samples were made using both fired glass microscope slides and lacquered G10.

Since the controlling property was the αs, achieving the lowest value was achieved using evaporated Ag and layers of Al_2O_3 and/or SiO_2 which were applied over the Ag to achieve an α_s/ε ratio of 1. Approximately 100 Å of Al_2O_3 was applied to all substrates prior to coating the Ag for adhesion. For the matte finish, the following layers were applied: Al_2O_3 (100Å)/Ag (700 Å) /Al_2O_3 (2000 Å) /SiO_2 (6000 Å). The α was 0.18 and ε (300° K) was 0.18. For the specular surface, we applied Al_2O_3 (100 Å) /Ag (700 Å) /Al_2O_3 (1000 Å) which had an α of 0.07, and 0.07 for ε (300 K).

THE PATCH COATING

With more latitude on the cooler design, only specular substrates were considered. Initially we applied the dielectric layers over evaporated Al, and then replaced the Al with Ag, which resulted in a decrease in α_s while exhibiting no change in the IR properties. The intent was to have a high spectral emittance at 30 micrometers (100 K BB peak) and minimize the spectral absorptance at 10 micrometers (300 K BB peak).

The following candidate coatings were developed and tested:

- SSC2-AL -- Al applied to the substrate with 16,000Å aluminum oxide, 6,000 Å titanium dioxide, and another 16000 Å layer of Al_2O_3.
- SSC3-Al -- Al / Al_2O_3(14,000 Å) / TiO_2(6,000 Å) / Al_2O_3(14,000 Å)
- SSC4-Al -- SSC3 / MgO (6,000 Å).
- SSC2-Ag -- same as SSC2-Al except Ag replaces the Al.
- SSC3-Ag -- same as SSC3-Al except Ag replaces the Al.
- SSC4-Ag -- same as SSC4-Al except Ag replaces the Al.

Table 1 presents the results of the thermal radiative measurements performed on several of the candidate materials. All the thermal radiative properties shown in Table 1 are for coatings applied to glass microscope slides. On polished aluminum substrates, the α was 0.005 higher than the coatings on glass, but the IR properties were unchanged.

TABLE 1. Thermal Radiative Properties for Several Candidate Patch Coatings on Glass.

Coating	α_s (Al /Ag)	α_{IR} (300 K)	α_{IR} (200 K)	ε (100 K)
SSC2- Al/Ag	0.15/0.085	0.69	0.79	0.82
SSC3- Al/Ag	0.15/0.085	0.65	0.78	0.80
SSC4- Al/Ag	0.15/0.085	0.60	0.68	0.76

For all of the reported coatings, the metal films, Al and Ag, were deposited by resistive heating and the dielectric films, metal oxides, were electron beam deposited.

The thermal radiative properties were acquired by measurements of spectral reflectance over the wavelength range from 0.25 to 50 micrometers and then integrating spectrally with the solar (0.25 to 2.4 micrometers) and blackbody (5 to 50 micrometers) distribution functions.

Eventually, the SSC3-Ag coating was chosen as the most suitable candidate for the patch coating application (see following section for details). Figure 3 presents the spectral reflectance characterization of the SSC3-Al coating.

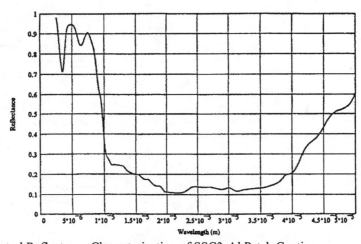

FIGURE 3. Spectral Reflectance Characterization of SSC3-Al Patch Coating.

TESTING SCENARIOS AND RESULTS

The baseline design for the Astromast consisted of no coating over the bare fiberglass pieces. The $\alpha s/\varepsilon$ for fiberglass is 0.8/0.8 which results in the Astromast components running at room temperature. The large emissivity of 0.8 results in the thermal IR backloading to the patch. Initial analysis investigated changes in coatings (vapor deposited aluminum, silver Teflon, and silver aluminum oxide), shape (square battens verses round), and tilting the Astromast relative to the spacecraft. This paper focuses on the use of silver aluminum oxide which showed the most promise of all these changes. Three variations of silver/aluminum oxide were investigated. The assumptions for this investigation are shown in Table 2 and the results of the analyses are shown in Table 3. The silver aluminum oxide (smooth cases and regular case), reduced the IR energy by reducing the emissivity to approximately 0.1. By keeping the $\alpha s/\varepsilon$ ratio to 1, the Astromast runs at room temperature. The reduced absorptance however puts more solar energy on the patch since the Astromast would now reflect 90% (smooth case) of the solar energy that strikes it. The difference between the smooth case and the regular case is that the coating is applied on a smooth lacquered fiberglass part in the smooth case, while in the regular case, the coating is applied to an unmodified fiberglass part. The lowest flux (partial case), was obtained by coating the longerons and battens with silver/aluminum oxide, and leaving the diagonals bare.

TABLE 2. Parametric Cases of Astromast Coating.

CASE	Baseline	Smooth	Regular	Partial
COATING	bare	Ag/Al_2O_3	Ag/Al_2O_3	**
FINISH	none	smooth*	none	**
PROPERTIES				
(solar absorptance)	0.8	0.07	0.18	**
(emittance)	0.8	0.07	0.18	**
% specular (Solar)	0.0	0.99	0.8	**
% specular (IR)	0.0	0.99	0.99	**

Notes: * Smooth lacquer finish on fiberglass part
** Minimum total flux case, longerons and battens are Ag/Al_2O_3 on fiberglass, while the diagonal are from bare fiberglass.

TABLE 3. Parametric Cases Of Patch Energy Incident from Astromast (in milliwatts).

CASE	Baseline	Smooth	Regular	Partial
Longeron				
IR	9.5	0.8	2.1	0.8
Solar	4.6	0.1	2.9	0.1
Total	14.1	0.9	5.0	0.9
Batten				
IR	13.4	1.2	3.3	1.2
Solar	1.4	2.4	2.6	2.4
Total	14.8	3.6	5.9	3.6
Diagonal				
IR	12.9	1.2	3.0	12.9
Solar	2.2	31.6	24.1	2.2
Total	15.1	32.8	27.1	15.1
Joints	2.9	2.9	2.9	2.9
Sail	-0.1	-0.2	-0.2	-0.1
SUMMATION				
IR	38.4	5.7	10.9	17.5
Solar	8.4	34.3	29.8	4.9
TOTAL	46.8	40.0	40.7	22.4

It was desired to coat the patch in such a manner that the patch rejects some of the radiated energy from the Astromast and patch for Summer Solstice with EOL properties, the Astromast and sunshield contribute 80% of the incoming energy. If the coating could reject a large portion of that energy without severely sacrificing its radiation to space, substantial savings can result. Absorptance is the ability of a surface to absorb energy as compared to a perfect black absorber. This value is a function of properties of the absorbing surface and the temperature of the emitting surface. The Astromast is at room temperature (300 K), but the sunshield varies in temperature as a function of season.

For the patch coating, the candidates listed in Table 1 were analyzed. The silver coatings were cooler than their aluminum counterparts due to the lower α_s. The difference between SSC2, 3, and 4 was just a fraction of a degree. However, for manufacturing and coating reliability reasons, the SSC3 was chosen as the most appropriate candidate. The assumptions for the selective patch coating are shown in Table 4.

TABLE 4. Patch Selective Coating Properties SSC3- Ag.

Season	Sunshield Temp K	ε Patch	α_s Solar	α Sunshield	α Astromast
SS EOL	250	0.8	0.09	0.72	0.65
SS BOL	230	0.8	0.09	0.74	0.65
WS	170	0.8	0.09	0.82	0.65

The Geometric Math Model (GMM) was built using TSS. TSS uses a ray tracing technique to calculate radiation couplings and absorbed powers for specular as well as diffuse surfaces. The Thermal Math Model (TMM) was built using SINDA85 to determine Astromast/sail temperatures. The fluxes and radiation couplings calculated in TSS were input into this model. All non Astromast spacecraft and instrument nodes (except MLI) were set as boundary nodes at their SS EOL temperatures. A spreadsheet program was then used to calculate the contribution from the Astromast/sail to the radiator and patch. The IR power is calculated as follows: $Q_{IR} = \sigma \times G_{A/P} \times (T^4_A - T^4_P)$

The sun reflected off the Astromast/sail is calculated by TSS and put into the spreadsheet. The energy contributed by the Astromast/sail to the radiator and patch is totaled by the spreadsheet and input into the thermal balance equations for the radiator and patch [Ref. 6]. The overall results of this investigation are summarized in Table 5. The patch temperature is shown for three seasons/conditions: Summer Solstice End Of Life Properties (SS EOL), Summer Solstice Beginning Of Life Properties (SS BOL), and Winter Solstice (WS).

TABLE 5. Patch Temperature Reduction Due To Selective Coating Of Patch and/or Astromast.

Name	Patch Coating	Astromast Coating	Astromast Flux mW	SS EOL Temp K	SS EOL ΔT(K)	SS BOL Temp K	SS BOL ΔT (K)	WS Temp K	WS ΔT (K)
Baseline	As is	Baseline	8.4+38.4= 46.8	99.6	0.0	96.2	0.0	88.6	0.0
Astromast only	As is	Case g	4.9+17.5= 22.4	95.3	4.3	91.5	4.7	81.8	6.8
Astromast only	As is	Case a	34.3+5.7=40.0	98.5	1.1	95.0	1.2	86.9	1.7
Patch only	Selective	Baseline	8.4+38.4=46.8	97.4	2.2	94.1	2.1	86.5	2.1
Both	Selective	Case g	4.9+17.5=22.4	94.2	5.4	90.6	5.6	81.6	7.0
Both	Selective	Case a	34.3+5.7=40.0	92.9	6.7	89.3	6.9	79.7	8.9

Notes:
Solar + IR = Total Astromast flux
SS EOL = Summer Solstice, End Of Life
WS = Winter Solstice
SS BOL = Summer Solstice, Beginning Of Life

The partial coating case was analyzed since it contributed the minimum energy to the patch. The smooth case is also shown and results in a greater reduction in patch temperature. Although the total energy contributed by the smooth case is higher, it is mostly in the form of solar energy.

Since the low solar absorptance (0.09) of the patch rejects most of the solar input, the silver / aluminum oxide on the entire Astromast yields the best results. This, combined with the ease of manufacturing, and coating reliability issues led the team to the conclusion that this coating would be the best suited for this application.

CONCLUSIONS

Because of the complicated and specific nature of this GOES coating problem, a comprehensive testing and analysis program was developed and implemented. A large number of candidate coatings for both the Astromast and the patch were investigated. Computer analyses of the various configurations and coating combinations were completed in order to fully assess the differences amongst all the coatings possibilities.

Based on the results of this testing, the option that includes coatings modifications to both the patch and the Astromast, clearly offer the largest benefits for the GOES mission. The recommendation to the GOES project was to incorporate the selective coating (SSC3/Ag) for the patch in conjunction with coating all Astromast components with silver/aluminum oxide.

With the incorporation of these changes, the patch temperature will be reduced as much as 9 K. The decreased patch temperature results in an overall improvement in the GOES instrument performance, thus helping to ensure the success of the GOES mission.

Acknowledgments

The authors wish to acknowledge the support and expertise of Mr. Lon Kauder, of the NASA Goddard Space Flight Center.

References

Annable, R. (1995) "GOES-M Imager Radiant Cooler: Patch, Radiator, Shield/Housing."

Bradford, A.P., G. Hass, J.B. Heaney, and J.J. Triolo, (1970) *Applied Optics 9*: 339.

Hass, G., J.B. Ramsey, J.B. Heaney, and J.J. Triolo, (1971) *Applied Optics 8:* 1296.

Hass, G., J.B. Ramsey, J.B. Heaney, and J.J. Triolo, (1971) *Applied Optics 10:* 1089.

Hass, G., J.B. Ramsey, J.J. Triolo, H.T. Albright, *Progress in Astronautics and Aeronautics*, G.B. Heller, Ed. (Academic Press, New York, 1966) 18:47-60.

Triolo, J.J., J.B. Heaney, and G. Hass, (1977) *"Optics in Adverse Environment."* SPIE 121.

Nomenclature

Q_{IR} IR power
σ Stephen Boltzman constant
$G_{A/P}$ Radiation coupling between the Astromast and the patch
T_A Astromast temperature (in K)
T_P Patch temperature (in K)

Design, Fabrication and Testing of an SP-100-like Phase Separator in a Microgravity Environment

Shannon Bragg, De Leah Lockridge, Dan Dorsey
James Fuller, Mike Ellis and Frederick Best
Nuclear Engineering Department
Texas A&M University
College Station, Texas 77843
409-845-4108

Abstract

The project encompassed the conception, construction, and testing of an SP-100 like design for the separation of gas-liquid phases in a microgravity environment. The system was successful in obtaining a stable column of air within a circulating liquid water volume in a fixed cylinder; however, some difficulties were noted. Air bubbles did not readily coalesce with the main vapor column within the short microgravity period established inside the KC-135A. Potential mechanisms to eliminate this problem in subsequent tests are described.

INTRODUCTION: TECHNICAL GOALS

The primary goal of the phase separation design was to achieve two-phase separation based on flow induced, centripetally driven buoyancy forces. The intrinsic momentum of a flowing two-phase stream was used to produce a rotating body of liquid in a stationary, cylindrical vessel. During microgravity, density differences between the air and water allowed the air to migrate to the center while the denser water was forced to the extremities of the separation chamber. Separate extraction ports were constructed for the removal of each phase from the vessel. For purposes of experimental investigation, the separated air and water were recirculated through separate liquid and vapor pumps and then re-injected into the system. The proposed design was expected to provide sufficient information to quantify and model the factors determining optimal phase separation in microgravity. To optimize operational efficiency and to meet the low power requirements stated in the initial design criteria, an acceptably low pressure drop was desired to minimize the pumping power required to provide the necessary flow rate in the cylinder.

THEORY

The initial stage in the experiment was to demonstrate the validity of using centripetally driven buoyance forces to obtain a defined region of air and a separate, well defined region of water. A vortex was created by tangential inlet of the water or two phase mixture, providing sufficient fluid momentum and a symmetrical flow within the cylinder. Figure 1 in a side view illustrates the geometry chosen for the separation vessel. The material chosen for this application was clear acrylic (Plexiglas). This polymer provided a transparent cylinder wall while maintaining relatively low surface roughness. (The approximate Manning roughness coefficient for Plexiglas is 0.01.) Modeling of the air-water system begins with modeling the motion of a bubble in a rotating liquid body.

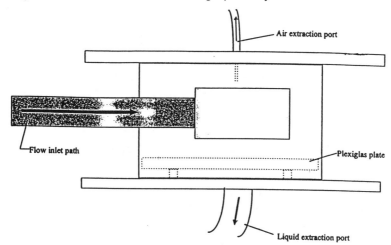

FIGURE 1. Basic Configuration of Separation Chamber.

The buoyancy force, F_B, is given by Eq. 1:

$$F_B = \rho \cdot g \cdot V \tag{1}$$

This equation is applied easily to centripetally driven buoyancy forces existing in a microgravity environment by using centripetal acceleration, as given by Eq. 2:

$$g = \frac{v^2}{r} \tag{2}$$

While centripetally driven buoyancy forces were responsible for bubble rise to the central vapor column, drag forces acted opposing bubble migration. The drag correlation is given by Eq. 3, as it applies to air bubbles entrained in the liquid water.

$$F_D = C_D \cdot \frac{1}{2} \rho \cdot v^2 \cdot A \tag{3}$$

The forces acting on the entrained air bubbles are shown in Figure 2.

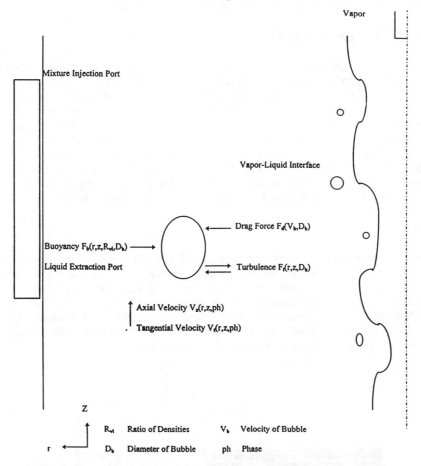

FIGURE 2. Parameters Affecting Hydraulics and Phase Separation in a Cylinder Under Microgravity

BUBBLE COALESCENCE WITH THE CENTRAL VAPOR CORE

Having described the phenomena driving bubble migration to the central core, bubble coalescence with the central vapor core is now considered. The ratio of centripetal force to the surface tension of a stable column was utilized by Dr. Fred Leslie of the Marshall Space Flight Center in a 1985, (Leslie 1985), investigation to measure bubble shapes in a low-gravity environment. This ratio is given in Eq. 4:

$$F = (\rho_{liq} - \rho_{bubble}) \cdot \frac{\omega^2 \cdot r_{bubble}^3}{\sigma} \tag{4}$$

Knowledge of the two-phase flow state is fundamental for two-phase flow systems. This is especially true in the microgravity acceleration environment of an orbital spacecraft. To a first order (or assuming a homogeneous flow exists), the void fraction is directly related to the energy carried by the two-phase mixture. Therefore, the void fraction is a key parameter in monitoring the operating state of a two-phase flow system. Furthermore, two phase flows in microgravity are known to be different from 1-g flows. Therefore, it is essential to calibrate the void fraction sensors under a microgravity environment.

This program was conducted jointly with JSC, GSFC, Lewis Research Center (LeRC), Creare Inc., the Interphase Transport Phenomena (ITP) group of Texas A&M University (TAMU), and the Center for Space Power (CSP) at TAMU.

EXPERIMENTAL EQUIPMENT

This experiment was conducted on the Flow Regime In Microgravity (FRIM) two-phase flow test loop. The FRIM is a modification of the breadboard experiment of GSFC described by Benner et al. (1995). The breadboard was modified to include two void fraction sensors, a clear sight tube, and quick acting valves. Figure 1 is a schematic diagram of the modified FRIM. A detailed description of the capacitance sensors is given by Crowley et al. (1996).

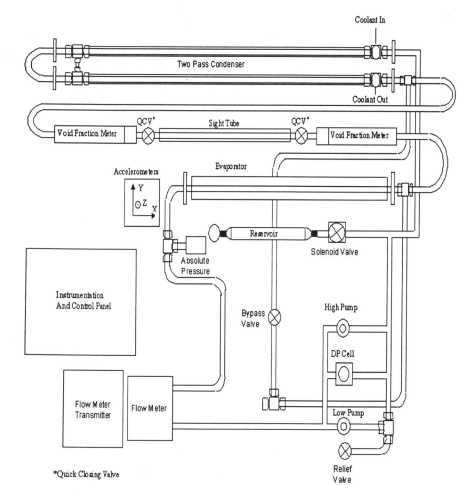

FIGURE 1. FRIM Schematic.

Vapor-liquid interfaces tend to form structures having the lease surface area; thus, a spherical shape is observed. As bubbles come in contact, the surface tension should break down between bubbles.

SEPARATOR DESIGN

The phase separation system incorporated a right circular cylinder designed to contain the two-phase mixture. The cylinder was constructed of one-half inch thick Plexiglas, having dimensions of approximately eight inches in height and ten and one-half inches interior diameter. The two phase mixture, water and air, was cycled through separate piping loops driven by individual pumps.

The separate liquid and air loops were mixed into a two-phase flow. The mixture entered a manifold designed to create a "sheet" of fluid, inlet tangentially. This manifold maximized the flow velocity while evenly distributing the flow along the cylinder height.

Upon entering the cylinder, the fluid experienced centripetally driven buoyancy forces as a result of the intrinsic momentum of the flow. Thus, the lighter gaseous phase migrated inward, forming a stable, central column of air. The gaseous phase was extracted from the central vapor column by a centrally located port extending approximately on and one-half inches into the cylinder interior. The liquid phase was extracted from the periphery of the bottom of the separation chamber. A Plexiglas plate was mounted in the interior of the cylinder approximately one inch above the bottom. A gap between the plate and the cylinder wall corresponding to the area of the inlet port was the path for liquid extraction.

TESTING PROCEDURES AND RESULTS

The in-flight trials were conducted during two separate flights, flying forty parabolic maneuvers each day over the course of approximately 2 1/2 hours.

Trials completed on day one yielded excellent results for the development of a stable vapor column at an approximate fill level of 5.5 - 6 inches. Higher fill levels were initially expected to produce the preferred results; however, these levels did not allow a "column" to form. Instead, a large "bubble was observed. Thus, the fill level chosen for the second day of experimental trials was approximately 5.5 inches. Each investigated fill level was observed at water flow rates of approximately 8 gpm and the maximum available flow rate (~ 17 gpm), and air flow speeds of zero and 5 (as indicated on the peristaltic pump control knob) at each water flow rate.

Study of the hydraulic flow in microgravity without injection of air into the inlet flow was highly successful. When only a single phase was injected into the cylinder containing a given void fraction, the centripetal motion that was created succeeded in separating the existing air and liquid volumes in the cylinder.

Upon addition of air to the inlet flow, creating a two-phased fluid, some difficulties were noted in the development of a separated vapor column. Though the column that was formed remained stable and the extraction lines for air and water maintained isolated (no air was noted in the water line, and vice versa), coalescence of the additional air bubbles with the central column was slow at any flow rate that was tested.

The second day of in-flight trials began with the optimal fill level determined in the initial trial. Although the fill was adjusted from this level slightly, the average level maintained was approximately 5 inches of liquid. Again, the trails produced excellent column stability with no air injection (single-phase fluid injection). The ideal configuration was noted at approximately 12 to 15 gpm liquid flow with a 5 inch fill level. However, problems similar to those noted on day one regarding bubble coalescence were noted with air injection at all flow rates.

DISCUSSION

As described in the experimental results, an optimal liquid fill level and approximate liquid flow rates were determined. However, difficulty was noted in bubble coalescence throughout the test matrix. Some suggestions to alleviate this problem are currently being considered for subsequent application. Some of these include the addition of a non-foaming soap or similar substance to the flowing body of fluid, or the use of distilled water (tap water was used in the experimental trials, which was highly mineralized, hard water). However, this difficulty also suggests that the design would not be directly applicable to any generic two phase system. Surface tension of the fluids must meet the necessary criteria for the separation to occur.

CONCLUSION

Overall, the design was successful in obtaining a stable column of air within a circulating liquid volume; however, some difficulties were noted with coalescence of air in the cylinder. Thus, the separation system was successful in part, and offered a significant basis for further study of hydraulics in microgravity. Some problems remain to be resolved; further ground study and analysis must be completed to resolve the issues prior to the next KC-135 flight test.

Acknowledgment

The separator team gratefully acknowledges the creation of the flight program by the Texas Space Grant Consortium and the NASA Johnson Manned Space Flight Center. Additional thanks are due to Dr. John Poston, Head of the Nuclear Engineering Department, Dr. B. Don Russell, Associate Director of the Texas Engineering Experiment Station, and the Center for Space Power for their financial support.

References

Hung, R. J., Lee, C. C., Leslie, F. W. (1991) "Response of Gravity Level Fluctuations on the Gravity Probe Spacecraft Propellant System," from *J. Propulsion*, , July-Aug. 1991, 7: 269- 279.

Leslie, F. W. (1985) "Measurements of Rotating Bubble Shapes in a Low-Gravity Environment," from *J. Fluid Mechanics*, 161: 556- 564.

---------------------Nomenclature---------------------

ρ = liquid density (approx. 1kg/m^3)
V = volume of immersed object
v = radial velocity of bubble rise
ρ_{liq} = liquid density
r_{bubble} = average bubble radius

g = acceleration of gravity
r = cylinder radius
r_b = bubble radius
ρ_{bubble} = air density
A = total surface area in contact with fluid = $4 \pi r_b^2$

v = inlet fluid velocity
C_D = drag coefficient
σ = surface tension

VOID FRACTION MEASUREMENTS BY QUICK ACTING VALVES AND CAPACITANCE MEASUREMENTS

Jae H. Chang and Frederick R. Best
Department of Nuclear Engineering
Texas A&M University
College Station, Texas 77843-3133
(409) 845-4108

Abstract

Two-phase flow systems are widely estimated to have superior capability in comparison with single phase thermal management systems for spacecraft. However, microgravity two-phase flow technology insufficiently advanced to allow development with acceptable risk levels. A capacitance effect, void fraction measurement sensor has been developed by Creare Inc. to begin to satisfy microgravity technology needs. Under a NASA Johnson Space Center grant, microgravity tests of the capacitance void fraction sensors were performed aboard the NASA KC-135. Twelve KC-135 flights were conducted in three series. Test points were collected over a wide range of void fractions (0 % - 90 %). Data were collected from stratified, slug, and annular flow regimes. Void fraction measurements from the capacitance sensors were compared with the void fractions from a trapped volume in the test section between two quick acting valves. Under the annular flow regime, void fractions measured by the capacitance sensors compared well with values from the trapped volume. In slug flow regime, some discrepancies between the sensors and trapped volumes were found. However, when the working fluid (Suva) mass flow rate increased from 0.00314 kg/s to 0.007756 kg/s, the void fraction measurements between the capacitance sensors and the trapped volume had better agreement. Overall, the FR experimental package produced satisfactory test conditions in the microgravity conditions of the KC-1 aircraft, to validate and calibrate the Creare capacitance void fraction sensors.

INTRODUCTION

Currently, the space program uses gravity independent, single-phase flow for high power thermal management purposes. Since technology for space applications is sensitive to volume, mass, and power, future space missions should take advantage of two phase flow systems (co-current vapor and liquid flow). This technology has the ability to carry more energy per unit mass than single-phase flow while simultaneously operating at a constant temperature. A two-phase system also requires less pumping power per unit thermal energy carried and has better heat transfer characteristics than single phase systems. All of these features lead directly to improved performance and smaller mass requirement than a single-phase system.

In traditional two-phase flow studies, the flow regime refers to the physical location of the gas and liquid in a conduit. The flow configuration is important for engineering data correlation such as heat and mass transfer, pressure drop, and wall shear. However, it is somewhat subjective since it is mostly defined the experimenter's eye, which results in an approximate definition. Thus, there is a need for better discretizing instrumentation. In designing a thermal management system, one can usually estimate phasic flow rates. However, the topological configurations of the phases are unknown. Hence, many investigations performed on this subject have tried to determine, through experimental and empirical means, when each flow regime configuration is likely to occur given the system parameters.

Several tools have been developed over the years to identify the flow regime in two-phase flows. Some approaches include the study of pressure drop measurements, gamma densitometry, resistive capacitance sensors as well as intrusive probes. Recently, with the advent of powerful personal computers and the development of technologies based on non-destructive evaluation, new technologies have been developed for non-intrusive monitoring. One such technology is the capacitance method developed by Creare Inc. for the purpose of void fraction measurements.

The FRIM uses R-134a, also called Suva, an environmentally benign tetrafluroethane, as the working fluid. The description begins at the pump discharge manifold. From the outlet of the Suva pumps, subcooled Suva enters the evaporator. The evaporator is a bored rod evaporator, which can provide up to 2 kW of electrical power input. The function of the evaporator is to produce a two-phase mixture from the inlet subcooled liquid. The evaporator is insulated with Nomex cloth to limit ambient losses. The two-phase flow from the evaporator passes through the test section. The test section consists of an upstream void fraction meter, a flow visualization tube, and a downstream void fraction meter.

After passing through the downstream void fraction meter, the flow enters the condenser section. The condenser is a two pass, tube-in-tube heat exchanger. Cold water was used as the coolant. The ultimate heat sink for the condenser coolant is an ice bath contained in a 10 gallon reservoir located on the bolt-down package (BDP). The subcooled liquid leaving the condenser returns to the pumps and begins the cycle again.

The flow rate is measured by a Coriolis mass flow meter, located on the outlet of the Suva pump manifold. A vertical, cylindrical vessel, located on the flow line after the condenser, is used to control the pressure in the test system. This accumulator is nominally filled half and half with liquid and vapor Suva. Pressure is regulated through external electrical heaters located on the outer walls of the accumulator. Temperature controllers, located on the instrumentation panel, regulate the temperatures of the accumulator heaters to produce the desired saturation temperature, thereby controlling pressure. A solenoid valve permits the isolation of the accumulator from the rest to the test loop when desired. Absolute and differential pressure transducers provided absolute pressure reading for the system and differential pressure measurement across the pump. A bypass line was also included in the FRIM. When the quick acting valves shut, the bypass line opens, permitting the flow to bypass the test section and continue from the evaporator to the condenser. This helps to preserve constant system conditions and prevents dryout of the evaporator walls.

The quick acting valves, located at each end of the clear sight tube, allow the experimenters to isolate the two-phase flow and measure the liquid level trapped in the sight tube. The test section was constructed to be rotated vertically for the mechanical measurements. The test section lies in the horizontal position during the microgravity periods of the flight. It will be raised into the vertical position after the quick acting valves are closed. The liquid level in the sight tube is measured during 2 g or level flight conditions.

Furthermore, 30 type T thermocouples were placed on the outer walls of the test system piping to monitor the temperatures in the system and one thermocouple was used to record ambient temperatures during the flight testing. Tri-axial accelerometers were mounted on the FRIM co-axially with the orientation of the test system piping. The accelerometers provide vertical, transverse, and axial acceleration profiles experienced by the test package during the flight test. Void fraction data, test system parameters, and accelerations were recorded by a personal computer.

FLIGHT PROCEDURE

The following is a description of the flight procedure.
- The operating conditions for the system were set prior to entering the first microgravity period. These conditions included the evaporator power, the liquid Suva mass flow rate, condenser coolant flow rate and the accumulator temperature setting. The goal was to set the test system under a steady operating condition prior to isolation of the sight tube.
- Data collection of the housekeeping data and the instrumentation recorder for the void fraction sensors was initiated prior to entering the first microgravity period. The data collection continued until the end of the parabola set.
- During the last 4 to 8 seconds of the microgravity period of the fourth parabola, an imager shot of the sight tube was taken. The sight tube was then isolated with quick acting valves prior to exiting the microgravity period. The test section was raised to the vertical position during the fifth microgravity period, with the liquid level being measured during the ensuing 2-g period.

- While the sight tube was isolated, the system settings were adjusted to the next test point conditions.
- A second imager shot and isolation were taken on the tenth microgravity period of the parabola set. After the tenth microgravity period, the plane returned to level flight (1 g) for several minutes. The test section was then raised to the vertical position and measurement of the liquid level in the sight tube was made. Also, during this time, housekeeping data stored in RAM was written to a floppy disk.

RESULTS

Three flight series were conducted under this project. Figure 2 is a plot of the void fraction data from the upstream and downstream capacitance sensor. The void fraction measurements from the last eight seconds of the microgravity period were averaged and compared to the void fraction measurement from the trapped volume between the two quick acting valves.

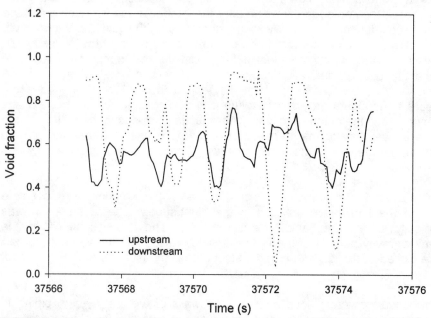

FIGURE 2. Time Trace of Upstream and Downstream Capacitance Void Fraction Sensors.

The first flight series of four flights were conducted during July 1995. Test points in slug and annular points were collected. Figure 3 is a plot of void fraction measurements from the trapped volume between the quick acting valves and the two capacitance void fraction sensors. Comparison of averaged void fraction measurements of the upstream and downstream void fraction sensors showed good agreement with the trapped volume void fraction within + 0.10 of the void fraction.

The second flight series of four flights were conducted during January 1996. The aim of this flight series was to use lower power to produce void fractions less than 0.50 in the slug or bubbly flow regimes. Figure 4 is plot of void fraction measurements from the trapped volume between the quick acting valves and the two capacitance void fraction sensors for this flight series. It was difficult to draw conclusions from the results of this flight series with the difficulty obtaining steady flow conditions with low evaporator powers used during this flight series. Transient behavior and non-uniform distribution of vapor and liquid across the test tube during the microgravity periods were observed throughout the flight series. Due to the transient behavior observed during the microgravity periods, it is difficult to conclude the significance of the correlation between the averaged void fraction measurements of the two sensors and the trapped volume from the quick acting valves. An analysis of the transit time for the liquid/vapor front from the inlet of the evaporator to the end of the downstream sensor was completed. This analysis concluded that with the Suva mass flow rate and evaporator power used on this flight series, steady state conditions in the test section were not possible. Therefore, the Suva mass flow rate was increased and higher evaporator power was used on the third flight series.

The third flight series was conducted during May 1996. A new Coriolis flow meter and transmitter were installed into the test system. With this flow meter system, the maximum measurable Suva mass flow rate was increased from 0.00314 kg/s to 0.00756 kg/s. The higher Suva flow rate coupled with the higher evaporator power shortened the transit time of the two-phase flow and increased the velocity of the fluid in the system. Figure 5 is a comparison between the upstream and downstream void fraction sensor with the void fraction measurements from the quick acting valves for the May 1996, flight series. The upstream and the downstream void fraction measurements are averages of the last 8 seconds prior to the end of the microgravity period in the parabola. There was good agreement between the capacitance sensors and the void fraction measurements from the quick acting valves. It can be seen from Figure 5, that the upstream void fraction sensor agrees well with the void fraction from the trapped volume for the entire range tested (20% to 80% void). The downstream void fraction sensor shows good agreement with the trapped volume for void fractions above 35%.

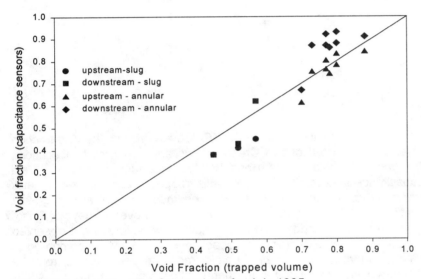

FIGURE 3. Void Fraction Comparison for July 1995.

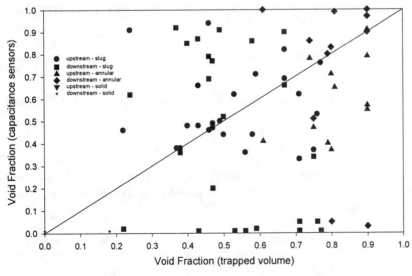

FIGURE 4. Void Fraction Comparison for January 1996.

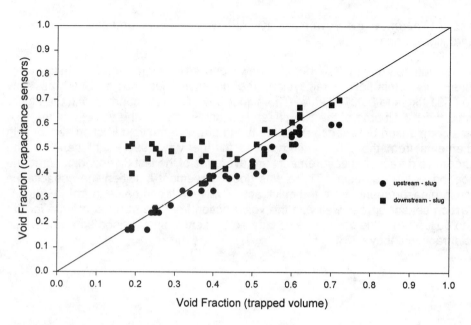

FIGURE 5. Void Fraction Comparison for May 1996.

CONCLUSIONS

Three flight series were conducted in reduced gravity on the NASA KC-135 aircraft with the void fraction experimental package. The test package consisted of two Creare capacitance void fraction sensors, GSFC FRIM package, and the ITP data acquisition system and KC-135 test support system. Void fraction measurements from the capacitance void fraction sensors and trapped volume from the quick acting valves were collected. Furthermore, digital imager shots were recorded during the microgravity periods just prior to trapping of the sight tube. Over 70 test points were collected over 12 days of flying. Void fractions ranging from 0 % to 90% were measured. Slug and annular flow regimes were observed during the flights. Average void fraction measurements from the capacitance sensors were compared to the void fraction measurement from trapped volume. Overall, the void fraction measurements from the sensors compared well with the trapped volume, especially in the annular flow regime. However, some discrepancies existed in the slug regime. These discrepancies were due to the lack of steady state conditions in the test system with the low liquid Suva mass flow rates and low evaporator powers. With the higher Suva mass flow rates and evaporator powers, void fraction measurements from the sensors and the trapped volume compared well. Overall, the Creare void fraction sensors where found to produce the correct result whenever the test system achieved equilibrium.

Acknowledgements

This work was funded by a NASA JSC Grant No. NAG3-941 and the Center for Space Power (CSP) at Texas A&M University.

Reference

Benner, S., G. Durback, K. Kolos, and R. Bayt (1995)"A Breadboard Flight Experiment for Two-Phase Flow Visualization in Microgravity," *Proc. 33rd Aerospace Sciences Meeting & Exhibit*, held in Reno, Nevada, January 9-12, 1995. AIAA-95-0696: 1-10.

Crowley, C. J., P. J. Magari, C. M. Martin, and M. E. Hill (1996) "A Void Fraction Instrument for Two-Phase Flow in Dielectric Liquids," *Proc. 34th Aerospace Sciences Meeting & Exhibit*, held in Reno, Nevada, January 15-18, 1996. AIAA-96-0925: 1-10.

LOOP HEAT PIPE FLIGHT EXPERIMENT

Walter B. Bienert
Dynatherm Corporation
1 Beaver Court
Cockeysville, MD 21030
(410) 584-7500

Extended Summary

The Loop Heat Pipe (LHP) is a passive two-phase thermal control system that uses the latent heat of vaporization of an internal working fluid to transfer heat from an evaporator (the heat source) to a condenser (the heat sink). The circulation of the working fluid is accomplished by capillary pressure gradients in a wick with very small pores. The main components of the LHP are the evaporator, vapor and liquid transport lines, the condenser and a fluid reservoir (compensation chamber). Like a conventional heat pipe, the LHP is a sealed device that is completely passive and requires no external power. But unlike the heat pipe, which has a wick along its entire length, the LHP's wick is confined to the evaporator/reservoir assembly; the transport lines which comprise the major length of the device, are simple unobstructed tubes.

LHPs are rapidly gaining acceptance in the aerospace community and several terrestrial applications are emerging as well. The intrinsic advantages of LHPs over conventional heat pipes are:

- high heat transport capability over long distances
- insensitivity to adverse tilts up to several meters
- tolerance of complicated layouts because the transport lines are small diameter tubes
- ease of incorporating flexible sections into the transport lines

The LHP is related to, and sometimes considered a derivative, of the Capillary Pumped Loop (CPL) which shares most of the above features with the LHP. However, the CPL requires extensive preconditioning of the reservoir prior to start-up and is susceptible to depriming during certain conditions. The LHP, on the other hand, is self-starting and largely immune to depriming problems.

LHP's are currently baselined as integral thermal control components for the next generation of large communication satellites. They are an enabling technology for deployable thermal radiators. Their ability to function in almost any orientation facilitates ground testing and the small diameter transport lines permit easy routing around equipment within the spacecraft. At least one major domestic spacecraft manufacturer is committed to LHPs for the next series of satellites; most manufacturers, both domestic and abroad, are actively evaluating the technology. Applications for LHPs in aircraft are under development. The high capillary pumping capability of the wicks used in LHP's permits the operation under high accelerations typical in aircraft. Specific applications under development include deicing of control surfaces using waste engine heat and cooling of actuators. In the commercial field, LHPs are being considered for roof top solar installations because of their ability to transport heat against gravity. Another interesting application is the cooling of remote communication sheds in a hot desert environment by transporting the heat into the ground.

The LHP has been originally developed in the former Soviet Union. The Russians have incorporated the LHP in several of their satellites and have demonstrated reliable, long term operation in micro-gravity; however, only limited test data are available in the open literature. Dynatherm, through its parent organization DTX, transferred the Russian technology to the US through a cooperative agreement with a Russian firm. In order to qualify the LHP for use in the US, the ability to duplicate the Russian technology must be demonstrated and a US made device must be flight tested on a US spacecraft. This is the overall objective of the Loop Heat Pipe Flight experiment. A second objective is to provide a data base for correlating and verifying analytical models.

The experiment was conceived and developed by Dynatherm/DTX who is also the program manager and principal investigator. The planned flight on STS-87 in the form of a Hitchhiker experiment was sponsored by the Center for Space Power of Texas A&M university. Several government laboratories (NASA, BMDO, U.S. Air Force - Phillips Lab & Wright Lab, Naval Research Lab, National Aerospace Laboratory in Netherlands) and industry (Hughes) contributed either through direct funding or by providing technical support.

The specific LHP that is being tested in this experiment is a direct derivative of a design that will be used with a deployable radiator in a communication satellite. Pertinent parameters of the LHP are:

Evaporator:	Stainless steel tube (300 mm by 25 mm diameter) with aluminum saddle
	Sintered nickel wick with 2.2 micron pores
Transport Lines:	Stainless steel tubing, 4.5 m long, 4.5 mm ID
Condenser:	Flanged aluminum extrusion, 3.7 m long, 4.0 mm ID
Fluid:	Ammonia

The LHP is capable of transporting over 800 Watts over the range from -40 to +65 ^{0}C, however, the available experiment power will limit operation to about 400 Watts. Its overall conductance is approximately 50 Watts/^{0}C when the condenser is fully active. The experiment is located in a standard, 5 cft NASA Hitchhiker canister. In order to package the device inside the compact canister, the 4.5 m long transport lines are coiled. The flanged condenser is bonded to the upper lid of the canister which serves as the thermal radiator. In order to maximize the heat rejection capability, the basic radiator is augmented by a "visor". The radiator and the visor are covered with silver-teflon tape, all other external surfaces are insulated. The Hitchhiker canister is sidewall mounted on the starboard side in Bay 6 of the shuttle cargo bay. Figure 1 shows the layout of the experiment inside the Hitchhiker canister.

The experiment is instrumented with 35 temperature sensors. Tape heaters on the evaporator allow the input power to be varied in steps of 12.5 Watts from zero to 388 Watts. Standard Hitchhiker services for power, data and command of the experiment are used. The command and telemetry interface between the experiment and the Hitchhiker electronics is via a Data Acquisition and Control Unit (DACU) which is an electronics box mounted on the mid-plate between the canister and a 5 inch extension cylinder. Real time monitoring and command of the experiment is accomplished from the Hitchhiker control center at NASA Goddard.

The test program includes start-ups, power cycling, steady state operation, and temperature control tests. Start-up tests will be conducted at low initial temperature (-55 to -40 ^{0}C), intermediate temperature (-15 to 0 ^{0}C) and at high temperature (+20 to +35 ^{0}C). At each temperature, the power will be cycled frequently from low-to-high and from high-to-low. Multi-orbit steady state operation (typically 8 hours duration) will be performed at low and high operating temperature. In addition to these heat transport tests, the ability of the LHP to control the source (evaporator) temperature will be evaluated. For this purpose, a small auxiliary heater and thermostat is installed on the compensation chamber. It will control the evaporator temperature at +40 ^{0}C while the power and sink temperatures vary.

The experiment underwent an extensive thermal test program, first under ambient laboratory conditions and later in thermal vacuum. For the ambient test, a liquid cooled cold plate was attached to the radiator. Sink temperatures were set to -40 ^{0}C, 0 ^{0}C, and +35 ^{0}C; orientation of the LHP was vertical (evaporator above condenser) and horizontal, and the power was varied from 20 to 800 Watts. For the thermal vacuum tests at NASA Goddard, a liquid cooled radiation sink was located adjacent to the radiator. The TV tests were conducted with the evaporator above the condenser. The sink temperature was varied between -55 ^{0}C and ambient and the power from 25 to 380 Watts. Steady state performance was demonstrated at operating temperatures ranging from +5 ^{0}C to +55 ^{0}C and transient performance up to +78 ^{0}C. The ground test program showed that the experiment met all design requirements. The test data obtained under ambient conditions were correlated successfully with Dynatherm's analytical LHP model, the TV data with a Sinda model of the entire experiment which includes a simplified algorithm for the performance of the LHP.

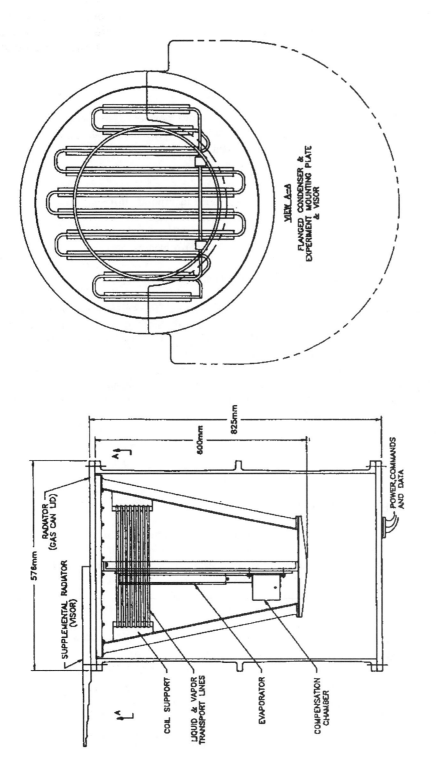

Figure 1
LHPFX Layout

SENSORS AND COMPONENTS FOR AEROSPACE THERMAL CONTROL, LIFE SCIENCES AND PROPELLANT SYSTEMS

A.A.M. Delil and A. Pauw
National Aerospace Laboratory NLR
P.O. Box 153, 8300 AD Emmeloord, Netherlands
phone +31 527 24 8229, fax +31 527 24 8210
E-Mail adelil@nlr.nl

R.G.H.M. Voeten and P. van Put
Bradford Engineering
De Wijpert 26, 4726 TG Heerle, Netherlands
phone +31 165 304 440, fax +31 165 304 442
E-Mail bradford@concepts.com.nl

Abstract

Various aspects of different sensors and components, (being) developed or fine-tuned for aerospace thermal control and propellant systems, are discussed: rotatable radial heat pipe joints, vapour quality sensors, controllable valves, condensers, flow metering assemblies and propellant gauges.

INTRODUCTION

Aerospace heat transport, life sciences and propellant systems incorporate different sensors and components. Some are spatialised versions of existing terrestrial hardware, others are dedicated direct developments for space. Various aspects of different sensors and components (being) developed or fine-tuned will be discussed in detail: background, applications foreseen, design, and terrestrial and (low-gravity) performance data.

The sensors and components, developed for ESA and the Netherlands Agency for Aerospace Programmes, are:

- Rotatable Thermal Joints, to be used to minimise the temperature drop between e.g. a spacecraft thermal bus and a deployable or steerable heat pipe radiator, the latter resulting in minimum radiator size and mass.
- Vapour Quality Sensors, to measure or control the vapour quality (vapour mass flow ratio) of a flowing vapour/liquid mixture in mechanically/capillary pumped two-phase heat transport and propellant system lines.
- From the Vapour Quality Sensor derived Propellant Gauges to accurately measure the remaining level of fuel (Mono Methyl Hydrazine) or oxidizer (Mixed Oxides of Nitrides) in the tanks of spin-stabilised spacecraft.
- Flow Metering Assemblies, to measure and control the flow rate in aerospace life science systems and single or two-phase thermal control loops and to assess, by integrating the consumption during lifetime, the remaining level of fuel or oxidizer in the propellant tanks of three-axes stabilised spacecraft.
- High Efficiency Low Pressure Drop Condensers for direct or indirect condensation spacecraft radiators.
- Controllable (motorised, three-ways) Valves, to be used in single or two-phase loops to control fluid temperature setpoint, flow rates, pressure drop, vapour quality, etcetera.

ROTATABLE RADIAL HEAT PIPE JOINT

Recalling earlier discussions (Delil 1987), it is remarked that dedicated heat pipe radiators will be used to reject (spacecraft) waste heat into space. Such a radiator, stowed during launch, will be deployed in orbit. The radiator may even be chosen to be steerable to achieve maximum performance, hence minimum radiator size/mass. In such radiator systems, the coupling to the central (two-phase) loop or heat pipe has to incorporate a rotatable/ flexible thermal joint. Drivers for the design of such joints are low thermal resistance, hence limited temperature drop across the joint, and small deployment/retraction or steering torque. A quantitative discussion on movable joint concepts identified the rotatable radial heat pipe as a promising solution for steerable radiators (Delil 1987). Figure 1 shows a schematic of a rotatable radial heat pipe. An essential component is the wick to provide the capillary head to return the condensate from condenser to evaporator and to distribute the liquid properly over the evaporator surface. Therefore the fine gauze wick structure should be uniformly fixed to the evaporator surface (inner tube outer surface). In this way burn-out, caused by blockage due to vapour bubbles generated, is prevented. Since the outer tube must be rotatable with respect to the inner tube there must be a clearance between porous structure and tube wall. This clearance is located at the condenser, the less critical side of the heat pipe, where the condensate has to be collected only (relatively easy, especially for a slightly overfilled heat pipe). An accurate design combines proper condensate collection and transport, hence good heat pipe performance, and low rotation torque, hence long lifetime for a steerable radiator. It is obvious that the end caps of a rotatable radial heat pipe must be leak-tight. This problem must be solved using appropriate seals. The thermal performance of a radial heat pipe is hard to predict. A rough estimate follows from flat plate vapour chamber data: heat transfer coefficient ≈ 4000 W/m^2·K, for methanol as working fluid, between 250 and 305 K. For a radial joint with an external diameter of 40 mm, this means a conductance of 500 W/K per meter joint length.

> The contents of this document have been initially prepared for publication as "Sensors and Components for Aerospace Thermal Control, Life Sciences, and Propellant Systems" in *Space Technology and Applications International Forum - 1998*, January 1998 by The American Institute of Physics.

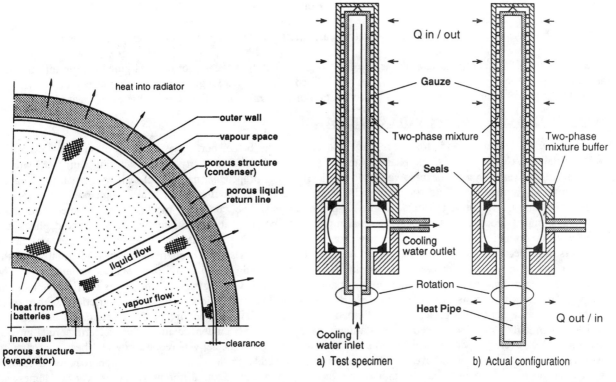

FIGURE 1. Cross-section of a Radial Heat Pipe Joint.

FIGURE 2. Rotatable Radial Heat Pipe Joint.

To prove the feasibility of the concept, a simple 10 cm long test specimen (Figure 2a) has been manufactured. It simulates the realistic configuration shown in figure 2b, and consists of: A 10/12 mm inner tube, cooled by liquid flow, simulating the heat pipe. A 13/15 mm outer tube, heated by a heater (simulating the heat source: a condensing two-phase mixture). A rotatable section (ball valve) allowing the outer and inner tube to rotate with respect to each other. The 0.5 mm clearance between the tubes contains a wick simulating metal gauze and working fluid R114.

Figure 3 shows the results of a test to determine the optimum working fluid content. Starting with pure liquid, the temperature drop across the joint is ≈ 13 K. By stepwise blow-off R114, this temperature drop is reduced to 7 K at the optimum mixture quality. Continuing blow-off causes increase of the temperature drop up to 26 K (pure vapour conduction/solid conduction of the gauze). The optimum joint conductance is 3 W/K for this 0.1 m long, 13 mm OD R114 joint, or 600 W/K for the mentioned 4 cm diameter, 1 m long methanol filled joint. This is ≈ 1500 W/K if ammonia is the working fluid. Figure 4 presents the (optimally filled) joint performance during rotation (at 17

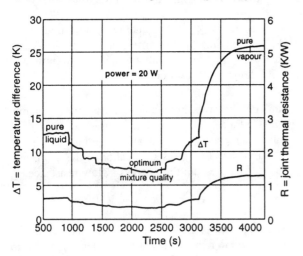

FIGURE 3. Determination of Optimum Filling.

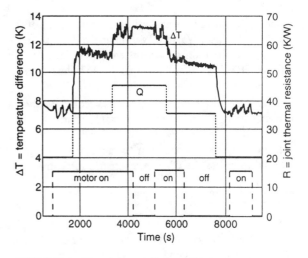

FIGURE 4. Power Dependence of Joint Resistance (non-rotating and rotating at 17.5 rpm).

revolutions per hour) and in non- rotating periods, for different power values (45, 35 and 20 W). The figure clearly confirms the aforementioned joint conductance value both for the rotating and non-rotating case. This conductance, showing the more stable values in the non-rotating case, increases slightly with power (temperature) level.

Summarizing, the concept is feasible as the joint did not leak and the performance figures are promising. Seal improvements and the use of a buffer volume (also filled with the working fluid at approximately the same, heat pipe, temperature) are expected to lead to the realisation of a mature long lifetime rotatable radial heat pipe joint.

VAPOUR QUALITY SENSOR AND CONTROLLABLE VALVE

Thermal management systems for future large spacecraft have to transport large amounts dissipated power over large distances. Conventional single-phase systems (based on the heat capacity of the working fluid) are simple, well understood, easy to test, inexpensive and low risk. But for a proper thermal control task with small temperature drops from equipment to radiator (to limit radiator size/mass), they require noisy, heavy, high power pumps and large solar arrays. As an alternative for single-phase systems one considers mechanically pumped two-phase systems: pumped loops, accepting heat by working fluid evaporation at heat dissipating stations (cold plates and heat exchangers) and releasing heat by condensation at heat demanding stations (hot plates, heat exchangers) and at radiators, for rejection to space. Such a system relies on heat of vaporisation: it operates nearly isothermally and the pumping power is reduced by orders of magnitude, thus minimising the sizes of radiators and solar arrays. The stations can be arranged in a pure series, a pure parallel or a hybrid configuration. The series configuration is the simplest, it offers the possibility of heat load sharing between different stations, with some restrictions with respect to their sequence in the loop. But it has limited growth potential and the higher flow resistance. In the low resistance modular parallel concept, the stations operate relatively independently, thus offering full growth capability. However, the parallel configuration is more complicated, especially if redundancy and heat load sharing (some cold plates operating in reverse mode) is foreseen. In addition, a parallel configuration requires a control system consisting of various sensors, monitoring the loop performance at different locations, control logic and actuators to adjust pump speed, fluid reservoir content and throughputs of valves. Sensors necessary for control are pressure gauges, flowmeters, temperature gauges and vapour quality sensors, measuring the relative vapour mass content of the flowing mixture (Delil 1988). An important application for Vapour Quality Sensors (VQS) is at cold plate exits, as a part of a control system adjusting the liquid fed into a cold plate to prevent evaporator dry-out or to maintain a prescribed quality value at evaporator exits, independent from transient heat sources and sink conditions.

As two-phase flow and heat transfer characteristics are different in 1-g and low-g, the technology of two-phase heat-transport systems and their components was to be demonstrated in orbit. Therefore, a Two-Phase eXperiment (TPX) has been developed within the ESA In-orbit Technology Demonstration Programme, by NLR (prime), Fokker Space, SABCA, Bradford and SPE. TPX is a scaled-down capillary pumped two-phase ammonia system together with scaled-down components of a mechanically pumped loop: multichannel condensers, vapour quality sensors and a controllable three-way valve. TPX has successfully flown in the Get Away Special G557 canister (5 ft^3, Nitrogen gas filled), aboard Space Shuttle STS-60, February 1994 (Delil et al. 1995). The experiment has run autonomously, using own power supply, data handling and experiment control after a switch-on command from the Shuttle crew.

One TPX objective was the in orbit calibration of the VQS, by setting the VQS vapour quality by adjusting the 3-ways Controllable Valve (CV), mixing the by-passed vapour and liquid leaving the condenser. In-orbit test data and terrestrial calibration curves differed considerably (Figure 5). This VQS (Figure 5), with updated electronics, will refly in TPX II (TPX follow-up) as G467 aboard STS-90, April 1998. Figure 6 shows the TPX II schematic.

Control exercises with VQS and CV could not be carried out in TPX, because the vapour quality control setpoint chosen turned out to be in the unstable churn flow pattern regime (Figure 5). The control excercises will be redone in TPX II (with the TPX II CV depicted in figure 7). Control will be done around two quality setpoints, one in the slug flow regime, the other in the annular flow regime. The NLR ammonia test rig, used to calibrate the VQS, will be used to assess the CV characteristics to control both vapour quality and pressure drop across a flow resistance.

CONDENSERS

High Efficiency Low Pressure Drop condensers/radiators are crucial for two-phase systems. Two radiator solutions can be distinguished: A direct condensing radiator: condenser attached to the radiator, radiating condensation heat to space, and a hybrid condenser radiator, where the condenser is not an integrated radiator part (condensation heat is transported from condenser via central heat pipe to heat pipes distributing the heat over the radiator).

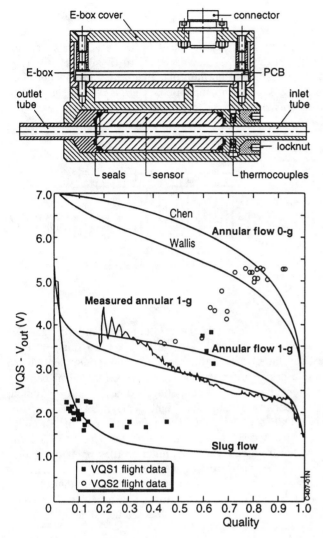

FIGURE 5. VQS, plus Theoretical Response and Test Data.

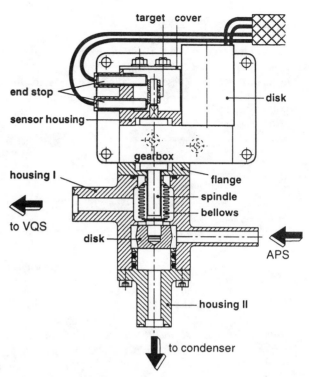

FIGURE 7. TPX II Controllable Valve.

APS	Absolute Pressure Sensor
DPS	Differential Pressure Sensor
CV	Controllable 3-way Valve
T_V	Vapour Temperature
T_ℓ	Liquid Temperature
EMP	Experiment Mounting Plate (GAS canister lid)
GAS	Get Away Special
LFM	Liquid Flow Meter
$\dot{m}$	Mass flow
$\dot{m}_c$	Flow Rate in Condenser Branch
Q_{E_F}	Flat Evaporator Heater Power
Q_{E_C}	Cylindrical Evaporator Heater Power
Q_{COND}	Condenser Imbalancing Power
VQS	Vapour Quality Sensor

FIGURE 6. TPX II Schematic.

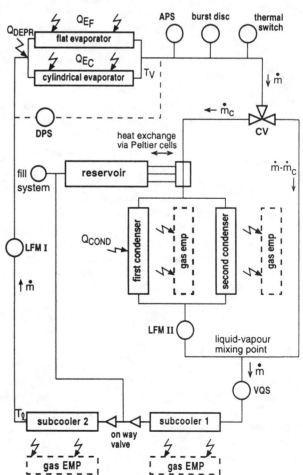

Direct condensing radiator:

Two direct condensing radiators have been designed and manufactured for the ATLID Laser Head Thermal Control Breadboard (Figure 8), developed for ESA by MSS-UK (prime), NLR and Bradford (Dunbar 1996). They are configured to represent the allowable areas for the ATLID instrument on the Polar Platform. One radiator, 1.05 m high by 1.0 m wide (radiator A), is fixed to the instrument baseplate and supported by struts. The other radiator (B), 0.8 m high and 1.45 m wide, is deployable and fixed only along its edge by cantilever support beams. The

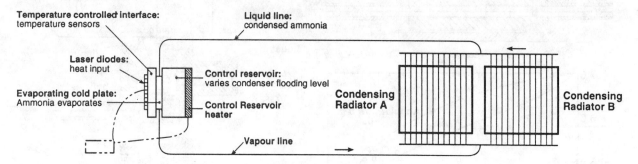

FIGURE 8. ATLID Laser Head Thermal Control Schematic.

struts for radiator A are constructed from filament wound carbon fibre tubes with aluminium end fittings. The cantilever beams of radiator B are 100 mm deep to provide adequate stiffness. The radiators are too small to reject the heat load in steady state conditions. They are only just capable of meeting the heat rejection requirements when a full orbital cycle is considered. The radiator B deployable design incorporates a unique torsion/helical bending configuration to minimise pipe strain and allow multiple repetition of the deployment. Although the instrument requires deployment only for ground access, the design is equally suitable for flight deployment. The radiator surfaces would for flight be covered with advanced glass OSR's to give low beginning and end of life solar absorptivities. For the breadboard tests sunlight has been simulated by altering heat sink temperatures, and the radiators are simply black painted. The radiators are constructed from extruded aluminium profiles rivetted together to form a continuous surface. Each profile section contains one 2 mm internal diameter condensing pipe, clamped into good thermal contact with a channel in the extrusion. Isolators in each liquid line and one at the liquid header outlet ensure even vapour distribution and prevent differential dryout. The rivetted construction provides stiffness in two axes, and the remaining axis is stiffened by the addition of a beam crossing all the profiles. The ATLID test programme conclusions (Dunbar 1996) report the following achievements:

- The two-phase system is treated as just another thermal tool able conform to installation, accommodation and structural requirements imposed by the overall instrument.
- The system has successfully completed severe sine and random vibration tests to qualification levels.
- The deployable radiator concept has been demonstrated.
- The tests demonstrated that the ATLID breadboard meets or exceeds nearly all performance requirements. In particular the principal requirement to maintain the laserdiode interface to within 1 °C of nominal temperature during simulated low earth orbits was met with a significant margin. Due to restrictions on radiator area the end of life heat rejection performance only just meets the requirements. Some extra margin is recommended.
- Some modifications will be necessary if the 125 Hz first radiator frequency is to be met for the actual flight units, but these have been identified and are not considered critical.

Detailed information can be found in the aforementioned reference.

Hybrid condenser:

For ESA, a high efficiency low pressure drop condenser (Figure 9) for a hybrid (heat pipe) radiator has been successfully developed and brought up to pre-qualification level, by NLR (prime), Bradford and DASA RI (Delil et al. 1996). This condenser has been subjected to tests in the test rig under conditions reflecting realistic in-orbit conditions: vapour temperature between 263 and 313 K, for a condensed power up to 300 W. The tested hybrid condenser design (Figure 9) consists of a concentric tube around a liquid cooled inner tube, simulating the heat pipe. Vapour entering the condenser is uniformly distributed by a cone. The condensing part is an annulus with ID 25 mm and OD 28 mm, hence a gap width 1.5 mm. Six wires with a diameter of 1.5 mm subdivide the annulus

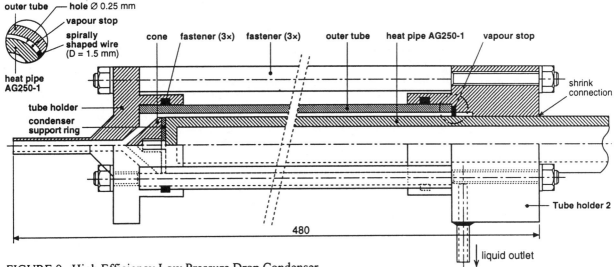

FIGURE 9. High Efficiency Low Pressure Drop Condenser.

into six parts. The wires are coiled around the central tube, leading to helical condensation channels, providing a swirling to improve performance. The tests proved the quality of the design, being a good compromise between high-efficient thermal performance and low pressure drop (Figure 10): for 300 W a temperature drop below 7.5 K at a pressure drop below 400 Pa (the latter can be considerably reduced by increasing the number of condenser outlet vapour stops). The tests proved that there is no significant difference in performance for vertical and horizontal orientation. Furthermore the condenser design satisfied all other requirements. Three of these condensers in series, equipped with 25 mm OD central heat pipes, are part of the ESA Capillary-pumped Loop Engineering Model CLEM currently developed by MMS-UK (prime), MMS-F, Bradford and NLR.

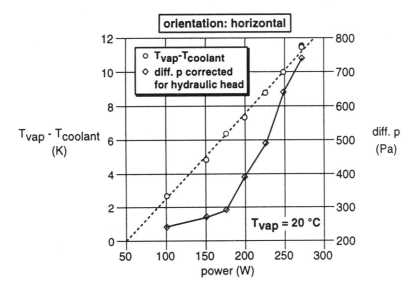

FIGURE 10. Temperature and Pressure Drops as a Function of Power for the Horizontal Condenser.

FLOW METER ASSEMBLIES

Activities to spatialise commercial Flow Metering Assemblies (FMA) for ESA (by NLR, prime, SPPS/Bradford and SABCA), started with selecting and screening more than 80 commercial meters. Trade-offs identified possible candidates for applications in spacecraft thermal, life sciences and propellant systems. The most promising ones were subjected to a test programme that included qualification level functional performance/environmental tests. A dedicated test facility has been developed for performance testing and calibration. The test rig accomodates all type of working fluids: thermal (ammonia, freons), life science (water) and propellants (MMH, MON). Tests can be executed between 233 and 358 K, for flows up to 3 g/s for ammonia, to 200 g/s for water and propellants. Test bench accuracy: 0.025 % Full Scale, flow meter accuracy 0.1 % Full Scale. The system can be pressurised up to 2.5 Mpa. Results of extensive testing led to the choice of two meters to be spatialised: ITT Barton 7182 turbine meter for water, ITT Barton 7506 pelton wheel meter for ammonia. The activities will be concluded December 1997.

Other flow meter activities of Bradford/SPPS (prime), NLR and Panametrics focus on the development of a non-intrusive ultrasonic meter for applications in propellant systems of three-axes stabilised spacecraft. The accuracy has to be better than 0.1 % Full Scale (design goal 0.05 % FS) as the meters are to be used to assess remaining propellant mass by totalising the propellant consumption.

PROPELLANT GAUGES

Since the introduction of the de-orbit requirement for geostationary spacecraft, increased attention has been paid to accurate and reliable on-board gauging of propellants. Various gauging systems have been developed and are in use with a wide spreading of accuracy and complexity (Hufenbach et al. 1997). A from the VQS derived capacitive gauge is currently being developed to very accurately determine the remaining fuel (MMH) and oxidiser (MON) in propellant tanks of spin-stabilised spacecraft. The gauges will be integrated parts of the tanks of the Meteosat Second Generation Unified Propulsion System (Hufenbach et al. 1997). This Gauging Sensor Unit (GSU), depicted in figure 11, consists of a measuring unit and electronics. The measurement principle is schematically shown in figure 12. The platinum covered central glass rod is the inner electrode of the GSU, the titanium holder is the outer one. The combination of a segmented inner electrode, intelligent electronics and a dedicated ground handling protocol, yields level accuracies ranging from 0.015 mm for chemically stable liquids (Freons) to 0.3 mm for less stable liquids (MON). The GSU breadboard model has been subjected to extensive testing. Figure 13 shows the

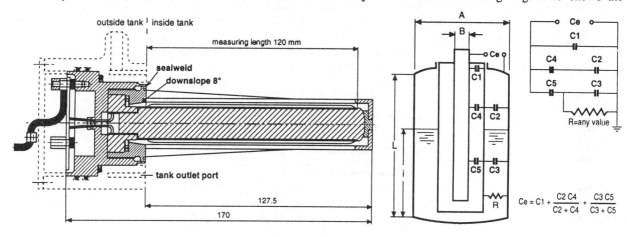

FIGURE 11. Propellant Gauging Sensor Unit.

FIGURE 12. Coated Capacitance Probe (Schematic).

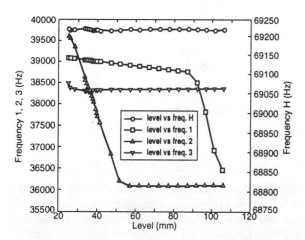

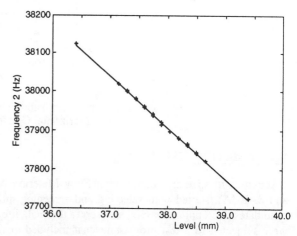

FIGURE 13. GSU Levels versus Frequency Test Data (MON-1).

FIGURE 14. GSU Resolution Test Data (MON-1).

measured MON level curves for the different segments. Figure 14 presents the results of the resolution/accuracy verification tests with MON. To correctly interpret the resolution/accuracy data, the liquid properties (dielectric permittivity and electric conductivity) must be known. As this knowledge could not be found in literature, a test rig has been built to characterise dielectric properties of propellants. Successful testing has been recently carried out on MMH and MON (Delil et al. 1997).

Acknowledgement

We express our appreciation for the numerous efforts of our colleagues in the different projects: J. Heemskerk, O. Mastenbroek, G. van Donk, A. Monkel and M. Versteeg (all NLR), R. Jacobs (Bradford), M. Dubois (SABCA), N. Dunbar (MMS-UK), R. Müller (Daimler Benz Aerospace RI) and S. Costa (SPPS).

References

Delil, A.A.M. (1987) "Moveable Thermal Joints for Deployable or Steerable Spacecraft Radiator Systems", *NLR MP 87016 U, SAE 871460, 17th Intersociety Conference on Environmental Systems*, Seattle, WA.

Delil. A.A.M. (1988) "A Sensor for High-Quality Two-Phase Flow", *NLR MP 88025 U, Proc. 16th International symposium on Space Technology and Science*, Sapporo, 957-966.

Delil, A.A.M. et al. (1996) "TPX for In-Orbit Demonstration of Two-Phase Heat Transport Technology - Evaluation of Flight & Postflight Experiment Results", *NLR TP 95192 U, SAE 95150, 25th International Conference on Environmental Systems*, San Diego, CA.

Delil, A.A.M., et al. (1997) "High Efficiency Low Pressure Drop Two-Phase Condenser for Space", *NLR TP 96380 U, SAE 961562, 26th International Conference on Environmental Systems*, Monterey, CA.

Delil, A.A.M. et al. (1997) "Characterisation of the Dielectric Properties of Mixed Oxydes of Nitrides and Mono Methyl Hydrazine, *NLR CR 97268 L.*

Dunbar, N.W. (1996) "ATLID Laser Head Thermal Control-Design and Development of a Two-Phase Heat Transport System for Practical Application", *SAE 961561, 26th International Conference on Environmental Systems*, Monterey, CA.

Hufenbach, B. et al. (1997) "Comparative Assessment of Gauging Systems and Description of a Liquid Gauging Concept for a Spin-Stabilised Spacecraft", *ESA SP398, 2nd European Spacecraft Propulsion Conference*, Noordwijk, Netherlands, 561-570.

Author Index

(**Bold** Page Numbers Indicate Senior Authorship)

A

Abshire, J. B., **33**
Ackerman, N., 495
Afzal, R. S., **107**
Agan, M., **236**
Ageev, V. P., 666
Aguilar, L., 563, 654, **693**, 698
Allen, J. S., **471**
An, Y., 1087
Anderson, D. M., 1033
Anderson, G. A., **616**
Annaballi, A. L., 80
Arnowitz, L., **703**
Arrington, L., 295
Aslam, S., 61
Atchison, D., **816**
Atsumi, M., 1066
Audet, N., 769
Ayers, R., 627

B

Bac, N., 544
Badi, N., **666**
Bailey, P. G., **1527**
Bajic, D. M., 729
Balance, J., 354
Balasubramaniam, R., **429**
Banks, S., 199
Barkan, A., 1542
Barnes, J. C., **95**
Barrett, T. W., 1449
Barthel, C. A., 1020
Barthelmy, S., 55
Bass, R. W., 1125
Bateman, T., 627
Bates, S. C., **711**
Batty, J. C., 276
Baumgartner, R. I., **867**
Bayard, D. S., 134, 147
Beauchamp, P. M., 159
Beering, D. R., 235
Begg, L. L., 318
Bennett, G. L., **1269**
Bensaoula, A., 648, 666, 724
Benson, J. W., **995**
Berichev, I., 648
Berryman, J., 803
Berte, M., 1196, **1200**
Best, F. R., 501, 505

Bhasin, K., 235
Bienert, W. B., **511**
Bilbro, J. W., **185**
Binot, R., 399
Biswas, A., 243
Blandino, J. J., 295
Bochove, E. J., **112**
Booher, R. A., 1167
Borkowski, C. A., 1479, 1491, **1542**
Borowski, S. K., 1301
Borshchevsky, A., 1647
Botello, A. M., 609
Boyce, W. D., 1056
Bozich, W. F., 926
Bragg, S., **501**
Brito, H. H., **1509**
Britt, R. G., **569**
Bromaghim, D. R., 295, **302**
Bromley, B., 1373
Brooks, D. E., 235
Brower, T. L., **452**
Brown, M. A., 282
Brown, R., 633
Brown, W., 1092
Bubenheim, D., 835
Bugby, D. C., **1119**
Bula, R. J., 593
Bult, R. P., 769
Bunn, T. L., **21**
Burchfield, J., 370
Burgess, A., 45
Burleson, D., **921**
Bush, H., 1007
Bussard, R. W., 1289, **1344**
Butler, J. R., 769
Butz, J. R., 852
Byer, R. L., **77**
Byrd, R. J., 962
Byrne, K., 205

C

Caillat, T., **1647**
Campbell, J. W., **1212**
Capell, B., **1196**, 1200
Carlson, M. E., **1486**
Carrington, H., **199**, 205
Cassenti, B. N., **1125**
Castles, S., 199
Cataldo, R. L., **1301**
Chacon, L., 1373

Author Index

(**Bold** Page Numbers Indicate Senior Authorship)

Chakrabarti, S., 1365
Chan, S. H., **440**
Chang, J. H., **505**
Chang, K., 1225
Charache, G. W., 1400
Charette, R. O., **926, 969**
Chato, D. J., **276**
Chen, I. Y., 479
Chen, X., 672
Cheng, R., 822
Cheng, R. K., **858**
Cheung, K., 464
Chin, K., 698
Christopherson, E., **574**
Chubb, D. L., **1410**
Cierpik, K., 633
Clark, K., 647
Cockfield, R. D., 1157
Coleman, J. R., **1423**
Connolly, J. C., 1394, 1400
Conrad, D., 647
Criswell, D. R., **1219**
Croft, J., **141**
Croop, H. C., **1020**
Crowe, D. A., 173
Curcio, M., 159
Curreri, P. A., **389**

D

Darling, T. W., 1628
Das, N. C., 61
Dashevsky, Z., **1634**
Davis, E. W., **1502**
Davis, L., 85
Deamer, D. A., 926
Deane, N. A., 1275
DeHainaut, L., 91
Deily, J., 141
Delaney, A., 693, **698**
Delil, A. A. M., **258, 514**
DeMora, J., 1373
Denman, C., 91
DePoy, D. M., 1400
Dessiatoun, S. V., 464
Deuser, M. S., **793**
Devereaux, A., 236
De Volder, B., 1138
Dhar, H. P., **679**
Dietrich, F. J., **229**
Difilippo, F. C., **1281**

Doarn, C. R., **829**
Donet, C., 211
Dorsey, D., 501
Drabkin, I., 1634
Draeger, N. A., **586**
Dundore, B., 1365
Dyer, J., 179

E

Eckmann, J. B., **270**
Egorov, V. S., 1621
Eguchi, K., 364
Eipers-Smith, K., 724
El-Genk, M. S., 308, **1461**, 1471, 1552, 1576, **1586**, 1595
Ellis, M., 501
Emrich, Jr., W. J., 1145, 1189, 1338
Ender, A. Ya., **1565**
Eryomin, S. A., 318
Escher, W. J. D., **999**
Espinoza, J., 1307
Ewin, A., **61**

F

Fatemi, N. S., 1417
Fattaey, H. K., 622
Faulconer, W., **779**
Ferguson, V., 627
Fisher, C., 556
Fleurial, J-P., 1647
Flynn, M., **835**
Folta, D. C., **119**
Foltyn, E. M., 1163, 1314
Ford, R. N., **1045**
Forouhar, S., **85**
Fortin, M., 61
Frappier, G., **534**
Freibert, F., 1628
Freundlich, A., 563, 654, **660**, 693, 698
Froning, Jr., H. D., **1289**, 1344, **1449**
Frye, P. E., 348
Fujii, T., 364
Fukui, T., **937**
Fuller, J., 501
Fulmer, J., 1365

AUTHOR INDEX

(**Bold** Page Numbers Indicate Senior Authorship)

G

Gaidos, G., **1365**
Galati, T., 881
Galli, G., 458
Garbuzov, D. Z., 1394, **1400**
Garnov, S. V., 666
Garton, D. J., 724
Garverick, L. M., **1417**
Gaubatz, W. A., 887
Gentry, B. M., **101**
George, T. G., **1163**, 1429
Giglio, J. C., 1486
Giuliano, C., **80**
Glass, J. F., **979**
Godfrey, J., 45
Godfroy, T., 1377
Goracke, B. D., 979
Greenberg, J. S., **949**
Gregg, M., 91
Gregory, P., 534
Gresser, J. D., 749
Grotz, T., 1527
Grymes, R. A., 816
Gu, Y., 1373
Guerrero, A., 1429
Gugliermetti, F., 422
Gully, W. J., 199, **205**
Gunn, S., **1234**
Guray, I., 544
Gusev, V. V., 1641

H

Ha, C. T., **1275**
Haberman, E., 270
Hachkowski, M. R., 188
Hahn, S. E., 1020
Haisch, B., **1443**
Hallinan, K. P., 471
Hardaway, L. R., 188
Hardin, J. R., 793
Harper, J., 603
Harris, G., 495
Hart, D., **773**
Hartman, K., 141
Harwick, W. T., **956**
Hastings, L. J., **331**
Hatfield, M., 1092
Hauber, B. K., **1027**
Hauser, Jr., R. L., **893**

Havelka, D., 685
Hawkins, J., **1092**
Hayes, W. A., 1056
Hemmati, H., 243
Hendricks, T. J., 1479, **1491**, 1542
Herrera, A., 1329
Herring, R., 534
Hertzfeld, H. R., **810**
Heslop, J. M., 1314
Heyenga, G., 578
Hikida, S., **127**
Hill, C., 211
Hill, R. C., **535**
Hinckley, J. E., 1321
Hinkal, S., 76
Hinkle, J. D., 188
Hoder, D. J., **235**
Hoehn, A., 578
Hoffman, A., 1352
Holt, A. C., **1295**
Homer, M. L., 1607
Horne, W. E., **1385**
Hoskins, W. A., 295
Houle, J. M., 729
Houston, S., 1092
Houts, M. G., **1189**
Howe, S. D., **1138**
Howell, E. I., 1167
Hsieh, K. C., 404
Hsu, Y. Y., 749
Hu, K., 45
Huang, C., 1491
Huang, J., **446**
Huang, L., **1471**, 1613
Huang, Z. D., **39**
Huber, F. M., 1435
Hurtak, J. J., 1527

I

Igarashi, T., 364
Ignatiev, A., 660, 672, 698
Itagaki, H., 127
Itoh, K., 364
Izhvanov, O. L., 318

J

Jackson, K., 786
James, E., 199

Author Index

(**Bold** Page Numbers Indicate Senior Authorship)

James, E. L., 359
Jarvinen, G. D., 1307
Jeganathan, M., 243
Jenkins, P. P., 1417
Jenstrom, D., 76
Jhabvala, M., 67
Johnson, L. **354**
Johnson, L. K., 302
Johnson, M. R., 858
Johnson, T. C., 622
Jones, J. M., **1056**
Josef, J. F., 1045
Joshi, A. M., **67, 153**
Jurczyk, B., 1373

K

Kammash, T., **1145, 1338, 1377**
Kampas, F. J., **1263**
Karthikeyan, M., 446
Kascak, A., 685
Kasl, E. P., **173**, 179
Kaukler, W. F., 389
Kaya, N., 1225
Kearney, M. E., **786**
Kee, R. J., **840**
Keener, D., **211**
Keller, P. C., 1045
Kennedy, F. G., **1185**
Kennel, E. B., **435**
Kenny, A., 685
Keo, S., 85
Khalfin, V., 1400
Kiehl, W. K., 199, 205
Kim, E., 648
Kim, J. H., 464
Kirchman, F., 76
Kishimoto, K., 1066
Klaus, D., 616, **633**
Klevatt, P. L., **887**
Klimentov, S. M., 666
Knight, K. S., 711
Koehler, F. A., 1167
Kolyshkin, I. N., 1565
Konno, A., **1066**
Korolev, V. U., 318
Korotaev, V., 1634
Korotkiy, Y. G., **943**
Kos, L., **1206**
Kostiuk, L. W., 858
Kotecki, C., 45

Kramer, D. P., **1167**
Krebs, D., 45
Krishnan, A., 718
Kronberg, B., **531**
Krumweide, G. C., **179**
Kubricht, T., 660
Kudija, C. T., **348**
Kufner, E., 399
Kulik, J., 666
Kumar, V., 1107
Kuznetsov, V. I., 1565

L

Laiho, J., 1365
Lake, M. S., **188**
Landau, A., 609, 736
Lapochkin, N. V., 318
Lara, L., 1607
Larson, A. A., **1535**
Latini, G., **458**
Lechtenberg, T. A., 318
LeDuc, J. R., 295, 302
Lee, H., 1394, 1400
Lee, S., 666
Legros, J-C., 773
Lei, S., 685
Lekan, J., 471
Leonard, B. G., 926
Lesh, J., **243**
Levack, D. J. H., 979
Lewinski, K. A., 679
Lewis, M. L., 793
Lewis, R. A., 1365, **1435**
Li, J., **381**
Li, M., 685
Lin, C.-H., 603
Lindley, C. A., 905
Linne, M., **556**
Linton, S., 822
Liou, S. G., **479**
Lisitsa, V. G., 1621
Little, F. E., **1225**
Liu, D., 672
Lockridge, D. L., 501
Lopez, B., 1329
Loughin, S., 1263
Lowndes III, H. B., 1020
Lowry, S. A., 718
Luchau, D., 324
Ludewig, H., 1131

Author Index

(**Bold** Page Numbers Indicate Senior Authorship)

Luke, R., 308
Lundquist, C., 551
Lynch, C., 1329, 1429

M

MacCallum, T. K., 616
Maise, G., 1131
Maker, P., 85
Malak, J., 295
Mallasch, P. G., 235
Malloy, J., 370
Maltsev, V. G., 1621
Manthripragada, S., 45
Marceau, M., 544
Marchin, G. L., **765**
Marshall, A. C., **317**
Martin, J. A., **985**
Martin, J. J., 331
Martineau, R. J., **45**
Martinelli, R. U., **1394**, 1400
Marvin, D., 211
Mayberry, C., 211
McAdoo, J., 61
McBirney, T. R., 1119
McCabe, D., 534
McCarson, T. D., 324
McCormick, E. D., **1173**
McCulloch, W. H., 1423
McDougal, J. R., 1167
McDougall, F., 881
McKinnon, J. T., 852
McNeal, S. R., 166
McNeil, D. C., 1167
McQuillen, J. B., 413
McSpadden, J. O., 1225
Meckel, N. J., 295
Medelci, N., 648, 666
Mehle, G. V., 179
Mehra, R. K., **15, 134, 147**
Merrill, J. M., **1613**
Messerschmid, E. W., 1435
Metzger, J. D., **1151**
Meyer, G. A., 729
Middleton, N., 556
Migliori, A., **1628**
Miho, K., 937
Miles, B. J., 375
Miley, G. H., 1081, **1373**
Millar, P. S., **27**
Miller, J., 1607

Miller, K., **1352**
Millis, M. G., **3**
Minami, Y., **1516**
Minick, S. D., 1062
Minster, O., 399
Minton, T., 724
Miskolczy, G., **370**
Mo, B., **464**
Momozaki, Y., 308
Monacos, S., 243
Mondt, J. F., **1098**
Mongan, P., 786
Monier, C., 563, 654, 693, 698
Moniz, P., 1329, 1429
Montel, F., 773
Montgomery, K., 822
Moore, B., 295
Moore, J. J., 755
Morgan, M. D., 1385
Morgenthaler, G. W., **609, 736, 803**
Morris, N. A., 1394, 1400
Moshopoulou, E., 1628
Motaffaf, F., 622
Motakef, S., 743
Mott, D. B., 39, 45
Moyer, M. W., 1429
Mukunda, M., **1245**
Muller, C., **91**
Muller, R., 85

N

Nakano, E., 937
Nasmyth, P. W., 769
Naumann, R., 551
Nesmith, B. J., 1098
Neuberger, D., **495**
Newman, F., 563, **654**, 693, 698
Newquist, C. W., **1033**
Nikolaev, Y. V., **318**
Noblett, P. M., 21
Noojin, S., 798
Noravian, H., 1107
Nuñez, G. R., 609

O

O'Connor, E. A., 899
Odubanjo, T., 1394
Ohadi, M. M., 464

Author Index
(**Bold** Page Numbers Indicate Senior Authorship)

Okamoto, K-i, 364
O'Leary, R., 257
Olenchenko, O. A., 1621
Olsen, A. D., **342**
O'Neill, M. J., **288**
Ono, R. M., 166
Or, C. T., 1107, **1257**
Otting, W. D., 21
Overmyer, C. M., **638**, 786

P

Pacas, **D. A., 755**
Padilla, M., 1429
Page, N., 243
Palazzolo, **A., 685**
Paniagua, J., 1131
Pankin, M. I., 1641
Pantolin, J. E., 1479
Paramonov, D. V., 1565
Parsons, A., 55
Passerini, G., **422**, 458
Pauw, A., 514
Peery, S. D., 1062
Pei, S. S., 603
Pencil, E., 295
Penn, J. P., **905**
Perry, S. S., 666
Peters, F., 45
Peterson, L., **647**
Peterson, L. D., 188
Peterson, S. W., 1571, **1652**
Peterson, T., **295**
Petra, M., **1081**
Phillips, J. M., 1179
Phillips, T. H., **257**
Phillips, W. M., 1607
Phipps, C., **1073**
Pipes, III, W. E., 1045
Piszczor, M. F., 288
Placr, A., 1429
Plawsky, J., 446
Pollard, J. E., **264**
Polonara, F., 422, 458
Portianoy, A. G., 1621
Poston, D. I., 1189
Powell, J., **1131**
Powers, D. F., 488
Poynter, J. E., 616
Preedy, M. W., 1033
Proffitt, G. D., 440

Purdy, G. M., 1307
Pustovalov, A. A., **1641**

Q

Quinn, D. A., 119
Quinton, **T. M., 622**

R

Rabinovich, D., 1634
Radzykewycz, D., 211
Rago, C., 134
Raja, L., 840
Ramsey, K. B., 1307, **1314**
Ramsey, W. D., 359
Randolph, D., 779
Rasmussen, J. R., 1479, 1542
Ravichandran, B., 15
Redden, R. F., **769**
Reichert, II, K. W., 729
Reimus, M. A. H., **1321, 1329, 1429**
Reinhardt, K., 211
Reinstrom, R. M., **1157**
Rhim, W-K, **397**
Rhodes, R. E., **962**
Riley, M. W., **798**
Rinehart, G. H., 1307, 1329
Ritums, D. L., **672**
Rodgers, D. H., 159
Ross, M. D., **822**
Rueda, A., 1443
Ruhkamp, J. D., 1167
Ryan, M. A., 1607

S

Saban, S. B., 1385
Sacco, Jr., A., **544**
Sacco, T. L., 544
Sack, W. F., 1056
Salasovich, R. M., 302
Samson, J., 647
Sandusky, J., 243
Sarrao, J. L., 1628
Sasso, S. E., **888**
Sato, H., 364
Sawyer, J. W., **1007**
Schick, S., 276

Author Index

(**Bold** Page Numbers Indicate Senior Authorship)

Schmidt, M. H., 729
Schock, A., **1107**
Schowengerdt, F., 755
Schuiling, R. L., **899**
Schuller, M. J., 1613
Schulte, L. D., **1307**
Schultz, J. A., 666, **724**
Schultz, S., 724
Schutz, B. E., 33
Schwalm, M., **159**
Seereeram, S., 134, 147, **223**
Sen, S., 389
Serdiukova, I., 693
Serdun, E. N., 1621
Seward, C., **1455**
Shen, T. R., 440
Sheu, J. S., 479
Shi, J. Z., 45, 55
Shimizu, M., **364**
Sholtis, Jr., J. A., 1423
Shorey, M. W., 1033
Shrairman, R., 609, 736
Shu, P. K., 39, 45, 55, 61, 67
Sievers, R. K., **1479**, 1486, 1542
Silver, G. L., 1307
Simberg, R., **991**
Simske, S., 627
Singh, B., 440
Sirota, J. M., 27
Skinner, D., **1179**
Skrabek, E. A., 1257
Slough, J., 1352
Smiljanic, R. R., 969
Smith, G. A., 847, 1365, 1435
Smith, J., 822
Smith, J. C., 33
Smith, Jr., J. E., 798
Smith, W., 61
Smitherman, Jr., D. V., **538**
Snodgrass, S., 55
Sokolsky, I., **282**
Sorokin, A. P., **1621**
Soto, J., 609
Sprenger, H. J., **525**
Srivastava, R., **852**
Stahle, C., 55
Stapleton, D., **881**
Starikov, D., **648**, 666, 724
Steadman, K. B., 1039
Steinmeyer, D. A., 969
Sterling, M., 724
Stickler, P. B., **1045**

Stodieck, L. S., **578, 627**
Stout, P. J., **718**
Strauss, A. M., **1571**, 1652
Street, S., 693, 698
Strength, V., **643**
Stubbers, R., 1373
Summerford, R., 875
Sutton, A. M., **1062**

T

Takahashi, H., **1087, 1359**
Talbot, G. J., **359**
Tanck, P. A., **1039**
Tang, C., 435
Taylor, C. R., 1212
Taylor, G. C., 1394, 1400
Tempez, A., 666
Thode, L., 1138
Thompson, C., 638
Tobin, S. M., **875**
Todosow, M., 1131
Tokarev, V. N., 666
Tolson, J., **488**
Tournier, J.-M., 1461, **1552, 1576**, 1586, **1595**
Trantolo, D. J., **749**
Trevgoda, M. M., 1621
Tripathi, V. K., 679
Trugman, S. A., 1628
Tsetshladze, D. L., 318
Tucker, D. S., **847**

U

Ugarov, M. V., 666
Underwood, M. L., 1098
Urban, J. E., **761**

V

Van Dyke, M., 276
van Put, P., 514
Velela, M., 698
Vellinger, J. C., 793
Vilela, M. F., **563**, 654, 693
Voeten, R. G. H. M., 514
Voisinet, L.-A., 236
Vujisic, L., **743**

AUTHOR INDEX
(**Bold** Page Numbers Indicate Senior Authorship)

W

Walter, H. U., **399**
Wang, A. W., **249**
Wang, D., 159
Wang, Y., 648
Warzywoda, J., 544
Waters, K., 666, 724
Watson, C., **551**
Wayner, Jr., P. C., 446
Weidow, D., 141
Weimer, M. B., 603
Weislogel, M. M., **404, 413**
Wen, J. T., 147, 223
Wessling, F., 551
Westerman, K. O., **375**
Whalen, L. M., 544
Whelan, H. T., **729**
White, R. L., 749
Whitten, R., 829
Widman, F., 257
Williams, B. E., **166**
Williams, R. M., 1607
Wilson, K., 243
Wilson, M., 76
Wilt, D. M., **1410**
Wise, D. L., 749
Wiswell, R. L., 270
Wolanski, T. W., 755
Wood, H. J., **76**
Workman, G. L., 847
Wu, N. J., 672
Wyant, F. J., **324**

Y

Ya. Kucherov, R., 318
Yakovlev, E. V., 1565
Yamazaki, Y., 1087
Yang, B. H., 603
Yang, R. Q., 603
Yoel, D. W., 711
Young, M. G., 85
Yu, A., 1359

Z

Zee, R. H., 381
Zerkle, D., 1138
Zernic, M. J., 235
Zhang, D., **603**
Zhou, W., **593**
Zomorrodian, V., 666